Quantity	Unit	SI Symbol
(Base Units)		
Length	meter°	m
Mass	kilogram	kg
Time	second	s
(Derived Units)		
Acceleration, linear	meter/second²	m/s²
Acceleration, angular	radian/second²	rad/s²
Area	meter²	m²
Density	kilogram/meter³	kg/m³
Force	newton	N (= kg·m/s²)
Frequency	hertz	Hz (= 1/s)
Impulse, linear	newton-second	N·s
Impulse, angular	newton-meter-second	N·m·s
Moment of force	newton-meter	N·m
Moment of inertia, area	meter⁴	m⁴
Moment of inertia, mass	kilogram-meter²	kg·m²
Momentum, linear	kilogram-meter/second	kg·m/s (= N·s)
Momentum, angular	kilogram-meter²/second	kg·m²/s (= N·m·s)
Power	watt	W (= J/s = N·m/s)
Pressure, stress	pascal	Pa (= N/m²)
Product of inertia, area	meter⁴	m⁴
Product of inertia, mass	kilogram-meter²	kg·m²
Spring constant	newton/meter	N/m
Velocity, linear	meter/second	m/s
Velocity, angular	radian/second	rad/s
Volume	meter³	m³
Work, energy	joule	J (= N·m)
(Supplementary and Other Acceptable Units)		
Distance (navigation)	nautical mile	(= 1.852 km)
Mass	ton (metric)	t (= 1000 kg)
Plane angle	degrees (decimal)	°
Plane angle	radian	
Speed	knot	(1.852 km/h)
Time	day	d
Time	hour	h
Time	minute	min

° Also spelled *metre*.

SI UNIT PREFIXES

Multiplication Factor		Prefix	Symbol
1 000 000 000 000	= 10¹²	tera	T
1 000 000 000	= 10⁹	giga	G
1 000 000	= 10⁶	mega	M
1 000	= 10³	kilo	k
100	= 10²	hecto	h
10	= 10	deka	da
0.1	= 10⁻¹	deci	d
0.01	= 10⁻²	centi	c
0.001	= 10⁻³	milli	m
0.000 001	= 10⁻⁶	micro	μ
0.000 000 001	= 10⁻⁹	nano	n
0.000 000 000 001	= 10⁻¹²	pico	p

SELECTED RULES FOR WRITING METRIC QUANTITIES

1. (a) Use prefixes to keep numerical values generally between 0.1 and 1000.
   (b) Use of the prefixes hecto, decka, deci, and centi should be generally avoided except for certain areas or volumes where the numbers would be otherwise awkward.
   (c) Use prefixes only in the numerator of unit combinations. The one exception is the base unit kilogram. (*Example:* write kN/m not N/mm; J/kg not mJ/g)
   (d) Avoid double prefixes. (*Example:* write GN not kMN)
2. Unit designations
   (a) Use a dot for multiplication of units. (*Example:* write N·m not Nm)
   (b) Avoid ambiguous double solidus. (*Example:* write N/m² not N/m/m)
   (c) Exponents refer to entire unit. (*Example:* mm² means (mm)²)
3. Number grouping
   Use a space rather than a comma to separate numbers in groups of three, counting from the decimal point in both directions. (*Example:* 4 607 321.048 72)
   Space may be omitted for numbers of four digits. (*Example:* 4296 or 0.0476)

Mele Kai—Hawaiian for *Song of the Sea*—is the name Professor Meriam gave to his 24-foot cruising sloop, which he designed and built at his house over a period of three years. *Mele Kai* was constructed with full-length cedar planking bronze-fastened to oak frames, teak trim, lead ballast keel, and custom-made bronze fittings. A unique structural element in the cabin is a post turned from camphor wood salvaged by the builder from the remains of an ancient Chinese pirate ship built in 1773. Professor Meriam has cruised extensively in *Mele Kai* among the islands off the Southern California coast and in 1955 sailed across the Pacific to the Hawaiian Islands in a similar 25-foot sloop.

ENGINEERING MECHANICS

VOLUME 2

DYNAMICS

FOURTH EDITION

J.L. MERIAM
*University of California
Santa Barbara*

L.G. KRAIGE
*Virginia Polytechnic Institute and
State University*

JOHN WILEY & SONS, INC.

New York • Chichester • Brisbane • Weinheim • Toronto • Singapore

SI VERSION

Acquisitions Editor	Joseph Hayton
Assistant Editor	Sharon Smith
Marketing Manager	Harper Mooy
Senior Production Editor	Nancy Prinz
Photo Editor	Hilary Newman
Illustration Editor	Sigmund Malinowski
Cover Art	Julian Baum/Science Photo Library
Electronic Illustrations	Precision Graphics
Outside Production Management	Ingrao Associates

This book was set in Century Schoolbook by Ruttle, Shaw & Wetherill, Inc. and printed and bound by Von Hoffmann Press, Inc. The cover was printed by The Lehigh Press, Inc.

Recognizing the importance of preserving what has been written, it is a policy of John Wiley & Sons, Inc. to have books of enduring value published in the United States printed on acid-free paper, and we exert our best efforts to that end.

The paper in this book was manufactured by a mill whose forest management programs include sustained yield harvesting of its timberlands. Sustained yield harvesting principles ensure that the number of trees cut each year does not exceed the amount of new growth.

Meriam, J. L. (James L.)
    Engineering mechanics / J. L. Meriam, L. G. Kraige.—4th ed.
      p.   cm.
    "SI version."
    Includes index.
    Contents: v. 2. Dynamics.
    ISBN 0-471-24167-9 (alk. paper)
    1. Mechanics, Applied.   I. Kraige, L. G. (L. Glenn) II. Title.

TA350.M458 1998
620.1—dc20
                                96-22505
                                  CIP

Printed in the United States of America

10 9 8 7 6 5 4

This book is dedicated to the

Engineering Profession

in recognition of the pivotal role played by
mechanics in the analysis and design of engineering systems

FOREWORD

The innovations and contributions of Dr. James L. Meriam to the field of engineering mechanics cannot be overstated. He has undoubtedly had as much influence on instruction in mechanics during the last forty-five years as any one individual. His first books on mechanics in 1951 literally reconstructed undergraduate mechanics and became the definitive textbooks for the decades that followed. His texts were logically organized, easy to read, directed to the average engineering undergraduate, and were packed with exciting examples of real-life engineering problems which were superbly illustrated. These books became the model for other engineering mechanics texts in the 1950s and beyond.

Dr. Meriam began his work in mechanical engineering at Yale University where he earned his B.E., M. Eng., and Ph.D. degrees. He had early industrial experience with Pratt and Whitney Aircraft and the General Electric Company, which stimulated his first contributions to mechanics in mathematical and experimental stress analysis. During the Second World War he served in the U.S. Coast Guard.

Dr. Meriam was a member of the faculty of the University of California, Berkeley, for twenty-one years where he served as Professor of Engineering Mechanics, Assistant Dean of Graduate Studies, and Chairman of the Division of Mechanics and Design. In 1963 he became Dean of Engineering at Duke University where he devoted his full energies to the development of its School of Engineering. In 1972 Professor Meriam followed his desire to return to full-time teaching and served as Professor of Mechanical Engineering at California Polytechnic State University, where he retired in 1980. During the following ten years he served as visiting professor at the University of California, Santa Barbara, retiring for the second time in 1990. Professor Meriam has always placed great emphasis on teaching, and this trait has been recognized by his students wherever he has taught. At Berkeley in 1963 he was the first recipient of the Outstanding Faculty Award of Tau Beta Pi, given primarily for excellence in teaching. In 1978 he received the Distinguished Educator Award for Outstanding Service to Engineering Mechanics Education from the American Society for Engineering Education, and in 1992 was the Society's recipient of the Benjamin Garver Lamme Award, which is their highest annual national award.

Professor Meriam was the first author to show clearly how the method of virtual work in statics can be employed to solve a class of problems largely neglected by previous authors. In dynamics, plane motion became understandable, and in his later editions, three-dimensional kinematics and kinetics received the same treatment. He is credited with original developments in the theory of variable-mass dynamics, which are contained in his *Dynamics, Second Edition* (1971). Professor Meriam has also been a leader in promoting the use of SI units, and his *SI Versions* of *Statics* and *Dynamics* (1975) were the first mechanics textbooks in SI units in this country.

Dr. L. Glenn Kraige, coauthor for the third time in the *Engineering Mechanics* series, has also made significant contributions to mechanics education. Dr. Kraige earned his B.S., M.S., and Ph.D. degrees at the University of Virginia, principally in aerospace engineering, and he currently serves as Professor of Engineering Science and Mechanics at Virginia Polytechnic Institute and State University.

In addition to his recognized research and publications in the field of spacecraft dynamics, Professor Kraige has devoted his attention to the teaching of mechanics at both introductory and advanced levels. His outstanding teaching has been widely recognized and has earned him teaching awards at the departmental, college, university, state, regional, and national levels. These include the Francis J. Maher Award for excellence in education in the Department of Engineering Science and Mechanics, an AT&T award for outstanding teaching in the Southeastern Section of the American Society for Engineering Education, and the Outstanding Educator Award from the State Council of Higher Education for the Commonwealth of Virginia. In 1996, the Mechanics Division of ASEE bestowed upon him the Archie Higdon Distinguished Educator Award. In his teaching, Professor Kraige stresses the development of analytical capabilities along with the strengthening of physical insight and engineering judgment. Since the early 1980s, he has worked on personal-computer software designed to enhance the teaching/learning process in statics and dynamics.

The Fourth Edition of *Engineering Mechanics* continues the same high standards set by previous editions and adds new features of help and interest to students. It contains a vast collection of interesting and instructive problems. Analysis and applications are the cornerstones of a successful learning experience in engineering mechanics, and J. L. Meriam and L. G. Kraige have shown again that they are the best at melding these essential characteristics.

Robert F. Steidel, Jr.
Professor Emeritus of Mechanical Engineering
University of California, Berkeley

PREFACE
To the Student

As you undertake the study of engineering mechanics, first statics and then dynamics, you will be building a foundation of analytical capability for the solution of a great variety of engineering problems. Modern engineering design demands a high level of analytical capability, and you will find that your study of mechanics will help you immensely in developing this capability.

In engineering mechanics we learn to construct and to solve mathematical models which describe the effects of force and motion on a variety of structures and machines that are of concern to engineers. In applying our principles of mechanics we formulate these models by incorporating appropriate physical assumptions and mathematical approximations. In both the formulation and solution of mechanics problems you will have frequent occasion to use your background in plane and solid geometry, scalar and vector algebra, trigonometry, analytic geometry, and calculus. Indeed, you are likely to discover new significance to these mathematical tools as you make them work for you in mechanics.

Your success in mechanics (and throughout engineering) will be highly contingent on developing a well-disciplined method of attack from hypothesis to conclusion in which the applicable principles are applied rigorously. Years of experience in teaching and engineering disclose the importance of developing the ability to represent one's work in a clear, logical, and concise manner. Mechanics is an excellent place in which to develop these habits of logical thinking and effective communication.

Engineering Mechanics contains a large number of sample problems in which the solutions are presented in detail. Also included in these examples are helpful observations that mention common errors and pitfalls to be avoided. In addition, the book contains a large selection of simple, introductory problems and problems of intermediate difficulty to help you gain initial confidence and understanding of each new topic. Also included are many problems that illustrate significant and contemporary engineering design situations to stimulate your interest and help you to develop an appreciation for the many applications of mechanics in engineering.

We are pleased to extend our encouragement to you as a student of mechanics. We hope this book will provide both help and stimulation as you develop your background in engineering.

J. L. Meriam

L. Glenn Kraige

Santa Barbara, California
January 1998

Blacksburg, Virginia
January 1998

PREFACE
To the Instructor

The primary purpose of the study of engineering mechanics is to develop the capacity to predict the effects of force and motion in the course of carrying out the creative design function of engineering. Successful prediction requires more than a mere knowledge of the physical and mathematical principles of mechanics. Prediction also requires the ability to visualize physical configurations in terms of real materials, actual constraints, and the practical limitations that govern the behavior of machines and structures. One of our primary objectives in teaching mechanics is to help the student develop this ability to visualize, which is so vital to problem formulation. Indeed, the construction of a meaningful mathematical model is often a more important experience than its solution. Maximum progress is made when the principles and their limitations are learned together within the context of engineering application.

Courses in mechanics are often regarded by students as a difficult requirement and frequently as an uninteresting academic hurdle as well. The difficulty stems from the extent to which reasoning from fundamentals, as distinguished from rote learning, is required. The lack of interest that is frequently experienced is due primarily to the extent to which mechanics is presented as an academic discipline often lacking in engineering purpose and challenge. This attitude is traceable to the frequent tendency in the presentation of mechanics to use problems mainly as a vehicle to illustrate theory rather than to develop theory for the purpose of solving problems. When the first view is allowed to predominate, problems tend to become overly idealized and unrelated to engineering with the result that the exercise becomes dull, academic, and uninteresting. This approach deprives the student of much of the valuable experience in formulating problems and thus of discovering the need for and meaning of theory. The second view provides by far the stronger motive for learning theory and leads to a better balance between theory and application. The crucial role of interest and purpose in providing the strongest possible motive for learning cannot be overemphasized. Furthermore, we should stress the view that, at best, theory can only approximate the real world of mechanics rather than the view that the real world approximates the theory. This difference in philosophy is indeed basic and distinguishes the *engineering* of mechanics from the *science* of mechanics.

During the past thirty years there has been a strong trend in engineering education to increase the extent and level of theory in the engineering-science courses. Nowhere has this trend been more evident than in mechanics courses. To the extent that students are prepared to handle the accelerated treatment, the trend is beneficial. There is evidence and justifiable concern, however, that a significant disparity has more recently appeared between coverage and comprehension. Among the contributing factors we note three trends. First, emphasis on the geometric and physical meanings of prerequisite mathematics appears to have diminished. Second, there has been a significant reduction and even elimination of instruction in graphics, which in the past enhanced the visualization and representation of mechanics problems. Third, in advancing the mathematical level of our treatment of mechanics, there has been a tendency to allow the notational manipulation of vector operations to mask or replace geometric visualization. Mechanics is inherently a subject that depends on geometric and physical perception, and we should increase our efforts to develop this ability.

One of our responsibilities as teachers of mechanics is to use the mathematics that is most appropriate for the problem at hand. The use of vector notation for one-dimensional problems is usually trivial; for two-dimensional problems it is often optional; but for three-dimensional problems it is usually essential. As we introduce vector operations in two-dimensional problems, it is especially important that their geometric meaning be emphasized. A vector equation is brought to life by a sketch of the corresponding vector polygon, which often discloses through its geometry the shortest solution. There are, of course, many mechanics problems where the complexity of variable independence is beyond the normal powers of visualization and physical perception, and reliance on analysis is essential. Nevertheless, our students become better engineers when their abilities to perceive, visualize, and represent are developed to the fullest.

As teachers of engineering mechanics, we have the strongest obligation to the engineering profession to set reasonable standards of performance and to uphold them. In addition, we have a serious responsibility to encourage our students to think for themselves. Too much help with details that students should be reasonably able to handle from prerequisite subjects can be as bad as too little help and can easily condition them to become overly dependent on others rather than to exercise their own initiative and ability. Also, when mechanics is subdivided into an excessive number of small compartments, each with detailed and repetitious instructions, students can have difficulty seeing the "forest for the trees" and, consequently, will fail to perceive the unity of mechanics and the far-reaching applicability of its few basic principles and methods.

This Fourth Edition of *Engineering Mechanics*, as with previous editions, is written with the foregoing philosophy in mind. It is intended primarily for the first engineering course in mechanics, generally taught in the second year of study. *Engineering Mechanics* is written in a style that is both concise and friendly. The major emphasis is on basic principles and methods rather than on a multitude of special cases. Strong effort has been made to show both the cohesiveness of the relatively few fundamental ideas and the great variety of problems that these few ideas will solve. A major feature of the book is the extensive treatment of sample problems, which are presented in a single-page format for convenient self-study. In addition to presenting the solution in detail, each sample problem also contains comments and cautions keyed to salient points in the solution and printed in blue type.

In order to bring the greatest possible degree of realism and clarity to the illustrations, the Fourth Edition is being produced in full color. It is important to note that color is used consistently for the identification of like quantities—red for forces and moments, green for velocity and acceleration arrows, and orange dashes for selected trajectories of moving points. Subdued colors are used for those parts of an illustration which are not central to the problem at hand. Whenever possible, mechanisms or objects which commonly have a certain color will be portrayed in that color. All of the fundamental elements of technical illustration which have been an essential part of this *Engineering Mechanics* series of textbooks have been retained. The authors wish to restate the conviction that a high standard of illustration is critical to any written work in the field of mechanics.

Volume 2, Dynamics, contains 114 sample problems and 1482 unsolved problems from which a wide choice of assignments can be made. Of these problems, approximately 50 percent are new to the Fourth Edition. SI units are used throughout the book.

A notable feature of the Fourth Edition, as with all previous editions, is the wealth of interesting and important problems which apply to engineering design. Whether directly stated or not, virtually all of the problems deal with principles and procedures inherent in the design and analysis of engineering structures and mechanical systems.

In the Fourth Edition the problem sets are divided into two sections entitled *Introductory Problems* and *Representative Problems*. In the first section are simple, uncomplicated problems designed to help students gain confidence with the new topic, while most of the problems in the second section are of average difficulty and length. The problems are arranged generally in order of increasing difficulty; near the end of the *Representative Problems* are more difficult exercises, which are marked with the symbol ▶. Computer-oriented problems are in a special section at the conclusion of the Review Problems at the end of each chapter. The answers to all odd-numbered problems and to all difficult problems have been provided. Simple numerical values have been used throughout so as not to complicate the solutions and divert attention from the principles. All numerical solutions have been carried out and checked with an electronic calculator without rounding intermediate values. Consequently, the final answers should be correct to within the number of significant figures cited.

A special note on the use of computers is in order. We wish to emphasize that the experience of formulating problems, where reason and judgment are developed, is vastly more important for the student than is the manipulative exercise in carrying out the solution. For this reason, we believe that computer usage must be carefully controlled. At the present time, the processes of constructing free-body diagrams and formulating governing equations are best done with pencil and paper. On the other hand, there are instances in which the *solution* to the governing equations can best be carried out and displayed via the computer. Computer-oriented problems should be genuine in the sense that there is a condition of design or criticality to be found, rather than "makework" problems in which some parameter is varied for no apparent reason other than to force artificial use of the computer. These thoughts have been in mind as we have designed the computer-oriented problems in the Fourth Edition. To conserve adequate time for problem formulation, it is suggested that the student be assigned only a limited number of the computer problems.

The logical division between particle dynamics and rigid-body dynamics has been preserved, with each part treating the kinematics prior to the kinetics. This arrangement greatly facilitates a more thorough and rapid excursion in rigid-body dynamics with the prior benefit of a comprehensive introduction to particle dynamics in Chapters 2 and 3.

Chapter 3 on particle kinetics focuses on the three basic methods, force-mass-acceleration, work-energy, and impulse-momentum. The special topics of impact, central-force motion, and relative motion are grouped together in Section D of Chapter 3 on special applications and serve as optional material to be assigned according to instructor preference and available time. With this arrangement, the attention of the student is focused more strongly on the three basic approaches to kinetics, which are developed in Sections A, B, and C of the chapter.

Chapter 4 on systems of particles is an extension of the principles of motion for a single particle and develops the general relationships that are so basic to the modern comprehension of dynamics. The chapter also includes the topics of steady mass flow and variable mass, which may be considered as optional material depending on the time available.

In Chapter 5 on the kinematics of rigid bodies in plane motion where the equations of relative velocity and relative acceleration are encountered, emphasis is placed jointly on solution by vector geometry and solution by vector algebra. This dual approach serves the purpose of reinforcing the meaning of vector mathematics.

In Chapter 6 on the kinetics of rigid bodies, we place great emphasis on the basic equations which govern all categories of plane motion. Strong dependence is placed on the identification of knowns and unknowns and on the necessary and sufficient motion equations that guarantee solution. Special emphasis is also placed on forming the direct equivalence between the actual applied forces and couples and their $m\bar{a}$ and $\bar{I}\alpha$ resultants. In this way the versatility of the moment principle is emphasized, and the student is encouraged to think directly in terms of resultant dynamics effects.

Chapter 7, which may be treated as optional, provides a basic introduction to three-dimensional dynamics which is sufficient to solve many of the more common space-motion problems. For students who later pursue more advanced work in dynamics, Chapter 7 will provide a solid foundation. Gyroscopic motion with steady precession is treated in two ways. The first approach makes use of the analogy between the relation of force and linear-momentum vectors and the relation of moment and angular-momentum vectors. With this treatment the student can understand the gyroscopic phenomenon of steady precession and can handle most of the engineering problems on gyros without a detailed study of three-dimensional dynamics. The second approach makes use of the more general momentum equations for three-dimensional rotation where all components of momentum are accounted for.

Chapter 8 is devoted to the topic of vibrations. This full chapter coverage will be especially useful for engineering students whose only exposure to vibrations is acquired in the basic dynamics course.

Moments and products of inertia of mass are presented in Appendix B. Appendix C contains a summary review of selected topics of elementary mathematics as well as several numerical techniques that the student should be prepared to use in computer-solved problems. Useful tables of physical constants, centroids, and moments of inertia are contained in Appendix D.

We wish to specially cite the outstanding contributions to this series of mechanics books over a period of twenty-five years by illustrator John Balbalis, who died in October 1991. His dedication to high standards of illustrative achievement greatly increased the educational potential of the *Engineering Mechanics* series by providing clarity, reality, and interest to the many thousands of students who have been challenged by his efforts. Building upon and adding to the work of Mr. Balbalis, the team of illustrators at Precision Graphics is recognized for their work in bringing full color to this series.

Special recognition is again due Dr. A. L. Hale, formerly of the Bell Telephone Laboratories, for his continuing contribution in the form of invaluable suggestions and accurate checking of the manuscript. Dr. Hale has rendered similar service to all previous versions in this entire series of mechanics books, and his input has been a great asset. A number of members of the Department of Engineering Science and Mechanics at Virginia Polytechnic Institute and State University have offered helpful suggestions, including Professors Norman E. Dowling, J. Wallace Grant, D. P. H. Hasselman, Scott L. Hendricks, Robert M. Jones, Don H. Morris, and Saad A. Ragab. The contribution by the staff of John Wiley & Sons, Inc., including Editors Charity Robey, Senior Production Editor Nancy Prinz, and Illustration Editor Sigmund Malinowski, reflects a high degree of professional competence and is duly recognized. We wish to specially cite the valuable coordinating work of Editor Cliff Robichaud, who died in 1997. The authors also wish to acknowledge Boris Starosta for his suggestions on the illustrations and for coordinating artwork activities with Precision Graphics. Finally, we acknowledge the patience, support, and assistance of our wives, Julia and Dale, during the many hours required to prepare this manuscript.

J. L. Meriam

Santa Barbara, California
January 1998

L. Glenn Kraige

Blacksburg, Virginia
January 1998

CONTENTS

ENGINEERING MECHANICS

DYNAMICS

PART I

DYNAMICS OF PARTICLES

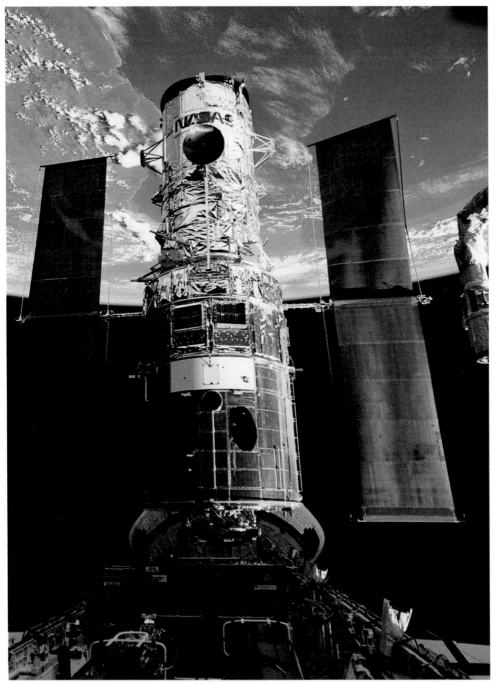

The Hubble Space Telescope is shown in the cargo bay of the space shuttle *Endeavour* after its recapture in space in 1993. This spectacular engineering achievement relied heavily upon the use of the principles of dynamics as applied to particle and rigid-body motion. Dynamics also played a key role in the subsequent successful relaunch of the telescope, which is continuing to expand our knowledge of deep space.

INTRODUCTION TO DYNAMICS

1

1/1 HISTORY AND MODERN APPLICATIONS

Dynamics is that branch of mechanics which deals with the motion of bodies under the action of forces. The study of dynamics in engineering usually follows the study of statics, which deals with the action of forces on bodies at rest. Dynamics has two distinct parts: *kinematics*, which is the study of motion without reference to the forces which cause motion, and *kinetics*, which relates the action of forces on bodies to their resulting motions. The student of engineering will find that a thorough comprehension of dynamics will provide one of his or her most useful and powerful tools for analysis in engineering.

Historically, dynamics is a relatively recent subject compared with statics. The beginning of a rational understanding of dynamics is credited to Galileo (1564–1642), who made careful observations concerning bodies in free fall, motion on an inclined plane, and motion of the pendulum. He was largely responsible for bringing a scientific approach to the investigation of physical problems. Galileo was continually under severe criticism for refusing to accept the established beliefs of his day, such as the philosophies of Aristotle which held, for example, that heavy bodies fall more rapidly than light bodies. The lack of accurate means for the measurement of time was a severe handicap to Galileo, and further significant development in dynamics awaited the invention of the pendulum clock by Huygens in 1657. Newton (1642–1727), guided by Galileo's work, was able to make an accurate formulation of the laws of motion and, hence, to place dynamics on a sound basis. Newton's famous work was published in the first edition of his *Principia*,* which is generally recognized as one of the greatest of all recorded contributions to knowledge. In addition to stating the laws governing the motion of a particle, Newton was the first to correctly for-

---

*The original formulations of Sir Isaac Newton may be found in the translation of his *Principia* (1687), revised by F. Cajori, University of California Press, 1934.

mulate the law of universal gravitation. Although his mathematical description was accurate, he felt that the concept of remote transmission of gravitational force without a supporting medium was an absurd notion. Following Newton's time, important contributions to mechanics were made by Euler, D'Alembert, Lagrange, Laplace, Poinsot, Coriolis, Einstein, and others.

In terms of engineering application, dynamics is an even more recent science. Only since machines and structures have operated with high speeds and appreciable accelerations has it been necessary to make calculations based on the principles of dynamics rather than on the principles of statics. The rapid technological developments of the present day require increasing application of the principles of mechanics, particularly dynamics. These principles are basic to the analysis and design of moving structures, to fixed structures subject to shock loads, to robotic devices, to automatic control systems, to rockets, missiles, and spacecraft, to ground and air transportation vehicles, to electron ballistics of electrical devices, and to machinery of all types such as turbines, pumps, reciprocating engines, hoists, machine tools, etc. Students whose interests lead them into one or more of these and many other activities will find a constant need for applying the fundamentals of dynamics.

1/2 BASIC CONCEPTS

The concepts basic to mechanics were set forth in Art. 1/2 of *Vol. 1 Statics*. They are summarized here along with additional comments of special relevance to the study of dynamics.

Space is the geometric region occupied by bodies. Position in space is determined relative to some geometric reference system by means of linear and angular measurements. The basic frame of reference for the laws of Newtonian mechanics is the *primary inertial system* or *astronomical frame of reference*, which is an imaginary set of rectangular axes assumed to have no translation or rotation in space. Measurements show that the laws of Newtonian mechanics are valid for this reference system as long as any velocities involved are negligible compared with the speed of light, which is 300 000 km/s or 186,000 mi/sec. Measurements made with respect to this reference are said to be *absolute*, and this reference system may be considered "fixed" in space. A reference frame attached to the surface of the earth has a somewhat complicated motion in the primary system, and a correction to the basic equations of mechanics must be applied for measurements made relative to the earth's reference frame. In the calculation of rocket and spaceflight trajectories, for example, the absolute motion of the earth becomes an important parameter. For most engineering problems of machines and structures which remain on the earth's surface, the corrections are extremely small and may be neglected. For these problems the laws of mechanics may be applied directly for measurements made relative to the earth, and in a practical sense such measurements will be referred to as *absolute*.

Time is a measure of the succession of events and is considered an absolute quantity in Newtonian mechanics.

Mass is the quantitative measure of the inertia or resistance to change in motion of a body. Mass can also be considered as the quantity of matter in a body as well as the property that gives rise to gravitational attraction.

Force is the vector action of one body on another. The properties of forces have been thoroughly treated in *Vol. 1 Statics*.

A **particle** is a body of negligible dimensions. Also, when the dimensions of a body are irrelevant to the description of its motion or the action of forces on it, the body may be treated as a particle. An airplane, for example, may be treated as a particle for the description of its flight path.

A **rigid body** is a body whose changes in shape are negligible compared with the overall dimensions of the body or with the changes in position of the body as a whole. As an example of the assumption of rigidity, the small flexural movement of the wing tip of an airplane flying through turbulent air is clearly of no consequence to the description of the motion of the airplane as a whole along its flight path. For this purpose, then, the treatment of the airplane as a rigid body offers no complication. On the other hand, if the problem is one of examining the internal stresses in the wing structure due to changing dynamic loads, then the deformation characteristics of the structure would have to be examined, and for this purpose the airplane could no longer be considered a rigid body.

Vector and **scalar** quantities have been treated extensively in *Vol. 1 Statics*, and their distinction should be perfectly clear by now. Scalar quantities are printed in lightface italic type, and vectors are shown in boldface type. Thus, V is the scalar magnitude of the vector **V**. It is important that we use an identifying mark, such as an underline $\underline{V}$, for all handwritten vectors to take the place of the boldface designation in print. For two nonparallel vectors recall, for example, that $\mathbf{V}_1 + \mathbf{V}_2$ and $V_1 + V_2$ have two entirely different meanings.

It is assumed that the reader is familiar with the geometry and algebra of vectors through previous study of statics and mathematics. Students who need to review these topics will find a brief summary of them in Appendix C along with other mathematical relations that find frequent use in mechanics. Experience has shown that the geometry of mechanics is often a source of difficulty. Mechanics by its very nature is geometrical, and students should bear this in mind as they review their mathematics. In addition to vector algebra, dynamics requires the use of vector calculus, and the essentials of this topic will be developed in the text as they are needed.

Dynamics involves the frequent use of time derivatives of both vectors and scalars. As a notational shorthand, a dot over a quantity will frequently be used to indicate a derivative with respect to time. Thus, $\dot{x}$ means dx/dt and $\ddot{x}$ stands for d^2x/dt^2.

1/3 NEWTON'S LAWS

Newton's three laws of motion, stated in Art. 1/4 of *Vol. 1 Statics*, are restated here because of their special significance to dynamics. In modern terminology they are

Law I. A particle remains at rest or continues to move in a straight line with a constant velocity if there is no unbalanced force acting on it.

Law II. The acceleration of a particle is proportional to the resultant force acting on it and is in the direction of this force.*

Law III. The forces of action and reaction between interacting bodies are equal in magnitude, opposite in direction, and collinear.

These laws have been verified by countless physical measurements. The first two laws hold for measurements made in an absolute frame of reference, but are subject to some correction when the motion is measured relative to a reference system having acceleration, such as one attached to the earth's surface.

Newton's second law forms the basis for most of the analysis in dynamics. For a particle of mass m subjected to a resultant force $\mathbf{F}$, the law may be stated as

$$\boxed{\mathbf{F} = m\mathbf{a}} \tag{1/1}$$

where $\mathbf{a}$ is the resulting acceleration measured in a nonaccelerating frame of reference. Newton's first law is a consequence of the second law since there is no acceleration when the force is zero, and the particle is either at rest or is moving with constant velocity. The third law constitutes the principle of action and reaction with which we should be thoroughly familiar from our work in statics.

1/4 UNITS

The International System of metric units (SI) is defined and used in *Vol. 2 Dynamics*. To become familiar with this system, it is

---

*To some it is preferable to interpret Newton's second law as meaning that the resultant force acting on a particle is proportional to the time rate of change of momentum of the particle and that this change is in the direction of the force. Both formulations are equally correct when applied to a particle of constant mass.

necessary to think directly in SI units. Although in engineering practice numerical conversion from the U.S. customary or British system to SI must be accomplished for some years, it would be a mistake to assume that familiarity with SI units can be achieved simply by the conversion of numerical results from the old system. Tables defining the SI units and giving numerical conversions between U.S. customary and SI units are included inside the front cover of the book. Charts comparing selected quantities in SI and U.S. customary units are included inside the back cover of the book to facilitate conversion and to help establish a feel for the relative size of units in both systems.

The four fundamental quantities of mechanics and their units and symbols for the two systems are summarized in the following table:

QUANTITY	DIMENSIONAL SYMBOL	SI UNITS		U.S. CUSTOMARY UNITS	
		UNIT	SYMBOL	UNIT	SYMBOL
Mass	M	Base units { kilogram	kg	Base units { slug	—
Length	L	meter*	m	foot	ft
Time	T	second	s	second	sec
Force	F	newton	N	pound	lb

*Also spelled *metre*.

In SI the units for mass, length, and time are taken as base units and the units for force are derived from Newton's second law of motion, Eq. 1/1. In the U.S. customary system the units for force, length, and time are base units and the units for mass are derived from the second law. The SI system is termed an *absolute* system since mass is taken to be an absolute or base quantity. The U.S. customary system is termed a *gravitational* system since force (as measured from gravitational pull) is taken as a base quantity. This distinction is a fundamental difference between the two systems.

In SI units by definition one newton is that force which will give a one-kilogram mass an acceleration of one meter per second squared. In the U.S. customary system a 32.1740-pound mass (1 slug) will have an acceleration of one foot per second squared when acted on by a force of one pound. Thus, for each system we have from Eq. 1/1

SI UNITS	U.S. CUSTOMARY UNITS
$(1\ N) = (1\ kg)(1\ m/s^2)$	$(1\ lb) = (1\ slug)(1\ ft/sec^2)$
$N = kg \cdot m/s^2$	$slug = lb\text{-}sec^2/ft$

In SI units each unit and symbol stands for only one quantity. The kilogram is to be used *only* as a unit of mass and *never* as a unit of force. On the other hand, in U.S. customary units the *pound* is used both as a unit of force (lbf) and a unit of mass (lbm). The distinction between the meaning as force and the meaning as mass is usually clear from the situation at hand. The symbol for pound force is most frequently written merely lb.

Additional quantities used in mechanics and their equivalent base units will be defined as they are introduced in the chapters which follow. However, for convenient reference these quantities along with their SI base units are listed in one place in the first table inside the front cover of the book.

Detailed guidelines have been established for the consistent use of SI units, and these guidelines have been followed throughout this book. The most essential ones are summarized inside the front cover, and the student should observe these rules carefully.

1/5   GRAVITATION

Newton's law of gravitation that governs the mutual attraction between bodies is

$$F = G \frac{m_1 m_2}{r^2} \qquad\qquad (1/2)$$

where

F = the mutual force of attraction between two particles
G = a universal constant called the constant of gravitation
m_1, m_2 = the masses of the two particles
r = the distance between the centers of the particles

The value of the gravitational constant from experimental data is $G = 6.673(10^{-11})$ m^3/(kg·s^2). The only gravitational force of appreciable magnitude is the force due to the attraction of the earth. It was shown in *Vol. 1 Statics*, for example, that each of two iron spheres 100 mm in diameter is attracted to the earth with a gravitational force of 37.1 N, which is called its *weight*, but the force of mutual attraction between them if they are just touching is only 0.000 000 095 1 N.

Since the gravitational attraction or weight of a body is a force, it should always be expressed in force units, newtons (N) in SI units and pounds force (lb) in U.S. customary units. Unfortunately, the mass unit kilogram (kg) has been used widely in common practice as a measure of weight. When expressed in kilograms, the word "weight" technically means mass. To avoid confusion, the word "weight" in this book shall be restricted to mean the force of gravitational attraction, and it will always be expressed in newtons or pounds force.

The force of gravitational attraction of the earth on a body depends on the position of the body relative to the earth. If the earth were a perfect sphere of the same mass and volume, a body with a mass of exactly 1 kg would be attracted to the earth by a force of 9.825 N on the earth's surface, 9.822 N at an altitude of 1 km, 9.523 N at an altitude of 100 km, 7.340 N at an altitude of 1000 km, and 2.456 N at an altitude equal to the mean radius 6371 km of the earth. It is at once apparent that the variation in gravitational attraction of high-altitude rockets and spacecraft becomes a major consideration.

Every object which is allowed to fall in a vacuum at a given position near the earth's surface will have the same acceleration g, as can be seen by combining Eqs. 1/1 and 1/2 and canceling the term representing the mass of the falling object. This combination gives

$$g = \frac{Gm_e}{R^2}$$

where m_e is the mass of the earth and R is the radius of the earth.* The mass m_e and the mean radius R of the earth have been found through experimental measurements to be $5.976(10^{24})$ kg and $6.371(10^6)$ m, respectively. These values, together with the value of G already cited, when substituted into the expression for g, give a mean value of $g = 9.825$ m/s^2.

The acceleration due to gravity as determined from the gravitational law is the acceleration which would be measured from a set of axes whose origin is at the center of the earth but which does not rotate with the earth. With respect to these "fixed" axes, then, this value may be termed the absolute value of g. Because the earth rotates, the acceleration of a freely falling body as measured from a position attached to the earth's surface is slightly less than the absolute value. Accurate values of the gravitational acceleration as measured relative to the earth's surface account for the fact that the earth is a rotating oblate spheroid with flattening at the poles. These values may be calculated to a high degree of accuracy from the 1980 International Gravity Formula, which is

$$g = 9.780\ 327(1 + 0.005\ 279 \sin^2 \gamma + 0.000\ 023 \sin^4 \gamma + \cdots)$$

where γ is the latitude and g is expressed in meters per second squared. The formula is based on an ellipsoidal model of the earth and also accounts for the effect of the earth's rotation. The absolute acceleration due to gravity as determined for a nonrotating earth may be computed from the relative values to a close approximation

---

*It can be proved that the earth, when taken as a sphere with a symmetrical distribution of mass about its center, may be considered a particle with its entire mass concentrated at its center.

by adding $3.382(10^{-2}) \cos^2 \gamma$ m/s^2, which removes the effect of the earth's rotation. The variation of both the absolute and the relative values of g with latitude is shown in Fig. 1/1 for sea-level conditions.*

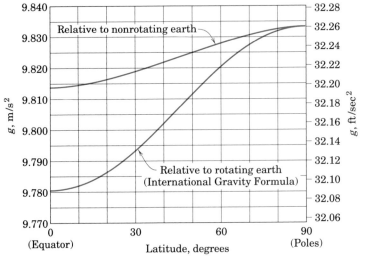

Figure 1/1

The standard value which has been adopted internationally for the gravitational acceleration relative to the rotating earth at sea level and at a latitude of 45° is 9.806 65 m/s^2 or 32.1740 ft/sec^2. This value differs very slightly from that obtained by evaluating the International Gravity Formula for $\gamma = 45°$. The reason for the small difference is that the earth is not exactly ellipsoidal, as assumed in the formulation of the International Gravity Formula.

The proximity of large land masses and the variations in the density of the earth's crust also influence the local value of g by a small but detectable amount. In almost all engineering problems where measurements are made on the surface of the earth, the difference between the absolute and relative values of the gravitational acceleration and the effect of local variations are neglected, and 9.81 m/s^2 in SI units and 32.2 ft/sec^2 in U.S. customary units are used for the sea-level value of g.

The variation of g with altitude is easily determined by the gravitational law. If g_0 represents the absolute acceleration due to gravity at sea level, the absolute value at an altitude h is

$$g = g_0 \frac{R^2}{(R + h)^2}$$

---

*Students will be able to derive these relations for a spherical earth following their study of relative motion in Chapter 3.

where R is the radius of the earth.

The earth's gravitational attraction on a body may be calculated from the results of the simple gravitational experiment. If the gravitational force of attraction or true weight of a body is **W**, then, since the body will fall with an absolute acceleration **g** in a vacuum, Eq. 1/1 gives

$$\boxed{\mathbf{W} = m\mathbf{g}} \tag{1/3}$$

The apparent weight of a body as determined by a spring balance, calibrated to read the correct force and attached to the surface of the earth, will be slightly less than its true weight. The difference is due to the rotation of the earth. The ratio of the apparent weight to the apparent or relative acceleration due to gravity still gives the correct value of mass. The apparent weight and the relative acceleration due to gravity are, of course, the quantities that are measured in experiments conducted on the surface of the earth.

1/6   DIMENSIONS

A given dimension such as length can be expressed in a number of different units such as meters, millimeters, or kilometers. Thus, the word *dimension* is distinguished from the word *unit*. Physical relations must always be dimensionally homogeneous, that is, the dimensions of all terms in an equation must be the same. It is customary to use the symbols L, M, T, and F to stand for length, mass, time, and force, respectively. In SI units force is a derived quantity and from Eq. 1/1 has the dimensions of mass times acceleration or

$$F = ML/T^2$$

One important use of the theory of dimensions is found in checking the dimensional correctness of some derived physical relation. The following expression for the velocity v of a body of mass m which is moved from rest a horizontal distance x by a force F may be derived:

$$Fx = \tfrac{1}{2}mv^2$$

where the $\tfrac{1}{2}$ is a dimensionless coefficient resulting from integration. This equation is dimensionally correct since substitution of L, M, and T gives

$$[MLT^{-2}][L] = [M][LT^{-1}]^2$$

Dimensional homogeneity is a necessary condition for correctness, but it is not sufficient since the correctness of dimensionless coefficients cannot be checked in this way.

1/7   *FORMULATION AND SOLUTION OF DYNAMICS PROBLEMS*

The study of dynamics is directed toward the understanding and description of the various quantities involved in the motions of bodies. This description, which is largely mathematical, enables predictions of dynamical behavior to be made. A dual thought process is necessary in formulating this description. It is necessary to think both in terms of the physical situation and in terms of the corresponding mathematical description. Analysis of every problem will require this repeated transition of thought between the physical and the mathematical. Without question, one of the greatest difficulties encountered by students is the inability to make this transition of thought freely. They should recognize that the mathematical formulation of a physical problem represents an ideal and limiting description, or model, that approximates but never quite matches the actual physical situation.

In the course of constructing the idealized mathematical model for any given engineering problem, certain approximations will always be involved. Some of these approximations may be mathematical, whereas others will be physical. For instance, it is often necessary to neglect small distances, angles, or forces compared with large distances, angles, or forces. If the change in velocity of a body with time is nearly uniform, then an assumption of constant acceleration may be justified. An interval of motion that cannot be easily described in its entirety is often divided into small increments, each of which can be approximated. The retarding effect of bearing friction on the motion acquired by a machine as the result of applied forces or moments may often be neglected if the friction forces are small. However, these same friction forces cannot be neglected if the purpose of the inquiry is a determination of the drop in efficiency of the machine due to the friction process. Thus, the degree of assumption involved depends on what information is desired and on the accuracy required. The student should be constantly alert to the various assumptions called for in the formulation of real problems. The ability to understand and make use of the appropriate assumptions in the course of the formulation and solution of engineering problems is certainly one of the most important characteristics of a successful engineer. Along with the development of the principles and analytical tools needed for modern dynamics, one of the major aims of this book is to provide a maximum of opportunity to develop ability in formulating good mathematical models. Strong emphasis is placed on a wide range of practical problems that not only require the full exercise of theory but also force consideration of the decisions that must be made concerning relevant assumptions.

An effective method of attack on dynamics problems, as in all engineering problems, is essential. The development of good habits in formulating problems and in representing their solutions will prove to be an invaluable asset. Each solution should proceed with a logical sequence of steps from hypothesis to conclusion, and its representation should include a clear statement of the following parts, each clearly identified:

1. Given data
2. Results desired
3. Necessary diagrams
4. Calculations
5. Answers and conclusions

In addition, it is well to incorporate a series of checks on the calculations at intermediate points in the solution. The reasonableness of numerical magnitudes should be observed, and the accuracy and dimensional homogeneity of terms should be checked frequently. It is also important that the arrangement of work be neat and orderly. Careless solutions that cannot be read easily by others are of little or no value. It will be found that the discipline involved in adherence to good form will in itself be an invaluable aid to the development of the abilities for formulation and analysis. Many problems which at first may seem difficult and complicated become clear and straightforward once they are begun with a logical and disciplined method of attack.

The subject of dynamics is based on a surprisingly few fundamental concepts and principles which, however, are extended and applied over an exceedingly wide range of conditions. One of the most valuable aspects of the study of dynamics is the experience afforded in reasoning from fundamentals. This experience cannot be obtained merely by memorizing the kinematical and dynamical equations which describe various motions. It must be obtained through exposure to a wide variety of problem situations which force the choice, use, and extension of basic principles to meet the given conditions.

In describing the relations between forces and the motions they produce, it is essential that the system to which a principle is applied be clearly defined. At times a single particle or a rigid body is the system to be isolated, whereas at other times two or more bodies taken together constitute the system. The definition of the system to be analyzed is made clear by constructing its *free-body diagram*. This diagram consists of a closed outline of the external boundary of the system defined. All bodies which contact and exert forces on the system but are not a part of it are removed and replaced by vectors representing the forces they exert *on* the system isolated.

In this way, we make a clear distinction between the action and reaction of each force, and account is taken of *all* forces on and external to the system. It is assumed that all students are familiar with the technique of drawing free-body diagrams from their prior work in statics.

In applying the laws of dynamics, numerical values of the quantities may be used directly in proceeding toward the solution, or algebraic symbols may be used to represent the quantities involved and the answer left as a formula. With numerical substitution the magnitudes of all quantities expressed in their particular units are evident at each stage of the calculation. This approach offers advantage when the practical significance of the magnitude of each term is important. The symbolic solution, however, has several advantages over the numerical solution. First, the abbreviation achieved by the use of symbols aids in focusing attention on the interconnection between the physical situation and its related mathematical description. Second, a symbolic solution permits a dimensional check to be made at every step, whereas dimensional homogeneity may not be checked when only numerical values are used. Third, a symbolic solution may be used repeatedly for obtaining answers to the same problem when different sets and sizes of units are used. Facility with both forms of solution is essential, and ample practice with each should be sought in the problem work.

Students will find that solutions to the various equations of dynamics may be obtained in one of three ways. First, a direct mathematical solution by hand calculation may be carried out, with answers appearing either as algebraic symbols or as numerical results. The large majority of the problems come under this category. Second, certain problems may be readily handled by graphical solutions, such as with the determination of velocities and accelerations in two-dimensional relative motion of rigid bodies. Third, there are a number of problems in *Vol. 2 Dynamics* which are designated as *computer-oriented problems*. They appear at the end of the Review Problem sets and are selected to illustrate the type of problem for which solution by computer offers a distinct advantage. The choice of the most expedient method of solution is an important aspect of the experience to be gained from the problem work. We emphasize, however, that the most important experience in learning mechanics lies in the formulation of problems, as distinct from their solution per se.

PROBLEMS

(Refer to Table D/2 in Appendix D for relevant solar-system values.)

1/1 The mass of one dozen apples is 2 kg. Determine the average weight of one apple in both SI and U.S. units. In the present case, how applicable is the "rule of thumb" that an average apple weighs 1 N?

Ans. $W = 1.635$ N, $W = 0.368$ lb

1/2 If a man's mass is 80 kg, determine his weight in newtons and calculate his corresponding mass in slugs.

1/3 The three 100-mm-diameter spheres constructed of different metals are located at the vertices of an equilateral triangle in deep space. Determine the resultant **R** of the gravitational forces which the aluminum and cast-iron spheres exert on the copper sphere.

Ans. $\mathbf{R} = (3.68\mathbf{i} + 13.98\mathbf{j})10^{-10}$ N

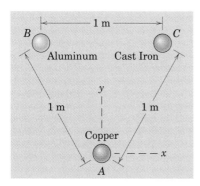

Problem 1/3

1/4 For the given vectors $\mathbf{V}_1$ and $\mathbf{V}_2$, determine $\mathbf{V}_1 + \mathbf{V}_2$, $\mathbf{V}_1 - \mathbf{V}_2$, $\mathbf{V}_1 \times \mathbf{V}_2$, $\mathbf{V}_2 \times \mathbf{V}_1$, and $\mathbf{V}_1 \cdot \mathbf{V}_2$. Consider the vectors to be nondimensional.

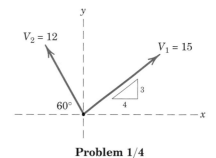

Problem 1/4

1/5 A space shuttle is in a circular orbit at an altitude of 250 km. Calculate the absolute value of g at this altitude and determine the corresponding true weight (gravitational force) of a shuttle passenger whose measured weight is 880 N when standing on the surface of the earth at a latitude of 45°. Are the terms "zero-g" and "weightless," which are sometimes used to describe conditions aboard orbiting spacecraft, correct in the absolute sense?

Ans. $g_h = 9.10$ m/s^2, $W_h = 816$ N

1/6 At what distance h above the earth should a spacecraft orbit the earth so that its gravitational attraction to the earth is equal to its true weight as measured on the surface of Mars?

1/7 Determine the absolute weight and the weight relative to the rotating earth of a 90-kg man if he is standing on the surface of the earth at a latitude of 40°.

Ans. $W_{abs} = 883.9$ N, $W_{rel} = 882.2$ N

1/8 Calculate the distance d from the center of the sun at which a particle experiences equal attractions from the earth and the sun. The particle is restricted to the line that joins the centers of the earth and the sun. Justify the two solutions physically.

Problem 1/8

1/9 Determine the ratio R_A of the force exerted by the sun on the moon to that exerted by the earth on the moon for position A of the moon. Repeat for moon position B.

Ans. $R_A = 2.19, R_B = 2.21$

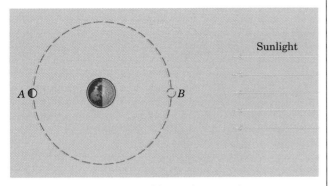

Problem 1/9

1/10 The drag coefficient C_D of an automobile is determined from the expression

$$C_D = \frac{D}{\frac{1}{2}\rho v^2 S}$$

where D is the drag force experimentally determined in a wind tunnel, ρ is the air density, v is the speed of the air in the wind tunnel, and S is the cross-sectional area of the car presented to the air flow. Determine the dimensions of C_D.

1/11 A mountain climber has a mass of 80 kg. Determine his loss of absolute weight in going from the foot of Mount Everest at an altitude of 2440 meters to its top at an altitude of 8848 m. Mount Everest has a latitude of 28° N, and the mean radius of the earth is 6371 km. Consult Fig. 1/1 as needed.

Ans. $\Delta W = 1.576$ N

The inside loop executed by a team of precision fliers is a graphic example of particle motion. Each airplane is treated as a particle whose dimensions are negligible compared with the dimensions of its path. Analysis of the kinematics of particle motion (relationships among displacement, velocity, acceleration, and time) is involved in virtually all dynamics problems in engineering.

KINEMATICS OF PARTICLES

2

2/1 INTRODUCTION

Kinematics is that branch of dynamics which describes the motion of bodies without reference to the forces which either cause the motion or are generated as a result of the motion. Kinematics is often referred to as the "geometry of motion." The design of cams, gears, linkages, and other machine elements to control or produce certain desired motions and the calculation of flight trajectories for aircraft, rockets, and spacecraft are a few examples of kinematic problems which engage the attention of engineers. A thorough working knowledge of kinematics is an absolute prerequisite to kinetics, which is the study of the relationships between motion and the corresponding forces that cause or accompany the motion.

We start our study of kinematics by first discussing in this chapter the motions of points or particles. A particle is a body whose physical dimensions are so small compared with the radius of curvature of its path that we can treat the motion of the particle as that of a point. For example, the wingspan of a jet transport flying between Los Angeles and New York is of no consequence compared with the radius of curvature of its flight path, and the treatment of the airplane as a particle or point should raise no question.

There are a number of ways in which the motion of a particle can be described, and the choice of the most convenient or appropriate method depends a great deal on experience and on how the data are given. Let us get an overview of the several methods developed in this chapter by referring to Fig. 2/1, which shows a particle P moving along some general path in space. If the particle is confined to a specified path, as with a bead sliding along a fixed wire, its motion is said to be *constrained*. If there are no physical guides, the motion is said to be *unconstrained*. A small rock tied to the end of a string and whirled in a circle undergoes constrained motion until the string breaks, after which instant its motion is unconstrained.

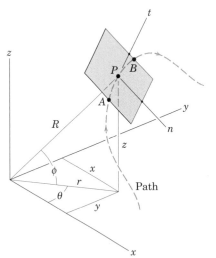

Figure 2/1

The position of particle P at any time t can be described by specifying its rectangular coordinates* x, y, z, its cylindrical coordinates r, θ, z, or its spherical coordinates R, θ, ϕ. The motion of P may also be described by measurements along the tangent t and normal n to the curve. The direction of n lies in the local plane of the curve.† These last measurements are known as *path variables*.

The motion of particles (or rigid bodies) may be described by using coordinates measured from fixed reference axes (*absolute-motion* analysis) or by using coordinates measured from moving reference axes (*relative-motion* analysis). Both descriptions will be developed and applied in the articles which follow.

With this conceptual picture of the description of particle motion in mind, we will now restrict our attention in the first part of this chapter to the case of *plane motion* where all movement occurs in or can be represented as occurring in a single plane. A large proportion of the motions of machines and structures in engineering can be represented as plane motion. Later, in Chapter 7 an introduction to three-dimensional motion is presented. We will begin our discussion of plane motion with *rectilinear motion*, which is motion along a straight line, and follow it with a description of motion along a plane curve.

2/2 RECTILINEAR MOTION

Consider a particle P moving along a straight line, Fig. 2/2. The position of P at any instant of time t may be specified by its distance s measured from some convenient reference point O fixed on the line. At time $t + \Delta t$ the particle has moved to P' and its coordinate becomes $s + \Delta s$. The change in the position coordinate during the interval Δt is called the *displacement* Δs of the particle. The displacement would be negative if the particle moved in the negative s-direction.

Figure 2/2

The average velocity of the particle during the interval Δt is the displacement divided by the time interval or $v_{\text{av}} = \Delta s/\Delta t$. As Δt becomes smaller and approaches zero in the limit, the average velocity approaches the *instantaneous velocity* of the particle, which is $v = \lim\limits_{\Delta t \to 0} \dfrac{\Delta s}{\Delta t}$ or

$$v = \frac{ds}{dt} = \dot{s} \qquad\qquad (2/1)$$

*Often called *Cartesian* coordinates, named after René Descartes (1596–1650), a French mathematician who was one of the inventors of analytic geometry.

†This plane is known as the *osculating* plane, which comes from the Latin word *osculari* meaning "to kiss." The plane that contains P and the two points A and B, one on either side of P, becomes the osculating plane as the distances between the points approach zero.

Thus, the velocity is the time rate of change of the position coordinate s. The velocity is positive or negative depending on whether the corresponding displacement is positive or negative.

The average acceleration of the particle during the interval Δt is the change in its velocity divided by the time interval or $a_{av} = \Delta v/\Delta t$. As Δt becomes smaller and approaches zero in the limit, the average acceleration approaches the *instantaneous acceleration* of the particle, which is $a = \lim\limits_{\Delta t \to 0} \dfrac{\Delta v}{\Delta t}$ or

$$a = \frac{dv}{dt} = \dot{v} \qquad \text{or} \qquad a = \frac{d^2s}{dt^2} = \ddot{s} \qquad (2/2)$$

The acceleration is positive or negative depending on whether the velocity is increasing or decreasing. Note that the acceleration would be positive if the particle had a negative velocity which was becoming less negative. If the particle is slowing down, the particle is said to be *decelerating*.

Velocity and acceleration are actually vector quantities, as we will see for curvilinear motion beginning with Art. 2/3. For rectilinear motion in the present article, where the direction of the motion is that of the given straight-line path, the sense of the vector along the path is described by a plus or minus sign. In our treatment of curvilinear motion, we will account for the changes in direction of the velocity and acceleration vectors as well as their changes in magnitude.

By eliminating the time dt between Eq. 2/1 and the first of Eqs. 2/2, a differential equation relating displacement, velocity, and acceleration results* which is

$$v\,dv = a\,ds \qquad \text{or} \qquad \dot{s}\,d\dot{s} = \ddot{s}\,ds \qquad (2/3)$$

Equations 2/1, 2/2, and 2/3 are the differential equations for the rectilinear motion of a particle. Problems in rectilinear motion involving finite changes in the motion variables are solved by integration of these basic differential relations. The position coordinate s, the velocity v, and the acceleration a are algebraic quantities, so that their signs, positive or negative, must be carefully observed. Note that the positive directions for v and a are the same as the positive direction for s.

Interpretation of the differential equations governing rectilinear motion is considerably clarified by representing the relationships among s, v, a, and t graphically. Figure 2/3a represents a schematic plot of the variation of s with t from time t_1 to time t_2 for some given rectilinear motion. By constructing the tangent to the curve at any time t, we obtain the slope, which is the velocity $v =$

*Differential quantities can be multiplied and divided in exactly the same way as other algebraic quantities.

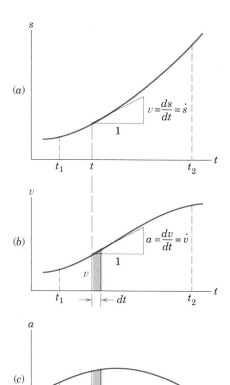

Figure 2/3

ds/dt. Thus, the velocity may be determined at all points on the curve and plotted against the corresponding time as shown in Fig. 2/3b. Similarly, the slope dv/dt of the v-t curve at any instant gives the acceleration at that instant, and the a-t curve can therefore be plotted as in Fig. 2/3c.

We now see from Fig. 2/3b that the area under the v-t curve during time dt is $v\,dt$, which from Eq. 2/1 is the displacement ds. Consequently, the net displacement of the particle during the interval from t_1 to t_2 is the corresponding area under the curve, which is

$$\int_{s_1}^{s_2} ds = \int_{t_1}^{t_2} v\,dt \quad \text{or} \quad s_2 - s_1 = \text{(area under } v\text{-}t \text{ curve)}$$

Similarly, from Fig. 2/3c we see that the area under the a-t curve during time dt is $a\,dt$, which, from the first of Eqs. 2/2, is dv. Thus, the net change in velocity between t_1 and t_2 is the corresponding area under the curve, which is

$$\int_{v_1}^{v_2} dv = \int_{t_1}^{t_2} a\,dt \quad \text{or} \quad v_2 - v_1 = \text{(area under } a\text{-}t \text{ curve)}$$

Two additional graphical relations are noted. When the acceleration a is plotted as a function of the position coordinate s, Fig. 2/4a, the area under the curve during a displacement ds is $a\,ds$, which, from Eq. 2/3, is $v\,dv = d(v^2/2)$. Thus, the net area under the curve between position coordinates s_1 and s_2 is

$$\int_{v_1}^{v_2} v\,dv = \int_{s_1}^{s_2} a\,ds \quad \text{or} \quad \tfrac{1}{2}(v_2{}^2 - v_1{}^2) = \text{(area under } a\text{-}s \text{ curve)}$$

When the velocity v is plotted as a function of the position coordinate s, Fig. 2/4b, the slope of the curve at any point A is dv/ds. By constructing the normal AB to the curve at this point, we see from the similar triangles that $\overline{CB}/v = dv/ds$. Thus, from Eq. 2/3, $\overline{CB} = v(dv/ds) = a$, the acceleration. It is necessary that the velocity and position coordinate axes have the same numerical scales so that the acceleration read on the position coordinate scale in meters (or feet), say, will represent the actual acceleration in meters (or feet) per second squared.

The graphical representations described are useful not only in visualizing the relationships among the several motion quantities but also in approximating results by graphical integration or differentiation when a lack of knowledge of the mathematical relationship prevents its expression as an explicit mathematical function. Experimental data and motions that involve discontinuous relationships between the variables are frequently analyzed graphically.

If the position coordinate s is known for all values of the time t, then successive mathematical or graphical differentiation with

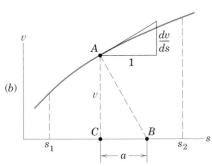

Figure 2/4

respect to t gives the velocity v and acceleration a. In many problems, however, the functional relationship between position coordinate and time is unknown, and we must determine it by successive integration from the acceleration. Acceleration is determined by the forces that act on moving bodies and is computed from the equations of kinetics discussed in subsequent chapters. Depending on the nature of the forces, the acceleration may be specified as a function of time, velocity, or position coordinate, or as a combined function of these quantities. The procedure for integrating the differential equation in each case is indicated as follows.

(a) Constant acceleration. When a is constant, the first of Eqs. 2/2 and 2/3 may be integrated directly. For simplicity with $s = s_0$, $v = v_0$, and $t = 0$ designated at the beginning of the interval, then for a lapse of time t the integrated equations become

$$\int_{v_0}^{v} dv = a \int_{0}^{t} dt \qquad \text{or} \qquad v = v_0 + at$$

$$\int_{v_0}^{v} v \, dv = a \int_{s_0}^{s} ds \qquad \text{or} \qquad v^2 = v_0^2 + 2a(s - s_0)$$

Substitution of the integrated expression for v into Eq. 2/1 and integration with respect to t give

$$\int_{s_0}^{s} ds = \int_{0}^{t} (v_0 + at) \, dt \qquad \text{or} \qquad s = s_0 + v_0 t + \tfrac{1}{2}at^2$$

These relations are necessarily restricted to the special case where the acceleration is constant. The integration limits depend on the initial and final conditions and for a given problem may be different from those used here. It may be more convenient, for instance, to begin the integration at some specified time t_1 rather than at time $t = 0$.

Caution: One of the most frequent mistakes made by students is the attempt to use the foregoing equations for problems of variable acceleration where they do not apply since they have been integrated for constant acceleration only.

(b) Acceleration given as a function of time, a = f(t). Substitution of the function into the first of Eqs. 2/2 gives $f(t) = dv/dt$. Multiplying by dt separates the variables and permits integration. Thus,

$$\int_{v_0}^{v} dv = \int_{0}^{t} f(t) \, dt \qquad \text{or} \qquad v = v_0 + \int_{0}^{t} f(t) \, dt$$

From this integrated expression for v as a function of t, the position coordinate s is obtained by integrating Eq. 2/1, which, in form, would be

$$\int_{s_0}^{s} ds = \int_{0}^{t} v \, dt \qquad \text{or} \qquad s = s_0 + \int_{0}^{t} v \, dt$$

If the indefinite integral is employed, the end conditions are used to establish the constants of integration with results that are identical with those obtained by using the definite integral.

If desired, the displacement s may be obtained by a direct solution of the second-order differential equation $\ddot{s} = f(t)$ obtained by substitution of $f(t)$ into the second of Eqs. 2/2.

(c) Acceleration given as a function of velocity, $a = f(v)$. Substitution of the function into the first of Eqs. 2/2 gives $f(v) = dv/dt$, which permits separating the variables and integrating. Thus,

$$t = \int_{0}^{t} dt = \int_{v_0}^{v} \frac{dv}{f(v)}$$

This result gives t as a function of v. Then it would be necessary to solve for v as a function of t so that Eq. 2/1 can be integrated to obtain the position coordinate s as a function of t.

Alternatively, the function $a = f(v)$ can be substituted into the first of Eqs. 2/3, giving $v \, dv = f(v) \, ds$. The variables are now separated and the equation integrated in the form

$$\int_{v_0}^{v} \frac{v \, dv}{f(v)} = \int_{s_0}^{s} ds \qquad \text{or} \qquad s = s_0 + \int_{v_0}^{v} \frac{v \, dv}{f(v)}$$

Note that this equation gives s in terms of v without explicit reference to t.

(d) Acceleration given as a function of displacement, $a = f(s)$. Substitution of the function into Eq. 2/3 and integrating give the form

$$\int_{v_0}^{v} v \, dv = \int_{s_0}^{s} f(s) \, ds \qquad \text{or} \qquad v^2 = v_0^2 + 2 \int_{s_0}^{s} f(s) \, ds$$

Next we solve for v to give $v = g(s)$, a function of s. Now we can substitute ds/dt for v, separate variables, and integrate in the form

$$\int_{s_0}^{s} \frac{ds}{g(s)} = \int_{0}^{t} dt \qquad \text{or} \qquad t = \int_{s_0}^{s} \frac{ds}{g(s)}$$

which gives t as a function of s. Finally, we can rearrange to obtain s as a function of t.

In each of the foregoing cases when the acceleration varies according to some functional relationship, the ability to solve the equations by direct mathematical integration will depend on the form of the function. In cases where the integration is excessively awkward or difficult, integration by graphical, numerical, or computer methods may be utilized.

Sample Problem   2/1

The position coordinate of a particle which is confined to move along a straight line is given by $s = 2t^3 - 24t + 6$, where s is measured in meters from a convenient origin and t is in seconds. Determine (a) the time required for the particle to reach a velocity of 72 m/s from its initial condition at $t = 0$, (b) the acceleration of the particle when $v = 30$ m/s, and (c) the net displacement of the particle during the interval from $t = 1$ s to $t = 4$ s.

Solution.   The velocity and acceleration are obtained by successive differentiation of s with respect to the time. Thus,

$$[v = \dot{s}] \qquad\qquad v = 6t^2 - 24 \text{ m/s}$$

$$[a = \dot{v}] \qquad\qquad a = 12t \text{ m/s}^2$$

(a)   Substituting $v = 72$ m/s into the expression for v gives us $72 = 6t^2 - 24$, from which $t = \pm 4$ s. The negative root describes a mathematical solution for t before the initiation of motion, so this root is of no physical interest. Thus, the desired result is

$$t = 4 \text{ s} \qquad\qquad Ans.$$

(b)   Substituting $v = 30$ m/s into the expression for v gives $30 = 6t^2 - 24$, from which the positive root is $t = 3$ s, and the corresponding acceleration is

$$a = 12(3) = 36 \text{ m/s}^2 \qquad\qquad Ans.$$

(c)   The net displacement during the specified interval is

$$\Delta s = s_4 - s_1 \qquad \text{or}$$

$$\Delta s = [2(4^3) - 24(4) + 6] - [2(1^3) - 24(1) + 6]$$

$$= 54 \text{ m} \qquad\qquad Ans.$$

which represents the net advancement of the particle along the s-axis from the position it occupied at $t = 1$ s to its position at $t = 4$ s.

To help visualize the motion, the values of s, v, and a are plotted against the time t as shown. Because the area under the v-t curve represents displacement, we see that the net displacement from $t = 1$ s to $t = 4$ s is the positive area Δs_{2-4} less the negative area Δs_{1-2}.

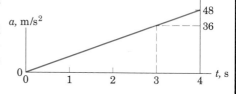

① Be alert to the proper choice of sign when taking a square root. When the situation calls for only one answer, the positive root is not always the one you may need.

② Note carefully the distinction between italic s for the position coordinate and the vertical s for seconds.

③ Note from the graphs that the values for v are the slopes ($\dot{s}$) of the s-t curve and that the values for a are the slopes ($\dot{v}$) of the v-t curve. *Suggestion:* Integrate $v\ dt$ for each of the two intervals and check the answer for Δs. Show that the total distance traveled during the interval $t = 1$ s to $t = 4$ s is 74 m.

Sample Problem 2/2

A particle moves along the x-axis with an initial velocity $v_x = 50$ m/s at the origin when $t = 0$. For the first 4 seconds it has no acceleration, and thereafter it is acted on by a retarding force which gives it a constant acceleration $a_x = -10$ m/s^2. Calculate the velocity and the x-coordinate

① of the particle for the conditions of $t = 8$ s and $t = 12$ s and find the maximum positive x-coordinate reached by the particle.

① Learn to be flexible with symbols. The position coordinate x is just as valid as s.

Solution. The velocity of the particle after $t = 4$ s is computed from

② $$\left[\int dv = \int a\, dt \right] \qquad \int_{50}^{v_x} dv_x = -10 \int_{4}^{t} dt \qquad v_x = 90 - 10t \text{ m/s}$$

② Note that we integrate to a general time t and then substitute specific values.

and is plotted as shown. At the specified times, the velocities are

$$t = 8 \text{ s}, \qquad v_x = 90 - 10(8) = 10 \text{ m/s}$$

$$t = 12 \text{ s}, \qquad v_x = 90 - 10(12) = -30 \text{ m/s} \qquad Ans.$$

The x-coordinate of the particle at any time greater than 4 seconds is the distance traveled during the first 4 seconds plus the distance traveled after the discontinuity in acceleration occurred. Thus,

$$\left[\int ds = \int v\, dt \right] \quad x = 50(4) + \int_{4}^{t} (90 - 10t)\, dt = -5t^2 + 90t - 80 \text{ m}$$

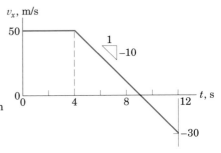

For the two specified times,

$$t = 8 \text{ s}, \qquad x = -5(8^2) + 90(8) - 80 = 320 \text{ m}$$

$$t = 12 \text{ s}, \qquad x = -5(12^2) + 90(12) - 80 = 280 \text{ m} \qquad Ans.$$

The x-coordinate for $t = 12$ s is less than that for $t = 8$ s since the motion is in the negative x-direction after $t = 9$ s. The maximum positive x-coordinate is, then, the value of x for $t = 9$ s which is

$$x_{\max} = -5(9^2) + 90(9) - 80 = 325 \text{ m} \qquad Ans.$$

③ These displacements are seen to be the net positive areas under the v-t graph up to the values of t in question.

③ Show that the total distance traveled by the particle in the 12 seconds is 370 m.

Sample Problem   2/3

The spring-mounted slider moves in the horizontal guide with negligible friction and has a velocity v_0 in the s-direction as it crosses the mid-position where $s = 0$ and $t = 0$. The two springs together exert a retarding force to the motion of the slider, which gives it an acceleration proportional to the displacement but oppositely directed and equal to $a = -k^2 s$, where k is constant. (The constant is arbitrarily squared for later convenience in the form of the expressions.) Determine the expressions for the displacement s and velocity v as functions of the time t.

Solution I.   Since the acceleration is specified in terms of the displacement, the differential relation $v\,dv = a\,ds$ may be integrated. Thus,

① $$\int v\,dv = \int -k^2 s\,ds + C_1 \text{ a constant}, \quad \text{or} \quad \frac{v^2}{2} = -\frac{k^2 s^2}{2} + C_1$$

① We have used an indefinite integral here and evaluated the constant of integration. For practice, obtain the same results by using the definite integral with the appropriate limits.

When $s = 0$, $v = v_0$, so that $C_1 = v_0^2/2$, and the velocity becomes

$$v = +\sqrt{v_0^2 - k^2 s^2}$$

The plus sign of the radical is taken when v is positive (in the plus s-direction). This last expression may be integrated by substituting $v = ds/dt$. Thus,

② $$\int \frac{ds}{\sqrt{v_0^2 - k^2 s^2}} = \int dt + C_2 \text{ a constant}, \quad \text{or} \quad \frac{1}{k}\sin^{-1}\frac{ks}{v_0} = t + C_2$$

② Again try the definite integral here as above.

With the requirement of $t = 0$ when $s = 0$, the constant of integration becomes $C_2 = 0$, and we may solve the equation for s so that

$$s = \frac{v_0}{k}\sin kt \qquad\qquad Ans.$$

The velocity is $v = \dot{s}$, which gives

$$v = v_0 \cos kt \qquad\qquad Ans.$$

Solution II.   Since $a = \ddot{s}$, the given relation may be written at once as

$$\ddot{s} + k^2 s = 0$$

This is an ordinary linear differential equation of second order for which the solution is well known and is

$$s = A \sin Kt + B \cos Kt$$

where A, B, and K are constants. Substitution of this expression into the differential equation shows that it satisfies the equation, provided that $K = k$. The velocity is $v = \dot{s}$, which becomes

$$v = Ak \cos kt - Bk \sin kt$$

The initial condition $v = v_0$ when $t = 0$ requires that $A = v_0/k$, and the condition $s = 0$ when $t = 0$ gives $B = 0$. Thus, the solution is

③ $$s = \frac{v_0}{k}\sin kt \quad \text{and} \quad v = v_0 \cos kt \qquad Ans.$$

③ This motion is called *simple harmonic motion* and is characteristic of all oscillations where the restoring force, and hence the acceleration, is proportional to the displacement but opposite in sign.

Sample Problem 2/4

A freighter is moving at a speed of 8 knots when its engines are suddenly stopped. If it takes 10 minutes for the freighter to reduce its speed to 4 knots, determine and plot the distance s in nautical miles moved by the ship and its speed v in knots as functions of the time t during this interval. The deceleration of the ship is proportional to the square of its speed, so that $a = -kv^2$.

① Recall that one knot is the speed of one nautical mile (1852 m) per hour. Work directly in the units of nautical miles and hours.

Solution. The speeds and the time are given, so we may substitute the expression for acceleration directly into the basic definition $a = dv/dt$ and integrate. Thus,

$$-kv^2 = \frac{dv}{dt} \qquad \frac{dv}{v^2} = -k\,dt \qquad \int_8^v \frac{dv}{v^2} = -k \int_0^t dt$$

$$-\frac{1}{v} + \frac{1}{8} = -kt \qquad v = \frac{8}{1 + 8kt}$$

② We choose to integrate to a general value of v and its corresponding time t so that we may obtain the variation of v with t.

Now we substitute the end limits of $v = 4$ knots and $t = \frac{10}{60} = \frac{1}{6}$ hour and get

$$4 = \frac{8}{1 + 8k(1/6)} \qquad k = \frac{3}{4}\,\text{mi}^{-1} \qquad v = \frac{8}{1 + 6t} \qquad \textit{Ans.}$$

The speed is plotted against the time as shown.

The distance is obtained by substituting the expression for v into the definition $v = ds/dt$ and integrating. Thus,

$$\frac{8}{1 + 6t} = \frac{ds}{dt} \qquad \int_0^t \frac{8\,dt}{1 + 6t} = \int_0^s ds \qquad s = \frac{4}{3}\ln(1 + 6t) \qquad \textit{Ans.}$$

The distance s is also plotted against the time as shown, and we see that the ship has moved through a distance $s = \frac{4}{3}\ln(1 + \frac{6}{6}) = \frac{4}{3}\ln 2 = 0.924$ mi (nautical) during the 10 minutes.

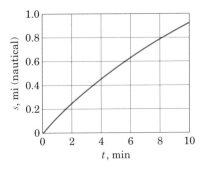

PROBLEMS

Introductory problems

Problems 2/1 through 2/7 treat the motion of a particle which moves along the *s*-axis shown in the figure.

Problems 2/1–2/7

2/1 The velocity of a particle is given by $v = 25t^2 - 80t - 200$, where v is in meters per second and t is in seconds. Plot the velocity v and acceleration a versus time for the first 6 seconds of motion and evaluate the velocity when a is zero.

Ans. $v = -264$ m/s

2/2 The position of a particle is given by $s = 2t^3 - 40t^2 + 200t - 50$, where s is in meters and t is in seconds. Plot the position, velocity, and acceleration as functions of time for the first 12 seconds of motion. Determine the time at which the velocity is zero.

2/3 The velocity of a particle which moves along the *s*-axis is given by $v = 2 - 4t + 5t^{3/2}$, where t is in seconds and v is in meters per second. Evaluate the position s, velocity v, and acceleration a when $t = 3$ s. The particle is at the position $s_0 = 3$ m when $t = 0$.

Ans. $s = 22.2$ m, $v = 15.98$ m/s, $a = 8.99$ m/s^2

2/4 The displacement of a particle which moves along the *s*-axis is given by $s = (-2 + 3t)e^{-0.5t}$, where s is in meters and t is in seconds. Plot the displacement, velocity, and acceleration versus time for the first 20 seconds of motion. Determine the time at which the acceleration is zero.

2/5 The acceleration of a particle is given by $a = 4t - 30$, where a is in meters per second squared and t is in seconds. Determine the velocity and displacement as functions of time. The initial displacement at $t = 0$ is $s_0 = -5$ m, and the initial velocity is $v_0 = 3$ m/s.

Ans. $v = 3 - 30t + 2t^2$ m/s
$s = -5 + 3t - 15t^2 + \frac{2}{3}t^3$ m

2/6 The velocity of a particle which moves along the *s*-axis is given by $\dot{s} = 40 - 3t^2$ m/s, where t is in seconds. Calculate the displacement Δs of the particle during the interval from $t = 2$ s to $t = 4$ s.

2/7 A particle moves along the *s*-direction with constant acceleration. The displacement, measured from a convenient position, is 2 m at time $t = 0$ and is zero when $t = 10$ s. If the velocity of the particle is momentarily zero when $t = 6$ s, determine the acceleration a and the velocity v when $t = 10$ s.

Ans. $a = 0.2$ m/s^2, $v = 0.8$ m/s

2/8 A ball is thrown vertically up with a velocity of 30 m/s at the edge of a 60-m cliff. Calculate the height h to which the ball rises and the total time t after release for the ball to reach the bottom of the cliff. Neglect air resistance and take the downward acceleration to be 9.81 m/s^2.

Problem 2/8

2/9 The velocity of a particle moving in a straight line is decreasing at the rate of 3 m/s per meter of displacement at an instant when the velocity is 10 m/s. Determine the acceleration a of the particle at this instant.

Ans. $a = -30$ m/s^2

2/10 A rocket is fired vertically up from rest. If it is designed to maintain a constant upward acceleration of $1.5g$, calculate the time t required for it to reach an altitude of 30 km and its velocity at that position.

2/11 Calculate the constant acceleration a in g's which the catapult of an aircraft carrier must provide to produce a launch velocity of 300 km/h in a distance of 100 m. Assume that the carrier is at anchor.

Ans. $a = 3.54g$

2/12 In the final stages of a moon landing, the lunar module descends under retrothrust of its descent engine to within $h = 5$ m of the lunar surface where it has a downward velocity of 2 m/s. If the descent engine is cut off abruptly at this point, compute the impact velocity of the landing gear with the moon. Lunar gravity is $\frac{1}{6}$ of the earth's gravity.

"Zero-g" test facility

Problem 2/13

2/14 The main elevator A of the CN Tower in Toronto rises about 350 m and for most of its run has a constant speed of 22 km/h. Assume that both the acceleration and deceleration have a constant magnitude of $\frac{1}{4}g$ and determine the time duration t of the elevator run.

Problem 2/12

2/13 To test the effects of "weightlessness" for short periods of time, a test facility is designed which accelerates a test package vertically up from A to B by means of a gas-activated piston and allows it to ascend and descend from B to C to B under free-fall conditions. The test chamber consists of a deep well and is evacuated to eliminate any appreciable air resistance. If a constant acceleration of $40g$ from A to B is provided by the piston and if the total test time for the "weightless" condition from B to C to B is 10 s, calculate the required working height h of the chamber. Upon returning to B, the test package is recovered in a basket filled with polystyrene pellets inserted in the line of fall.

Ans. $h = 125.7$ m

Problem 2/14

Representative problems

2/15 A projectile is fired vertically with an initial velocity of 200 m/s. Calculate the maximum altitude h reached by the projectile and the time t after firing for it to return to the ground. Neglect air resistance and take the gravitational acceleration to be constant at 9.81 m/s^2.

Ans. $h = 2040$ m, $t = 40.8$ s

2/16 Small steel balls fall from rest through the opening at A at the steady rate of two per second. Find the vertical separation h of two consecutive balls when the lower one has dropped 3 meters. Neglect air resistance.

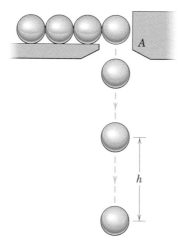

Problem 2/16

2/17 A particle traveling in a straight line encounters a retarding force which causes its velocity to decrease according to $v = 20e^{-t/10}$ m/s, where t is the time in seconds during which the force acts. Determine the acceleration a of the particle when $t = 10$ s and find the corresponding distance s which the particle has moved during the 10-s interval. Plot v as a function of t for the first 10 s.

Ans. $a = -0.736$ m/s^2, $s = 126.4$ m

2/18 A particle oscillates along a straight line with a sinusoidally varying velocity in millimeters per second given by $v = 16 \sin \pi t/6$, where t is in seconds. If the displacement of the particle is 8 mm when $t = 0$, determine its maximum displacement s_{max} and plot s versus t for one complete cycle.

2/19 During an 8-second interval the velocity of a particle moving in a straight line varies with the time as shown. Within reasonable limits of accuracy, determine the amount Δa by which the acceleration at $t = 4$ s exceeds the average acceleration during the interval. What is the displacement Δs during the interval?

Ans. $\Delta a = 0.50$ m/s^2, $\Delta s = 64$ m

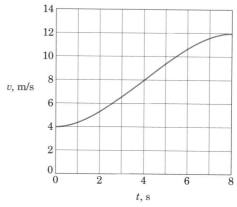

Problem 2/19

2/20 A vehicle enters a test section of straight road at $s = 0$ with a speed of 40 km/h. It then undergoes an acceleration which varies with displacement as shown. Determine the velocity v of the vehicle as it passes the position $s = 0.2$ km.

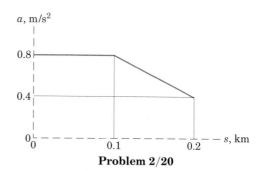

Problem 2/20

2/21 A retarding force acts on a particle moving initially with a velocity of 100 m/s and gives it a deceleration as recorded by the oscilloscope record shown. Approximate the velocity of the particle at $t = 4$ s and at $t = 8$ s.

Ans. $v_4 = 60.8$ m/s, $v_8 = -17.72$ m/s

Problem 2/21

2/22 A girl rolls a ball up an incline and allows it to return to her. For the angle θ and ball involved, the acceleration of the ball along the incline is constant at $0.25g$, directed down the incline. If the ball is released with a speed of 4 m/s, determine the distance s it moves up the incline before reversing its direction and the total time t required for the ball to return to the child's hand.

Problem 2/22

2/23 The car is traveling at a constant speed $v_0 = 100$ km/h on the level portion of the road. When the 6 percent ($\tan \theta = 6/100$) incline is encountered, the driver does not change the throttle setting and consequently the car decelerates at the constant rate $g \sin \theta$. Determine the speed of the car (a) 10 seconds after passing point A and (b) when $s = 100$ m.

Ans. (a) $v = 21.9$ m/s, (b) $v = 25.6$ m/s

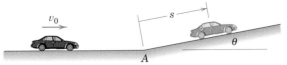

Problem 2/23

2/24 A body moves in a straight line with a velocity whose square decreases linearly with the displacement between two points A and B, which are 300 m apart as shown. Determine the displacement Δs of the body during the last 2 s before arrival at B.

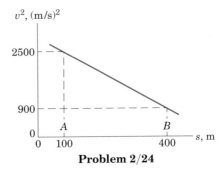

Problem 2/24

2/25 A motorcycle patrolman starts from rest at A two seconds after a car, speeding at the constant rate of 120 km/h, passes point A. If the patrolman accelerates at the rate of 6 m/s^2 until he reaches his maximum permissible speed of 150 km/h, which he maintains, calculate the distance s from point A to the point at which he overtakes the car.

Ans. $s = 912$ m

Problem 2/25

2/26 The deceleration of the mass center G of a car during a crash test is measured by an accelerometer with the results shown, where the distance x moved by G after impact is 0.8 m. Obtain a close approximation to the impact velocity v from the data given.

Problem 2/26

2/27 If the velocity v of a particle moving along a straight line decreases linearly with its displacement s from 20 m/s to a value approaching zero at $s = 30$ m, determine the acceleration a of the particle when $s = 15$ m and show that the particle never reaches the 30-m displacement.

Ans. $a = -6.67$ m/s^2

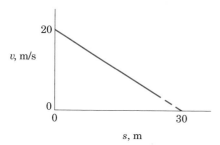

Problem 2/27

2/28 A sprinter reaches his maximum speed v_{max} in 2.5 s from rest with constant acceleration. He then maintains that speed and finishes the 100 meters in the overall time of 10.40 s. Determine his maximum speed v_{max}.

Problem 2/28

2/29 In an archery test, the acceleration of the arrow decreases linearly with distance s from its initial value of 4800 m/s^2 at A upon release to zero at B after a travel of 600 mm. Calculate the maximum velocity v of the arrow.

Ans. $v = 53.7$ m/s

Problem 2/29

2/30 A particle moving along the s-axis has a velocity given by $v = 18 - 2t^2$ m/s, where t is in seconds. When $t = 0$, the position of the particle is given by $s_0 = -3$ m. For the first 5 seconds of motion, determine the total distance D traveled, the net displacement Δs, and the value of s at the end of the interval.

2/31 A vacuum-propelled capsule for a high-speed tube transportation system of the future is being designed for operation between two stations A and B, which are 10 km apart. If the acceleration and deceleration are to have a limiting magnitude of $0.6g$ and if velocities are to be limited to 400 km/h, determine the minimum time t for the capsule to make the 10-km trip.

Ans. $t = 108.9$ s

Problem 2/31

2/32 A motorcycle starts from rest with an initial acceleration of 3 m/s², and the acceleration then changes with distance s as shown. Determine the velocity v of the motorcycle when $s = 200$ m. At this point also determine the value of the derivative $\frac{dv}{ds}$.

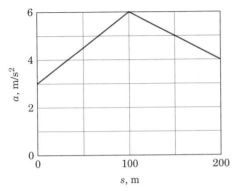

s, m

Problem 2/32

2/33 A particle moving along the positive x-direction with an initial velocity of 12 m/s is subjected to a retarding force that gives it a negative acceleration which varies linearly with time for the first 4 seconds as shown. For the next 5 seconds the force is constant and the acceleration remains constant. Plot the velocity of the particle during the 9 seconds and specify its value at $t = 4$ s. Also find the distance Δx traveled by the particle from its position at $t = 0$ to the point where it reverses its direction.

Ans. $v_4 = 6$ m/s, $\Delta x = 46$ m

Problem 2/33

2/34 Particle 1 is subjected to an acceleration $a = -kv$, particle 2 is subjected to $a = -kt$, and particle 3 is subjected to $a = -ks$. All three particles start at the origin $s = 0$ with an initial velocity $v_0 = 10$ m/s at time $t = 0$, and the magnitude of k is 0.1

for all three particles (note that the units of k vary from case to case). Plot the position, velocity, and acceleration versus time for each particle over the range $0 \leq t \leq 10$ s.

2/35 A car starts from rest with an acceleration of 6 m/s² which decreases linearly with time to zero in 10 seconds, after which the car continues at a constant speed. Determine the time t required for the car to travel 400 m from the start.

Ans. $t = 16.67$ s

2/36 The aerodynamic resistance to motion of a car is nearly proportional to the square of its velocity. Additional frictional resistance is constant, so that the acceleration of the car when coasting may be written $a = -C_1 - C_2v^2$, where C_1 and C_2 are constants which depend on the mechanical configuration of the car. If the car has an initial velocity v_0 when the engine is disengaged, derive an expression for the distance D required for the car to coast to a stop.

v

Problem 2/36

2/37 Compute the impact speed of a body released from rest at an altitude $h = 800$ km. (*a*) Assume a constant gravitational acceleration $g_0 = 9.81$ m/s² and (*b*) account for the variation of g with altitude (refer to Art. 1/5). Neglect the effects of atmospheric drag.

Ans. (*a*) $v = 3960$ m/s, (*b*) $v = 3730$ m/s

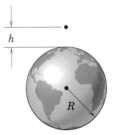

Problem 2/37

2/38 The acceleration a_x of the piston in a small recip- rocating engine is given in the following table in terms of the position x of the piston measured from the top of its stroke. From a plot of the data, de- termine to within two-significant-figure accuracy the maximum velocity v_{max} reached by the piston.

x, m	a_x, m/s^2	x, m	a_x, m/s^2
0	4950	0.075	-450
0.0075	4340	0.090	-1265
0.015	3740	0.105	-1960
0.030	2580	0.120	-2510
0.045	1490	0.135	-2910
0.060	476	0.150	-3150

Problem 2/38

2/39 In the design of the valve mechanism for an inter- nal-combustion engine, it is necessary to deter- mine the velocity and the acceleration of the valve during engagement by the rotating cam illustrated here. From the profile of the cam, the valve dis- placement x from the closed position can be deter- mined as a function of the angle of rotation of the cam. For the particular cam with symmetrical pro- file shown, the valve displacement x as a function of engagement angle θ is given in the following ta- ble for the 50° lifting part of the cycle. From an accurate plot of x versus θ, determine the maxi- mum velocity v_{max} of the valve during its lift and its acceleration a at $\theta = 50°$. The cam rotates at

the rate of 4000 rev/min. (Note that the cam ro- tates at the rate of $4000(360)/60 = 24\,000$ degrees per second. Also, account for the fact that $\dot{x} = 0$ at both $\theta = 0$ and $\theta = 50°$.)

θ, degrees	x, mm	θ, degrees	x, mm
0	0	30	3.43
5	0.20	35	4.03
10	0.67	40	4.44
15	1.30	45	4.73
20	2.01	50	4.80
25	2.78		

Ans. $v_{max} = 3.6$ m/s, $a \cong -4100$ m/s^2

Problem 2/39

2/40 A particle moving along a straight line decelerates according to $a = -kv$, where k is a constant and v is velocity. If its initial velocity at time $t = 0$ is $v_0 = 4$ m/s and its velocity at time $t = 2$ s is $v = 1$ m/s, determine the time T and corresponding distance D for the particle speed to be reduced to one-tenth of its initial value.

2/41 On its takeoff roll, the airplane starts from rest and accelerates according to $a = a_0 - kv^2$, where a_0 is the constant acceleration resulting from the engine thrust and $-kv^2$ is the acceleration due to aerodynamic drag. If $a_0 = 2$ m/s^2, $k = 0.00004$ m^{-1}, and v is in meters per second, determine the design length of runway required to reach the takeoff speed of 250 km/h if the drag term is (*a*) excluded and (*b*) included.

Ans. (*a*) $s = 1206$ m, (*b*) $s = 1268$ m

Problem 2/41

2/42 The steel ball A of diameter D slides freely on the horizontal rod which leads to the pole face of the electromagnet. The force of attraction obeys an inverse-square law, and the resulting acceleration of the ball is $a = K/(L - x)^2$, where K is a measure of the strength of the magnetic field. If the ball is released from rest at $x = 0$, determine the velocity v with which it strikes the pole face.

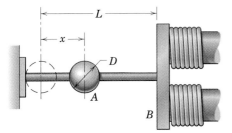

Problem 2/42

2/43 The cone falling with a speed v_0 strikes and penetrates the block of packing material. The acceleration of the cone after impact is $a = g - cy^2$, where c is a positive constant and y is the penetration distance. If the maximum penetration depth is observed to be y_m, determine the constant c.

Ans. $c = \dfrac{3v_0{}^2 + 6gy_m}{2y_m{}^3}$

Problem 2/43

2/44 The electronic throttle control of a model train is programmed so that the train speed varies with position as shown in the plot. Determine the time t required for the train to complete one lap.

Problem 2/44

2/45 A test projectile is fired horizontally into a viscous liquid with a velocity v_0. The retarding force is proportional to the square of the velocity, so that the acceleration becomes $a = -kv^2$. Derive expressions for the distance D traveled in the liquid and the corresponding time t required to reduce the velocity to $v_0/2$. Neglect any vertical motion.

Ans. $D = 0.693/k,\ t = 1/(kv_0)$

Problem 2/45

2/46 Compute the impact speed of body A which is released from rest at an altitude $h = 1200$ km above the surface of the moon. (*a*) First assume a constant gravitational acceleration $g_{m_0} = 1.620$ m/s^2 and (*b*) then account for the variation of g_m with altitude (refer to Art. 1/5).

— 3476 km — ⊢ h ⊣

•A

Problem 2/46

2/47 The driver of a car, which is initially at rest at the top A of the grade, releases the brakes and coasts down the grade with an acceleration in meters per second squared given by $a = 0.981 - 0.013v^2$, where v is the velocity in meters per second. Determine the velocity v_B at the bottom B of the grade.

Ans. $v_B = 8.66$ m/s

A

200 m

v

B

Problem 2/47

2/48 A small object is released from rest in a tank of oil. The downward acceleration of the object is $g - kv$, where g is the constant acceleration due to gravity, k is a constant which depends on the viscosity of the oil and shape of the object, and v is the downward velocity of the object. Derive expressions for the velocity v and vertical drop y as functions of the time t after release.

2/49 When the effect of aerodynamic drag is included, the y-acceleration of a baseball moving vertically upward is $a_u = -g - kv^2$, while the acceleration when the ball is moving downward is $a_d = -g + kv^2$, where k is a positive constant and v is the

speed in meters per second. If the ball is thrown upward at 30 m/s from essentially ground level, compute its maximum height h and its speed v_f upon impact with the ground. Take k to be 0.006 m^{-1} and assume that g is constant.

Ans. $h = 36.5$ m, $v_f = 24.1$ m/s

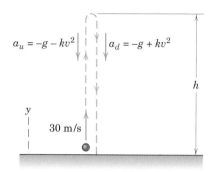

$a_u = -g - kv^2$ $a_d = -g + kv^2$

h

y

30 m/s

Problem 2/49

2/50 For the baseball of Prob. 2/49 thrown upward with an initial speed of 30 m/s, determine the time t_u from ground to apex and the time t_d from apex to ground.

2/51 Car A travels at a constant speed of 100 km/h. When in the position shown at time $t = 0$, car B has a speed of 40 km/h and accelerates at a constant rate of $0.1g$ along its path until it reaches a speed of 100 km/h, after which it travels at that constant speed. What is the steady-state position of car A with respect to car B?

Ans. A is ahead of B by 198.7 m

A

100 m

B

Problem 2/51

2/52 Repeat Prob. 2/51, except that car *B*, rather than possessing a constant acceleration, now accelerates as shown in the accompanying plot. Time t_2 is the time at which the speed of car *B* reaches 100 km/h. After time t_2, the speed remains constant. Compare your result with that stated for Prob. 2/51.

Problem 2/52

2/53 A particle which moves along the *x*-axis is subjected to an accelerating force which increases linearly with time and a retarding force which increases directly with displacement. The resulting acceleration is $a = Kt - k^2x$, where *K* and *k* are positive constants and where both *x* and $v = \dot{x}$ are zero when the time $t = 0$. Determine the displacement *x* as a function of *t*.

$$Ans. \quad x = \frac{K}{k^3}(kt - \sin kt)$$

2/54 The fuel of a model rocket is burned so quickly that one may assume that the rocket acquires its burn-out velocity of 120 m/s while essentially still at ground level. The rocket then coasts vertically upward to the trajectory apex. With the inclusion of aerodynamic drag, the *y*-acceleration is $a_y = -g - 0.0005v^2$ during this motion, where the units are meters and seconds. At apex a parachute pops out of the nose cone, and the rocket quickly acquires a constant downward speed of 4 m/s. Estimate the flight time *t*.

Problem 2/54

2/55 The stories of a tall building are uniformly 3 meters in height. A ball *A* is dropped from the rooftop position shown. Determine the times required for it to pass the 3 meters of the first, tenth, and one-hundredth stories (counted from the top). Neglect aerodynamic drag.

$$Ans. \quad t_1 = 0.782 \text{ s}, \ t_{10} = 0.1269 \text{ s}$$
$$t_{100} = 0.0392 \text{ s}$$

Problem 2/55

2/56 Repeat Prob. 2/55, except now include the effects of aerodynamic drag. The drag force causes an acceleration component in m/s² of $0.016v^2$ in the direction opposite the velocity vector, where *v* is in m/s.

▶2/57 The vertical acceleration of a certain solid-fuel rocket is given by $a = ke^{-bt} - cv - g$, where k, b, and c are constants, v is the vertical velocity acquired, and g is the gravitational acceleration, essentially constant for atmospheric flight. The exponential term represents the effect of a decaying thrust as fuel is burned, and the term $-cv$ approximates the retardation due to atmospheric resistance. Determine the expression for the vertical velocity of the rocket t seconds after firing.

$$Ans. \quad v = \frac{g}{c}(e^{-ct} - 1) + \frac{k}{c - b}(e^{-bt} - e^{-ct})$$

▶2/58 The preliminary design for a rapid-transit system calls for the train velocity to vary with time as shown in the plot as the train runs the 3.2 km between stations A and B. The slopes of the cubic transition curves (which are of form $a + bt + ct^2 + dt^3$) are zero at the end points. Determine the total run time t between the stations and the maximum acceleration.

$$Ans. \quad t = 103.6 \text{ s}, \quad a_{max} = 3.61 \text{ m/s}^2$$

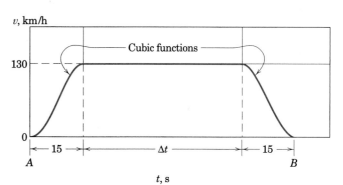

Problem 2/58

2/3 PLANE CURVILINEAR MOTION

We now begin our description of the motion of a particle along a curved path which lies in a single plane. This motion is seen to be a special case of the more general three-dimensional motion introduced in Art. 2/1 and illustrated in Fig. 2/1. If we let the plane of motion be the x-y plane, for instance, then the coordinates z and ϕ of Fig. 2/1 are both zero and R becomes the same as r. As mentioned previously, the vast majority of the motions of points or particles encountered in engineering practice may be represented as plane motion.

Before pursuing the description of plane curvilinear motion in any specific set of coordinates, we will first describe the motion with the aid of vector analysis, the results of which are independent of any particular coordinate system. What follows in this article will constitute one of the most basic of all treatments in dynamics, namely, development of the *time derivative of a vector*. A large fraction of the entire subject of dynamics hinges directly on the time rates of change of vector quantities, and all students of dynamics are well advised to master this topic at the outset because they will have frequent occasion to use it.

Consider now the continuous motion of a particle along a plane curve as represented in Fig. 2/5. At time t the particle is at position A, which is located by the *position vector* **r** measured from some convenient fixed origin O. If both the magnitude and direction of **r** are known at time t, then the position of the particle is completely specified. At time $t + \Delta t$, the particle is at A', located by the position vector $\mathbf{r} + \Delta \mathbf{r}$. We note, of course, that this combination is vector addition and not scalar addition. The *displacement* of the particle during time Δt is the vector $\Delta \mathbf{r}$ that represents the vector change of position and is clearly independent of the choice of origin. If an origin were chosen at some different location, the position vector **r** would be changed, but $\Delta \mathbf{r}$ would be unchanged. The *distance* actually traveled by the particle as it moves along the path from A to A' is the scalar length Δs measured along the path. Thus, we

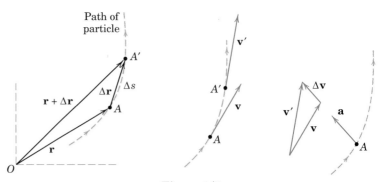

Figure 2/5

distinguish between the vector displacement $\Delta\mathbf{r}$ and the scalar distance Δs.

The *average velocity* of the particle between A and A' is defined as $\mathbf{v}_{av} = \Delta\mathbf{r}/\Delta t$, which is a vector whose direction is that of $\Delta\mathbf{r}$ and whose magnitude is the magnitude of $\Delta\mathbf{r}$ divided by Δt. The average *speed* of the particle between A and A' is the scalar quotient $\Delta s/\Delta t$. Clearly, the magnitude of the average velocity and the speed approach one another as the interval Δt decreases and A and A' become closer together.

The *instantaneous velocity* $\mathbf{v}$ of the particle is defined as the limiting value of the average velocity as the time interval approaches zero. Thus,

$$\mathbf{v} = \lim_{\Delta t \to 0} \frac{\Delta\mathbf{r}}{\Delta t}$$

We observe that the direction of $\Delta\mathbf{r}$ approaches that of the tangent to the path as Δt approaches zero and, hence, the velocity $\mathbf{v}$ is always a vector tangent to the path. We now extend the basic definition of the derivative of a scalar quantity to include a vector quantity and write

$$\mathbf{v} = \frac{d\mathbf{r}}{dt} = \dot{\mathbf{r}} \qquad\qquad (2/4)$$

The derivative of a vector is itself a vector having both a magnitude and a direction. The magnitude of $\mathbf{v}$ is called the *speed* and is the scalar

$$v = |\mathbf{v}| = \frac{ds}{dt} = \dot{s}$$

At this point we make a careful distinction between the *magnitude of the derivative* and the *derivative of the magnitude*. The magnitude of the derivative may be written in any one of the several ways $|d\mathbf{r}/dt| = |\dot{\mathbf{r}}| = \dot{s} = |\mathbf{v}| = v$ and represents the magnitude of the velocity or the speed of the particle. On the other hand, the derivative of the magnitude is written $d|\mathbf{r}|/dt = dr/dt = \dot{r}$, and represents the rate at which the length of the position vector $\mathbf{r}$ is changing. Thus, these two derivatives have two entirely different meanings, and we must be extremely careful to distinguish between them in our thinking and in our notation. For this and other reasons, students are urged to adopt a consistent notation for their handwritten work for all vector quantities to distinguish them from scalar quantities. For simplicity the underline $\underline{v}$ is recommended. Other handwritten symbols such as $\vec{v}$, $\underset{\sim}{v}$, and $\hat{v}$ are sometimes used.

With the concept of velocity as a vector established, we return to Fig. 2/5 and denote the velocity of the particle at A by the tangent vector **v** and the velocity at A' by the tangent vector **v'**. Clearly, there is a vector change in the velocity during the time Δt. The velocity **v** at A plus (vectorially) the change Δ**v** must equal the velocity at A', so we may write $\mathbf{v'} - \mathbf{v} = \Delta\mathbf{v}$. Inspection of the vector diagram shows that Δ**v** depends both on the change in magnitude (length) of **v** and on the change in direction of **v**. These two changes are fundamental characteristics of the derivative of a vector.

The *average acceleration* of the particle between A and A' is defined as $\Delta\mathbf{v}/\Delta t$, which is a vector whose direction is that of Δ**v**. The magnitude of this average acceleration is the magnitude of Δ**v** divided by Δt.

The *instantaneous acceleration* **a** of the particle is defined as the limiting value of the average acceleration as the time interval approaches zero. Thus,

$$\mathbf{a} = \lim_{\Delta t \to 0} \frac{\Delta\mathbf{v}}{\Delta t}$$

By definition of the derivative, then, we write

$$\mathbf{a} = \frac{d\mathbf{v}}{dt} = \dot{\mathbf{v}} \qquad (2/5)$$

As the interval Δt becomes smaller and approaches zero, the direction of the change Δ**v** approaches that of the differential change d**v** and, hence, **a**. The acceleration **a**, then, includes the effect of change in magnitude of **v** and the change of direction of **v**. It is apparent, in general, that the direction of the acceleration of a particle in curvilinear motion is neither tangent to the path nor normal to the path. We do observe, however, that the acceleration component which is normal to the path points toward the center of curvature of the path.

A further approach to the visualization of acceleration is shown in Fig. 2/6, where the position vectors to three arbitrary positions

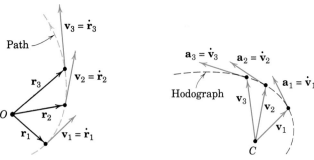

Figure 2/6

on the path of the particle are shown for illustrative purpose. There is a velocity vector tangent to the path corresponding to each position vector, and the relation is $\mathbf{v} = \dot{\mathbf{r}}$. Now if these velocity vectors are plotted from some arbitrary point C, then a curve, known as the *hodograph*, is formed. The derivatives of these velocity vectors will be the acceleration vectors $\mathbf{a} = \dot{\mathbf{v}}$ that are tangent to the hodograph. We see that the acceleration bears the same relation to the velocity as the velocity bears to the position vector.

The geometric portrayal of the derivatives of the position vector $\mathbf{r}$ and velocity vector $\mathbf{v}$ in Fig. 2/5 can be used to describe the derivative of any vector quantity with respect to t or with respect to any other scalar variable. With the introduction of the derivative of a vector in the definitions of velocity and acceleration, it is important at this point to establish the rules under which the differentiation of vector quantities may be carried out. These rules are the same as for the differentiation of scalar quantities, except for the case of the cross product where the order of the terms must be preserved. These rules are covered in Art. C/7 of Appendix C and should be reviewed at this point.

For curvilinear motion of a particle in a plane, there are three different coordinate systems that are in common use to describe the vector relationships developed in this article. An important lesson to be learned from the study of these coordinate systems is the proper choice of a reference system for a given problem. This choice is usually revealed by the manner in which the motion is generated or by the form in which the data are specified. Each of the three coordinate systems will now be developed and illustrated.

2/4   RECTANGULAR COORDINATES *(x-y)*

This system of coordinates is particularly useful for describing motions where the x- and y-components of acceleration are independently generated or determined. The resulting curvilinear motion is, then, obtained by a vector combination of the x- and y-components of the position vector, the velocity, and the acceleration.

The particle path of Fig. 2/5 is shown again in Fig. 2/7 along with x- and y-axes. The position vector $\mathbf{r}$, the velocity $\mathbf{v}$, and the acceleration $\mathbf{a}$ of the particle as developed in Art. 2/3 are represented in Fig. 2/7 together with their x- and y-components. With the aid of the unit vectors $\mathbf{i}$ and $\mathbf{j}$, we may write the vectors $\mathbf{r}$, $\mathbf{v}$, and $\mathbf{a}$ in terms of their x- and y-components. Thus,

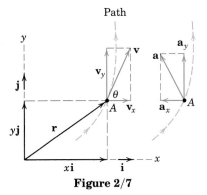

Figure 2/7

$$\boxed{\begin{aligned} \mathbf{r} &= x\mathbf{i} + y\mathbf{j} \\ \mathbf{v} &= \dot{\mathbf{r}} = \dot{x}\mathbf{i} + \dot{y}\mathbf{j} \\ \mathbf{a} &= \dot{\mathbf{v}} = \ddot{\mathbf{r}} = \ddot{x}\mathbf{i} + \ddot{y}\mathbf{j} \end{aligned}}$$ **(2/6)**

As we differentiate with respect to time, we observe that the unit vectors have no time derivatives since their magnitudes and directions remain constant. The scalar values of the components of $\mathbf{v}$ and $\mathbf{a}$ are merely $v_x = \dot{x}$, $v_y = \dot{y}$ and $a_x = \dot{v}_x = \ddot{x}$, $a_y = \dot{v}_y = \ddot{y}$. (As drawn in Fig. 2/7, a_x is in the negative x-direction, so that $\ddot{x}$ would be a negative number.)

As observed previously, the direction of the velocity is always tangent to the path, and from the figure it is clear that

$$v^2 = v_x{}^2 + v_y{}^2 \qquad v = \sqrt{v_x{}^2 + v_y{}^2} \qquad \tan\theta = \frac{v_y}{v_x}$$

$$a^2 = a_x{}^2 + a_y{}^2 \qquad a = \sqrt{a_x{}^2 + a_y{}^2}$$

If the angle θ is measured counterclockwise from the x-axis to $\mathbf{v}$ for the configuration of axes shown, then we may also observe that $dy/dx = \tan\theta = v_y/v_x$.

If the coordinates x and y are known independently as functions of time, $x = f_1(t)$ and $y = f_2(t)$, then for any value of the time we may combine them to obtain $\mathbf{r}$. Similarly, we combine their first derivatives $\dot{x}$ and $\dot{y}$ to obtain $\mathbf{v}$ and their second derivatives $\ddot{x}$ and $\ddot{y}$ to obtain $\mathbf{a}$. Or, if the acceleration components a_x and a_y are given as functions of the time, we may integrate each one separately with respect to time, once to obtain v_x and v_y and again to obtain $x = f_1(t)$ and $y = f_2(t)$. Elimination of the time t between these last two parametric equations gives the equation of the curved path $y = f(x)$.

From the foregoing discussion it should be recognized that the rectangular-coordinate representation of curvilinear motion is merely the superposition of the components of two simultaneous rectilinear motions in the x- and y-directions. Therefore, everything covered in Art. 2/2 on rectilinear motion may be applied separately to the x-motion and to the y-motion.

Projectile motion. An important application of two-dimensional kinematic theory is the problem of projectile motion. For a first treatment of the subject, we neglect aerodynamic drag and the earth's curvature and rotation, and we assume that the altitude range is small enough so that the acceleration due to gravity can be considered constant. With these assumptions, rectangular coordinates are useful for the trajectory analysis. For the axes shown in Fig. 2/8, the acceleration components are

$$a_x = 0 \qquad a_y = -g$$

Integration of these accelerations follows the results obtained previously in Art. 2/2a for constant acceleration and yields

$$v_x = (v_x)_0 \qquad\qquad v_y = (v_y)_0 - gt$$
$$x = x_0 + (v_x)_0 t \qquad y = y_0 + (v_y)_0 t - \tfrac{1}{2}gt^2$$
$$v_y{}^2 = (v_y)_0{}^2 - 2g(y - y_0)$$

In all these expressions, the subscript zero denotes initial conditions, frequently taken as those at launch where, for the case illustrated, $x_0 = y_0 = 0$. Note that the quantity g is taken to be positive throughout this text.

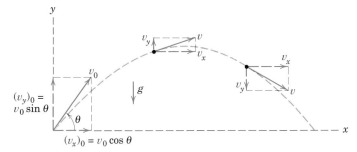

Figure 2/8

We can see that the x- and y-motions are independent for the simple projectile conditions under consideration. Elimination of the time t between the x- and y-displacement equations shows the path to be parabolic (see Sample Problem 2/6). If we were to introduce a drag force which depends on the speed squared (for example), then the x- and y-motions would be coupled (interdependent), and the trajectory would be nonparabolic.

When extreme accuracy is to be achieved in describing projectile motion when large velocities and high altitudes are involved, the shape of the projectile, the variation of g with altitude, the variation of the air density with altitude, and the rotation of the earth must all be accounted for. These factors introduce considerable complexity in the motion equations, and numerical integration of the acceleration equations is usually necessary.

Sample Problem 2/5

The curvilinear motion of a particle is defined by $v_x = 50 - 16t$ and $y = 100 - 4t^2$, where v_x is in meters per second, y is in meters, and t is in seconds. It is also known that $x = 0$ when $t = 0$. Plot the path of the particle and determine its velocity and acceleration when the position $y = 0$ is reached.

Solution. The x-coordinate is obtained by integrating the expression for v_x, and the x-component of the acceleration is obtained by differentiating v_x. Thus,

$$\left[\int dx = \int v_x \, dt\right] \qquad \int_0^x dx = \int_0^t (50 - 16t) \, dt \qquad x = 50t - 8t^2 \text{ m}$$

$$[a_x = \dot{v}_x] \qquad a_x = \frac{d}{dt}(50 - 16t) \qquad a_x = -16 \text{ m/s}^2$$

The y-components of velocity and acceleration are

$$[v_y = \dot{y}] \qquad v_y = \frac{d}{dt}(100 - 4t^2) \qquad v_y = -8t \text{ m/s}$$

$$[a_y = \dot{v}_y] \qquad a_y = \frac{d}{dt}(-8t) \qquad a_y = -8 \text{ m/s}^2$$

We now calculate corresponding values of x and y for various values of t and plot x against y to obtain the path as shown.

When $y = 0$, $0 = 100 - 4t^2$, so $t = 5$ s. For this value of the time, we have

$$v_x = 50 - 16(5) = -30 \text{ m/s}$$

$$v_y = -8(5) = -40 \text{ m/s}$$

$$v = \sqrt{(-30)^2 + (-40)^2} = 50 \text{ m/s}$$

$$a = \sqrt{(-16)^2 + (-8)^2} = 17.89 \text{ m/s}^2$$

The velocity and acceleration components and their resultants are shown on the separate diagrams for point A, where $y = 0$. Thus, for this condition we may write

$$\mathbf{v} = -30\mathbf{i} - 40\mathbf{j} \text{ m/s} \qquad \qquad Ans.$$

$$\mathbf{a} = -16\mathbf{i} - 8\mathbf{j} \text{ m/s}^2 \qquad \qquad Ans.$$

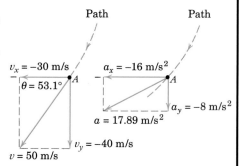

We observe that the velocity vector lies along the tangent to the path as it should, but that the acceleration vector is not tangent to the path. Note especially that the acceleration vector has a component that points toward the inside of the curved path. We concluded from our diagram in Fig. 2/5 that it is impossible for the acceleration to have a component that points toward the outside of the curve.

Sample Problem 2/6

A rocket has expended all its fuel when it reaches position A, where it has a velocity $\mathbf{u}$ at an angle θ with respect to the horizontal. It then begins unpowered flight and attains a maximum added height h at position B after traveling a horizontal distance s from A. Determine the expressions for h and s, the time t of flight from A to B, and the equation of the path. For the interval concerned, assume a flat earth with a constant gravitational acceleration g and neglect any atmospheric resistance.

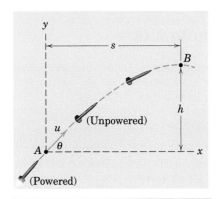

Solution. Since all motion components are directly expressible in terms of horizontal and vertical coordinates, a rectangular set of axes x-y ① will be employed. With the neglect of atmospheric resistance, $a_x = 0$ and $a_y = -g$, and the resulting motion is a direct superposition of two rectilinear motions with constant acceleration. Thus,

$$[dx = v_x \, dt] \qquad x = \int_0^t u \cos \theta \, dt \qquad x = ut \cos \theta$$

$$[dv_y = a_y \, dt] \qquad \int_{u \sin \theta}^{v_y} dv_y = \int_0^t (-g) \, dt \qquad v_y = u \sin \theta - gt$$

$$[dy = v_y \, dt] \qquad y = \int_0^t (u \sin \theta - gt) \, dt \qquad y = ut \sin \theta - \tfrac{1}{2}gt^2$$

Position B is reached when $v_y = 0$, which occurs for $0 = u \sin \theta - gt$ or

$$t = (u \sin \theta)/g \qquad \textit{Ans.}$$

Substitution of this value for the time into the expression for y gives the maximum added altitude

$$h = u \left(\frac{u \sin \theta}{g}\right) \sin \theta - \frac{1}{2} g \left(\frac{u \sin \theta}{g}\right)^2 \qquad h = \frac{u^2 \sin^2 \theta}{2g} \qquad \textit{Ans.}$$

The horizontal distance is seen to be

$$s = u \left(\frac{u \sin \theta}{g}\right) \cos \theta \qquad s = \frac{u^2 \sin 2\theta}{2g} \qquad \textit{Ans.}$$

which is clearly a maximum when $\theta = 45°$. The equation of the path is obtained by eliminating t from the expressions for x and y, which gives

$$y = x \tan \theta - \frac{gx^2}{2u^2} \sec^2 \theta \qquad \textit{Ans.}$$

③ This equation describes a vertical parabola as indicated in the figure.

① Note that this problem is simply the description of projectile motion neglecting atmospheric resistance.

② We see that the total range and time of flight for a projectile fired above a horizontal plane would be twice the respective values of s and t given here.

③ If atmospheric resistance were to be accounted for, the dependency of the acceleration components on the velocity would have to be established before an integration of the equations could be carried out. This becomes a much more difficult problem.

PROBLEMS

(In the following problems where motion as a projectile in air is involved, neglect air resistance unless otherwise stated and use $g = 9.81$ m/s^2.)

Introductory problems

2/59 At time $t = 0$, the position vector of a particle moving in the x-y plane is $\mathbf{r} = 5\mathbf{i}$ m. By time $t = 0.02$ s, its position vector has become $5.1\mathbf{i} + 0.4\mathbf{j}$ m. Determine the magnitude v_{av} of its average velocity during this interval and the angle θ made by the average velocity with the positive x-axis.

$\qquad$ *Ans.* $v_{av} = 20.6$ m/s, $\theta = 76.0°$

2/60 At time $t = 10$ s, the velocity of a particle moving in the x-y plane is $\mathbf{v} = +0.1\mathbf{i} + 2\mathbf{j}$ m/s. By time $t = 10.1$ s, its velocity has become $-0.1\mathbf{i} + 1.8\mathbf{j}$ m/s. Determine the magnitude a_{av} of its average acceleration during this interval and the angle θ made by the average acceleration with the positive x-axis.

2/61 A particle which moves in two-dimensional motion has coordinates given in millimeters by $x = t^2 - 4t + 20$ and $y = 3 \sin 2t$, where the time t is in seconds. Determine the magnitudes of the velocity $\mathbf{v}$ and the acceleration $\mathbf{a}$ and the angle θ between these two vectors at time $t = 3$ s.

$\qquad$ *Ans.* $v = 6.10$ mm/s
$\qquad\qquad$ $a = 3.90$ mm/s^2, $\theta = 11.67°$

2/62 The x- and y-motions of guides A and B with right-angle slots control the curvilinear motion of the connecting pin P, which slides in both slots. For a short interval, the motions are governed by $x = 20 + \frac{1}{4}t^2$ and $y = 15 - \frac{1}{6}t^3$, where x and y are in millimeters and t is in seconds. Calculate the magnitudes of the velocity $\mathbf{v}$ and acceleration $\mathbf{a}$ of the pin for $t = 2$ s. Sketch the direction of the path and indicate its curvature for this instant.

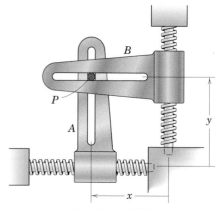

Problem 2/62

2/63 The particle P moves along the curved slot, a portion of which is shown. Its distance in meters measured along the slot is given by $s = t^2/4$, where t is in seconds. The particle is at A when $t = 2.00$ s and at B when $t = 2.20$ s. Determine the magnitude a_{av} of the average acceleration of P between A and B. Also express the acceleration as a vector $\mathbf{a}_{av}$ using unit vectors $\mathbf{i}$ and $\mathbf{j}$.

$\qquad$ *Ans.* $a_{av} = 2.76$ m/s^2
$\qquad\qquad$ $\mathbf{a}_{av} = 2.26\mathbf{i} - 1.580\mathbf{j}$ m/s^2

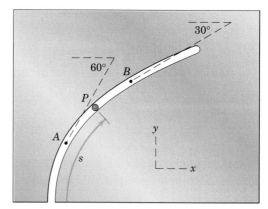

Problem 2/63

2/64 The rectangular coordinates of a particle moving in the x-y plane are given by $x = 3 \cos 4t$ and $y = 2 \sin 4t$, where the time t is in seconds and x and y are in meters. Sketch the position $\mathbf{r}$, velocity $\mathbf{v}$, and acceleration $\mathbf{a}$ at time $t = 1.4$ s and determine the angles θ_1 between $\mathbf{v}$ and $\mathbf{a}$ and θ_2 between $\mathbf{r}$ and $\mathbf{a}$.

2/65 The x-coordinate of a particle in curvilinear motion is given by $x = 2t^3 - 3t$, where x is in meters and t is in seconds. The y-component of acceleration in meters per second squared is given by $a_y = 4t$. If the particle has y-components $y = 0$ and $\dot{y} = 4$ m/s when $t = 0$, find the magnitudes of the velocity $\mathbf{v}$ and acceleration $\mathbf{a}$ when $t = 2$ s. Sketch the path for the first 2 seconds of motion, and show the velocity and acceleration vectors for $t = 2$ s.

Ans. $v = 24.2$ m/s, $a = 25.3$ m/s^2

2/66 Use the notation and conditions in Sample Problem 2/6 to show that the horizontal range $R = 2s$ of the projectile can be expressed in terms of the angle θ and the maximum altitude h only.

2/67 A broadjumper approaches his takeoff board A with a horizontal velocity of 10 m/s. Determine the vertical component v_y of the velocity of his center of gravity at takeoff for him to make the jump shown. What is the vertical rise h of his center of gravity?

Ans. $v_y = 3.68$ m/s, $h = 0.690$ m

Problem 2/67

2/68 The center of mass G of a high jumper follows the trajectory shown. Determine the component v_0, measured in the vertical plane of the figure, of his take-off velocity and angle θ if the apex of the trajectory just clears the bar at A. (In general, must the mass center G of the jumper clear the bar during a successful jump?)

Problem 2/68

2/69 With what minimum horizontal velocity u can a boy throw a rock at A and have it just clear the obstruction at B?

Ans. $u = 28.0$ m/s

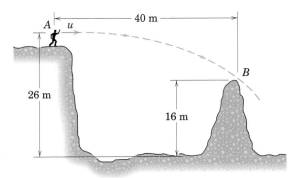

Problem 2/69

Representative problems

2/70 The rifle is aimed at point A and fired. Calculate the distance b below A to the point B where the bullet strikes. The muzzle velocity is 800 m/s.

Problem 2/70

2/71 A projectile is fired from the vertical tube mounted on the vehicle which is traveling at the constant speed $u = 30$ km/h. The projectile leaves the tube with a velocity $v_r = 20$ m/s relative to the tube. If air resistance is neglected, show that the projectile will land on the vehicle at the tube location and calculate the distance s traveled by the vehicle during the flight of the projectile.

Ans. $s = 34.0$ m

Problem 2/71

2/72 A rocket is released at point A from a jet aircraft flying horizontally at 1000 km/h at an altitude of 800 m. If the rocket thrust remains horizontal and gives the rocket a horizontal acceleration of $0.5g$, determine the angle θ from the horizontal to the line of sight to the target.

Problem 2/72

2/73 Water issues from the nozzle at A, which is 1.5 m above the ground. Determine the coordinates of the point of impact of the stream if the initial water speed is (*a*) $v_0 = 14$ m/s and (*b*) $v_0 = 18$ m/s.

Ans. (*a*) $x = 9$ m, $y = 8.98$ m
(*b*) $x = 29.4$ m, $y = 0$

Problem 2/73

2/74 Prove the well-known result that, for a given launch speed v_0, the launch angle $\theta = 45°$ yields the maximum horizontal range R. Determine the maximum range. (Note that this result does not hold when aerodynamic drag is included in the analysis.)

2/75 The pilot of an airplane carrying a package of mail to a remote outpost wishes to release the package at the right moment to hit the recovery location A. What angle θ with the horizontal should the pilot's line of sight to the target make at the instant of release? The airplane is flying horizontally at an altitude of 100 m with a velocity of 200 km/h.

Ans. $\theta = 21.7°$

Problem 2/75

2/76 A small airplane flying horizontally with a speed of 300 km/h at an altitude of 120 m above a remote valley drops an emergency medical package at A. The package has a parachute which deploys at B and allows the package to descend vertically at the constant rate of 1.8 m/s. If the drop is designed so that the package is to reach the ground 37 seconds after release at A, determine the horizontal lead L so that the package hits the target. Neglect atmospheric resistance from A to B.

300 km/h

120 m

Target

L

Problem 2/76

2/77 An outfielder experiments with two different trajectories for throwing to home plate from the position shown: (a) $v_0 = 42$ m/s with $\theta = 8°$ and (b) $v_0 = 36$ m/s with $\theta = 12°$. For each set of initial conditions, determine the time t required for the baseball to reach home plate and the altitude h as the ball crosses the plate.

Ans. (a) $t = 1.443$ s, $h = 0.525$ m
(b) $t = 1.704$ s, $h = 0.813$ m

A v_0 θ

2.3 m

h

B

60 m

Problem 2/77

2/78 A projectile is launched from point A with the initial conditions shown in the figure. Determine the slant distance s which locates the point B of impact. Calculate the time of flight t.

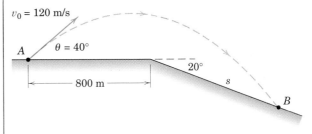

$v_0 = 120$ m/s

A $\theta = 40°$

800 m

20°

s

B

Problem 2/78

2/79 A projectile is launched with a speed $v_0 = 25$ m/s from the floor of a 5-m-high tunnel as shown. Determine the maximum horizontal range R of the projectile and the corresponding launch angle θ.

Ans. $R = 46.4$ m, $\theta = 23.3°$

5 m

$v_0 = 25$ m/s

θ

A

Problem 2/79

2/80 If the tennis player serves the ball horizontally ($\theta = 0$), calculate its velocity v if the center of the ball clears the 0.9-m net by 150 mm. Also find the distance s from the net to the point where the ball hits the court surface. Neglect air resistance and the effect of ball spin.

θ

A

v

2.55 m

0.9 m

s

11.7 m

Problem 2/80

2/81 A long-range artillery rifle at A is aimed at an angle of 45° with the horizontal, and its shell is just able to clear the mountain peak at the top of its trajectory. Determine the magnitude u of the muzzle velocity, the height H of the mountain above sea level, and the range R to the sea.

Ans. $u = 396$ m/s, $H = 4600$ m, $R = 16.58$ km

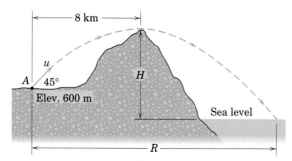

Problem 2/81

2/82 The muzzle velocity of a long-range rifle at A is $u = 400$ m/s. Determine the two angles of elevation θ which will permit the projectile to hit the mountain target B.

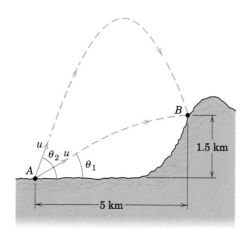

Problem 2/82

2/83 A projectile is launched with an initial speed of 200 m/s at an angle of 60° with respect to the horizontal. Compute the range R as measured up the incline.

Ans. $R = 2970$ m

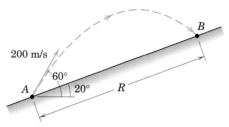

Problem 2/83

2/84 A team of engineering students is designing a catapult to launch a small ball at A so that it lands in the box. If it is known that the initial velocity vector makes a 30° angle with the horizontal, determine the range of launch speeds v_0 for which the ball will land inside the box.

Problem 2/84

2/85 Ball bearings leave the horizontal trough with a velocity of magnitude u and fall through the 70-mm-diameter hole as shown. Calculate the permissible range of u which will enable the balls to enter the hole. Take the dotted positions to represent the limiting conditions.

Ans. $u_{max} = 1.135$ m/s, $u_{min} = 0.744$ m/s

Problem 2/85

2/86 A horseshoe player releases the horseshoe at A with an initial speed $v_0 = 11$ m/s. Determine the range for the launch angle θ for which the shoe will strike the 350-mm vertical stake.

Problem 2/86

2/87 A boy throws a ball upward with a speed $v_0 = 12$ m/s. The wind imparts a horizontal acceleration of 0.4 m/s^2 to the left. At what angle θ must the ball be thrown so that it returns to the point of release? Assume that the wind does not affect the vertical motion.

$$Ans. \quad \theta = 2.33°$$

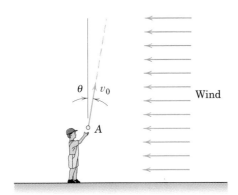

Problem 2/87

2/88 In the cathode-ray tube, electrons traveling horizontally from their source with the velocity v_0 are deflected by an electric field E due to the voltage gradient across the plates P. The deflecting force causes an acceleration in the vertical direction on the sketch equal to eE/m, where e is the electron charge and m is its mass. When clear of the plates, the electrons travel in straight lines. Determine the expression for the deflection δ for the tube and plate dimensions shown.

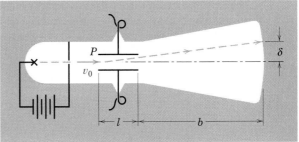

Problem 2/88

2/89 To meet design criteria, small ball bearings must bounce through an opening of limited size at the top of their trajectory when rebounding from a heavy plate as shown. Calculate the angle θ made by the rebound velocity with the horizontal and the velocity v of the balls as they pass through the opening.

$$Ans. \quad \theta = 68.2°, v = 1.253 \text{ m/s}$$

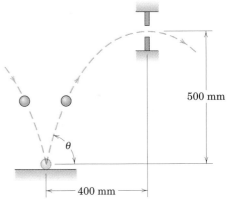

Problem 2/89

2/90 Determine the location h of the spot toward which the pitcher must throw if the ball is to hit the catcher's mitt. The ball is released with a speed of 40 m/s.

Problem 2/90

2/91 A projectile is fired with a velocity u at right angles to the slope, which is inclined at an angle θ with the horizontal. Derive an expression for the distance R to the point of impact.

$$\text{Ans. } R = \frac{2u^2}{g} \tan \theta \sec \theta$$

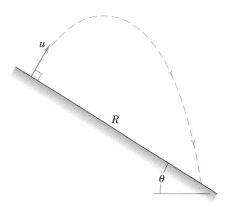

Problem 2/91

2/92 The basketball player likes to release his foul shots at an angle $\theta = 50°$ to the horizontal as shown. What initial speed v_0 will cause the ball to pass through the center of the rim?

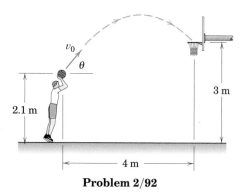

Problem 2/92

2/93 A projectile is launched from point A with the initial conditions shown in the figure. Determine the x- and y-coordinates of the point of impact.

$$\text{Ans. } x = 373 \text{ m}, y = 18.75 \text{ m}$$

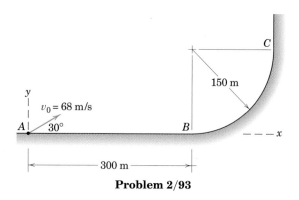

Problem 2/93

2/94 An object which is released from rest from the top A of a tower of height h will appear not to fall straight down due to the effect of the earth's rotation. It may be shown that the object has an eastward horizontal acceleration relative to the horizontal surface of the earth equal to $2v_y\omega \cos \gamma$, where v_y is the free-fall downward velocity, ω is the angular velocity of the earth, and γ is the latitude, north or south. Determine the deflection b if $h = 300$ m and $\gamma = 30°$ north. From Table D/3, $\omega = 0.7292(10^{-4})$ rad/s and from Fig. 1/1, $g = 9.793$ m/s^2.

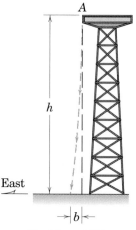

Problem 2/94

2/95 A projectile is ejected into an experimental fluid at time $t = 0$. The initial speed is v_0 and the angle to the horizontal is θ. The drag on the projectile results in an acceleration term $\mathbf{a}_D = -k\mathbf{v}$, where k is a constant and $\mathbf{v}$ is the velocity of the projectile. Determine the x- and y-components of both the velocity and displacement as functions of time. What is the terminal velocity? Include the effects of gravitational acceleration.

Ans. $v_x = (v_0 \cos \theta)e^{-kt}$, $x = \dfrac{v_0 \cos \theta}{k}(1 - e^{-kt})$

$$v_y = \left(v_0 \sin \theta + \frac{g}{k}\right)e^{-kt} - \frac{g}{k}$$

$$y = \frac{1}{k}\left(v_0 \sin \theta + \frac{g}{k}\right)(1 - e^{-kt}) - \frac{g}{k}t$$

$$v_x \to 0, \ v_y \to -\frac{g}{k}$$

Problem 2/95

▶**2/96** A projectile is launched with speed v_0 from point A. Determine the launch angle θ that results in the maximum range R up the incline of angle α (where $0 \le \alpha \le 90°$). Evaluate your results for $\alpha = 0, 30°$, and 45°.

Ans. $\theta = \dfrac{90° + \alpha}{2}$, $\theta = 45°, 60°, 67.5°$

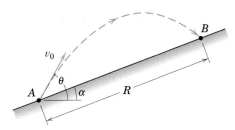

Problem 2/96

2/5 NORMAL AND TANGENTIAL COORDINATES (n-t)

In the introduction to this chapter, Art. 2/1, it was mentioned that one of the common descriptions of curvilinear motion involves the use of path variables, which are measurements made along the tangent t and normal n to the path of the particle. These coordinates provide a very natural description for curvilinear motion and are frequently the most direct and convenient coordinates to use. The n- and t-coordinates are considered to move along the path with the particle, as seen in Fig. 2/9 where the particle advances from A to B to C. The positive direction for n at any position is always taken toward the center of curvature of the path. As seen from Fig. 2/9, the positive n-direction will shift from one side of the curve to the other side if the curvature changes direction.

Figure 2/9

The coordinates n and t will now be used to describe the velocity $\mathbf{v}$ and acceleration $\mathbf{a}$ that were introduced in Art. 2/3 for the curvilinear motion of a particle. For this purpose, we introduce unit vectors $\mathbf{e}_n$ in the n-direction and $\mathbf{e}_t$ in the t-direction, as shown in Fig. 2/10a for the position of the particle at point A on its path. During a differential increment of time dt, the particle moves a differential distance ds along the curve from A to A'. With the radius of curvature of the path at this position designated by ρ, we see that $ds = \rho\,d\beta$, where β is in radians. We see that it is unnecessary to consider the differential change in ρ between A and A' since a higher-order term would be introduced that disappears in the limit. Thus, the magnitude of the velocity may be written $v = ds/dt = \rho\,d\beta/dt$, and we may write the velocity as the vector

$$\mathbf{v} = v\mathbf{e}_t = \rho\dot{\beta}\mathbf{e}_t \qquad (2/7)$$

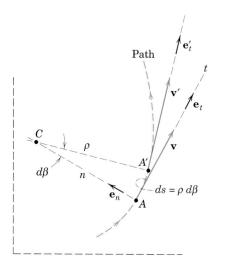

(a)

The acceleration $\mathbf{a}$ of the particle was defined in Art. 2/3 as $\mathbf{a} = d\mathbf{v}/dt$, and we observed from Fig. 2/5 that the acceleration is a vector which reflects both the change in magnitude and the change in direction of $\mathbf{v}$. We now differentiate $\mathbf{v}$ in Eq. 2/7 by applying the ordinary rule for the differentiation of the product of a scalar and a vector* and get

$$\mathbf{a} = \frac{d\mathbf{v}}{dt} = \frac{d(v\mathbf{e}_t)}{dt} = v\dot{\mathbf{e}}_t + \dot{v}\mathbf{e}_t \qquad (2/8)$$

where the unit vector $\mathbf{e}_t$ now has a derivative because its direction changes. To find $\dot{\mathbf{e}}_t$ we analyze the change in $\mathbf{e}_t$ during a differential increment of motion as the particle moves from A to A' in Fig. 2/10a. The unit vector $\mathbf{e}_t$ correspondingly changes to $\mathbf{e}_t'$, and the vector difference $d\mathbf{e}_t$ is shown in the b-part of the figure. The vector $d\mathbf{e}_t$ in the limit has a magnitude equal to the length of the arc

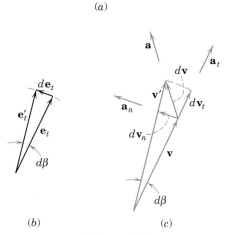

(b)　　　(c)

Figure 2/10

*See Art. C/7 of Appendix C.

$|\mathbf{e}_t|\, d\beta = d\beta$ obtained by swinging the unit vector $\mathbf{e}_t$ through the angle $d\beta$ expressed in radians. The direction of $d\mathbf{e}_t$ is given by $\mathbf{e}_n$. Thus, we may write $d\mathbf{e}_t = \mathbf{e}_n\, d\beta$. Dividing by $d\beta$ gives

$$\frac{d\mathbf{e}_t}{d\beta} = \mathbf{e}_n$$

or dividing by dt gives $d\mathbf{e}_t/dt = (d\beta/dt)\mathbf{e}_n$, which may be written

$$\dot{\mathbf{e}}_t = \dot{\beta}\mathbf{e}_n \qquad\qquad (2/9)$$

With the substitution of Eq. 2/9 and $\dot{\beta}$ from the relation $v = \rho\dot{\beta}$, Eq. 2/8 for the acceleration becomes

$$\mathbf{a} = \frac{v^2}{\rho}\mathbf{e}_n + \dot{v}\mathbf{e}_t \qquad\qquad (2/10)$$

where

$$a_n = \frac{v^2}{\rho} = \rho\dot{\beta}^2 = v\dot{\beta}$$

$$a_t = \dot{v} = \ddot{s}$$

$$a = \sqrt{a_n{}^2 + a_t{}^2}$$

We may also note that $a_t = \dot{v} = d(\rho\dot{\beta})/dt = \rho\ddot{\beta} + \dot{\rho}\dot{\beta}$. However, this relation finds little use since we seldom have reason to compute $\dot{\rho}$.

A clear understanding of Eq. 2/10 comes only when the geometry of the physical changes that it describes can be clearly seen. Figure 2/10c shows the velocity vector $\mathbf{v}$ when the particle is at A and $\mathbf{v}'$ when it is at A'. The vector change in the velocity is $d\mathbf{v}$, which establishes the direction of the acceleration $\mathbf{a}$. The n-component of $d\mathbf{v}$ is labeled $d\mathbf{v}_n$, and in the limit its magnitude equals the length of the arc generated by swinging the vector $\mathbf{v}$ as a radius through the angle $d\beta$. Thus, $|d\mathbf{v}_n| = v\, d\beta$ and the n-component of acceleration is $a_n = |d\mathbf{v}_n|/dt = v(d\beta/dt) = v\dot{\beta}$ as before. The t-component of $d\mathbf{v}$ is labeled $d\mathbf{v}_t$, and its magnitude is simply the change dv in the magnitude or length of the velocity vector. Therefore, the t-component of acceleration is $a_t = dv/dt = \dot{v} = \ddot{s}$ as before. The acceleration vectors resulting from the corresponding vector changes in velocity are shown in Fig. 2/10c.

It is especially important to observe that the normal component of acceleration a_n is *always directed toward the center of curvature* C. The tangential component of acceleration, on the other hand, will be in the positive t-direction of motion if the speed v is increasing and in the negative t-direction if the speed is decreasing. In Fig. 2/11 are shown schematic representations of the variation

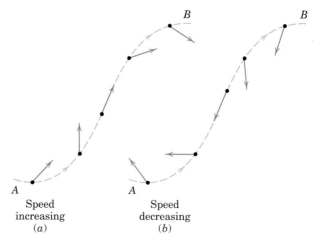

Speed
increasing
(a)

Speed
decreasing
(b)

Acceleration vectors for
particle moving from A to B

Figure 2/11

in the acceleration vector for a particle moving from A to B with
(a) increasing speed and (b) decreasing speed. At an inflection point
on the curve, the normal acceleration v^2/ρ goes to zero since ρ becomes infinite.

Circular motion is an important special case of plane curvilinear motion where the radius of curvature ρ becomes the constant
radius r of the circle and the angle β is replaced by the angle θ
measured from any convenient radial reference to OP, Fig. 2/12.
The velocity and the acceleration components for the circular motion of the particle P become

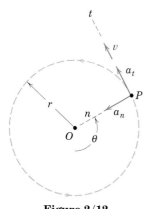

Figure 2/12

$$
\begin{aligned}
v &= r\dot\theta \\
a_n &= v^2/r = r\dot\theta^2 = v\dot\theta \\
a_t &= \dot v = r\ddot\theta
\end{aligned}
\tag{2/11}
$$

We find repeated use for Eqs. 2/10 and 2/11 in dynamics, and
these relations and the principles behind them should be *mastered
thoroughly*.

Sample Problem 2/7

To anticipate the dip and hump in the road, the driver of a car applies her brakes to produce a uniform deceleration. Her speed is 100 km/h at the bottom A of the dip and 50 km/h at the top C of the hump, which is 120 m along the road from A. If the passengers experience a total acceleration of 3 m/s^2 at A and if the radius of curvature of the hump at C is 150 m, calculate (a) the radius of curvature ρ at A, (b) the acceleration at the inflection point B, and (c) the total acceleration at C.

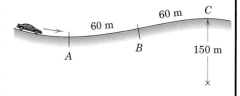

(1) **Solution.** The dimensions of the car are small compared with those of the path, so we will treat the car as a particle. The velocities are

$$v_A = \left(100 \frac{\text{km}}{\text{h}}\right)\left(\frac{1 \text{ h}}{3600 \text{ s}}\right)\left(1000 \frac{\text{m}}{\text{km}}\right) = 27.8 \text{ m/s}$$

$$v_C = 50 \frac{1000}{3600} = 13.89 \text{ m/s}$$

We find the constant deceleration along the path from

$$\left[\int v \, dv = \int a_t \, ds\right] \qquad \int_{v_A}^{v_C} v \, dv = a_t \int_0^s ds$$

$$a_t = \frac{1}{2s}(v_C^2 - v_A^2) = \frac{(13.89)^2 - (27.8)^2}{2(120)} = -2.41 \text{ m/s}^2$$

(a) Condition at A. With the total acceleration given and a_t determined, we can easily compute a_n and hence ρ from

$$[a^2 = a_n^2 + a_t^2] \qquad a_n^2 = 3^2 - (2.41)^2 = 3.19 \qquad a_n = 1.785 \text{ m/s}^2$$

$$[a_n = v^2/\rho] \qquad \rho = v^2/a_n = (27.8)^2/1.785 = 432 \text{ m} \qquad \textit{Ans.}$$

(b) Condition at B. Since the radius of curvature is infinite at the inflection point, $a_n = 0$ and

$$a = a_t = -2.41 \text{ m/s}^2 \qquad \textit{Ans.}$$

(c) Condition at C. The normal acceleration becomes

$$[a_n = v^2/\rho] \qquad a_n = (13.89)^2/150 = 1.286 \text{ m/s}^2$$

With unit vectors $\mathbf{e}_n$ and $\mathbf{e}_t$ in the n- and t-directions, the acceleration may be written

$$\mathbf{a} = 1.286\mathbf{e}_n - 2.41\mathbf{e}_t \text{ m/s}^2$$

where the magnitude of $\mathbf{a}$ is

$$[a = \sqrt{a_n^2 + a_t^2}] \qquad a = \sqrt{(1.286)^2 + (-2.41)^2} = 2.73 \text{ m/s}^2 \qquad \textit{Ans.}$$

The acceleration vectors representing the conditions at each of the three points are shown for clarification.

(1) Actually, the radius of curvature to the road differs by about 1 m from that to the path followed by the center of mass of the passengers, but we have neglected this relatively small difference.

Sample Problem 2/8

A certain rocket maintains a horizontal attitude of its axis during the powered phase of its flight at high altitude. The thrust imparts a horizontal component of acceleration of 6 m/s^2, and the downward acceleration component is the acceleration due to gravity at that altitude, which is $g = 9$ m/s^2. At the instant represented, the velocity of the mass center G of the rocket along the 15° direction of its trajectory is $20(10^3)$ km/h. For this position determine (a) the radius of curvature of the flight trajectory, (b) the rate at which the speed v is increasing, (c) the angular velocity $\dot{\beta}$ of the radial line from G to the center of curvature C, and (d) the vector expression for the total acceleration $\mathbf{a}$ of the rocket.

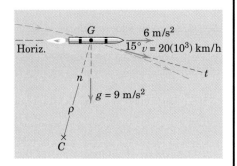

Solution. We observe that the radius of curvature appears in the expression for the normal component of acceleration, so we use n- and t-coordinates to describe the motion of G. The n- and t-components of the total acceleration are obtained by resolving the given horizontal and vertical accelerations into their n- and t-components and then combining. From the figure we get

$$a_n = 9 \cos 15° - 6 \sin 15° = 7.14 \text{ m/s}^2$$

$$a_t = 9 \sin 15° + 6 \cos 15° = 8.12 \text{ m/s}^2$$

(①) Alternatively, we could find the resultant acceleration and then resolve it into n- and t-components.

(a)   We may now compute the radius of curvature from

(②) $[a_n = v^2/\rho]$ $\rho = \dfrac{v^2}{a_n} = \dfrac{[20(10^3)/3.6]^2}{7.14} = 4.32(10^6)$ m   *Ans.*

(②) To convert from km/h to m/s, multiply by $\dfrac{1000 \text{ km/m}}{3600 \text{ s/h}}$ or divide by 3.6, which is easily remembered.

(b)   The rate at which v is increasing is simply the t-component of acceleration.

$[\dot{v} = a_t]$ $\dot{v} = 8.12$ m/s^2 *Ans.*

(c)   The angular rate $\dot{\beta}$ of line GC depends on v and ρ and is given by

$[v = \rho\dot{\beta}]$ $\dot{\beta} = v/\rho = \dfrac{20(10^3)/3.6}{4.32(10^6)} = 12.85(10^{-4})$ rad/s   *Ans.*

(d)   With unit vectors $\mathbf{e}_n$ and $\mathbf{e}_t$ for the n- and t-directions, respectively, the total acceleration becomes

$$\mathbf{a} = 7.14\mathbf{e}_n + 8.12\mathbf{e}_t \text{ m/s}^2$$ *Ans.*

PROBLEMS

Introductory problems

2/97 A test car starts from rest on a horizontal circular track of 80-m radius and increases its speed at a uniform rate to reach 100 km/h in 10 seconds. Determine the magnitude a of the total acceleration of the car 8 seconds after the start.

Ans. $a = 6.77 \text{ m/s}^2$

2/98 Six acceleration vectors are shown for the car whose velocity vector is directed forward. For each acceleration vector describe in words the instantaneous motion of the car.

Problem 2/98

2/99 The particle P moves in the circular path shown. Sketch the acceleration vector **a** and determine its magnitude a for the following cases: (a) speed v is 1.2 m/s and is constant, (b) speed is 1.2 m/s and is increasing at the rate of 2.4 m/s each second, and (c) speed is 1.2 m/s and is decreasing at the rate of 4.8 m/s each second. In each case the particle is in the position shown in the figure.

Ans. (a) $a = 2.4 \text{ m/s}^2$, (b) $a = 3.39 \text{ m/s}^2$
(c) $a = 5.37 \text{ m/s}^2$

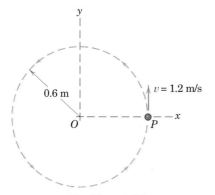

Problem 2/99

2/100 A particle moves along the curved path shown. The particle has a speed $v_A = 4$ m/s at time t_A and a speed $v_B = 4.2$ m/s at time t_B. Determine the average values of the normal and tangential accelerations of the particle between points A and B.

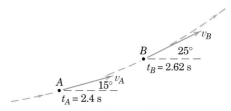

Problem 2/100

2/101 A train enters a curved horizontal section of track at a speed of 100 km/h and slows down with constant deceleration to 50 km/h in 12 seconds. An accelerometer mounted inside the train records a horizontal acceleration of 2 m/s² when the train is 6 seconds into the curve. Calculate the radius of curvature ρ of the track for this instant.

Ans. $\rho = 266$ m

2/102 The car travels at a constant speed from the bottom A of the dip to the top B of the hump. If the radius of curvature of the road at A is $\rho_A = 120$ m and the car acceleration at A is $0.4g$, determine the car speed v. If the acceleration at B must be limited to $0.25g$, determine the minimum radius of curvature ρ_B of the road at B.

Problem 2/102

2/103 The speed of a car increases uniformly with time from 50 km/h at A to 100 km/h at B during 10 seconds. The radius of curvature of the hump at A is 40 m. If the magnitude of the total acceleration of the car's mass center is the same at B as at A, compute the radius of curvature ρ_B of the dip in the road at B. The mass center of the car is 0.6 m from the road.

Ans. $\rho_B = 163.0$ m

Problem 2/103

2/104 A particle moves on a circular path of radius $r = 0.8$ m with a constant speed of 2 m/s. The velocity undergoes a vector change $\Delta\mathbf{v}$ from A to B. Express the magnitude of $\Delta\mathbf{v}$ in terms of v and $\Delta\theta$ and divide it by the time interval Δt between A and B and obtain the magnitude of the average acceleration of the particle for (*a*) $\Delta\theta = 30°$, (*b*) $\Delta\theta = 15°$, and (*c*) $\Delta\theta = 5°$. In each case, determine the percentage difference from the instantaneous value of acceleration.

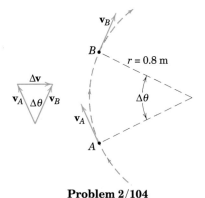

Problem 2/104

2/105 The figure shows two possible paths for negotiating an unbanked turn on a horizontal portion of a race course. Path AA follows the center line of the road and has a radius of curvature $\rho_A = 85$ m, while path BB uses the width of the road to good advantage in increasing the radius of curvature to $\rho_B = 200$ m. If the drivers limit their speeds in their curves so that the lateral acceleration does not exceed $0.8g$, determine the maximum speed for each path.

Ans. $v_A = 25.8$ m/s, $v_B = 39.6$ m/s

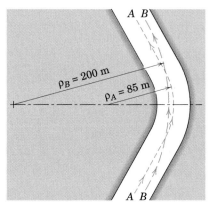

Problem 2/105

2/106 A minivan starts from rest on the road whose constant radius of curvature is 40 m and whose bank angle is 10°. The motion occurs in a horizontal plane. If the constant forward acceleration of the minivan is 1.8 m/s², determine the magnitude a of its total acceleration 5 s after starting.

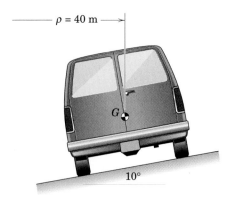

Problem 2/106

Representative problems

2/107 Consider the polar axis of the earth to be fixed in space and compute the magnitude of the acceleration **a** of a point P on the earth's surface at latitude 40° north. The mean diameter of the earth is 12 742 km and its angular velocity is $0.729(10^{-4})$ rad/s.

Ans. $a = 0.0259$ m/s^2

Problem 2/107

2/108 A space shuttle that moves in a circular orbit around the earth at a height $h = 240$ km above its surface must have a speed of 27 955 km/h. Calculate the gravitational acceleration g for this altitude. The mean radius of the earth is 6371 km. (Check your answer by computing g from the gravitational law $g = g_0 \left(\dfrac{R}{R + h}\right)^2$, where $g_0 = 9.821$ m/s^2 from Table D/2 in Appendix D.)

$h = 240$ km

Problem 2/108

2/109 A spacecraft S is orbiting Jupiter in a circular path 1000 km above the surface with a constant speed. Using the gravitational law, calculate the magnitude v of its orbital velocity with respect to Jupiter. The diameter of Jupiter is 142 984 km and its surface-level gravitational acceleration is 24.85 m/s^2.

Ans. $v = 41\ 900$ m/s

Problem 2/109

2/110 In the design of a tape transfer mechanism, magnetic tape is being transferred from reel A to reel B and passes around idler pulleys C and D. At a certain instant, point P_1 on the tape is in contact with pulley C and point P_2 is in contact with pulley D. If the normal component of acceleration of P_1 is 40 m/s^2 and the tangential component of acceleration of P_2 is 30 m/s^2 at this instant, compute the corresponding speed v of the tape, the magnitude of the total acceleration of P_1, and the magnitude of the total acceleration of P_2.

Problem 2/110

2/111 To simulate a condition of "weightlessness" in its cabin, a jet transport traveling at 800 km/h moves on a sustained vertical curve as shown. At what rate $\dot\beta$ in degrees per second should the pilot drop his longitudinal line of sight to effect the desired condition? The maneuver takes place at a mean altitude of 8 km, and the gravitational acceleration may be taken as 9.79 m/s^2.

Ans. $\dot\beta = 2.52$ deg/s

Problem 2/111

2/112 The preliminary design for a "small" space station to orbit the earth in a circular path consists of a ring (torus) with a circular cross section as shown. The living space within the torus is shown in section A, where the "ground level" is 6 m from the center of the section. Calculate the angular speed N in revolutions per minute required to simulate standard gravity at the surface of the earth (9.81 m/s^2). Recall that you would be unaware of a gravitational field if you were in a non-rotating spacecraft in a circular orbit around the earth.

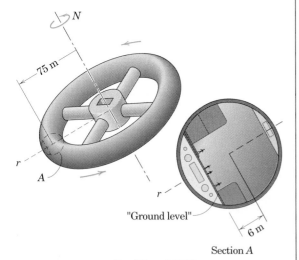

Section A

Problem 2/112

2/113 The design of a camshaft-drive system of a four-cylinder automobile engine is shown. As the engine is revved up, the belt speed v changes uniformly from 3 m/s to 6 m/s over a two-second interval. Calculate the magnitudes of the accelerations of points P_1 and P_2 halfway through this time interval.

Ans. $a_{P_1} = 338$ m/s^2, $a_{P_2} = 1.5$ m/s^2

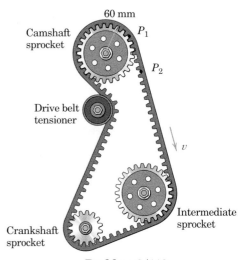

Problem 2/113

2/114 A baseball player releases a ball with the initial conditions shown in the figure. Determine the radius of curvature of the trajectory (*a*) just after release and (*b*) at the apex. For each case, compute the time rate of change of the speed.

Problem 2/114

2/115 For the baseball of Prob. 2/114, determine the radius of curvature ρ of the path and the time rate of change $\dot{v}$ of the speed at times $t = 1$ s and $t = 2.5$ s, where $t = 0$ is the time of release from the player's hand.

> *Ans.* (a) $\rho = 73.0$ m, $\dot{v} = -1.922$ m/s^2
> (b) $\rho = 83.1$ m, $\dot{v} = 3.38$ m/s^2

2/116 A car travels along the level curved road with a speed that is decreasing at the constant rate of 0.6 m/s each second. The speed of the car as it passes point A is 16 m/s. Calculate the magnitude of the total acceleration of the car as it passes point B which is 120 m along the road from A. The radius of curvature of the road at B is 60 m.

Problem 2/116

2/117 In the design of a timing mechanism, the motion of the pin A in the fixed circular slot is controlled by the guide B, which is being elevated by its lead screw with a constant upward velocity $v_0 = 2$ m/s for an interval of its motion. Calculate both the normal and tangential components of acceleration of pin A as it passes the position for which $\theta = 30°$.

> *Ans.* $a_n = 21.3$ m/s^2, $a_t = -12.32$ m/s^2

Problem 2/117

2/118 At a certain point in the reentry of the space shuttle into the earth's atmosphere, the total acceleration of the shuttle may be represented by two components. One component is the gravitational acceleration $g = 9.66$ m/s^2 at this altitude. The second component equals 12.90 m/s^2 due to atmospheric resistance and is directed opposite to the velocity. The shuttle is at an altitude of 48.2 km and has reduced its orbital velocity of 28 300 km/h to 15 450 km/h in the direction $\theta = 1.50°$. For this instant, calculate the radius of curvature ρ of the path and the rate $\dot{v}$ at which the speed is changing.

Problem 2/118

2/119 Race car A follows path a-a while race car B follows path b-b on the unbanked track. If each car has a constant speed limited to that corresponding to a lateral (normal) acceleration of $0.8g$, determine the times t_A and t_B for both cars to negotiate the turn as delimited by the line C-C.

Ans. $t_A = 10.52$ s, $t_B = 10.86$ s

Problem 2/119

2/120 An earth satellite which moves in the elliptical equatorial orbit shown has a velocity v in space of 17 970 km/h when it passes the end of the semiminor axis at A. The earth has an absolute surface value of g of 9.821 m/s^2 and has a radius of 6371 km. Determine the radius of curvature ρ of the orbit at A.

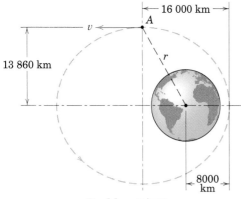

Problem 2/120

2/121 A small particle P starts from point O with a negligible speed and increases its speed to a value $v = \sqrt{2gy}$, where y is the vertical drop from O. When $x = 15$ m, determine the n-component of acceleration of the particle. (See Art. C/10 for the radius of curvature.)

Ans. $a_n = 0.520$ m/s^2

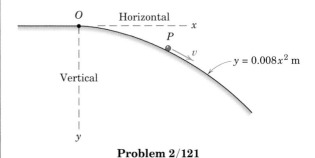

Problem 2/121

2/122 The mine skip is being hauled to the surface over the curved track by the cable wound around the 750-mm drum which turns at the constant clockwise speed of 120 rev/min. The shape of the track is designed so that $y = x^2/16$, where x and y are in meters. Calculate the magnitude of the total acceleration of the skip as it reaches a level of 1 m below the top. Neglect the dimensions of the skip compared with those of the path. Recall that the radius of curvature is given by

$$\rho = \frac{\left[1 + \left(\dfrac{dy}{dx}\right)^2\right]^{3/2}}{\dfrac{d^2y}{dx^2}}$$

Problem 2/122

2/123 Magnetic tape runs over the idler pulley in a computer as shown. If the total acceleration of a point P on the tape in contact with the pulley makes an angle of 4° with the tangent to the tape at time $t = 0$ when the velocity v of the tape is 4 m/s, determine the time t required to bring the pulley to a stop with constant deceleration. Assume no slipping between the pulley and the tape.

Ans. $t = 2.10(10^{-3})$ s

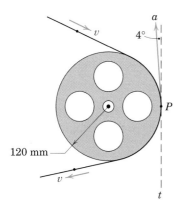

Problem 2/123

2/124 In the design of a control mechanism, the vertical slotted guide is moving with a constant velocity $\dot{x} = 150$ mm/s during the interval of motion from $x = -80$ mm to $x = +80$ mm. For the instant when $x = 60$ mm, calculate the n- and t-components of acceleration of the pin P, which is confined to move in the parabolic groove. From these results, determine the radius of curvature ρ of the path at this position. Verify your result by computing ρ from the expression cited in Appendix C/10.

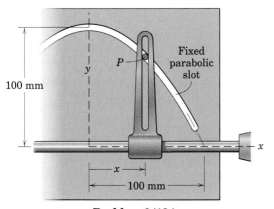

Problem 2/124

2/125 A ball is thrown horizontally from the top of a 50-m cliff at A with a speed of 15 m/s and lands at point C. Because of a strong horizontal wind the ball has a constant acceleration in the negative x-direction. Determine the radius of curvature ρ of the path of the ball at B where its trajectory makes an angle of 45° with the horizontal. Neglect any effect of air resistance in the vertical direction.

Ans. $\rho = 41.8$ m

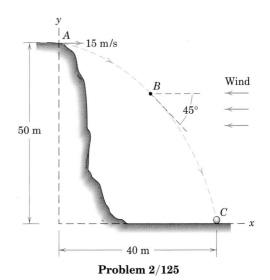

Problem 2/125

2/126 A race driver traveling at a speed of 250 km/h on the straightaway applies his brakes at point A and reduces his speed at a uniform rate to 200 km/h at C in a distance of $150 + 150 = 300$ m. Calculate the magnitude of the total acceleration of the race car an instant after it passes point B.

Problem 2/126

2/127 A particle moving in the x-y plane has a position vector $\mathbf{r} = [-4t^2 + 14t + 4]\mathbf{i} + [6t + 1]\mathbf{j}$, where $\mathbf{r}$ is in meters and t is in seconds. Determine and sketch the unit vectors $\mathbf{e}_n$ and $\mathbf{e}_t$ in terms of $\mathbf{i}$ and $\mathbf{j}$ at time $t = 1$ s. Also calculate the corresponding values of v, a_t, a_n, and ρ.

> *Ans.* $\mathbf{e}_t = 0.707\mathbf{i} + 0.707\mathbf{j}$
> $\mathbf{e}_n = -0.707\mathbf{i} + 0.707\mathbf{j}$, $v = 8.49$ m/s
> $a_t = -5.66$ m/s^2, $a_n = 5.66$ m/s^2
> $\rho = 12.73$ m

2/128 During a short interval the slotted guides are designed to move according to $x = 16 - 12t + 4t^2$ and $y = 2 + 15t - 3t^2$, where x and y are in millimeters and t is in seconds. At the instant when $t = 2$ s, determine the radius of curvature ρ of the path of the constrained pin P.

Problem 2/128

▶2/129 As a handling test, a car is driven through the slalom course shown. It is assumed that the car path is sinusoidal and that the maximum lateral acceleration is $0.7g$. If the testers wish to design a slalom through which the maximum speed is 80 km/h, what cone spacing L should be used?

> *Ans.* $L = 46.1$ m

Problem 2/129

▶2/130 A particle which moves with curvilinear motion has coordinates in meters which vary with time t in seconds according to $x = 2t^2 + 3t - 1$ and $y = 5t - 2$. Determine the coordinates of the center of curvature C at time $t = 1$ s.

> *Ans.* $x_C = 22.5$ m, $y_C = -22.9$ m

2/6 *POLAR COORDINATES (r-θ)*

We now consider the third description of plane curvilinear motion, namely, polar coordinates where the particle is located by the radial distance r from a fixed pole and by an angular measurement θ to the radial line. Polar coordinates are particularly useful when a motion is constrained through the control of a radial distance and an angular position or when an unconstrained motion is observed by measurements of a radial distance and an angular position.

Figure 2/13a shows the polar coordinates r and θ that locate a particle traveling on a curved path. An arbitrary fixed line, such as the x-axis, is used as a reference for the measurement of θ. Unit vectors $\mathbf{e}_r$ and $\mathbf{e}_\theta$ are established in the positive r- and θ-directions, respectively. The position vector $\mathbf{r}$ to the particle at A has a magnitude equal to the radial distance r and a direction specified by the unit vector $\mathbf{e}_r$. Thus, we express the location of the particle at A by the vector

$$\mathbf{r} = r\mathbf{e}_r$$

(a)

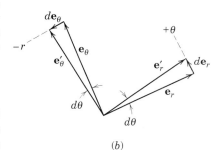

(b)

Figure 2/13

As we differentiate this relation with respect to time to obtain $\mathbf{v} = \dot{\mathbf{r}}$ and $\mathbf{a} = \dot{\mathbf{v}}$, we will need expressions for the time derivatives of both unit vectors $\mathbf{e}_r$ and $\mathbf{e}_\theta$. We obtain $\dot{\mathbf{e}}_r$ and $\dot{\mathbf{e}}_\theta$ in exactly the same way that we derived $\dot{\mathbf{e}}_t$ in the preceding article. During time dt the coordinate directions rotate through the angle $d\theta$, and the unit vectors also rotate through the same angle from $\mathbf{e}_r$ and $\mathbf{e}_\theta$ to $\mathbf{e}_r'$ and $\mathbf{e}_\theta'$, as shown in Fig. 2/13b. We note that the vector change $d\mathbf{e}_r$ is in the plus θ-direction and that $d\mathbf{e}_\theta$ is in the minus r-direction. Since their magnitudes in the limit are equal to the unit vector as radius times the angle $d\theta$ in radians, we may write them as $d\mathbf{e}_r = \mathbf{e}_\theta \, d\theta$ and $d\mathbf{e}_\theta = -\mathbf{e}_r \, d\theta$. If we divide these equations by $d\theta$, we have

$$\frac{d\mathbf{e}_r}{d\theta} = \mathbf{e}_\theta \qquad \text{and} \qquad \frac{d\mathbf{e}_\theta}{d\theta} = -\mathbf{e}_r$$

Or, if we divide them by dt, we have $d\mathbf{e}_r/dt = (d\theta/dt)\mathbf{e}_\theta$ and $d\mathbf{e}_\theta/dt = -(d\theta/dt)\mathbf{e}_r$, or simply

$$\boxed{\dot{\mathbf{e}}_r = \dot{\theta}\mathbf{e}_\theta \qquad \text{and} \qquad \dot{\mathbf{e}}_\theta = -\dot{\theta}\mathbf{e}_r} \qquad (2/12)$$

We are now ready to differentiate $\mathbf{r} = r\mathbf{e}_r$ with respect to time. Using the rule for differentiating the product of a scalar and a vector gives

$$\mathbf{v} = \dot{\mathbf{r}} = \dot{r}\mathbf{e}_r + r\dot{\mathbf{e}}_r$$

With the substitution of $\dot{\mathbf{e}}_r$ from Eq. 2/12, the vector expression for the velocity becomes

$$\boxed{\mathbf{v} = \dot{r}\mathbf{e}_r + r\dot{\theta}\mathbf{e}_\theta} \qquad (2/13)$$

where
$$v_r = \dot{r}$$
$$v_\theta = r\dot{\theta}$$
$$v = \sqrt{v_r^2 + v_\theta^2}$$

The r-component of $\mathbf{v}$ is merely the rate at which the vector $\mathbf{r}$ stretches. The θ-component of $\mathbf{v}$ is due to the rotation of $\mathbf{r}$.

We now differentiate the expression for $\mathbf{v}$ to obtain the acceleration $\mathbf{a} = \dot{\mathbf{v}}$. Note that the derivative of $r\dot{\theta}\mathbf{e}_\theta$ will produce three terms since all three factors are variable. Thus,

$$\mathbf{a} = \dot{\mathbf{v}} = (\ddot{r}\mathbf{e}_r + \dot{r}\dot{\mathbf{e}}_r) + (\dot{r}\dot{\theta}\mathbf{e}_\theta + r\ddot{\theta}\mathbf{e}_\theta + r\dot{\theta}\dot{\mathbf{e}}_\theta)$$

Substitution of $\dot{\mathbf{e}}_r$ and $\dot{\mathbf{e}}_\theta$ from Eq. 2/12 and collecting terms give

$$\boxed{\mathbf{a} = (\ddot{r} - r\dot{\theta}^2)\mathbf{e}_r + (r\ddot{\theta} + 2\dot{r}\dot{\theta})\mathbf{e}_\theta} \qquad (2/14)$$

where
$$a_r = \ddot{r} - r\dot{\theta}^2$$
$$a_\theta = r\ddot{\theta} + 2\dot{r}\dot{\theta}$$
$$a = \sqrt{a_r^2 + a_\theta^2}$$

We may write the θ-component alternatively as

$$a_\theta = \frac{1}{r}\frac{d}{dt}(r^2\dot{\theta})$$

which can be verified easily by carrying out the differentiation. This form for a_θ will be found useful when we treat the angular momentum of particles in the next chapter.

An adequate appreciation of the terms in Eq. 2/14 comes only when the geometry of the physical changes can be clearly seen. For this purpose, Fig. 2/14a is developed to show the velocity vectors and their r- and θ-components at position A and at position A' after an infinitesimal movement. Each of these components undergoes a change in magnitude and direction as shown in Fig. 2/14b. In this figure we see the following changes:

Magnitude change of $\mathbf{v}_r$. This change is simply the increase in length of v_r or $dv_r = d\dot{r}$, and the corresponding acceleration term is $d\dot{r}/dt = \ddot{r}$ in the positive r-direction.

Direction change of $\mathbf{v}_r$. The magnitude of this change is seen from the figure to be $v_r\, d\theta = \dot{r}\, d\theta$, and its contribution to the acceleration becomes $\dot{r}\, d\theta/dt = \dot{r}\dot{\theta}$ which is in the positive θ-direction.

Magnitude change of $\mathbf{v}_\theta$. This term is the change in length of $\mathbf{v}_\theta$ or $d(r\dot{\theta})$, and its contribution to the acceleration is $d(r\dot{\theta})/dt = r\ddot{\theta} + \dot{r}\dot{\theta}$ and is in the positive θ-direction.

Path

(a)

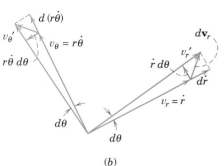

(b)

Figure 2/14

Direction change of v_θ. The magnitude of this change is $v_\theta \, d\theta = r\dot\theta \, d\theta$, and the corresponding acceleration term is observed to be $r\dot\theta(d\theta/dt) = r\dot\theta^2$ in the negative r-direction.

Collecting terms gives $a_r = \ddot r - r\dot\theta^2$ and $a_\theta = r\ddot\theta + 2\dot r\dot\theta$ as obtained previously. We see that the term $\ddot r$ is the acceleration that the particle would have along the radius in the absence of a change in θ. The term $-r\dot\theta^2$ is the normal component of acceleration if r were constant, as in circular motion. The term $r\ddot\theta$ is the tangential acceleration that the particle would have if r were constant, but is only a part of the acceleration due to the change in magnitude of $\mathbf v_\theta$ when r is variable. Finally, the term $2\dot r\dot\theta$ is composed of two effects. The first effect comes from that portion of the change in magnitude $d(r\dot\theta)$ of v_θ due to the change in r, and the second effect comes from the change in direction of $\mathbf v_r$. The term $2\dot r\dot\theta$ represents, therefore, a combination of changes and is not so easily perceived as are the other acceleration terms.

We note carefully the difference between the vector change $d\mathbf v_r$ in $\mathbf v_r$ and the change dv_r in the magnitude of v_r. Similarly, the vector change $d\mathbf v_\theta$ is not the same as the change dv_θ in the magnitude of v_θ. When we divide these changes by dt to obtain expressions for the derivatives, we see clearly that the magnitude of the derivative $|d\mathbf v_r/dt|$ and the derivative of the magnitude dv_r/dt are *not* the same. We also note carefully that a_r is not $\dot v_r$ and that a_θ is not $\dot v_\theta$.

The total acceleration $\mathbf a$ and its components are represented in Fig. 2/15. If $\mathbf a$ has a component normal to the path, we know from our analysis of n- and t-components in Art. 2/5 that the sense of the n-component *must* be toward the center of curvature.

For motion in a circular path with r constant, the components of Eqs. 2/13 and 2/14 become simply

$$v_r = 0 \qquad v_\theta = r\dot\theta$$
$$a_r = -r\dot\theta^2 \qquad a_\theta = r\ddot\theta$$

This description is the same as that obtained with n- and t-components, where the θ- and t-directions coincide but the positive r-direction is in the negative n-direction. Hence, $a_r = -a_n$ for circular motion centered at the origin of the polar coordinates.

The expressions for a_r and a_θ in scalar form may also be obtained by direct differentiation of the coordinate relations $x = r\cos\theta$ and $y = r\sin\theta$ to get $a_x = \ddot x$ and $a_y = \ddot y$. Each of these rectangular components of acceleration may then be resolved into r- and θ-components which, when combined, will yield the expressions of Eq. 2/14.

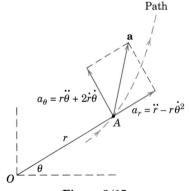

Figure 2/15

Sample Problem 2/9

Rotation of the radially slotted arm is governed by $\theta = 0.2t + 0.02t^3$, where θ is in radians and t is in seconds. Simultaneously, the power screw in the arm engages the slider B and controls its distance from O according to $r = 0.2 + 0.04t^2$, where r is in meters and t is in seconds. Calculate the magnitudes of the velocity and acceleration of the slider for the instant when $t = 3$ s.

Solution. The coordinates and their time derivatives which appear in the expressions for velocity and acceleration in polar coordinates are obtained first and evaluated for $t = 3$ s.

$r = 0.2 + 0.04t^2 \qquad r_3 = 0.2 + 0.04(3^2) = 0.56$ m

$\dot{r} = 0.08t \qquad \dot{r}_3 = 0.08(3) = 0.24$ m/s

$\ddot{r} = 0.08 \qquad \ddot{r}_3 = 0.08$ m/s^2

$\theta = 0.2t + 0.02t^3 \qquad \theta_3 = 0.2(3) + 0.02(3^3) = 1.14$ rad

$\qquad\qquad$ or $\theta_3 = 1.14(180/\pi) = 65.3°$

$\dot{\theta} = 0.2 + 0.06t^2 \qquad \dot{\theta}_3 = 0.2 + 0.06(3^2) = 0.74$ rad/s

$\ddot{\theta} = 0.12t \qquad \ddot{\theta}_3 = 0.12(3) = 0.36$ rad/s^2

The velocity components are obtained from Eq. 2/13 and for $t = 3$ s are

$[v_r = \dot{r}] \qquad v_r = 0.24$ m/s

$[v_\theta = r\dot{\theta}] \qquad v_\theta = 0.56(0.74) = 0.414$ m/s

$[v = \sqrt{v_r^2 + v_\theta^2}] \qquad v = \sqrt{(0.24)^2 + (0.414)^2} = 0.479$ m/s $\qquad$ *Ans.*

The velocity and its components are shown for the specified position of the arm.

The acceleration components are obtained from Eq. 2/14 and for $t = 3$ s are

$[a_r = \ddot{r} - r\dot{\theta}^2] \qquad a_r = 0.08 - 0.56(0.74)^2 = -0.227$ m/s^2

$[a_\theta = r\ddot{\theta} + 2\dot{r}\dot{\theta}] \qquad a_\theta = 0.56(0.36) + 2(0.24)(0.74) = 0.557$ m/s^2

$[a = \sqrt{a_r^2 + a_\theta^2}] \qquad a = \sqrt{(-0.227)^2 + (0.557)^2} = 0.601$ m/s^2 $\qquad$ *Ans.*

The acceleration and its components are also shown for the 65.3° position of the arm.

① We see that this problem is an example of constrained motion where the center B of the slider is mechanically constrained by the rotation of the slotted arm and by engagement with the turning screw.

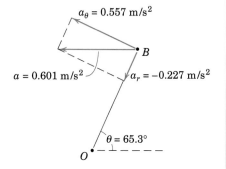

Sample Problem 2/10

A tracking radar lies in the vertical plane of the path of a rocket which is coasting in unpowered flight above the atmosphere. For the instant when $\theta = 30°$, the tracking data give $r = 8(10^4)$ m, $\dot{r} = 1200$ m/s, and $\dot{\theta} = 0.80$ deg/s. The acceleration of the rocket is due only to gravitational attraction and for its particular altitude is 9.20 m/s² vertically down. For these conditions determine the velocity v of the rocket and the values of $\ddot{r}$ and $\ddot{\theta}$.

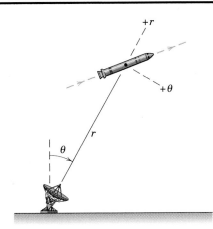

Solution. The components of velocity from Eq. 2/13 are

$[v_r = \dot{r}]$ $\qquad$ $v_r = 1200$ m/s

① $[v_\theta = r\dot{\theta}]$ $\qquad$ $v_\theta = 8(10^4)(0.80)\left(\dfrac{\pi}{180}\right) = 1117$ m/s

$[v = \sqrt{v_r^2 + v_\theta^2}]$ $\qquad$ $v = \sqrt{(1200)^2 + (1117)^2} = 1639$ m/s $\qquad$ *Ans.*

Since the total acceleration of the rocket is $g = 9.20$ m/s² down, we can easily find its r- and θ-components for the given position. As shown in the figure, they are

② $\qquad a_r = -9.20 \cos 30° = -7.97$ m/s²

$\qquad a_\theta = 9.20 \sin 30° = 4.60$ m/s²

We now equate these values to the polar-coordinate expressions for a_r and a_θ that contain the unknowns $\ddot{r}$ and $\ddot{\theta}$. Thus, from Eq. 2/14

③ $[a_r = \ddot{r} - r\dot{\theta}^2]$ $\qquad -7.97 = \ddot{r} - 8(10^4)\left(0.80\,\dfrac{\pi}{180}\right)^2$

$\qquad\qquad\qquad\quad \ddot{r} = 7.63$ m/s² $\qquad$ *Ans.*

$[a_\theta = r\ddot{\theta} + 2\dot{r}\dot{\theta}]$ $\qquad 4.60 = 8(10^4)\ddot{\theta} + 2(1200)\left(0.80\,\dfrac{\pi}{180}\right)$

$\qquad\qquad\qquad\quad \ddot{\theta} = -3.61(10^{-4})$ rad/s² $\qquad$ *Ans.*

① We observe that the angle θ in polar coordinates need not always be taken positive in a counterclockwise sense.

② Note that the r-component of acceleration is in the negative r-direction, so it carries a minus sign.

③ We must be careful to convert $\dot{\theta}$ from deg/s to rad/s.

PROBLEMS

Introductory problems

2/131 Motion of the sliding block P in the rotating radial slot is controlled by the power screw as shown. For the instant represented, $\dot{\theta} = 0.1$ rad/s, $\ddot{\theta} = -0.04$ rad/s², and $r = 300$ mm. Also, the screw turns at a constant speed giving $\dot{r} = 40$ mm/s. For this instant, determine the magnitudes of the velocity $\mathbf{v}$ and acceleration $\mathbf{a}$ of P. Sketch $\mathbf{v}$ and $\mathbf{a}$ if $\theta = 120°$.

Ans. $v = 50$ mm/s, $a = 5$ mm/s²

Problem 2/131

2/132 The boom OAB pivots about point O, while section AB simultaneously extends from within section OA. Determine the velocity and acceleration of the center B of the pulley for the following conditions: $\theta = 20°$, $\dot{\theta} = 5$ deg/s, $\ddot{\theta} = 2$ deg/s², $l = 2$ m, $\dot{l} = 0.5$ m/s, $\ddot{l} = -1.2$ m/s². The quantities $\dot{l}$ and $\ddot{l}$ are the first and second time derivatives, respectively, of the length l of section AB.

Problem 2/132

2/133 The ladder of a fire truck is designed to be extended at the constant rate $\dot{l} = 150$ mm/s and to be elevated at the constant rate $\dot{\theta} = 2$ deg/s. As the position $\theta = 50°$ and $l = 4$ m is reached, determine the magnitudes of the velocity $\mathbf{v}$ and the acceleration $\mathbf{a}$ of the fireman at A.

Ans. $v = 0.380$ m/s, $a = 0.01607$ m/s²

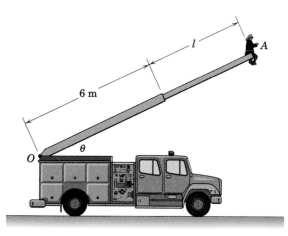

Problem 2/133

2/134 A model airplane flies over an observer O with constant speed in a straight line as shown. Determine the signs (plus, minus, or zero) for r, $\dot{r}$, $\ddot{r}$, θ, $\dot{\theta}$, and $\ddot{\theta}$ for each position A, B, and C.

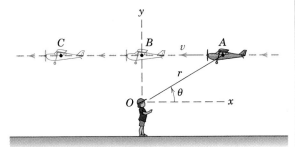

Problem 2/134

2/135 The nozzle shown rotates with constant angular speed Ω about a fixed horizontal axis through point O. Because of the change in diameter by a factor of 2, the water speed relative to the nozzle at A is v, while that at B is $4v$. The water speeds at both A and B are constant. Determine the velocity and acceleration of a water particle as it passes (*a*) point A and (*b*) point B.

$$\text{Ans. } (a) \ \mathbf{v}_A = v\mathbf{e}_r + l\Omega\mathbf{e}_\theta$$
$$\mathbf{a}_A = -l\Omega^2\mathbf{e}_r + 2v\Omega\mathbf{e}_\theta$$
$$(b) \ \mathbf{v}_B = 4v\mathbf{e}_r + 2l\Omega\mathbf{e}_\theta$$
$$\mathbf{a}_B = -2l\Omega^2\mathbf{e}_r + 8v\Omega\mathbf{e}_\theta$$

Problem 2/135

2/136 The plane curvilinear motion of a particle is given by the polar coordinates $r = t^3/3$ and $\theta = 4\cos(\pi t/6)$, where r is in meters, θ is in radians (measured counterclockwise from the positive x-axis), and t is in seconds. Determine and sketch the unit vectors $\mathbf{e}_r$ and $\mathbf{e}_\theta$ in terms of $\mathbf{i}$ and $\mathbf{j}$ at time $t = 2$ s.

2/137 An internal mechanism is used to maintain a constant angular rate $\Omega = 0.05$ rad/s about the z-axis of the spacecraft as the telescopic booms are extended at a constant rate. The length l is varied from essentially zero to 3 m. The maximum acceleration to which the sensitive experiment modules P may be subjected is 0.011 m/s². Determine the maximum allowable boom extension rate $\dot{l}$. *Ans.* $\dot{l} = 32.8$ mm/s

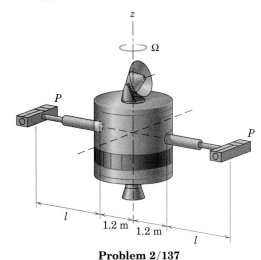

Problem 2/137

2/138 The rocket is fired vertically and tracked by the radar shown. When θ reaches 60°, other corresponding measurements give the values $r = 9$ km, $\ddot{r} = 21$ m/s², and $\dot{\theta} = 0.02$ rad/s. Calculate the magnitudes of the velocity and acceleration of the rocket at this position.

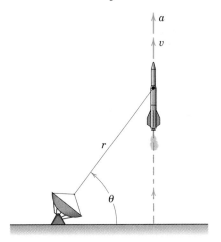

Problem 2/138

2/139 A jet plane flying at a constant speed v at an altitude $h = 10$ km is being tracked by radar located at O directly below the line of flight. If the angle θ is decreasing at the rate of 0.020 rad/s when $\theta = 60°$, determine the value of $\ddot{r}$ at this instant and the magnitude of the velocity $\mathbf{v}$ of the plane.

Ans. $\ddot{r} = 4.62$ m/s^2, $v = 960$ km/h

Problem 2/139

Representative problems

2/140 A particle moving along a plane curve has a position vector $\mathbf{r}$, a velocity $\mathbf{v}$, and an acceleration $\mathbf{a}$. Unit vectors in the r- and θ-directions are $\mathbf{e}_r$ and $\mathbf{e}_\theta$, respectively, and both r and θ are changing with time. Explain why each of the following statements is correctly marked as an inequality.

$\dot{\mathbf{r}} \neq v \qquad \ddot{\mathbf{r}} \neq a \qquad \dot{r} \neq \dot{r}\mathbf{e}_r$

$\dot{r} \neq v \qquad \ddot{r} \neq a \qquad \ddot{r} \neq \ddot{r}\mathbf{e}_r$

$\dot{r} \neq \mathbf{v} \qquad \ddot{r} \neq \mathbf{a} \qquad \dot{\mathbf{r}} \neq r\dot{\theta}\mathbf{e}_\theta$

2/141 At the bottom of a loop in the vertical $(r\text{-}\theta)$ plane at an altitude of 400 m, the airplane P has a horizontal velocity of 600 km/h and no horizontal acceleration. The radius of curvature of the loop is 1200 m. For the radar tracking at O, determine the recorded values of $\ddot{r}$ and $\ddot{\theta}$ for this instant.

Ans. $\ddot{r} = 12.15$ m/s^2, $\ddot{\theta} = 0.0365$ rad/s^2

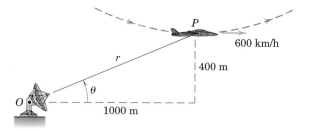

Problem 2/141

2/142 The slider P can be moved inward by means of the string S as the bar OA rotates about the pivot O. The angular position of the bar is given by $\theta = 0.4 + 0.12t + 0.06t^3$, where θ is in radians and t is in seconds. The position of the slider is given by $r = 0.8 - 0.1t - 0.05t^2$, where r is in meters and t is in seconds. Determine and sketch the velocity and acceleration of the slider at time $t = 2$ s. Find the angles α and β which $\mathbf{v}$ and $\mathbf{a}$ make with the positive x-axis.

Problem 2/142

2/143 Car A is moving with constant speed v on the straight and level highway. The police officer in the stationary car P attempts to measure the speed v with radar. If the radar measures "line-of-sight" velocity, what velocity v' will the officer observe? Evaluate your general expression for the values $v = 115$ km/h, $L = 150$ m, and $D = 6$ m, and draw any appropriate conclusions.

Ans. $v' = \dfrac{vL}{\sqrt{L^2 + D^2}}$, $v' = 114.9$ km/h

Problem 2/143

2/144 The radial position of a fluid particle P in a certain centrifugal pump with radial vanes is approximated by $r = r_0 \cosh Kt$, where t is time and $K = \dot{\theta}$ is the constant angular rate at which the impeller turns. Determine the expression for the magnitude of the total acceleration of the particle just prior to leaving the vane in terms of r_0, R, and K.

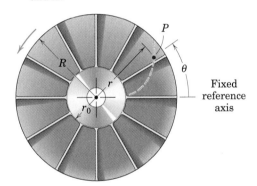

Problem 2/144

2/145 At the instant depicted in the figure, the radar station at O measures the range rate of the space shuttle P to be $\dot{r} = -3742$ m/s, with O considered fixed. If it is known that the shuttle is in a circular orbit at an altitude $h = 240$ km, determine the orbital speed of the shuttle from this information.   *Ans.* $v = 7766$ m/s

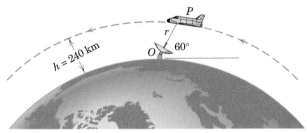

Problem 2/145

2/146 A satellite m moves in an elliptical orbit around the earth. There is no force on the satellite in the θ-direction, so that $a_\theta = 0$. Prove Kepler's second law of planetary motion, which says that the radial line r sweeps through equal areas in equal times. The area dA swept by the radial line during time dt is shaded in the figure.

Problem 2/146

2/147 The cam is designed so that the center of the roller A which follows the contour moves on a limaçon defined by $r = b - c \cos \theta$, where $b > c$. If the cam does not rotate, determine the magnitude of the total acceleration of A in terms of θ if the slotted arm revolves with a constant counterclockwise angular rate $\dot{\theta} = \omega$.

$$\text{Ans. } a = \omega^2 \sqrt{4c^2 - 4bc \cos \theta + b^2}$$

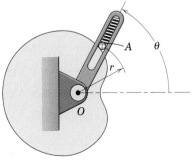

Problem 2/147

2/148 A locomotive is traveling on the straight and level track with a speed $v = 90$ km/h and a deceleration $a = 0.5$ m/s^2 as shown. Relative to the fixed observer at O, determine the quantities $\dot{r}$, $\ddot{r}$, $\dot{\theta}$, and $\ddot{\theta}$ at the instant when $\theta = 60°$ and $r = 400$ m.

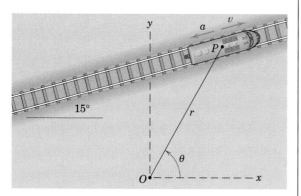

Problem 2/148

2/149 The robot arm is elevating and extending simultaneously. At a given instant, $\theta = 30°$, $\dot{\theta} = 10$ deg/s = constant, $l = 0.5$ m, $\dot{l} = 0.2$ m/s, and $\ddot{l} = -0.3$ m/s^2. Compute the magnitudes of the velocity $\mathbf{v}$ and acceleration $\mathbf{a}$ of the gripped part P. In addition, express $\mathbf{v}$ and $\mathbf{a}$ in terms of the unit vectors $\mathbf{i}$ and $\mathbf{j}$.

Ans. $v = 0.296$ m/s, $a = 0.345$ m/s^2
$\mathbf{v} = 0.064\mathbf{i} + 0.289\mathbf{j}$ m/s
$\mathbf{a} = -0.328\mathbf{i} - 0.1086\mathbf{j}$ m/s^2

Problem 2/149

2/150 The fixed horizontal guide carries a slider and pin P whose motion is controlled by the rotating slotted arm OA. If the arm is revolving about O at the constant rate $\dot{\theta} = 2$ rad/s for an interval of its designed motion, determine the magnitudes of the velocity and acceleration of the slider in the slot for the instant when $\theta = 60°$. Also find the r-components of the velocity and acceleration.

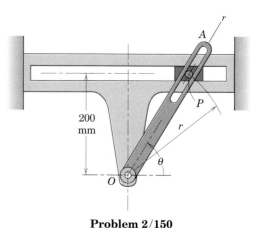

Problem 2/150

2/151 An aircraft flying in a straight line at a climb angle β to the horizontal is tracked by radar located directly below the line of flight. At a certain instant, the following data are recorded: $r = 3600$ m, $\dot{r} = 110$ m/s, $\ddot{r} = 6$ m/s^2, $\theta = 30°$, and $\dot{\theta} = 2.20$ deg/s. For this instant determine the aircraft altitude h, velocity v, angle of climb β, $\ddot{\theta}$, and acceleration a.

Ans. $h = 3120$ m, $v = 176.7$ m/s, $\beta = 8.5°$
$\ddot{\theta} = -2.10(10^{-3})$ rad/s^2, $a = 1.112$ m/s^2

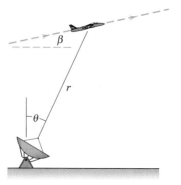

Problem 2/151

2/152 The hydraulic cylinder gives pin A a constant velocity $v = 2$ m/s along its axis for an interval of motion and, in turn, causes the slotted arm to rotate about O. Determine the values of $\dot{r}$, $\ddot{r}$, and $\ddot{\theta}$ for the instant when $\theta = 30°$. (*Hint:* Recognize that all acceleration components are zero when the velocity is constant.)

300 mm

Problem 2/152

2/153 The small block P starts from rest at time $t = 0$ at point A and moves up the incline with constant acceleration a. Determine $\dot{r}$ as a function of time.

$$Ans. \ \dot{r} = \frac{\frac{1}{2}at \ (2R \ \cos \ \alpha \ + \ at^2)}{\sqrt{R^2 \ + \ Rat^2 \ \cos \ \alpha \ + \ \frac{1}{4}a^2t^4}}$$

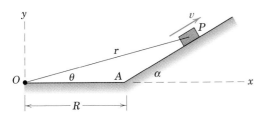

Problem 2/153

2/154 The paint-spraying robot is programmed to paint a production line of curved surfaces A (seen on edge). The length of the telescoping arm is controlled according to $b = 0.3 \sin (\pi t/2)$, where b is in meters and t is in seconds. Simultaneously, the arm is programmed to rotate according to $\theta = \pi/4 + (\pi/8) \sin (\pi t/2)$ radians. Calculate the magnitude v of the velocity of the nozzle N and the magnitude a of the acceleration of N for $t = 1$ s and for $t = 2$ s.

Problem 2/154

2/155 A rocket follows a trajectory in the vertical plane and is tracked by radar from point A. At a certain instant, the radar measurements give $r = 10.5$ km, $\dot{r} = 480$ m/s, $\dot{\theta} = 0$, and $\ddot{\theta} = -0.00720$ rad/s². Sketch the position of the rocket for this instant and determine the radius of curvature ρ of the trajectory at this position of the rocket.

$$Ans. \ \rho = 3.05 \ km$$

Problem 2/155

2/156 The slotted arm is pivoted at O and carries the slider C. The position of C in the slot is governed by the cord which is fastened at D and remains taut. The arm turns counterclockwise with a constant angular rate $\dot{\theta} = 4$ rad/s during an interval of its motion. The length DBC of the cord equals R, which makes $r = 0$ when $\theta = 0$. Determine the magnitude a of the acceleration of the slider at the position for which $\theta = 30°$. The distance R is 375 mm.

Problem 2/156

2/157 As a result of an overhead serve, the flight of a volleyball has the initial conditions shown in the figure. Determine the quantities $\dot{r}$, $\ddot{r}$, $\dot{\theta}$, and $\ddot{\theta}$, all relative to the x-y axes shown, just after the ball is struck at $\theta = 90°$. Neglect aerodynamic drag.

Ans. $\dot{r} = 6.84$ m/s, $\ddot{r} = 158.4$ m/s^2
$\dot{\theta} = -8.95$ rad/s, $\ddot{\theta} = 58.3$ rad/s^2

Problem 2/157

2/158 During a portion of a vertical loop, an airplane flies in an arc of radius $\rho = 600$ m with a constant speed $v = 400$ km/h. When the airplane is at A, the angle made by $\mathbf{v}$ with the horizontal is $\beta = 30°$, and radar tracking gives $r = 800$ m and $\theta = 30°$. Calculate v_r, v_θ, a_r, and $\ddot{\theta}$ for this instant.

Problem 2/158

2/159 A meteor P is tracked by a radar observatory on the earth at O. When the meteor is directly overhead ($\theta = 90°$), the following observations are recorded: $r = 80$ km, $\dot{r} = -20$ km/s, and $\dot{\theta} = 0.4$ rad/s. (*a*) Determine the speed v of the meteor and the angle β which its velocity vector makes with the horizontal. Neglect any effects due to the earth's rotation. (*b*) Repeat with all given quantities remaining the same, except that $\theta = 75°$.

Ans. (*a*) $v = 37.7$ km/s, $\beta = 32.0°$
(*b*) $v = 37.7$ km/s, $\beta = 17.01°$

Problem 2/159

2/160 A fireworks shell P fired in a vertical trajectory has a y-acceleration given by $a_y = -g - kv^2$, where the latter term is due to aerodynamic drag. If the speed of the shell is 15 m/s at the instant shown, determine the corresponding values of r, $\dot{r}$, $\ddot{r}$, θ, $\dot{\theta}$, and $\ddot{\theta}$. The drag parameter k has a constant value of 0.01 m^{-1}.

100 m

200 m

θ

O

Problem 2/160

2/161 In the design of a control device, the spring-loaded pin P in the slotted arm follows the contour of the four-lobe fixed cam as the arm rotates about O. The path of the center of the pin is governed by $r = 80 + 12 \cos 4\theta$, where r is in millimeters and θ is in radians. If the arm rotates at the constant rate of 2 revolutions per second, find the acceleration of the pin when (a) $\theta = 0$, (b) $\theta = \pi/8$, and (c) $\theta = \pi/4$. Express your results in vector notation using unit vectors $\mathbf{e}_r$ and $\mathbf{e}_\theta$.

$$Ans. \ (a) \ \mathbf{a} = -44.8\mathbf{e}_r \text{ m/s}^2$$
$$(b) \ \mathbf{a} = -12.63\mathbf{e}_r - 15.16\mathbf{e}_\theta \text{ m/s}^2$$
$$(c) \ \mathbf{a} = 19.58\mathbf{e}_r \text{ m/s}^2$$

Problem 2/161

2/162 The piston of the hydraulic cylinder gives pin A a constant velocity $v = 1.5$ m/s in the direction shown for an interval of its motion. For the instant when $\theta = 60°$, determine $\dot{r}$, $\ddot{r}$, $\dot{\theta}$, and $\ddot{\theta}$ where $r = \overline{OA}$.

150 mm

Problem 2/162

2/163 An earth satellite traveling in the elliptical orbit shown has a velocity $v = 17\,970$ km/h as it passes the end of the semiminor axis at A. The acceleration of the satellite at A is $a = a_r = -1.556$ m/s^2 as calculated from the gravitational law. For position A calculate the values of $\dot{v}$ and $\ddot{r}$.

$$Ans. \ \dot{v} = -0.778 \text{ m/s}^2, \ \ddot{r} = -0.388 \text{ m/s}^2$$

16 000 km

A

v

13 800 km

r

θ

O

P

8000 km

Problem 2/163

▶**2/164** The baseball player of Prob. 2/114 is repeated here with additional information supplied. At time $t = 0$, the ball is thrown with an initial speed of 30 m/s at an angle of 30° to the horizontal. Determine the quantities r, $\dot{r}$, $\ddot{r}$, θ, $\dot{\theta}$, and $\ddot{\theta}$, all relative to the x-y coordinate system shown, at time $t = 0.5$ s.

> *Ans.* $r = 15.40$ m, $\dot{r} = 27.3$ m/s
> $\ddot{r} = -3.35$ m/s², $\theta = 32.5°$
> $\dot{\theta} = -0.353$ rad/s, $\ddot{\theta} = 0.717$ rad/s²

Problem 2/164

▶**2/165** If the slotted arm of Prob. 2/147, shown again here, is revolving counterclockwise at the constant rate of 40 rev/min and the cam is revolving clockwise at the constant rate of 30 rev/min, determine the magnitude of the acceleration of the center of the roller A when the cam and arm are in the relative position for which $\theta = 30°$. The limaçon has the dimensions $b = 100$ mm and $c = 75$ mm. (*Caution:* Redefine the coordinates as necessary after noting that the θ in the expression $r = b - c \cos \theta$ is not the absolute angle appearing in Eq. 2/14.)

> *Ans.* $a = 3.68$ m/s²

Problem 2/165

▶**2/166** For a limited range of motion, crank CP causes the slotted link OA to rotate. If β is increasing at the constant rate of 4 rad/s when $\beta = \pi/4$, determine the r- and θ-components of the acceleration of pin P for this position and specify the corresponding values of $\dot{r}$ and $\ddot{r}$. Solve first by computing the components of Eq. 2/14 and second by resolving the acceleration of P in its circular motion into r- and θ-components.

> *Ans.* $a_r = -1774$ mm/s², $a_\theta = -735$ mm/s²
> $\dot{r} = -183.7$ mm/s, $\ddot{r} = -887$ mm/s²

Problem 2/166

2/7 *SPACE CURVILINEAR MOTION*

The general case of three-dimensional motion of a particle along a space curve was introduced in Art. 2/1 and illustrated in Fig. 2/1. Mention was made of the three coordinate systems, rectangular (x-y-z), cylindrical (r-θ-z), and spherical (R-θ-ϕ), which are commonly used to describe this motion. These systems are indicated in Fig. 2/16, which also shows the unit vectors for the three coordinate systems.*

Before describing the use of these coordinate systems, we may observe that a path-variable description, using n- and t-coordinates, which we developed in Art. 2/5, may be applied in the osculating plane shown in Fig. 2/1. We defined this plane earlier as the plane which contains the curve at the location in question. We see that the velocity **v**, which is along the tangent t to the curve, lies in the osculating plane. The acceleration **a** also lies in the osculating plane and, as in the case of plane motion, has a component $a_t = \dot{v}$ tangent to the path due to the change in magnitude of the velocity and a component $a_n = v^2/\rho$ normal to the curve due to the change in direction of the velocity. As before, ρ is the radius of curvature of the path at the point in question and would be measured in the osculating plane. This description of motion, which we found to be natural and direct for many plane-motion problems, finds little use for space motion because the osculating plane continually shifts its orientation, thus making its use as a reference awkward. We shall confine our attention, therefore, to the three fixed coordinate systems shown in Fig. 2/16.

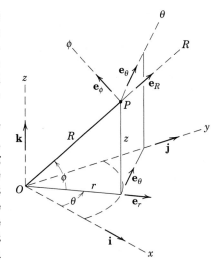

Figure 2/16

(a) Rectangular coordinates (x-y-z). The extension from two to three dimensions offers no particular difficulty. We merely add the z-coordinate and its two time derivatives to the two-dimensional expressions of Eqs. 2/6 so that the position vector **R**, the velocity **v**, and the acceleration **a** become

$$\mathbf{R} = x\mathbf{i} + y\mathbf{j} + z\mathbf{k}$$
$$\mathbf{v} = \dot{\mathbf{R}} = \dot{x}\mathbf{i} + \dot{y}\mathbf{j} + \dot{z}\mathbf{k} \qquad (2/15)$$
$$\mathbf{a} = \dot{\mathbf{v}} = \ddot{\mathbf{R}} = \ddot{x}\mathbf{i} + \ddot{y}\mathbf{j} + \ddot{z}\mathbf{k}$$

Note that in three dimensions we are using **R** in place of **r** for the position vector.

(b) Cylindrical coordinates (r-θ-z). If we understand the polar-coordinate description of plane motion, then there should be

*In a variation of spherical coordinates commonly used, angle ϕ is replaced by its complement.

no difficulty with cylindrical coordinates because all that is required is the addition of the z-coordinate and its two time derivatives. The position vector $\mathbf{R}$ to the particle for cylindrical coordinates is simply

$$\mathbf{R} = r\mathbf{e}_r + z\mathbf{k}$$

In place of Eq. 2/13 for plane motion, we may write the velocity as

$$\mathbf{v} = \dot{r}\mathbf{e}_r + r\dot{\theta}\mathbf{e}_\theta + \dot{z}\mathbf{k} \tag{2/16}$$

where

$$v_r = \dot{r}$$

$$v_\theta = r\dot{\theta}$$

$$v_z = \dot{z}$$

$$v = \sqrt{v_r^2 + v_\theta^2 + v_z^2}$$

Similarly, the acceleration is written by adding the z-component to Eq. 2/14, which gives us

$$\mathbf{a} = (\ddot{r} - r\dot{\theta}^2)\mathbf{e}_r + (r\ddot{\theta} + 2\dot{r}\dot{\theta})\mathbf{e}_\theta + \ddot{z}\mathbf{k} \tag{2/17}$$

where

$$a_r = \ddot{r} - r\dot{\theta}^2$$

$$a_\theta = r\ddot{\theta} + 2\dot{r}\dot{\theta} = \frac{1}{r}\frac{d}{dt}(r^2\dot{\theta})$$

$$a_z = \ddot{z}$$

$$a = \sqrt{a_r^2 + a_\theta^2 + a_z^2}$$

Whereas the unit vectors $\mathbf{e}_r$ and $\mathbf{e}_\theta$ have time derivatives due to the changes in their directions, we note that the unit vector $\mathbf{k}$ in the z-direction remains fixed in direction and therefore has no time derivative.

(c) Spherical coordinates (R-θ-φ). When a radial distance and two angles are used to locate the position of a particle, as in the case of radar measurements, for example, spherical coordinates R, θ, ϕ are used. Derivation of the expression for the velocity $\mathbf{v}$ is easily obtained, but the expression for the acceleration $\mathbf{a}$ is more complex because of the added geometry. Consequently, only the results will be cited here.* First we designate unit vectors $\mathbf{e}_R$, $\mathbf{e}_\theta$, $\mathbf{e}_\phi$ as shown in Fig. 2/16. Note that the unit vector $\mathbf{e}_R$ is in the direction that the

*For a complete derivation of $\mathbf{v}$ and $\mathbf{a}$ in spherical coordinates, see the first author's book *Dynamics*, 2nd Edition, 1971 or SI Version, 1975 (John Wiley & Sons, Inc.).

particle P would move if R increases but θ and ϕ are held constant. Also the unit vector $\mathbf{e}_\theta$ is in the direction that P would move if θ increases while R and ϕ are held constant. Lastly, the unit vector $\mathbf{e}_\phi$ is in the direction in which P would move if ϕ increases while R and θ are held constant. The resulting expressions for $\mathbf{v}$ and $\mathbf{a}$ are

$$\boxed{\mathbf{v} = v_R\mathbf{e}_R + v_\theta\mathbf{e}_\theta + v_\phi\mathbf{e}_\phi} \qquad (2/18)$$

where

$$v_R = \dot{R}$$
$$v_\theta = R\dot{\theta}\cos\phi$$
$$v_\phi = R\dot{\phi}$$

and

$$\boxed{\mathbf{a} = a_R\mathbf{e}_R + a_\theta\mathbf{e}_\theta + a_\phi\mathbf{e}_\phi} \qquad (2/19)$$

where

$$a_R = \ddot{R} - R\dot{\phi}^2 - R\dot{\theta}^2\cos^2\phi$$

$$a_\theta = \frac{\cos\phi}{R}\frac{d}{dt}(R^2\dot{\theta}) - 2R\dot{\theta}\dot{\phi}\sin\phi$$

$$a_\phi = \frac{1}{R}\frac{d}{dt}(R^2\dot{\phi}) + R\dot{\theta}^2\sin\phi\cos\phi$$

We should mention that linear algebraic transformations between any two of the three coordinate-system expressions for velocity or acceleration may be developed. These transformations make it possible to express the motion components in rectangular coordinates, for example, if the components are known in spherical coordinates, or vice versa.* These transformations are easily handled with the aid of matrix algebra and a simple computer program.

---

*These coordinate transformations are developed and illustrated in the first author's book *Dynamics*, 2nd Edition, 1971 or SI Version, 1975 (John Wiley & Sons, Inc.).

Sample Problem   2/11

The power screw starts from rest and is given a rotational speed $\dot{\theta}$ that increases uniformly with time t according to $\dot{\theta} = kt$, where k is a constant. Determine the expressions for the velocity v and acceleration a of the center of ball A when the screw has turned through one complete revolution from rest. The lead of the screw (advancement per revolution) is L.

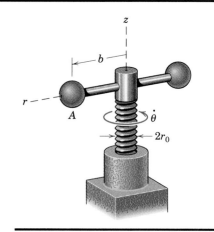

Solution.   The center of ball A moves in a helix on the cylindrical surface of radius b, and the cylindrical coordinates r, θ, z are clearly indicated.

Integrating the given relation for $\dot{\theta}$ gives $\theta = \Delta\theta = \int \dot{\theta}\,dt = \frac{1}{2}kt^2$. For one revolution from rest we have

$$2\pi = \tfrac{1}{2}kt^2$$

giving

$$t = 2\sqrt{\pi/k}$$

Thus, the angular rate at one revolution is

$$\dot{\theta} = kt = k(2\sqrt{\pi/k}) = 2\sqrt{\pi k}$$

① The helix angle γ of the path followed by the center of the ball governs the relation between the θ- and z-components of velocity and is given by $\tan\gamma = L/(2\pi b)$. Now from the figure we see that $v_\theta = $ ② $v\cos\gamma$. Substituting $v_\theta = r\dot{\theta} = b\dot{\theta}$ from Eq. 2/16 gives $v = v_\theta/\cos\gamma = b\dot{\theta}/\cos\gamma$. With $\cos\gamma$ obtained from $\tan\gamma$ and with $\dot{\theta} = 2\sqrt{\pi k}$, we have for the one-revolution position

$$v = 2b\sqrt{\pi k}\,\frac{\sqrt{L^2 + 4\pi^2 b^2}}{2\pi b} = \sqrt{\frac{k}{\pi}}\sqrt{L^2 + 4\pi^2 b^2} \qquad Ans.$$

The acceleration components from Eq. 2/17 become

③ $[a_r = \ddot{r} - r\dot{\theta}^2]$ $\qquad a_r = 0 - b(2\sqrt{\pi k})^2 = -4b\pi k$

$[a_\theta = r\ddot{\theta} + 2\dot{r}\dot{\theta}]$ $\qquad a_\theta = bk + 2(0)(2\sqrt{\pi k}) = bk$

$[a_z = \ddot{z} = \dot{v}_z]$ $\qquad a_z = \dfrac{d}{dt}(v_z) = \dfrac{d}{dt}(v_\theta \tan\gamma) = \dfrac{d}{dt}(b\dot{\theta}\tan\gamma)$

$$= (b\tan\gamma)\ddot{\theta} = b\,\frac{L}{2\pi b}\,k = \frac{kL}{2\pi}$$

Now we combine the components to give the magnitude of the total acceleration, which becomes

$$a = \sqrt{(-4b\pi k)^2 + (bk)^2 + \left(\frac{kL}{2\pi}\right)^2}$$

$$= bk\sqrt{(1 + 16\pi^2) + L^2/(4\pi^2 b^2)} \qquad Ans.$$

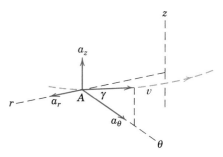

① Unless we recognize that the helix angle of the path governs the z-motion in relation to the rotation of the screw, we cannot complete the problem. We must also be careful to divide the lead L by the circumference $2\pi b$ and not the diameter $2b$ to obtain $\tan\gamma$. If in doubt, unwrap one turn of the helix traced by the center of the ball.

② Sketch a right triangle and recall that for $\tan\beta = a/b$ the cosine of β becomes $b/\sqrt{a^2 + b^2}$.

③ The negative sign for a_r is consistent with our previous knowledge that the normal component of acceleration is directed toward the center of curvature.

PROBLEMS

Introductory problems

2/167 The particle P moves along the space curve and has a velocity $\mathbf{v} = 4\mathbf{i} - 2\mathbf{j} - \mathbf{k}$ m/s for the instant shown. At the same instant the particle has an acceleration $\mathbf{a}$ whose magnitude is 8 m/s². Calculate the radius of curvature ρ of the path for this position and the rate $\dot{v}$ at which the magnitude of the velocity is increasing.

Ans. $\rho = 7.67$ m, $\dot{v} = 7.52$ m/s²

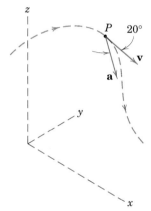

Problem 2/167

2/168 A projectile is launched from point O with an initial speed $v_0 = 150$ m/s directed as shown in the figure. Compute the x-, y-, and z-components of position, velocity, and acceleration 20 seconds after launch. Neglect aerodynamic drag.

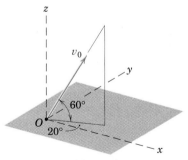

Problem 2/168

2/169 The rectangular coordinates of a particle are given in millimeters as functions of time t in seconds by $x = 30 \cos 2t$, $y = 40 \sin 2t$, and $z = 20t + 3t^2$. Determine the angle θ_1 between the position vector $\mathbf{r}$ and the velocity $\mathbf{v}$ and the angle θ_2 between the position vector $\mathbf{r}$ and the acceleration $\mathbf{a}$, both at time $t = 2$ s.

Ans. $\theta_1 = 60.8°$, $\theta_2 = 122.4°$

2/170 The velocity and acceleration of a particle are given for a certain instant by $\mathbf{v} = 6\mathbf{i} - 3\mathbf{j} + 2\mathbf{k}$ m/s and $\mathbf{a} = 3\mathbf{i} - \mathbf{j} - 5\mathbf{k}$ m/s². Determine the angle β between $\mathbf{v}$ and $\mathbf{a}$, $\dot{v}$, and the radius of curvature ρ in the osculating plane.

2/171 An amusement ride called the "corkscrew" takes the passengers through the upside-down curve of a horizontal cylindrical helix. The velocity of the cars as they pass position A is 15 m/s, and the component of their acceleration measured along the tangent to the path is $g \cos \gamma$ at this point. The effective radius of the cylindrical helix is 5 m, and the helix angle is $\gamma = 40°$. Compute the magnitude of the acceleration of the passengers as they pass position A.

Ans. $a = 27.5$ m/s²

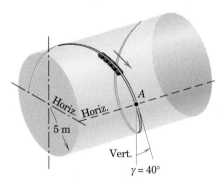

Problem 2/171

2/172 The rotating element in a mixing chamber is given a periodic axial movement $z = z_0 \sin 2\pi nt$ while it is rotating at the constant angular velocity $\dot{\theta} = \omega$. Determine the expression for the maximum magnitude of the acceleration of a point A on the rim of radius r. The frequency n of vertical oscillation is constant.

Problem 2/172

Representative problems

2/173 The rotating nozzle sprays a large circular area and turns with the constant angular rate $\dot{\theta} = K$. Particles of water move along the tube at the constant rate $\dot{l} = c$ relative to the tube. Write expressions for the magnitudes of the velocity and acceleration of a water particle P for a given position l in the rotating tube.

$$Ans.\ v = \sqrt{c^2 + K^2 l^2 \sin^2 \beta}$$
$$a = K \sin \beta \sqrt{K^2 l^2 + 4c^2}$$

Problem 2/173

2/174 The car A is ascending a parking-garage ramp in the form of a cylindrical helix of 7.2-m radius rising 3 m for each half turn. At the position shown the car has a speed of 25 km/h, which is decreasing at the rate of 3 km/h per second. Determine the r-, θ-, and z-components of the acceleration of the car.

Problem 2/174

2/175 An aircraft takes off at A and climbs at a steady angle with slope of 1 to 2 in the vertical y-z plane at a constant speed $v = 400$ km/h. The aircraft is tracked by radar at O. For the position B, determine the value of $\dot{R}$, $\dot{\theta}$, and $\dot{\phi}$.

$$Ans.\ \dot{R} = 92.0 \text{ km/h}, \ \dot{\theta} = 0.1988 \text{ rad/s}$$
$$\dot{\phi} = 0.0731 \text{ rad/s}$$

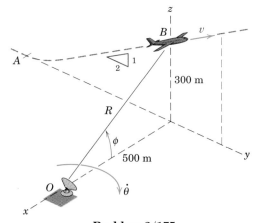

Problem 2/175

2/176 For the conditions of Prob. 2/175, find the values of $\ddot{R}$, $\ddot{\theta}$, and $\ddot{\phi}$ for the radar tracking coordinates as the aircraft passes point B. Use the results cited for Prob. 2/175.

2/177 The robotic device of Prob. 2/149 now rotates about a fixed vertical axis while its arm extends and elevates. At a given instant, $\phi = 30°$, $\dot{\phi} = 10$ deg/s = constant, $l = 0.5$ m, $\dot{l} = 0.2$ m/s, $\ddot{l} = -0.3$ m/s^2, and $\Omega = 20$ deg/s = constant. Determine the magnitudes of the velocity $\mathbf{v}$ and the acceleration $\mathbf{a}$ of the gripped part P.

                      Ans. $v = 0.480$ m/s, $a = 0.474$ m/s^2

2/178 The base structure of the firetruck ladder rotates about a vertical axis through O with a constant angular velocity $\Omega = 10$ deg/s. At the same time, the ladder unit OB elevates at a constant rate $\dot{\phi} = 7$ deg/s, and section AB of the ladder extends from within section OA at the constant rate of 0.5 m/s. At the instant under consideration, $\phi = 30°$, $\overline{OA} = 9$ m, and $\overline{AB} = 6$ m. Determine the magnitudes of the velocity and acceleration of the end B of the ladder.

Problem 2/177

Problem 2/178

2/179 The revolving crane has a boom of length $\overline{OP} =$ 24 m and is turning about its vertical axis at a constant rate of 2 rev/min. At the same time, the boom is being lowered at the constant rate $\dot{\beta} =$ 0.10 rad/s. Calculate the magnitude of the velocity and acceleration of the end P of the boom at the instant when the position $\beta = 30°$ is passed.

Ans. $v = 3.48$ m/s, $a = 1.104$ m/s^2

2/180 In a design test of the actuating mechanism for a telescoping antenna on a spacecraft, the supporting shaft rotates about the fixed z-axis with an angular rate $\dot{\theta}$. Determine the R-, θ-, and ϕ-components of the acceleration $\mathbf{a}$ of the end of the antenna at the instant when $L = 1.2$ m and $\beta = 45°$ if the rates $\dot{\theta} = 2$ rad/s, $\dot{\beta} = \frac{3}{2}$ rad/s, and $\dot{L} = 0.9$ m/s are constant during the motion.

Problem 2/179

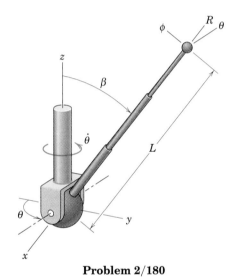

Problem 2/180

2/181 At the bottom of a vertical loop in the *x*-*y* plane at an altitude of 400 m, the airplane has a speed of 600 km/h with no horizontal *x*-acceleration. The radius of curvature of the loop at the bottom is 1200 m. For the radar tracking at O, determine the recorded values of $\ddot{R}$ and $\ddot{\phi}$ for this instant.

Ans. $\ddot{R} = 34.4 \text{ m/s}^2$, $\ddot{\phi} = 0.01038 \text{ rad/s}^2$

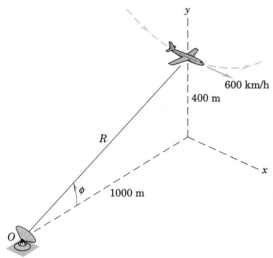

Problem 2/181

2/182 The rod OA is held at the constant angle $\beta = 30°$ while it rotates about the vertical with a constant angular rate $\dot{\theta} = 120$ rev/min. Simultaneously, the sliding ball P oscillates along the rod with its distance in millimeters from the fixed pivot O given by $R = 200 + 50 \sin 2\pi nt$, where the frequency n of oscillation along the rod is a constant 2 cycles per second and where t is the time in seconds. Calculate the magnitude of the acceleration of P for an instant when its velocity along the rod from O toward A is a maximum.

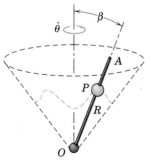

Problem 2/182

▶**2/183** The disk A rotates about the vertical *z*-axis with a constant speed $\omega = \dot{\theta} = \pi/3$ rad/s. Simultaneously, the hinged arm OB is elevated at the constant rate $\dot{\phi} = 2\pi/3$ rad/s. At time $t = 0$, both $\theta = 0$ and $\phi = 0$. The angle θ is measured from the fixed reference *x*-axis. The small sphere P slides out along the rod according to $R = 50 + 200t^2$, where R is in millimeters and t is in seconds. Determine the magnitude of the total acceleration **a** of P when $t = \frac{1}{2}$ s.

Ans. $a = 0.904 \text{ m/s}^2$

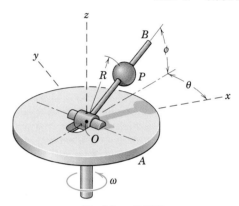

Problem 2/183

▶**2/184** In the design of an amusement-park ride the cars are attached to arms of length R that are hinged to a central rotating collar which drives the assembly about the vertical axis with a constant angular rate $\omega = \dot{\theta}$. The cars rise and fall with the track according to the relation $z = (h/2)(1 - \cos 2\theta)$. Find the *R*-, *θ*-, and *φ*-components of the velocity **v** of each car as it passes the position $\theta = \pi/4$ rad.

Ans. $v_R = 0$, $v_\theta = R\omega\sqrt{1 - (h/2R)^2}$
$v_\phi = h\omega/\sqrt{1 - (h/2R)^2}$

Problem 2/184

2/8 RELATIVE MOTION (TRANSLATING AXES)

In the previous articles of this chapter, we have described particle motion using coordinates that were referred to fixed reference axes. The displacements, velocities, and accelerations so determined are termed *absolute*. But it is not always possible or convenient to use a fixed set of axes for the observation of motion, and there are many engineering problems for which the analysis of motion is simplified by using measurements made with respect to a moving reference system. These measurements, when combined with the absolute motion of the moving coordinate system, permit us to determine the absolute motion in question. This approach is known as a *relative-motion* analysis.

The motion of the moving coordinate system is specified with respect to a fixed coordinate system. Strictly speaking, this fixed system in Newtonian mechanics is the primary inertial system that is assumed to have no motion in space. For engineering purposes, the fixed system may be taken as any system whose absolute motion is negligible for the problem at hand. For most earthbound engineering problems, it is sufficiently precise to take for the fixed reference system a set of axes attached to the earth, in which case we neglect the motion of the earth. For the motion of satellites around the earth, a nonrotating coordinate system with origin on the earth's axis of rotation is chosen. For interplanetary travel, a nonrotating coordinate system fixed to the sun would be used. Hence, the choice of the fixed system depends on the type of problem involved.

We will confine our attention in this article to moving reference systems which translate but do not rotate. Motion measured in rotating systems will be discussed in Art. 5/7 of Chapter 5 on rigid-body kinematics, where this approach finds special but important application. We will also confine our attention here to relative motion analysis for plane motion.

Consider now two particles A and B that may have separate curvilinear motions in a given plane or in parallel planes, Fig. 2/17. We will arbitrarily attach the origin of a set of translating (non-rotating) axes x-y to particle B and observe the motion of A from our moving position on B. The position vector of A as measured relative to the frame x-y is $\mathbf{r}_{A/B} = x\mathbf{i} + y\mathbf{j}$, where the subscript notation "A/B" means "A relative to B" or "A with respect to B." The unit vectors along the x- and y-axes are $\mathbf{i}$ and $\mathbf{j}$, and x and y are the coordinates of A measured in x-y. The absolute position of B is defined by the vector $\mathbf{r}_B$ measured from the origin of the fixed axes X-Y. The absolute position of A is seen, therefore, to be determined by the vector

$$\mathbf{r}_A = \mathbf{r}_B + \mathbf{r}_{A/B}$$

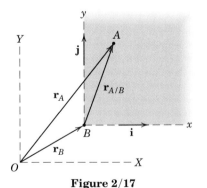

Figure 2/17

We now differentiate this vector equation once with respect to time to obtain velocities and twice to obtain accelerations. Thus,

$$\dot{\mathbf{r}}_A = \dot{\mathbf{r}}_B + \dot{\mathbf{r}}_{A/B} \quad \text{or} \quad \boxed{\mathbf{v}_A = \mathbf{v}_B + \mathbf{v}_{A/B}} \quad \textbf{(2/20)}$$

$$\ddot{\mathbf{r}}_A = \ddot{\mathbf{r}}_B + \ddot{\mathbf{r}}_{A/B} \quad \text{or} \quad \boxed{\mathbf{a}_A = \mathbf{a}_B + \mathbf{a}_{A/B}} \quad \textbf{(2/21)}$$

In Eq. 2/20 the velocity which we observe A to have from our position at B attached to the moving axes x-y is $\dot{\mathbf{r}}_{A/B} = \mathbf{v}_{A/B} = \dot{x}\mathbf{i} + \dot{y}\mathbf{j}$. This term is the velocity of A with respect to B. Similarly, in Eq. 2/21 the acceleration that we observe A to have from our nonrotating position on B is $\ddot{\mathbf{r}}_{A/B} = \dot{\mathbf{v}}_{A/B} = \ddot{x}\mathbf{i} + \ddot{y}\mathbf{j}$. This term is the acceleration of A with respect to B. We note carefully that the unit vectors $\mathbf{i}$ and $\mathbf{j}$ have no derivatives because their directions as well as their magnitudes remain unchanged. (Later when we discuss rotating reference axes, we will be obliged to account for the derivatives of the unit vectors when they change direction.)

In words, Eq. 2/20 (or 2/21) states that the absolute velocity (or acceleration) of A equals the absolute velocity (or acceleration) of B plus, vectorially, the velocity (or acceleration) of A relative to B. The relative term is the velocity (or acceleration) measurement which an observer attached to the moving coordinate system x-y would make. We may express the relative motion terms in whatever coordinate system is convenient—rectangular, normal and tangential, or polar—and the formulations in the preceding articles may be used for this purpose. The appropriate fixed system of the previous articles becomes the moving system in the present article.

The selection of the moving point B for attachment of the reference coordinate system is arbitrary. As shown in Fig. 2/18, point A could be used just as well for the attachment of the moving system, in which case the three corresponding relative-motion equations for position, velocity, and acceleration are

$$\mathbf{r}_B = \mathbf{r}_A + \mathbf{r}_{B/A} \qquad \mathbf{v}_B = \mathbf{v}_A + \mathbf{v}_{B/A} \qquad \mathbf{a}_B = \mathbf{a}_A + \mathbf{a}_{B/A}$$

It is seen, therefore, that $\mathbf{r}_{B/A} = -\mathbf{r}_{A/B}$, $\mathbf{v}_{B/A} = -\mathbf{v}_{A/B}$, and $\mathbf{a}_{B/A} = -\mathbf{a}_{A/B}$.

An important observation to be made in relative-motion analysis is that the acceleration of a particle as observed in a translating system x-y is the same as that observed in a fixed system X-Y if the moving system has a constant velocity. This conclusion broadens the application of Newton's second law of motion, treated in Chapter 3. We conclude, consequently, that a set of axes that has a constant absolute velocity may be used in place of a "fixed" system for the determination of accelerations. A translating reference system which has no acceleration is known as an *inertial system*.

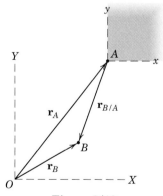

Figure 2/18

Sample Problem 2/12

Passengers in the jet transport A flying east at a speed of 800 km/h observe a second jet plane B that passes under the transport in horizontal flight. Although the nose of B is pointed in the 45° northeast direction, plane B appears to the passengers in A to be moving away from the transport at the 60° angle as shown. Determine the true velocity of B.

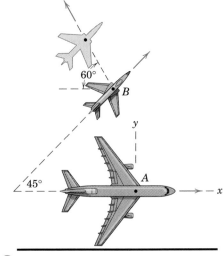

---

① *Solution.* The moving reference axes x-y are attached to A, from which the relative observations are made. We write, therefore,

$$\mathbf{v}_B = \mathbf{v}_A + \mathbf{v}_{B/A}$$

② ③ Next we identify the knowns and unknowns. The velocity $\mathbf{v}_A$ is given in both magnitude and direction. The 60° direction of $\mathbf{v}_{B/A}$, the velocity that B appears to have to the moving observers in A, is known, and the true velocity of B is in the 45° direction in which it is heading. The two remaining unknowns are the magnitudes of $\mathbf{v}_B$ and $\mathbf{v}_{B/A}$. We may solve the vector equation in any one of three ways.

(I) Graphical. We start the vector sum at some point P by drawing $\mathbf{v}_A$ to a convenient scale and then construct a line through the tip of $\mathbf{v}_A$ with the known direction of $\mathbf{v}_{B/A}$. The known direction of $\mathbf{v}_B$ is then drawn through P, and the intersection C yields the unique solution enabling us to complete the vector triangle and scale off the unknown magnitudes, which are found to be

$$v_{B/A} = 586 \text{ km/h} \quad \text{and} \quad v_B = 717 \text{ km/h} \quad \textit{Ans.}$$

(II) Trigonometric. A sketch of the vector triangle is made to reveal the trigonometry, which gives

④
$$\frac{v_B}{\sin 60°} = \frac{v_A}{\sin 75°} \qquad v_B = 800 \frac{\sin 60°}{\sin 75°} = 717 \text{ km/h} \quad \textit{Ans.}$$

(III) Vector algebra. Using unit vectors $\mathbf{i}$ and $\mathbf{j}$, we express the velocities in vector form as

$$\mathbf{v}_A = 800\mathbf{i} \text{ km/h} \qquad \mathbf{v}_B = (v_B \cos 45°)\mathbf{i} + (v_B \sin 45°)\mathbf{j}$$

$$\mathbf{v}_{B/A} = (v_{B/A} \cos 60°)(-\mathbf{i}) + (v_{B/A} \sin 60°)\mathbf{j}$$

Substituting these relations into the relative-velocity equation and solving separately for the $\mathbf{i}$ and $\mathbf{j}$ terms give

$$\textbf{(i-terms)} \qquad v_B \cos 45° = 800 - v_{B/A} \cos 60°$$

$$\textbf{(j-terms)} \qquad v_B \sin 45° = v_{B/A} \sin 60°$$

⑤ Solving simultaneously yields the unknown velocity magnitudes

$$v_{B/A} = 586 \text{ km/h} \quad \text{and} \quad v_B = 717 \text{ km/h} \quad \textit{Ans.}$$

It is worth noting the solution of this problem from the viewpoint of an observer in B. With reference axes attached to B, we would write $\mathbf{v}_A = \mathbf{v}_B + \mathbf{v}_{A/B}$. The apparent velocity of A as observed by B is then $\mathbf{v}_{A/B}$, which is the negative of $\mathbf{v}_{B/A}$.

① We treat each airplane as a particle.

② We assume no side slip due to cross wind.

③ Students should become familiar with all three solutions.

④ We must be prepared to recognize the appropriate trigonometric relation, which here is the law of sines.

⑤ We can see that the graphical or trigonometric solution is shorter than the vector algebra solution in this particular problem.

Sample Problem 2/13

Car A is accelerating in the direction of its motion at the rate of 1.2 m/s^2. Car B is rounding a curve of 150-m radius at a constant speed of 54 km/h. Determine the velocity and acceleration that car B appears to have to an observer in car A if car A has reached a speed of 72 km/h for the positions represented.

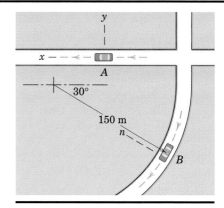

Solution. We choose nonrotating reference axes attached to car A since the motion of B with respect to A is desired.

Velocity. The relative-velocity equation is

$$\mathbf{v}_B = \mathbf{v}_A + \mathbf{v}_{B/A}$$

and the velocities of A and B for the position considered have the magnitudes

$$v_A = 72/3.6 = 20 \text{ m/s} \qquad v_B = 54/3.6 = 15 \text{ m/s}$$

The triangle of velocity vectors is drawn in the sequence required by the equation, and application of the law of cosines and the law of sines gives

$$v_{B/A} = 18.03 \text{ m/s} \qquad \theta = 46.1° \qquad \textit{Ans.}$$

Acceleration. The relative-acceleration equation is

$$\mathbf{a}_B = \mathbf{a}_A + \mathbf{a}_{B/A}$$

The acceleration of A is given, and the acceleration of B is normal to the curve in the n-direction and has the magnitude

$$[a_n = v^2/\rho] \qquad a_B = (15)^2/150 = 1.5 \text{ m/s}^2$$

The triangle of acceleration vectors is drawn in the sequence required by the equation as illustrated. Solving for the x- and y-components of $\mathbf{a}_{B/A}$ gives us

$$(a_{B/A})_x = 1.5 \cos 30° - 1.2 = 0.0990 \text{ m/s}^2$$

$$(a_{B/A})_y = 1.5 \sin 30° = 0.750 \text{ m/s}^2$$

$$\text{from which } a_{B/A} = \sqrt{(0.0990)^2 + (0.750)^2} = 0.757 \text{ m/s}^2 \quad \textit{Ans.}$$

The direction of $\mathbf{a}_{B/A}$ may be specified by the angle β which, by the law of sines, becomes

$$\frac{1.5}{\sin \beta} = \frac{0.757}{\sin 30°} \qquad \beta = \sin^{-1}\left(\frac{1.5}{0.757} 0.5\right) = 97.5° \qquad \textit{Ans.}$$

① Alternatively, we could use either a graphical or a vector algebraic solution.

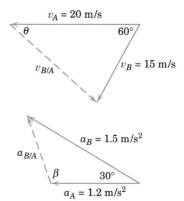

② Be careful to choose between the two values 82.5° and 180 − 82.5 = 97.5°.

Suggestion: To gain familiarity with the manipulation of vector equations, it is suggested that the student rewrite the relative-motion equations in the form $\mathbf{v}_{B/A} = \mathbf{v}_B - \mathbf{v}_A$ and $\mathbf{a}_{B/A} = \mathbf{a}_B - \mathbf{a}_A$ and redraw the vector polygons to conform with these alternative relations.

Caution: So far we are only prepared to handle motion relative to *nonrotating* axes. If we had attached the reference axes rigidly to car B, they would rotate with the car, and we would find that the velocity and acceleration terms relative to the rotating axes are *not* the negative of those measured from the nonrotating axes moving with A. Rotating axes are treated in Art. 5/7.

PROBLEMS

Introductory problems

2/185 Rapid-transit trains A and B travel on parallel tracks. Train A has a speed of 80 km/h and is slowing at the rate of 2 m/s^2, while train B has a constant speed of 40 km/h. Determine the velocity and acceleration of train B relative to train A.

Ans. $\mathbf{v}_{B/A} = 120\mathbf{i}$ km/h, $\mathbf{a}_{B/A} = -2\mathbf{i}$ m/s^2

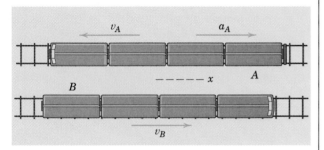

Problem 2/185

2/186 A woman P walks on an east-west street at a speed of 6 km/h. The wind blows out of the northwest as shown at a speed of 4 km/h. Determine the velocity of the wind relative to the woman if she (*a*) walks west and (*b*) walks east on the street. Express your results both in terms of unit vectors $\mathbf{i}$ and $\mathbf{j}$ and as magnitudes and compass directions.

Problem 2/186

2/187 The passenger aircraft B is flying east with a velocity $v_B = 800$ km/h. A military jet traveling south with a velocity $v_A = 1200$ km/h passes under B at a slightly lower altitude. What velocity does A appear to have to a passenger in B, and what is the direction of that apparent velocity?

Ans. $v_{A/B} = 1442$ km/h

$\beta = 33.7°$ west of south

Problem 2/187

2/188 The jet transport B is flying north with a velocity $v_B = 600$ km/h when a smaller aircraft A passes underneath the transport headed in the 60° direction shown. To passengers in B, however, A appears to be flying sideways and moving east. Determine the actual velocity of A and the velocity which A appears to have relative to B.

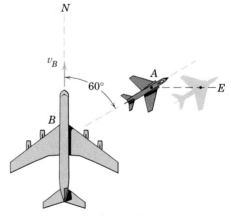

Problem 2/188

2/189 The small airplane *A* initially flying north with a ground speed of 225 km/h encounters a 75-km/h west wind (blowing east). Airplane *B* flying west with an airspeed of 270 km/h passes *A* at nearly the same altitude. Determine the magnitude and direction of the velocity which *A* appears to have to the pilot of *B*.

Ans. $v_{A/B}$ = 351 km/h
50.2° east of north

Problem 2/189

2/190 Train *A* travels with a constant speed v_A = 120 km/h along the straight and level track. The driver of car *B*, anticipating the railway grade crossing *C*, decreases the car speed of 90 km/h at the rate of 3 m/s². Determine the velocity and acceleration of the train relative to the car.

Problem 2/190

Representative problems

2/191 A sailboat moving in the direction shown is tacking to windward against a north wind. The log registers a hull speed of 6.5 knots. A "telltale" (light string tied to the rigging) indicates that the direction of the apparent wind is 35° from the centerline of the boat. What is the true wind velocity v_w?

Ans. v_w = 14.40 knots

Problem 2/191

2/192 Two ships *A* and *B* are moving with constant speeds v_A and v_B, respectively, along straight intersecting courses. The navigator of ship *B* notes the time rates of change of the separation distance *r* between the ships and the bearing angle θ. Show that $\ddot{\theta} = -2\dot{r}\dot{\theta}/r$ and $\ddot{r} = r\dot{\theta}^2$.

Problem 2/192

2/193 To increase his speed, the water skier A cuts across the wake of the tow boat B, which has a velocity of 60 km/h. At the instant when $\theta = 30°$, the actual path of the skier makes an angle $\beta = 50°$ with the tow rope. For this position determine the velocity v_A of the skier and the value of $\dot{\theta}$.

Ans. $v_A = 80.8$ km/h, $\dot{\theta} = 0.887$ rad/s

2/195 Hockey player A carries the puck on his stick and moves in the direction shown with a speed $v_A = 4$ m/s. In passing the puck to his stationary teammate B, by what shot angle α should the direction of his shot trail the line of sight if he launches the puck with a speed of 7 m/s relative to himself.?

Ans. $\alpha = 23.8°$

Problem 2/195

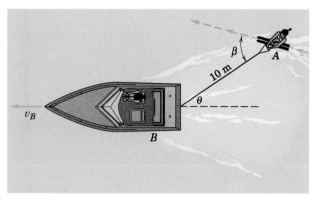

Problem 2/193

2/194 The car A has a forward speed of 18 km/h and is accelerating at 3 m/s². Determine the velocity and acceleration of the car relative to observer B, who rides in a nonrotating chair on the ferris wheel. The angular rate $\Omega = 3$ rev/min of the ferris wheel is constant.

2/196 A drop of water falls with no initial speed from point A of a highway overpass. After dropping 6 m, it strikes the windshield at point B of a car which is traveling at a speed of 100 km/h on the horizontal road. If the windshield is inclined 50° from the vertical as shown, determine the angle θ relative to the normal n to the windshield at which the water drop strikes.

Problem 2/194

Problem 2/196

2/197 The shuttle orbiter A is in a circular orbit of altitude 320 km, while spacecraft B is in a geosynchronous circular orbit of altitude 36 000 km. Determine the acceleration of B relative to a nonrotating observer in the shuttle A. Use $g_0 = 9.823$ m/s^2 for the surface-level gravitational acceleration and $R = 6371$ km for the radius of the earth.

$$\textit{Ans. } \mathbf{a}_{B/A} = 0.222\mathbf{i} + 8.91\mathbf{j} \text{ m/s}^2$$

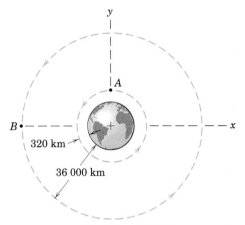

Problem 2/197

2/198 The captain of a small ship capable of making a speed of 6 knots through still water desires to set a course which will take the boat due east from A to B a distance of 10 nautical miles. To allow for a steady 2-knot current running northeast, determine his necessary compass heading H, measured clockwise from the north to the nearest degree. Also determine the time t of the trip. (Recall that 1 knot is 1 nautical mile per hour.)

Problem 2/198

2/199 The spacecraft S approaches the planet Mars along a trajectory b–b in the orbital plane of Mars with an absolute velocity of 19 km/s. Mars has a velocity of 24.1 km/s along its trajectory a–a. Determine the angle β between the line of sight S–M and the trajectory b–b when Mars appears from the spacecraft to be approaching it head on.

$$\textit{Ans. } \beta = 55.6°$$

Problem 2/199

2/200 Airplane A is flying horizontally with a constant speed of 200 km/h and is towing the glider B, which is gaining altitude. If the tow cable has a length $r = 60$ m and θ is increasing at the constant rate of 5 degrees per second, determine the magnitudes of the velocity $\mathbf{v}$ and acceleration $\mathbf{a}$ of the glider for the instant when $\theta = 15°$.

Problem 2/200

2/201 Car A is traveling at the constant speed of 60 km/h as it rounds the circular curve of 300-m radius and at the instant represented is at the position $\theta = 45°$. Car B is traveling at the constant speed of 80 km/h and passes the center of the circle at this same instant. Car A is located with respect to car B by polar coordinates r and θ with the pole moving with B. For this instant determine $v_{A/B}$ and the values of $\dot{r}$ and $\dot{\theta}$ as measured by an observer in car B.

　　Ans. $v_{A/B} = 36.0$ m/s
　　　　$\dot{r} = -15.71$ m/s, $\dot{\theta} = 0.1079$ rad/s

Problem 2/201

2/202 For the conditions of Prob. 2/201, determine the values of $\ddot{r}$ and $\ddot{\theta}$ as measured by an observer in car B at the instant represented. Use the results for $\dot{r}$ and $\dot{\theta}$ cited in the answers for that problem.

2/203 After starting from the position marked with the "x", a football receiver B runs the slant-in pattern shown, making a cut at P and thereafter running with a constant speed $v_B = 7$ m/s in the direction shown. The quarterback releases the ball with a horizontal velocity of 30 m/s at the instant the receiver passes point P. Determine the angle α at which the quarterback must throw the ball, and the velocity of the ball relative to the receiver when the ball is caught. Neglect any vertical motion of the ball.

　　Ans. $\alpha = 32.0°$, $\mathbf{v}_{A/B} = 21.9\mathbf{i} + 21.9\mathbf{j}$ m/s

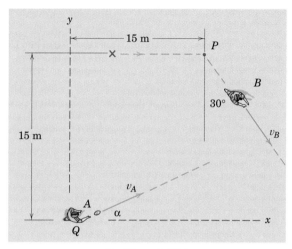

Problem 2/203

▶**2/204** A batter hits the baseball A with an initial velocity $v_0 = 30$ m/s directly toward fielder B at an angle of 30° to the horizontal; the initial position of the ball is 0.9 m above ground level. Fielder B requires $\frac{1}{4}$ s to judge where the ball should be caught and begins moving to that position with constant speed. Because of great experience, fielder B chooses his running speed so that he arrives at the "catch position" simultaneously with the baseball. The catch position is the field location at which the ball altitude is 2.1 m. Determine the velocity of the ball relative to the fielder at the instant the catch is made.

　　Ans. $\mathbf{v}_{A/B} = 21.5\mathbf{i} - 14.19\mathbf{j}$ m/s

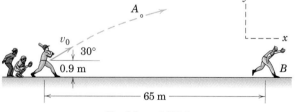

Problem 2/204

▶2/205 The aircraft A with radar detection equipment is flying horizontally at an altitude of 12 km and is increasing its speed at the rate of 1.2 m/s each second. Its radar locks onto an aircraft flying in the same direction and in the same vertical plane at an altitude of 18 km. If A has a speed of 1000 km/h at the instant when $\theta = 30°$, determine the values of $\ddot{r}$ and $\ddot{\theta}$ at this same instant if B has a constant speed of 1500 km/h.

$$\text{Ans. } \ddot{r} = -0.637 \text{ m/s}^2$$
$$\ddot{\theta} = 1.660(10^{-4}) \text{ rad/s}^2$$

Problem 2/205

▶2/206 At a certain instant after jumping from the airplane A, a skydiver B is in the position shown and has reached a terminal (constant) speed $v_B = 50$ m/s. The airplane has the same constant speed $v_A = 50$ m/s, and after a period of level flight is just beginning to follow the circular path shown of radius $\rho_A = 2000$ m. (a) Determine the velocity and acceleration of the airplane relative to the skydiver. (b) Determine the time rate of change of the speed v_r of the airplane and the radius of curvature ρ_r of its path, both as observed by the nonrotating skydiver.

$$\text{Ans. } (a) \ \mathbf{v}_{A/B} = 50\mathbf{i} + 50\mathbf{j} \text{ m/s}$$
$$\mathbf{a}_{A/B} = 1.250\mathbf{j} \text{ m/s}^2$$
$$(b) \ \dot{v}_r = 0.884 \text{ m/s}^2, \ \rho_r = 5660 \text{ m}$$

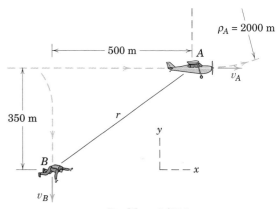

Problem 2/206

2/9   CONSTRAINED MOTION OF CONNECTED PARTICLES

There are occasions when the motions of particles are interrelated by virtue of the constraints of interconnecting members, and it becomes necessary to account for these constraints in order to determine the respective motions of the particles.

Consider first the very simple system of two interconnected particles A and B shown in Fig. 2/19. Although it should be quite evident by inspection that the horizontal motion of A is twice the vertical motion of B, we will use this example to illustrate the method of analysis which will carry over to more complex situations where the results cannot be easily perceived by inspection. Clearly, the motion of B is the same as that of the center of its pulley, so we establish position coordinates y and x measured from a convenient fixed datum. The total length of the cable is

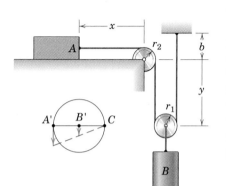

Figure 2/19

$$L = x + \frac{\pi r_2}{2} + 2y + \pi r_1 + b$$

With L, r_2, r_1, and b all constant, the first and second time derivatives of the equation give

$$0 = \dot{x} + 2\dot{y} \quad \text{or} \quad 0 = v_A + 2v_B$$

$$0 = \ddot{x} + 2\ddot{y} \quad \text{or} \quad 0 = a_A + 2a_B$$

The velocity and acceleration constraint equations indicate that, for the coordinates selected, the velocity of A must have a sign which is opposite to that of the velocity of B, and similarly for the accelerations. The constraint equations are valid for the motion of the system in either direction. We emphasize that $v_A = \dot{x}$ is positive to the left and that $v_B = \dot{y}$ is positive down.

Since the results do not depend on the lengths or pulley radii, we should be able to analyze the motion without considering them. In the lower-left portion of Fig. 2/19 is shown an enlarged view of the horizontal diameter $A'B'C$ of the lower pulley at an instant of time. Clearly, A' and A have the same motion magnitudes, as do B and B'. During an infinitesimal motion of A', it is easy to see that B' moves half as far since point C as a point on the fixed portion of the cable momentarily has no motion. Thus, with differentiation by time in mind, we can see the velocity and acceleration magnitude relationships by inspection. The pulley, in effect, is a wheel that rolls on the fixed vertical cable. (The kinematics of a rolling wheel will be treated more extensively in Chapter 5 on rigid-body motion.) The system of Fig. 2/19 is said to have *one degree of freedom* since only one variable, either x or y, is needed to specify the positions of all parts of the system.

A system with *two degrees of freedom* is shown in Fig. 2/20. Here the positions of the lower cylinder and pulley C depend on the separate specifications of the two coordinates y_A and y_B. The lengths of the cables attached to cylinders A and B may be written, respectively, as

$$L_A = y_A + 2y_D + \text{constant}$$

$$L_B = y_B + y_C + (y_C - y_D) + \text{constant}$$

and their time derivatives are

$$0 = \dot{y}_A + 2\dot{y}_D \quad \text{and} \quad 0 = \dot{y}_B + 2\dot{y}_C - \dot{y}_D$$

$$0 = \ddot{y}_A + 2\ddot{y}_D \quad \text{and} \quad 0 = \ddot{y}_B + 2\ddot{y}_C - \ddot{y}_D$$

Eliminating the terms in $\dot{y}_D$ and $\ddot{y}_D$ gives

$$\dot{y}_A + 2\dot{y}_B + 4\dot{y}_C = 0 \quad \text{or} \quad v_A + 2v_B + 4v_C = 0$$

$$\ddot{y}_A + 2\ddot{y}_B + 4\ddot{y}_C = 0 \quad \text{or} \quad a_A + 2a_B + 4a_C = 0$$

Clearly, it is impossible for the signs of all three terms to be positive simultaneously. So, for example, if both A and B have downward (positive) velocities, then C will have an upward (negative) velocity.

These results may also be found by inspection of the motions of the two pulleys at C and D. For an increment dy_A (with y_B held fixed), the center of D moves up an amount $dy_A/2$, which causes an upward movement $dy_A/4$ of the center of C. For an increment dy_B (with y_A held fixed), the center of C moves up a distance $dy_B/2$. A combination of the two movements gives an upward movement

$$-dy_C = \frac{dy_A}{4} + \frac{dy_B}{2}$$

so that $-v_C = v_A/4 + v_B/2$ as before. Visualization of the actual geometry of the motion is an important ability.

A second type of constraint where the direction of the connecting member changes with the motion is illustrated in the second of the two sample problems which follow.

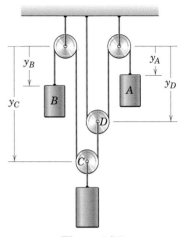

Figure 2/20

Sample Problem 2/14

In the pulley configuration shown, cylinder A has a downward velocity of 0.3 m/s. Determine the velocity of B. Solve in two ways.

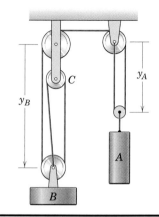

Solution (I). The centers of the pulleys at A and B are located by the coordinates y_A and y_B measured from fixed positions. The total constant length of cable in the pulley system is

$$L = 3y_B + 2y_A + \text{constants}$$

① where the constants account for the fixed lengths of cable in contact with the circumferences of the pulleys and the constant vertical separation between the two upper left-hand pulleys. Differentiation with time gives

$$0 = 3\dot{y}_B + 2\dot{y}_A$$

Substitution of $v_A = \dot{y}_A = 0.3$ m/s and $v_B = \dot{y}_B$ gives

$$0 = 3(v_B) + 2(0.3) \qquad \text{or} \qquad v_B = -0.2 \text{ m/s} \qquad \textit{Ans.}$$

Solution (II). An enlarged diagram of the pulleys at A, B, and C is shown. During a differential movement ds_A of the center of pulley A, the left end of its horizontal diameter has no motion since it is attached to the fixed part of the cable. Therefore, the right-hand end has a movement of $2ds_A$ as shown. This movement is transmitted to the left-hand end of the horizontal diameter of the pulley at B. Further, from pulley C with its fixed center, we see that the displacements on each side are equal and opposite. Thus, for pulley B, the right-hand end of the diameter has a downward displacement equal to the upward displacement ds_B of its center. By inspection of the geometry, we conclude that

②
$$2ds_A = 3ds_B \qquad \text{or} \qquad ds_B = \tfrac{2}{3}ds_A$$

Dividing by dt gives

$$|v_B| = \tfrac{2}{3}v_A = \tfrac{2}{3}(0.3) = 0.2 \text{ m/s (upward)} \qquad \textit{Ans.}$$

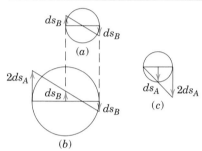

① We neglect the small angularity of the cables between B and C.

② The negative sign indicates that the velocity of B is *upward*.

Sample Problem 2/15

The tractor A is used to hoist the bale B with the pulley arrangement shown. If A has a forward velocity v_A, determine an expression for the upward velocity v_B of the bale in terms of x.

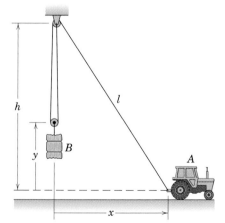

Solution. We designate the position of the tractor by the coordinate x and the position of the bale by the coordinate y, both measured from a fixed reference. The total constant length of the cable is

$$L = 2(h - y) + l = 2(h - y) + \sqrt{h^2 + x^2}$$

① Differentiation with time yields

$$0 = -2\dot{y} + \frac{x\dot{x}}{\sqrt{h^2 + x^2}}$$

Substituting $v_A = \dot{x}$ and $v_B = \dot{y}$ gives

$$v_B = \frac{1}{2}\frac{xv_A}{\sqrt{h^2 + x^2}} \qquad \textit{Ans.}$$

① Differentiation of the relation for a right triangle occurs frequently in mechanics.

PROBLEMS

Introductory problems

2/207 If block *A* has a velocity of 0.6 m/s to the right, determine the velocity of cylinder *B*.

Ans. $v_B = 1.8$ m/s down

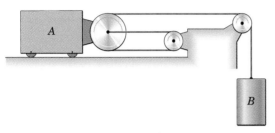

Problem 2/207

2/208 If the velocity $\dot{x}$ of block *A* up the incline is increasing at the rate of 0.044 m/s each second, determine the acceleration of *B*.

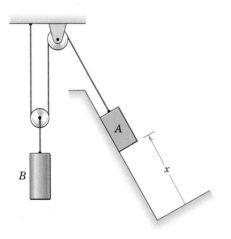

Problem 2/208

2/209 A truck equipped with a power winch on its front end pulls itself up a steep incline with the cable and pulley arrangement shown. If the cable is wound up on the drum at the constant rate of 40 mm/s, how long does it take for the truck to move 4 m up the incline?

Ans. $t = 3$ min 20 s

Problem 2/209

2/210 Cylinder *B* has a downward velocity of 0.6 m/s and an upward acceleration of 0.15 m/s². Calculate the velocity and acceleration of block *A*.

Problem 2/210

2/211 At a certain instant, cylinder A has a downward velocity of 0.8 m/s and an upward acceleration of 2 m/s². Determine the corresponding velocity and acceleration of cylinder B.

$$\text{Ans. } v_B = 1.2 \text{ m/s up}$$
$$a_B = 3 \text{ m/s}^2 \text{ down}$$

Problem 2/211

2/212 Determine the constraint equation which relates the accelerations of bodies A and B. Assume that the upper surface of A remains horizontal.

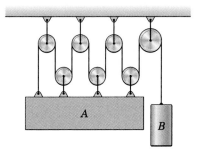

Problem 2/212

Representative problems

2/213 For the pulley system shown, each of the cables at A and B is given a velocity of 2 m/s in the direction of the arrow. Determine the upward velocity v of the load m.

$$\text{Ans. } v = 1.5 \text{ m/s}$$

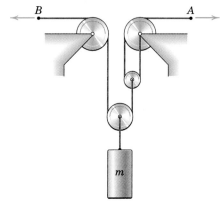

Problem 2/213

2/214 Neglect the diameters of the small pulleys and establish the relationship between the velocity of A and the velocity of B for a given value of y.

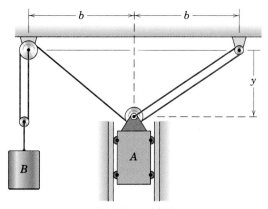

Problem 2/214

2/215 Determine an expression for the velocity v_A of the cart A down the incline in terms of the upward velocity v_B of cylinder B.

$$Ans. \ v_A = \frac{2v_B\sqrt{x^2 + h^2}}{x}$$

Problem 2/215

2/216 Under the action of force P, the constant acceleration of block B is 2 m/s² up the incline. For the instant when the velocity of B is 1.2 m/s up the incline, determine the velocity of B relative to A, the acceleration of B relative to A, and the absolute velocity of point C of the cable.

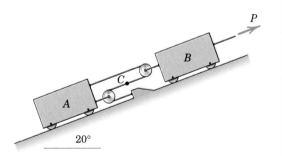

Problem 2/216

2/217 Determine the vertical rise h of the load W during 10 seconds if the hoisting drum draws in cable at the constant rate of 180 mm/s.

$$Ans. \ h = 300 \text{ mm}$$

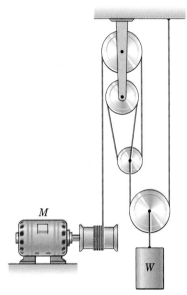

Problem 2/217

2/218 The power winches on the industrial scaffold enable it to be raised or lowered. For rotation in the sense indicated, the scaffold is being raised. If each drum has a diameter of 200 mm and turns at the rate of 40 rev/min, determine the upward velocity v of the scaffold.

Problem 2/218

2/219 Collars A and B slide along the fixed right-angled rods and are connected by a cord of length L. Determine the acceleration a_x of collar B as a function of y if collar A is given a constant upward velocity v_A.

$$\text{Ans. } a_x = -\frac{L^2 v_A{}^2}{(L^2 - y^2)^{3/2}}$$

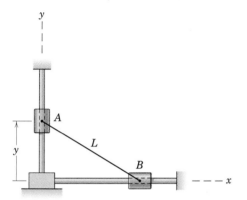

Problem 2/219

2/220 If load B has a downward velocity v_B, determine the upward component $(v_A)_y$ of the velocity of A in terms of b, the boom length l, and the angle θ. Assume that the cable supporting A remains vertical.

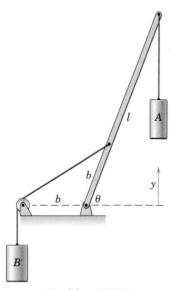

Problem 2/220

▶**2/221** Under the action of force P, the constant acceleration of block B is 3 m/s² to the right. At the instant when the velocity of B is 2 m/s to the right, determine the velocity of B relative to A, the acceleration of B relative to A, and the absolute velocity of point C of the cable.

$$\text{Ans. } v_{B/A} = 0.5 \text{ m/s}, \ a_{B/A} = 0.75 \text{ m/s}^2$$
$$v_C = 1 \text{ m/s, all to the right}$$

Problem 2/221

▶**2/222** Neglect the diameter of the small pulley attached to body A and determine the magnitude of the total velocity of B in terms of the velocity v_A which body A has to the right. Assume that the cable between B and the pulley remains vertical and solve for a given value of x.

$$\text{Ans. } v_B = v_A \sqrt{\frac{2x^2 + h^2}{x^2 + h^2}}$$

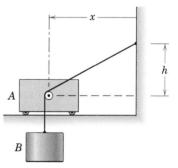

Problem 2/222

2/10 *PROBLEM FORMULATION AND REVIEW*

In Chapter 2 we have developed and illustrated the basic elements of theory and their use for the description of particle motion. It is important for the student to review this material before proceeding to consolidate and enlarge his or her perspective of particle kinematics. The concepts and procedures developed and illustrated in this chapter form the basis for building much of the entire subject of dynamics, and satisfactory progress in the topics ahead will be contingent on a firm grasp of Chapter 2.

By far the most important concept in Chapter 2 is the time derivative of a vector. The independent contributions of magnitude change and direction change to the time derivative are, indeed, basic. As we proceed in our study of dynamics, we will have further occasion to examine the time derivatives of vectors other than position and velocity vectors, and the principles and procedures developed in Chapter 2 will be utilized directly.

When one concentrates primarily on the individual parts of a subject, the relationships among its parts are not always clearly grasped. When guided by a specific topical assignment, the student automatically recognizes the problem category and the indicated method of solution. But without this guide, the choice of method is up to the student. The ability to choose the most appropriate method is certainly a key to the successful formulation and solution of engineering problems. One of the best ways to acquire this ability is to compare alternative methods after each of them is studied.

To facilitate this comparison, it will be helpful if we recognize the following breakdowns:

(a) Type of motion.   The three conspicuous categories are

Rectilinear motion (one coordinate)

Plane curvilinear motion (two coordinates)

Space curvilinear motion (three coordinates)

Generally, the geometry of a given problem permits this identification to be made readily. One exception to this categorization is encountered when only the magnitudes of the motion quantities measured along the path are of interest. In this event, we may use the single distance coordinate measured along the curved path, together with its scalar time derivatives giving the speed $|\dot{s}|$ and the tangential acceleration $\ddot{s}$.

Plane motion is easier to generate and control, particularly in machinery, than space motion, so that a large fraction of our motion problems come under the plane-curvilinear or rectilinear categories.

(b) Reference fixity.   We commonly make motion measurements with respect to fixed reference axes (absolute motion) and moving axes (relative motion). The acceptable choice of the fixed

axes depends on the problem. Axes attached to the surface of the earth are sufficiently "fixed" for most engineering problems, although important exceptions include earth–satellite and interplanetary motion, accurate projectile trajectories, navigation, and other problems. The equations of relative motion discussed in Chapter 2 are restricted to translating reference axes.

(c) Coordinates. Of prime importance is the choice of coordinates. We have developed the description of motion using

Rectangular (Cartesian) coordinates (x-y) and (x-y-z)

Normal and tangential coordinates (n-t)

Polar coordinates (r-θ)

Cylindrical coordinates (r-θ-z)

Spherical coordinates (R-θ-ϕ)

When the coordinates are not specified, the choice usually depends on how the motion is generated or measured. Thus, for a particle which slides radially along a rotating rod, polar coordinates are the natural ones to use. Radar tracking calls for polar or spherical coordinates. When measurements are made along a curved path, normal and tangential coordinates are indicated. An x-y plotter clearly involves rectangular coordinates.

Figure 2/21 is a composite representation of the x-y, n-t, and r-θ coordinate descriptions of the velocity **v** and acceleration **a** for curvilinear motion. It is frequently essential to transpose motion description from one set of coordinates to another, and Fig. 2/21 contains the information necessary for that transition.

(d) Approximations. One of the most important abilities to acquire is that of making appropriate approximations. The assumption of constant acceleration is valid when the forces that cause the acceleration do not vary appreciably. When motion data are acquired experimentally, we must utilize the nonexact data to acquire the best possible description, often with the aid of graphical or numerical approximations.

(e) Mathematical method. We frequently have a choice of solution using scalar algebra, vector algebra, trigonometric geometry, or graphical geometry. All of these methods have been illustrated and all are important to learn. The choice of method will depend on the geometry of the problem, how the motion data are given, and the accuracy desired. Inasmuch as mechanics by its very nature is geometrical, the student is encouraged to develop facility in sketching vector relationships, both as an aid to the disclosure of appropriate geometrical and trigonometrical relations and as a means of solving vector equations graphically. Geometric portrayal is the most direct representation of the vast majority of mechanics problems.

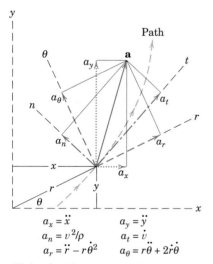

$$v_x = \dot{x} \qquad v_y = \dot{y}$$
$$v_n = 0 \qquad v_t = v$$
$$v_r = \dot{r} \qquad v_\theta = r\dot{\theta}$$

(a) Velocity components

$$a_x = \ddot{x} \qquad a_y = \ddot{y}$$
$$a_n = v^2/\rho \qquad a_t = \dot{v}$$
$$a_r = \ddot{r} - r\dot{\theta}^2 \qquad a_\theta = r\ddot{\theta} + 2\dot{r}\dot{\theta}$$

(b) Acceleration components

Figure 2/21

REVIEW PROBLEMS

2/223 In a test of vertical leaping ability, a basketball player (*a*) crouches just before jumping, (*b*) has given his mass center *G* a vertical velocity v_0 at the instant his feet leave the surface, and (*c*) reaches the maximum height. If the player can raise his mass center 1 m as shown, estimate the initial velocity v_0 of his mass center in position (*b*).

Ans. $v_0 = 4.43$ m/s

(a) (b) (c)

Problem 2/223

2/224 At time $t = 2$ s, the position vector of a particle moving in the x-y plane is $\mathbf{r} = 1.040\mathbf{i} + 0.600\mathbf{j}$ m. At time $t = 2.05$ s, the position vector has become $\mathbf{r}' = 1.060\mathbf{i} + 0.740\mathbf{j}$ m. Plot these two position vectors along with the change $\Delta\mathbf{r} = \mathbf{r}' - \mathbf{r}$ and determine the average velocity $\mathbf{v}$ of the particle over this time interval. Express $\mathbf{v}$ both in $\mathbf{i}$-$\mathbf{j}$ form and in terms of its magnitude and direction θ with respect to the x-axis.

2/225 At time $t = 0$ a small ball is projected from point A with a velocity of 60 m/s at the 60° angle. Neglect atmospheric resistance and determine the two times t_1 and t_2 when the velocity of the ball makes an angle of 45° with the horizontal x-axis.

Ans. $t_1 = 2.24$ s, $t_2 = 8.35$ s

Problem 2/225

2/226 For a certain interval of motion, the pin P is forced to move in the fixed parabolic slot by the vertical slotted guide, which moves in the x-direction at the constant rate of 20 mm/s. All measurements are in millimeters and seconds. Calculate the magnitudes of the velocity $\mathbf{v}$ and acceleration $\mathbf{a}$ of pin P when $x = 60$ mm.

Problem 2/226

2/227 A stone is thrown down the slope as shown. Determine the magnitude u and direction θ of its initial velocity so that the stone will rise 12 m and still have a range of 50 m down the slope.

Ans. $u = 17.74$ m/s, $\theta = 96.7°$

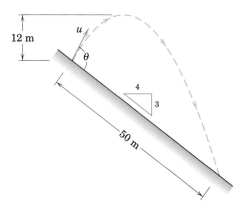

Problem 2/227

2/228 The angular displacement of the centrifuge is given by $\theta = 4[t + 30e^{-0.03t} - 30]$ rad, where t is in seconds and $t = 0$ is the startup time. If the person loses consciousness at an acceleration level of $10g$, determine the time t at which this would occur. Verify that the tangential acceleration is negligible as the normal acceleration approaches $10g$.

Problem 2/228

2/229 A rocket is fired vertically and tracked by the radar as shown. At the instant when $\theta = 60°$, measurements give $\dot{\theta} = 0.03$ rad/s and $r = 7500$ m, and the vertical acceleration of the rocket is found to be $a = 20$ m/s^2. For this instant determine the values of $\ddot{r}$ and $\ddot{\theta}$.

Ans. $\ddot{r} = 24.1$ m/s^2
$\ddot{\theta} = -1.784(10^{-3})$ rad/s^2

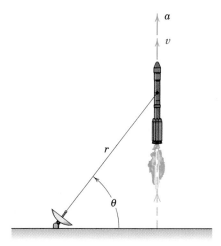

Problem 2/229

2/230 As part of a training exercise, the pilot of aircraft A adjusts her airspeed (speed relative to the wind) to 220 km/h while in the level portion of the approach path and thereafter holds her absolute speed constant as she negotiates the 10° glide path. The absolute speed of the aircraft carrier is 30 km/h and that of the wind is 48 km/h. What will be the angle β of the glide path with respect to the horizontal as seen by an observer on the ship?

Problem 2/230

2/231 The vertical displacement of cylinder A in meters is given by $y = t^2/4$ where t is in seconds. Calculate the downward acceleration a_B of cylinder B. Identify the number of degrees of freedom.

Ans. $a_B = 4$ m/s^2, one degree of freedom

Problem 2/231

2/232 A jet aircraft pulls up into a vertical curve as shown. As it passes the position where $\theta = 30°$, its speed is 1000 km/h and is decreasing at the rate of 15 km/h per second. If the radius of curvature ρ of the flight path is 1.5 km at this point, calculate the corresponding horizontal and vertical components, $\ddot{x}$ and $\ddot{y}$, of the acceleration of the aircraft.

Problem 2/232

2/233 The launching catapult of the aircraft carrier gives the jet fighter a constant acceleration of 50 m/s^2 from rest relative to the flight deck and launches the aircraft in a distance of 100 m measured along the angled takeoff ramp. If the carrier is moving at a steady 30 knots (1 knot = 1.852 km/h), determine the magnitude v of the actual velocity of the fighter when it is launched.

Ans. $v = 414$ km/h

Problem 2/233

2/234 At the instant depicted, assume that the particle P, which moves on a curved path, is 80 m from the pole O and has the velocity v and acceleration a as indicated. Determine the instantaneous values of $\dot{r}$, $\ddot{r}$, $\dot{\theta}$, $\ddot{\theta}$, the n- and t-components of acceleration, and the radius of curvature ρ.

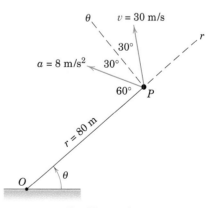

Problem 2/234

2/235 A rocket fired vertically up from the north pole achieves a velocity of 27 000 km/h at an altitude of 350 km when its fuel is exhausted. Calculate the additional vertical height h reached by the rocket before it starts its descent back to the earth. The coasting phase of its flight occurs above the atmosphere. Consult Fig. 1/1 in choosing the appropriate value of gravitational acceleration and use the mean radius of the earth from Table D/2. (*Note:* Launching from the earth's pole avoids considering the effect of the earth's rotation.)

Ans. $h = 6048$ km

2/236 Cylinder A has a constant downward speed of 1 m/s. Compute the velocity of cylinder B for (*a*) $\theta = 45°$, (*b*) $\theta = 30°$, and (*c*) $\theta = 15°$. The spring is in tension throughout the motion range of interest, and the pulleys are connected by the cable of fixed length.

Problem 2/236

2/237 The motion of pin P is controlled by the two moving slots A and B in which the pin slides. If B has a velocity $v_B = 3$ m/s to the right while A has an upward velocity $v_A = 2$ m/s, determine the magnitude v_P of the velocity of the pin.

Ans. $v_P = 2.72$ m/s

Problem 2/237

2/238 Particle P moves along the curved path shown. At the instant represented $r = 2$ m, $\theta = 30°$, and the velocity **v** makes an angle $\beta = 60°$ with the horizontal x-axis and has a magnitude of 3.2 m/s. If the y- and r-components of the acceleration of P are 5 m/s^2 and -1.83 m/s^2, respectively, at this position, determine the corresponding radius of curvature ρ of the path and the x-component of the acceleration of the particle. Solve graphically or analytically.

Problem 2/238

2/239 Just after being struck by the club, a golf ball has a velocity of 38 m/s directed at 35° to the horizontal as shown. Determine the location of the point of impact.

Ans. $R = 138.3$ m

Problem 2/239

2/240 A particle has the following position, velocity, and acceleration components: $x = 50$ m, $y = 25$ m, $\dot{x} = -10$ m/s, $\dot{y} = 10$ m/s, $\ddot{x} = -10$ m/s², and $\ddot{y} = 5$ m/s². Determine the following quantities: v, a, $\mathbf{e}_t$, $\mathbf{e}_n$, a_t, $\mathbf{a}_t$, a_n, $\mathbf{a}_n$, ρ, $\mathbf{e}_r$, $\mathbf{e}_\theta$, v_r, $\mathbf{v}_r$, v_θ, $\mathbf{v}_\theta$, a_r, $\mathbf{a}_r$, a_θ, $\mathbf{a}_\theta$, r, $\dot{r}$, $\ddot{r}$, θ, $\dot{\theta}$, and $\ddot{\theta}$. Express all vectors in terms of $\mathbf{i}$ and $\mathbf{j}$, and graph all vectors on one set of x-y axes as you proceed.

▶**2/241** The radar tracking antenna oscillates about its vertical axis according to $\theta = \theta_0 \cos \omega t$, where ω is the constant circular frequency and $2\theta_0$ is the double amplitude of oscillation. Simultaneously, the angle of elevation ϕ is increasing at the constant rate $\dot{\phi} = K$. Determine the expression for the magnitude a of the acceleration of the signal horn (a) as it passes position A and (b) as it passes the top position B, assuming that $\theta = 0$ at this instant.

Ans. $(a)\ a = b\sqrt{K^4 + \omega^4\theta_0^2 \cos^2 \phi}$

$(b)\ a = bK\sqrt{K^2 + 4\omega^2\theta_0^2}$

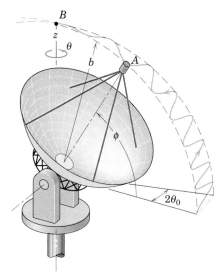

Problem 2/241

Computer-oriented problems

*2/242 The figure of Prob. 2/161 is repeated here. For this position, the slotted arm is fixed and the four-lobe cam rotates counterclockwise at the constant speed of 2 revolutions per second, and as previously, $r = 80 + 12 \cos 4\theta$, where r is in millimeters and θ is in radians. Plot the radial velocity v_r and the radial acceleration a_r of pin P versus θ from $\theta = 0$ to $\theta = \pi/2$. State the acceleration of pin P for $(a)\ \theta = 0$, $(b)\ \theta = \pi/8$, and $(c)\ \theta = \pi/4$, and compare these results with the corresponding results from Prob. 2/161.

Problem 2/242

*2/243 A baseball is dropped from an altitude $h = 60$ m and is found to be traveling at 26 m/s when it strikes the ground. In addition to gravitational acceleration, which may be assumed constant, air resistance causes a deceleration component of magnitude kv^2, where v is the speed and k is a constant. Determine the value of the coefficient k. Plot the speed of the baseball as a function of altitude y. If the baseball were dropped from a high altitude, but one at which g may still be assumed constant, what would be the terminal velocity v_t? (The *terminal velocity* is that speed at which the acceleration of gravity and that due to air resistance are equal and opposite, so that the baseball drops at a constant speed.) If the baseball were dropped from $h = 60$ m, at what speed v' would it strike the ground if air resistance were neglected?

Ans. $k = 0.01029$ m^{-1}, $v_t = 30.9$ m/s

$v' = 34.3$ m/s

*2/244 If all frictional effects are neglected, the expression for the angular acceleration of the simple pendulum is $\ddot{\theta} = \dfrac{g}{l} \cos \theta$, where g is the acceleration of gravity and l is the length of the rod OA. If the pendulum has a clockwise angular velocity $\dot{\theta} = 2$ rad/s when $\theta = 0$ at $t = 0$, determine the time t' at which the pendulum passes the vertical position $\theta = 90°$. The pendulum length is $l = 0.6$ m. Also plot the time t versus the angle θ.

Problem 2/244

*2/245 The acceleration of the drag racer is modeled by $a = c_1 - c_2 v^2$, where the v^2-term accounts for aerodynamic drag and where c_1 and c_2 are positive constants. If c_1 is known to be 9.14 m/s^2, determine c_2 if the racer completes the run in 9.40 s. Then plot the velocity and displacement as functions of time. The drag race is a 402-m straight run from a standing start.

Ans. $c_2 = 33.2(10^{-6})$ m^{-1}

Problem 2/245

*2/246 A ship with a total displacement of 16 000 metric tons (1 metric ton = 1000 kg) starts from rest in still water under a constant propeller thrust $T = 250$ kN. The ship develops a total resistance to motion through the water given by $R = 4.50v^2$, where R is in kilonewtons and v is in meters per second. The acceleration of the ship is $a = (T - R)/m$, where m equals the mass of the ship in metric tons. Plot the speed v of the ship in knots as a function of the distance s in nautical miles which the ship goes for the first 5 nautical miles from rest. Find the speed after the ship has gone 1 nautical mile. What is the maximum speed which the ship can reach?

*2/247 By means of the control unit M, the pendulum OA is given an oscillatory motion about the vertical given by $\theta = \theta_0 \sin \sqrt{\dfrac{g}{l}}\,t$, where θ_0 is the maximum angular displacement in radians, g is the acceleration of gravity, l is the pendulum length, and t is the time in seconds measured from an instant when OA is vertical. Determine and plot the magnitude a of the acceleration of A as a function of time and as a function of θ over the first quarter cycle of motion. Determine the minimum and maximum values of a and the corresponding values of t and θ. Use the values $\theta_0 = \pi/3$ radians, $l = 0.8$ m, and $g = 9.81$ m/s^2. (*Note:* The prescribed motion is not precisely that of a freely swinging pendulum for large amplitudes.)

  Ans. $a_{\min} = 9.03$ m/s^2 at $\theta = 44.3°$, $t = 0.237$ s

  $a_{\max} = 10.76$ m/s^2 at $\theta = 0$, $t = 0$

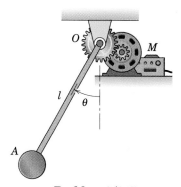

Problem 2/247

*2/248 A bullet with a muzzle velocity of 600 m/s is fired vertically upward and reaches a maximum height of 1600 m. Air resistance causes an additional component of downward acceleration kv^2 proportional to the square of the velocity v. Take g to be constant at 9.81 m/s^2 and calculate the coefficient k.

This cork-screw amusement-park ride is an example of particle kinetics. The forces required to support each car, treated as a particle, must be determined by applying Newton's second law of motion. The safe design of the structure is completely dependent on this analysis.

KINETICS OF PARTICLES

3

3/1 INTRODUCTION

According to Newton's second law, a particle will accelerate when it is subjected to unbalanced forces. Kinetics is the study of the relations between unbalanced forces and the changes in motion that they produce. In Chapter 3 we shall study the kinetics of particles. This topic requires that we combine our knowledge of two previously learned parts of mechanics, namely, the properties of forces that we developed in our earlier study of statics, and the kinematics of particle motion that we have just covered in Chapter 2. With the aid of Newton's second law of motion, we are now ready to combine these two topics and prepare for the solution of engineering problems involving force, mass, and motion.

There are three general approaches to the solution of kinetics problems: (A) direct application of Newton's second law (called the force-mass-acceleration method), (B) use of work and energy principles, and (C) solution by impulse and momentum methods. Each approach has its special characteristics and advantages, and Chapter 3 is subdivided into sections according to these three methods of solution. In addition, a fourth section treats special applications and combinations of the three basic approaches. Before proceeding, the student is strongly advised to review carefully the definitions and concepts of Chapter 1, as they are fundamental to the developments which follow.

SECTION A. FORCE, MASS, AND ACCELERATION

3/2 NEWTON'S SECOND LAW

The basic relation between force and acceleration is found in Newton's second law of motion, Eq. 1/1, the verification of which is entirely experimental. We will describe the fundamental meaning of this law by an ideal experiment in which force and acceleration are assumed to be measured without error. A mass particle is isolated in the primary inertial system* and is subjected to the action of a single force $\mathbf{F}_1$. The acceleration $\mathbf{a}_1$ of the particle is measured, and the ratio F_1/a_1 of the magnitudes of the force and the acceleration will be some number C_1 whose value depends on the units used for measurement of force and acceleration. The experiment is now repeated by subjecting the same particle to a different force $\mathbf{F}_2$ and measuring the corresponding acceleration $\mathbf{a}_2$. Again the ratio F_2/a_2 of the magnitudes will produce a number C_2. The experiment is repeated as many times as desired. We draw two important conclusions from the results. First, the ratios of applied force to corresponding acceleration all equal the *same* number, provided the units used for measurement are not changed in the experiments. Thus,

$$\frac{F_1}{a_1} = \frac{F_2}{a_2} = \cdots = \frac{F}{a} = C, \quad \text{a constant}$$

We conclude that the constant C is a measure of some property of the particle that does not change. This property is the *inertia* of the particle which is its *resistance to rate of change of velocity*. For a particle of high inertia (large C), the acceleration will be small for a given force F. On the other hand, if the inertia is small, the acceleration will be large. The mass m is used as a quantitative measure of inertia, and therefore the expression $C = km$ may be written, where k is a constant to account for the units used. Thus, the experimental relation becomes

$$F = kma \tag{3/1}$$

where F is the magnitude of the resultant force acting on the particle of mass m, and a is the magnitude of the resulting acceleration of the particle.

The second conclusion we draw from the ideal experiment is that the acceleration is always in the direction of the applied force. Thus, Eq. 3/1 becomes a *vector* relation and may be written

$$\mathbf{F} = km\mathbf{a} \tag{3/2}$$

Although an actual experiment cannot be performed in the ideal manner described, the conclusions are inferred from the measurements of countless accurately performed experiments where the results are correctly predicted from the hypothesis of the ideal experiment. One of the most accurate checks lies in the precise prediction of the motions of planets based on Eq. 3/2.

---

*The primary inertial system or astronomical frame of reference is an imaginary set of reference axes that are assumed to have no translation or rotation in space. See Art. 1/2, Chapter 1.

(a) Inertial system. Whereas the results of the ideal experiment are obtained for measurements made relative to the "fixed" primary inertial system, they are equally valid for measurements made with respect to any nonrotating reference system that translates with a constant velocity with respect to the primary system. From our study of relative motion in Art. 2/8, we saw that the acceleration measured in a system translating with no acceleration is the same as that measured in the primary system. Thus, Newton's second law holds equally well in a nonaccelerating system, so that we may define an *inertial system* as any system in which Eq. 3/2 is valid.

If the ideal experiment described were performed on the surface of the earth and all measurements were made relative to a reference system attached to the earth, the measured results would show a slight discrepancy upon substitution into Eq. 3/2. This discrepancy would be due to the fact that the measured acceleration would not be the correct absolute acceleration. The discrepancy would disappear when we introduced the corrections due to the acceleration components of the earth. These corrections are negligible for most engineering problems which involve the motions of structures and machines on the surface of the earth. In such cases, the accelerations measured with respect to reference axes attached to the surface of the earth may be treated as "absolute," and Eq. 3/2 may be applied with negligible error to experimental measurements made on the surface of the earth.*

There is an increasing number of problems, particularly in the fields of rocket and spacecraft design, where the acceleration components of the earth are of primary concern. For this work it is essential that the fundamental basis of Newton's law be thoroughly understood and that the appropriate absolute acceleration components be employed.

Before 1905 the laws of Newtonian mechanics had been verified by innumerable physical experiments and were considered the final description of the motion of bodies. The concept of *time*, considered an absolute quantity in the Newtonian theory, received a basically

---

*An example of the magnitude of the error introduced by neglect of the motion of the earth may be cited for the case of a particle that is allowed to fall from rest (relative to the earth) at a height h above the ground. We can show that the rotation of the earth gives rise to an eastward acceleration (Coriolis acceleration) relative to the earth and, neglecting air resistance, that the particle falls to the ground a distance

$$x = \frac{2}{3} \omega \sqrt{\frac{2h^3}{g}} \cos \gamma$$

east of the point on the ground directly under that from which it was dropped. The angular velocity of the earth is $\omega = 0.729(10^{-4})$ rad/s, and the latitude, north or south, is γ. At a latitude of 45° and from a height of 200 m, this eastward deflection would be $x = 43.9$ mm.

different interpretation in the theory of relativity announced by Einstein in 1905. The new concept called for a complete reformulation of the accepted laws of mechanics. The theory of relativity was subjected to early ridicule, but has had experimental check and is now universally accepted by scientists the world over. Although the difference between the mechanics of Newton and that of Einstein is basic, there is a practical difference in the results given by the two theories only when velocities of the order of the speed of light (300×10^6 m/s) are encountered.* Important problems dealing with atomic and nuclear particles, for example, involve calculations based on the theory of relativity and are of basic concern to both scientists and engineers.

(b) Units. It is customary to take k equal to unity in Eq. 3/2, thus putting the relation in the usual form of Newton's second law

$$\boxed{\mathbf{F} = m\mathbf{a}} \tag{1/1}$$

A system of units for which k is unity is known as a *kinetic* system. Thus, for a kinetic system the units of force, mass, and acceleration are not independent. In SI units, as explained in Art. 1/4, the units of force (newtons, N) are derived by Newton's second law from the base units of mass (kilograms, kg) times acceleration (meters per second squared, m/s^2). Thus, N = kg·m/s^2. This system is known as an *absolute* system since the unit for force is dependent on the absolute value of mass.

In U.S. customary units, on the other hand, the units of mass (slugs) are derived from the units of force (pounds force, lb) divided by acceleration (feet per second squared, ft/sec^2). Thus, the mass units are slugs = lb-sec^2/ft. This system is known as a *gravitational* system since mass is derived from force as determined from gravitational attraction.

For measurements made relative to the rotating earth, the relative value of g should be used. The internationally accepted value of g relative to the earth at sea level and at a latitude of 45° is 9.806 65 m/s^2. Except where greater precision is required, the value of 9.81 m/s^2 will be used for g. For measurements relative to a nonrotating earth, the absolute value of g should be used. At a latitude of 45° and at sea level, the absolute value is 9.8236 m/s^2. The

---

*The theory of relativity demonstrates that there is no such thing as a preferred primary inertial system and that measurements of time which are made in two coordinate systems which have a velocity relative to one another are different. On this basis, for example, the principles of relativity show that a clock carried by the pilot of a spacecraft traveling around the earth in a circular polar orbit of 644 km altitude at a velocity of 27 080 km/h would be slow compared with a clock at the pole by 0.000 001 85 s for each orbit.

sea-level variation in both the absolute and relative values of g with latitude is shown in Fig. 1/1 of Art. 1/5.

In the U.S. customary system, the standard value of g relative to the rotating earth at sea level and at a latitude of 45° is 32.1740 ft/sec^2. The corresponding value relative to a nonrotating earth is 32.2230 ft/sec^2.

By virtue of our need to use both SI units and U.S. customary units, we must be absolutely sure that we have a clear understanding of the correct force and mass units in each system. These units were explained in Art. 1/4, but it will be helpful to illustrate them here using simple numbers before moving ahead to the applications of Newton's second law. Consider, first, the free-fall experiment as depicted in Fig. 3/1a where we release an object from rest near the earth's surface and allow it to fall freely under the influence of the force of gravitational attraction W on the body, which we call its *weight*. In SI units for a mass $m = 1$ kg, the weight is $W = 9.81$ N, and the corresponding downward acceleration a is $g = 9.81$ m/s^2. In U.S. customary units for a mass $m = 1$ lbm $(1/32.2$ slug), the weight is $W = 1$ lbf and the resulting gravitational acceleration is $g = 32.2$ ft/sec^2. For a mass $m = 1$ slug $(32.2$ lbm), the weight is $W = 32.2$ lbf and the acceleration, of course, is also $g = 32.2$ ft/sec^2.

(a) Gravitational Free-Fall

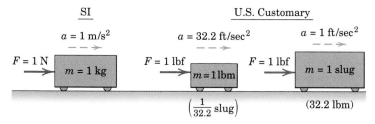

(b) Newton's Second Law

Figure 3/1

In Fig. 3/1*b* we illustrate the proper units with the simplest example where we accelerate an object of mass m along the horizontal with a force F. In the coherent system of SI units, a force $F = 1$ N causes a mass $m = 1$ kg to accelerate at the rate $a = 1$ m/s². Thus, 1 N $= 1$ kg·m/s². In the U.S. customary system (which is a gravitational, noncoherent system), a force $F = 1$ lbf causes a mass $m = 1$ lbm (1/32.2 slug) to accelerate at the rate $a = 32.2$ ft/sec², whereas a force $F = 1$ lbf causes a mass $m = 1$ slug (32.2 lbm) to accelerate at the rate $a = 1$ ft/sec².

We note that in SI units where the mass is expressed in kilograms (kg), the weight W of the body in newtons (N) is given by $W = mg$, where $g = 9.81$ m/s². In U.S. customary units, the weight W of a body is expressed in pounds force (lbf), and the mass in slugs (lbf-sec²/ft) is given by $m = W/g$, where $g = 32.2$ ft/sec². In U.S. customary units, we frequently speak of the weight of a body when we really mean mass. It is entirely proper to specify the mass of a body in pounds (lbm) which must be converted to mass in slugs before substituting into Newton's second law. Unless otherwise stated, the pound (lb) is normally used as the unit of force (lbf).

3/3 EQUATION OF MOTION AND SOLUTION OF PROBLEMS

When a particle of mass m is subjected to the action of concurrent forces $\mathbf{F}_1, \mathbf{F}_2, \mathbf{F}_3, \ldots$ whose vector sum is $\Sigma\mathbf{F}$, Eq. 1/1 becomes

$$\boxed{\Sigma\mathbf{F} = m\mathbf{a}} \qquad\qquad (3/3)$$

In the solution of problems, Eq. 3/3 is usually expressed in scalar component form with the use of one of the coordinate systems developed in Chapter 2. The choice of the appropriate coordinate system is dictated by the type of motion involved and is a vital step in the formulation of any problem. Equation 3/3, or any one of the component forms of the force-mass-acceleration equation, is usually referred to as the *equation of motion*. The equation of motion gives the instantaneous value of the acceleration corresponding to the instantaneous values of the forces which are acting.

(a) The two problems of dynamics. We encounter two types of problems when applying Eq. 3/3. In the first type, the acceleration is either specified or can be determined directly from known kinematic conditions. The corresponding forces which act on the particle whose motion is specified are then determined by direct substitution into Eq. 3/3. This problem is generally quite straightforward.

In the second type of problem, the forces are specified and the resulting motion is to be determined. If the forces are constant, the acceleration is constant and is easily found from Eq. 3/3.

When the forces are functions of time, position, velocity, or acceleration, Eq. 3/3 becomes a differential equation which must be integrated to determine the velocity and displacement. Problems of this second type are often more formidable, as the integration may be difficult to carry out, particularly when the force is a mixed function of two or more motion variables. In practice, it is frequently necessary to resort to approximate integration techniques, either numerical or graphical, particularly when experimental data are involved. The procedures for a mathematical integration of the acceleration when it is a function of the motion variables were developed in Art. 2/2, and these same procedures apply when the force is a specified function of these same parameters, since force and acceleration differ only by the constant factor of the mass.

(b) Constrained and unconstrained motion.   There are two physically distinct types of motion, both described by Eq. 3/3. The first type is *unconstrained* motion where the particle is free of mechanical guides and follows a path determined by its initial motion and by the forces which are applied to it from external sources. An airplane or rocket in flight and an electron moving in a charged field are examples of unconstrained motion. The second type is *constrained* motion where the path of the particle is partially or totally determined by restraining guides. An ice-hockey puck moves with the partial constraint of the ice. A train moving along its track and a collar sliding along a fixed shaft are examples of more fully constrained motion. Some of the forces acting on a particle during constrained motion may be applied from outside sources and others may be the reactions on the particle from the constraining guides. *All forces*, both applied and reactive, that act *on* the particle must be accounted for in applying Eq. 3/3.

The choice of a coordinate system is frequently indicated by the number and geometry of the constraints. Thus, if a particle is free to move in space, as is the center of mass of the airplane or rocket in free flight, the particle is said to have *three degrees of freedom* since three independent coordinates are required to specify the position of the particle at any instant. All three of the scalar components of the equation of motion would have to be applied and integrated to obtain the space coordinates as a function of time. If a particle is constrained to move along a surface, as is the hockey puck or a marble sliding on the curved surface of a bowl, only two coordinates are needed to specify its position, and in this case it is said to have *two degrees of freedom*. If a particle is constrained to move along a fixed linear path, as is the collar sliding along a fixed shaft, its position may be specified by the coordinate measured along the shaft. In this case, the particle would have only *one degree of freedom*.

(c) Free-body diagram. In the application of any of the force-mass-acceleration equations of motion, it is absolutely necessary to account correctly for *all* forces acting on the particle. The only forces that we may neglect are those whose magnitudes are negligible compared with other forces acting, such as the forces of mutual attraction between two particles compared with their attraction to a celestial body such as the earth. The vector sum $\Sigma\mathbf{F}$ of Eq. 3/3 means the vector sum of *all* forces acting *on* the particle in question. Likewise, the corresponding scalar force summation in any one of the component directions means the sum of the components of *all* forces acting *on* the particle in that particular direction. The only reliable way to account accurately and consistently for every force is to *isolate* the particle under consideration from *all* contacting and influencing bodies and replace the bodies removed by the forces they exert on the particle isolated. The resulting *free-body diagram* is the means by which every force, known and unknown, which acts on the particle is represented and hence accounted for. Only after this vital step has been completed should the appropriate equation or equations of motion be written. The free-body diagram serves the same key purpose in dynamics as it does in statics. This purpose is simply to establish a *thoroughly reliable method* for the correct evaluation of the resultant of all actual forces acting on the particle or body in question. In statics this resultant equals zero, whereas in dynamics it is equated to the product of mass and acceleration. If we recognize that the equations of motion must be interpreted literally and exactly, and if in so doing, we respect the full scalar and vector meaning of the equals sign in the motion equation, then a minimum of difficulty will be experienced. Every experienced student of engineering mechanics recognizes that careful and consistent observance of the *free-body method* is the *most important single lesson* to be learned in the study of engineering mechanics. As a part of the drawing of a free-body diagram, the coordinate axes and their positive directions should be clearly indicated. When the equations of motion are written, all force summations should be consistent with the choice of these positive directions. Also, as an aid to the identification of external forces which act on the body in question, these forces are shown as heavy red vectors in the illustrations throughout the remainder of the book. Sample Problems 3/1 through 3/5 in the next article contain five examples of free-body diagrams which may be easily reviewed as a reminder of their construction.

In solving problems, the student often wonders how to get started and what sequence of steps to follow in arriving at the solution. This difficulty may be minimized by forming the habit of first recognizing some relationship between the desired unknown quantity in the problem and other quantities, known and unknown. Additional relationships between these unknowns and other quantities, known and unknown, are then perceived. Finally, the dependence on the original data is established, and the procedure for the analysis and computation is indicated. A few minutes spent organizing the plan of attack through recognition of the dependence of one quantity on another will be time well spent and will usually prevent groping for the answer with irrelevant calculations.

3/4 RECTILINEAR MOTION

We now apply the concepts discussed in Arts. 3/2 and 3/3 to problems in particle motion, starting with rectilinear motion in this article and treating curvilinear motion in Art. 3/5. In both articles, we shall analyze the motions of bodies which can be treated as particles. This simplification is permitted as long as we are interested only in the motion of the mass center of the body, in which case, the forces may be treated as concurrent through the mass center. When we discuss the kinetics of rigid bodies in Chapter 6, we shall account for the action of nonconcurrent forces on the motions of bodies.

If we choose the x-direction, for example, as the direction of the rectilinear motion of a particle of mass m, the acceleration in the y- and z-directions will be zero and the scalar components of Eq. 3/3 become

$$\Sigma F_x = ma_x$$
$$\Sigma F_y = 0 \qquad\qquad (3/4)$$
$$\Sigma F_z = 0$$

For cases where we are not free to choose a coordinate direction along the motion, we would have in the general case all three component equations

$$\Sigma F_x = ma_x$$
$$\Sigma F_y = ma_y \qquad\qquad (3/5)$$
$$\Sigma F_z = ma_z$$

where the acceleration and resultant force are given by

$$\mathbf{a} = a_x\mathbf{i} + a_y\mathbf{j} + a_z\mathbf{k}$$
$$a = \sqrt{a_x^2 + a_y^2 + a_z^2}$$
$$\Sigma\mathbf{F} = \Sigma F_x\mathbf{i} + \Sigma F_y\mathbf{j} + \Sigma F_z\mathbf{k}$$
$$|\Sigma\mathbf{F}| = \sqrt{(\Sigma F_x)^2 + (\Sigma F_y)^2 + (\Sigma F_z)^2}$$

Sample Problem 3/1

A 75-kg man stands on a spring scale in an elevator. During the first 3 seconds of motion from rest, the tension T in the hoisting cable is 8300 N. Find the reading R of the scale in newtons during this interval and the upward velocity v of the elevator at the end of the 3 seconds. The total mass of the elevator, man, and scale is 750 kg.

Solution. The force registered by the scale and the velocity both depend on the acceleration of the elevator, which is constant during the interval for which the forces are constant. From the free-body diagram of the elevator, scale, and man taken together, the acceleration is found to be

$$[\Sigma F_y = ma_y] \qquad 8300 - 7360 = 750a_y \qquad a_y = 1.257 \text{ m/s}^2$$

The scale reads the downward force exerted on it by the man's feet. The equal and opposite reaction R to this action is shown on the free-body diagram of the man alone together with his weight, and the equation of motion for him gives

① $[\Sigma F_y = ma_y] \qquad R - 736 = 75(1.257) \qquad R = 830 \text{ N} \qquad$ *Ans.*

The velocity reached at the end of the 3 seconds is

$$[\Delta v = \int a \, dt] \qquad v - 0 = \int_0^3 1.257 \, dt \qquad v = 3.77 \text{ m/s} \qquad \textit{Ans.}$$

① If the scale is calibrated in kilograms it would read $830/9.81 = 84.6$ kg which, of course, is not his true mass since the measurement was made in a noninertial (accelerating) system. *Suggestion:* Rework this problem in U.S. customary units.

Sample Problem 3/2

A small inspection car with a mass of 200 kg runs along the fixed overhead cable and is controlled by the attached cable at A. Determine the acceleration of the car when the control cable is horizontal and under a tension $T = 2.4$ kN. Also find the total force P exerted by the supporting cable on the wheels.

Solution. The free-body diagram of the car and wheels taken together and treated as a particle discloses the 2.4-kN tension T, the weight $W = mg = 200(9.81) = 1962$ N, and the force P exerted on the wheel assembly by the cable.

The car is in equilibrium in the y-direction since there is no acceleration in this direction. Thus,

$$[\Sigma F_y = 0] \qquad P - 2.4(\tfrac{5}{13}) - 1.962(\tfrac{12}{13}) = 0 \qquad P = 2.73 \text{ kN} \qquad \textit{Ans.}$$

① In the x-direction the equation of motion gives

$$[\Sigma F_x = ma_x] \qquad 2400(\tfrac{12}{13}) - 1962(\tfrac{5}{13}) = 200a \qquad a = 7.30 \text{ m/s}^2 \qquad \textit{Ans.}$$

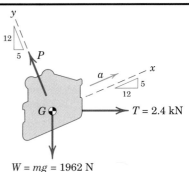

① By choosing our coordinate axes along and normal to the direction of the acceleration, we are able to solve the two equations independently. Would this be so if x and y were chosen as horizontal and vertical?

Sample Problem 3/3

The 125-kg concrete block A is released from rest in the position shown and pulls the 200-kg log up the 30° ramp. If the coefficient of kinetic friction between the log and the ramp is 0.5, determine the velocity of the block as it hits the ground at B.

200 kg

$\mu_k = 0.5$

30°

A

125 kg

6 m

B

Solution. The motions of the log and the block A are clearly dependent. Although by now it should be evident that the acceleration of the log up the incline is half the downward acceleration of A, we may prove it formally. The constant total length of the cable is $L = 2s_C + s_A +$ constant, where the constant accounts for the cable portions

① wrapped around the pulleys. Differentiating twice with respect to time gives $0 = 2\ddot{s}_C + \ddot{s}_A$, or

$$0 = 2a_C + a_A$$

We assume here that the masses of the pulleys are negligible and that they turn with negligible friction. With these assumptions the free-body diagram of the pulley C discloses force and moment equilibrium. Hence, the tension in the cable attached to the log is twice that applied to the block. Note that the accelerations of the log and the center of pulley C are identical.

The free-body diagram of the log shows the friction force $\mu_k N$ for motion up the plane. Equilibrium of the log in the y-direction gives

② $[\Sigma F_y = 0]$ $N - 200(9.81) \cos 30° = 0$ $N = 1699$ N

and its equation of motion in the x-direction gives

$[\Sigma F_x = ma_x]$ $0.5(1699) - 2T + 200(9.81) \sin 30° = 200 a_C$

For the block in the positive downward direction, we have

③ $[+\downarrow \Sigma F = ma]$ $125(9.81) - T = 125 a_A$

Solving the three equations in a_C, a_A, and T gives us

$a_A = 1.777$ m/s^2 $a_C = -0.888$ m/s^2 $T = 1004$ N

④ For the 6-m drop with constant acceleration, the block acquires a velocity

$[v^2 = 2ax]$ $v_A = \sqrt{2(1.777)(6)} = 4.62$ m/s   *Ans.*

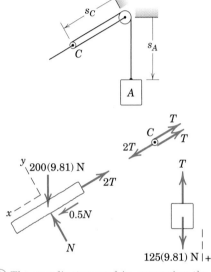

① The coordinates used in expressing the final kinematic constraint relationship must be consistent with those used for the kinetic equations of motion.

② We can verify that the log will indeed move up the ramp by calculating the cable force necessary to initiate motion from the equilibrium condition. This force is $2T = 0.5N + 200(9.81) \sin 30° = 1831$ N or $T = 915$ N, which is less than the 1226-N weight of block A. Hence, the log will move up.

③ Note the serious error in assuming that $T = 125(9.81)$ N, in which case, block A would not accelerate.

④ Because the forces on this system remain constant, the resulting accelerations also remain constant.

Sample Problem 3/4

The design model for a new ship has a mass of 10 kg and is tested in an experimental towing tank to determine its resistance to motion through the water at various speeds. The test results are plotted on the accompanying graph, and the resistance R may be closely approximated by the dotted parabolic curve shown. If the model is released when it has a speed of 2 m/s, determine the time t required for it to reduce its speed to 1 m/s and the corresponding travel distance x.

Solution. We approximate the resistance-velocity relation by $R = kv^2$ and find k by substituting $R = 8$ N and $v = 2$ m/s into the equation, which gives $k = 8/2^2 = 2$ N·s^2/m^2. Thus, $R = 2v^2$.

The only horizontal force on the model is R, so that

① $[\Sigma F_x = ma_x]$ $\qquad -R = ma_x \qquad$ or $\qquad -2v^2 = 10\dfrac{dv}{dt}$

We separate the variables and integrate to obtain

$$\int_0^t dt = -5 \int_2^v \frac{dv}{v^2} \qquad t = 5\left(\frac{1}{v} - \frac{1}{2}\right) \text{ sec}$$

Thus, when $v = v_0/2 = 1$ m/s, the time is $t = 5(\frac{1}{1} - \frac{1}{2}) = 2.5$ s $\qquad$ *Ans.*

The distance traveled during the 2.5 seconds is obtained by integrating $v = dx/dt$. Thus, $v = 10/(5 + 2t)$ so that

② $\displaystyle\int_0^x dx = \int_0^{2.5} \frac{10}{5 + 2t}\, dt \qquad x = \frac{10}{2}\ln(5 + 2t)\Big|_0^{2.5} = 3.47 \text{ m} \qquad$ *Ans.*

① Be careful to observe the minus sign for R.

② *Suggestion:* Express the distance x after release in terms of the velocity v and see if you agree with the resulting relation $x = 5\ln(v_0/v)$.

Sample Problem 3/5

The collar of mass m slides up the vertical shaft under the action of a force F of constant magnitude but variable direction. If $\theta = kt$ where k is a constant and if the collar starts from rest with $\theta = 0$, determine the magnitude F of the force that will result in the collar coming to rest as θ reaches $\pi/2$. The coefficient of kinetic friction between the collar and shaft is μ_k.

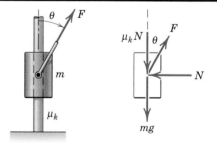

Solution. After drawing the free-body diagram, we apply the equation of motion in the y-direction to get

① $[\Sigma F_y = ma_y] \qquad F\cos\theta - \mu_k N - mg = m\dfrac{dv}{dt}$

where equilibrium in the horizontal direction requires $N = F\sin\theta$. Substituting $\theta = kt$ and integrating first between general limits give

$$\int_0^t (F\cos kt - \mu_k F\sin kt - mg)\, dt = m\int_0^v dv$$

which becomes

$$\frac{F}{k}[\sin kt + \mu_k(\cos kt - 1)] - mgt = mv$$

For $\theta = \pi/2$ the time becomes $t = \pi/2k$, and $v = 0$ so that

② $\dfrac{F}{k}[1 + \mu_k(0 - 1)] - \dfrac{mg\pi}{2k} = 0 \qquad$ and $\qquad F = \dfrac{mg\pi}{2(1 - \mu_k)} \qquad$ *Ans.*

① If θ were expressed as a function of the vertical displacement y instead of the time t, the acceleration would become a function of the displacement and we would use $v\, dv = a\, dy$.

② We see that the results do not depend on k, the rate at which the force changes direction.

PROBLEMS

Introductory problems

3/1 During a brake test, the rear-engine car is stopped from an initial speed of 100 km/h in a distance of 50 m. If it is known that all four wheels contribute equally to the braking force, determine the braking force F at each wheel. Assume a constant deceleration for the 1500-kg car.

Ans. $F = 2890$ N

$v_1 = 100$ km/h $v_2 = 0$

Problem 3/1

3/2 The 100-kg crate is carefully placed with zero velocity on the incline. Describe what happens if (*a*) $\theta = 15°$ and (*b*) $\theta = 20°$.

$\mu_s = 0.30$
$\mu_k = 0.25$

100 kg

θ

Problem 3/2

3/3 At a certain instant, the 40-kg crate has a velocity of 10 m/s up the 20° incline. Calculate the time t required for the crate to come to rest and the corresponding distance d traveled. Also, determine the distance d' traveled when the crate speed has been reduced to 5 m/s.

Ans. $t = 1.767$ s, $d = 8.83$ m, $d' = 6.63$ m

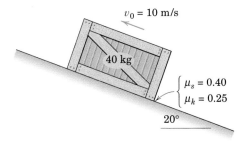

$v_0 = 10$ m/s

40 kg

$\mu_s = 0.40$
$\mu_k = 0.25$

20°

Problem 3/3

3/4 The 50-kg crate is stationary when the force P is applied. Determine the resulting acceleration of the crate if (*a*) $P = 0$, (*b*) $P = 150$ N, and (*c*) $P = 300$ N.

$\mu_s = 0.20$
$\mu_k = 0.15$

50 kg

P

15°

Problem 3/4

3/5 The 10-Mg truck hauls the 20-Mg trailer. If the unit starts from rest on a level road with a tractive force of 20 kN between the driving wheels of the truck and the road, compute the tension T in the horizontal drawbar and the acceleration a of the rig.

Ans. $T = 13.33$ kN, $a = 0.667$ m/s^2

20 Mg   10 Mg

Problem 3/5

3/6 The 300-Mg jet airliner has four engines, each of which produces a nearly constant thrust of 180 kN during the takeoff roll. Determine the length s of runway required if the takeoff speed is 220 km/h. Compute s first for an uphill takeoff direction from A to B and second for a downhill takeoff from B to A on the slightly inclined runway. Neglect air and rolling resistance.

0.5°

A

B

Horizontal

Problem 3/6

3/7 The 340-Mg jetliner A has four engines, each of which produces a nearly constant thrust of 200 kN during the takeoff roll. A small commuter aircraft B taxis toward the end of the runway at a constant speed $v_B = 25$ km/h. Determine the velocity and acceleration which A appears to have relative to an observer in B 10 seconds after A begins its takeoff roll. Neglect air and rolling resistance.

Ans. $\mathbf{v}_{A/B} = -29.5\mathbf{i} - 3.47\mathbf{j}$ m/s

$\mathbf{a}_{A/B} = -2.35\mathbf{i}$ m/s^2

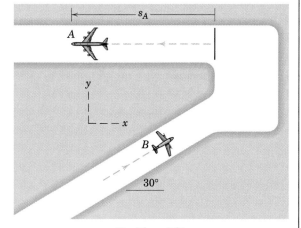

s_A

A

y

x

B

30°

Problem 3/7

3/8 The 80-kg man in the bosun's chair exerts a pull of 270 N on the rope for a short interval. Find his acceleration. Neglect the mass of the chair, rope, and pulleys.

Problem 3/8

3/9 A man pulls himself up the 15° incline by the method shown. If the combined mass of the man and cart is 100 kg, determine the acceleration of the cart if the man exerts a pull of 250 N on the rope. Neglect all friction and the mass of the rope, pulleys, and wheels.

Ans. $a = 4.96$ m/s^2

15°

Problem 3/9

3/10 The tractor-trailer unit is moving down the incline with a speed of 8 km/h when the driver brakes the tractor to a stop in a distance of 1.2 m. Estimate the percent increase n in the hitch-force component which is parallel to the incline, compared with the force present at steady speed. The cart and its load have a combined mass of 220 kg. State any assumptions.

Problem 3/10

3/11 A skier starts from rest on the 40° slope at time $t = 0$ and is clocked at $t = 2.58$ s as he passes a speed checkpoint 20 m down the slope. Determine the coefficient of kinetic friction between the snow and the skis. Neglect wind resistance.

Ans. $\mu_k = 0.0395$

Problem 3/11

3/12 Determine the tension P in the cable which will give the 50-kg block a steady acceleration of 2 m/s² up the incline.

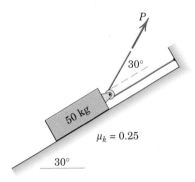

Problem 3/12

Representative problems

3/13 The collar A is free to slide along the smooth shaft B mounted in the frame. The plane of the frame is vertical. Determine the horizontal acceleration a of the frame necessary to maintain the collar in a fixed position on the shaft.

Ans. $a = 5.66$ m/s²

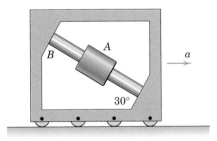

Problem 3/13

...

3/14 The coefficient of static friction between the flat bed of the truck and the crate it carries is 0.30. Determine the minimum stopping distance s which the truck can have from a speed of 70 km/h with constant deceleration if the crate is not to slip forward.

Problem 3/14

3/15 A car is descending the hill of slope θ_1 with the brakes slightly applied so that the speed v is constant. The slope decreases abruptly to θ_2 at point A. If the driver does not change the braking force, determine the acceleration a of the car after it passes point A. Evaluate your expression for $\theta_1 = 6°$ and $\theta_2 = 2°$.

Ans. $a = g(\sin \theta_2 - \sin \theta_1)$, $a = -0.0696g$

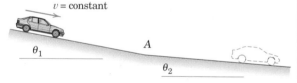

Problem 3/15

3/16 A small package is deposited by the conveyor belt onto the 30° ramp at A with a velocity of 0.8 m/s. Calculate the distance s on the level surface BC at which the package comes to rest. The coefficient of kinetic friction for the package and supporting surface from A to C is 0.3.

Problem 3/16

3/17 Small objects are delivered to the 2-m inclined chute by a conveyor belt A which moves at a speed $v_1 = 0.4$ m/s. If the conveyor belt B has a speed $v_2 = 0.9$ m/s and the objects are delivered to this belt with no slipping, calculate the coefficient of friction μ_k between the objects and the chute.

Ans. $\mu_k = 0.558$

Problem 3/17

3/18 The 10-kg steel sphere is suspended from the 15-kg frame which slides down the 20° incline. If the coefficient of kinetic friction between the frame and incline is 0.15, compute the tension in each of the supporting wires A and B.

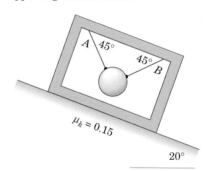

Problem 3/18

3/19 A train consists of a 180-Mg locomotive and one hundred 90-Mg hopper cars. If the locomotive exerts a friction force of 180 kN on the rails in starting the train from rest, compute the forces in couplers 1 and 100. Assume no slack in the couplers and neglect friction.

Ans. $T_1 = 176.5$ kN, $T_{100} = 1765$ N

Problem 3/19

3/20 A three-car subway train is traveling down a 5-percent grade when the individual car brakes are simultaneously applied. Each 10-Mg car can generate a braking force of 0.5 times the normal force exerted on it by the tracks. Determine the train deceleration a and the forces T_1 and T_2 in the couplings 1 and 2 for the cases when (*a*) all brakes function normally, (*b*) the brakes of car A fail, (*c*) the brakes of car B fail, and (*d*) the brakes of car C fail.

Problem 3/20

3/21 During a reliability test, a circuit board of mass m is attached to an electromagnetic shaker and subjected to a harmonic displacement $x = X \sin \omega t$, where X is the motion amplitude, ω is the motion frequency in radians per second, and t is time. Determine the magnitude F_{max} of the maximum horizontal force which the shaker exerts on the circuit board.

Ans. $F_{max} = mX\omega^2$

Problem 3/21

3/22 Determine the acceleration of the slider relative to the frame, which has an acceleration $a = 6$ m/s^2. Friction is to be neglected.

Problem 3/22

3/23 Determine the vertical acceleration of the 30-kg cylinder for each of the two cases. Neglect friction and the mass of the pulleys.

Ans. (*a*) $a = 3.27$ m/s^2 up
(*b*) $a = 0.892$ m/s^2 up

Problem 3/23

3/24 If the coefficients of static and kinetic friction between the 20-kg block A and the 100-kg cart B are both essentially the same value of 0.50, determine the acceleration of each part for (*a*) $P = 60$ N and (*b*) $P = 40$ N.

Problem 3/24

3/25 The system of Prob. 2/210 is repeated here with additional mass information specified. Neglect all friction and the mass of the pulleys and determine the accelerations of bodies A and B upon release from rest.

$$\text{Ans. } a_A = 1.024 \text{ m/s}^2 \text{ down the incline}$$
$$a_B = 0.682 \text{ m/s}^2 \text{ up}$$

Problem 3/25

3/26 The system is released from rest with the cable taut. For the friction coefficients $\mu_s = 0.25$ and $\mu_k = 0.20$, calculate the acceleration of each body and the tension T in the cable. Neglect the small mass and friction of the pulleys.

Problem 3/26

3/27 A player pitches a baseball horizontally toward a speed-sensing radar gun. The baseball has a mass of 146 g and has a circumference of 232 mm. If the speed at $x = 0$ is $v_0 = 150$ km/h, estimate the speed as a function of x. Assume that the horizontal aerodynamic drag on the baseball is given by $D = C_D(\frac{1}{2}\rho v^2)S$, where C_D is the drag coefficient, ρ is the air density, v is the speed, and S is the cross-sectional area of the baseball. Use a value of 0.3 for C_D. Neglect the y-component of the motion but comment on the validity of this assumption. Evaluate your answer for $x = 18$ m, which is the approximate distance between a pitcher's hand and home plate.

$$\text{Ans. } v = v_0 e^{(-0.5C_D\rho Sx/m)} = v_0 e^{-5.31(10^{-3})x}$$
$$v_{18} = 136.3 \text{ km/h}$$

Problem 3/27

3/28 During its final approach to the runway, the aircraft speed is reduced from 300 km/h at A to 200 km/h at B. Determine the net external aerodynamic force R which acts on the 200-Mg aircraft during this interval, and find the components of this force which are parallel to and normal to the flight path.

Problem 3/28

3/29 In a test to determine the crushing characteristics of polystyrene packing material, a steel cone of mass m is dropped so that it falls a distance h and then penetrates the material. The resistance R of polystyrene to penetration depends on the cross-sectional area of the penetrating object and thus is proportional to the square of the cone penetration distance x, or $R = kx^2$. If the cone comes to rest at a distance $x = d$, determine the constant k in terms of the test conditions and results.

$$Ans. \ k = \frac{3mg}{d^3}(h + d)$$

Problem 3/29

3/30 A small block is given an initial velocity v measured along the horizontal floor of an elevator moving with a downward acceleration a. Because of friction the block moves only a distance s measured along the floor before it stops sliding. The experiment is repeated with the same initial velocity relative to the floor when the elevator has an upward acceleration of the same magnitude a, and the block slides a shorter distance s_2. Determine the elevator acceleration a.

3/31 A heavy chain with a mass ρ per unit length is pulled by the constant force P along a horizontal surface consisting of a smooth section and a rough section. The chain is initially at rest on the rough surface with $x = 0$. If the coefficient of kinetic friction between the chain and the rough surface is μ_k, determine the velocity v of the chain when $x = L$. The force P is greater than $\mu_k \rho g L$ in order to initiate motion.

$$Ans. \ v = \sqrt{\frac{2P}{\rho} - \mu_k g L}$$

Problem 3/31

3/32 A force P is applied to the initially stationary cart. Determine the velocity and displacement at time $t = 5$ s for each of the force histories P_1 and P_2. Neglect friction.

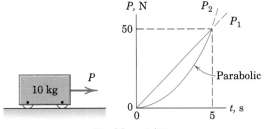

Problem 3/32

3/33 Determine the accelerations of bodies A and B and the tension in the cable due to the application of the 300-N force. Neglect all friction and the masses of the pulleys.

$$Ans. \ a_A = 2.34 \ \text{m/s}^2$$
$$a_B = 1.558 \ \text{m/s}^2$$
$$T = 81.8 \ \text{N}$$

Problem 3/33

3/34 The design of a lunar mission calls for a 1200-kg spacecraft to lift off from the surface of the moon and travel in a straight line from point A and pass point B. If the spacecraft motor has a constant thrust of 2500 N, determine its speed as it passes point B. Use Table D/2 and the gravitational law from Chapter 1 as needed.

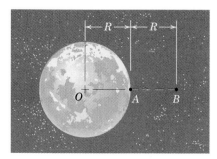

Problem 3/34

3/35 The nonlinear spring has a tensile force-deflection relationship given by $F_s = 150x + 400x^2$, where x is in meters and F_s is in newtons. Determine the acceleration of the 6-kg block if it is released from rest at (*a*) $x = 50$ mm and (*b*) $x = 100$ mm.

Ans. (*a*) $a = 0$
(*b*) $a = -0.714$ m/s^2

Problem 3/35

3/36 The spring of constant $k = 200$ N/m is attached to both the support and the 2-kg cylinder, which slides freely on the horizontal guide. If a constant 10-N force is applied to the cylinder at time $t = 0$ when the spring is undeformed and the system is at rest, determine the velocity of the cylinder when $x = 40$ mm and the maximum displacement of the cylinder.

Problem 3/36

3/37 In a test of resistance to motion in an oil bath, a small steel ball of mass m is released from rest at the surface $(y = 0)$. If the resistance to motion is given by $R = kv$ where k is a constant, derive an expression for the depth h required for the ball to reach a velocity v.

Ans. $h = \dfrac{m^2g}{k^2} \ln \left(\dfrac{1}{1 - kv/(mg)} \right) - \dfrac{mv}{k}$

Problem 3/37

3/38 If the steel ball of Prob. 3/37 is released from rest at the surface of a liquid in which the resistance to motion is $R = cv^2$, where c is a constant and v is the downward velocity of the ball, determine the depth h required to reach a velocity v.

3/39 A shock absorber is a mechanical device which provides resistance to compression or extension given by $R = cv$, where c is a constant and v is the time rate of change of the length of the absorber. An absorber of constant $c = 3000$ N·s/m is shown being tested with a 100-kg cylinder suspended from it. The system is released with the cable taut at $y = 0$ and allowed to extend. Determine (a) the steady-state velocity v_s of the lower end of the absorber and (b) the time t and displacement y of the lower end when the cylinder has reached 90 percent of its steady-state speed. Neglect the mass of the piston and attached rod.

> *Ans.* (a) $v_s = 0.327$ m/s
>
> (b) $t = 0.0768$ s, $y = 15.29$ mm

Problem 3/39

3/40 The sliders A and B are connected by a light rigid bar of length $l = 0.5$ m and move with negligible friction in the horizontal slots shown. For the position where $x_A = 0.4$ m, the velocity of A is $v_A = 0.9$ m/s to the right. Determine the acceleration of each slider and the force in the bar at this instant.

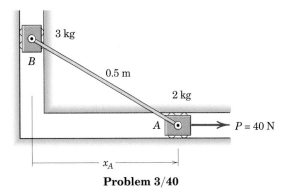

Problem 3/40

3/41 The sliders A and B are connected by a light rigid bar and move with negligible friction in the slots, both of which lie in a horizontal plane. For the position shown, the velocity of A is 0.4 m/s to the right. Determine the acceleration of each slider and the force in the bar at this instant.

> *Ans.* $a_A = 7.95$ m/s^2 right
>
> $a_B = 8.04$ m/s^2 down
>
> $T = 25.0$ N

Problem 3/41

3/42 For what value(s) of the angle θ will the acceleration of the 35-kg block be 9 m/s² to the right?

$\mu_s = 0.6, \mu_k = 0.5$

Problem 3/42

3/43 Compute the acceleration of block A for the instant
depicted. Neglect the mass of the pulley.

Ans. $a = 1.406$ m/s²

Problem 3/43

3/44 Determine the range of applied force P over which
the block of mass m_2 will not slip on the wedge-
shaped block of mass m_1. Neglect friction associated with the wheels of the tapered block.

Problem 3/44

▶**3/45** The chain is released from rest with the length b
of overhanging links just sufficient to initiate motion. The coefficients of static and kinetic friction
between the links and the horizontal surface have
essentially the same value μ. Determine the velocity v of the chain when the last link leaves the edge.
Neglect any friction at the corner.

$$\text{Ans. } v = \sqrt{\frac{gL}{1 + \mu}}$$

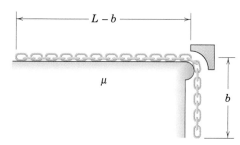

Problem 3/45

▶**3/46** Two iron spheres, each of which is 100 mm in diameter, are released from rest with a center-to-
center separation of 1 m. Assume an environment
in space with no forces other than the force of mutual gravitational attraction and calculate the time
t required for the spheres to contact each other and
the absolute speed v of each sphere upon contact.

$$\text{Ans. } t = 13 \text{ h } 33 \text{ min}$$
$$v = 4.76(10^{-5}) \text{ m/s}$$

Problem 3/46

▶3/47 An object projected vertically up from the surface of the earth with a sufficiently high velocity v_0 can escape from the earth's gravitational field. Calculate this velocity on the basis of the absence of an atmosphere to offer resistance due to air friction. To eliminate the effect of the earth's rotation on the velocity measurement, consider the launch to be from the north or south pole. Use the mean radius of the earth and the absolute value of g as cited in Art. 1/5 and compare your answer with the value cited in Table D/2.

Ans. $v_0 = 11.19$ km/s

▶3/48 The motorized drum turns clockwise at constant speed causing the vertical cable to have a constant downward velocity v. As part of the design of this system, determine the tension T in the cable in terms of the y-coordinate of the cylinder of mass m. Neglect the diameter and mass of the small pulleys.

Ans. $T = \dfrac{m}{2y}\sqrt{b^2 + y^2}\left(g + \dfrac{b^2v^2}{4y^3}\right)$

Problem 3/48

3/5 CURVILINEAR MOTION

We turn our attention now to the kinetics of particles which move along plane curvilinear paths. In applying Newton's second law, Eq. 3/3, we shall make use of the three coordinate descriptions of acceleration in curvilinear motion which we developed and applied in Arts. 2/4, 2/5, and 2/6.

The choice of coordinate system depends on the conditions of the problem and is one of the basic decisions to be made in solving curvilinear-motion problems. We now rewrite Eq. 3/3 in three ways, the choice of which depends on which coordinate system is most appropriate.

Rectangular coordinates (Art. 2/4, Fig. 2/7)

$$\begin{aligned} \Sigma F_x &= ma_x \\ \Sigma F_y &= ma_y \end{aligned} \tag{3/6}$$

where $\qquad a_x = \ddot{x} \qquad$ and $\qquad a_y = \ddot{y}$

Normal and tangential coordinates (Art. 2/5, Fig. 2/10)

$$\begin{aligned} \Sigma F_n &= ma_n \\ \Sigma F_t &= ma_t \end{aligned} \tag{3/7}$$

where

$$a_n = \rho\dot{\beta}^2 = v^2/\rho = v\dot{\beta}, \qquad a_t = \dot{v}, \qquad \text{and} \qquad v = \rho\dot{\beta}$$

Polar coordinates (Art. 2/6, Fig. 2/15)

$$\begin{aligned} \Sigma F_r &= ma_r \\ \Sigma F_\theta &= ma_\theta \end{aligned} \tag{3/8}$$

where $\qquad a_r = \ddot{r} - r\dot{\theta}^2 \qquad$ and $\qquad a_\theta = r\ddot{\theta} + 2\dot{r}\dot{\theta}$

In the application of these motion equations, the general procedure established in the previous article on rectilinear motion should be followed. After the motion is identified and the coordinate system is designated, the free-body diagram of the body to be treated as a particle should be drawn. The appropriate force summations are then obtained from this diagram in the usual way. The free-body diagram should be complete so that no mistakes are made in figuring the force summations. Once an assignment of reference axes is made, the expressions for both the forces and the acceleration must be consistent with that assignment. In the first of Eqs. 3/7, for example, the positive sense of the n-axis is *toward* the center of curvature, and so the positive sense of our force summation ΣF_n must also be *toward* the center of curvature to agree with the positive sense of the acceleration $a_n = v^2/\rho$.

Sample Problem   3/6

Determine the maximum speed v which the sliding block may have as it passes point A without losing contact with the surface.

Solution.   The condition for loss of contact is that the normal force N which the surface exerts on the block goes to zero. Summing forces in the normal direction gives

$$[\Sigma F_n = ma_n] \qquad\qquad mg = m\frac{v^2}{\rho} \qquad v = \sqrt{g\rho} \qquad\qquad Ans.$$

If the speed at A were less than $\sqrt{g\rho}$, then an upward normal force exerted by the surface on the block would exist. In order for the block to have a speed at A that is greater than $\sqrt{g\rho}$, some type of constraint, such as a second curved surface above the block, would have to be introduced to provide additional downward force.

Sample Problem   3/7

Small objects are released from rest at A and slide down the smooth circular surface of radius R to a conveyor B. Determine the expression for the normal contact force N between the guide and each object in terms of θ and specify the correct angular velocity ω of the conveyor pulley of radius r to prevent any sliding on the belt as the objects transfer to the conveyor.

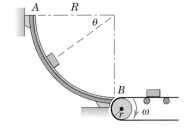

①

Solution.   The free-body diagram of the object is shown together with the coordinate directions n and t. The normal force N depends on the n-component of the acceleration which, in turn, depends on the velocity. The velocity will be cumulative according to the tangential acceleration a_t. Hence, we will find a_t first for any general position.

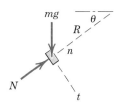

$$[\Sigma F_t = ma_t] \qquad\qquad mg\cos\theta = ma_t \qquad a_t = g\cos\theta$$

Now we can find the velocity by integrating

$$[v\,dv = a_t\,ds] \qquad \int_0^v v\,dv = \int_0^\theta g\cos\theta\,d(R\theta) \qquad v^2 = 2gR\sin\theta$$

① It is essential here that we recognize the need to express the tangential acceleration as a function of position so that v may be found by integrating the kinematical relation $v\,dv = a_t\,ds$, in which all quantities are measured along the path.

We obtain the normal force by summing forces in the positive n-direction, which is the direction of the n-component of acceleration.

$$[\Sigma F_n = ma_n] \quad N - mg\sin\theta = m\frac{v^2}{R} \qquad N = 3mg\sin\theta \qquad Ans.$$

The conveyor pulley must turn at the rate $v = r\omega$ for $\theta = \pi/2$, so that

$$\omega = \sqrt{2gR}/r \qquad\qquad Ans.$$

Sample Problem 3/8

A 1500-kg car enters a section of curved road in the horizontal plane and slows down at a uniform rate from a speed of 100 km/h at A to a speed of 50 km/h as it passes C. The radius of curvature ρ of the road at A is 400 m and at C is 80 m. Determine the total horizontal force exerted by the road on the tires at positions A, B, and C. Point B is the inflection point where the curvature changes direction.

Solution. The car will be treated as a particle so that the effect of all forces exerted by the road on the tires will be treated as a single force. Since the motion is described along the direction of the road, normal and tangential coordinates will be used to specify the acceleration of the car. We will then determine the forces from the accelerations.

The constant tangential acceleration is in the negative t-direction and its magnitude is given by

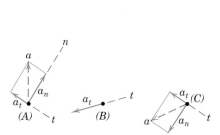

① $[v_C^2 = v_A^2 + 2a_t \,\Delta s]$ $\quad a_t = \left| \dfrac{(50/3.6)^2 - (100/3.6)^2}{2(200)} \right| = 1.447 \text{ m/s}^2$ ① Recognize the numerical value of the conversion factor from km/h to m/s as 1000/3600 or 1/3.6.

The normal components of acceleration at A, B, and C are

② $[a_n = v^2/\rho]$ $\quad$ At A, $\quad a_n = \dfrac{(100/3.6)^2}{400} = 1.929 \text{ m/s}^2$ ② Note that a_n is always directed toward the center of curvature.

$\quad$ At B, $\quad a_n = 0$

$\quad$ At C, $\quad a_n = \dfrac{(50/3.6)^2}{80} = 2.41 \text{ m/s}^2$

Application of Newton's second law in both the n- and t-directions to the free-body diagrams of the car gives

$[\Sigma F_t = ma_t]$ $\qquad\qquad F_t = 1500(1.447) = 2170 \text{ N}$

③ $[\Sigma F_n = ma_n]$ $\quad$ At A, $\quad F_n = 1500(1.929) = 2890 \text{ N}$

$\quad$ At B, $\quad F_n = 0$

$\quad$ At C, $\quad F_n = 1500(2.41) = 3620 \text{ N}$ ③ Note that the direction of F_n must agree with that of a_n.

Thus, the total horizontal force acting on the tires becomes

At A, $\quad F = \sqrt{F_n^2 + F_t^2} = \sqrt{(2890)^2 + (2170)^2} = 3620 \text{ N}$ $\quad$ *Ans.*

At B, $\quad F = F_t = 2170 \text{ N}$ $\quad$ *Ans.*

④ At C, $\quad F = \sqrt{F_n^2 + F_t^2} = \sqrt{(3620)^2 + (2170)^2} = 4220 \text{ N}$ $\quad$ *Ans.* ④ The angle made by **a** and **F** with the direction of the path can be computed if desired.

Sample Problem 3/9

Compute the magnitude v of the velocity required for the spacecraft S to maintain a circular orbit of altitude 320 km above the surface of the earth.

①

Solution. The only external force acting on the spacecraft is the force of gravitational attraction to the earth (i.e., its weight), as shown in the free-body diagram. Summing forces in the normal direction yields

$$[\Sigma F_n = ma_n] \quad G\frac{mm_e}{(R+h)^2} = m\frac{v^2}{(R+h)}, \quad v = \sqrt{\frac{Gm_e}{(R+h)}} = R\sqrt{\frac{g}{(R+h)}}$$

where the substitution $gR^2 = Gm_e$ has been made. Substitution of numbers gives

$$v = (6371)(1000)\sqrt{\frac{9.825}{(6371+320)(1000)}} = 7720 \text{ m/s} \quad Ans.$$

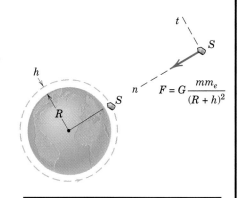

① Note that, for observations made within an inertial frame of reference, there is no such quantity as "centrifugal force" acting in the minus n-direction. Note also that neither the spacecraft nor its occupants are "weightless," because the weight in each case is given by Newton's law of gravitation. For this altitude, the weights are only about 10 percent less than the earth-surface values. Finally, the term "zero-g" is also misleading. It is only when we make our observations with respect to a coordinate system which has an acceleration equal to the gravitational acceleration (such as in an orbiting spacecraft) that we appear to be in a "zero-g" environment. The quantity which does go to zero aboard orbiting spacecraft is the familiar normal force associated with, for example, an object in contact with a horizontal surface within the spacecraft.

Sample Problem 3/10

Tube A rotates about the vertical O-axis with a constant angular rate $\dot\theta = \omega$ and contains a small cylindrical plug B of mass m whose radial position is controlled by the cord which passes freely through the tube and shaft and is wound around the drum of radius b. Determine the tension T in the cord and the horizontal component F_θ of force exerted by the tube on the plug if the constant angular rate of rotation of the drum is ω_0 first in the direction for case (a) and second in the direction for case (b). Neglect friction.

Solution. With r a variable, we use the polar-coordinate form of the equations of motion, Eqs. 3/8. The free-body diagram of B is shown in the horizontal plane and discloses only T and F_θ. The equations of motion are

$$[\Sigma F_r = ma_r] \qquad -T = m(\ddot r - r\dot\theta^2)$$

$$[\Sigma F_\theta = ma_\theta] \qquad F_\theta = m(r\ddot\theta + 2\dot r\dot\theta)$$

Case (a). With $\dot r = +b\omega_0$, $\ddot r = 0$, and $\ddot\theta = 0$, the forces become

$$T = mr\omega^2 \qquad F_\theta = 2mb\omega_0\omega \qquad Ans.$$

Case (b). With $\dot r = -b\omega_0$, $\ddot r = 0$, and $\ddot\theta = 0$, the forces become

$$T = mr\omega^2 \qquad F_\theta = -2mb\omega_0\omega \qquad Ans.$$

① The minus sign shows that F_θ is in the direction opposite to that shown on the free-body diagram.

PROBLEMS

Introductory problems

3/49 The small 0.6-kg block slides with a small amount of friction on the circular path of radius 3 m in the vertical plane. If the speed of the block is 5 m/s as it passes point A and 4 m/s as it passes point B, determine the normal force exerted on the block by the surface at each of these two locations.

Ans. $N_A = 10.89$ N, $N_B = 8.30$ N

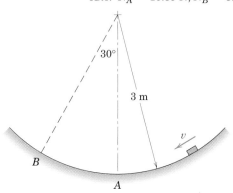

Problem 3/49

3/50 The position vector of a 1.5-kg particle is given by $\mathbf{r} = 8t^2\mathbf{i} + 3t^3\mathbf{j} - \frac{1}{2}t^4\mathbf{k}$, where $\mathbf{r}$ is in meters and t is in seconds. Determine the magnitude F of the resultant of all forces on the particle when $t = 2$ s.

3/51 The 120-g slider has a speed $v = 1.4$ m/s as it passes point A of the smooth guide, which lies in a horizontal plane. Determine the magnitude R of the force which the guide exerts on the slider (*a*) just before it passes point A of the guide and (*b*) as it passes point B.

Ans. (*a*) $R = 1.177$ N, (*b*) $R = 1.664$ N

Problem 3/51

3/52 If the 2-kg block passes over the top B of the circular portion of the path with a speed of 3.5 m/s, calculate the magnitude N_B of the normal force exerted by the path on the block. Determine the maximum speed v which the block can have at A without losing contact with the path.

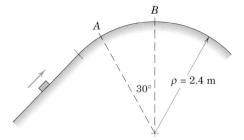

Problem 3/52

3/53 If the speed of the block shown with Prob. 3/52 is 4.5 m/s as it passes point A of the smooth track, determine the corresponding normal force exerted on the block by the track and the time rate of change of the speed.

Ans. $N_A = 0.1164$ N, $\dot{v} = -4.90$ m/s^2

3/54 The standard test to determine the maximum lateral acceleration of a car is to drive it around a 60-m-diameter circle painted on a level asphalt surface. The driver slowly increases the vehicle speed until he is no longer able to keep both wheel pairs straddling the line. If this maximum speed is 55 km/h for a 1400-kg car, determine its lateral acceleration capability a_n in g's and compute the magnitude F of the total friction force exerted by the pavement on the car tires.

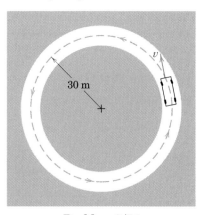

Problem 3/54

3/55 The car of Prob. 3/54 is traveling at 40 km/h when the driver applies the brakes, and the car continues to move along the circular path. What is the maximum deceleration possible if the tires are limited to a total horizontal friction force of 10.6 kN?

Ans. $a_t = -6.36$ m/s^2

3/56 In the design of a space station to operate outside the earth's gravitational field, it is desired to give the structure a rotational speed N which will simulate the effect of the earth's gravity for members of the crew. If the centers of the crew's quarters are to be located 12 m from the axis of rotation, calculate the necessary rotational speed N of the space station in revolutions per minute.

Problem 3/56

3/57 Determine the proper bank angle θ for the airplane flying at 600 km/h and making a turn of 3-km radius. Note that the force exerted by the air is normal to the supporting wing surface.

Ans. $\theta = 43.3°$

Problem 3/57

3/58 The car passes over the top of a vertical curve at A with a speed of 60 km/h and then passes through the bottom of a dip at B. The radii of curvature of the road at A and B are both 100 m. Find the speed of the car at B if the normal force between the road and the tires at B is twice that at A. The mass center of the car is 1 meter from the road.

Problem 3/58

3/59 In order to simulate a condition of apparent "weightlessness" experienced by astronauts in an orbiting spacecraft, a jet transport can change its direction at the top of its flight path by dropping its flight-path direction at a prescribed rate $\dot{\theta}$ for a short interval of time. Specify $\dot{\theta}$ if the aircraft has a speed $v = 600$ km/h.

Ans. $\dot{\theta} = 3.37$ deg/s

Problem 3/59

3/60 The weight of an object as measured by a spring scale at rest on the ground is 20 N. The object is weighed by the same spring scale placed on the cabin floor of an aircraft which is at the top of a vertical circular arc as shown in the figure for Prob. 3/59. If the airplane is flying with a speed $v = 800$ km/h at an altitude of 6000 m and is dropping its forward line of sight at the rate of 1 degree per second, what reading W' is registered by the scale?

3/61 The small spheres are free to move on the inner surface of the rotating spherical chambers shown in section with radius $R = 200$ mm. If the spheres reach a steady-state angular position $\beta = 45°$, determine the angular velocity Ω of the device.

Ans. $\Omega = 3.64$ rad/s

Problem 3/61

3/62 A child twirls a small 50-g ball attached to the end of a 1-m string so that the ball traces a circle in a vertical plane as shown. What is the minimum speed v which the ball must have when in position 1? If this speed is maintained throughout the circle, calculate the tension T in the string when the ball is in position 2. Neglect any small motion of the child's hand.

Problem 3/62

3/63 A pilot flies an airplane at a constant speed of 600 km/h in the vertical circle of radius 1000 m. Calculate the force exerted by the seat on the 90-kg pilot at point A and at point B.

Ans. $N_A = 3380$ N, $N_B = 1617$ N

Problem 3/63

Representative problems

3/64 Determine the altitude h (in kilometers) above the surface of the earth at which a satellite in a circular orbit has the same period, 23.9344 h, as the earth's absolute rotation. If such an orbit lies in the equatorial plane of the earth, it is said to be geosynchronous, because the satellite does not appear to move relative to an earth-fixed observer.

3/65 A small object is given an initial horizontal velocity v_0 at the bottom of a smooth slope. The angle θ made by the slope with the horizontal varies according to $\sin \theta = ks$, where k is a constant and s is the distance measured along the slope from the bottom. Determine the maximum distance s which the object slides up the slope.

Ans. $s = v_0/\sqrt{kg}$

3/66 A small bead of mass m is carried by a circular hoop of radius r which rotates about a fixed vertical axis. Show how one might determine the angular speed ω of the hoop by observing the angle θ which locates the bead. Neglect friction in your analysis, but assume that a small amount of friction is present to damp out any motion of the bead relative to the hoop once a constant angular speed has been established. Note any restrictions on your solution.

Problem 3/66

3/67 As the skateboarder negotiates the surface shown, his mass-center speeds at $\theta = 0$, 45°, and 90° are 8.5 m/s, 6 m/s, and 0, respectively. Determine the normal force between the surface and the skateboard wheels if the combined mass of the person and the skateboard is 70 kg and his center of mass is 750 mm from the surface.

Ans. $N_0 = 2040$ N, $N_{45°} = 1158$ N, $N_{90°} = 0$

Problem 3/67

3/68 A small object A is held against the vertical side of the rotating cylindrical container of radius r by centrifugal action. If the coefficient of static friction between the object and the container is μ_s, determine the expression for the minimum rotational rate $\dot{\theta} = \omega$ of the container which will keep the object from slipping down the vertical side.

Problem 3/68

3/69 The 30-Mg aircraft is climbing at the angle $\theta = 15°$ under a jet thrust T of 180 kN. At the instant shown, its speed is 300 km/h and is increasing at the rate of 1.96 m/s². Also θ is decreasing as the aircraft begins to level off. If the radius of curvature of the path at this instant is 20 km, compute the lift L and drag D. (Lift L and drag D are the aerodynamic forces normal to and opposite to the flight direction, respectively.)

Ans. $D = 45.0$ kN, $L = 274$ kN

Problem 3/69

3/70 Explain how to utilize the graduated pendulum to measure the speed of a vehicle traveling in a horizontal circular arc of known radius r.

Problem 3/70

3/71 Beginning from rest when $\theta = 20°$, a 35-kg child slides with negligible friction down the sliding board which is in the shape of a 2.5-m circular arc. Determine the tangential acceleration and speed of the child, and the normal force exerted on her (a) when $\theta = 30°$ and (b) when $\theta = 90°$.

Ans. (a) $a_t = 8.50$ m/s^2, $v = 2.78$ m/s
$N = 280$ N
(b) $a_t = 0$, $v = 5.68$ m/s
$N = 795$ N

Problem 3/71

3/72 Determine the speed v at which the racecar will have no tendency to slip sideways on the banked track, i.e., the speed at which there is no reliance on friction. In addition, determine the minimum and maximum speeds, using the coefficient of static friction $\mu_s = 0.90$. State any assumptions.

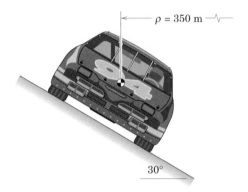

Problem 3/72

3/73 The small ball of mass m is attached to a light cord of length L and moves as a conical pendulum in a horizontal circle with a tangential velocity v. Locate the plane of motion by determining h, and find the tension T in the cord. (*Note:* Use the relation $v = r\dot{\theta} = r\omega$, where ω is the angular velocity about the vertical axis.)

Ans. $h = g/\omega^2$, $T = mL\omega^2$

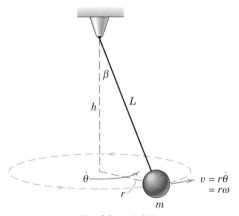

Problem 3/73

3/74 The cars of an amusement park ride have a speed $v_A = 22$ m/s at A and a speed $v_B = 12$ m/s at B. If a 75-kg rider sits on a spring scale (which registers the normal force exerted on it), determine the scale readings as the car passes points A and B. Assume that the person's arms and legs do not support appreciable force.

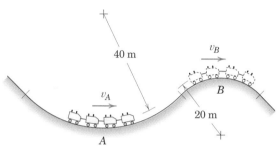

40 m

v_B

B

v_A

20 m

A

Problem 3/74

3/75 The slotted arm revolves in the horizontal plane about the fixed vertical axis through point O. The 2-kg slider C is drawn toward O at the constant rate of 50 mm/s by pulling the cord S. At the instant for which $r = 225$ mm, the arm has a counterclockwise angular velocity $\omega = 6$ rad/s and is slowing down at the rate of 2 rad/s^2. For this instant, determine the tension T in the cord and the magnitude N of the force exerted on the slider by the sides of the smooth radial slot. Indicate which side, A or B, of the slot contacts the slider.

Ans. $T = 16.20$ N, $N = 2.10$ N, side B

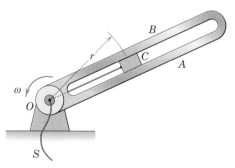

B

r

C

A

ω

O

S

Problem 3/75

3/76 The rotating drum of a clothes dryer is shown in the figure. Determine the angular velocity Ω of the drum which results in loss of contact between the clothes and the drum at $\theta = 50°$. Assume that the small vanes prevent slipping until loss of contact.

Ω

A

θ

330 mm

Problem 3/76

3/77 The amusement-park ride pivots about the fixed point O. A mechanism (not shown) drives the unit according to $\theta = (\pi/3) \sin 0.950t$, where θ is in radians and t is in seconds. Determine the maximum normal force N exerted by the seat on a rider of mass m, and state which riders are subjected to the maximum force.

Ans. $N = 20.7m$

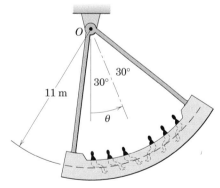

O

30°

30°

11 m

θ

Problem 3/77

3/78 The small object is placed on the inner surface of the conical dish at the radius shown. If the coefficient of static friction between the object and the conical surface is 0.30, for what range of angular velocities ω about the vertical axis will the block remain on the dish without slipping? Assume that speed changes are made slowly so that any angular acceleration may be neglected.

Problem 3/78

3/79 The small object of mass m is placed on the rotating conical surface at the radius shown. If the coefficient of static friction between the object and the rotating surface is 0.8, calculate the maximum angular velocity ω of the cone about the vertical axis for which the object will not slip. Assume very gradual angular-velocity changes.

Ans. $\omega = 2.73$ rad/s

Problem 3/79

3/80 The robot arm of Prob. 2/149 is repeated here, with additional information supplied. At a given instant, $\theta = 30°$, $\dot{\theta} = 40$ deg/s, $\ddot{\theta} = 120$ deg/s^2, $l = 0.5$ m, $\dot{l} = 0.4$ m/s, and $\ddot{l} = -0.3$ m/s^2. Compute the radial and transverse forces F_r and F_θ that the arm must exert on the gripped part P, which has a mass of 1.2 kg. Compare with the case of static equilibrium in the same position.

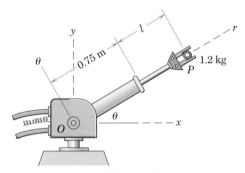

Problem 3/80

3/81 The flatbed truck starts from rest on a road whose constant radius of curvature is 30 m and whose bank angle is 10°. If the constant forward acceleration of the truck is 2 m/s^2, determine the time t after the start of motion at which the crate on the bed begins to slide. The coefficient of static friction between the crate and truck bed is $\mu_s = 0.3$, and the truck motion occurs in a horizontal plane.

Ans. $t = 5.58$ s

Problem 3/81

3/82 The 1500-kg car is traveling at 100 km/h on the straight portion of the road, and then its speed is reduced uniformly from A to C, at which point it comes to rest. Compute the magnitude F of the total friction force exerted by the road on the car (*a*) just before it passes point B, (*b*) just after it passes point B, and (*c*) just before it stops at point C.

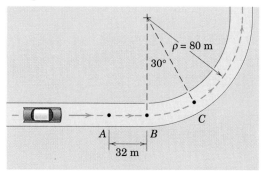

Problem 3/82

3/83 The flat circular disk rotates about a vertical axis through O with a slowly increasing angular velocity ω. Prior to rotation, each of the 0.5-kg sliding blocks has the position $x = 25$ mm with no force in its attached spring. Each spring has a stiffness of 400 N/m. Determine the value of x for each spring for a steady speed of 240 rev/min. Also calculate the normal force N exerted by the side of the slot on the block. Neglect any friction between the blocks and the slots, and neglect the mass of the springs. (*Hint:* Sum forces along and normal to the slot.)

Ans. $x = 118.8$ mm, $N = 25.3$ N

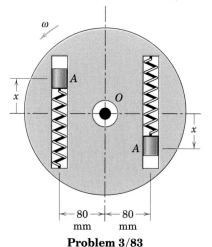

Problem 3/83

3/84 The disk with the circular groove rotates about the vertical axis with a constant speed of 30 rev/min and carries the two 4-kg spheres. Calculate the larger of the two forces of contact between the disk and each sphere. (Can this result be reached by using only one force equation?)

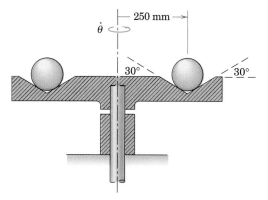

Problem 3/84

3/85 The spring-mounted 0.8-kg collar A oscillates along the horizontal rod which is rotating at the constant angular rate $\dot{\theta} = 6$ rad/s. At a certain instant, r is increasing at the rate of 800 mm/s. If the coefficient of kinetic friction between the collar and the rod is 0.40, calculate the friction force F exerted by the rod on the collar at this instant.

Ans. $F = 4.39$ N

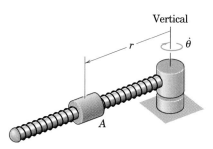

Problem 3/85

3/86 A small coin is placed on the horizontal surface of the rotating disk. If the disk starts from rest and is given a constant angular acceleration $\ddot{\theta} = \alpha$, determine an expression for the number of revolutions N through which the disk turns before the coin slips. The coefficient of static friction between the coin and the disk is μ_s.

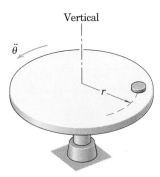

Vertical

Problem 3/86

3/87 A small vehicle enters the top A of the circular path with a horizontal velocity v_0 and gathers speed as it moves down the path. Determine an expression for the angle β to the position where the vehicle leaves the path and becomes a projectile. Evaluate your expression for $v_0 = 0$. Neglect friction and treat the vehicle as a particle.

Ans. $\beta = \cos^{-1}\left(\dfrac{2}{3} + \dfrac{v_0^2}{3gR}\right), \beta = 48.2°$

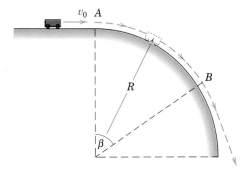

Problem 3/87

3/88 The car has a speed of 70 km/h at the bottom of the dip when the driver applies the brakes, causing a deceleration of 0.5g. What is the minimum seat cushion angle θ for which a package will not slide forward? The coefficient of static friction between the package and seat cushion is (a) 0.2 and (b) 0.4.

$\rho = 80$ m

Horizontal

Seat cushion

Problem 3/88

3/89 The small 180-g slider A moves without appreciable friction in the hollow tube, which rotates in a horizontal plane with a constant speed $\Omega = 7$ rad/s. The slider is launched with an initial speed $\dot{r}_0 = 20$ m/s relative to the tube at the inertial coordinates $x = 150$ mm and $y = 0$. Determine the magnitude P of the horizontal force exerted on the slider by the tube just before the slider exits the tube.

Ans. $P = 53.3$ N

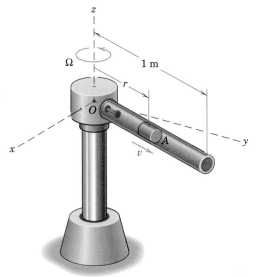

Problem 3/89

3/90 For the conditions given in Prob. 3/89, determine the inertial x- and y- components of the horizontal force **P** exerted on the slider by the tube just before the slider exits the tube.

3/91 A right-handed baseball pitcher throws a curve ball initially aimed at the right edge of home plate B. It curves so as to "break" 150 mm as shown. Assume that the horizontal velocity component is constant at $v = 38$ m/s, neglect vertical motion, and estimate (a) the average radius of curvature ρ of the baseball path and (b) the normal force R acting on the 146-g baseball.

Ans. $\rho = 1080$ m, $R = 0.1952$ N

Problem 3/91

3/92 The particle P is released at time $t = 0$ from the position $r = r_0$ inside the smooth tube with no velocity relative to the tube, which is driven at the constant angular velocity ω_0 about a vertical axis. Determine the radial velocity v_r, the radial position r, and the transverse velocity v_θ as functions of time t. Explain why the radial velocity increases with time in the absence of radial forces. Plot the absolute path of the particle during the time it is inside the tube for $r_0 = 0.1$ m, $l = 1$ m, and $\omega_0 = 1$ rad/s.

Problem 3/92

3/93 The tangential or hoop stress in the rim of a flywheel can be approximated by simulating the rim as a metal hoop rotating in its plane about its geometric axis. Derive an expression for the hoop stress σ_t in terms of the rim speed v, the mass ρ per unit length of the rim, and the cross-sectional area A of the hoop rim. Apply the equation of motion to the free-body diagram of a differential element of the hoop, treated as a particle.

Ans. $\sigma_t = \rho v^2 / A$

Problem 3/93

3/94 The mechanism of Prob. 2/117 is repeated here. At the instant when $\theta = 30°$, the horizontal guide is given a constant upward velocity $v_0 = 2$ m/s. For this instant calculate the force N exerted by the circular slot and the force P exerted by the horizontal slot on the 0.5-kg pin A. The width of the slots is slightly greater than the diameter of the pin, and friction is negligible.

Problem 3/94

3/95 The pin P is forced to move in the fixed parabolic slot by the moving vertical guide. The center line of the slot is described by $y = 0.015x^2$, where x and y are in millimeters. When $x = 40$ mm, the speed v of the guide is 100 mm/s and is not changing at this instant. Determine the magnitudes N and R of the forces which the guide and fixed vertical plate, respectively, exert on the 50-g pin.

Ans. $R = 0.790$ N, $N = 0.607$ N

Problem 3/95

3/96 The amusement-park ride consists of a fixed support near O, the 6-m arm OA, which rotates about the pivot at O, and the compartment, which remains horizontal by means of a mechanism at A (not shown). At a certain instant, $\beta = 45°$, $\dot{\beta} = 0.8$ rad/s, and $\ddot{\beta} = 0.4$ rad/s^2, all clockwise. Determine the horizontal and vertical forces (F and N) exerted by the bench on the 80-kg rider at P. Compare your results with the static values of these forces. Note that every rider moves in a circle of radius 6 m.

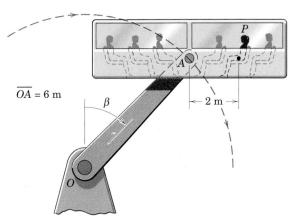

Problem 3/96

3/97 When a V-belt drives a pulley at high speed, the centrifugal action tends to lessen its contact with the pulley and hence to reduce the capacity to transmit torque. A device to compensate for this effect is shown and consists of a cage and four balls which rotate with the pulley. The balls press against the two 30° conical surfaces and force the inner side A to slide to the left toward the opposite side B, thus tightening the belt. Part A is splined to B along the hub, so it rotates with the remainder of the pulley but is free to slide on B. Compute the axial force F on A caused by the action of the balls for a speed of 600 rev/min, if the mass of each of the four balls is 2.5 kg.

Ans. $F = 5130$ N

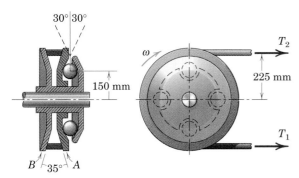

Problem 3/97

3/98 A fire-fighting helicopter hovers over a lake while its water bucket is immersed and filled. It then rises slightly, clears the bucket from the lake, and starts its flight essentially from rest with a horizontal acceleration a_0. Obtain an expression for the angle θ for which $\dot{\theta}$ is a maximum. Also determine the tension T in the cable as a function of θ and the mass m of the filled bucket.

Problem 3/98

3/99 The centrifugal pump with smooth radial vanes rotates about its vertical axis with a constant angular velocity $\dot{\theta} = \omega$. Find the magnitude N of the force exerted by a vane on a particle P of mass m as it moves out along the vane. The particle is introduced at $r = r_0$ without radial velocity. Assume that the particle contacts the side of the vane only.

$$Ans. \ N = 2m\omega^2\sqrt{r^2 - r_0{}^2}$$

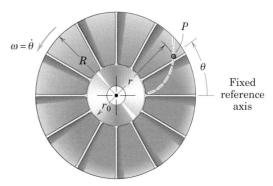

Problem 3/99

3/100 The spacecraft P is in the elliptical orbit shown. At the instant represented, its speed is $v = 4230$ m/s. Determine the corresponding values of $\dot{r}$, $\dot{\theta}$, $\ddot{r}$, and $\ddot{\theta}$. Use $g = 9.825$ m/s^2 as the acceleration of gravity on the surface of the earth and $R = 6371$ km as the radius of the earth.

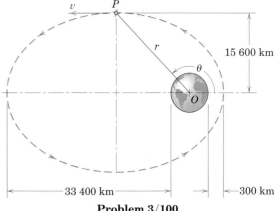

Problem 3/100

▶3/101 Each tire on the 1350-kg car can support a maximum friction force parallel to the road surface of 2500 N. This force limit is nearly constant over all possible rectilinear and curvilinear car motions and is attainable only if the car does not skid. Under this maximum braking, determine the total stopping distance s if the brakes are first applied at point A when the car speed is 25 m/s and if the car follows the center line of the road.

Ans. $s = 47.4$ m

▶3/102 A small collar of mass m is given an initial velocity of magnitude v_0 on the horizontal circular track fabricated from a slender rod. If the coefficient of kinetic friction is μ_k, determine the distance traveled before the collar comes to rest. (*Hint:* Recognize that the friction force depends on the net normal force.)

$$Ans. \quad s = \frac{r}{2\mu_k} \ln \left[\frac{v_0^2 + \sqrt{v_0^4 + r^2 g^2}}{rg} \right]$$

Problem 3/102

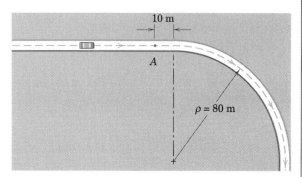

Problem 3/101

SECTION B. WORK AND ENERGY

3/6   *WORK AND KINETIC ENERGY*

In the previous two articles, we applied Newton's second law $\mathbf{F} = m\mathbf{a}$ to various problems of particle motion to establish the instantaneous relationship between the net force acting on a particle and the resulting acceleration of the particle. When intervals of motion were involved where the change in velocity or the corresponding displacement of the particle was required, we integrated the computed acceleration over the interval by using the appropriate kinematical equations.

There are two general classes of problems in which the cumulative effects of unbalanced forces acting on a particle over an interval of motion are of interest to us. These cases involve, respectively, integration of the forces with respect to the displacement of the particle and integration of the forces with respect to the time they are applied. We may incorporate the results of these integrations directly into the governing equations of motion so that it becomes unnecessary to solve directly for the acceleration. Integration with respect to displacement leads to the equations of work and energy, which are the subject of this article. Integration with respect to time leads to the equations of impulse and momentum, discussed in Section C.

(a) Work.   The quantitative meaning of the term work will now be developed.* Figure 3/2*a* shows a force $\mathbf{F}$ acting on a particle at A which moves along the path shown. The position vector $\mathbf{r}$ measured from some convenient origin O locates the particle as it passes point A, and $d\mathbf{r}$ is the differential displacement associated with an infinitesimal movement from A to A'. The work done by the force $\mathbf{F}$ during the displacement $d\mathbf{r}$ is defined as

$$dU = \mathbf{F} \cdot d\mathbf{r}$$

The magnitude of this dot product is $dU = F\,ds\,\cos\alpha$, where α is the angle between $\mathbf{F}$ and $d\mathbf{r}$ and where ds is the magnitude of $d\mathbf{r}$. This expression may be interpreted as the displacement multiplied by the force component $F_t = F\cos\alpha$ in the direction of the displacement, as represented by the dotted lines in Fig. 3/2*b*. Alternatively, the work dU may be interpreted as the force multiplied by the displacement component $ds\cos\alpha$ in the direction of the force, as represented by the full lines in Fig. 3/2*b*. With this definition of work, it should be noted that the component $F_n = F\sin\alpha$ normal to the

(a)

(b)

Figure 3/2

---

*The concept of work was also developed in the study of virtual work in Chapter 7 of *Vol. 1 Statics.*

displacement does no work. Hence, the work dU may be written as

$$dU = F_t \, ds$$

Work is positive if the working component F_t is in the direction of the displacement and negative if it is in the opposite direction. Forces that do work are termed *active forces*. Constraint forces that do no work are termed *reactive forces*.

In SI units work has the units of force (N) times displacement (m) or N·m. This unit is given the special name *joule* (J), which is defined as the work done by a force of 1 N moving through a distance of 1 m in the direction of the force. Consistent use of the joule for work (and energy) rather than the units N·m will avoid possible ambiguity with the units of moment of a force or torque, which are also written N·m.

It should be noted that work is a scalar as given by the dot product and involves the product of a force and a distance, both measured along the same line. Moment, on the other hand, is a vector as given by the cross product and involves the product of force and distance measured at right angles to the force.

During a finite movement of the point of application of a force, the force does an amount of work equal to

$$U = \int \mathbf{F} \cdot d\mathbf{r} = \int (F_x \, dx + F_y \, dy + F_z \, dz)$$

or

$$U = \int F_t \, ds$$

In order to carry out this integration, it is necessary to know the relations between the force components and their respective coordinates or the relation between F_t and s. If the functional relationship is not known as a mathematical expression which can be integrated but is specified in the form of approximate or experimental data, then the work may be evaluated by carrying out a numerical or graphical integration which would be represented by the area under the curve of F_t versus s, as shown in Fig. 3/3.

A common example of the work done on a body by a variable force is found in the action of a spring on a movable body to which it is attached. We consider here the common linear spring of stiffness k, where the force F in the spring, tension or compression, is proportional to its deformation x, so that $F = kx$. Figure 3/4 shows the two cases where the body is moved by a force P so as to stretch the spring a distance x or to compress the spring a distance x. The

Figure 3/3

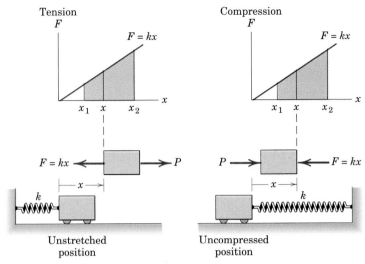

Figure 3/4

force exerted by the spring on the body in each case is in the sense *opposite* to the displacement so it does *negative* work on the body. Thus, for both stretching and compressing the spring, the work done *on* the body is *negative* and is given by

$$U_{1\text{-}2} = -\int_{x_1}^{x_2} F \, dx = -\int_{x_1}^{x_2} kx \, dx = -\tfrac{1}{2}k(x_2^2 - x_1^2)$$

When a spring under tension is being relaxed or when a spring under compression is being relaxed, then we see from both examples in Fig. 3/4 that the deformation changes from x_2 to a lesser deformation x_1. For this condition the force exerted on the body by the spring in both cases is in the *same* sense as the displacement, and, therefore, the work done *on* the body is *positive*.

The magnitude of the work, positive or negative, is seen to be equal to the shaded trapezoidal area shown for both cases in Fig. 3/4. In calculating the work done by a spring force, care must be taken to see that the units for k and x are consistent. Thus, if x is in meters, k must be in N/m.

The expression $F = kx$ is actually a static relationship which is true only when elements of the spring have no acceleration. Dynamic behavior of a spring when its mass is accounted for is a fairly complex problem which will not be treated here. We shall assume that the mass of the spring is small compared with the masses of other accelerating parts of the system, in which case, the linear static relationship will not involve appreciable error.

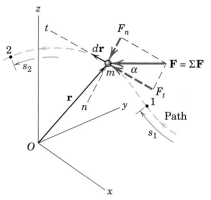

Figure 3/5

We now consider the work done on a particle of mass m, Fig. 3/5, moving along a curved path under the action of the force **F**, which stands for the resultant $\Sigma\mathbf{F}$ of all forces acting on the particle. The position of m is established by the position vector **r**, and its displacement along its path during time dt is represented by the change $d\mathbf{r}$ in its position vector. The work done by **F** during a finite movement of the particle from point 1 to point 2 is

$$U_{1\text{-}2} = \int_1^2 \mathbf{F}\cdot d\mathbf{r} = \int_{s_1}^{s_2} F_t\, ds$$

where the limits specify the initial and final end points of the interval of motion involved. When we substitute Newton's second law $\mathbf{F} = m\mathbf{a}$, the expression for the work of all forces becomes

$$U_{1\text{-}2} = \int_1^2 \mathbf{F}\cdot d\mathbf{r} = \int_1^2 m\mathbf{a}\cdot d\mathbf{r}$$

But $\mathbf{a}\cdot d\mathbf{r} = a_t\, ds$, where a_t is the tangential component of the acceleration of m. In terms of the velocity v of the particle, Eq. 2/3 gives $a_t\, ds = v\, dv$. Thus, the expression for the work of **F** becomes

$$U_{1\text{-}2} = \int_1^2 \mathbf{F}\cdot d\mathbf{r} = \int_{v_1}^{v_2} mv\, dv = \tfrac{1}{2}m(v_2{}^2 - v_1{}^2) \qquad (3/9)$$

where the integration is carried out between points 1 and 2 along the curve, at which points the velocities have the magnitudes v_1 and v_2, respectively.

(b) Kinetic energy.   The *kinetic energy* T of the particle is defined as

$$\boxed{T = \tfrac{1}{2}mv^2} \qquad (3/10)$$

and is the total work which must be done on the particle to bring it from a state of rest to a velocity v. Kinetic energy T is a scalar quantity with the units of N·m or *joules* (J). Kinetic energy is *always* positive, regardless of the direction of the velocity. Equation 3/9 may be restated as

$$\boxed{U_{1\text{-}2} = T_2 - T_1 = \Delta T} \qquad (3/11)$$

which is the *work-energy equation* for a particle. The equation states that the *total work done* by all forces acting on a particle during an interval of its motion from condition 1 to condition 2 equals the corresponding *change in kinetic energy* of the particle. Although T is always positive, the change ΔT may be positive, negative, or zero.

When written in this concise form, Eq. 3/11 tells us that the work always results in a *change* of kinetic energy.

Alternatively, the work-energy relation may be expressed as the initial kinetic energy T_1 plus the work done $U_{1\text{-}2}$ equals the final kinetic energy T_2 or

$$\boxed{T_1 + U_{1\text{-}2} = T_2}$$ **(3/11a)**

When written in this form, the terms correspond to the natural sequence of events. Clearly, the two forms 3/11 and 3/11a are equivalent.

We now see from Eq. 3/11 that a major advantage of the method of work and energy is that it avoids the necessity of computing the acceleration and leads directly to the velocity changes as functions of the forces which do work. Further, the work-energy equation involves only those forces which do work and thus give rise to changes in the magnitude of the velocities.

We consider now two particles joined together by a connection which is frictionless and incapable of any deformation. The forces in the connection constitute a pair of equal and opposite forces, and the points of application of these forces necessarily have identical displacement components in the direction of the forces. Hence, the net work done by these internal forces is zero during any movement of the system of the two connected particles. Thus, Eq. 3/11 is applicable to the entire system, where $U_{1\text{-}2}$ is the total or net work done on the system by forces external to it and ΔT is the change, $T_2 - T_1$, in the total kinetic energy of the system. The total kinetic energy is the sum of the kinetic energies of both elements of the system. It may now be observed that a further advantage of the work-energy method is that it permits the analysis of a system of particles joined in the manner described without dismembering the system.

Application of the work-energy method calls for an isolation of the particle or system under consideration. For a single particle a *free-body diagram* showing all externally applied forces should be drawn. For a system of particles rigidly connected without springs, an *active-force diagram* that shows only those external forces which do work (active forces) on the entire system may be drawn.*

(c) Power. The capacity of a machine is measured by the time rate at which it can do work or deliver energy. The total work or energy output is not a measure of this capacity since a motor, no matter how small, can deliver a large amount of energy if given

---

*The active-force diagram was introduced in the method of virtual work in statics. See Chapter 7 of *Vol. 1 Statics*.

sufficient time. On the other hand, a large and powerful machine is required to deliver a large amount of energy in a short period of time. Thus, the capacity of a machine is rated by its *power*, which is defined as the *time rate of doing work*.

Accordingly, the power P developed by a force $\mathbf{F}$ which does an amount of work U is $P = dU/dt = \mathbf{F} \cdot d\mathbf{r}/dt$. Since $d\mathbf{r}/dt$ is the velocity $\mathbf{v}$ of the point of application of the force, we have

$$\boxed{P = \mathbf{F} \cdot \mathbf{v}} \qquad (3/12)$$

Power is clearly a scalar quantity, and in SI units it has the units of $N \cdot m/s = J/s$. The special unit for power is the *watt* (W), which equals one joule per second (J/s).

(d) Efficiency. The ratio of the work done *by* a machine to the work done *on* the machine during the same time interval is called the *mechanical efficiency* e_m of the machine. This definition assumes that the machine operates uniformly so that there is no accumulation or depletion of energy within it. Efficiency is always less than unity since every device operates with some loss of energy and since energy cannot be created within the machine. In mechanical devices that involve moving parts, there will always be some loss of energy due to the negative work of kinetic friction forces. This work is converted to heat energy which, in turn, is dissipated to the surroundings. The mechanical efficiency at any instant of time may be expressed in terms of mechanical power P by

$$e_m = \frac{P_{\text{output}}}{P_{\text{input}}} \qquad (3/13)$$

In addition to energy loss by mechanical friction, there may also be electrical and thermal energy loss, in which case, the *electrical efficiency* e_e and *thermal efficiency* e_t are also involved. The *overall efficiency* e in such instances would be

$$e = e_m e_e e_t$$

Sample Problem 3/11

Calculate the velocity v of the 50-kg crate when it reaches the bottom of the chute at B if it is given an initial velocity of 4 m/s down the chute at A. The coefficient of kinetic friction is 0.30.

Solution. The free-body diagram of the crate is drawn and includes the normal force R and the kinetic friction force F calculated in the usual manner. The work done by the component of the weight down the plane is positive, whereas that done by the friction force is negative. The total work done on the crate during the motion is

① $[U = Fs]$ $U_{1\text{-}2} = [50(9.81) \sin 15° - 142.1]10 = -151.9$ J

The change in kinetic energy is $T_2 - T_1 = \Delta T$

$[T = \tfrac{1}{2}mv^2]$ $\Delta T = \tfrac{1}{2}(50)(v^2 - 4^2)$

The work-energy equation gives

$[U_{1\text{-}2} = \Delta T]$ $-151.9 = 25(v^2 - 16)$

$v^2 = 9.93 \text{ (m/s)}^2$ $v = 3.15$ m/s *Ans.*

① Since the net work done is negative, we obtain a decrease in the kinetic energy.

Sample Problem 3/12

The flatbed truck, which carries an 80-kg crate, starts from rest and attains a speed of 72 km/h in a distance of 75 m on a level road with constant acceleration. Calculate the work done by the friction force acting on the crate during this interval if the static and kinetic coefficients of friction between the crate and the truck bed are (a) 0.30 and 0.28, respectively, or (b) 0.25 and 0.20, respectively.

Solution. If the crate does not slip on the bed, its acceleration will be that of the truck, which is

$[v^2 = 2as]$ $a = \dfrac{v^2}{2s} = \dfrac{(72/3.6)^2}{2(75)} = 2.67 \text{ m/s}^2$

Case (a). This acceleration requires a friction force on the block of

$[F = ma]$ $F = 80(2.67) = 213$ N

which is less than the maximum possible value of $\mu_s N = 0.30(80)$ (9.81) $= 235$ N. Therefore, the crate does not slip and the work done by the actual static friction force of 213 N is

① $[U = Fs]$ $U_{1\text{-}2} = 213(75) = 16\ 000$ J or 16.0 kJ

Case (b). For $\mu_s = 0.25$, the maximum possible friction force is $0.25(80)(9.81) = 196.2$ N, which is slightly less than the value of 213 N required for no slipping. Therefore, we conclude that the crate slips, and the friction force is governed by the kinetic coefficient and is $F = 0.20(80)(9.81) = 157.0$ N. The acceleration becomes

$[F = ma]$ $a = F/m = 157.0/80 = 1.962 \text{ m/s}^2$

The distances traveled by the crate and the truck are in proportion to their accelerations. Thus, the crate has a displacement of $(1.962/2.67)75 = 55.2$ m, and the work done by kinetic friction is

② $[U = Fs]$ $U_{1\text{-}2} = 157.0(55.2) = 8660$ J or 8.66 kJ *Ans.*

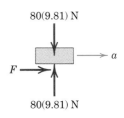

① We note that static friction forces do no work when the contacting surfaces are both at rest. When they are in motion, however, as in this problem, the static friction force acting on the crate does positive work and that acting on the truck bed does negative work.

② This problem shows that a kinetic friction force can do positive work when the surface that supports the object and generates the friction force is in motion. If the supporting surface is at rest, then the kinetic friction force acting on the moving part always does negative work.

Sample Problem 3/13

The 50-kg block at A is mounted on rollers so that it moves along the fixed horizontal rail with negligible friction under the action of the constant 300-N force in the cable. The block is released from rest at A, with the spring to which it is attached extended an initial amount $x_1 = 0.233$ m. The spring has a stiffness $k = 80$ N/m. Calculate the velocity v of the block as it reaches position B.

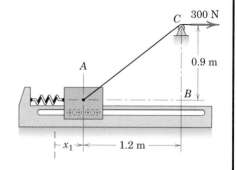

Solution. It will be assumed initially that the stiffness of the spring is small enough to allow the block to reach position B. The active-force diagram for the system composed of both block and cable is shown for a general position. The spring force $80x$ and the 300-N tension are the only forces external to this system that do work on the system. The force exerted on the block by the rail, the weight of the block, and the reaction of the small pulley on the cable do no work on the system and are not included on the active-force diagram.

As the block moves from $x = 0.233$ m to $x = 0.233 + 1.2 = 1.433$ m, the work done by the spring force acting on the block is negative and equals

① $[U = \int F\,dx]$ $\quad U_{1\text{-}2} = -\int_{0.233}^{1.433} 80x\,dx = -40x^2 \Big|_{0.233}^{1.433} = -80.0$ J

The work done on the system by the constant 300-N force in the cable is the force times the net horizontal movement of the cable over pulley C, which is $\sqrt{(1.2)^2 + (0.9)^2} - 0.9 = 0.6$ m. Thus, the work done is $300(0.6) = 180$ J. We now apply the work-energy equation to the system and get

$[U_{1\text{-}2} = \Delta T]$ $\quad -80.0 + 180 = \tfrac{1}{2}(50)(v^2 - 0)$ $\quad v = 2.0$ m/s $\quad$ *Ans.*

We take special note of the advantage to our choice of system. If the block alone had constituted the system, the horizontal component of the 300-N cable tension acting on the block would have to be integrated over the 1.2-m displacement. This step would require considerably more effort than was needed in the solution as presented. If there had been appreciable friction between the block and its guiding rail, we would have found it necessary to isolate the block alone in order to compute the variable normal force and, hence, the variable friction force. Integration of the friction force over the displacement would then be required to evaluate the negative work that it would do.

① If the variable x had been measured from the starting position A, the spring force would be $80(0.233 + x)$, and the limits of integration would be 0 and 1.2 m.

Sample Problem 3/14

The power winch A hoists the 360-kg log up the 30° incline at a constant speed of 1.2 m/s. If the power output of the winch is 4 kW, compute the coefficient of kinetic friction μ_k between the log and the incline. If the power is suddenly increased to 6 kW, what is the corresponding instantaneous acceleration a of the log?

Solution. From the free-body diagram of the log, we get $N = 360(9.81)\cos 30° = 3060$ N, and the kinetic friction force becomes $3060\mu_k$. For constant speed, the forces are in equilibrium so that

$[\Sigma F_x = 0]$ $T - 3060\mu_k - 360(9.81)\sin 30° = 0$

$T = 3060\mu_k + 1766$

① The power output of the winch gives the tension in the cable

$[P = Tv]$ $T = P/v = 4000/1.2 = 3330$ N

Substituting T gives

$3330 = 3060\mu_k + 1766$ $\mu_k = 0.513$ *Ans.*

When the power is increased, the tension momentarily becomes

$[P = Tv]$ $T = P/v = 6000/1.2 = 5000$ N

and the corresponding acceleration is given by

$[\Sigma F_x = ma_x]$ $5000 - 3060(0.513) - 360(9.81)\sin 30° = 360\,a$

$a = 4.63$ m/s^2 *Ans.*

① Note the conversion from kilowatts to watts. Also remember to use J/s rather than N·m/s.

② As the speed increases, the acceleration will drop until the speed stabilizes at a value higher than 1.2 m/s.

Sample Problem 3/15

A satellite of mass m is put into an elliptical orbit around the earth. At point A, its distance from the earth is $h_1 = 500$ km and it has a velocity $v_1 = 30\,000$ km/h. Determine the velocity v_2 of the satellite as it reaches point B, a distance $h_2 = 1200$ km from the earth.

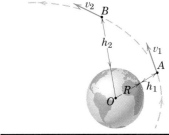

Solution. The satellite is moving outside of the earth's atmosphere so that the only force acting on it is the gravitational attraction of the earth. With the mass and radius of the earth expressed by m_e and R, respectively, the gravitational law of Eq. 1/2 gives $F = Gmm_e/r^2 = gR^2m/r^2$ when the substitution $Gm_e = gR^2$ is made for the surface values $F = mg$ and $r = R$. The work done by F is due only to the radial component of motion along the line of action of F and is negative for increasing r.

$$U_{1\text{-}2} = -\int_{r_1}^{r_2} F\,dr = -mgR^2 \int_{r_1}^{r_2} \frac{dr}{r^2} = mgR^2\left(\frac{1}{r_2} - \frac{1}{r_1}\right)$$

The work-energy equation $U_{1\text{-}2} = \Delta T$ gives

① ②

$$mgR^2\left(\frac{1}{r_2} - \frac{1}{r_1}\right) = \tfrac{1}{2}m(v_2^2 - v_1^2) \qquad v_2^2 = v_1^2 + 2gR^2\left(\frac{1}{r_2} - \frac{1}{r_1}\right)$$

Substituting the numerical values gives

$$v_2^2 = \left(\frac{30\,000}{3.6}\right)^2 + 2(9.81)[(6371)(10^3)]^2\left(\frac{10^{-3}}{6371 + 1200} - \frac{10^{-3}}{6371 + 500}\right)$$

$$= 69.44(10^6) - 10.72(10^6) = 58.73(10^6)\ (\text{m/s})^2$$

$v_2 = 7663$ m/s   or $v_2 = 7663(3.6) = 27\,590$ km/h   *Ans.*

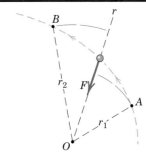

① Note that the result is independent of the mass of the satellite.

② Consult Table D/2, Appendix D, to find the radius R of the earth.

PROBLEMS

Introductory problems

3/103 The spring is unstretched at the position $x = 0$. Under the action of a force P, the cart moves from the initial position $x_1 = -150$ mm to the final position $x_2 = 80$ mm. Determine (a) the work done on the cart by the spring and (b) the work done on the cart by its weight.

$$Ans.\ (a)\ U_{1\text{-}2} = 4.03\ J$$
$$(b)\ U_{1\text{-}2} = -3.50\ J$$

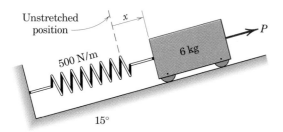

Problem 3/103

3/104 The small cart has a speed $v_A = 4$ m/s as it passes point A. It moves without appreciable friction and passes over the top hump of the track. Determine the cart speed as it passes point B. Is knowledge of the shape of the track necessary?

Problem 3/104

3/105 Refer to the figure of Prob. 3/104. If it is known that the 3-kg cart passes over the top of the track and arrives at B with a speed $v_B = 6$ m/s, determine the work done by friction between A and B.

$$Ans.\ U_f = -23.0\ J$$

3/106 A two-engine jet transport has a loaded mass of 40 Mg and a forward thrust of 44 kN per engine during takeoff. If the transport requires 1450 m of level runway starting from rest to become airborne at a speed of 260 km/h, determine the average resistance R to motion over the runway length due to drag (air resistance) and mechanical retardation by the landing gear.

3/107 The 0.5-kg collar C starts from rest at A and slides with negligible friction on the fixed rod in the vertical plane. Determine the velocity v with which the collar strikes end B when acted upon by the 5-N force, which is constant in direction. Neglect the small dimensions of the collar.

$$Ans.\ v = 2.32\ m/s$$

Problem 3/107

3/108 The crawler wrecking crane is moving with a constant speed of 3 km/h when it is suddenly brought to a stop. Compute the maximum angle θ through which the cable of the wrecking ball swings.

6 m

θ

3 km/h

Problem 3/108

3/109 The 15-kg collar A is released from rest in the position shown and slides with negligible friction up the fixed rod inclined 30° from the horizontal under the action of a constant force $P = 200$ N applied to the cable. Calculate the required stiffness k of the spring so that its maximum deflection equals 180 mm. The position of the small pulley at B is fixed.

Ans. $k = 1957$ N/m

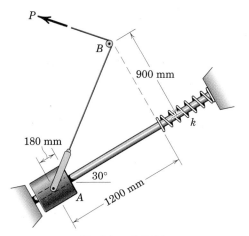

P

B

900 mm

k

180 mm

30°

A

1200 mm

Problem 3/109

3/110 In the design of a spring bumper for a 1500-kg car, it is desired to bring the car to a stop from a speed of 8 km/h in a distance equal to 150 mm of spring deformation. Specify the required stiffness k for each of the two springs behind the bumper. The springs are undeformed at the start of impact.

8 km/h

Problem 3/110

3/111 The small collar of mass m is released from rest at A and slides down the curved rod in the vertical plane with negligible friction. Express the velocity v of the collar as it strikes the base at B in terms of the given conditions.

Ans. $v = \sqrt{2gh}$

b

A

h

B

Problem 3/111

3/112 For the sliding collar of Prob. 3/111, if $m = 0.5$ kg, $b = 0.8$ m, and $h = 1.5$ m, and if the velocity of the collar as it strikes the base B is 4.70 m/s after release of the collar from rest at A, calculate the work Q of friction. What happens to the energy that is lost?

3/113 The 2-kg collar is released from rest at A and slides down the inclined fixed rod in the vertical plane. The coefficient of kinetic friction is 0.4. Calculate (*a*) the velocity v of the collar as it strikes the spring and (*b*) the maximum deflection x of the spring.

Ans. (*a*) $v = 2.56$ m/s, (*b*) $x = 98.9$ mm

2 kg

A

0.5 m

$\mu_k = 0.4$

60° $k = 1.6$ kN/m

Problem 3/113

3/114 The car is moving with a speed $v_0 = 105$ km/h up the 6-percent grade and the driver applies the brakes at point A, causing all wheels to skid. The coefficient of kinetic friction for the rain-slicked road is $\mu_k = 0.6$. Determine the stopping distance s_{AB}. Repeat your calculations for the case when the car is moving downhill from B to A.

A s B

v_0

6

100

Problem 3/114

3/115 A 40-kg boy starts from rest at the bottom A of a 10-percent incline and increases his speed at a constant rate to 8 km/h as he passes B, 15 m along the incline from A. Determine his power output as he approaches B.

Ans. $P = 101.4$ W

15 m

B

A

10

100

Problem 3/115

3/116 The 54-kg woman jogs up the flight of stairs in 5 seconds. Determine her average power output.

2.75 m

Problem 3/116

Representative problems

3/117 In the design of a conveyor-belt system, small metal blocks are discharged with a velocity of 0.4 m/s onto a ramp by the upper conveyor belt shown. If the kinetic coefficient of friction between the blocks and the ramp is 0.30, calculate the angle θ which the ramp must make with the horizontal so that the blocks will transfer without slipping to the lower conveyor belt moving at the speed of 0.14 m/s.

Ans. $\theta = 16.62°$

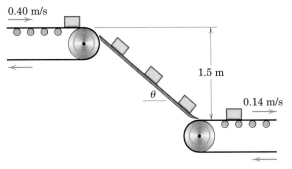

Problem 3/117

3/118 A department-store escalator handles a steady load of 30 people per minute in elevating them from the first to the second floor through a vertical rise of 7 m. The average person has a mass of 65 kg. If the motor that drives the unit delivers 3 kW, calculate the mechanical efficiency e of the system.

Problem 3/118

3/119 A car with a mass of 1500 kg starts from rest at the bottom of a 10-percent grade and acquires a speed of 50 km/h in a distance of 100 m with constant acceleration up the grade. What is the power P delivered to the drive wheels by the engine when the car reaches this speed?

Ans. $P = 40.4$ kW

3/120 The motor unit A is used to elevate the 300-kg cylinder at a constant rate of 2 m/s. If the power meter B registers an electrical input of 2.20 kW, calculate the combined electrical and mechanical efficiency e of the system.

Problem 3/120

3/121 The 6-kg cylindrical collar is released from rest in the position shown and drops onto the spring. Calculate the velocity v of the cylinder when the spring has been compressed 50 mm.

Ans. $v = 2.41$ m/s

Problem 3/121

3/122 The resistance R to penetration x of a 0.25-kg projectile fired with a velocity of 600 m/s into a certain block of fibrous material is shown in the graph. Represent this resistance by the dashed line and compute the velocity v of the projectile for the instant when $x = 25$ mm if the projectile is brought to rest after a total penetration of 75 mm.

Problem 3/122

3/123 The 0.5-kg collar slides with negligible friction along the fixed spiral rod which lies in the vertical plane. The rod has the shape of the spiral $r = 0.3\theta$, where r is in meters and θ is in radians. The collar is released from rest at A and slides to B under the action of a constant radial force $T = 10$ N. Calculate the velocity v of the slider as it reaches B.

Ans. $v = 5.30$ m/s

Problem 3/123

3/124 Each of the two systems is released from rest. Calculate the velocity v of each 25-kg cylinder after the 20-kg cylinder has dropped 2 m. The 10-kg cylinder of case (a) is replaced by a 10(9.81)-N force in case (b).

Problem 3/124

3/125 The small slider of mass m is released from rest while in position A and then slides along the vertical-plane track. The track is smooth from A to D and rough (coefficient of kinetic friction μ_k) from point D on. Determine (a) the normal force N_B exerted by the track on the slider just after it passes point B, (b) the normal force N_C exerted by the track on the slider as it passes the bottom point C, and (c) the distance s traveled along the incline past point D before the slider stops.

Ans. (a) $N_B = 4mg$, (b) $N_C = 7mg$

$$(c)\ s = \frac{4R}{1 + \mu_k\sqrt{3}}$$

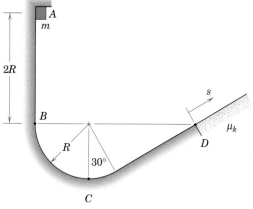

Problem 3/125

3/126 In a design test of penetration resistance, a 12-g bullet is fired through a 400-mm stack of fibrous plates with an entering velocity of 600 m/s. If the bullet emerges with a velocity of 300 m/s, calculate the average resistance R to penetration. What is the loss ΔQ of energy and where does it go?

Problem 3/126

3/127 In the structural design of the upper floors of an industrial building, allowance must be made for the accidental dropping of heavy machinery through a small distance. For a machine of mass m dropped through a very small distance onto a floor that acts elastically, determine the maximum force F supported by the floor. (The problem is modeled by the mass m mounted on supports a negligible distance above a spring of stiffness k, with the action occurring when the supports are suddenly removed.)

Ans. $F = 2mg$

Problem 3/127

3/128 The ball is released from position A with a velocity of 3 m/s and swings in a vertical plane. At the bottom position, the cord strikes the fixed bar at B, and the ball continues to swing in the dashed arc. Calculate the velocity v_C of the ball as it passes position C.

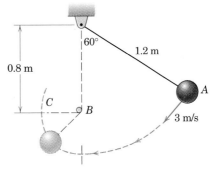

Problem 3/128

3/129 The 150-kg carriage has an initial velocity of 3 m/s down the incline at A, when a constant force of 550 N is applied to the hoisting cable as shown. Calculate the velocity of the carriage when it reaches B. Show that in the absence of friction this velocity is independent of whether the initial velocity of the carriage at A was up or down the incline.

Ans. $v = 5.51$ m/s

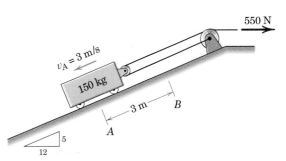

Problem 3/129

3/130 Once under way at a steady speed, the 1000-kg elevator A rises at the rate of 1 story (3 m) per second. Determine the power input P_{in} into the motor unit M if the combined mechanical and electrical efficiency of the system is $e = 0.8$.

Problem 3/130

3/131 It is experimentally determined that the drive wheels of a car must exert a tractive force of 560 N on the road surface in order to maintain a steady vehicle speed of 90 km/h on a horizontal road. If it is known that the overall drivetrain efficiency is $e_m = 0.70$, determine the required motor power output P.

Ans. $P = 20$ kW

3/132 A 1400-kg car is traveling at a speed $v_A = 100$ km/h as it passes point A, and then the car goes down the 6-percent incline. The driver applies her brakes so as to bring the car speed at B to $v_B = 20$ km/h. Calculate the energy Q dissipated from the brakes in the form of heat. Neglect friction losses from other causes such as air resistance.

Problem 3/132

3/133 The 6-kg cylinder is released from rest in the position shown and falls on the spring that has been initially precompressed 50 mm by the light strap and restraining wires. If the stiffness of the spring is 4 kN/m, compute the additional deflection δ of the spring produced by the falling cylinder before it rebounds.

Ans. $\delta = 29.4$ mm

Problem 3/133

3/134 The nest of two springs is used to bring the 0.5-kg plunger A to a stop from a speed of 5 m/s and reverse its direction of motion. The inner spring increases the deceleration, and the adjustment of its position is used to control the exact point at which the reversal takes place. If this point is to correspond to a maximum deflection $\delta = 200$ mm for the outer spring, specify the adjustment of the inner spring by determining the distance s. The outer spring has a stiffness of 300 N/m and the inner one a stiffness of 150 N/m.

Problem 3/134

3/135 The 7-kg collar A slides with negligible friction on the fixed vertical shaft. When the collar is released from rest at the bottom position shown, it moves up the shaft under the action of the constant force $F = 200$ N applied to the cable. Calculate the stiffness k which the spring must have if its maximum compression is to be limited to 75 mm. The position of the small pulley at B is fixed.

Ans. $k = 8.79$ kN/m

60 N

12 mm

12 mm

12 mm

100 mm

Problem 3/136

Problem 3/135

3/136 In a design test of piston-ring pressure, the special 100-mm-diameter aluminum piston with a mass of 2.7 kg is released from rest in the vertical cylinder under the action of the constant 60-N force. The piston reaches a velocity of 2.5 m/s in 250 mm of travel. The coefficient of kinetic friction between the cast-iron rings and the cylinder is 0.15. The piston diameter is slightly smaller than the cylinder diameter so that all frictional resistance to motion is due to piston-ring friction. Calculate the average pressure p between the rings and the cylinder wall. Each of the two rings of 12-mm width is free to expand in its piston groove.

3/137 The 0.60-kg collar slides on the curved rod in the vertical plane with negligible friction under the action of a constant force F in the cord guided by the small pulleys at D. If the collar is released from rest at A, determine the force F which will result in the collar striking the stop at B with a velocity of 4 m/s.

Ans. $F = 13.21$ N

Problem 3/137

3/138 Extensive testing of an experimental 900-kg automobile reveals the aerodynamic drag force F_D and the total nonaerodynamic rolling-resistance force F_R to be as shown in the plot. Determine (a) the power required for steady speeds of 50 and 100 km/h on a level road, (b) the power required for a steady speed of 100 km/h both up and down a 6-percent incline, and (c) the steady speed at which no power is required going down the 6-percent incline.

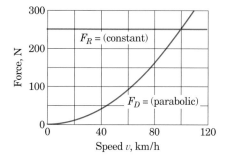

Problem 3/138

3/139 The 25-kg slider in the position shown has an initial velocity $v_0 = 0.6$ m/s on the inclined rail and slides under the influence of gravity and friction. The coefficient of kinetic friction between the slider and the rail is 0.5. Calculate the velocity of the slider as it passes the position for which the spring is compressed a distance $x = 100$ mm. The spring offers a compressive resistance C and is known as a "hardening" spring, since its stiffness increases with deflection as shown in the accompanying graph.

Ans. $v = 2.19$ m/s

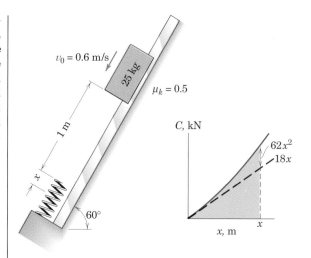

Problem 3/139

3/140 The 10-kg block is released from rest on the horizontal surface at point B, where the spring has been stretched a distance of 0.5 m from its neutral position A. The coefficient of kinetic friction between the block and the plane is 0.30. Calculate (a) the velocity v of the block as it passes point A and (b) the maximum distance x to the left of A which the block goes.

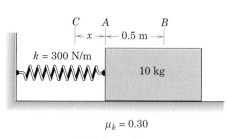

Problem 3/140

3/141 The spring attached to the 10-kg mass is non-linear, having the force–deflection relationship shown in the figure, and is unstretched when $x = 0$. If the mass is moved to the position $x = 100$ mm and released from rest, determine its velocity v when $x = 0$. Determine the corresponding velocity v' if the spring were linear according to $F = 4x$, where x is in meters and the force F is in kilonewtons.

Ans. $v' = 1.899$ m/s

Force F, kN

Linear, $F = 4x$

Nonlinear, $F = 4x - 120x^3$

Deflection x, m

x

10 kg

$\mu_s = 0.25$ $\mu_k = 0.20$

Problem 3/141

3/142 The car of mass m accelerates on a level road under the action of the driving force F from a speed v_1 to a higher speed v_2 in a distance s. If the engine develops a constant power output P, determine v_2. Treat the car as a particle under the action of the single horizontal force F.

F

Problem 3/142

3/7 POTENTIAL ENERGY

In the previous article on work and kinetic energy, a particle or a combination of joined particles was isolated, and the work done by gravity forces, spring forces, and other externally applied forces acting on the particle or system was determined to evaluate U in the work-energy equation. In the present article we shall treat the work done by gravity forces and by spring forces by introducing the concept of *potential energy*. This concept will simplify the analysis of many problems.

$V_g = mgh$

h

mg

$V_g = 0$

(a)

(b)

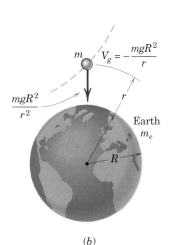

$V_g = -\dfrac{mgR^2}{r}$

$\dfrac{mgR^2}{r^2}$

r

Earth
m_e

R

Figure 3/6

(a) Gravitational potential energy. We consider first the motion of a particle of mass m in close proximity to the earth's surface, where the gravitational attraction (weight) mg is essentially constant, Fig. 3/6a. The *gravitational potential energy* V_g of the particle is defined as the work mgh done *against* the gravitational field to elevate the particle a distance h above some arbitrary reference plane, where V_g is taken to be zero. Thus, we write the potential energy as

$$V_g = mgh \qquad (3/14)$$

This work is called potential energy because it may be converted into energy if the particle is allowed to do work on a supporting body while it returns to its lower original datum plane. In going from one level at $h = h_1$ to a higher level at $h = h_2$, the *change* in potential energy becomes

$$\Delta V_g = mg(h_2 - h_1) = mg\,\Delta h$$

The corresponding work done *by* the gravitational force on the particle is $-mg\,\Delta h$. Thus, the work done by the gravitational force is the negative of the change in potential energy.

When large changes in altitude in the earth's field are encountered, Fig. 3/6b, the gravitational force $Gmm_e/r^2 = mgR^2/r^2$ is no longer constant. The work done *against* this force to change the radial position of the particle from r to r' is the change $V_g{}' - V_g$ in gravitational potential energy, which is

$$\int_r^{r'} mgR^2\frac{dr}{r^2} = mgR^2\left(\frac{1}{r} - \frac{1}{r'}\right) = V_g{}' - V_g$$

It is customary to take $V_g{}' = 0$ when $r' = \infty$, so that with this datum we have

$$V_g = -\frac{mgR^2}{r} \qquad (3/15)$$

In going from r_1 to r_2, the corresponding change in potential energy is

$$\Delta V_g = mgR^2 \left(\frac{1}{r_1} - \frac{1}{r_2} \right)$$

which, again, is the *negative* of the work done *by* the gravitational force. We note that the potential energy of a given particle depends only on its position, h or r, and not on the particular path it followed in reaching that position.

(b) Elastic potential energy.   The second example of potential energy is found in the deformation of an elastic body, such as a spring. The work which is done on the spring to deform it is stored in the spring and is called its *elastic potential energy V_e*. This energy is recoverable in the form of work done by the spring on the body attached to its movable end during the release of the spring's deformation. For the one-dimensional linear spring of stiffness k, which we discussed in Art. 3/6 and illustrated in Fig. 3/4, the force supported by the spring at any deformation x, tensile or compressive, from its undeformed position is $F = kx$. Thus, we define the elastic potential energy of the spring as the work done on it to deform it an amount x, and we have

$$V_e = \int_0^x kx \, dx = \tfrac{1}{2}kx^2 \qquad\qquad (3/16)$$

If the deformation, either tensile or compressive, of a spring increases from x_1 to x_2 during the motion interval, then the change in potential energy of the spring is its final value minus its initial value or

$$\Delta V_e = \tfrac{1}{2}k(x_2{}^2 - x_1{}^2)$$

which is positive. Conversely, if the deformation of a spring decreases during the motion interval, then the change in potential energy of the spring becomes negative. The magnitude of these changes is represented by the shaded trapezoidal area in the *F-x* diagram of Fig. 3/4.

Since the force exerted *on* the spring *by* the moving body is equal and opposite to the force F exerted *by* the spring *on* the body (Fig. 3/4), it follows that the work done on the spring is the negative of the work done on the body. Therefore, we may replace the work U done by the spring on the body by $-\Delta V_e$, the negative of the potential energy change for the spring, provided the spring is now included within the system.

(c) Work-energy equation.   With the elastic member included in the system, we now modify the work-energy equation to account for the potential energy terms. If $U'_{1\text{-}2}$ stands for the work

of all external forces *other than* gravitational forces and spring forces, we may write Eq. 3/11 as $U'_{1\text{-}2} + (-\Delta V_g) + (-\Delta V_e) = \Delta T$ or

$$U'_{1\text{-}2} = \Delta T + \Delta V_g + \Delta V_e \qquad (3/17)$$

This alternative form of the work-energy equation is often far more convenient to use than Eq. 3/11, since the work of both gravity and spring forces is accounted for by focusing attention on the end-point positions of the particle and on the end-point lengths of the elastic spring. The path followed between these end-point positions is of no consequence in the evaluation of ΔV_g and ΔV_e.

Note that Eq. 3/17 may be rewritten in the equivalent form

$$T_1 + V_{g_1} + V_{e_1} + U'_{1\text{-}2} = T_2 + V_{g_2} + V_{e_2} \qquad (3/17a)$$

To help clarify the difference between the use of Eqs. 3/11 and 3/17, Fig. 3/7 shows schematically a particle of mass m constrained to move along a fixed path under the action of forces F_1 and F_2, the gravitational force $W = mg$, the spring force F, and the normal

(a)

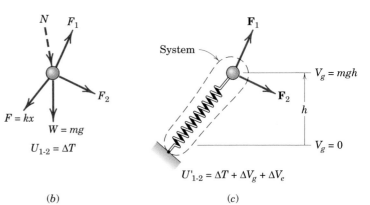

(b)

(c)

Figure 3/7

reaction N. In the b-part of the figure, the particle is isolated with its free-body diagram, and the work done by each of the forces F_1, F_2, W, and the spring force $F = kx$ is evaluated for the interval of motion in question, say, from A to B, and equated to the change ΔT in kinetic energy using Eq. 3/11. The constraint reaction N, if normal to the path, will do no work. In the c-part of the figure for the alternative approach, the spring is included as a part of the isolated system. The work done during the interval by F_1 and F_2 constitutes the $U'_{1\text{-}2}$-term of Eq. 3/17 with the changes in elastic and gravitational potential energies included on the energy side of the equation. We note with the first approach that the work done by $F = kx$ could require a somewhat awkward integration to account for the changes in magnitude and direction of F as the particle moved from A to B. With the second approach, however, only the initial and final lengths of the spring would be required to evaluate ΔV_e, which would greatly simplify the calculation.

We may rewrite the alternative work-energy relation, Eq. 3/17, for a particle-and-spring system as

$$U'_{1\text{-}2} = \Delta(T + V_g + V_e) = \Delta E \qquad (3/17b)$$

where $E = T + V_g + V_e$ is the total mechanical energy of the particle and its attached linear spring. Equation 3/17b states that the net work done on the system by all forces other than gravitational forces and elastic forces equals the change in the total mechanical energy of the system. For problems where the only forces are gravitational, elastic, and nonworking constraint forces, the U'-term is zero, and the energy equation becomes merely

$$\Delta E = 0 \qquad \text{or} \qquad E = \text{constant} \qquad (3/18)$$

When E is constant, we see that transfers of energy between kinetic and potential may take place as long as the total mechanical energy $T + V_g + V_e$ does not change. Equation 3/18 expresses the *law of conservation of dynamical energy.*

(d) Conservative force fields. We have observed that the work done against a gravitational or an elastic force depends only on the net change of position and not on the particular path followed in reaching the new position. Forces with this characteristic are associated with *conservative force fields*, which possess an important mathematical property. Consider a force field where the force $\mathbf{F}$ is a function of the coordinates, Fig. 3/8. The work done by $\mathbf{F}$ during a displacement $d\mathbf{r}$ of its point of application is $dU = \mathbf{F} \cdot d\mathbf{r}$. The total work done along its path from 1 to 2 is

$$U = \int \mathbf{F} \cdot d\mathbf{r} = \int (F_x \, dx + F_y \, dy + F_z \, dz)$$

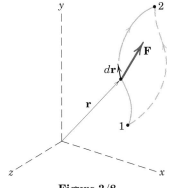

Figure 3/8

*Optional.

The integral $\int \mathbf{F} \cdot d\mathbf{r}$ is a line integral dependent, in general, on the particular path followed between any two points 1 and 2 in space. If, however, $\mathbf{F} \cdot d\mathbf{r}$ is an *exact differential** $-dV$ of some scalar function V of the coordinates, then

$$U_{1\text{-}2} = \int_{V_1}^{V_2} -dV = -(V_2 - V_1) \qquad (3/19)$$

which depends only on the end points of the motion and which is thus *independent* of the path followed. The minus sign before dV is arbitrary but is chosen to agree with the customary designation of the sign of potential energy change in the earth's gravity field. If V exists, the differential change in V becomes

$$dV = \frac{\partial V}{\partial x}\, dx + \frac{\partial V}{\partial y}\, dy + \frac{\partial V}{\partial z}\, dz$$

Comparison with $-dV = \mathbf{F} \cdot d\mathbf{r} = F_x\, dx + F_y\, dy + F_z\, dz$ gives us

$$F_x = -\frac{\partial V}{\partial x} \qquad F_y = -\frac{\partial V}{\partial y} \qquad F_z = -\frac{\partial V}{\partial z}$$

The force may also be written as the vector

$$\mathbf{F} = -\boldsymbol{\nabla} V \qquad (3/20)$$

where the symbol $\boldsymbol{\nabla}$ stands for the vector operator "del", which is

$$\boldsymbol{\nabla} = \mathbf{i}\,\frac{\partial}{\partial x} + \mathbf{j}\,\frac{\partial}{\partial y} + \mathbf{k}\,\frac{\partial}{\partial z}$$

The quantity V is known as the *potential function*, and the expression $\boldsymbol{\nabla} V$ is known as the *gradient of the potential function*.

When force components are derivable from a potential as described, the force is said to be *conservative*, and the work done by $\mathbf{F}$ between any two points is independent of the path followed.

---

*Recall that a function $d\phi = P\, dx + Q\, dy + R\, dz$ is an exact differential in the coordinates x-y-z if

$$\frac{\partial P}{\partial y} = \frac{\partial Q}{\partial x} \qquad \frac{\partial P}{\partial z} = \frac{\partial R}{\partial x} \qquad \frac{\partial Q}{\partial z} = \frac{\partial R}{\partial y}$$

Sample Problem 3/16

The 10-kg slider A moves with negligible friction up the inclined guide. The attached spring has a stiffness of 60 N/m and is stretched 0.6 m in position A, where the slider is released from rest. The 250-N force is constant and the pulley offers negligible resistance to the motion of the cord. Calculate the velocity v of the slider as it passes point C.

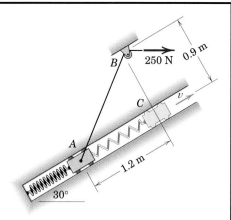

Solution. The slider and inextensible cord together with the attached spring will be analyzed as a system, which permits the use of Eq. 3/17. The only nonpotential force doing work on this system is the 250-N tension applied to the cord. While the slider moves from A to C, the point of application of the 250-N force moves a distance of $\overline{AB} - \overline{BC}$ or $1.5 - 0.9 = 0.6$ m.

① $$U'_{1\text{-}2} = 250(0.6) = 150 \text{ J}$$

The change in kinetic energy of the slider is

$$\Delta T = \tfrac{1}{2}m(v^2 - v_0^2) = \tfrac{1}{2}(10)v^2$$

where the initial velocity v_0 is zero. The change in gravitational potential energy is

② $$\Delta V_g = mg(\Delta h) = 10(9.81)(1.2 \sin 30°) = 58.9 \text{ J}$$

The change in elastic potential energy is

③ $$\Delta V_e = \tfrac{1}{2}k(x_2{}^2 - x_1{}^2) = \tfrac{1}{2}(60)([1.2 + 0.6]^2 - [0.6]^2) = 86.4 \text{ J}$$

Substitution into the alternative work-energy equation gives

$$[U'_{1\text{-}2} = \Delta T + \Delta V_g + \Delta V_e] \quad 150 = \tfrac{1}{2}(10)v^2 + 58.9 + 86.4$$

$$v = 0.974 \text{ m/s} \qquad\qquad Ans.$$

① The reactions of the guides on the slider are normal to the direction of motion and do no work.

② Since the center of mass of the slider has an upward component of displacement, ΔV_g is positive.

③ Be very careful not to make the mistake of using $\tfrac{1}{2}k(x_2 - x_1)^2$ for ΔV_e. We want the difference of the squares and not the square of the difference.

Sample Problem 3/17

The 3-kg slider is released from rest at point A and slides with negligible friction in a vertical plane along the circular rod. The attached spring has a stiffness of 350 N/m and has an unstretched length of 0.6 m. Determine the velocity of the slider as it passes position B.

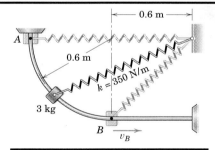

Solution. The work done by the weight and the spring force on the slider will be treated as changes in the potential energies, and the reaction of the rod on the slider is normal to the motion and does no work. Hence, $U'_{1\text{-}2} = 0$. The changes in the potential and kinetic energies for the system of slider and spring are

① $$\Delta V_e = \tfrac{1}{2}k(x_B{}^2 - x_A{}^2) = \tfrac{1}{2}(350)\{(0.6[\sqrt{2} - 1])^2 - (0.6)^2\} = -52.2 \text{ J}$$

$$\Delta V_g = W\Delta h = 3(9.81)(-0.6) = -17.66 \text{ J}$$

$$\Delta T = \tfrac{1}{2}m(v_B{}^2 - v_A{}^2) = \tfrac{1}{2}(3)(v_B{}^2 - 0) = 1.5v_B{}^2$$

$$[\Delta T + \Delta V_g + \Delta V_e = 0] \quad 1.5v_B{}^2 - 17.66 - 52.2 = 0$$

$$v_B = 6.82 \text{ m/s} \qquad\qquad Ans.$$

① Note that if we evaluated the work done by the spring force acting on the slider by means of the integral $\int \mathbf{F} \cdot d\mathbf{r}$, it would necessitate a lengthy computation to account for the change in the magnitude of the force, along with the change in the angle between the force and the tangent to the path. Note further that v_B depends only on the end conditions of the motion and does not require knowledge of the shape of the path.

PROBLEMS

Introductory problems

3/143 The spring has an unstretched length of 0.4 m and a stiffness of 200 N/m. The 3-kg slider and attached spring are released from rest at A and move in the vertical plane. Calculate the velocity v of the slider as it reaches B in the absence of friction.

Ans. $v = 1.537$ m/s

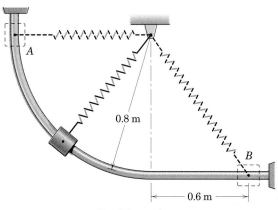

Problem 3/143

3/144 The 1.2-kg slider is released from rest in position A and slides without friction along the vertical-plane guide shown. Determine (a) the speed v_B of the slider as it passes position B and (b) the maximum deflection δ of the spring.

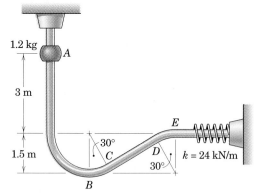

Problem 3/144

3/145 The 1.2-kg slider of the system of Prob. 3/144 is released from rest in position A and slides without friction along the vertical-plane guide. Determine the normal force exerted by the guide on the slider (a) just before it passes point C, (b) just after it passes point C, and (c) just before it passes point E.

Ans. (a) $N_C = 77.7$ N, (b) $N_C = 10.19$ N
(c) $N_E = 35.3$ N (down)

3/146 The two particles of equal mass are joined by a rod of negligible mass. If they are released from rest in the position shown and slide on the smooth guide in the vertical plane, calculate their velocity v when A reaches B's position and B is at B'.

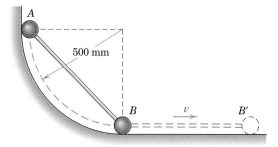

Problem 3/146

3/147 The 2-kg plunger is released from rest in the position shown where the spring of stiffness $k = 500$ N/m has been compressed to one half its uncompressed length of 200 mm. Calculate the maximum height h above the starting position reached by the plunger.

Ans. $h = 95.6$ mm

Problem 3/147

3/148 The spring of constant k is unstretched when the slider of mass m passes position B. If the slider is released from rest in position A, determine its speed as it passes points B and C. What is the normal force exerted by the guide on the slider at position C? Neglect friction between the mass and the circular guide, which lies in a vertical plane.

Problem 3/148

3/149 The light rod is pivoted at O and carries the 2- and 4-kg particles. If the rod is released from rest at $\theta = 60°$ and swings in the vertical plane, calculate (a) the velocity v of the 2-kg particle just before it hits the spring in the dashed position and (b) the maximum compression x of the spring. Assume that x is small so that the position of the rod when the spring is compressed is essentially horizontal.

Ans. (a) $v = 1.162$ m/s, (b) $x = 12.07$ mm

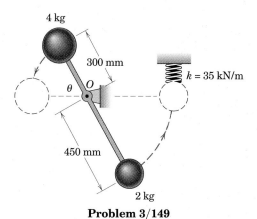

Problem 3/149

3/150 The springs are undeformed in the position shown. If the 6-kg collar is released from rest in the position where the lower spring is compressed 125 mm, determine the maximum compression x_B of the upper spring.

Problem 3/150

Representative problems

3/151 The system is released from rest with the spring initially stretched 75 mm. Calculate the velocity v of the cylinder after it has dropped 12 mm. The spring has a stiffness of 1050 N/m. Neglect the mass of the small pulley.

Ans. $v = 0.371$ m/s

Problem 3/151

3/152 The 10-kg collar slides on the smooth vertical rod and has a velocity $v_1 = 2$ m/s in position A where each spring is stretched 0.1 m. Calculate the velocity v_2 of the collar as it passes point B.

Problem 3/152

3/153 When the mechanism is released from rest in the position where $\theta = 60°$, the 4-kg carriage drops and the 6-kg sphere rises. Determine the velocity v of the sphere when $\theta = 180°$. Neglect the mass of the links and treat the sphere as a particle.

Ans. $v = 0.990$ m/s

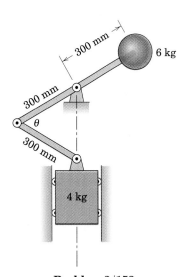

Problem 3/153

3/154 The system is released from rest in the position shown. The 6-kg cylinder passes through the hole in the bracket, but the 4-kg collar does not. Determine the maximum height h which the 8-kg cylinder rises. Explain what happens to the kinetic energy of the collar. Neglect the mass of the cable and small pulleys.

Problem 3/154

3/155 The two springs, each of stiffness $k = 1.2$ kN/m, are of equal length and undeformed when $\theta = 0$. If the mechanism is released from rest in the position $\theta = 20°$, determine its angular velocity $\dot{\theta}$ when $\theta = 0$. The mass m of each sphere is 3 kg. Treat the spheres as particles and neglect the masses of the light rods and springs.

Ans. $\dot{\theta} = 4.22$ rad/s

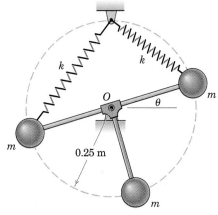

Problem 3/155

3/156 The 0.9-kg collar is released from rest at *A* and slides freely up the inclined rod, striking the stop at *B* with a velocity *v*. The spring of stiffness *k* = 24 N/m has an unstretched length of 375 mm. Calculate *v*.

Problem 3/156

3/157 If the system is released from rest, determine the speeds of both masses after *B* has moved 1 m. Neglect friction and the masses of the pulleys.
Ans. v_A = 0.616 m/s, v_B = 0.924 m/s

Problem 3/157

3/158 The collar has a mass of 2 kg and is attached to the light spring which has a stiffness of 30 N/m and an unstretched length of 1.5 m. The collar is released from rest at *A* and slides up the smooth rod under the action of the constant 50-N force. Calculate the velocity *v* of the collar as it passes position *B*.

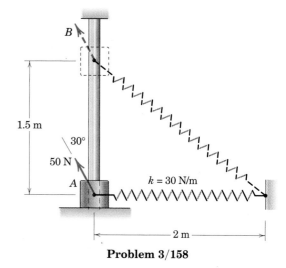

Problem 3/158

3/159 The two 1.5-kg spheres are released from rest and gently nudged outward from the position $\theta = 0$ and then rotate in a vertical plane about the fixed centers of their attached gears, thus maintaining the same angle θ for both rods. Determine the velocity *v* of each sphere as the rods pass the position $\theta = 30°$. The spring is unstretched when $\theta = 0$, and the masses of the two identical rods and the two gear wheels may be neglected.
Ans. v = 0.331 m/s

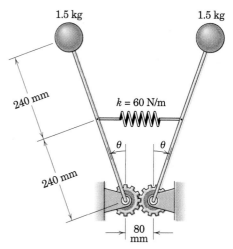

Problem 3/159

3/160 The 2-kg sliding collar C with attached spring moves with friction from A to B along the fixed rod. If the collar has a velocity of 3 m/s at A and a velocity of 5 m/s at B, determine the loss U_f of energy due to friction. The spring has a stiffness of 30 N/m and an unstretched length of 0.5 m. The x-y plane is horizontal. Also determine the average friction force F during the motion from A to B.

Problem 3/160

3/161 A spacecraft m is heading toward the center of the moon with a velocity of 3000 km/h at a distance from the moon's surface equal to the radius R of the moon. Compute the impact velocity v with the surface of the moon if the spacecraft is unable to fire its retro-rockets. Consider the moon fixed in space. The radius R of the moon is 1738 km, and the acceleration due to gravity at its surface is 1.62 m/s².

Ans. $v = 6740$ km/h

Problem 3/161

3/162 The two right-angled rods with attached spheres are released from rest in the position $\theta = 0$. If the system is observed to momentarily come to rest when $\theta = 45°$, determine the spring constant k. The spring is unstretched when $\theta = 0$. Treat the spheres as particles and neglect friction.

Problem 3/162

3/163 When $\theta = 90°$, the spring of stiffness $k = 2$ kN/m is uncompressed. If the balls are released from rest in their top position, calculate their velocity v as they strike the horizontal surface with $\theta = 0$. Treat each ball as a particle and neglect the mass of the right-angled members and any friction.

Ans. $v = 1.368$ m/s

Problem 3/163

3/164 The 5-kg cylinder is released from rest in the position shown and compresses the spring of stiffness $k = 1.8$ kN/m. Determine the maximum compression x_{max} of the spring and the maximum velocity v_{max} of the cylinder along with the corresponding deflection x of the spring.

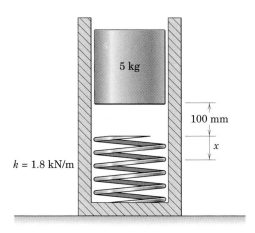

5 kg

100 mm

x

$k = 1.8$ kN/m

Problem 3/164

3/165 The mass of Prob. 3/127, shown again here, is released from rest onto the top of the uncompressed spring of stiffness k by removal of the supports. Determine the maximum velocity v acquired by the mass.

Ans. $v = g\sqrt{m/k}$

m

Problem 3/165

3/166 An 80-kg pole vaulter carrying a uniform 4.9-m, 4.5-kg pole approaches the jump with a velocity v and manages to barely clear the bar set at a height of 5.5 m. As he clears the bar, his velocity and that of the pole are essentially zero. Calculate the minimum possible value of v required for him to make the jump. Both the horizontal pole and the center of mass of the vaulter are 1.1 m above the ground during the approach.

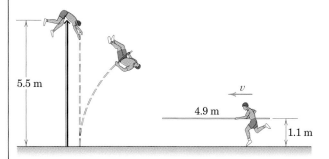

5.5 m

v

4.9 m

1.1 m

Problem 3/166

3/167 When the 5-kg plunger is released from rest in its vertical guide at $\theta = 0$, each spring of stiffness $k = 3.5$ kN/m is uncompressed. The links are free to slide through their pivoted collars and compress their springs. Calculate the velocity v of the plunger when the position $\theta = 30°$ is passed.

Ans. $v = 0.972$ m/s

5 kg

k θ θ k

$\leftarrow$ 150 mm $\rightarrow$$\leftarrow$ 150 mm $\rightarrow$

Problem 3/167

3/168 The cars of an amusement-park ride have a speed $v_1 = 90$ km/h at the lowest part of the track. Determine their speed v_2 at the highest part of the track. Neglect energy loss due to friction. (*Caution:* Give careful thought to the change in potential energy of the system of cars.)

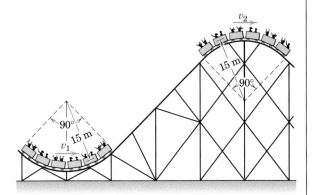

Problem 3/168

3/169 An artificial satellite moving in an elliptical orbit has a velocity of 25 000 km/h at an altitude of 2200 km at point A. Determine its velocity v_B at point B where the altitude is 2500 km. Treat the earth as a sphere of radius $R = 6371$ km and use $g = 9.825$ m/s^2 for the acceleration of gravity at the earth's surface.

Ans. $v_B = 24\ 170$ km/h

Problem 3/169

3/170 A satellite is put into an elliptical orbit around the earth and has a velocity v_P at the perigee position P. Determine the expression for the velocity v_A at the apogee position A. The radii to A and P are, respectively, r_A and r_P. Note that the total energy remains constant.

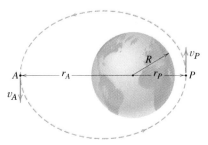

Problem 3/170

3/171 By "pumping" as he swings, the boy increases the swing amplitude from θ_0 to θ_1 by abruptly changing from the sitting to the supine position at the start of each forward swing and reversing positions at the start of each back swing. Treat the boy as a particle in each of the two configurations where the path of his mass center is shown by the dashed trajectory. Any loss of mechanical energy $(T + V_g)$ between G_1 and G_3 may be assumed to be negligible. Express θ_1 in terms of θ_0, R, and h.

$$Ans.\ \theta_1 = \cos^{-1}\left(\frac{R}{R+h}\cos\theta_0\right)$$

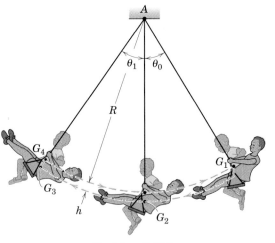

Problem 3/171

3/172 The rope shown in the *a*-part of the figure is released from rest in the smooth semicircular guide and acquires a velocity *v* when all of it is clear of the guide as seen in the *b*-part of the figure. Find *v*. (*Hint:* In finding the potential energy change, imagine that the semicircular portion of the rope in the initial position becomes the bottom part of the rope of equal length in the final position.)

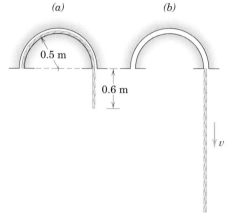

(a)   (b)

0.5 m

0.6 m

v

Problem 3/172

3/173 The flexible bicycle-type chain of length $\pi r/2$ and mass per unit length ρ is released from rest with $\theta = 0$ in the smooth circular channel and falls through the hole in the supporting surface. Determine the velocity v of the chain as the last link leaves the slot.

$$Ans.\ v = \sqrt{gr\left(\frac{\pi}{2} + \frac{4}{\pi}\right)}$$

Problem 3/173

▶**3/174** The free end of a flexible rope of length L and mass ρ per unit length is released from rest in the position shown in the *a*-part of the figure. Determine the velocity v of the moving portion of the rope in terms of y. Discuss the meaning of v when $y = L$ and account for all energy changes.

$$Ans.\ v = \sqrt{2gy\,\frac{L - y/2}{L - y}}$$

(a)   (b)   (c)

$L/2$

y

v

Problem 3/174

▶3/175 The cable railway consists of two passenger gondolas, each of mass m, one on each end of the cable of total length L and mass ρ per unit length. The system is operated by applying a torque M to the drum of radius r at the top of the railway. Several turns of cable around the drum prevent slipping, and the length of cable around the drum may be neglected compared with L. If the gondolas start from rest at $x = 0$ with M constant, derive an expression for the velocity v of each gondola for a given value of x. Neglect the mass of the drum and all friction.

$$Ans.\ v = \sqrt{\frac{2}{2m + \rho L}} \sqrt{\frac{Mx}{r} - \rho gx(L - x)\sin\theta}$$

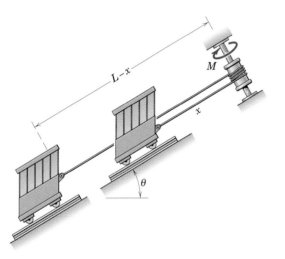

Problem 3/175

▶3/176 The two particles of mass m and $2m$, respectively, are connected by a rigid rod of negligible mass and slide with negligible friction in a circular path of radius r on the inside of the vertical circular ring. If the unit is released from rest at $\theta = 0$, determine (a) the velocity v of the particles when the rod passes the horizontal position, (b) the maximum velocity v_{max} of the particles, and (c) the maximum value of θ.

$$Ans.\ (a)\ v_{45°} = 0.865\sqrt{gr}$$
$$(b)\ v_{max} = 0.908\sqrt{gr}$$
$$(c)\ \theta_{max} = 126.9°$$

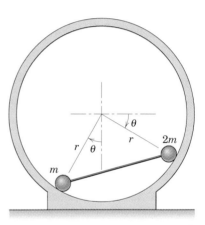

Problem 3/176

SECTION C. IMPULSE AND MOMENTUM

3/8 *INTRODUCTION*

In the previous two articles, we focused attention on the equations of work and energy which are obtained by integrating the equation of motion $\mathbf{F} = m\mathbf{a}$ with respect to the displacement of the particle. In consequence, we found that the velocity changes could be expressed directly in terms of the work done or in terms of the overall changes in energy. In the next two articles, we will direct our attention to the integration of the equation of motion with respect to time rather than displacement. This approach leads to the equations of impulse and momentum. We will find that these equations greatly facilitate the solution of many problems in which the applied forces act during extremely short periods of time (as in impact problems) or over specified intervals of time.

3/9 *LINEAR IMPULSE AND LINEAR MOMENTUM*

Consider again the general curvilinear motion in space of a particle of mass m, Fig. 3/9, where the particle is located by its position vector $\mathbf{r}$ measured from a fixed origin O. The velocity of the particle is $\mathbf{v} = \dot{\mathbf{r}}$ and is tangent to its path (shown as a dashed line). The resultant $\Sigma\mathbf{F}$ of all forces on m is in the direction of its acceleration $\dot{\mathbf{v}}$. We may now write the basic equation of motion for the particle, Eq. 3/3, as

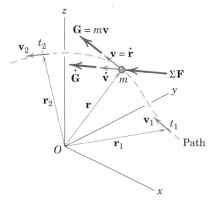

Figure 3/9

$$\Sigma\mathbf{F} = m\dot{\mathbf{v}} = \frac{d}{dt}(m\mathbf{v}) \quad \text{or} \quad \boxed{\Sigma\mathbf{F} = \dot{\mathbf{G}}} \qquad (3/21)$$

where the product of the mass and velocity is defined as the *linear momentum* $\mathbf{G} = m\mathbf{v}$ of the particle. Equation 3/21 states that *the resultant of all forces acting on a particle equals its time rate of change of linear momentum*. In SI the units of linear momentum $m\mathbf{v}$ are seen to be kg·m/s, which also equals N·s.

Since Eq. 3/21 is a vector equation, we recognize that, in addition to the equality of the magnitudes of $\Sigma\mathbf{F}$ and $\dot{\mathbf{G}}$, the direction of the resultant force coincides with the direction of the rate of change in linear momentum, which is the direction of the rate of change in velocity. Equation 3/21 is one of the most useful and important relationships in dynamics, and it is valid as long as the mass m of the particle is not changing with time. The case where m changes with time is discussed in Art. 4/7 of Chapter 4. We now write the three scalar components of Eq. 3/21 as

$$\Sigma F_x = \dot{G}_x \qquad \Sigma F_y = \dot{G}_y \qquad \Sigma F_z = \dot{G}_z \qquad (3/22)$$

These equations may be applied independently of one another.

All that we have done so far in this article is to rewrite Newton's second law in an alternative form in terms of momentum. But we are now able to describe the effect of the resultant force $\Sigma\mathbf{F}$ on the linear momentum of the particle over a finite period of time simply by integrating Eq. 3/21 with respect to the time t. Multiplying the equation by dt gives $\Sigma\mathbf{F}\,dt = d\mathbf{G}$, which we integrate from time t_1 to time t_2 to obtain

$$\int_{t_1}^{t_2} \Sigma\mathbf{F}\,dt = \mathbf{G}_2 - \mathbf{G}_1 = \Delta\mathbf{G}$$

(3/23)

Here the linear momentum at time t_2 is $\mathbf{G}_2 = m\mathbf{v}_2$ and the linear momentum at time t_1 is $\mathbf{G}_1 = m\mathbf{v}_1$. The product of force and time is defined as *linear impulse* of the force, and Eq. 3/23 states that *the total linear impulse on m equals the corresponding change in linear momentum of m.*

Alternatively, we may write Eq. 3/23 as

$$\mathbf{G}_1 + \int_{t_1}^{t_2} \Sigma\mathbf{F}\,dt = \mathbf{G}_2$$

(3/23*a*)

which says that the initial linear momentum of the body plus the linear impulse applied to it equals its final linear momentum.

The impulse integral is a vector which, in general, may involve changes in both magnitude and direction during the time interval. Under these conditions, it will be necessary to express $\Sigma\mathbf{F}$ and $\mathbf{G}$ in component form and then combine the integrated components. The components of Eq. 3/23 become the scalar equations

$$\int_{t_1}^{t_2} \Sigma F_x\,dt = (mv_x)_2 - (mv_x)_1$$

$$\int_{t_1}^{t_2} \Sigma F_y\,dt = (mv_y)_2 - (mv_y)_1$$

$$\int_{t_1}^{t_2} \Sigma F_z\,dt = (mv_z)_2 - (mv_z)_1$$

These three scalar impulse-momentum equations are completely independent. The scalar expressions corresponding to the vector equations 3/23*a* are simply rearrangements of these equations.

There are times when a force acting on a particle varies with the time in a manner determined by experimental measurements or by other approximate means. In this case, a graphical or numerical integration must be accomplished. If, for example, a force F

acting on a particle in a given direction varies with the time t as indicated in Fig. 3/10, then the impulse, $\int_{t_1}^{t_2} F \, dt$, of this force from t_1 to t_2 becomes the shaded area under the curve.

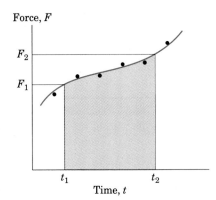

Force, F

Figure 3/10

In evaluating the resultant impulse, it is necessary to include the effect of *all* forces acting on m, except those whose magnitudes are negligible. The student should be fully aware at this point that the only reliable method of accounting for the effects of *all* forces is to isolate the particle in question by drawing its *free-body diagram*.

Conservation of linear momentum. If the resultant force on a particle is zero during an interval of time, we see that Eq. 3/21 requires that its linear momentum **G** remain constant. In this case, the linear momentum of the particle is said to be *conserved*. Linear momentum may be conserved in one coordinate direction, such as x, but not necessarily in the y- or z-direction. A careful examination of the free-body diagram of the particle will disclose whether the total linear impulse on the particle in a particular direction is zero. If it is, the corresponding linear momentum is unchanged (conserved) in that direction.

Consider now the motion of two particles a and b that interact during an interval of time. If the interactive forces **F** and $-$**F** between them are the only unbalanced forces acting on the particles during the interval, it follows that the linear impulse on particle a is the negative of the linear impulse on particle b. Therefore, from Eq. 3/23, the change in linear momentum $\Delta\mathbf{G}_a$ of particle a is the negative of the change $\Delta\mathbf{G}_b$ in linear momentum of particle b. So we have $\Delta\mathbf{G}_a = -\Delta\mathbf{G}_b$ or $\Delta(\mathbf{G}_a + \mathbf{G}_b) = \mathbf{0}$. Thus, the total linear momentum $\mathbf{G} = \mathbf{G}_a + \mathbf{G}_b$ for the system of the two particles remains constant during the interval, and we write

$$\boxed{\Delta\mathbf{G} = \mathbf{0} \qquad \text{or} \qquad \mathbf{G}_1 = \mathbf{G}_2} \qquad (3/24)$$

Equation 3/24 expresses the *principle of conservation of linear momentum*.

Sample Problem 3/18

A 0.2-kg particle moves in the vertical y-z plane (z up, y horizontal) under the action of its weight and a force $\mathbf{F}$ that varies with the time. The linear momentum of the particle in N·s is given by the expression $\mathbf{G} = \frac{3}{2}(t^2 + 3)\mathbf{j} - \frac{2}{3}(t^3 - 4)\mathbf{k}$, where t is the time in seconds. Determine $\mathbf{F}$ for the instant when $t = 2$ s.

Solution. The weight expressed as a vector is $-0.2(9.81)\mathbf{k}$ N. Thus, the force-momentum equation becomes

① $[\Sigma \mathbf{F} = \dot{\mathbf{G}}]$ $\qquad \mathbf{F} - 0.2(9.81)\mathbf{k} = \dfrac{d}{dt}[\frac{3}{2}(t^2 + 3)\mathbf{j} - \frac{2}{3}(t^3 - 4)\mathbf{k}]$

$\qquad\qquad\qquad\qquad\qquad = 3t\mathbf{j} - 2t^2\mathbf{k}$

For $t = 2$ s, $\qquad \mathbf{F} = 0.2(9.81)\mathbf{k} + 3(2)\mathbf{j} - 2(2^2)\mathbf{k} = 6\mathbf{j} - 6.04\mathbf{k}$ N

Thus, $\qquad\qquad F = \sqrt{6^2 + 6.04^2} = 8.51$ N $\qquad\qquad$ *Ans.*

① Don't forget that $\Sigma\mathbf{F}$ includes *all* forces on the particle, of which the weight is one.

Sample Problem 3/19

A particle with a mass of 0.5 kg has a velocity $u = 10$ m/s in the x-direction at time $t = 0$. Forces $\mathbf{F}_1$ and $\mathbf{F}_2$ act on the particle, and their magnitudes change with time according to the graphical schedule shown. Determine the velocity $\mathbf{v}$ of the particle at the end of the 3 s.

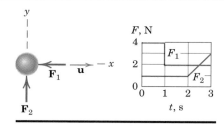

Solution. The impulse-momentum equation is applied in component form and gives for the x- and y-directions, respectively,

① $\left[\int \Sigma F_x \, dt = m\Delta v_x\right]$ $\qquad -[4(1) + 2(3 - 1)] = 0.5(v_x - 10)$

$\qquad\qquad\qquad\qquad\qquad\qquad v_x = -6$ m/s

$\left[\int \Sigma F_y \, dt = m\Delta v_y\right]$ $\qquad [1(2) + 2(3 - 2)] = 0.5(v_y - 0)$

$\qquad\qquad\qquad\qquad\qquad\qquad v_y = 8$ m/s

Thus,

$\qquad \mathbf{v} = -6\mathbf{i} + 8\mathbf{j}$ m/s $\qquad$ and $\qquad v = \sqrt{6^2 + 8^2} = 10$ m/s

$\qquad\qquad \theta_x = \tan^{-1}\dfrac{8}{-6} = 126.9°$ $\qquad\qquad$ *Ans.*

② $\qquad$ Although not called for, the path of the particle for the first 3 s is plotted in the figure. The velocity at $t = 3$ s is shown together with its components.

① The impulse in each direction is the corresponding area under the force-time graph. Note that F_1 is in the negative x-direction so its impulse is negative.

② It is important to note that the algebraic signs must be carefully respected in applying the momentum equations. Also we must recognize that impulse and momentum are vector quantities in contrast to work and energy, which are scalars.

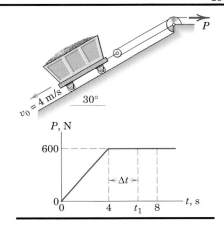

Sample Problem 3/20

The loaded 150-kg skip is rolling down the incline at 4 m/s when a force P is applied to the cable as shown at time $t = 0$. The force P is increased uniformly with the time until it reaches 600 N at $t = 4$ s, after which time it remains constant at this value. Calculate (a) the time t_1 at which the skip reverses its direction and (b) the velocity v of the skip at $t = 8$ s. Treat the skip as a particle.

Solution. The stated variation of P with the time is plotted, and the free-body diagram of the skip is drawn.

Part (a). The skip reverses direction when its velocity becomes zero. We will assume that this condition occurs at $t = 4 + \Delta t$ s. The impulse-momentum equation applied consistently in the positive x-direction gives

$$\left[\int \Sigma F_x \, dt = m \Delta v_x \right]$$

①

$$\tfrac{1}{2}(4)(2)(600) + 2(600)\Delta t - 150(9.81) \sin 30°(4 + \Delta t) = 150(0 - [-4])$$

$$464 \Delta t = 1143 \qquad \Delta t = 2.46 \text{ s} \qquad t = 4 + 2.46 = 6.46 \text{ s} \qquad Ans.$$

Part (b). Applying the impulse-momentum equation to the entire interval gives

$$\left[\int \Sigma F_x \, dt = m \Delta v_x \right]$$

$$\tfrac{1}{2}(4)(2)(600) + 4(2)(600) - 150(9.81) \sin 30°(8) = 150(v - [-4])$$

$$150v = 714 \qquad v = 4.76 \text{ m/s} \qquad Ans.$$

The same result is obtained by analyzing the interval from t_1 to 8 s.

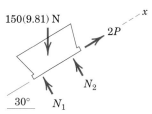

① The free-body diagram keeps us from making the error of using the impulse of P rather than $2P$ or of forgetting the impulse of the component of the weight. The first term in the equation is the triangular area of the P-t relation for the first 4 s, doubled for the force of $2P$.

Sample Problem 3/21

The 50-g bullet traveling at 600 m/s strikes the 4-kg block centrally and is embedded within it. If the block slides on a smooth horizontal plane with a velocity of 12 m/s in the direction shown prior to impact, determine the velocity $\mathbf{v}$ of the block and embedded bullet immediately after impact.

Solution. Since the force of impact is internal to the system composed of the block and bullet and since there are no other external forces acting on the system in the plane of motion, it follows that the linear momentum of the system is conserved. Thus,

① $[\mathbf{G}_1 = \mathbf{G}_2]$ $\quad 0.050(600\mathbf{j}) + 4(12)(\cos 30°\mathbf{i} + \sin 30°\mathbf{j}) = (4 + 0.050)\mathbf{v}$

$$\mathbf{v} = 10.26\mathbf{i} + 13.33\mathbf{j} \text{ m/s} \qquad Ans.$$

The final velocity and its direction are given by

$$[v = \sqrt{v_x^2 + v_y^2}] \qquad v = \sqrt{(10.26)^2 + (13.33)^2} = 16.83 \text{ m/s} \qquad Ans.$$

$$[\tan \theta = v_y/v_x] \qquad \tan \theta = \frac{13.33}{10.26} = 1.299 \qquad \theta = 52.4° \qquad Ans.$$

① Working with the vector form of the principle of conservation of linear momentum is clearly equivalent to working with the component form.

PROBLEMS

Introductory problems

3/177 The rocket engine of a 30-Mg spacecraft traveling at a speed of 24 000 km/h is fired and produces a thrust of 20 kN in the direction of its circular path for a period of 3 min. Determine the new speed of the spacecraft. The loss of mass due to fuel burned is negligibly small.

Ans. $v = 24\ 432$ km/h

3/178 The 1500-kg car has a velocity of 30 km/h up the 10-percent grade when the driver applies more power for 8 s to bring the car up to a speed of 60 km/h. Calculate the time average F of the total force tangent to the road exerted on the tires during the 8 s. Treat the car as a particle and neglect air resistance.

Problem 3/178

3/179 The jet fighter has a mass of 6450 kg and requires 10 seconds from rest to reach its takeoff speed of 250 km/h under the constant jet thrust $T = 48$ kN. Compute the time average R of the combined air and ground resistance during takeoff.

Ans. $R = 3.21$ kN

Problem 3/179

3/180 A 75-g projectile traveling at 600 m/s strikes and becomes embedded in the 50-kg block, which is initially stationary. Compute the energy lost during the impact. Express your answer as an absolute value $|\Delta E|$ and as a percentage n of the original system energy E.

75 g 600 m/s 50 kg

Problem 3/180

3/181 Freight car A with a total mass of 80 Mg is moving along the horizontal track in a switching yard at 3 km/h. Freight car B with a total mass of 60 Mg and moving at 5 km/h overtakes car A and is coupled to it. Determine (*a*) the common velocity v of the two cars as they move together after being coupled and (*b*) the loss of energy $|\Delta E|$ due to the impact.

Ans. (*a*) $v = 3.86$ km/h, (*b*) $|\Delta E| = 5290$ J

5 km/h   3 km/h

B A

Problem 3/181

3/182 A 60-g bullet is fired horizontally with a velocity $v_1 = 600$ m/s into the 3-kg block of soft wood initially at rest on the horizontal surface. The bullet emerges from the block with the velocity $v_2 = 400$ m/s, and the block is observed to slide a distance of 2.70 m before coming to rest. Determine the coefficient of kinetic friction μ_k between the block and the supporting surface.

400 m/s   3 kg   600 m/s

60 g

2.70 m

Problem 3/182

3/183 The two orbital maneuvering engines of the space shuttle develop 26 kN of thrust each. If the shuttle is traveling in orbit at a speed of 28 000 km/h, how long would it take to reach a speed of 28 100 km/h after the two engines are fired? The mass of the shuttle is 90 Mg.

Ans. t = 48.1 s

3/184 A 45-kg boy runs and jumps on his 10-kg sled with a horizontal velocity of 4.6 m/s. If the sled and boy coast 25 m on the level snow before coming to rest, compute the coefficient of kinetic friction μ_k between the snow and the runners of the sled.

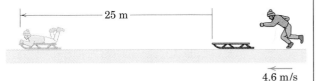

Problem 3/184

3/185 The 200-kg lunar lander is descending onto the moon's surface with a velocity of 6 m/s when its retro-engine is fired. If the engine produces a thrust T for 4 s which varies with the time as shown and then cuts off, calculate the velocity of the lander when $t = 5$ s, assuming that it has not yet landed. Gravitational acceleration at the moon's surface is 1.62 m/s^2.

Ans. v = 2.10 m/s

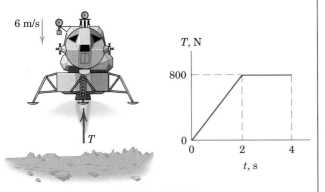

Problem 3/185

3/186 The pilot of a 40-Mg airplane which is originally flying horizontally at a speed of 650 km/h cuts off all engine power and enters a 5° glide path as shown. After 120 s the airspeed is 600 km/h. Calculate the time-average drag force D (air resistance to motion along the flight path).

Problem 3/186

3/187 The inspection gondola for a cableway is being drawn up the sloping cable at the speed of 4 m/s. If the control cable at A suddenly breaks, calculate the time t after the break occurs for the gondola to reach a speed of 8 m/s down the inclined cable. Neglect friction and treat the gondola as a particle.

Ans. t = 3.18 s

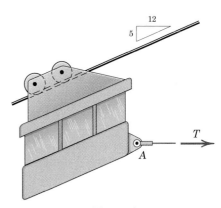

Problem 3/187

Representative problems

3/188 The third and fourth stages of a rocket are coasting in space with a velocity of 18 000 km/h when a small explosive charge between the stages separates them. Immediately after separation the fourth stage has increased its velocity to $v_4 =$ 18 060 km/h. What is the corresponding velocity v_3 of the third stage? At separation the third and fourth stages have masses of 400 and 200 kg, respectively.

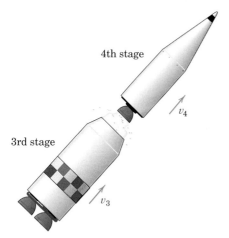

Problem 3/188

3/189 The space shuttle launches an 800-kg satellite by ejecting it from the cargo bay as shown. The ejection mechanism is activated and is in contact with the satellite for 4 s to give it a velocity of 0.3 m/s in the z-direction relative to the shuttle. The mass of the shuttle is 90 Mg. Determine the component of velocity v_f of the shuttle in the minus z-direction resulting from the ejection. Also find the time average F_{av} of the ejection force.

  Ans. $v_f = 0.00264$ m/s, $F_{av} = 59.5$ N

Problem 3/189

3/190 The projectile-shaped body of mass m has a washer of mass $m/5$ resting on its shoulder. As the combined body passes downward through an opening with speed v, the washer strikes the solid surface and is left behind. If the duration time of the impact is Δt, determine the total force R exerted on the washer by the surface and the percent loss n of system kinetic energy.

Problem 3/190

3/191 The cart of mass m is subjected to the exponentially decreasing force F, which represents a shock or blast loading. If the cart is stationary at time $t = 0$, determine its velocity v and displacement s as functions of time. What is the value of v for large values of t?

$$Ans. \quad v = \frac{F_0}{mb}(1 - e^{-bt}), \quad v = \frac{F_0}{mb} \text{ as } t \to \infty$$

$$s = \frac{F_0}{mb}\left[t - \frac{1}{b}(1 - e^{-bt})\right]$$

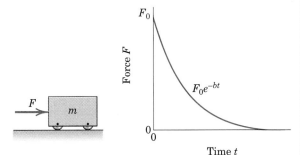

Problem 3/191

3/192 The 140-g projectile is fired with a velocity of 600 m/s and picks up three washers, each with a mass of 100 g. Find the common velocity v of the projectile and washers. Determine also the loss $|\Delta E|$ of energy during the interaction.

Problem 3/192

3/193 An emergency evacuation system at the launch tower for astronauts consists of a long slide-wire cable, down which the escape cage travels to a safe distance from the tower. The cage, together with its two occupants, has a mass of 320 kg and approaches the netting horizontally at a speed of 28 m/s. The netting is held to the cable by a breakaway lashing and is attached to 20 m of heavy chain with a mass of 18 kg/m. The coefficient of kinetic friction between the chain and the ground is 0.70. Determine the initial velocity v of the chain when the cage has engaged the net, and find the time t to bring the cage to a stop after engagement. Assume all links of the chain remain in contact with the ground.

$$Ans. \quad v = 13.18 \text{ m/s}, \quad t = 3.62 \text{ s}$$

Problem 3/193

3/194 The tow truck with attached 1200-kg car accelerates uniformly from 30 km/h to 70 km/h over a 15-s interval. The average rolling resistance for the car over this speed interval is 500 N. Assume that the 60° angle shown represents the time average configuration and determine the average tension in the tow cable.

Problem 3/194

3/195 The 10-kg block is moving to the left with a speed of 1.2 m/s at time $t = 0$, at which time the force P is applied as shown on the graph. After 0.2 s the force continues at the 10-N level. If the coefficient of kinetic friction is $\mu_k = 0.2$, determine the time t at which the block comes to a stop.

Ans. $t = 0.338$ s

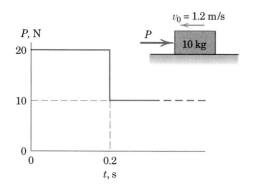

Problem 3/195

3/196 A supertanker with a total displacement (mass) of $150(10^3)$ metric tons (one ton equals 10^3 kg) is moving forward at a speed of 2 knots when the engines are reversed to produce a rearward propeller thrust of 400 kN. How long would it take the tanker to acquire a speed of 2 knots in the reverse direction? Can you justify neglecting the impulse of water resistance of the hull? (Recall 1 knot = 1.852 km/h.)

3/197 The 2.4-kg particle moves in the horizontal x-y plane and has the velocity shown at time $t = 0$. If the force $F = 2 + 3t^2/4$ newtons, where t is time in seconds, is applied to the particle in the y-direction beginning at time $t = 0$, determine the velocity v of the particle 4 seconds after F is applied and specify the corresponding angle θ measured counterclockwise from the x-axis to the direction of the velocity.

Ans. $v = 8.06$ m/s, $\theta = 60.3°$

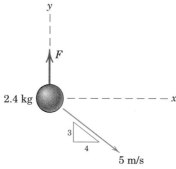

Problem 3/197

3/198 Car B is initially stationary and is struck by car A moving with initial speed $v_1 = 30$ km/h. The cars become entangled and move together with speed v' after the collision. If the time duration of the collision is 0.1 sec, determine (*a*) the common final speed v', (*b*) the average acceleration of each car during the collision, and (*c*) the magnitude R of the average force exerted by each car on the other car during the impact. All brakes are released during the collision.

Problem 3/198

3/199 The vertical motion of the 3-kg cylinder is controlled by the forces P applied to the end rollers of the extensible framework shown. If the upward velocity v of the cylinder is increased from 2 m/s to 4 m/s in 2 seconds, calculate the average force R_{av} under each of the two rollers during the 2-s interval. Analyze the entire system as a unit and neglect the small mass of the framework.

Ans. $R_{av} = 16.22$ N

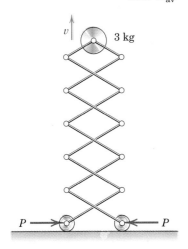

Problem 3/199

3/200 Car B (1500 kg) traveling west at 48 km/h collides with car A (1600 kg) traveling north at 32 km/h as shown. If the two cars become entangled and move together as a unit after the crash, compute the magnitude v of their common velocity immediately after the impact and the angle θ made by the velocity vector with the north direction.

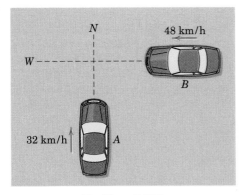

Problem 3/200

3/201 A 1000-kg spacecraft is traveling in deep space with a speed of $v_s = 2000$ m/s when a 10-kg meteor moving with a velocity $\mathbf{v}_m$ of magnitude 5000 m/s in the direction shown strikes and becomes embedded in the spacecraft. Determine the final velocity $\mathbf{v}$ of the mass center G of the spacecraft. Calculate the angle β between $\mathbf{v}$ and the initial velocity $\mathbf{v}_s$ of the spacecraft.

Ans. $\mathbf{v} = 36.9\mathbf{i} + 1951\mathbf{j} - 14.76\mathbf{k}$ m/s
$\beta = 1.167°$

Problem 3/201

3/202 The force P, which is applied to the 10-kg block initially at rest, varies linearly with the time as indicated. If the coefficients of static and kinetic friction between the block and the horizontal surface are 0.6 and 0.4, respectively, determine the velocity of the block when $t = 4$ s.

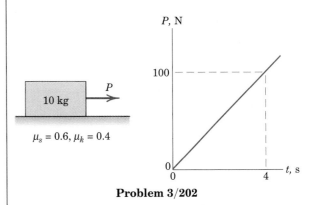

Problem 3/202

3/203 The 45.9-g golf ball is struck by the five-iron and acquires the velocity shown in a time period of 0.001 s. Determine the magnitude R of the average force exerted by the club on the ball. What acceleration magnitude a does this force cause, and what is the distance d over which the launch velocity is achieved, assuming constant acceleration?

$$\text{Ans.}\ R = 2110\ \text{N},\ a = 46\,000\ \text{m/s}^2$$
$$d = 23\ \text{mm}$$

Problem 3/203

3/204 The cart is moving down the incline with a velocity $v_0 = 20$ m/s at $t = 0$, at which time the force P begins to act as shown. After 5 seconds the force continues at the 50-N level. Determine the velocity of the cart at time $t = 8$ s and calculate the time t at which the cart velocity is zero.

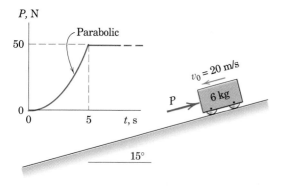

Problem 3/204

3/205 The 30-kg carriage moves with negligible friction along the horizontal rail under the action of the sinusoidal force $F = 40 \sin \pi t$ in newtons measured positive to the left as shown. If the carriage has a velocity of 1.5 m/s to the right at $t = 0$, determine the velocity v_1 of the carriage when $t = 1$ s. By inspection, what is the velocity v_2 of the carriage when $t = 2$ s?

$$\text{Ans.}\ v_1 = 0.651\ \text{m/s},\ v_2 = 1.5\ \text{m/s}$$

Problem 3/205

3/206 The 10-kg cart is stationary at time $t = 0$ and thereafter is subjected to the sinusoidal force $F = b + 45 \sin 6t$, where F and b are in newtons and time t is in seconds. (*a*) If $b = 22$ N, determine the velocity v of the cart at $t = 1.5$ s. (*b*) Determine the value of b for which the velocity of the cart would be zero after the first complete cycle of force application. Neglect friction.

Problem 3/206

3/207 The 900-kg motorized unit A is designed to raise and lower the 600-kg bucket B of concrete. Determine the average force R which supports unit A during the 6 seconds required to slow the descent of the bucket from 3 m/s to 0.5 m/s. Analyze the entire system as a unit without finding the tension in the cable.

Ans. $R = 14.96$ kN

Problem 3/207

3/208 The ballistic pendulum is a simple device to measure projectile velocity v by observing the maximum angle θ to which the box of sand with embedded projectile swings. Calculate the angle θ if the 60-g projectile is fired horizontally into the suspended 20-kg box of sand with a velocity $v = 600$ m/s. Also find the percentage of energy lost during impact.

Problem 3/208

3/209 The ice-hockey puck with a mass of 0.20 kg has a velocity of 12 m/s before being struck by the hockey stick. After the impact the puck moves in the new direction shown with a velocity of 18 m/s. If the stick is in contact with the puck for 0.04 s, compute the magnitude of the average force $\mathbf{F}$ exerted by the stick on the puck during contact, and find the angle β made by $\mathbf{F}$ with the x-direction.

Ans. $F = 147.8$ N, $\beta = 12.02°$

Problem 3/209

3/210 The baseball is traveling with a horizontal velocity of 135 km/h just before impact with the bat. Just after the impact, the velocity of the 146-g ball is 210 km/h directed at 35° to the horizontal as shown. Determine the x- and y-components of the average force $\mathbf{R}$ exerted by the bat on the baseball during the 0.005-s impact. Comment on treatment of the weight of the baseball (a) during the impact and (b) over the first few seconds after impact.

Problem 3/210

3/211 A tennis player strikes the tennis ball with her racket while the ball is still rising. The ball speed before impact with the racket is $v_1 = 15$ m/s and after impact its speed is $v_2 = 22$ m/s, with directions as shown in the figure. If the 60-g ball is in contact with the racket for 0.05 s, determine the magnitude of the average force **R** exerted by the racket on the ball. Find the angle β made by **R** with the horizontal. Comment on the treatment of the ball weight during impact.

Ans. $R = 43.0$ N, $\beta = 8.68°$

Problem 3/211

3/212 The small marble is projected with a velocity of 3 m/s in a direction 15° from the horizontal y-direction on the smooth inclined plane. Calculate the magnitude v of its velocity after 2 seconds.

Problem 3/212

3/213 The 500-g sphere is moving in the horizontal x-y plane with a velocity of 3 m/s in the direction shown and encounters a steady flow of air in the x-direction. If the air stream exerts an essentially constant force of 0.9 N on the sphere in the x-direction, determine the time t required for the sphere to cross the y-axis again.

Ans. $t = 1.667$ s

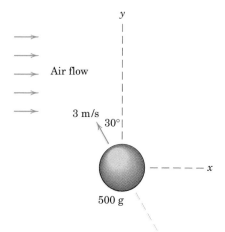

Problem 3/213

3/214 The 40-kg boy has taken a running jump from the upper surface and lands on his 5-kg skateboard with a velocity of 5 m/s in the plane of the figure as shown. If his impact with the skateboard has a time duration of 0.05 s, determine the final speed v along the horizontal surface and the total normal force N exerted by the surface on the skateboard wheels during the impact.

Problem 3/214

3/215 The 10-kg block is resting on the horizontal surface when the force T is applied to it for 7 seconds. The variation of T with time is shown. Calculate the maximum velocity reached by the block and the total time Δt during which the block is in motion. The coefficients of static and kinetic friction are both 0.50.

Ans. $v_{\text{max}} = 5.19$ m/s, $\Delta t = 5.54$ s

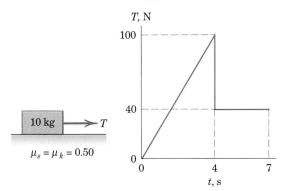

Problem 3/215

3/216 A smooth 3-kg particle traveling at 10 m/s encounters a circular guide which changes its direction by 45° in the horizontal plane as indicated. If the radius of the circular path is 25 m, determine by the principle of impulse and momentum the average force R_{av} exerted on the particle during the motion from A to B. Compare your result with the exact instantaneous value obtained from $\Sigma F = ma$.

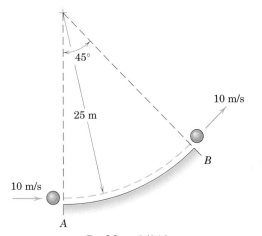

Problem 3/216

3/217 A spacecraft with a mass of 260 kg is moving with a velocity $u = 30\,000$ km/h in the fixed x-direction remote from any attracting celestial body. The spacecraft is spin-stabilized and rotates about the z-axis at the constant rate $\dot{\theta} = \pi/10$ rad/s. During a quarter of a revolution from $\theta = 0$ to $\theta = \pi/2$, a jet is activated which produces a thrust $T = 600$ N of constant magnitude. Determine the y-component of the velocity of the spacecraft when $\theta = \pi/2$. Neglect the small change in mass due to the loss of exhaust gas through the control nozzle and treat the spacecraft as a particle.

Ans. $v_y = 7.35$ m/s

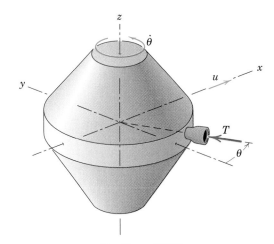

Problem 3/217

3/218 The loaded mine skip has a mass of 3 Mg. The hoisting drum produces a tension T in the cable according to the time schedule shown. If the skip is at rest against A when the drum is activated, determine the speed v of the skip when $t = 6$ s. Friction loss may be neglected.

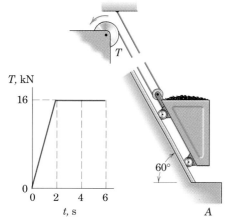

Problem 3/218

▶3/219 The 400-kg ram of a pile driver is designed to fall 1.5 m from rest and strike the top of a 300-kg pile partially embedded in the ground. The deeper the penetration, the greater is the tendency for the ram to rebound as a result of the impact. Calculate the velocity v of the pile immediately after impact for the following three conditions: (*a*) initial resistance to penetration is small at the outset, and the ram is observed to move with the pile immediately after impact; (*b*) resistance to penetration has increased, and the ram is seen to have zero velocity immediately after impact; (*c*) resistance to penetration is high, and the ram is seen to rebound to a height of 100 mm above the point of impact. Why is it permissible to neglect the impulse of the weight of the ram during impact?

$\qquad$ *Ans.* (*a*) $v = 3.10$ m/s, (*b*) $v = 7.23$ m/s
$\qquad\quad$ (*c*) $v = 9.10$ m/s

Problem 3/219

▶3/220 Two barges, each with a displacement (mass) of 500 metric tons, are loosely moored in calm water. A stunt driver starts his 1500-kg car from rest at A, drives along the deck, and leaves the end of the 15° ramp at a speed of 50 km/h relative to the barge and ramp. The driver successfully jumps the gap and brings his car to rest relative to barge 2 at B. Calculate the velocity v_2 imparted to barge 2 just after the car has come to rest on the barge. Neglect the resistance of the water to motion at the low velocities involved.

$\qquad\qquad\qquad$ *Ans.* $v_2 = 40.0$ mm/s

Problem 3/220

3/10 ANGULAR IMPULSE AND ANGULAR MOMENTUM

In addition to the equations of linear impulse and linear momentum, there exists a parallel set of equations for angular impulse and angular momentum. First, we define the term angular momentum. In Fig. 3/11a is shown a particle P of mass m moving along a curve in space. The particle is located by its position vector $\mathbf{r}$ with respect to a convenient origin O of fixed coordinates x-y-z. The velocity of the particle is $\mathbf{v} = \dot{\mathbf{r}}$, and its linear momentum is $\mathbf{G} = m\mathbf{v}$. The *moment* of the *linear momentum vector* $m\mathbf{v}$ about the origin O is defined as the *angular momentum* $\mathbf{H}_O$ of P about O and is given by the cross-product relation for the moment of a vector

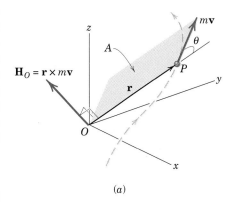

$$\boxed{\mathbf{H}_O = \mathbf{r} \times m\mathbf{v}} \qquad (3/25)$$

(a)

The angular momentum then is a vector perpendicular to plane A defined by $\mathbf{r}$ and $\mathbf{v}$. The sense of $\mathbf{H}_O$ is clearly defined by the right-hand rule for cross products.

The scalar components of angular momentum may be obtained from the expansion

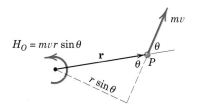

$$\mathbf{H}_O = \mathbf{r} \times m\mathbf{v} = m(v_z y - v_y z)\mathbf{i} + m(v_x z - v_z x)\mathbf{j} + m(v_y x - v_x y)\mathbf{k}$$

$$\mathbf{H}_O = m \begin{vmatrix} \mathbf{i} & \mathbf{j} & \mathbf{k} \\ x & y & z \\ v_x & v_y & v_z \end{vmatrix} \qquad (3/26)$$

View in plane A

(b)

Figure 3/11

so that

$$H_x = m(v_z y - v_y z) \qquad H_y = m(v_x z - v_z x) \qquad H_z = m(v_y x - v_x y)$$

Each of these expressions for angular momentum may be checked easily from Fig. 3/12, which shows the three linear-momentum components, by taking the moments of these components about the respective axes.

To help visualize angular momentum, we show in Fig. 3/11b a two-dimensional representation in plane A of the vectors shown in the a-part of the figure. The motion is viewed in plane A defined by $\mathbf{r}$ and $\mathbf{v}$. The magnitude of the moment of $m\mathbf{v}$ about O is simply the linear momentum mv times the moment arm $r \sin \theta$ or $mvr \sin \theta$, which is the magnitude of the cross product $\mathbf{H}_O = \mathbf{r} \times m\mathbf{v}$.

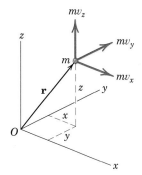

Figure 3/12

Angular momentum is the moment of linear momentum and must not be confused with linear momentum. In SI units, angular momentum has the units $\text{kg} \cdot (\text{m/s}) \cdot \text{m} = \text{kg} \cdot \text{m}^2/\text{s} = \text{N} \cdot \text{m} \cdot \text{s}$. We are now ready to relate the moment of the forces acting on the particle P to its angular momentum. If $\Sigma \mathbf{F}$ represents the resultant of *all* forces acting on the particle P of Fig. 3/11, the moment $\mathbf{M}_O$ about the origin O is the vector cross product

$$\Sigma \mathbf{M}_O = \mathbf{r} \times \Sigma \mathbf{F} = \mathbf{r} \times m\dot{\mathbf{v}}$$

where Newton's second law $\Sigma \mathbf{F} = m\dot{\mathbf{v}}$ has been substituted. We now differentiate Eq. 3/25 with time, using the rule for the differentiation of a cross product (see item 9, Art. C/7, Appendix C) and get

$$\dot{\mathbf{H}}_O = \dot{\mathbf{r}} \times m\mathbf{v} + \mathbf{r} \times m\dot{\mathbf{v}} = \mathbf{v} \times m\mathbf{v} + \mathbf{r} \times m\dot{\mathbf{v}}$$

The term $\mathbf{v} \times m\mathbf{v}$ is zero since the cross product of parallel vectors is identically zero. Substitution into the expression for $\Sigma \mathbf{M}_O$ gives

$$\boxed{\Sigma \mathbf{M}_O = \dot{\mathbf{H}}_O} \qquad (3/27)$$

Equation 3/27 states that the *moment about the fixed point O of all forces acting on m equals the time rate of change of angular momentum of m about O.* This relation, particularly when extended to a system of particles, rigid or nonrigid, constitutes one of the most powerful tools of analysis in all of dynamics. Equation 3/27 is a vector equation with scalar components

$$\Sigma M_{O_x} = \dot{H}_{O_x} \qquad \Sigma M_{O_y} = \dot{H}_{O_y} \qquad \Sigma M_{O_z} = \dot{H}_{O_z} \qquad (3/28)$$

Equation 3/27 gives the instantaneous relation between the moment and the time rate of change of angular momentum. To obtain the effect of the moment $\Sigma \mathbf{M}_O$ on the angular momentum of the particle over a finite period of time, Eq. 3/27 may be integrated from time t_1 to time t_2. Multiplying the equation by dt gives $\Sigma \mathbf{M}_O \, dt = d\mathbf{H}_O$, which we integrate to obtain

$$\boxed{\int_{t_1}^{t_2} \Sigma \mathbf{M}_O \, dt = \mathbf{H}_{O_2} - \mathbf{H}_{O_1} = \Delta \mathbf{H}_O} \qquad (3/29)$$

where $\mathbf{H}_{O_2} = \mathbf{r}_2 \times m\mathbf{v}_2$ and $\mathbf{H}_{O_1} = \mathbf{r}_1 \times m\mathbf{v}_1$. The product of moment and time is defined as *angular impulse*, and Eq. 3/29 states that the *total angular impulse on m about the fixed point O equals the corresponding change in angular momentum of m about O.*

Alternatively, we may write Eq. 3/29 as

$$\boxed{\mathbf{H}_{O_1} + \int_{t_1}^{t_2} \Sigma \mathbf{M}_O \, dt = \mathbf{H}_{O_2}} \qquad (3/29a)$$

which states that the initial angular momentum of the particle plus the angular impulse applied to it equals its final angular momentum. The units of angular impulse are clearly those of angular momentum, which are $N \cdot m \cdot s$ or $kg \cdot m^2/s$ in SI units and ft-lb-sec in U.S. customary units.

As in the case of linear impulse and linear momentum, the equation of angular impulse and angular momentum is a vector equation where changes in direction as well as magnitude may occur during the interval of integration. Under these conditions, it is necessary to express $\Sigma \mathbf{M}_O$ and $\mathbf{H}_O$ in component form and then combine the integrated components. Thus, the x-component of Eq. 3/29 becomes

$$\int_{t_1}^{t_2} \Sigma M_{O_x}\, dt = (H_{O_x})_2 - (H_{O_x})_1$$

$$= m[(v_z y - v_y z)_2 - (v_z y - v_y z)_1]$$

where the subscripts 1 and 2 refer to the values of the respective quantities at times t_1 and t_2. Similar expressions exist for the y- and z-components of the angular-momentum integral.

The foregoing angular-impulse and angular-momentum relations have been developed in their general three-dimensional sense. Most of the applications with which we shall be concerned, on the other hand, can be analyzed as plane-motion problems where moments are taken about a single axis normal to the plane of motion. In this event, the angular momentum may change magnitude and sense, but the direction of the vector remains unaltered. Thus, for a particle of mass m moving along a curved path in the x-y plane, Fig. 3/13, the angular momenta about O at points 1 and 2 have the magnitudes $H_{O_1} = |\mathbf{r}_1 \times m\mathbf{v}_1| = mv_1 d_1$ and $H_{O_2} = |\mathbf{r}_2 \times m\mathbf{v}_2| = mv_2 d_2$, respectively. In the illustration both H_{O_1} and H_{O_2} are represented in the counterclockwise sense according to the direction of

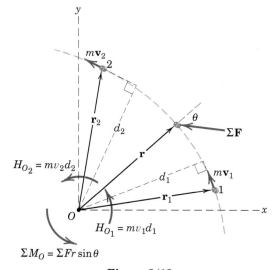

$$\Sigma M_O = \Sigma Fr \sin \theta$$

Figure 3/13

the moment of the linear momentum. The scalar form of Eq. 3/29 applied to the motion between points 1 and 2 during the time interval t_1 to t_2 becomes

$$\int_{t_1}^{t_2} \Sigma M_O \, dt = H_{O_2} - H_{O_1} \quad \text{or} \quad \int_{t_1}^{t_2} \Sigma Fr \sin \theta \, dt = mv_2 d_2 - mv_1 d_1$$

This example should help clarify the relation between the scalar and vector forms of the angular impulse-momentum relations.

Equations 3/21 and 3/27 add no new basic information since they are merely alternative forms of Newton's second law. However, we will discover in subsequent chapters that the motion equations expressed in terms of the time rate of change of momentum are applicable to the motion of rigid and nonrigid bodies and constitute a very general and powerful approach to many problems. The full generality of Eq. 3/27 is usually not required to describe the motion of a single particle or the plane motion of rigid bodies, but it does find important use in the analysis of the space motion of rigid bodies introduced in Chapter 7.

Conservation of angular momentum. If the resultant moment about a fixed point O of all forces acting on a particle is zero during an interval of time, Eq. 3/27 requires that its angular momentum $\mathbf{H}_O$ about that point remain constant. In this case, the angular momentum of the particle is said to be *conserved*. Angular momentum may be conserved about one axis but not about another axis. A careful examination of the free-body diagram of the particle will disclose whether the moment of the resultant force on the particle about a fixed point is zero, in which case, the angular momentum about that point is unchanged (conserved).

Consider now the motion of two particles a and b that interact during an interval of time. If the interactive forces $\mathbf{F}$ and $-\mathbf{F}$ between them are the only unbalanced forces acting on the particles during the interval, it follows that the moments of the equal and opposite forces about any fixed point O not on their line of action are equal and opposite. If we apply Eq. 3/29 to particle a and then to particle b and add the two equations, we obtain $\Delta \mathbf{H}_a + \Delta \mathbf{H}_b = \mathbf{0}$ (where all angular momenta are referred to point O). Thus, the total angular momentum for the system of the two particles remains constant during the interval, and we write

$$\boxed{\Delta \mathbf{H}_O = \mathbf{0} \quad \text{or} \quad \mathbf{H}_{O_1} = \mathbf{H}_{O_2}} \qquad (3/30)$$

which expresses the *principle of conservation of angular momentum.*

Sample Problem 3/22

The small 2-kg block slides on a smooth horizontal surface under the action of the force in the spring and a force F. The angular momentum of the block about O varies with time as shown in the graph. When $t = 6.5$ s, it is known that $r = 150$ mm and $\beta = 60°$. Determine F for this instant.

Solution. The only moment of the forces about O is due to F since the spring force passes through O. Thus, $\Sigma M_O = Fr \sin \beta$. From the graph the time rate of change of H_O for $t = 6.5$ s is very nearly $(8 - 4)/(7 - 6)$ or $\dot{H}_O = 4$ kg·m²/s. The moment-angular momentum relation gives

① $[\Sigma M_O = \dot{H}_O]$ $\qquad$ $F\,(0.150) \sin 60° = 4$ $\qquad$ $F = 30.8$ N $\qquad$ *Ans.*

① We do not need vector notation here since we have plane motion where the direction of the vector $\mathbf{H}_O$ does not change.

Sample Problem 3/23

A small mass particle is given an initial velocity $\mathbf{v}_0$ tangent to the horizontal rim of a smooth hemispherical bowl at a radius r_0 from the vertical centerline, as shown at point A. As the particle slides past point B, a distance h below A and a distance r from the vertical centerline, its velocity $\mathbf{v}$ makes an angle θ with the horizontal tangent to the bowl through B. Determine θ.

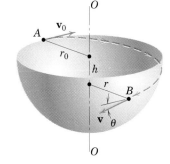

Solution. The forces on the particle are its weight and the normal reaction exerted by the smooth surface of the bowl. Neither force exerts a moment about the axis $O\text{-}O$, so that angular momentum is conserved about that axis. Thus,

① $[(H_O)_1 = (H_O)_2]$ $\qquad$ $mv_0 r_0 = mvr \cos \theta$

Also, energy is conserved so that $E_1 = E_2$. Thus

$[T_1 + V_{g_1} = T_2 + V_{g_2}]$ $\qquad$ $\tfrac{1}{2}mv_0^2 + mgh = \tfrac{1}{2}mv^2 + 0$

$$v = \sqrt{v_0^2 + 2gh}$$

Eliminating v and substituting $r^2 = r_0^2 - h^2$ give

$$v_0 r_0 = \sqrt{v_0^2 + 2gh}\sqrt{r_0^2 - h^2} \cos \theta$$

$$\theta = \cos^{-1} \frac{1}{\sqrt{1 + \dfrac{2gh}{v_0^2}}\sqrt{1 - \dfrac{h^2}{r_0^2}}} \qquad Ans.$$

① The angle θ is measured in the plane tangent to the hemispherical surface at B.

PROBLEMS

Introductory problems

3/221 Determine the magnitude H_O of the angular momentum of the 2-kg sphere about point O (a) by using the vector definition of angular momentum and (b) by using an equivalent scalar approach. The center of the sphere lies in the x-y plane.

Ans. $H_O = 128.7$ kg·m²/s

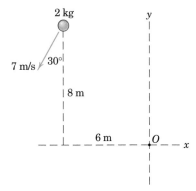

Problem 3/221

3/222 The 3-kg sphere moves in the x-y plane and has the indicated velocity at a particular instant. Determine its (a) linear momentum, (b) angular momentum about point O, and (c) kinetic energy.

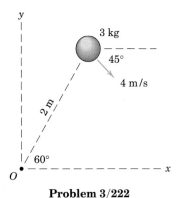

Problem 3/222

3/223 At a certain instant, the particle of mass m has the position and velocity shown in the figure, and it is acted upon by the force $\mathbf{F}$. Determine its angular momentum about point O and the time rate of change of this angular momentum.

Ans. $\mathbf{H}_O = mv(-c\mathbf{i} + a\mathbf{k})$, $\dot{\mathbf{H}}_O = F(b\mathbf{i} - a\mathbf{j})$

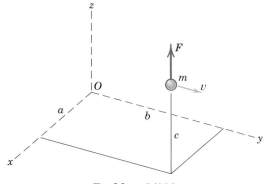

Problem 3/223

3/224 The small spheres, which have the masses and initial velocities shown in the figure, strike and become attached to the spiked ends of the rod which is freely pivoted at O and is initially at rest. Determine the angular velocity ω of the assembly after impact. Neglect the mass of the rod.

Problem 3/224

3/225 The assembly starts from rest and reaches an angular speed of 150 rev/min under the action of a 20-N force T applied to the string for t seconds. Determine t. Neglect friction and all masses except those of the four 3-kg spheres, which may be treated as particles.

Ans. $t = 15.08$ s

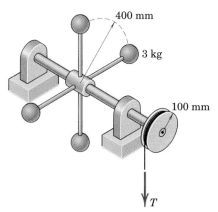

Problem 3/225

3/226 A particle of mass m moves with negligible friction on a horizontal surface and is connected to a light spring fastened at O. At position A the particle has the velocity $v_A = 4$ m/s. Determine the velocity v_B of the particle as it passes position B.

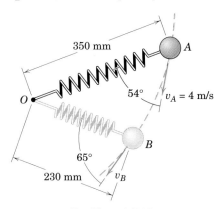

Problem 3/226

Representative problems

3/227 Each of the four spheres of mass m is treated as a particle. Spheres A and B are mounted on a light rod and are rotating initially with an an-

gular velocity ω_0 about a vertical axis through O. Spheres C and D are also mounted on a light rod which is pivoted independently about O and is initially at rest. Assembly AB contacts CD where slots in A and B allow engagement and latching to CD in the dashed position shown. Both units then rotate with a common angular velocity ω. Frictional resistance is negligible. Determine expressions for ω and the percentage loss n of kinetic energy.

Ans. $\omega = \omega_0/5$, $n = 80\%$

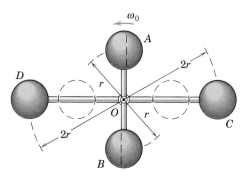

Problem 3/227

3/228 A small 110-g particle is projected with a horizontal velocity of 2 m/s into the top A of the smooth circular guide fixed in the vertical plane. Calculate the time rate of change $\dot{H}_B$ of angular momentum about point B when the particle passes the bottom of the guide at C.

Problem 3/228

3/229 The central attractive force F on an earth satellite can have no moment about the center O of the earth. For the particular elliptical orbit with major and minor axes as shown, a satellite will have a velocity of 33 880 km/h at the perigee altitude of 390 km. Determine the velocity of the satellite at point B and at apogee A. The radius of the earth is 6371 km.

$$Ans. \ v_B = 19\ 540 \text{ km/h}$$
$$v_A = 11\ 300 \text{ km/h}$$

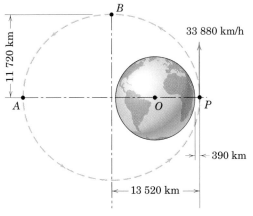

Problem 3/229

3/230 A small 0.1-kg particle is given a velocity of 2 m/s on the horizontal x-y plane and is guided by the fixed curved rail. Friction is negligible. As the particle crosses the y-axis at A, its velocity is in the x-direction, and as it crosses the x-axis at B, its velocity makes a 60° angle with the x-axis. The radius of curvature of the path at B is 500 mm. Determine the time rate of change of angular momentum H_O of the particle about the z-axis through O at both A and B.

Problem 3/230

3/231 The two spheres of equal mass m are able to slide along the horizontal rotating rod. If they are initially latched in position a distance r from the rotating axis with the assembly rotating freely with an angular velocity ω_0, determine the new angular velocity ω after the spheres are released and finally assume positions at the ends of the rod at a radial distance of $2r$. Also find the fraction n of the initial kinetic energy of the system which is lost. Neglect the small mass of the rod and shaft.

$$Ans. \ \omega = \omega_0/4, \ n = 3/4$$

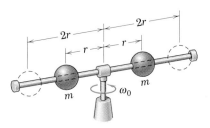

Problem 3/231

3/232 The 0.2-kg ball and its supporting cord are revolving about the vertical axis on the fixed smooth conical surface with an angular velocity of 4 rad/s. The ball is held in the position $b = 300$ mm by the tension T in the cord. If the distance b is reduced to the constant value of 200 mm by increasing the tension T in the cord, compute the new angular velocity ω and the work $U'_{1\text{-}2}$ done on the system by T.

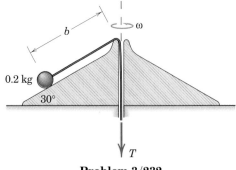

Problem 3/232

3/233 The particle of mass m is launched from point O with a horizontal velocity $\mathbf{u}$ at time $t = 0$. Determine its angular momentum $\mathbf{H}_O$ relative to point O as a function of time.

Ans. $\mathbf{H}_O = -\frac{1}{2}mgut^2\mathbf{k}$

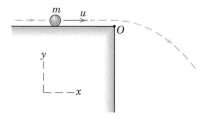

Problem 3/233

3/234 The 0.02-kg particle moves along the dashed trajectory shown and has indicated velocities at positions A and B. Calculate the time average of the moment about O of the resultant force P acting on the particle during the 0.5 second required for it to go from A to B.

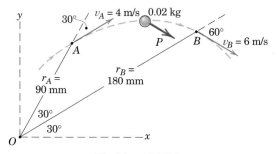

Problem 3/234

3/235 At the point A of closest approach to the sun, a comet has a velocity $v_A = 57.45(10^3)$ m/s. Determine the radial and transverse components of its velocity v_B at point B, where the radial distance from the sun is $120.7(10^6)$ km.

Ans. $v_r = 27.1(10^3)$ m/s, $v_\theta = 38.3(10^3)$ m/s

Problem 3/235

3/236 The simple pendulum of mass m and length l is released from rest at $\theta = 0$. Using only the principle of angular impulse and momentum, determine the expression for $\ddot{\theta}$ in terms of θ and the velocity v of the pendulum at $\theta = 90°$. Compare this approach with a solution by the work-energy principle.

Problem 3/236

3/237 A particle is released on the smooth inside wall of a cylindrical tank at A with a velocity v_0 that makes an angle β with the horizontal tangent. When the particle reaches a point B a distance h below A, determine the expression for the angle θ made by its velocity with the horizontal tangent at B.

$$Ans. \quad \theta = \cos^{-1} \frac{\cos \beta}{\sqrt{1 + \dfrac{2gh}{v_0^2}}}$$

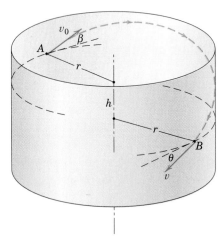

Problem 3/237

3/238 A pendulum consists of two 3.2-kg concentrated masses positioned as shown on a light but rigid bar. The pendulum is swinging through the vertical position with a clockwise angular velocity $\omega = 6$ rad/s when a 50-g bullet traveling with velocity $v = 300$ m/s in the direction shown strikes the lower mass and becomes embedded in it. Calculate the angular velocity ω' which the pendulum has immediately after impact and find the maximum angular deflection θ of the pendulum.

Problem 3/238

3/239 The small particle of mass m and its restraining cord are spinning with an angular velocity ω on the horizontal surface of a smooth disk, shown in section. As the force F is slightly relaxed, r increases and ω changes. Determine the rate of change of ω with respect to r and show that the work done by F during a movement dr equals the change in kinetic energy of the particle.

$$Ans. \quad \frac{d\omega}{dr} = -\frac{2\omega}{r}$$

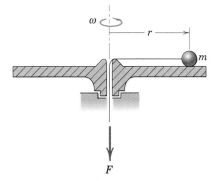

Problem 3/239

3/240 The 0.7-kg sphere moves in a horizontal plane and is controlled by a cord which is reeled in and out below the table in such a way that the center of the sphere is confined to the path given by

$(x^2/2.25) + (y^2/1.44) = 1$ where x and y are in meters. If the speed of the sphere is $v_A = 2$ m/s as it passes point A, determine the tension T_B in the cord as the sphere passes point B. Friction is negligible.

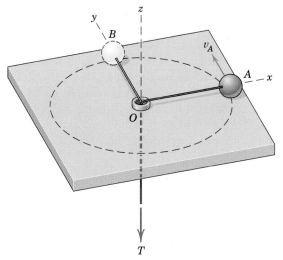

Problem 3/240

▶3/241 The assembly of two 5-kg spheres is rotating freely about the vertical axis at 40 rev/min with $\theta = 90°$. If the force F which maintains the given position is increased to raise the base collar and reduce θ to 60°, determine the new angular velocity ω. Also determine the work U done by F in changing the configuration of the system. Assume that the mass of the arms and collars is negligible.

Ans. $\omega = 3.00$ rad/s, $U = 5.34$ J

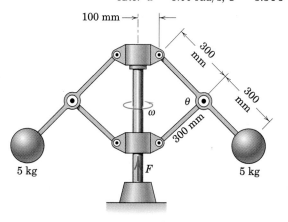

Problem 3/241

▶3/242 The small particle of mass m is given an initial high velocity in the horizontal plane and winds its cord around the fixed vertical shaft of radius a. All motion occurs essentially in the horizontal plane. If the angular velocity of the cord is ω_0 when the distance from the particle to the tangency point is r_0, determine the angular velocity ω of the cord and its tension T after it has turned through an angle θ. Does either of the principles of conservation of momentum apply?

$$Ans.\ \omega = \frac{\omega_0}{1 - a\theta/r_0},\ T = mr_0\omega_0\omega$$

Problem 3/242

SECTION D. SPECIAL APPLICATIONS

3/11 INTRODUCTION

The basic principles and methods of particle kinetics were developed and illustrated in the first three sections of this chapter. This treatment included the direct use of Newton's second law, the equations of work and energy, and the equations of impulse and momentum. Special attention was paid to the kind of problem for which each of the approaches was most appropriate.

There are several topics of specialized interest in particle kinetics that will be briefly treated in Section D. These topics involve further extension and application of the fundamental principles of dynamics, and their study will help to broaden one's background in mechanics.

3/12 IMPACT

The principles of impulse and momentum find important use in describing the behavior of colliding bodies. *Impact* refers to the collision between two bodies and is characterized by the generation of relatively large contact forces that act over a very short interval of time. Before we begin treatment of the subject, it should be stated that an impact is a very complex event involving material deformation and recovery and the generation of heat and sound. Small changes in the impact conditions may cause large changes in the impact process and thus in the conditions immediately following the impact. Hence, we must be careful not to place excessive reliance on the results of impact calculations.

(a) Direct central impact. As an introduction to impact, we consider the collinear motion of two spheres of masses m_1 and m_2, Fig. 3/14a, traveling with velocities v_1 and v_2. If v_1 is greater than v_2, collision occurs with the contact forces directed along the line of centers. This condition is called *direct central impact*. Following initial contact, a short period of increasing deformation takes place

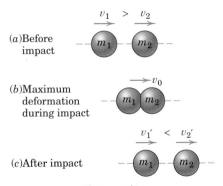

(a) Before impact

(b) Maximum deformation during impact

(c) After impact

Figure 3/14

until the contact area between the spheres ceases to increase. At this instant, both spheres, Fig. 3/14b, are moving with the same velocity v_0. During the remainder of contact, a period of restoration occurs during which the contact area decreases to zero. The final condition is shown in the c-part of the figure where the spheres now have new velocities v_1' and v_2', where of necessity v_1' is less than v_2'. All velocities are arbitrarily assumed positive to the right, so that with this scalar notation a velocity to the left would carry a negative sign. If the impact is not overly severe and if the spheres are highly elastic, they will regain their original shape following the restoration. With a more severe impact and with less elastic bodies, a permanent deformation may result.

Inasmuch as the contact forces are equal and opposite during impact, the linear momentum of the system remains unchanged, as discussed in Art. 3/9. Thus, we apply the law of conservation of linear momentum and write

$$m_1 v_1 + m_2 v_2 = m_1 v_1' + m_2 v_2' \tag{3/31}$$

It is assumed that any forces other than the large internal forces of contact that may act on the spheres during impact are relatively small and produce negligible impulses compared with the impulse associated with each internal impact force. In addition, we assume that no appreciable position change occurs during the short duration of the impact.

For given masses and initial conditions, the momentum equation contains two unknowns, v_1' and v_2'. Clearly, an additional relationship is required before the final velocities can be found. This relationship must reflect the capacity of the contacting bodies to recover from the impact and can be expressed by the ratio e of the magnitude of the restoration impulse to the magnitude of the deformation impulse. This ratio is called the *coefficient of restitution*. If F_r and F_d represent the magnitudes of the contact forces during the restoration and deformation periods, respectively, as shown in Fig. 3/15, for particle 1 the definition of e together with the impulse-momentum equation give us

$$e = \frac{\int_{t_0}^{t} F_r \, dt}{\int_{0}^{t_0} F_d \, dt} = \frac{m_1[-v_1' - (-v_0)]}{m_1[-v_0 - (-v_1)]} = \frac{v_0 - v_1'}{v_1 - v_0}$$

Similarly, for particle 2 we have

$$e = \frac{\int_{t_0}^{t} F_r \, dt}{\int_{0}^{t_0} F_d \, dt} = \frac{m_2(v_2' - v_0)}{m_2(v_0 - v_2)} = \frac{v_2' - v_0}{v_0 - v_2}$$

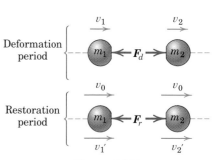

Deformation period

Restoration period

Figure 3/15

We are careful in these equations to express the change of momentum (and hence Δv) in the same direction as the impulse (and hence the force). The time for the deformation is taken as t_0 and the total time of contact is t. Eliminating v_0 between the two expressions for e gives us

$$e = \frac{v_2' - v_1'}{v_1 - v_2} = \frac{|\text{relative velocity of separation}|}{|\text{relative velocity of approach}|} \quad (3/32)$$

In addition to the initial conditions, if e is known for the impact condition at hand, then Eqs. 3/31 and 3/32 give us two equations in the two unknown final velocities v_1' and v_2'.

Impact phenomena are almost always accompanied by energy loss, which may be calculated by subtracting the kinetic energy of the system just after impact from that just before impact. Energy is lost through the generation of heat during the localized inelastic deformation of the material, through the generation and dissipation of elastic stress waves within the bodies, and through the generation of sound energy.

According to this classical theory of impact, the value $e = 1$ means that the capacity of the two particles to recover equals their tendency to deform. This condition is one of *elastic impact* with no energy loss. The value $e = 0$, on the other hand, describes *inelastic* or *plastic impact* where the particles cling together after collision and the loss of energy is a maximum. All impact conditions lie somewhere between these two extremes. Also it should be noted that a coefficient of restitution must be associated with a *pair* of contacting bodies.

The coefficient of restitution is frequently considered a constant for given geometries and a given combination of contacting materials. Actually, it depends on the impact velocity and approaches unity as the impact velocity approaches zero as shown schematically in Fig. 3/16. A handbook value for e is generally unreliable.

Figure 3/16

(b) Oblique central impact. We now extend the relationships developed for direct central impact to the case where the initial and final velocities are not parallel, Fig. 3/17. Here spherical particles of mass m_1 and m_2 have initial velocities $\mathbf{v}_1$ and $\mathbf{v}_2$ in the same plane and approach each other on a collision course, as shown in the *a*-part of the figure. The directions of the velocity vectors are measured from the direction tangent to the contacting surfaces, Fig. 3/17*b*. Thus, the initial velocity components along the *t*- and *n*-axes are $(v_1)_n = -v_1 \sin \theta_1$, $(v_1)_t = v_1 \cos \theta_1$, $(v_2)_n = v_2 \sin \theta_2$, and $(v_2)_t = v_2 \cos \theta_2$. Note that $(v_1)_n$ is a negative quantity for the particular coordinate system and initial velocities shown. The final rebound conditions are shown in the *c*-part of the figure. The impact forces are $\mathbf{F}$ and $-\mathbf{F}$, as seen in the *d*-part of the figure. They vary

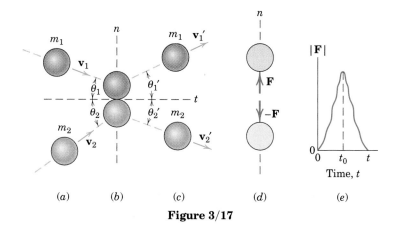

(a) (b) (c) (d) (e)

Figure 3/17

from zero to their peak value during the deformation portion of the impact and back again to zero during the restoration period, as indicated in the *e*-part of the figure where t is the duration of the impact interval.

For given initial conditions of m_1, m_2, $(v_1)_n$, $(v_1)_t$, $(v_2)_n$, and $(v_2)_t$, there will be four unknowns, namely, $(v_1')_n$, $(v_1')_t$, $(v_2')_n$, and $(v_2')_t$. The four needed equations are:

(1) Momentum of the system is conserved in the *n*-direction, giving

$$m_1(v_1)_n + m_2(v_2)_n = m_1(v_1')_n + m_2(v_2')_n$$

(2) and (3) The momentum for each particle is conserved in the *t*-direction since there is no impulse on either particle in the *t*-direction. Thus,

$$m_1(v_1)_t = m_1(v_1')_t$$

$$m_2(v_2)_t = m_2(v_2')_t$$

(4) The coefficient of restitution, as in the case of direct central impact, is the positive ratio of the recovery impulse to the deformation impulse. Equation 3/32 applies, then, to the velocity components in the *n*-direction, and for the notation adopted with Fig. 3/17, we have

$$e = \frac{(v_2')_n - (v_1')_n}{(v_1)_n - (v_2)_n}$$

Once the four final velocity components are found, the angles θ_1' and θ_2' of Fig. 3/17 may be easily determined.

Sample Problem 3/24

The ram of a pile driver has a mass of 800 kg and is released from rest 2 m above the top of the 2400-kg pile. If the ram rebounds to a height of 0.1 m after impact with the pile, calculate (a) the velocity v_p' of the pile immediately after impact, (b) the coefficient of restitution e, and (c) the percentage loss of energy due to the impact.

2 m drop

0.1 m rebound

Solution. Conservation of energy during free fall gives the initial and final velocities of the ram from $v = \sqrt{2gh}$. Thus,

$$v_r = \sqrt{2(9.81)(2)} = 6.26 \text{ m/s} \qquad v_r' = \sqrt{2(9.81)(0.1)} = 1.401 \text{ m/s}$$

(a) Conservation of momentum $(G_1 = G_2)$ for the system of the ram and pile gives ①

$$800(6.26) + 0 = 800(-1.40) + 2400v_p' \qquad v_p' = 2.55 \text{ m/s} \quad Ans.$$

(b) The coefficient of restitution yields

$$e = \frac{|\text{rel. vel. separation}|}{|\text{rel. vel. approach}|} \qquad e = \frac{2.55 + 1.401}{6.26 + 0} = 0.631 \quad Ans.$$

(c) The kinetic energy of the system just before impact is the same as the potential energy of the ram above the pile and is

$$T = V_g = mgh = 800(9.81)(2) = 15\ 700 \text{ J}$$

The kinetic energy T' just after impact is

$$T' = \tfrac{1}{2}(800)(1.40)^2 + \tfrac{1}{2}(2400)(2.55)^2 = 8620 \text{ J}$$

The percentage loss of energy is, therefore,

$$\frac{15\ 700 - 8620}{15\ 700}(100) = 45.1\% \qquad Ans.$$

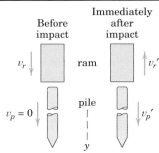

Immediately
Before   after
impact   impact

v_r ram v_r'

$v_p = 0$ pile v_p'

y

① The impulses of the weights of the ram and pile are very small compared with the impulses of the impact forces and thus are neglected during the impact.

Sample Problem 3/25

A ball is projected onto the heavy plate with a velocity of 16 m/s at the 30° angle shown. If the effective coefficient of restitution is 0.5, compute the rebound velocity v' and its angle θ'.

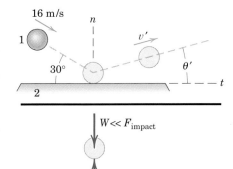

16 m/s

n

1

v'

30°

θ'

2

t

Solution. Let the ball be denoted body 1 and the plate body 2. The mass of the heavy plate may be considered infinite and its corresponding velocity zero after impact. The coefficient of restitution is applied to the velocity components normal to the plate in the direction of the impact force and gives

$$e = \frac{(v_2')_n - (v_1')_n}{(v_1)_n - (v_2)_n} \qquad 0.5 = \frac{0 - (v_1')_n}{-16 \sin 30° - 0} \qquad (v_1')_n = 4 \text{ m/s}$$ ①

Momentum of the ball in the t-direction is unchanged since, with assumed smooth surfaces, there is no force acting on the ball in that direction. Thus,

$$m(v_1)_t = m(v_1')_t \qquad (v_1')_t = (v_1)_t = 16 \cos 30° = 13.86 \text{ m/s}$$

The rebound velocity v' and its angle θ' are then

$$v' = \sqrt{(v_1')_n{}^2 + (v_1')_t{}^2} = \sqrt{4^2 + 13.86^2} = 14.42 \text{ m/s} \qquad Ans.$$

$$\theta' = \tan^{-1}\left(\frac{(v_1')_n}{(v_1')_t}\right) = \tan^{-1}\left(\frac{4}{13.86}\right) = 16.10° \qquad Ans.$$

$W \ll F_{\text{impact}}$

F_{impact}

① We observe here that for infinite mass there is no way of applying the principle of conservation of momentum for the system in the n-direction. From the free-body diagram of the ball during impact, we note that the impulse of the weight W is neglected since W is very small compared with the impact force.

Sample Problem 3/26

Spherical particle 1 has a velocity $v_1 = 6$ m/s in the direction shown and collides with spherical particle 2 of equal mass and diameter and initially at rest. If the coefficient of restitution for these conditions is $e = 0.6$, determine the resulting motion of each particle following impact. Also calculate the percentage loss of energy due to the impact.

Solution. The geometry at impact indicates that the normal n to the contacting surfaces makes an angle $\theta = 30°$ with the direction of $\mathbf{v}_1$, as indicated in the figure. Thus, the initial velocity components are ① $(v_1)_n = v_1 \cos 30° = 6 \cos 30° = 5.20$ m/s, $(v_1)_t = v_1 \sin 30° = 6 \sin 30° = 3$ m/s, and $(v_2)_n = (v_2)_t = 0$.

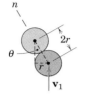

Momentum conservation for the two-particle system in the n-direction gives

$$m_1(v_1)_n + m_2(v_2)_n = m_1(v_1')_n + m_2(v_2')_n$$

or, with $m_1 = m_2$,

$$5.20 + 0 = (v_1')_n + (v_2')_n \qquad (a)$$

The coefficient-of-restitution relationship is

$$e = \frac{(v_2')_n - (v_1')_n}{(v_1)_n - (v_2)_n} \qquad 0.6 = \frac{(v_2')_n - (v_1')_n}{5.20 - 0} \qquad (b)$$

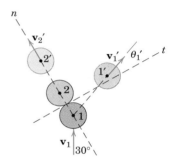

② Simultaneous solution of Eqs. a and b yields

$$(v_1')_n = 1.039 \text{ m/s} \qquad (v_2')_n = 4.16 \text{ m/s}$$

Conservation of momentum for each particle holds in the t-direction because, with assumed smooth surfaces, there is no force in the t-direction. Thus for particles 1 and 2, we have

$$m_1(v_1)_t = m_1(v_1')_t \qquad (v_1')_t = (v_1)_t = 3 \text{ m/s}$$

$$m_2(v_2)_t = m_2(v_2')_t \qquad (v_2')_t = (v_2)_t = 0$$

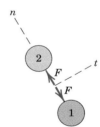

The final speeds of the particles are

$$v_1' = \sqrt{(v_1')_n{}^2 + (v_1')_t{}^2} = \sqrt{(1.039)^2 + 3^2} = 3.17 \text{ m/s} \quad Ans.$$

$$v_2' = \sqrt{(v_2')_n{}^2 + (v_2')_t{}^2} = \sqrt{(4.16)^2 + 0^2} = 4.16 \text{ m/s} \quad Ans.$$

The angle θ' that $\mathbf{v}_1'$ makes with the t-direction is

$$\theta' = \tan^{-1}\left(\frac{(v_1')_n}{(v_1')_t}\right) = \tan^{-1}\left(\frac{1.039}{3}\right) = 19.11° \quad Ans.$$

The kinetic energies just before and just after impact, with $m = m_1 = m_2$, are

$$T = \tfrac{1}{2}m_1v_1{}^2 + \tfrac{1}{2}m_2v_2{}^2 = \tfrac{1}{2}m(6)^2 + 0 = 18m$$

$$T' = \tfrac{1}{2}m_1v_1'{}^2 + \tfrac{1}{2}m_2v_2'{}^2 = \tfrac{1}{2}m(3.17)^2 + \tfrac{1}{2}m(4.16)^2 = 13.68m$$

The percentage energy loss is then

$$\frac{|\Delta E|}{E}(100) = \frac{T - T'}{T}(100) = \frac{18m - 13.68m}{18m}(100) = 24.0\% \quad Ans.$$

① Be sure to set up n- and t-coordinates that are, respectively, normal to and tangent to the contacting surfaces. Calculation of the 30° angle is critical to all that follows.

② Note that, even though there are four equations in four unknowns for the standard problem of oblique central impact, only one pair of the equations is coupled.

③ We note that particle 2 has no initial or final velocity component in the t-direction. Hence, its final velocity $\mathbf{v}_2'$ is restricted to the n-direction.

PROBLEMS

Introductory problems

3/243 As a check of the basketball before the start of a game, the referee releases the ball from the overhead position shown, and the ball rebounds to about waist level. Determine the coefficient of restitution e and the percentage n of the original energy lost during the impact.

Ans. $e = 0.724$, $n = 47.6\%$

Problem 3/243

3/244 Determine the final velocities v_1' and v_2' after the collision of the two cylinders which slide on the smooth horizontal shaft. The coefficient of restitution is $e = 0.6$. Also determine the percent n of the original energy lost during the impact.

Problem 3/244

3/245 For the system of Prob. 3/244, determine the initial velocity v_2 which results in cylinder 2 being motionless after the impact.

Ans. $v_2 = 12.44$ m/s to the left

3/246 Car B is initially stationary and is struck by car A, which is moving with speed v. The mass of car B is pm, where m is the mass of car A and p is a positive constant. If the coefficient of restitution is $e = 0.1$, express the speeds v_A' and v_B' of the two cars at the end of the impact in terms of p and v. Evaluate your expressions for $p = 0.5$.

Problem 3/246

3/247 The sphere of mass m_1 travels with an initial velocity v_1 directed as shown and strikes the stationary sphere of mass m_2. For a given coefficient of restitution e, what condition on the mass ratio m_1/m_2 ensures that the final velocity of m_2 is greater than v_1?

Ans. $\dfrac{m_1}{m_2} > \dfrac{1}{e}$

Problem 3/247

3/248 Freight car A of mass m_A is rolling to the right when it collides with freight car B of mass m_B initially at rest. If the two cars are coupled together at impact, show that the fractional loss of energy equals $m_B/(m_A + m_B)$.

Problem 3/248

3/249 If the center of the ping-pong ball is to clear the net as shown, at what height h should the ball be horizontally served? Also determine h_2. The coefficient of restitution for the impacts between ball and table is $e = 0.9$, and the radius of the ball is $r = 18.75$ mm.

Ans. $h = 273$ mm, $h_2 = 185.8$ mm

Problem 3/249

3/250 The steel ball strikes the heavy steel plate with a velocity $v_0 = 24$ m/s at an angle of $60°$ with the horizontal. If the coefficient of restitution is $e = 0.8$, compute the velocity v and its direction θ with which the ball rebounds from the plate.

Problem 3/250

Representative problems

3/251 Pool ball B is to be shot into the side pocket D by banking it off the cushion at C. Specify the location x of the cushion impact for coefficients of restitution (a) $e = 1$ and (b) $e = 0.8$.

Ans. (a) $x = \dfrac{d}{3}$, (b) $x = 0.286d$

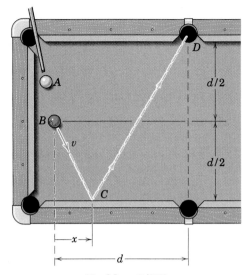

Problem 3/251

3/252 The 200-kg ram of a pile driver falls 1.2 m from rest and strikes the top of a 320-kg pile embedded 0.9 m in the ground. Immediately after impact the ram is seen to have no velocity. Determine the coefficient of restitution e and the velocity v' of the pile immediately after impact.

Problem 3/252

3/253 The figure shows n spheres of equal mass m suspended in a line by wires of equal length so that the spheres are almost touching each other. If sphere 1 is released from the dashed position and strikes sphere 2 with a velocity v_1, write an expression for the velocity v_n of the nth sphere immediately after being struck by the one adjacent to it. The common coefficient of restitution is e.

$$Ans. \ v_n = \left(\frac{1+e}{2}\right)^{n-1} v_1$$

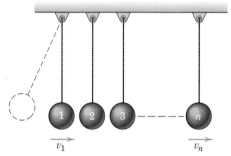

Problem 3/253

3/254 The ball is released from position A and drops 0.75 m to the incline. If the coefficient of restitution in the impact is $e = 0.85$, determine the slant range R.

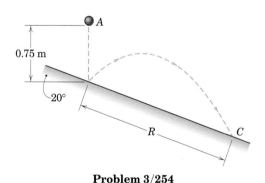

Problem 3/254

3/255 To pass inspection, steel balls designed for use in ball bearings must clear the fixed bar A at the top of their rebound when dropped from rest through the vertical distance $H = 900$ mm onto the heavy inclined steel plate. If balls which have a coeffi-

cient of restitution of less than 0.7 with the rebound plate are to be rejected, determine the position of the bar by specifying h and s. Neglect any friction during impact.

$$Ans. \ h = 379 \text{ mm}, \ s = 339 \text{ mm}$$

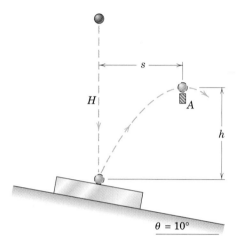

Problem 3/255

3/256 The two cars collide at right angles in the intersection of two icy roads. Car A has a mass of 1200 kg and car B has a mass of 1600 kg. The cars become entangled and move off together with a common velocity v' in the direction indicated. If car A was traveling 50 km/h at the instant of impact, compute the corresponding velocity of car B just before impact.

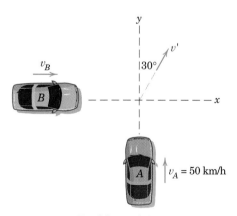

Problem 3/256

3/257 In a pool game the cue ball A must strike the eight ball in the position shown in order to send it to the pocket P with a velocity v_2'. The cue ball has a velocity v_1 before impact and a velocity v_1' after impact. The coefficient of restitution is 0.9. Both balls have the same mass and diameter. Calculate the rebound angle θ and the fraction n of the kinetic energy which is lost during the impact.

Ans. $\theta = 2.86°,\ n = 0.0475$

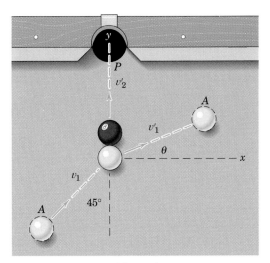

Problem 3/257

3/258 Two steel balls of the same diameter are connected by a rigid bar of negligible mass as shown and are dropped in the horizontal position from a height of 150 mm above the heavy steel and brass base plates. If the coefficient of restitution between the ball and the steel base is 0.6 and that between the other ball and the brass base is 0.4, determine the angular velocity ω of the bar immediately after impact. Assume that the two impacts are simultaneous.

Problem 3/258

3/259 A projectile is launched from point A and has a horizontal range L_1 as shown. If the coefficient of restitution at B is e, determine the distance L_2.

Ans. $L_2 = eL_1$

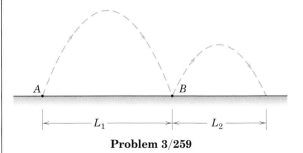

Problem 3/259

3/260 A basketball traveling with the velocity shown in the figure strikes the backboard at A. If the coefficient of restitution for this impact is $e = 0.84$, determine the required distance h above the hoop if the ball is to arrive at the center B of the hoop. Carry out two solutions: (a) an approximate solution obtained by neglecting the effects of gravity from A to B and (b) a solution which accounts for gravity from A to B. Neglect the diameter of the ball compared with h.

Problem 3/260

3/261 Two identical hockey pucks moving with initial velocities v_A annd v_B collide as shown. If the coefficient of restitution is $e = 0.75$, determine the velocity (magnitude and direction θ with respect to the positive x-axis) of each puck just after impact. Also calculate the percentage loss n of system kinetic energy.

$$Ans. \quad v_{A}' = 6.83 \text{ m/s at } \theta_A = 180°$$
$$v_{B}' = 6.51 \text{ m/s at } \theta_B = 50.2°$$
$$n = 34.6\%$$

Problem 3/261

3/262 Sphere A has a mass of 23 kg and a radius of 75 mm, while sphere B has a mass of 4 kg and a radius of 50 mm. If the spheres are traveling initially along the parallel paths with the speeds shown, determine the velocities of the spheres immediately after impact. Specify the angles θ_A and θ_B with respect to the x-axis made by the rebound velocity vectors. The coefficient of restitution is 0.4 and friction is neglected.

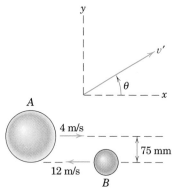

Problem 3/262

3/263 A miniature-golf shot from position A to the hole D is to be accomplished by "banking off" the 45° wall. Using the theory of this article, determine the location x for which the shot can be made. The coefficient of restitution associated with the wall collision is $e = 0.8$.

$$Ans. \quad x = 0.1088d$$

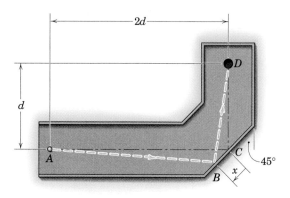

Problem 3/263

3/264 During a pregame warmup period, two basketballs collide above the hoop when in the positions shown. Just before impact, ball 1 has a velocity $\mathbf{v}_1$ which makes a 30° angle with the horizontal. If the velocity $\mathbf{v}_2$ of ball 2 just before impact has the same magnitude as $\mathbf{v}_1$, determine the two possible values of the angle θ, measured from the horizontal, which will cause ball 1 to go directly through the center of the basket. The coefficient of restitution is $e = 0.8$.

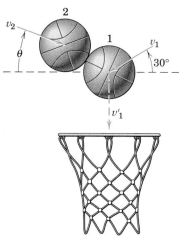

Problem 3/264

3/265 In a game of pool, the eight ball is to be struck by the cue ball A so that the eight ball enters the right corner pocket B. Specify the distance x from the center of the left corner pocket C to the point where the cue ball strikes the cushion after hitting the eight ball. The equal-mass balls are 50 mm in diameter, and the coefficient of restitution is $e = 0.9$.   *Ans.* $x = 410$ mm

Problem 3/265

3/266 The 3000-kg anvil A of the drop forge is mounted on a nest of heavy coil springs having a combined stiffness of $2.8(10^6)$ N/m. The 600-kg hammer B falls 500 mm from rest and strikes the anvil, which suffers a maximum downward deflection of 24 mm from its equilibrium position. Determine the height h of rebound of the hammer and the coefficient of restitution e which applies.

Problem 3/266

▶**3/267** A child throws a ball from point A with a speed of 15 m/s. It strikes the wall at point B and then returns exactly to point A. Determine the necessary angle α if the coefficient of restitution in the wall impact is $e = 0.5$.

Ans. $\alpha = 11.55°$ or $78.4°$

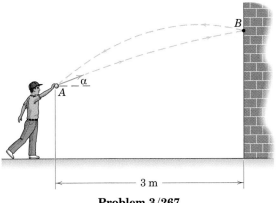

Problem 3/267

▶**3/268** The small smooth sphere is released from rest at position A and slides without friction down the inclined guide until it strikes the rigid horizontal surface at B. If the coefficient of restitution for the impact is e, determine the x-component of velocity of the sphere after impact and the fraction n of the energy lost during the impact. Compare your results with the case where the sharp corner is replaced by a rounded corner.

Ans. $v_x = \sqrt{2gh}\cos\theta$

$n = 1 - (\cos^2\theta + e^2\sin^2\theta)$

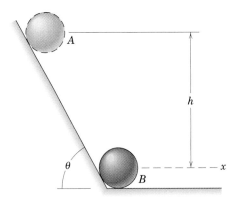

Problem 3/268

3/13 CENTRAL-FORCE MOTION

When a particle moves under the influence of a force directed toward a fixed center of attraction, the motion is called *central-force motion*. The most common example of central-force motion is found with the orbits of planets and satellites. The laws that govern this motion were deduced from observation of the motions of the planets by J. Kepler (1571–1630). The dynamics of central-force motion is basic to the design of high-altitude rockets, earth satellites, and space vehicles of all types.

Consider a particle of mass m, Fig. 3/18, moving under the action of the central gravitational attraction

$$F = G \frac{mm_0}{r^2}$$

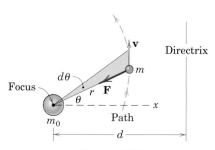

Figure 3/18

where m_0 is the mass of the attracting body assumed to be fixed, G is the universal gravitational constant, and r is the distance between the centers of the masses. The particle of mass m could represent the earth moving about the sun, the moon moving about the earth, or a satellite in its orbital motion about the earth above the atmosphere. The most convenient coordinate system to use is polar coordinates in the plane of motion since $\mathbf{F}$ will always be in the negative r-direction and there is no force in the θ-direction.

Equations 3/8 may be applied directly for the r- and θ-directions to give

$$-G \frac{mm_0}{r^2} = m(\ddot{r} - r\dot{\theta}^2)$$

$$0 = m(r\ddot{\theta} + 2\dot{r}\dot{\theta})$$

(3/33)

The second of the two equations when multiplied by r/m is seen to be the same as $d(r^2\dot{\theta})/dt = 0$, which is integrated to give

$$r^2\dot{\theta} = h, \quad \text{a constant} \qquad (3/34)$$

The physical significance of Eq. 3/34 is made clear when we note that the angular momentum $\mathbf{r} \times m\mathbf{v}$ of m about m_0 has the magnitude $mr^2\dot{\theta}$. Thus, Eq. 3/34 merely states that the angular momentum of m about m_0 remains constant, or is conserved. This statement is easily deduced from Eq. 3/27 where it is observed that the angular momentum $\mathbf{H}_O$ remains constant (is conserved) if there is no moment acting on the particle about a fixed point O.

We observe that during time dt, the radius vector sweeps out an area, shaded in Fig. 3/18, equal to $dA = (\frac{1}{2}r)(r\,d\theta)$. Therefore, the rate at which area is swept by the radius vector is $\dot{A} = \frac{1}{2}r^2\dot{\theta}$, which is constant according to Eq. 3/34. This conclusion is expressed in Kepler's *second law* of planetary motion which states that the areas swept through in equal times are equal.

The shape of the path followed by m may be obtained by solving the first of Eqs. 3/33, with the time t eliminated through combination with Eq. 3/34. To this end the mathematical substitution $r = 1/u$ is useful. Thus, $\dot{r} = -(1/u^2)\dot{u}$, which from Eq. 3/34 becomes $\dot{r} = -h(\dot{u}/\dot{\theta})$ or $\dot{r} = -h(du/d\theta)$. The second time derivative is $\ddot{r} = -h(d^2u/d\theta^2)\dot{\theta}$, which by combining with Eq. 3/34, becomes $\ddot{r} = -h^2u^2(d^2u/d\theta^2)$. Substitution into the first of Eqs. 3/33 now gives

$$-Gm_0u^2 = -h^2u^2\frac{d^2u}{d\theta^2} - \frac{1}{u}h^2u^4$$

or

$$\frac{d^2u}{d\theta^2} + u = \frac{Gm_0}{h^2} \qquad (3/35)$$

which is a nonhomogeneous linear differential equation. The solution of this familiar second-order equation may be verified by direct substitution and is

$$u = \frac{1}{r} = C\cos(\theta + \delta) + \frac{Gm_0}{h^2}$$

where C and δ are the two integration constants. The phase angle δ may be eliminated by choosing the x-axis so that r is a minimum when $\theta = 0$. Thus,

$$\frac{1}{r} = C\cos\theta + \frac{Gm_0}{h^2} \qquad (3/36)$$

The interpretation of Eq. 3/36 requires a knowledge of the equations for conic sections. We recall that a conic section is formed by the locus of a point which moves so that the ratio e of its distance from a point (focus) to a line (directrix) is constant. Thus, from Fig. 3/18, $e = r/(d - r\cos\theta)$, which may be rewritten as

$$\frac{1}{r} = \frac{1}{d}\cos\theta + \frac{1}{ed} \qquad (3/37)$$

which is the same form as Eq. 3/36. Hence, it is seen that the motion of m is along a conic section with $d = 1/C$ and $ed = h^2/(Gm_0)$ or

$$e = \frac{h^2C}{Gm_0} \qquad (3/38)$$

There are three cases to be investigated corresponding to $e < 1$ (ellipse), $e = 1$ (parabola), and $e > 1$ (hyperbola). The trajectory for each of these cases in shown in Fig. 3/19.

Case I: ellipse ($e < 1$). From Eq. 3/37 it is seen that r is a minimum when $\theta = 0$ and is a maximum when $\theta = \pi$. Thus,

$$2a = r_{min} + r_{max} = \frac{ed}{1 + e} + \frac{ed}{1 - e} \quad \text{or} \quad a = \frac{ed}{1 - e^2}$$

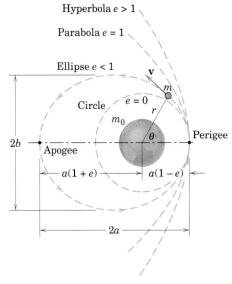

Figure 3/19

With the distance d expressed in terms of a, Eq. 3/37 and the maximum and minimum values of r may be written as

$$\frac{1}{r} = \frac{1 + e \cos \theta}{a(1 - e^2)}$$

(3/39)

$$r_{\text{min}} = a(1 - e) \qquad r_{\text{max}} = a(1 + e)$$

In addition, the relation $b = a\sqrt{1 - e^2}$, which comes from the geometry of the ellipse, gives the expression for the semiminor axis. It is seen that the ellipse becomes a circle with $r = a$ when $e = 0$. Equation 3/39 is an expression of Kepler's *first law* which says that the planets move in elliptical orbits around the sun as a focus.

The period τ for the elliptical orbit is the total area A of the ellipse divided by the constant rate $\dot{A}$ at which the area is swept through. Thus,

$$\tau = A/\dot{A} = \frac{\pi a b}{\frac{1}{2}r^2\dot{\theta}} \qquad \text{or} \qquad \tau = \frac{2\pi a b}{h}$$

by Eq. 3/34. Substitution of Eq. 3/38, the identity $d = 1/C$, the geometric relationships $a = ed/(1 - e^2)$ and $b = a\sqrt{1 - e^2}$ for the ellipse, and the equivalence $Gm_0 = gR^2$ yields upon simplification

$$\boxed{\tau = 2\pi \frac{a^{3/2}}{R\sqrt{g}}}$$

(3/40)

In this equation it is noted that R is the mean radius of the central attracting body and g is the absolute value of the acceleration due to gravity at the surface of the attracting body.

Equation 3/40 expresses Kepler's *third law* of planetary motion which states that the square of the period of motion is proportional to the cube of the semimajor axis of the orbit.

Case II: parabola ($e = 1$). Equations 3/37 and 3/38 become

$$\frac{1}{r} = \frac{1}{d}(1 + \cos \theta) \qquad \text{and} \qquad h^2 C = Gm_0$$

The radius vector becomes infinite as θ approaches π, so the dimension a is infinite.

Case III: hyperbola ($e > 1$). From Eq. 3/37 it is seen that the radial distance r becomes infinite for the two values of the polar angle θ_1 and $-\theta_1$ defined by $\cos \theta_1 = -1/e$. Only branch I corresponding to $-\theta_1 < \theta < \theta_1$, Fig. 3/20, represents a physically possible motion. Branch II corresponds to angles in the remaining sector (with r negative). For this branch positive r's may be used if θ is replaced by $\theta - \pi$ and $-r$ by r. Thus, Eq. 3/37 becomes

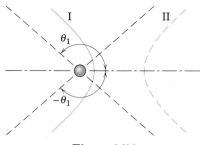

Figure 3/20

$$\frac{1}{-r} = \frac{1}{d} \cos(\theta - \pi) + \frac{1}{ed} \quad \text{or} \quad \frac{1}{r} = -\frac{1}{ed} + \frac{\cos\theta}{d}$$

But this expression contradicts the form of Eq. 3/36 where Gm_0/h^2 is necessarily positive. Hence, branch II does not exist (except for repulsive forces).

Now consider the energies of particle m. The system is conservative, and the constant energy E of m is the sum of its kinetic energy T and potential energy V. The kinetic energy is $T = \frac{1}{2}mv^2 = \frac{1}{2}m(\dot{r}^2 + r^2\dot{\theta}^2)$ and the potential energy from Eq. 3/15 is $V = -mgR^2/r$. Recall that g is the absolute acceleration due to gravity measured at the surface of the attracting body, R is the radius of the attracting body, and $Gm_0 = gR^2$. Thus,

$$E = \tfrac{1}{2}m(\dot{r}^2 + r^2\dot{\theta}^2) - \frac{mgR^2}{r}$$

This constant value of E can be determined from its value at $\theta = 0$, where $\dot{r} = 0$, $1/r = C + gR^2/h^2$ from Eq. 3/36, and $r\dot{\theta} = h/r$ from Eq. 3/34. Substitution into the expression for E and simplification yield

$$\frac{2E}{m} = h^2C^2 - \frac{g^2R^4}{h^2}$$

Now eliminate C by substitution of Eq. 3/38, which may be written as $h^2C = egR^2$, and obtain

$$e = +\sqrt{1 + \frac{2Eh^2}{mg^2R^4}} \tag{3/41}$$

The plus value of the radical is mandatory since by definition e is positive. It is now seen that for the

elliptical orbit $e < 1$, E is negative

parabolic orbit $e = 1$, E is zero

hyperbolic orbit $e > 1$, E is positive

These conclusions, of course, depend on the arbitrary selection of the datum condition for zero potential energy ($V = 0$ when $r = \infty$).

The expression for the velocity v of m may be found from the energy equation, which is

$$\tfrac{1}{2}mv^2 - \frac{mgR^2}{r} = E$$

The total energy E is obtained from Eq. 3/41 by combining Eq. 3.38 and $1/C = d = a(1 - e^2)/e$ to give for the elliptical orbit

$$E = -\frac{gR^2m}{2a} \tag{3/42}$$

Substitution into the energy equation yields

$$v^2 = 2gR^2 \left(\frac{1}{r} - \frac{1}{2a} \right) \qquad (3/43)$$

from which the magnitude of the velocity may be computed for a particular orbit in terms of the radial distance r. Combination of the expressions for r_{min} and r_{max} corresponding to perigee and apogee, Eq. 3/39, with Eq. 3/43 gives for the respective velocities at these two positions for the elliptical orbit

$$v_P = R\sqrt{\frac{g}{a}}\sqrt{\frac{1+e}{1-e}} = R\sqrt{\frac{g}{a}}\sqrt{\frac{r_{max}}{r_{min}}}$$

$$v_A = R\sqrt{\frac{g}{a}}\sqrt{\frac{1-e}{1+e}} = R\sqrt{\frac{g}{a}}\sqrt{\frac{r_{min}}{r_{max}}} \qquad (3/44)$$

Selected numerical data pertaining to the solar system are included in Appendix D and will be found useful in applying the foregoing relationships to problems in planetary motion.

The foregoing analysis is based on three assumptions:

1. The two bodies possess spherical mass symmetry so that they may be treated as if their masses were concentrated at their centers, that is, as if they were particles.

2. There are no forces present except the gravitational force which each mass exerts on the other.

3. Mass m_0 is fixed in space.

Assumption (1) is excellent for bodies that are distant from the central attracting body, which is the case for most heavenly bodies. A significant class of problems for which assumption (1) is poor is that of artificial satellites in the very near vicinity of oblate planets. As a comment on assumption (2), we note that aerodynamic drag on a low-altitude earth satellite is a force that usually cannot be ignored in the orbital analysis. For an artificial satellite in earth orbit, the error of assumption (3) is negligible because the ratio of the mass of the satellite to that of the earth is very small. On the other hand, for the earth–moon system, a small but significant error is introduced if assumption (3) is invoked—note that the lunar mass is about 1/81 times that of the earth.

Two-body problem. We now account for the motion of both masses and allow the presence of forces in addition to those of mutual attraction by considering the "perturbed two-body" problem.

Figure 3/21 depicts the major mass m_0, the minor mass m, their respective position vectors $\mathbf{r}_1$ and $\mathbf{r}_2$ measured relative to an inertial frame, the gravitational forces $\mathbf{F}$ and $-\mathbf{F}$, and a non-two-body force $\mathbf{P}$ that is exerted on mass m. The force $\mathbf{P}$ might be due to aerodynamic drag, solar pressure, the presence of a third body, on-board thrusting activities, a nonspherical gravitational field, or a combination of these and other sources. Application of Newton's second law to each mass results in

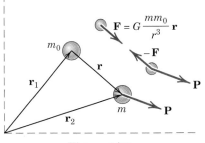

Figure 3/21

$$G \frac{mm_0}{r^3}\mathbf{r} = m_0\ddot{\mathbf{r}}_1 \qquad \text{and} \qquad -G\frac{mm_0}{r^3}\mathbf{r} + \mathbf{P} = m\ddot{\mathbf{r}}_2$$

Dividing the first equation by m_0, the second equation by m, and subtracting the first equation from the second give

$$-G\frac{(m_0 + m)}{r^3}\mathbf{r} + \frac{\mathbf{P}}{m} = \ddot{\mathbf{r}}_2 - \ddot{\mathbf{r}}_1 = \ddot{\mathbf{r}}$$

or

$$\ddot{\mathbf{r}} + G\frac{(m_0 + m)}{r^3}\mathbf{r} = \frac{\mathbf{P}}{m} \qquad (3/45)$$

Equation 3/45 is a second-order differential equation which, when solved, yields the relative position vector $\mathbf{r}$ as a function of time. Numerical techniques are usually required for the integration of the scalar differential equations which are equivalent to the vector equation 3/45, especially if $\mathbf{P}$ is nonzero. If $m_0 >> m$ and $\mathbf{P} = \mathbf{0}$, we have the restricted two-body problem, the equation of motion of which is

$$\ddot{\mathbf{r}} + G\frac{m_0}{r^3}\mathbf{r} = \mathbf{0} \qquad (3/45a)$$

With $\mathbf{r}$ and $\ddot{\mathbf{r}}$ expressed in polar coordinates, Eq. 3/45a becomes

$$(\ddot{r} - r\dot{\theta}^2)\mathbf{e}_r + (r\ddot{\theta} + 2\dot{r}\dot{\theta})\mathbf{e}_\theta + G\frac{m_0}{r^3}(r\mathbf{e}_r) = \mathbf{0}$$

When we equate coefficients of like unit vectors, we recover Eqs. 3/33.

Comparison of Eq. 3/45 (with $\mathbf{P} = \mathbf{0}$) and Eq. 3/45a allows us to relax the assumption that mass m_0 is fixed in space. If we replace m_0 by $(m_0 + m)$ in the expressions derived with the assumption of m_0 fixed, then we have expressions that account for the motion of m_0. For example, the corrected expression for the period of elliptical motion of m about m_0 is, from Eq. 3/40,

$$\tau = 2\pi\frac{a^{3/2}}{\sqrt{G(m_0 + m)}} \qquad (3/45b)$$

where the equality $R^2g = Gm_0$ has been used.

Sample Problem 3/27

An artificial satellite is launched from point B on the equator by its carrier rocket and inserted into an elliptical orbit with a perigee altitude of 2000 km. If the apogee altitude is to be 4000 km, compute (a) the necessary perigee velocity v_P and the corresponding apogee velocity v_A, (b) the velocity at point C where the altitude of the satellite is 2500 km, and (c) the period τ for a complete orbit.

Solution. **(a)** The perigee and apogee velocities for specified altitudes are given by Eqs. 3/44, where

$$r_{\text{max}} = 6371 + 4000 = 10\ 371 \text{ km}$$

$$r_{\text{min}} = 6371 + 2000 = 8371 \text{ km}$$

$$a = (r_{\text{min}} + r_{\text{max}})/2 = 9371 \text{ km}$$

Thus,

$$v_P = R\sqrt{\frac{g}{a}}\sqrt{\frac{r_{\text{max}}}{r_{\text{min}}}} = 6371(10^3)\sqrt{\frac{9.825}{9371(10^3)}}\sqrt{\frac{10\ 371}{8371}}$$

$$= 7261 \text{ m/s} \quad \text{or} \quad 26\ 140 \text{ km/h} \quad \textit{Ans.}$$

$$v_A = R\sqrt{\frac{g}{a}}\sqrt{\frac{r_{\text{min}}}{r_{\text{max}}}} = 6371(10^3)\sqrt{\frac{9.825}{9371(10^3)}}\sqrt{\frac{8371}{10\ 371}}$$

$$= 5861 \text{ m/s} \quad \text{or} \quad 21\ 099 \text{ km/h} \quad \textit{Ans.}$$

(b) For an altitude of 2500 km the radial distance from the earth's center is $r = 6371 + 2500 = 8871$ km. From Eq. 3/43 the velocity at point C becomes

$$v_C{}^2 = 2gR^2\left(\frac{1}{r} - \frac{1}{2a}\right) = 2(9.825)[(6371)(10^3)]^2\left(\frac{1}{8871} - \frac{1}{18\ 742}\right)\frac{1}{10^3}$$

$$= 47.353(10^6)(\text{m/s})^2$$

$$v_C = 6881 \text{ m/s} \quad \text{or} \quad 24\ 773 \text{ km/h} \quad \textit{Ans.}$$

(c) The period of the orbit is given by Eq. 3/40, which becomes

$$\tau = 2\pi\frac{a^{3/2}}{R\sqrt{g}} = 2\pi\frac{[(9371)(10^3)]^{3/2}}{(6371)(10^3)\sqrt{9.825}} = 9026 \text{ s}$$

$$\text{or} \quad \tau = 2.507 \text{ h} \quad \textit{Ans.}$$

① The mean radius of $12\ 742/2 = 6371$ km from Table D/2 in Appendix D is used. Also the absolute acceleration due to gravity $g = 9.825 \text{ m/s}^2$ from Art. 1/5 will be used.

② We must be careful with units. It is often safer to work in base units, meters in this case, and convert later.

③ We should observe here that the time interval between successive overhead transits of the satellite as recorded by an observer on the equator is longer than the period calculated here since the observer will have moved in space due to the counterclockwise rotation of the earth, as seen looking down on the north pole.

PROBLEMS

(Unless otherwise indicated, the velocities mentioned in the problems which follow are measured from a nonrotating reference frame moving with the center of the attracting body. Also, aerodynamic drag is to be neglected unless stated otherwise. Use $g = 9.825$ m/s^2 for the absolute gravitational acceleration at the surface of the earth and treat the earth as a sphere of radius $R = 6371$ km.)

Introductory problems

3/269 What velocity v must the space shuttle have in order to release the Hubble space telescope in a circular earth orbit 590 km above the earth?

Ans. $v = 7569$ m/s

590 km

Problem 3/269

3/270 Show that the path of the moon is concave toward the sun at the position shown. Assume that the sun, earth, and moon are in the same line.

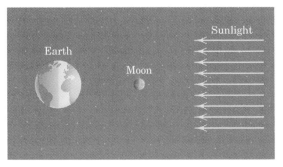

Problem 3/270

3/271 Calculate the velocity of a spacecraft which orbits the moon in a circular path of 80-km altitude.

Ans. $v = 1641$ m/s

80 km

Problem 3/271

3/272 A spacecraft is orbiting the earth in a circular orbit of altitude H. If its rocket engine is activated to produce a sudden burst of speed, determine the increase Δv necessary to allow the spacecraft to escape from the earth's gravity field. Calculate Δv if $H = 320$ km.

3/273 A satellite is in a circular earth orbit of radius $2R$, where R is the radius of the earth. What is the minimum velocity boost Δv necessary to reach point B, which is a distance $3R$ from the center of the earth? At what point in the original circular orbit should the velocity increment be added?

Ans. $\Delta v = 534$ m/s

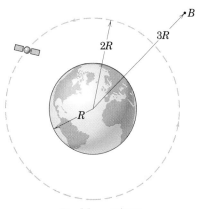

Problem 3/273

3/274 Determine the apparent velocity v_{rel} of a satellite moving in a circular equatorial orbit 300 km above the earth as measured by an observer on the equator (*a*) for a west-to-east orbit and (*b*) for an east-to-west orbit. Why is the west-to-east orbit more easily achieved?

3/275 A satellite is in a circular polar orbit of altitude 300 km. Determine the separation *d* at the equator between the ground tracks (shown dashed) associated with two successive overhead passes of the satellite.

Ans. $d = 2520$ km

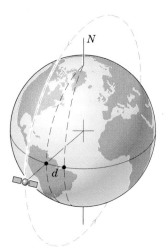

Problem 3/275

3/276 The Mars orbiter for the Viking mission was designed to make one complete trip around the planet in exactly the same time that it takes Mars to revolve once about its own axis. This time is 24 h, 37 min, 23 s. In this way, it is possible for the orbiter to pass over the landing site of the lander capsule at the same time in each Martian day at the orbiter's minimum (periapsis) altitude.

For the Viking I mission, the periapsis altitude of the orbiter was 1508 km. Make use of the data in Table D/2 in Appendix D and compute the maximum (apoapsis) altitude h_a for the orbiter in its elliptical path.

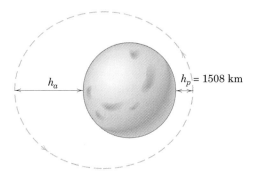

Problem 3/276

3/277 Determine the speed *v* required of an earth satellite at point *A* for (*a*) a circular orbit, (*b*) an elliptical orbit of eccentricity $e = 0.1$, (*c*) an elliptical orbit of eccentricity $e = 0.9$, and (*d*) a parabolic orbit. In cases (*b*), (*c*), and (*d*), *A* is the orbit perigee.

Ans. (*a*) $v = 7544$ m/s, (*b*) $v = 7912$ m/s
(*c*) $v = 10\ 398$ m/s, (*d*) $v = 10\ 668$ m/s

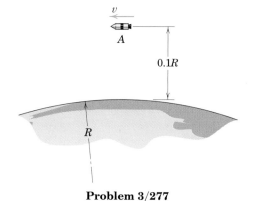

Problem 3/277

Representative problems

3/278 Satellite *A* moving in the circular orbit and satellite *B* moving in the elliptical orbit collide and become entangled at point *C*. If the masses of the satellites are equal, determine the maximum altitude h_{max} of the resulting orbit.

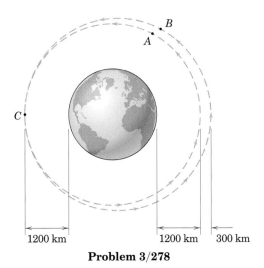

1200 km   1200 km   300 km

Problem 3/278

3/279 If the earth were suddenly deprived of its orbital velocity around the sun, find the time *t* that it would take for the earth to "fall" to the location of the center of the sun. (*Hint:* The time would be one half the period of a degenerate elliptical orbit around the sun with the semiminor axis approaching zero.) Refer to Table D/2 for the exact period of the earth around the sun.

Ans. $t = 64.6$ days

3/280 After launch from the earth, the 85 000-kg space-shuttle orbiter is in the elliptical orbit shown. If the orbit is to be circularized at the apogee altitude of 320 km, determine the necessary time duration Δt during which its two orbital maneuvering system (OMS) engines, each of which has a thrust of 30 kN, must be fired when the apogee position *C* is reached.

240 km →   ← → ← 320 km

Problem 3/280

3/281 A "drag-free" satellite is one that carries a small mass inside a chamber as shown. If the satellite speed decreases because of drag, the mass speed will not, and so the mass moves relative to the chamber as indicated. Sensors detect this change in the position of the mass within the chamber, and the satellite thruster is periodically fired to recenter the mass. In this manner, compensation is made for drag. If the satellite is in a circular earth orbit of 200-km altitude and a total thruster burn time of 300 seconds occurs during 10 orbits, determine the drag force *D* acting on the 100-kg satellite. The thruster force *T* is 2 N.

Ans. $D = 0.01132$ N

Problem 3/281

3/282 Determine the required velocity v_B in the direction indicated so that the spacecraft path will be tangent to the circular orbit at point C. What must be the distance b so that this path is possible?

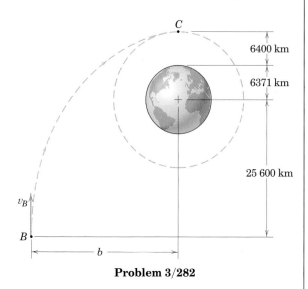

Problem 3/282

3/283 Two satellites B and C are in the same circular orbit of altitude 800 kilometers. Satellite B is 2000 km ahead of satellite C as indicated. Show that C can catch up to B by "putting on the brakes." Specifically, by what amount Δv should the circular orbit velocity of C be reduced so that it will rendezvous with B after one period in its new elliptical orbit? Check to see that C does not strike the earth in the elliptical orbit.

Ans. $\Delta v = 115.5$ m/s

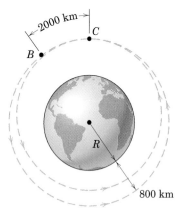

Problem 3/283

3/284 Determine the necessary amount Δv by which the circular orbit velocity of satellite C should be reduced if the catch-up maneuver of Prob. 3/283 is to be accomplished with not one but two periods in a new elliptical orbit.

3/285 A spacecraft is in a circular orbit of radius $3R$ around the moon. At point A, the spacecraft ejects a probe which is designed to arrive at the surface of the moon at point B. Determine the necessary velocity v_r of the probe relative to the spacecraft just after ejection. Also calculate the position θ of the spacecraft when the probe arrives at point B.

Ans. $v_r = 284$ m/s, $\theta = 98.0°$

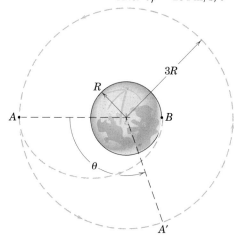

Problem 3/285

3/286 Prove that the velocity of a satellite traveling in an elliptical orbit equals $R\sqrt{g/a}$ when it reaches point C on the end of the semiminor axis. Also show that this velocity is the same as that for a circular orbit of radius a.

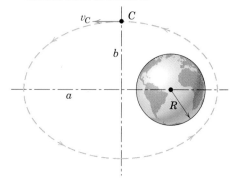

Problem 3/286

3/287 An earth satellite requires 1 h 36 min 22 s to make one complete space orbit and has a perigee altitude of 500 km. Determine the altitude H_A and velocity v_A at apogee and the eccentricity e of the orbit.

Ans. $H_A = 686$ km, $v_A = 26\ 882$ km/h
$e = 0.01333$

3/288 The 80-Mg space shuttle orbiter is in a circular orbit of altitude 320 km. The two orbital maneuvering system (OMS) engines, each of which has a thrust of 27 kN, are fired in retro-thrust for 150 seconds. Determine the angle β which locates the intersection of the shuttle trajectory with the earth's surface. Assume that the shuttle position B corresponds to the completion of the OMS burn and that no loss of altitude occurs during the burn.

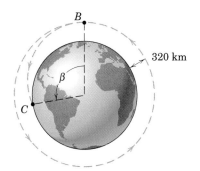

Problem 3/288

3/289 A satellite is placed in a circular polar orbit a distance H above the earth. As the satellite goes over the north pole at A, its retro-rocket is activated to produce a burst of negative thrust which reduces its velocity to a value that will ensure an equatorial landing. Derive the expression for the required reduction Δv_A of velocity at A. Note that A is the apogee of the elliptical path.

Ans. $\Delta v_A = R\sqrt{\dfrac{g}{R+H}}\left(1 - \sqrt{\dfrac{R}{R+H}}\right)$

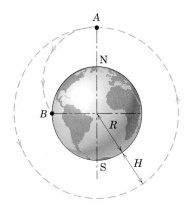

Problem 3/289

3/290 A projectile is launched from B with a speed of 2000 m/s at an angle α of 30° with the horizontal as shown. Determine the maximum altitude h_{max}.

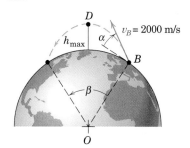

Problem 3/290

3/291 Compute the magnitude of the necessary launch velocity at B if the projectile trajectory is to intersect the earth's surface so that the angle β equals 90°. The altitude at the highest point of the trajectory is $0.5R$.

Ans. $v_B = 7560$ m/s

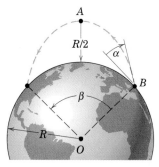

Problem 3/291

3/292 The perigee and apogee altitudes above the surface of the earth of an artificial satellite are h_p and h_a, respectively. Derive the expression for the radius of curvature ρ_p of the orbit at the perigee position. The radius of the earth is R.

3/293 A synchronous satellite is one whose velocity in its circular orbit allows it to remain above the same position on the rotating earth's surface. Calculate the required distance H above the earth's surface to the satellite. Locate the position of the orbital plane of the satellite and calculate the angular range β of longitude on the surface of the earth for which there is a direct line of sight to the satellite.

Ans. $H = 35\ 800$ km, $\beta = 162.6°$

3/294 Determine the angle β made by the velocity vector **v** with respect to the θ-direction for an earth satellite traveling in an elliptical orbit of eccentricity e. Express β in terms of the angle θ measured from perigee.

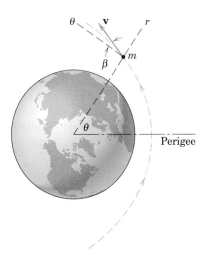

Problem 3/294

3/295 An earth satellite A is in a circular west-to-east equatorial orbit a distance 300 km above the surface of the earth as indicated. An observer B on the equator who sees the satellite directly overhead will see it directly overhead in the next orbit at position B' because of the rotation of the earth.

The radial line to the satellite will have rotated through the angle $2\pi + \theta$, and the observer will measure the apparent period τ' as a value slightly greater than the true period τ. Calculate τ' and $\tau' - \tau$.

Ans. $\tau' = 1$ h 36 min 25 s, $\tau' - \tau = 6$ min 4 s

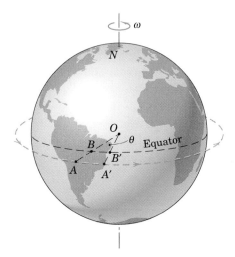

Problem 3/295

3/296 A spacecraft with a mass of 800 kg is traveling in a circular orbit 6000 km above the earth. It is desired to change the orbit to an elliptical one with a perigee altitude of 3000 km as shown. The transition is made by firing the retro-engine at A with a reverse thrust of 2000 N. Calculate the required time t for the engine to be activated.

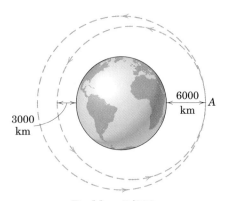

Problem 3/296

*3/297 In 1995 a spacecraft called the Solar and Helio-spheric Observatory (SOHO) was placed into a circular orbit inside that of the earth about the sun as shown. Determine the distance h so that the period of the spacecraft's orbit will match that of the earth, with the result that the spacecraft will remain between the earth and the sun in a "halo" orbit.

Ans. $h = 1.482(10^6)$ km

path 1? Compute each Δv if $r_1 = (6371 + 500)$ km and $r_2 = (6371 + 35\ 800)$ km. Note that r_2 has been chosen as the radius of a geosynchronous orbit.

$$\textit{Ans.} \ \ \Delta v_A = R\sqrt{\frac{g}{r_1}} \left(\sqrt{\frac{2r_2}{r_1 + r_2}} - 1 \right)$$

$$= 2370 \text{ m/s}$$

$$\Delta v_B = R\sqrt{\frac{g}{r_2}} \left(1 - \sqrt{\frac{2r_1}{r_1 + r_2}} \right)$$

$$= 1447 \text{ m/s}$$

Problem 3/298

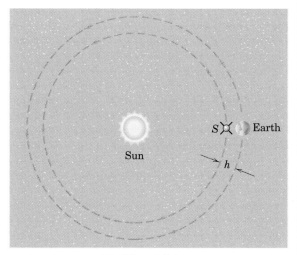

Problem 3/297

▶3/298 A space vehicle moving in a circular orbit of radius r_1 transfers to a larger circular orbit of radius r_2 by means of an elliptical path between A and B. (This transfer path is known as the Hohmann transfer ellipse.) The transfer is accomplished by a burst of speed Δv_A at A and a second burst of speed Δv_B at B. Write expressions for Δv_A and Δv_B in terms of the radii shown and the value g of the acceleration due to gravity at the earth's surface. If each Δv is positive, how can the velocity for path 2 be less than the velocity for

▶3/299 The satellite has a velocity at B of 3200 m/s in the direction indicated. Determine the angle β which locates the point C of impact with the earth.

Ans. $\beta = 109.1°$

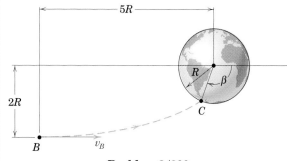

Problem 3/299

▶3/300 At the instant represented in the figure, a small
experimental satellite A is ejected from the shuttle orbiter with a velocity $v_r = 100$ m/s relative
to the shuttle, directed toward the center of the
earth. The shuttle is in a circular orbit of altitude
$h = 200$ km. For the resulting elliptical orbit of
the satellite, determine the semi-major axis a and
its orientation, the period τ, eccentricity e, apogee
speed v_a, perigee speed v_p, r_{max}, and r_{min}. Sketch
the satellite orbit.

> *Ans.* $a = 6572$ km (parallel to the x-axis)
> $\tau = 5301$ s, $e = 0.01284$
> $v_a = 7690$ m/s, $v_p = 7890$ m/s
> $r_{max} = 6.66(10^6)$ m, $r_{min} = 6.49(10^6)$ m

Problem 3/300

3/14 *RELATIVE MOTION*

Up to this point in our development of the kinetics of particle motion, we have applied Newton's second law and the equations of work-energy and impulse-momentum to problems where all measurements of motion were made with respect to a reference system which was considered fixed. The nearest we can come to a "fixed" reference system is the primary inertial system or astronomical frame of reference, which is an imaginary set of axes attached to the fixed stars. All other reference systems then are considered to have motion in space, including any reference system attached to the moving earth.

The accelerations of points attached to the earth as measured in the primary system are quite small, however, and we normally neglect them for most earth-surface measurements. For example, the acceleration of the center of the earth in its near-circular orbit around the sun considered fixed is 0.00593 m/s^2, and the acceleration of a point on the equator at sea level with respect to the center of the earth considered fixed is 0.0339 m/s^2. Clearly, these accelerations are small compared with g and with most other significant accelerations in engineering work. Thus, we make only a small error when we assume that our earth-attached reference axes are equivalent to a fixed reference system.

(a) Relative-motion equation. We now consider a particle A of mass m, Fig. 3/22, whose motion is observed from a set of axes x-y-z which have translatory motion with respect to a fixed reference frame X-Y-Z. Thus, the x-y-z directions always remain parallel to the X-Y-Z directions. A discussion of motion relative to a rotating reference system is reserved for a later treatment in Arts. 5/7 and 7/7. The acceleration of the origin B of x-y-z is $\mathbf{a}_B$. The acceleration of A as observed from or relative to x-y-z is $\mathbf{a}_{\text{rel}} = \mathbf{a}_{A/B} = \ddot{\mathbf{r}}_{A/B}$, and by the relative-motion principle of Art. 2/8, the absolute acceleration of A is

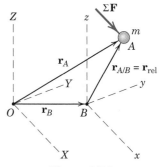

Figure 3/22

$$\mathbf{a}_A = \mathbf{a}_B + \mathbf{a}_{\text{rel}}$$

Thus, Newton's second law $\Sigma \mathbf{F} = m\mathbf{a}_A$ becomes

$$\Sigma \mathbf{F} = m(\mathbf{a}_B + \mathbf{a}_{\text{rel}}) \qquad (3/46)$$

The force sum $\Sigma \mathbf{F}$ is disclosed, as always, by a complete free-body diagram which will appear the same to an observer in x-y-z or to one in X-Y-Z as long as only the real forces acting on the particle are represented. We can conclude immediately that Newton's second law does not hold with respect to an accelerating system since $\Sigma \mathbf{F} \neq m\mathbf{a}_{\text{rel}}$.

(a)   *(b)*

Figure 3/23

(a)

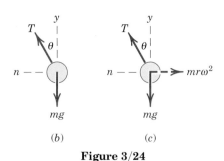

(b)   *(c)*

Figure 3/24

D'Alembert's principle. When a particle is observed from a fixed set of axes *X-Y-Z*, Fig. 3/23*a*, its absolute acceleration **a** is measured and the familiar relation $\Sigma\mathbf{F} = m\mathbf{a}$ is applied. When the particle is observed from a moving system *x-y-z* to which it is attached at the origin, Fig. 3/23*b*, the particle necessarily appears to be at rest or in equilibrium in *x-y-z*. Thus, the observer who is accelerating with *x-y-z* concludes that a force $-m\mathbf{a}$ acts on the particle to balance $\Sigma\mathbf{F}$. This point of view, which permits the treatment of a dynamics problem by the methods of statics, was an outgrowth of the work of D'Alembert contained in his *Traité de Dynamique* published in 1743. This approach merely amounts to rewriting the equation of motion as $\Sigma\mathbf{F} - m\mathbf{a} = \mathbf{0}$, which assumes the form of a zero force summation if $-m\mathbf{a}$ is treated as a force. This fictitious force is known as the *inertia force*, and the artificial state of equilibrium created is known as *dynamic equilibrium*. The apparent transformation of a problem in dynamics to one in statics has become known as *D'Alembert's principle.*

Opinion differs concerning the original interpretation of D'Alembert's principle, but the principle in the form in which it is generally known is regarded in this book as being mainly of historical interest. It was evolved during a time when understanding and experience with dynamics were extremely limited and was a means of explaining dynamics in terms of the principles of statics that during earlier times were more fully understood. This excuse for using an artificial situation to describe a real one is open to question, as there is today a wealth of knowledge and experience with the phenomena of dynamics to support strongly the direct approach of thinking in terms of dynamics rather than in terms of statics. It is somewhat difficult to justify the long persistence in the acceptance of statics as a way of understanding dynamics, particularly in view of the continued search for the understanding and description of physical phenomena in their most direct form.

Only one simple example of the method known as D'Alembert's principle will be cited. The conical pendulum of mass *m*, Fig. 3/24*a*, is swinging in a horizontal circle, with its radial line *r* having an angular velocity ω. In the straightforward application of the equation of motion $\Sigma\mathbf{F} = m\mathbf{a}_n$ in the direction *n* of the acceleration, the free-body diagram in the *b*-part of the figure shows that $T \sin\theta = mr\omega^2$. When the equilibrium requirement in the *y*-direction, $T \cos\theta - mg = 0$, is introduced, the unknowns *T* and θ can be found. But if the reference axes are attached to the particle, the particle will appear to be in equilibrium relative to these axes. Accordingly, the inertia force $-m\mathbf{a}$ must be added, which amounts to visualizing the application of $mr\omega^2$ in the direction opposite to the acceleration, as shown in the *c*-part of the figure. With this pseudo free-body diagram, a zero force summation in the *n*-direction gives $T \sin\theta - mr\omega^2 = 0$ which, of course, gives us the same result as before. We may conclude that no advantage results from this alternative formulation. The authors recommend against using it since it introduces no simplification and adds a nonexistent force to the diagram. In the case of a particle moving in a circular path, this hypothetical inertia force is known as the *centrifugal force* since it is directed away from the center and is opposite to the direction of the acceleration. The student is urged to recognize that there is no actual centrifugal force acting on the particle. The only actual force that may properly be called centrifugal is the horizontal component of the tension T exerted *by* the particle *on* the cord.

(b) Constant-velocity translating systems.   In discussing particle motion relative to moving reference systems, it is of importance for us to note the special case where the reference system has a constant translational velocity. If the *x-y-z* axes of Fig. 3/22 have a constant velocity, then $\mathbf{a}_B = \mathbf{0}$ and the acceleration of the particle is $\mathbf{a}_A = \mathbf{a}_{\text{rel}}$. Therefore, we may write Eq. 3/46 as

$$\boxed{\Sigma\mathbf{F} = m\mathbf{a}_{\text{rel}}} \qquad (3/47)$$

which tells us that Newton's second law holds for measurements made in a system moving with a constant velocity. Such a system is known as an inertial system or as a Newtonian frame of reference. Observers in the moving system and in the fixed system will also agree on the designation of the resultant force acting on the particle from their identical free-body diagrams, provided they avoid the use of any so-called "inertia forces."

We will now examine the parallel question concerning the validity of the work-energy equation and the impulse-momentum equation relative to a constant-velocity, nonrotating system. Again, we take the *x-y-z* axes of Fig. 3/22 to be moving with a constant velocity $\mathbf{v}_B = \dot{\mathbf{r}}_B$ relative to the fixed axes *X-Y-Z*. The path of the particle *A* relative to *x-y-z* is governed by $\mathbf{r}_{\text{rel}}$ and is represented schematically in Fig. 3/25. The work done by $\Sigma\mathbf{F}$ relative to *x-y-z* is $dU_{\text{rel}} = \Sigma\mathbf{F}\cdot d\mathbf{r}_{\text{rel}}$. But $\Sigma\mathbf{F} = m\mathbf{a}_A = m\mathbf{a}_{\text{rel}}$ since $\mathbf{a}_B = \mathbf{0}$. Also $\mathbf{a}_{\text{rel}}\cdot d\mathbf{r}_{\text{rel}} = \mathbf{v}_{\text{rel}}\cdot d\mathbf{v}_{\text{rel}}$ for the same reason that $a_t\, ds = v\, dv$ in Art. 2/5 on curvilinear motion. Thus, we have

$$dU_{\text{rel}} = m\mathbf{a}_{\text{rel}}\cdot d\mathbf{r}_{\text{rel}} = mv_{\text{rel}}\, dv_{\text{rel}} = d(\tfrac{1}{2}mv_{\text{rel}}{}^2)$$

We define the kinetic energy relative to *x-y-z* as $T_{\text{rel}} = \tfrac{1}{2}mv_{\text{rel}}{}^2$ so that we now have

$$\boxed{dU_{\text{rel}} = dT_{\text{rel}}} \quad \text{or} \quad \boxed{U_{\text{rel}} = \Delta T_{\text{rel}}} \qquad (3/48)$$

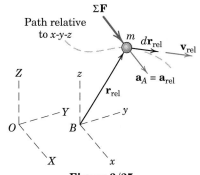

Figure 3/25

which shows that the work-energy equation holds for measurements made relative to a constant-velocity system.

Relative to *x-y-z*, the impulse on the particle during time dt is $\Sigma F\, dt = m\mathbf{a}_A\, dt = m\mathbf{a}_{\mathrm{rel}}\, dt$. But $m\mathbf{a}_{\mathrm{rel}}\, dt = m\, d\mathbf{v}_{\mathrm{rel}} = d(m\mathbf{v}_{\mathrm{rel}})$ so

$$\Sigma \mathbf{F}\, dt = d(m\mathbf{v}_{\mathrm{rel}})$$

We define the linear momentum of the particle relative to *x-y-z* as $\mathbf{G}_{\mathrm{rel}} = m\mathbf{v}_{\mathrm{rel}}$, which gives us $\Sigma \mathbf{F}\, dt = d\mathbf{G}_{\mathrm{rel}}$. Dividing by dt and integrating give

$$\boxed{\Sigma \mathbf{F} = \dot{\mathbf{G}}_{\mathrm{rel}}} \quad \text{and} \quad \boxed{\int \Sigma \mathbf{F}\, dt = \Delta \mathbf{G}_{\mathrm{rel}}} \qquad (3/49)$$

Thus, the impulse-momentum equations for a fixed reference system also hold for measurements made relative to a constant-velocity system.

Finally, we define the relative angular momentum of the particle about a point in *x-y-z*, such as the origin *B*, as the moment of the relative linear momentum. Thus, $\mathbf{H}_{B_{\mathrm{rel}}} = \mathbf{r}_{\mathrm{rel}} \times \mathbf{G}_{\mathrm{rel}}$. The time derivative gives $\dot{\mathbf{H}}_{B_{\mathrm{rel}}} = \dot{\mathbf{r}}_{\mathrm{rel}} \times \mathbf{G}_{\mathrm{rel}} + \mathbf{r}_{\mathrm{rel}} \times \dot{\mathbf{G}}_{\mathrm{rel}}$. The first term is nothing more than $\mathbf{v}_{\mathrm{rel}} \times m\mathbf{v}_{\mathrm{rel}} = \mathbf{0}$ and the second term becomes $\mathbf{r}_{\mathrm{rel}} \times \Sigma \mathbf{F} = \Sigma \mathbf{M}_B$, the sum of the moments about *B* of all forces on *m*. Thus, we have

$$\boxed{\Sigma \mathbf{M}_B = \dot{\mathbf{H}}_{B_{\mathrm{rel}}}} \qquad (3/50)$$

which shows that the moment-angular momentum relation holds with respect to a constant-velocity system.

Although the work-energy and impulse-momentum equations hold relative to a system translating with a constant velocity, the individual expressions for work, kinetic energy, and momentum differ between the fixed and the moving systems. Thus,

$$(dU = \Sigma \mathbf{F} \cdot d\mathbf{r}_A) \neq (dU_{\mathrm{rel}} = \Sigma \mathbf{F} \cdot d\mathbf{r}_{\mathrm{rel}})$$

$$(T = \tfrac{1}{2}mv_A{}^2) \neq (T_{\mathrm{rel}} = \tfrac{1}{2}mv_{\mathrm{rel}}{}^2)$$

$$(\mathbf{G} = m\mathbf{v}_A) \neq (\mathbf{G}_{\mathrm{rel}} = m\mathbf{v}_{\mathrm{rel}})$$

Equations 3/47 through 3/50 are formal proof of the validity of the Newtonian equations of kinetics in any constant-velocity, nonrotating system. We might have surmised these conclusions from the fact that $\Sigma \mathbf{F} = m\mathbf{a}$ depends on acceleration and not velocity. We are also ready to conclude that there is no experiment which can be conducted in and relative to a constant-velocity system (Newtonian frame of reference) which discloses its absolute velocity. Any mechanical experiment will achieve the same results in any Newtonian system.

Sample Problem   3/28

A simple pendulum of mass m and length r is mounted on the flatcar which has a constant horizontal acceleration a_0 as shown. If the pendulum is released from rest relative to the flatcar at the position $\theta = 0$, determine the expression for the tension T in the supporting light rod for any value of θ. Also find T for $\theta = \pi/2$ and $\theta = \pi$.

Solution. We attach our moving x-y coordinate system to the translating car with origin at O for convenience. Relative to this system, n- and t-coordinates are the natural ones to use since the motion is circular within x-y. The acceleration of m is given by the relative-acceleration equation

$$\mathbf{a} = \mathbf{a}_0 + \mathbf{a}_{\text{rel}}$$

where $\mathbf{a}_{\text{rel}}$ is the acceleration that would be measured by an observer riding with the car. He would measure an n-component equal to $r\dot{\theta}^2$ and a t-component equal to $r\ddot{\theta}$. The three components of the absolute acceleration of m are shown in the separate view.

First, we apply Newton's second law to the t-direction and get

① $[\Sigma F_t = ma_t]$ $\qquad mg \cos\theta = m(r\ddot{\theta} - a_0 \sin\theta)$

$$r\ddot{\theta} = g \cos\theta + a_0 \sin\theta$$

Integrating to obtain $\dot{\theta}$ as a function of θ yields

$[\dot{\theta}\,d\dot{\theta} = \ddot{\theta}\,d\theta]$ $\qquad \displaystyle\int_0^{\dot{\theta}} \dot{\theta}\,d\dot{\theta} = \int_0^{\theta} \frac{1}{r}(g \cos\theta + a_0 \sin\theta)\,d\theta$

$$\frac{\dot{\theta}^2}{2} = \frac{1}{r}[g \sin\theta + a_0(1 - \cos\theta)]$$

We now apply Newton's second law to the n-direction, noting that the n-component of the absolute acceleration is $r\dot{\theta}^2 - a_0 \cos\theta$.

② $[\Sigma F_n = ma_n]$ $\quad T - mg \sin\theta = m(r\dot{\theta}^2 - a_0 \cos\theta)$

$$= m[2g \sin\theta + 2a_0(1 - \cos\theta) - a_0 \cos\theta]$$

$$T = m[3g \sin\theta + a_0(2 - 3\cos\theta)] \qquad Ans.$$

For $\theta = \pi/2$ and $\theta = \pi$, we have

$$T_{\pi/2} = m[3g(1) + a_0(2 - 0)] = m(3g + 2a_0) \qquad Ans.$$

$$T_{\pi} = m[3g(0) + a_0(2 - 3[-1])] = 5ma_0 \qquad Ans.$$

Free-body diagram   Acceleration components

① We choose the t-direction first since the n-direction equation, which contains the unknown T, will involve $\dot{\theta}^2$, which, in turn, is obtained from an integration of $\ddot{\theta}$.

② Be sure to recognize that $\dot{\theta}\,d\dot{\theta} = \ddot{\theta}\,d\theta$ may be obtained from $v\,dv = a_t\,ds$ by dividing by r^2.

Sample Problem   3/29

The flatcar moves with a constant speed v_0 and carries a winch that produces a constant tension P in the cable attached to the small carriage. The carriage has a mass m and rolls freely on the horizontal surface starting from rest relative to the flatcar at $x = 0$, at which instant $X = x_0 = b$. Apply the work-energy equation to the carriage, first, as an observer moving with the frame of reference of the car and, second, as an observer on the ground. Show the compatibility of the two expressions.

Solution.   To the observer on the flatcar, the work done by P is

(1)
$$U_{\text{rel}} = \int_0^x P\, dx = Px \qquad \text{for constant } P$$

The change in kinetic energy relative to the car is

$$\Delta T_{\text{rel}} = \tfrac{1}{2}m(\dot{x}^2 - 0)$$

The work-energy equation for the moving observer becomes

$[U_{\text{rel}} = \Delta T_{\text{rel}}]$ $\qquad Px = \tfrac{1}{2}m\dot{x}^2$

To the observer on the ground, the work done by P is

$$U = \int_b^X P\, dX = P(X - b)$$

The change in kinetic energy for the ground measurement is

(2)
$$\Delta T = \tfrac{1}{2}m(\dot{X}^2 - v_0^2)$$

The work-energy equation for the fixed observer gives

$[U = \Delta T]$ $\qquad P(X - b) = \tfrac{1}{2}m(\dot{X}^2 - v_0^2)$

To reconcile this equation with that for the moving observer, we can make the following substitutions:

$$X = x_0 + x, \qquad \dot{X} = v_0 + \dot{x}, \qquad \ddot{X} = \ddot{x}$$

Thus,

(3)
$$P(X - b) = Px + P(x_0 - b) = Px + m\ddot{x}(x_0 - b)$$
$$= Px + m\ddot{x}v_0 t = Px + mv_0\dot{x}$$

and

$$\dot{X}^2 - v_0^2 = (v_0^2 + \dot{x}^2 + 2v_0\dot{x} - v_0^2) = \dot{x}^2 + 2v_0\dot{x}$$

The work-energy equation for the fixed observer now becomes

$$Px + mv_0\dot{x} = \tfrac{1}{2}m\dot{x}^2 + mv_0\dot{x}$$

which is merely $Px = \tfrac{1}{2}m\dot{x}^2$, as concluded by the moving observer. We see, therefore, that the difference between the two work-energy expressions is

$$U - U_{\text{rel}} = T - T_{\text{rel}} = mv_0\dot{x}$$

① The only coordinate that the moving observer can measure is x.

② To the ground observer, the initial velocity of the carriage is v_0 so its initial kinetic energy is $\tfrac{1}{2}mv_0^2$.

③ The symbol t stands for the time of motion from $x = 0$ to $x = x$. The displacement $x_0 - b$ of the carriage is its velocity v_0 times the time t or $x_0 - b = v_0 t$. Also, since the constant acceleration times the time equals the velocity change, $\ddot{x}t = \dot{x}$.

PROBLEMS

Introductory problems

3/301 The flatbed truck is traveling at the constant speed of 60 km/h up the 15-percent grade when the 100-kg crate which it carries is given a shove which imparts to it an initial relative velocity $\dot{x} = 3$ m/s toward the rear of the truck. If the crate slides a distance $x = 2$ m measured on the truck bed before coming to rest on the bed, compute the coefficient of kinetic friction μ_k between the crate and the truck bed.

Ans. $\mu_k = 0.382$

Problem 3/301

3/302 If the spring of constant k is compressed a distance δ as indicated, calculate the acceleration a_{rel} of the block of mass m_1 relative to the frame of mass m_2 upon release of the spring. The system is initially stationary.

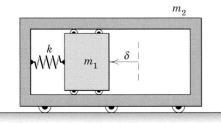

Problem 3/302

3/303 The cart with attached x-y axes moves with an absolute speed $v = 2$ m/s to the right. Simultaneously, the light arm of length $l = 0.5$ m rotates about point B of the cart with angular velocity $\dot{\theta} = 2$ rad/s. The mass of the sphere is $m = 3$ kg.

Determine the following quantities for the sphere when $\theta = 0$: $\mathbf{G}$, $\mathbf{G}_{\text{rel}}$, T, T_{rel}, $\mathbf{H}_O$, $\mathbf{H}_{B_{\text{rel}}}$, where the subscript "rel" indicates measurement relative to the x-y axes. Point O is an inertially fixed point coincident with point B at the instant under consideration.

Ans. $\mathbf{G} = 9\mathbf{i}$ kg·m/s, $\mathbf{G}_{\text{rel}} = 3\mathbf{i}$ kg·m/s
$$T = 13.5 \text{ J}, \ T_{\text{rel}} = 1.5 \text{ J}$$
$$\mathbf{H}_O = -4.5\mathbf{k} \ \frac{\text{kg·m}^2}{\text{s}}$$
$$\mathbf{H}_{B_{\text{rel}}} = -1.5\mathbf{k} \ \frac{\text{kg·m}^2}{\text{s}}$$

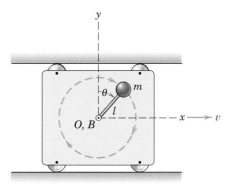

Problem 3/303

3/304 The aircraft carrier is moving at a constant speed and launches a jet plane with a mass of 3 Mg in a distance of 75 m along the deck by means of a steam-driven catapult. If the plane leaves the deck with a velocity of 240 km/h relative to the carrier and if the jet thrust is constant at 22 kN during takeoff, compute the constant force P exerted by the catapult on the airplane during the 75-m travel of the launch carriage.

Problem 3/304

3/305 The 2000-kg van is driven from position A to position B on the barge which is towed at a constant speed $v_0 = 16$ km/h. The van starts from rest relative to the barge at A, accelerates to $v = 24$ km/h relative to the barge over a distance of 25 m, and then stops with a deceleration of the same magnitude. Determine the magnitude of the net force F between the tires of the van and the barge during this maneuver.

Ans. $F = 1778$ N

Problem 3/305

Representative problems

3/306 The launch catapult of the aircraft carrier gives the 7-Mg jet airplane a constant acceleration and launches the airplane in a distance of 100 m measured along the angled takeoff ramp. The carrier is moving at a steady speed $v_C = 16$ m/s. If an absolute aircraft speed of 90 m/s is desired for takeoff, determine the net force F supplied by the catapult and the aircraft engines.

Problem 3/306

3/307 The coefficients of friction between the flat bed of the truck and crate are $\mu_s = 0.8$ and $\mu_k = 0.7$. The coefficient of kinetic friction between the truck tires and the road surface is 0.9. If the truck stops from an initial speed of 15 m/s with maximum braking (wheels skidding), determine where on the bed the crate finally comes to rest or the velocity v_{rel} relative to the truck with which the crate strikes the wall at the forward edge of the bed.

Ans. $v_{\text{rel}} = 2.46$ m/s

Problem 3/307

3/308 A boy of mass m is standing initially at rest relative to the moving walkway, which has a constant horizontal speed u. He decides to accelerate his progress and starts to walk from point A with a steadily increasing speed and reaches point B with a speed $\dot{x} = v$ relative to the walkway. During his acceleration he generates an average horizontal force F between his shoes and the walkway. Write the work-energy equations for his absolute and relative motions and explain the meaning of the term muv.

Problem 3/308

3/309 The block of mass m is attached to the frame by the spring of stiffness k and moves horizontally with negligible friction within the frame. The frame and block are initially at rest with $x = x_0$, the uncompressed length of the spring. If the frame is given a constant acceleration a_0, determine the maximum velocity $\dot{x}_{max} = (v_{rel})_{max}$ of the block relative to the frame.

Ans. $(v_{rel})_{max} = a_0 \sqrt{m/k}$

Problem 3/309

3/310 The slider A has a mass of 2 kg and moves with negligible friction in the 30° slot in the vertical sliding plate. What horizontal acceleration a_0 should be given to the plate so that the absolute acceleration of the slider will be vertically down? What is the value of the corresponding force R exerted on the slider by the slot?

Problem 3/310

3/311 The ball A of mass 10 kg is attached to the light rod of length $l = 0.8$ m. The mass of the carriage alone is 250 kg and it moves with an acceleration a_0 as shown. If $\dot{\theta} = 3$ rad/s when $\theta = 90°$, find the kinetic energy T of the system if the carriage has a velocity of 0.8 m/s (a) in the direction of a_0 and (b) in the direction opposite to a_0. Treat the ball as a particle.

Ans. (a) $T = 112$ J, (b) $T = 112$ J

Problem 3/311

3/312 Consider the system of Prob. 3/311 where the mass of the ball is $m = 10$ kg and the length of the light rod is $l = 0.8$ m. The ball–rod assembly is free to rotate about a vertical axis through O. The carriage, rod, and ball are initially at rest with $\theta = 0$ when the carriage is given a constant acceleration $a_0 = 3$ m/s². Write an expression for the tension T in the rod as a function of θ and calculate T for the position $\theta = \pi/2$.

3/313 A simple pendulum is placed on an elevator, which accelerates upward as shown. If the pendulum is displaced an amount θ_0 and released from rest relative to the elevator, find the tension T_0 in the supporting light rod when $\theta = 0$. Evaluate your result for $\theta_0 = \pi/2$.

$$Ans.\ \ T_0 = m(g + a_0)(3 - 2\cos\theta_0)$$

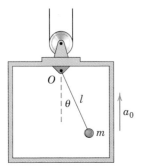

Problem 3/313

3/314 A boy of mass m is standing initially at rest relative to the moving walkway inclined at the angle θ and moving with a constant speed u. He decides to accelerate his progress and starts to walk from point A with a steadily increasing speed and reaches point B with a speed v_r relative to the walkway. During his acceleration he generates a constant average force F tangent to the walkway between his shoes and the walkway surface. Write the work-energy equations for the motion between A and B for his absolute motion and his relative motion and explain the meaning of the term muv_r. If the boy has a mass of 60 kg and if $u = 0.6$ m/s, $s = 10$ m, and $\theta = 10°$, calculate the power P_{rel} developed by the boy as he reaches the speed of 0.75 m/s relative to the walkway.

Problem 3/314

▶3/315 A ball is released from rest relative to the elevator at a distance h_1 above the floor. The speed of the elevator at the time of ball release is v_0. Determine the bounce height h_2 of the ball (a) if v_0 is constant and (b) if an upward elevator acceleration $a = g/4$ begins at the instant the ball is released. The coefficient of restitution for the impact is e.

$$Ans.\ \ (a)\ and\ (b)\ h_2 = e^2 h_1$$

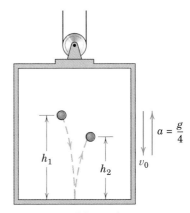

Problem 3/315

▶3/316 The small slider A moves with negligible friction down the tapered block, which moves to the right with constant speed $v = v_0$. Use the principle of work-energy to determine the magnitude v_A of the absolute velocity of the slider as it passes point C if it is released at point B with no velocity relative to the block. Apply the equation, both as an observer fixed to the block and as an observer fixed to the ground, and reconcile the two relations.

$$Ans.\ \ v_A = (v_0{}^2 + 2gl\sin\theta$$
$$+ 2v_0\cos\theta\sqrt{2gl\sin\theta})^{1/2}$$

Problem 3/316

▶3/317 When a particle is dropped from rest relative to the earth's surface at a latitude γ, the initial apparent acceleration is the relative acceleration due to gravity g_{rel}. The absolute acceleration due to gravity g is directed toward the center of the earth. Derive an expression for g_{rel} in terms of g, R, ω, and γ, where R is the radius of the earth treated as a sphere and ω is the constant angular velocity of the earth about the polar axis considered fixed. (Although axes x-y-z are attached to the earth and hence rotate, we may use Eq. 3/46 as long as the particle has no velocity relative to x-y-z.) (*Hint:* Use the first two terms of the binomial expansion for the approximation.)

$$Ans. \ g_{rel} = g - R\omega^2 \cos^2 \gamma \left(1 - \frac{R\omega^2}{2g} \right) + \cdots$$
$$= 9.825 - 0.03382 \cos^2 \gamma \ \text{m/s}^2$$

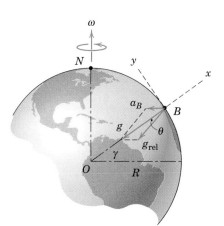

Problem 3/317

▶3/318 The figure represents the space shuttle S, which is (*a*) in a circular orbit about the earth and (*b*) in an elliptical orbit where P is its perigee position. The exploded views on the right represent the cabin space with its x-axis oriented in the direction of the orbit. The astronauts conduct an experiment by applying a known force F in the x-direction to a small mass m. Explain why $F = m\ddot{x}$ does or does not hold in each case, where x is measured within the spacecraft. Assume that the shuttle is between perigee and apogee in the elliptical orbit so that the orbital speed is changing with time. Note that the t- and x-axes are tangent to the path, and the θ-axis is normal to the radial r-direction.

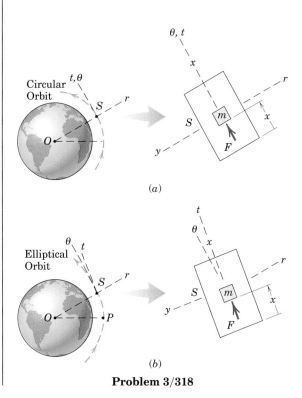

Problem 3/318

3/15 *PROBLEM FORMULATION AND REVIEW*

In Chapter 3 we have developed the three basic methods of solution to problems in particle kinetics. This experience is central to the study of dynamics and lays the foundation for the subsequent study of rigid-body and nonrigid-body dynamics. First, we applied Newton's second law $\Sigma \mathbf{F} = m\mathbf{a}$ to determine the instantaneous relation between forces and the acceleration they produce. With the background of Chapter 2 for identifying the kind of motion and with the aid of our familiar free-body diagram to be certain that all forces are accounted for, we were able to solve a large variety of problems using *x-y*, *n-t*, and *r-θ* coordinates for plane-motion problems and *x-y-z*, *r-θ-z*, and *R-θ-φ* coordinates for space problems.

Second, we integrated the basic equation of motion $\Sigma \mathbf{F} = m\mathbf{a}$ with respect to displacement and derived the scalar equations for work and energy. These equations enabled us to relate the initial and final velocities to the work done during the interval by forces external to our defined system. We expanded this approach to include potential energy, both elastic and gravitational. With these tools we discovered that the energy approach is especially valuable for conservative systems, that is, systems wherein the loss of energy due to friction or other forms of dissipation is negligible.

Third, we rewrote Newton's second law in the form of force equals time rate of change of linear momentum and moment equals time rate of change of angular momentum. Then we integrated these relations with respect to time and derived the impulse and momentum equations. These equations were then applied to motion intervals where the forces were functions of the time. We also investigated the interactions between particles under conditions where the linear momentum is conserved and where the angular momentum is conserved.

In the final section of Chapter 3, we employed the three basic methods in specific application areas. In particular, we noted that the impulse-momentum method is convenient in developing the relations governing particle impacts. We observed that the direct application of Newton's second law allows the determination of the trajectory properties associated with a particle under central-force attraction. And lastly, we have seen that any of the three basic methods may be applied to particle motion relative to a translating frame of reference.

It is clear that the successful solution of problems in particle kinetics depends on knowledge of the prerequisite particle kinematics. Analogously, the analytical developments associated with particle systems and rigid bodies, which are covered in the remainder of *Dynamics*, are heavily dependent on the fundamental principles of particle kinetics as treated in Chapter 3.

REVIEW PROBLEMS

3/319 Collar *A* is free to slide with negligible friction on the circular guide mounted in the vertical frame. Determine the angle θ assumed by the collar if the frame is given a constant horizontal acceleration *a* to the right.

$$Ans. \quad \theta = \tan^{-1}\frac{a}{g}$$

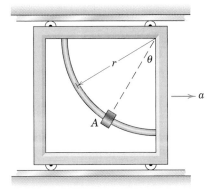

Problem 3/319

3/320 The block of mass *m* is given an initial velocity v_1 = 6 m/s up the 20° incline at point *A*. Calculate the velocity v_2 with which the block passes *A* as it slides back down.

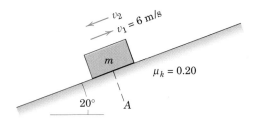

Problem 3/320

3/321 A 30-g tire-balance weight is attached to a vertical surface of the wheel rim by means of an adhesive backing. The tire-wheel unit is then given a final test on the tire-balance machine. If the adhesive can support a maximum shear force of 80 N, determine the maximum rotational speed *N* for which the weight remains fixed to the wheel. Assume very gradual speed changes.

$$Ans. \quad N = 1177 \text{ rev/min}$$

Problem 3/321

3/322 The 30-kg girl with mass center at *G* is in the lowest position in a swing at the instant represented. The effective length from *G* to the fixed support for the rope is 4.5 m, and the velocity of the girl's mass center is 3.6 m/s at this position. Neglect the mass of the seat and the ropes and calculate the tension *T* in the rope and the force *P* in the direction of her arms with which each of the girl's two hands pulls on the rope at this position. Also calculate the corresponding force *R* exerted on her by the seat.

Problem 3/322

3/323 The simple 2-kg pendulum is released from rest in the horizontal position. As it reaches the bottom position, the cord wraps around the smooth fixed pin at B and continues in the smaller arc in the vertical plane. Calculate the magnitude of the force R supported by the pin at B when the pendulum passes the position $\theta = 30°$.

Ans. $R = 46.7$ N

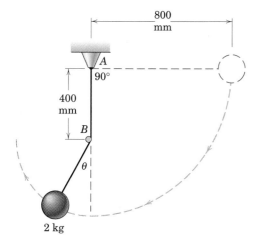

Problem 3/323

3/324 Six identical spheres are arranged as shown in the figure. The two spheres at the left end are released from the dotted positions and strike sphere 3 with speed v_1. Assuming that the common coefficient of restitution is $e = 1$, explain why two spheres leave the right end of the row with speed v_1 instead of one sphere leaving the right end with speed $2v_1$.

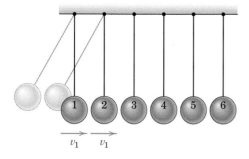

Problem 3/324

3/325 The spring of stiffness k is compressed and suddenly released, sending the particle of mass m sliding along the track. Determine the minimum spring compression δ for which the particle will not lose contact with the loop-the-loop track. The sliding surface is smooth except for the rough portion of length s equal to R, where the coefficient of kinetic friction is μ_k.

Ans. $\delta = \sqrt{\dfrac{mgR(5 + 2\mu_k)}{k}}$

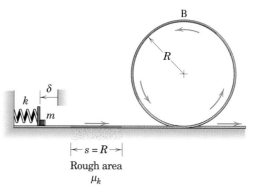

Problem 3/325

3/326 The drag force which acts on a body moving in a vertical plane through a fluid is accurately modeled by $\mathbf{D} = -C_D(\tfrac{1}{2}\rho v^2)S\mathbf{e}_v$, where $\mathbf{D}$ is the drag force as shown in the figure, C_D is the drag coefficient, ρ is the fluid density, $\mathbf{v}$ is the velocity of the body relative to the fluid, S is the cross-sectional area of the body presented to the flow, and $\mathbf{e}_v$ is a unit vector in the direction of $\mathbf{v}$. For a body of mass m, determine the x- and y-components of acceleration, and comment on the difficulty of integrating these two expressions. Assume a constant acceleration due to gravity.

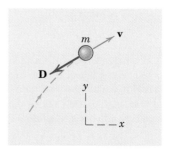

Problem 3/326

3/327 The 2-kg carriage moves up the 15° incline with negligible friction. The force P is applied in two separate impulses during a 2-second period as recorded by the instrumentation which measures the force as shown on the graph. The carriage has a velocity of 2 m/s at time $t = 0$. Approximate the P versus t relation and determine the velocity of the carriage when $t = 2$ s.

Ans. $v = 2.92$ m/s

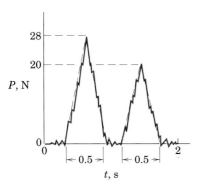

Problem 3/327

3/328 A ball is thrown from point O with a velocity of 10 m/s at a 60° angle with the horizontal and bounces on the inclined plane at A. If the coefficient of restitution is 0.6, calculate the magnitude v of the rebound velocity at A. Neglect air resistance.

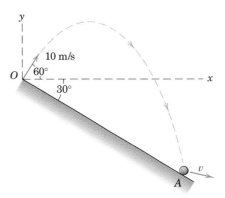

Problem 3/328

3/329 The figure shows a centrifugal clutch consisting in part of a rotating spider A which carries four plungers B. As the spider is made to rotate about its center with a speed ω, the plungers move outward and bear against the interior surface of the rim of wheel C, causing it to rotate. The wheel and spider are independent except for frictional contact. If each plunger has a mass of 2 kg with a center of mass at G, and if the coefficient of kinetic friction between the plungers and the wheel is 0.40, calculate the maximum moment M which can be transmitted to wheel C for a spider speed of 3000 rev/min.

Ans. $M = 18.96$ kN·m

Problem 3/329

3/330 The position of the small 0.5-kg blocks in the smooth radial slots in the disk which rotates about a vertical axis at O is used to activate a speed-control mechanism. If each block moves from $r = 150$ mm to $r = 175$ mm while the speed of the disk changes slowly from 300 rev/min to 400 rev/min, design the spring by calculating the spring constant k of each spring. The springs are attached to the inner ends of the slots and to the blocks.

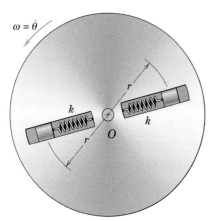

Problem 3/330

3/331 If the spring for the right-hand sliding block of Prob. 3/330 breaks, and the 0.5-kg block slides outward at the rate of $\dot{r} = 150$ mm/s as it passes the position $r = 180$ mm, compute the force exerted by the smooth side of the slot on the block at this instant. The disk turns at the steady rate of 400 rev/min. Also find $\ddot{r}$ at this instant.

Ans. $F = 6.28$ N, $\ddot{r} = 316$ m/s^2

3/332 A spacecraft is designed for Martian impact on a course heading directly toward the center of the planet. If the craft is 10 000 km from the surface of the planet and is traveling at a relative speed of 20 000 km/h, compute the impact velocity v. The gravitational acceleration at the surface of Mars is 3.73 m/s^2, and the planet has a diameter of approximately 6800 km. Neglect any atmospheric resistance.

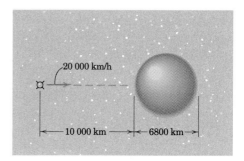

Problem 3/332

3/333 A person rolls a small ball with speed u along the floor from point A. If $x = 3R$, determine the required speed u so that the ball returns to A after rolling on the circular surface in the vertical plane from B to C and becoming a projectile at C. What is the minimum value of x for which the game could be played if contact must be maintained to point C? Neglect friction.

Ans. $u = \frac{5}{2}\sqrt{gR}$, $x_{min} = 2R$

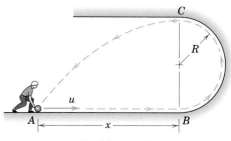

Problem 3/333

3/334 An automobile accident occurs as follows: The driver of a full-size car (vehicle A, 1800 kg) is traveling on a dry, level road and approaches a stationary compact car (vehicle B, 900 kg). Just 15 meters before collision, he applies the brakes, skidding all wheels. After impact, vehicle A skids an additional 15 m and vehicle B, whose driver had all brakes fully applied, skids 30 m. The final positions of the vehicles are shown in the figure. If the coefficient of kinetic friction is 0.9, was the driver of vehicle A exceeding the speed limit of 90 km/h before the initial application of his brakes?

Problem 3/334

3/335 The bungee jumper, an 80-kg man, falls from the bridge at A with the bungee cord secured to his ankles. He falls 20 m before the 17-m length of elastic bungee cord begins to stretch. The 3 m of rope above the elastic cord has no appreciable stretch. The man is observed to drop a total of 44 m before being projected upward. Neglect any energy loss and calculate (a) the stiffness k of the bungee cord per meter of its elongation, (b) the maximum velocity v_{max} of the man during his fall, and (c) his maximum acceleration a_{max}. Treat the man as a particle located at the end of the bungee cord.

> *Ans.* (a) $k = 119.9$ N/m
> (b) $v_{max} = 21.4$ m/s
> (c) $a_{max} = 26.2$ m/s^2

Problem 3/335

3/336 The 2-kg piece of putty is dropped 2 m onto the 18-kg block initially at rest on the two springs, each with a stiffness $k = 1.2$ kN/m. Calculate the additional deflection δ of the springs due to the impact of the putty which adheres to the block upon contact.

Problem 3/336

3/337 The 200-kg glider B is being towed by airplane A, which is flying horizontally with a constant speed of 220 km/h. The tow cable has a length $r = 60$ m and may be assumed to form a straight line. The glider is gaining altitude and when θ reaches 15°, the angle is increasing at the constant rate $\dot{\theta} = 5$ deg/s. At the same time the tension in the tow cable is 1520 N for this position. Calculate the aerodynamic lift L and drag D acting on the glider.

Ans. $L = 2540$ N, $D = 954$ N

Problem 3/337

3/338 Cars A and B of equal mass traveling in the directions shown collide at a blind intersection. They become entangled and move together, in the direction making the angle θ with the y-axis, with a common velocity v immediately after impact. Just prior to impact their velocities are $v_A = 60$ km/h and $v_B = 80$ km/h. Determine v and θ and the percentage n of energy loss. What is the coefficient of restitution e?

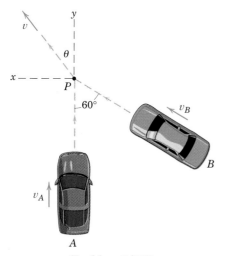

Problem 3/338

3/339 A baseball pitcher delivers a fast ball with a near-horizontal velocity of 145 km/h. The batter hits a home run over the center-field fence. The 146-g ball travels a horizontal distance of 110 m, with an initial velocity in the 45° direction shown. Determine the magnitude F_{av} of the average force exerted by the bat on the ball during the 0.005 s of contact between the bat and the ball. Neglect air resistance during the flight of the ball.

Ans. $F_{av} = 1974$ N

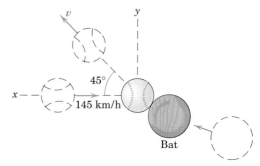

Problem 3/339

3/340 The system is released from rest with $\theta = 0$. The cord to the 1.5-kg cylinder is securely wound around the light 50-mm-diameter pulley at O, to which are attached the light arms and their 2-kg spheres. The centers of the spheres are 250 mm and 375 mm from the axis at O. Determine the downward velocity of the 1.5-kg cylinder when $\theta = 30°$.

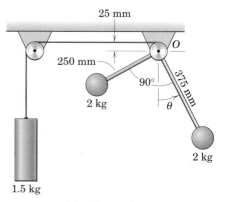

Problem 3/340

3/341 The 3-kg block A is released from rest in the 60° position shown and subsequently strikes the 1-kg cart B. If the coefficient of restitution for the collision is $e = 0.7$, determine the maximum displacement s of cart B beyond point C. Neglect friction.

Ans. $s = 2.28$ m

Problem 3/341

3/342 The crash-tested car of Prob. 2/26 is treated again here, where the impact velocity was found to be 38.9 km/h. If the car has a mass of 1100 kg and the front end of the car is crushed an amount $x = 0.8$ m, calculate the average force F_{av} exerted by the car on the barrier during the impact. Because the barrier does not move, F_{av} does no work. As a first approximation to the analysis, we consider the car to be composed of two parts as shown. The undamaged portion of mass 650 kg travels 0.8 m during the crash, with an average deceleration of $8g$. The accordioned portion of the car has a mass of 450 kg whose mass center has a deceleration of $4g$. Determine F_{av} by analyzing each part separately and find the loss of energy ΔE during the crash.

Problem 3/342

3/343 A short train consists of a 200-Mg locomotive and three 100-Mg hopper cars. The locomotive exerts a constant friction force of 200 kN on the rails as the train starts from rest. (*a*) If there is 300 mm of slack in each of the three couplers before the train begins moving, estimate the speed v of the train just after car C begins to move. Slack removal is a plastic short-duration impact. Neglect all friction except that of the locomotive tractive force and neglect the tractive force during the short time duration of the impacts associated with the slack removal. (*b*) If there is no slack in the train couplers, determine the speed v' which is acquired when the train has moved 900 mm.

Ans. (*a*) $v = 0.657$ m/s, (*b*) $v' = 0.849$ m/s

Problem 3/343

3/344 A long fly ball strikes the wall at point A (where $e_1 = 0.5$) and then hits the ground at B (where $e_2 = 0.3$). The outfielder likes to catch the ball when it is 1.2 m above the ground and 0.6 m in front of him as shown. Determine the distance x from the wall where he can catch the ball as described. Note the two possible solutions.

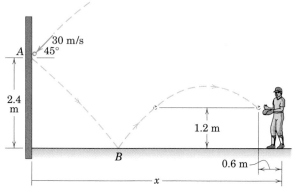

Problem 3/344

3/345 The satellite of Sample Problem 3/27 has a perigee velocity of 26 140 km/h at the perigee altitude of 2000 km. What is the minimum increase Δv in velocity required of its rocket motor at this position to allow the satellite to escape from the earth's gravity field?

Ans. $\Delta v = 9000$ km/h

3/346 Determine and plot the velocity v in kilometers per hour of a satellite in circular earth orbit as a function of altitude H ranging from $H = R$ to $H = 10R$. (Neglect atmospheric resistance for values of H close to R, the radius of the earth.)

3/347 One of the functions of the space shuttle is to release communications satellites at low altitude. A booster rocket is fired at B, placing the satellite in an elliptical transfer orbit, the apogee of which is the altitude necessary for a geosynchronous orbit. (A geosynchronous orbit is an equatorial-plane circular orbit whose period is equal to the absolute rotational period of the earth. A satellite in such an orbit appears to remain stationary to an earth-fixed observer.) A second booster rocket is then fired at C, and the final circular orbit is achieved. On one of the early space-shuttle missions, a 700-kg satellite was released from the shuttle at B, where $h_1 = 275$ km. The booster rocket was to fire for $t = 90$ seconds, forming a transfer orbit with $h_2 = 35\,900$ km. The rocket failed during its burn. Radar observations determined the apogee altitude of the transfer orbit to be only 1125 km. Determine the actual time t' that the rocket motor operated before failure. Assume negligible mass change during the booster rocket firing.

Ans. $t' = 8.47$ s

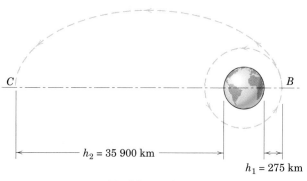

Problem 3/347

3/348 The vertical motion of the 60-kg cylinder is controlled by the two forces P applied to the bottom
rollers of the symmetrical frame. Determine the
constant force P which, if applied when the frame
is at rest with $\theta = 120°$, will give the cylinder an
upward velocity of 3 m/s when the position $\theta = 60°$ is passed. The links are very light so that their
mass may be neglected. Also, for the instant
when the upward acceleration of the cylinder is
20 m/s^2, find the corresponding force R under
each of the supporting rollers.

Problem 3/349

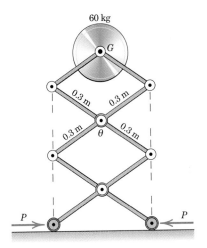

Problem 3/348

▶3/349 The fixed rod in the form of a spiral $r = k\theta$ lies
in the horizontal plane. A small collar of mass m
slides freely on the rod and is given a velocity v
along the rod starting at O. For negligible friction
the kinetic energy, and hence the magnitude of
the velocity, remains constant. Use the principle
established in Eq. 3/27 to determine an expression for the force P exerted on the rod in terms
of r.

$$\textit{Ans. } P = \frac{r^2 + 2k^2}{(r^2 + k^2)^{3/2}} mv^2$$

▶3/350 The retarding forces which act on the race car are
the drag force F_D and a nonaerodynamic force F_R.
The drag force is $F_D = C_D(\frac{1}{2}\rho v^2)S$, where C_D is
the drag coefficient, ρ is the air density, v is the
car speed, and $S = 2.8$ m^2 is the projected frontal
area of the car. The nonaerodynamic force F_R is
constant at 900 N. With its sheet metal in good
condition, the race car has a drag coefficient
$C_D = 0.3$ and it has a corresponding top speed
$v = 320$ km/h. After a minor collision, the dam-
aged front-end sheet metal causes the drag coefficient to be $C_D' = 0.4$. What is the corresponding
top speed v' of the race car?

$$\textit{Ans. } v' = 293 \text{ km/h}$$

Problem 3/350

▶3/351 The hollow tube rotates with a constant angular velocity ω_0 about a horizontal axis through end O. At time $t = 0$ the tube passes the vertical position $\theta = 0$, at which instant the small ball of mass m is released with r essentially zero. Determine r as a function of θ.

$$Ans. \quad r = \frac{g}{2\omega_0{}^2}(\cosh\theta - \cos\theta)$$

Problem 3/351

Problem 3/352

Computer-oriented problems

*3/353 The figure for Prob. 3/319 is repeated here. If the vertical frame starts from rest with a constant acceleration a and the smooth sliding collar A is initially at rest in the bottom position $\theta = 0$, plot $\dot{\theta}$ as a function of θ and find the maximum position angle θ_{max} reached by the collar. Use the values $a = g/2$ and $r = 0.3$ m.

$$Ans. \quad \dot{\theta} = 8.09\sqrt{0.5\sin\theta + \cos\theta - 1} \text{ rad/s}$$
$$\theta_{max} = 53.1°$$

▶3/352 Extensive wind-tunnel and coast-down studies of a 1000-kg automobile reveal the aerodynamic drag force F_D and the total nonaerodynamic rolling resistance force F_R to vary with speed as shown in the plot. Determine (*a*) the power P required for steady speeds of 45 km/h and 90 km/h and (*b*) the time t and the distance s required for the car to coast down to a speed of 5 km/h from an initial speed of 90 km/h. Assume a straight, level road and no wind.

$$Ans. \quad P_{45} = 2.11 \text{ kW}, \ P_{90} = 11.25 \text{ kW}$$
$$t = 250 \text{ s}, \ s = 1775 \text{ m}$$

Problem 3/353

*3/354 The 650-mm drum rotates about a horizontal axis with a constant angular velocity $\Omega = 7.5$ rad/s. The small block A has no motion relative to the drum surface as it passes the bottom position $\theta = 0$. Determine the coefficient of static friction μ_s that would result in block slippage at an angular position θ; plot your expression for $0 \le \theta \le 180°$. Determine the minimum required coefficient value $\mu_{\min}$ that would allow the block to remain fixed relative to the drum throughout a full revolution. For a friction coefficient slightly less than $\mu_{\min}$, at what angular position θ would slippage occur?

$\Omega = 7.5$ rad/s

$r = 325$ mm

O

A

θ

Problem 3/354

*3/355 The tennis player practices by hitting the ball against the wall at A. The ball bounces off the court surface at B and then up to its maximum height at C. For the conditions shown in the figure, plot the location of point C for values of the coefficient of restitution in the range $0.5 \le e \le 0.9$. (The value of e is common to both A and B.) For what value of e is $x = 0$ at point C, and what is the corresponding value of y?

Ans. $e = 0.617, y = 0.427$ m

y

$5°$ 24 m/s

0.9 m C

A

B

x

9 m

Problem 3/355

*3/356 The simple pendulum of length $l = 0.5$ m has an angular velocity $\dot\theta_0 = 0.2$ rad/s at time $t = 0$ when $\theta = 0$. Derive an integral expression for the time t required to reach an arbitrary angle θ. Plot t vs. θ for $0 \le \theta \le \dfrac{\pi}{2}$ and state the value of t for $\theta = \dfrac{\pi}{2}$.

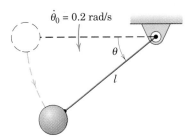

$\dot\theta_0 = 0.2$ rad/s

θ

l

Problem 3/356

*3/357 The 3-kg mass is released from rest in the position $x = 0$ where the spring of stiffness 1.8 kN/m has been compressed a horizontal distance $b = 100$ mm from its uncompressed position. Neglect any friction and plot the power P developed by the spring as a function of the recovery displacement x from its uncompressed position. Indicate the maximum value of P and the corresponding displacement x.

Ans. $P = \sqrt{38.9(10^8)}\,(0.1 - x)\sqrt{0.1x - 0.5x^2}$ W

$P_{\max} = 220$ W at $x = 0.0293$ m

100 mm

x

v

3 kg

$k = 1.8$ kN/m

Problem 3/357

*3/358 Wind-tunnel tests of the resistance of a certain 0.1-kg sphere in a moving airstream for low velocities give the plotted curve shown in the full line. If the sphere is released from rest in still air, use these data to predict the velocity v which it will acquire after dropping 10 m from rest. Multiply the equation of motion by dx and rewrite it as a finite-difference equation using 1-m intervals in your solution. Next, solve for v by approximating the data with the analytic expression $R = kv^2$, where agreement at $v = 6.5$ m/s represents a fair average with the experimental data over the region considered. (Read the curve and get $k = 0.0133$ N·s²/m².)

*3/359 The flexible bicycle-type chain of Prob. 3/173, reproduced again here, is released from rest in the smooth circular guide with $\theta = 0$. The length of the chain is $\pi r/2$. Plot the velocity of the chain in dimensionless form $v' = v/\sqrt{gr}$ as a function of θ from $\theta = 0$ to $\theta = \pi/2$.

$$Ans. \quad v' = \sqrt{2}\sqrt{\frac{1}{\pi}(2 + \theta^2) - \left(1 - \frac{2\theta}{\pi}\right)\frac{\sin^2\left(\frac{\pi}{4} - \frac{\theta}{2}\right)}{\frac{\pi}{4} - \frac{\theta}{2}}}$$

Problem 3/358

Problem 3/359

The dynamics of a group or system of masses is a generalization of the dynamics of a single-mass particle and leads to the equations of translation and rotation which describe the motion of nonrigid bodies as well as rigid bodies. The space shuttle, together with its detachable rocket boosters and its expendable fuel, is a good example of a nonrigid system which is losing mass.

KINETICS OF SYSTEMS OF PARTICLES

4

4/1 INTRODUCTION

In the previous two chapters, we have dealt with the principles of dynamics for the motion of a particle. Although attention was focused primarily on the kinetics of a single particle in Chapter 3, reference to the motion of two particles, considered together as a system, was made in the discussions of work-energy and impulse-momentum. Our next major step in the development of the subject of dynamics is to extend these principles, which we applied to a single particle, to describe the motion of a general system of particles. This extension provides a unity to the remaining sections of dynamics and enables us to treat the motion of rigid bodies and the motion of nonrigid systems as well. We recall that a rigid body is a solid system of particles, wherein the distances between particles remain essentially unchanged. The overall motions found with machines, land and air vehicles, rockets and spacecraft, and many moving structures provide examples of rigid-body problems. A nonrigid body, on the other hand, may be a solid body where the object of investigation is the time dependence of the changes in shape due to elastic or nonelastic deformations. Or a nonrigid body may be a defined mass of liquid or gaseous particles that are flowing at a specified rate. Examples would be the air and fuel flowing through the turbine of an aircraft engine, the burned gases issuing from the nozzle of a rocket motor, or the water passing through a rotary pump.

Although the extension of the equations for single-particle motion to a general system of particles is accomplished without undue difficulty, it cannot be expected that the generality and significance of these extended principles will be fully understood without considerable problem experience. For this reason, we strongly recommend that the general results obtained in the following articles be reviewed frequently during the remainder of the study of dynamics. In this way, the unity contained in these broader principles of dynamics will be brought into sharper focus and a more fundamental view of the subject will be acquired.

4/2 GENERALIZED NEWTON'S SECOND LAW

Figure 4/1

Newton's second law of motion will now be extended to cover a general mass system which we model by considering n mass particles bounded by a closed surface in space, Fig. 4/1. This bounding envelope, for example, might be the exterior surface of a given rigid body, the bounding surface of an arbitrary portion of the body, the exterior surface of a rocket containing both rigid and flowing particles, or a particular volume of fluid particles. In each case, the system considered is the mass within the envelope, and that mass must be clearly defined and isolated.

Figure 4/1 shows a representative particle of mass m_i of the system isolated with forces $\mathbf{F}_1, \mathbf{F}_2, \mathbf{F}_3, \ldots$ acting on m_i from sources *external* to the envelope, and forces $\mathbf{f}_1, \mathbf{f}_2, \mathbf{f}_3, \ldots$ acting on m_i from sources *internal* to the envelope. The external forces are due to contact with external bodies or to external gravitational, electric, or magnetic forces. The internal forces are forces of reaction with other mass particles within the envelope. A representative particle of mass m_i is located by its position vector $\mathbf{r}_i$ measured from the non-accelerating origin O of a Newtonian set of reference axes.* The center of mass G of the system of particles isolated is located by the position vector $\bar{\mathbf{r}}$ which, from the definition of the mass center as covered in statics, is given by

$$m\bar{\mathbf{r}} = \Sigma m_i \mathbf{r}_i$$

where the total system mass is $m = \Sigma m_i$. The summation sign Σ represents the summation $\Sigma_{i=1}^{n}$ over all n particles. Newton's second law, Eq. 3/3, when applied to m_i gives

$$\mathbf{F}_1 + \mathbf{F}_2 + \mathbf{F}_3 + \cdots + \mathbf{f}_1 + \mathbf{f}_2 + \mathbf{f}_3 + \cdots = m_i \ddot{\mathbf{r}}_i$$

where $\ddot{\mathbf{r}}_i$ is the acceleration of m_i. A similar equation may be written for each of the particles of the system. If these equations written for *all* particles of the system are added together, there results

$$\Sigma \mathbf{F} + \Sigma \mathbf{f} = \Sigma m_i \ddot{\mathbf{r}}_i$$

The term $\Sigma \mathbf{F}$ then becomes the vector sum of *all* forces acting on all particles of the isolated system from sources external to the system, and $\Sigma \mathbf{f}$ becomes the vector sum of all forces on all particles produced by the internal actions and reactions between particles. This last sum is identically zero since all internal forces occur in pairs of equal and opposite actions and reactions. By differentiating the equation defining $\bar{\mathbf{r}}$ twice with time, we have $m\ddot{\bar{\mathbf{r}}} = \Sigma m_i \ddot{\mathbf{r}}_i$ where m has no time derivative as long as mass is not entering or leaving

---

*It was shown in Art. 3/14 that any nonrotating and nonaccelerating set of axes constitutes a Newtonian reference system in which the principles of Newtonian mechanics are valid.

the system.* Substitution into the summation of the equations of motion gives

$$\Sigma \mathbf{F} = m\ddot{\bar{\mathbf{r}}} \quad \text{or} \quad \Sigma \mathbf{F} = m\bar{\mathbf{a}} \qquad (4/1)$$

where $\bar{\mathbf{a}}$ is the acceleration $\ddot{\bar{\mathbf{r}}}$ of the center of mass of the system.

Equation 4/1 is the generalized Newton's second law of motion for a mass system and is referred to as the *equation of motion of m*. The equation states that the resultant of the external forces on *any* system of mass equals the total mass of the system times the acceleration of the center of mass. This law expresses the so-called *principle of motion of the mass center*. It should be observed that $\bar{\mathbf{a}}$ is the acceleration of the mathematical point which represents instantaneously the position of the mass center for the given n particles. For a nonrigid body, this acceleration need not represent the acceleration of any particular particle. It should be noted further that Eq. 4/1 holds for each instant of time and is therefore an instantaneous relationship. Equation 4/1 for the mass system had to be proved as it may not be inferred directly from Eq. 3/3 for the single particle. Equation 4/1 may be expressed in component form using *x-y-z* coordinates or whatever coordinate system is most convenient for the problem at hand. Thus,

$$\Sigma F_x = m\bar{a}_x \qquad \Sigma F_y = m\bar{a}_y \qquad \Sigma F_z = m\bar{a}_z \qquad (4/1a)$$

Although Eq. 4/1, as a vector equation, requires that the acceleration vector $\bar{\mathbf{a}}$ must have the same direction as the resultant external force $\Sigma \mathbf{F}$, it does not follow that $\Sigma \mathbf{F}$ necessarily passes through G. In general, in fact, $\Sigma \mathbf{F}$ does not pass through G, as will be shown later.

4/3 WORK-ENERGY

In Art. 3/6 the work-energy relation for a single particle was developed. Applicability to a system of two joined particles was also noted. We now turn our attention to the general system of Fig. 4/1, where the work-energy relation for the representative particle of mass m_i is $(U_{1\text{-}2})_i = \Delta T_i$. Here $(U_{1\text{-}2})_i$ is the work done on m_i during an interval of motion by all forces $\mathbf{F}_i = \mathbf{F}_1 + \mathbf{F}_2 + \mathbf{F}_3 + \cdots$ applied from sources external to the system and by all forces $\mathbf{f}_i = \mathbf{f}_1 + \mathbf{f}_2 + \mathbf{f}_3 + \cdots$ applied from sources internal to the system. The kinetic energy of m_i is $T_i = \frac{1}{2}m_i v_i^2$, where v_i is the magnitude of the particle velocity $\mathbf{v}_i = \dot{\mathbf{r}}_i$.

For the entire system, the sum of the work-energy equations written for all particles is $\Sigma(U_{1\text{-}2})_i = \Sigma \Delta T_i$, which may be repre-

---

*If m is a function of time, a more complex situation develops; this situation is discussed in Art. 4/7 on variable mass.

sented by the same expression as Eq. 3/11 of Art. 3/6, namely,

$$U_{1\text{-}2} = \Delta T \quad \text{or} \quad T_1 + U_{1\text{-}2} = T_2 \qquad (4/2)$$

where $U_{1\text{-}2} = \Sigma(U_{1\text{-}2})_i$, the work done by all forces, external and internal, on all particles, and ΔT is the change in the total kinetic energy $T = \Sigma T_i$ of the system.

For a rigid body or a system of rigid bodies joined by ideal frictionless connections, no net work is done by the internal interacting forces or moments in the connections. We see that the work done by all pairs of internal forces, labeled here as $\mathbf{f}_i$ and $-\mathbf{f}_i$, at a typical connection, Fig. 4/2, in the system is zero since their points of application have identical displacement components while the forces are equal but opposite. For this situation $U_{1\text{-}2}$ becomes the work done on the system by the external forces only.

For a nonrigid mechanical system which includes elastic members capable of storing energy, a part of the work done by the external forces goes into changing the internal elastic potential energy V_e. Also, if the work done by the gravity forces is *excluded* from the work term and is accounted for instead by the changes in gravitational potential energy V_g, then we may equate the work $U'_{1\text{-}2}$ done on the system during an interval of motion to the change ΔE in the mechanical energy of the system. Thus, $U'_{1\text{-}2} = \Delta E$ or

Figure 4/2

$$U'_{1\text{-}2} = \Delta T + \Delta V_g + \Delta V_e \qquad (4/3)$$

or

$$T_1 + V_{g_1} + V_{e_1} + U'_{1\text{-}2} = T_2 + V_{g_2} + V_{e_2} \qquad (4/3a)$$

which are the same as Eqs. 3/17 and 3/17a.

We now examine the expression $T = \Sigma \frac{1}{2}m_i v_i^2$ for the kinetic energy of the mass system in more detail. By our principle of relative motion discussed in Art. 2/8, we may write the velocity of the representative particle as

$$\mathbf{v}_i = \bar{\mathbf{v}} + \dot{\boldsymbol{\rho}}_i$$

where $\bar{\mathbf{v}}$ is the velocity of the mass center G and $\dot{\boldsymbol{\rho}}_i$ is the velocity of m_i with respect to a translating reference frame moving with the mass center G. We recall the identity $v_i^2 = \mathbf{v}_i \cdot \mathbf{v}_i$ and write the kinetic energy of the system as

$$T = \Sigma \tfrac{1}{2}m_i \mathbf{v}_i \cdot \mathbf{v}_i = \Sigma \tfrac{1}{2}m_i (\bar{\mathbf{v}} + \dot{\boldsymbol{\rho}}_i) \cdot (\bar{\mathbf{v}} + \dot{\boldsymbol{\rho}}_i)$$
$$= \Sigma \tfrac{1}{2}m_i \bar{v}^2 + \Sigma \tfrac{1}{2}m_i |\dot{\boldsymbol{\rho}}_i|^2 + \Sigma m_i \bar{\mathbf{v}} \cdot \dot{\boldsymbol{\rho}}_i$$

Since ρ_i is measured from the mass center, $\Sigma m_i \rho_i = \mathbf{0}$ and the third term is $\overline{\mathbf{v}} \cdot \Sigma m_i \dot{\rho}_i = \overline{\mathbf{v}} \cdot \dfrac{d}{dt} \Sigma (m_i \rho_i) = 0$. Also $\Sigma \frac{1}{2} m_i \overline{v}^2 = \frac{1}{2} \overline{v}^2 \Sigma m_i = \frac{1}{2} m \overline{v}^2$. Therefore, the total kinetic energy becomes

$$\boxed{T = \tfrac{1}{2} m \overline{v}^2 + \Sigma \tfrac{1}{2} m_i |\dot{\rho}_i|^2} \tag{4/4}$$

This equation expresses the fact that the total kinetic energy of a mass system equals the energy of mass-center translation of the system as a whole plus the energy due to motion of all particles relative to the mass center.

4/4 IMPULSE-MOMENTUM

(a) *Linear momentum.* From our definition in Art. 3/8, the linear momentum of the representative particle of the system depicted in Fig. 4/1 is $\mathbf{G}_i = m_i \mathbf{v}_i$ where the velocity of m_i is $\mathbf{v}_i = \dot{\mathbf{r}}_i$. The linear momentum of the system is defined as the vector sum of the linear momenta of all of its particles or $\mathbf{G} = \Sigma m_i \mathbf{v}_i$. By substituting the relative-velocity relation $\mathbf{v}_i = \overline{\mathbf{v}} + \dot{\rho}_i$ and noting again that $\Sigma m_i \rho_i = m \overline{\rho} = \mathbf{0}$, we get

$$\mathbf{G} = \Sigma m_i (\overline{\mathbf{v}} + \dot{\rho}_i) = \Sigma m_i \overline{\mathbf{v}} + \frac{d}{dt} \Sigma m_i \rho_i$$

$$= \overline{\mathbf{v}} \Sigma m_i + \frac{d}{dt} (\mathbf{0})$$

or

$$\boxed{\mathbf{G} = m \overline{\mathbf{v}}} \tag{4/5}$$

Thus, the linear momentum of any mass system of constant mass is the product of the mass and the velocity of its center of mass.

The time derivative of $\mathbf{G}$ is $m \dot{\overline{\mathbf{v}}} = m \overline{\mathbf{a}}$, which by Eq. 4/1 is the resultant external force acting on the system. Thus, we have

$$\boxed{\Sigma \mathbf{F} = \dot{\mathbf{G}}} \tag{4/6}$$

which has the same form as Eq. 3/21 for a single particle. Equation 4/6 states that the resultant of the external forces on any mass system equals the time rate of change of the linear momentum of the system and is an alternative form of the generalized second law of motion, Eq. 4/1. As was noted at the end of the last article, $\Sigma \mathbf{F}$, in general, does not pass through the mass center G. In Eq. 4/6 the total mass was assumed to be constant during differentiation with time, so that the equation does not apply to systems whose mass changes with time.

(b) Angular momentum. The angular momentum of our general mass system will now be determined about the fixed point O, about the mass center G, and about an arbitrary point P, shown in Fig. 4/3, which may have an acceleration $\mathbf{a}_P = \ddot{\mathbf{r}}_P$.

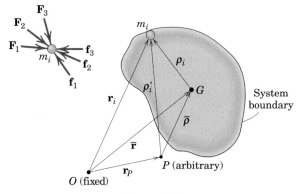

Figure 4/3

(About O). The angular momentum of the mass system about the point O, fixed in the Newtonian reference system, is defined as the vector sum of the moments of the linear momenta about O of all particles of the system and is

$$\mathbf{H}_O = \Sigma(\mathbf{r}_i \times m_i\mathbf{v}_i)$$

The time derivative of the vector product is $\dot{\mathbf{H}}_O = \Sigma(\dot{\mathbf{r}}_i \times m_i\mathbf{v}_i) + \Sigma(\mathbf{r}_i \times m_i\dot{\mathbf{v}}_i)$. The first summation vanishes since the cross product of two equal vectors $\dot{\mathbf{r}}_i$ and $\mathbf{v}_i$ is zero. The second summation is $\Sigma(\mathbf{r}_i \times m_i\mathbf{a}_i) = \Sigma(\mathbf{r}_i \times \mathbf{F}_i)$, which is the vector sum of the moments about O of all forces acting on all particles of the system. This moment sum $\Sigma\mathbf{M}_O$ represents only the moments of forces external to the system, since the internal forces cancel one another and their moments add up to zero. Thus, the moment sum is

$$\boxed{\Sigma\mathbf{M}_O = \dot{\mathbf{H}}_O} \tag{4/7}$$

which has the same form as Eq. 3/27 for a single particle. Equation 4/7 states that the resultant vector moment about any fixed point of all external forces on any system of mass equals the time rate of change of angular momentum of the system about the fixed point. As in the linear-momentum case, Eq. 4/7 does not apply if the total mass of the system is changing with time.

(About G). The angular momentum of the mass system about the mass center G is the sum of the moments of the linear momenta about G of all particles and is

$$\mathbf{H}_G = \Sigma \boldsymbol{\rho}_i \times m_i \dot{\mathbf{r}}_i \qquad (4/8)$$

We may write the absolute velocity $\dot{\mathbf{r}}_i$ as $(\dot{\mathbf{r}} + \dot{\boldsymbol{\rho}}_i)$ so that $\mathbf{H}_G$ becomes

$$\mathbf{H}_G = \Sigma \boldsymbol{\rho}_i \times m_i(\dot{\mathbf{r}} + \dot{\boldsymbol{\rho}}_i) = \Sigma \boldsymbol{\rho}_i \times m_i \dot{\mathbf{r}} + \Sigma \boldsymbol{\rho}_i \times m_i \dot{\boldsymbol{\rho}}_i$$

The first term on the right side of this equation may be rewritten as $-\dot{\mathbf{r}} \times \Sigma m_i \boldsymbol{\rho}_i$, which is zero because $\Sigma m_i \boldsymbol{\rho}_i = \mathbf{0}$ by definition of the mass center. Hence, we have

$$\mathbf{H}_G = \Sigma \boldsymbol{\rho}_i \times m_i \dot{\boldsymbol{\rho}}_i \qquad (4/8a)$$

The expression of Eq. 4/8 is called the *absolute* angular momentum because the absolute velocity $\dot{\mathbf{r}}_i$ is used. The expression of Eq. 4/8a is called the *relative* angular momentum because the relative velocity $\dot{\boldsymbol{\rho}}_i$ is used. With the mass center G as a reference, the absolute and relative angular momenta are seen to be identical. We will see that this identity does not hold for an arbitrary reference point P; there is no distinction for a fixed reference point O.

Differentiating Eq. 4/8 with respect to time gives

$$\dot{\mathbf{H}}_G = \Sigma \dot{\boldsymbol{\rho}}_i \times m_i(\dot{\mathbf{r}} + \dot{\boldsymbol{\rho}}_i) + \Sigma \boldsymbol{\rho}_i \times m_i \ddot{\mathbf{r}}_i$$

The first summation is expanded as $\Sigma \dot{\boldsymbol{\rho}}_i \times m_i \dot{\mathbf{r}} + \Sigma \dot{\boldsymbol{\rho}}_i \times m_i \dot{\boldsymbol{\rho}}_i$. The first term may be rewritten as $-\dot{\mathbf{r}} \times \Sigma m_i \dot{\boldsymbol{\rho}}_i = -\dot{\mathbf{r}} \times \dfrac{d}{dt} \Sigma m_i \boldsymbol{\rho}_i$ which is zero from the definition of the mass center. The second term is zero because the cross product of parallel vectors is zero. With $\mathbf{F}_i$ representing the sum of all external forces acting on m_i and $\mathbf{f}_i$ the sum of all internal forces acting on m_i, the second summation by Newton's second law becomes $\Sigma \boldsymbol{\rho}_i \times (\mathbf{F}_i + \mathbf{f}_i) = \Sigma \boldsymbol{\rho}_i \times \mathbf{F}_i = \Sigma \mathbf{M}_G$, the sum of all external moments about point G. Recall that the sum of all internal moments $\Sigma \boldsymbol{\rho}_i \times \mathbf{f}_i$ is zero. Thus, we are left with

$$\boxed{\Sigma \mathbf{M}_G = \dot{\mathbf{H}}_G} \qquad (4/9)$$

where we may use either the absolute or the relative angular momentum.

Equations 4/7 and 4/9 are among the most powerful of the governing equations in dynamics and apply to any defined system of constant mass—rigid or nonrigid.

(About P).   The angular momentum about an arbitrary point P (which may have an acceleration $\ddot{\mathbf{r}}_P$) will now be expressed with the notation of Fig. 4/3. Thus,

$$\mathbf{H}_P = \Sigma \boldsymbol{\rho}_i' \times m_i \dot{\mathbf{r}}_i = \Sigma(\overline{\boldsymbol{\rho}} + \boldsymbol{\rho}_i) \times m_i \dot{\mathbf{r}}_i$$

The first term may be written as $\bar{\boldsymbol{\rho}} \times \Sigma m_i \dot{\mathbf{r}}_i = \bar{\boldsymbol{\rho}} \times \Sigma m_i \mathbf{v}_i = \bar{\boldsymbol{\rho}} \times m\bar{\mathbf{v}}$. The second term is $\Sigma \boldsymbol{\rho}_i \times m_i \dot{\mathbf{r}}_i = \mathbf{H}_G$. Thus, rearranging gives

$$\boxed{\mathbf{H}_P = \mathbf{H}_G + \bar{\boldsymbol{\rho}} \times m\bar{\mathbf{v}}} \qquad (4/10)$$

Equation 4/10 states that the absolute angular momentum about any point P equals the angular momentum about G plus the moment about P of the linear momentum $m\bar{\mathbf{v}}$ of the system considered concentrated at G.

We now make use of the principle of moments developed in our study of statics where we represented a force system by a resultant force through any point, such as G, and a corresponding couple. Figure 4/4 represents the resultants of the external forces acting on the system expressed in terms of the resultant force $\Sigma\mathbf{F}$ through

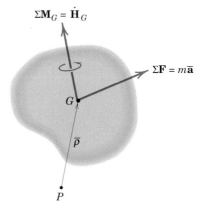

Figure 4/4

G and the corresponding couple $\Sigma\mathbf{M}_G$. We see that the sum of the moments about P of all forces external to the system must equal the moment of their resultants. Therefore, we may write

$$\Sigma\mathbf{M}_P = \Sigma\mathbf{M}_G + \bar{\boldsymbol{\rho}} \times \Sigma\mathbf{F}$$

which, by Eqs. 4/9 and 4/6, becomes

$$\boxed{\Sigma\mathbf{M}_P = \dot{\mathbf{H}}_G + \bar{\boldsymbol{\rho}} \times m\bar{\mathbf{a}}} \qquad (4/11)$$

Equation 4/11 enables us to write the moment equation about any convenient moment center P and is easily visualized with the aid of Fig. 4/4. This equation forms a rigorous basis for much of our treatment of planar rigid-body kinetics in Chapter 6.

We may also develop similar momentum relationships by using the momentum relative to P. Thus, from Fig. 4/3

$$(\mathbf{H}_P)_{\text{rel}} = \Sigma\boldsymbol{\rho}_i' \times m_i\dot{\boldsymbol{\rho}}_i'$$

where $\dot{\boldsymbol{\rho}}_i'$ is the velocity of m_i relative to P. With the substitution $\boldsymbol{\rho}_i' = \bar{\boldsymbol{\rho}} + \boldsymbol{\rho}_i$ and $\dot{\boldsymbol{\rho}}_i' = \dot{\bar{\boldsymbol{\rho}}} + \dot{\boldsymbol{\rho}}_i$ we may write

$$(\mathbf{H}_P)_{\text{rel}} = \Sigma\bar{\boldsymbol{\rho}} \times m_i\dot{\bar{\boldsymbol{\rho}}} + \Sigma\bar{\boldsymbol{\rho}} \times m_i\dot{\boldsymbol{\rho}}_i + \Sigma\boldsymbol{\rho}_i \times m_i\dot{\bar{\boldsymbol{\rho}}} + \Sigma\boldsymbol{\rho}_i \times m_i\dot{\boldsymbol{\rho}}_i$$

The first summation is $\bar{\boldsymbol{\rho}} \times m\bar{\mathbf{v}}_{\text{rel}}$. The second summation is $\bar{\boldsymbol{\rho}} \times \dfrac{d}{dt}\Sigma m_i\boldsymbol{\rho}_i$ and the third summation is $-\dot{\bar{\boldsymbol{\rho}}} \times \Sigma m_i\boldsymbol{\rho}_i$ where both are zero by definition of the mass center. The fourth summation is $(\mathbf{H}_G)_{\text{rel}}$. Rearranging gives us

$$(\mathbf{H}_P)_{\text{rel}} = (\mathbf{H}_G)_{\text{rel}} + \bar{\boldsymbol{\rho}} \times m\bar{\mathbf{v}}_{\text{rel}} \qquad (4/12)$$

where $(\mathbf{H}_G)_{\text{rel}}$ is the same as $\mathbf{H}_G$ (see Eqs. 4/8 and 4/8a). The similarity of Eqs. 4/12 and 4/10 should be noted.

The moment equation about P may now be expressed in terms of the angular momentum relative to P. We differentiate the definition $(\mathbf{H}_P)_{\text{rel}} = \Sigma\boldsymbol{\rho}_i' \times m_i\dot{\boldsymbol{\rho}}_i'$ with time and make the substitution $\ddot{\mathbf{r}}_i = \ddot{\mathbf{r}}_P + \ddot{\boldsymbol{\rho}}_i'$ to get

$$(\dot{\mathbf{H}}_P)_{\text{rel}} = \Sigma\dot{\boldsymbol{\rho}}_i' \times m_i\dot{\boldsymbol{\rho}}_i' + \Sigma\boldsymbol{\rho}_i' \times m_i\ddot{\mathbf{r}}_i - \Sigma\boldsymbol{\rho}_i' \times m_i\ddot{\mathbf{r}}_P$$

The first summation is identically zero, and the second summation is the sum $\Sigma\mathbf{M}_P$ of the moments of all external forces about P. The third summation becomes $\Sigma\boldsymbol{\rho}_i' \times m_i\mathbf{a}_P = -\mathbf{a}_P \times \Sigma m_i\boldsymbol{\rho}_i' = -\mathbf{a}_P \times m\bar{\boldsymbol{\rho}} = \bar{\boldsymbol{\rho}} \times m\mathbf{a}_P$. Substituting and rearranging terms give

$$\boxed{\;\Sigma\mathbf{M}_P = (\dot{\mathbf{H}}_P)_{\text{rel}} + \bar{\boldsymbol{\rho}} \times m\mathbf{a}_P\;} \qquad (4/13)$$

The form of Eq. 4/13 is convenient when a point P whose acceleration is known is used as a moment center. The equation reduces to the simpler form

$$\Sigma\mathbf{M}_P = (\dot{\mathbf{H}}_P)_{\text{rel}} \quad \text{if} \quad \begin{cases} \text{1. } \mathbf{a}_P = \mathbf{0} \text{ (equivalent to Eq. 4/7)} \\ \text{2. } \bar{\boldsymbol{\rho}} = \mathbf{0} \text{ (equivalent to Eq. 4/9)} \\ \text{3. } \bar{\boldsymbol{\rho}} \text{ and } \mathbf{a}_P \text{ are parallel } (\mathbf{a}_P \text{ directed} \\ \qquad \text{toward or away from } G) \end{cases}$$

4/5   CONSERVATION OF ENERGY AND MOMENTUM

Conditions frequently occur when there is no net change in the total mechanical energy of a system during an interval of motion. Other conditions exist when there is no net change in the momentum of a system. These conditions are treated separately as follows.

(a) Conservation of energy. A mass system is said to be *conservative* if it does not lose energy by virtue of internal friction forces that do negative work or by virtue of nonelastic members which dissipate energy upon cycling. If no work is done on a conservative system during an interval of motion by external forces (other than gravity or other potential forces), then no part of the energy of the system is lost. In this case, $\Delta E = 0$ or $E_{\text{initial}} = E_{\text{final}}$. Therefore, we may write Eq. 4/3 as

$$\Delta T + \Delta V_g + \Delta V_e = 0 \tag{4/14}$$

or
$$T_1 + V_{g_1} + V_{e_1} = T_2 + V_{g_2} + V_{e_2} \tag{4/14a}$$

which expresses the *law of conservation of dynamical energy*. This law holds only in the ideal case where internal kinetic friction is sufficiently small to be neglected.

(b) Conservation of momentum. If, for a certain interval of time, the resultant external force $\Sigma \mathbf{F}$ acting on a conservative or nonconservative mass system is zero, Eq. 4/6 requires that $\dot{\mathbf{G}} = \mathbf{0}$, so that during this interval

$$\mathbf{G}_1 = \mathbf{G}_2 \tag{4/15}$$

which expresses the *principle of conservation of linear momentum*. Thus, in the absence of an external impulse the linear momentum of a system remains unchanged.

Similarly, if the resultant moment about a fixed point O or about the mass center G of all external forces on any mass system is zero, Eq. 4/7 or 4/9 requires, respectively, that

$$(\mathbf{H}_O)_1 = (\mathbf{H}_O)_2 \quad \text{or} \quad (\mathbf{H}_G)_1 = (\mathbf{H}_G)_2 \tag{4/16}$$

These relations express the *principle of conservation of angular momentum* for a general mass system in the absence of an angular impulse. Thus, if there is no angular impulse about a fixed point (or about the mass center), the angular momentum of the system about the fixed point (or about the mass center) remains unchanged. Either equation may hold without the other.

It was proved in Art. 3/14 that the basic laws of Newtonian mechanics hold for measurements made relative to a set of axes that translate with a constant velocity. Thus, Eqs. 4/1 through 4/16 are valid provided all quantities are expressed relative to the translating axes.

Equations 4/1 through 4/16 are among the most important of the basic derived laws of mechanics. In this chapter these laws have been derived for the most general system of constant mass in order that the generality of the laws will be established. These laws find repeated use when applied to specific mass systems such as rigid and nonrigid solids and certain fluid systems which are discussed in the articles that follow. The reader is urged to study these laws carefully and to compare them with their more restricted forms encountered earlier in Chapter 3.

Sample Problem 4/1

Each of the three balls has a mass m and is welded to the rigid equi-angular frame of negligible mass. The assembly rests on a smooth horizontal surface. If a force $\mathbf{F}$ is suddenly applied to the one bar as shown, determine (a) the acceleration of point O and (b) the angular acceleration $\ddot{\theta}$ of the frame.

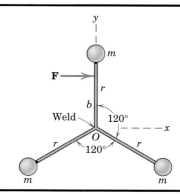

Solution. *(a)* Point O is the mass center of the system of the three balls, so that its acceleration is given by Eq. 4/1.

① $[\Sigma \mathbf{F} = m\bar{\mathbf{a}}]$ $\qquad F\mathbf{i} = 3m\bar{\mathbf{a}} \qquad \bar{\mathbf{a}} = \mathbf{a}_O = \dfrac{F}{3m}\mathbf{i}$ *Ans.*

(b) We determine $\ddot{\theta}$ from the moment principle, Eq. 4/9. To find $\mathbf{H}_G$ we note that the velocity of each ball relative to the mass center O as measured in the nonrotating axes x-y is $r\dot{\theta}$, where $\dot{\theta}$ is the common angular velocity of the spokes. The angular momentum of the system about O is the sum of the moments of the relative linear momenta as shown by Eq. 4/8, so it is expressed by

$$H_O = H_G = 3(mr\dot{\theta})r = 3mr^2\dot{\theta}$$

② Equation 4/9 now gives

$[\Sigma \mathbf{M}_G = \dot{\mathbf{H}}_G]$

$$Fb = \frac{d}{dt}(3mr^2\dot{\theta}) = 3mr^2\ddot{\theta} \qquad \text{so} \qquad \ddot{\theta} = \frac{Fb}{3mr^2} \quad \textit{Ans.}$$

① We note that the result depends only on the magnitude and direction of $\mathbf{F}$ and not on b, which locates the line of action of $\mathbf{F}$.

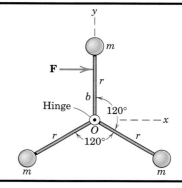

② Although $\dot{\theta}$ is initially zero, we need the expression for $H_O = H_G$ in order to get $\dot{H}_G$. We observe also that $\ddot{\theta}$ is independent of the motion of O.

Sample Problem 4/2

Consider the same conditions as for Sample Problem 4/1, except that the spokes are freely hinged at O and so do not constitute a rigid system. Explain the difference between the two problems.

Solution. The generalized Newton's second law holds for any mass system, so that the acceleration $\bar{\mathbf{a}}$ of the mass center G is the same as with Sample Problem 4/1, namely,

$$\bar{\mathbf{a}} = \frac{F}{3m}\mathbf{i} \qquad \textit{Ans.}$$

Although G coincides with O at the instant represented, the motion of the hinge O is not the same as the motion of G since O will not remain the center of mass as the angles between the spokes change.

① Both ΣM_G and $\dot{H}_G$ have the same values for the two problems at the instant represented. However, the angular motions of the spokes in this problem are all different and are not easily determined.

① This present system could be dismembered and the motion equations written for each of the parts, with the unknowns eliminated one by one. Or a more sophisticated method using the equations of Lagrange could be employed. (See the first author's *Dynamics, 2nd Edition SI Version,* 1975, for a discussion of this approach.)

Sample Problem 4/3

A shell with a mass of 20 kg is fired from point O, with a velocity $u = 300$ m/s in the vertical x-z plane at the inclination shown. When it reaches the top of its trajectory at P, it explodes into three fragments A, B, and C. Immediately after the explosion, fragment A is observed to rise vertically a distance of 500 m above P, and fragment B is seen to have a horizontal velocity $\mathbf{v}_B$ and eventually land at point Q. When recovered, the masses of the fragments A, B, and C are found to be 5, 9, and 6 kg, respectively. Calculate the velocity which fragment C has immediately after the explosion. Neglect atmospheric resistance.

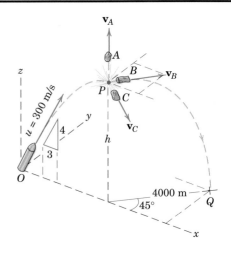

Solution. From our knowledge of projectile motion (Sample Problem 2/6), the time required for the shell to reach P and its vertical rise are

$$t = u_z/g = 300(4/5)/9.81 = 24.5 \text{ s}$$

$$h = \frac{u_z^2}{2g} = \frac{[(300)(4/5)]^2}{2(9.81)} = 2940 \text{ m}$$

The velocity of A has the magnitude

$$v_A = \sqrt{2gh_A} = \sqrt{2(9.81)(500)} = 99.0 \text{ m/s}$$

With no z-component of velocity initially, fragment B requires 24.5 s to return to the ground. Thus, its horizontal velocity, which remains constant, is

$$v_B = s/t = 4000/24.5 = 163.5 \text{ m/s}$$

Since the force of the explosion is internal to the system of the shell and its three fragments, the linear momentum of the system remains unchanged during the explosion. Thus,

① $[\mathbf{G}_1 = \mathbf{G}_2]$ $m\mathbf{v} = m_A\mathbf{v}_A + m_B\mathbf{v}_B + m_C\mathbf{v}_C$

$20(300)(\tfrac{3}{5})\mathbf{i} = 5(99.0\mathbf{k}) + 9(163.5)(\mathbf{i}\cos 45° + \mathbf{j}\sin 45°) + 6\mathbf{v}_C$

$6\mathbf{v}_C = 2560\mathbf{i} - 1040\mathbf{j} - 495\mathbf{k}$

② $\mathbf{v}_C = 427\mathbf{i} - 173.4\mathbf{j} - 82.5\mathbf{k}$ m/s

$v_C = \sqrt{(427)^2 + (173.4)^2 + (82.5)^2} = 468$ m/s   *Ans.*

① The velocity $\mathbf{v}$ of the shell at the top of its trajectory is, of course, the constant horizontal component of its initial velocity $\mathbf{u}$, which becomes $u(3/5)$.

② We note that the mass center of the three fragments while still in flight continues to follow the same trajectory that the shell would have followed if it had not exploded.

Sample Problem 4/4

The 16-kg carriage A moves horizontally in its guide with a speed of 1.2 m/s and carries two assemblies of balls and light rods which rotate about a shaft at O in the carriage. Each of the four balls has a mass of 1.6 kg. The assembly on the front face rotates counterclockwise at a speed of 80 rev/min, and the assembly on the back side rotates clockwise at a speed of 100 rev/min. For the entire system, calculate (a) the kinetic energy T, (b) the magnitude G of the linear momentum, and (c) the magnitude H_O of the angular momentum about point O.

Solution. (a) *Kinetic energy.* The velocity of the balls with respect to O is

$$[|\dot{\boldsymbol{\rho}}_i| = v_{\text{rel}} = r\dot{\theta}] \quad (v_{\text{rel}})_{1,2} = 0.450 \frac{80(2\pi)}{60} = 3.77 \text{ m/s}$$

$$(v_{\text{rel}})_{3,4} = 0.300 \frac{100(2\pi)}{60} = 3.14 \text{ m/s}$$

The kinetic energy of the system is given by Eq. 4/4. The translational part is

$$\tfrac{1}{2}m\bar{v}^2 = \tfrac{1}{2}[16 + 4(1.6)](1.2)^2 = 16.13 \text{ J}$$

The rotational part of the kinetic energy depends on the squares of the relative velocities and is

$$\Sigma\tfrac{1}{2}m_i|\dot{\boldsymbol{\rho}}_i|^2 = 2\,[(\tfrac{1}{2})(1.6)(3.77)^2]_{(1,2)} + 2\,[(\tfrac{1}{2})(1.6)(3.14)^2]_{(3,4)}$$

$$= 22.7 + 15.79 = 38.5 \text{ J}$$

The total kinetic energy is

$$T = \tfrac{1}{2}m\bar{v}^2 + \Sigma\tfrac{1}{2}m_i|\dot{\boldsymbol{\rho}}_i|^2 = 16.13 + 38.5 = 54.7 \text{ J} \qquad Ans.$$

(b) *Linear momentum.* The linear momentum of the system by Eq. 4/5 is the total mass times v_O, the velocity of the center of mass. Thus,

$$[\mathbf{G} = m\bar{\mathbf{v}}] \qquad G = [16 + 4(1.6)](1.2) = 26.9 \text{ kg·m/s}$$

(c) *Angular momentum about O.* The angular momentum about O is due to the moments of the linear momenta of the balls. Taking counterclockwise as positive, we have

$$H_O = \Sigma|\mathbf{r}_i \times m_i\mathbf{v}_i|$$

$$H_O = [2(1.6)(0.450)(3.77)]_{(1,2)} - [2(1.6)(0.300)(3.14)]_{(3,4)}$$

$$= 5.43 - 3.02 = 2.41 \text{ kg·m}^2/\text{s} \qquad Ans.$$

① Note that the mass m is the total mass, carriage plus the four balls, and that $\bar{v}$ is the velocity of the mass center O which is the carriage velocity.

② Note that the direction of rotation, clockwise or counterclockwise, makes no difference in the calculation of kinetic energy, which depends on the square of the velocity.

③ There is a temptation to overlook the contribution of the balls since their linear momenta relative to O in each pair are in opposite directions and cancel. However, each ball also has a velocity component $\bar{\mathbf{v}}$ and hence a momentum component $m_i\bar{\mathbf{v}}$.

④ Contrary to the case of kinetic energy where the direction of rotation was immaterial, angular momentum is a vector quantity and the direction of rotation must be accounted for.

PROBLEMS

Introductory problems

4/1 Each of the identical 4-kg steel balls is fastened to the other two by connecting bars of negligible mass and unequal length. In the absence of friction at the supporting horizontal surface, determine the initial acceleration $\bar{a}$ of the mass center of the assembly when it is subjected to the horizontal force $F = 200$ N applied to the supporting ball. The assembly is initially at rest in the vertical plane. Can you show that $\bar{a}$ is initially horizontal?

Ans. $\bar{a} = 16.67$ m/s^2

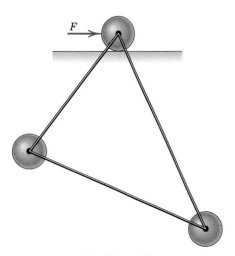

Problem 4/1

4/2 The three monkeys A, B, and C with masses of 10, 12, and 8 kg, respectively, are climbing up and down the rope suspended from D. At the instant represented, A is descending the rope with an acceleration of 1.6 m/s^2, and C is pulling himself up with an acceleration of 0.9 m/s^2. Monkey B is climbing up with a constant speed of 0.6 m/s. Treat the rope and the monkeys as a complete system and calculate the tension T in the rope at D.

Problem 4/2

4/3 The three identical bars, each with a mass of 8 kg, are connected by the two freely-pinned links of negligible mass and the assembly is resting on a smooth horizontal surface. Calculate the initial acceleration a of the center of the middle bar when the 100-N force is applied to the connecting link as shown.

Ans. $a = 4.17$ m/s^2

Problem 4/3

4/4 Calculate the acceleration of the center of mass of the system of the four 10-kg cylinders. Neglect friction and the mass of the pulleys and cables.

500 N   250 N

10 kg   10 kg

10 kg   10 kg

Problem 4/4

4/5 The two small spheres, each of mass m, are connected by a cord of length $2b$ (measured from the centers of the spheres) and are initially at rest on a smooth horizontal surface in the position shown. If a vertical force of constant magnitude F is applied to the center A of the cord, determine the velocity v of each sphere when they collide as θ approaches 90°. What is the maximum value of F for which the spheres do not lose contact with the surface? (Analyze the system without dismembering it.)

$$Ans. \ v = \sqrt{\frac{Fb}{m}}(1 - \sin\theta), \ F < 2mg$$

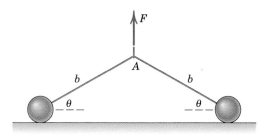

Problem 4/5

4/6 The two small spheres, each of mass m, are rigidly connected by a rod of negligible mass. The center C of the rod has a velocity v in the x-direction, and the rod is rotating counterclockwise at the constant rate $\dot{\theta}$. For a given value of θ, write the expressions for (a) the linear momentum of each sphere and (b) the linear momentum $\mathbf{G}$ of the system of the two spheres.

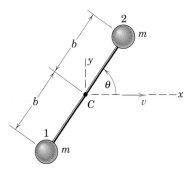

Problem 4/6

4/7 A 60-g bullet is fired horizontally with a velocity $v = 300$ m/s into the slender bar of a 1.5-kg pendulum initially at rest. If the bullet embeds itself in the bar, compute the resulting angular velocity of the pendulum immediately after the impact. Treat the sphere as a particle and neglect the mass of the rod. Why is the linear momentum of the system not conserved?

$$Ans. \ \omega = 11.88 \ rad/s$$

250 mm

60 g

v

250 mm

1.5 kg

Before   After

Problem 4/7

4/8 A centrifuge consists of four cylindrical containers, each of mass m, at a radial distance r from the rotation axis. Determine the time t required to bring the centrifuge to an angular velocity ω from rest under a constant torque M applied to the shaft. The diameter of each container is small compared with r, and the mass of the shaft and supporting arms is small compared with m.

Problem 4/8

Representative problems

4/9 The two small spheres, each of mass m, are rigidly connected by a rod of negligible mass and are released from rest in the position shown and slide down the smooth circular guide in the vertical plane. Determine their common velocity v as they reach the horizontal dashed position. Also find the force R between sphere 1 and the supporting surface an instant before the sphere reaches the bottom position A.

Ans. $v = 1.137gr$, $R = 2.29mg$

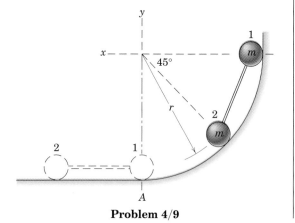

Problem 4/9

4/10 The four 3-kg balls are rigidly mounted to the rotating frame and shaft, which are initially rotating freely about the vertical z-axis at the angular rate of 20 rad/s clockwise when viewed from above. If a constant torque $M = 30$ N·m is applied to the shaft, calculate the time t to reverse the direction of rotation and reach an angular velocity $\dot{\theta} = 20$ rad/s in the same sense as M.

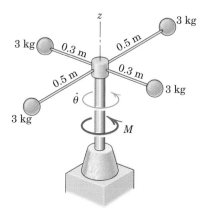

Problem 4/10

4/11 The 300-kg and 400-kg mine cars are rolling in opposite directions along the horizontal track with the respective speeds of 0.6 m/s and 0.3 m/s. Upon impact the cars become coupled together. Just prior to impact, a 100-kg boulder leaves the delivery chute with a velocity of 1.2 m/s in the direction shown and lands in the 300-kg car. Calculate the velocity v of the system after the boulder has come to rest relative to the car. Would the final velocity be the same if the cars were coupled before the boulder dropped?

Ans. $v = 0.205$ m/s

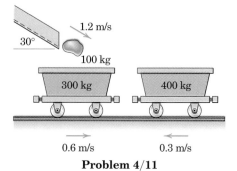

Problem 4/11

4/12 The three freight cars are rolling along the horizontal track with the velocities shown. After the impacts occur, the three cars become coupled together and move with a common velocity v. The loaded cars A, B, and C have masses of 65 Mg, 50 Mg, and 75 Mg, respectively. Determine v and calculate the percentage loss n of energy of the system due to coupling.

2 km/h   1 km/h   1.5 km/h

A B C

Problem 4/12

4/13 Billiard ball A is moving in the y-direction with a velocity of 2 m/s when it strikes ball B of identical size and mass initially at rest. Following the impact, the balls are observed to move in the directions shown. Calculate the velocities v_A and v_B which the balls have immediately after the impact. Treat the balls as particles and neglect any friction forces acting on the balls compared with the force of impact.

Ans. $v_A = 1.015$ m/s, $v_B = 1.556$ m/s

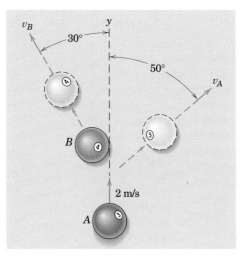

Problem 4/13

4/14 The man of mass m_1 and the woman of mass m_2 are standing on opposite ends of the platform of mass m_0 which moves with negligible friction and is initially at rest with $s = 0$. The man and woman begin to approach each other. Derive an expression for the displacement s of the platform when the two meet in terms of the displacement x_1 of the man relative to the platform.

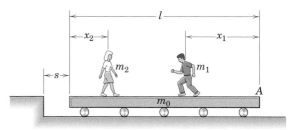

Problem 4/14

4/15 The 60-kg young woman A, the 90-kg captain B, and the 80-kg sailor C are sitting in the 150-kg skiff which is gliding through the water with a speed of 1 knot. If the three people change their positions as shown in the second figure, find the distance x from the skiff to the position where it would have been if the people had not moved. Neglect any resistance to motion afforded by the water. Does the sequence or timing of the change in positions affect the final result?

Ans. $x = 0.0947$ m

Problem 4/15

4/16 The two small spheres, each of mass m, and their connecting rod of negligible mass are rotating about their mass center G with an angular velocity ω. At the same instant the mass center has a velocity v in the x-direction. Determine the angular momentum $\mathbf{H}_O$ of the assembly at the instant when G has coordinates x and y.

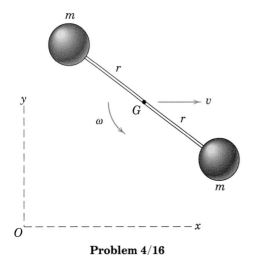

Problem 4/16

4/17 The two balls are attached to the light rigid rod, which is suspended by a cord from the support above it. If the balls and rod, initially at rest, are struck by the force $F = 60$ N, calculate the corresponding acceleration $\bar{a}$ of the mass center and the rate $\ddot{\theta}$ at which the angular velocity of the bar is changing.

Ans. $\bar{a} = 20$ m/s^2, $\ddot{\theta} = 336$ rad/s^2

Problem 4/17

4/18 The two spheres are rigidly connected to the rod of the negligible mass and are initially at rest on the smooth horizontal surface. A force F is suddenly applied to the one sphere in the y-direction and imparts an impulse of 10 N·s during a negligibly short period of time. As the spheres pass the dashed position, calculate the velocity of each one.

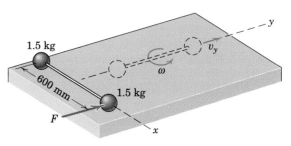

Problem 4/18

4/19 The three small steel balls, each of mass 2.75 kg, are connected by the hinged links of negligible mass and equal length. They are released from rest in the positions shown and slide down the quarter-circular guide in the vertical plane. When the upper sphere reaches the bottom position, the spheres have a horizontal velocity of 1.560 m/s. Calculate the energy loss ΔQ due to friction and the total impulse I_x on the system of three spheres during this interval.

Ans. $\Delta Q = 2.52$ J, $I_x = 12.87$ N·s

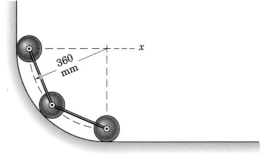

Problem 4/19

4/20   The flexible nonextensible rope of length $\pi r/2$ and mass ρ per unit length is attached at A to the fixed quarter-circular guide and allowed to fall from rest in the horizontal position. When the rope comes to rest in the dashed position, the system will have lost energy. Determine the loss ΔQ and explain what becomes of the lost energy.

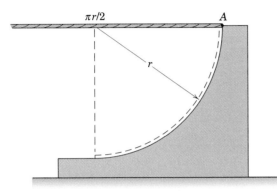

Problem 4/20

4/21   Two steel balls, each of mass m, are welded to a light rod of length L and negligible mass and are initially at rest on a smooth horizontal surface. A horizontal force of magnitude F is suddenly applied to the rod as shown. Determine (a) the instantaneous acceleration $\bar{a}$ of the mass center G and (b) the corresponding rate $\ddot{\theta}$ at which the angular velocity of the assembly about G is changing with time.

$$Ans.\ (a)\ \bar{a} = \frac{F}{2m},\ (b)\ \ddot{\theta} = \frac{2Fb}{mL^2}$$

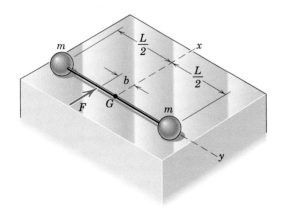

Problem 4/21

4/22   The small car, which has a mass of 20 kg, rolls freely on the horizontal track and carries the 5-kg sphere mounted on the light rotating rod with $r = 0.4$ m. A geared motor drive maintains a constant angular speed $\dot{\theta} = 4$ rad/s of the rod. If the car has a velocity $v = 0.6$ m/s when $\theta = 0$, calculate v when $\theta = 60°$. Neglect the mass of the wheels and any friction.

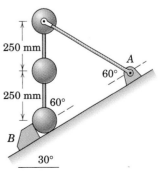

Problem 4/22

4/23   The three identical spheres, each of mass m, are supported in the vertical plane on the 30° incline. The spheres are welded to the two connecting rods of negligible mass. The upper rod, also of negligible mass, is pivoted freely to the upper sphere and to the bracket at A. If the stop at B is suddenly removed, determine the velocity v with which the upper sphere hits the incline. (Note that the corresponding velocity of the middle sphere is $v/2$.) Explain the loss of energy which has occurred after all motion has ceased.

$$Ans.\ v = 4.20\ m/s$$

Problem 4/23

4/24 The cars of a roller-coaster ride have a speed of 30 km/h as they pass over the top of the circular track. Neglect any friction and calculate their speed v when they reach the horizontal bottom position. At the top position, the radius of the circular path of their mass centers is 18 m, and all six cars have the same mass.

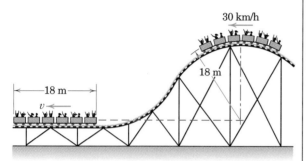

Problem 4/24

4/25 The two small spheres, each of mass m, are connected by a cord of length $2b$ (measured to the centers of the spheres) and are initially at rest on a smooth horizontal surface. A projectile of mass m_0 with a velocity v_0 perpendicular to the cord hits it in the middle, causing the deflection shown in the b-part of the figure. Determine the velocity v of m_0 as the two spheres near contact, with θ approaching 90° as indicated in the c-part of the figure. Also find $\dot{\theta}$ for this condition.

$$Ans. \ v = \frac{m_0}{m_0 + 2m} v_0$$

$$\dot{\theta} = \frac{v_0}{b} \sqrt{\frac{m_0}{m_0 + 2m}}$$

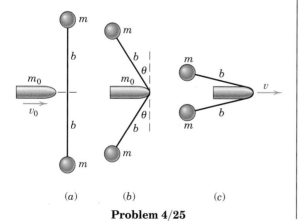

Problem 4/25

4/26 The three identical 2-kg spheres are welded to the connecting rods of negligible mass and are hanging by a cord from point A. The spheres are initially at rest when a horizontal force $F = 16$ N is applied to the upper sphere. Calculate the initial acceleration $\bar{a}$ of the mass center of the spheres, the rate $\ddot{\theta}$ at which the angular velocity is increasing, and the initial acceleration a of the top sphere.

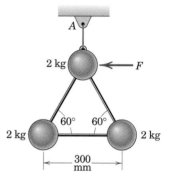

Problem 4/26

4/27 The system of eight symmetrically located spheres, each of mass 2 kg, is rotating freely about the vertical z-axis at the rate $N = 120$ rev/min. The outer spheres are welded to the ends of the rods of negligible mass. The inner spheres are initially latched in the positions shown but are able to slide outward along the rods when the latches (not shown) are released. Determine the new angular rate N' after release and after the spheres have moved 250 mm and come to rest against the outer spheres. Also calculate the loss ΔT in kinetic energy and account for the loss. The spheres may be treated as particles.

$$Ans. \ N' = 90.4 \ \text{rev/min}, \ \Delta T = 124.8 \ \text{J loss}$$

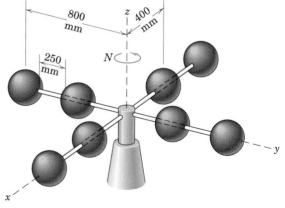

Problem 4/27

4/28 The carriage of mass $2m$ is free to roll along the horizontal rails and carries the two spheres, each of mass m, mounted on rods of length l and negligible mass. The shaft to which the rods are secured is mounted in the carriage and is free to rotate. If the system is released from rest with the rods in the vertical position where $\theta = 0$, determine the velocity v_x of the carriage and the angular velocity $\dot{\theta}$ of the rods for the instant when $\theta = 180°$. Treat the carriage and the spheres as particles and neglect any friction.

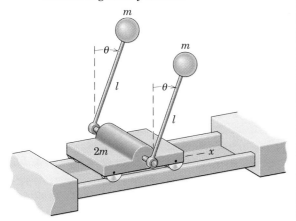

Problem 4/28

▶**4/29** In the unstretched position the coils of the 1.5-kg spring are just touching one another, as shown in the a-part of the figure. In the stretched position the force P, proportional to x, equals 900 N when $x = 500$ mm. If end A of the spring is suddenly released, determine the velocity v_A of the coil end A, measured positive to the left, as it approaches its unstretched position at $x = 0$. What happens to the kinetic energy of the spring?

Ans. $v_A = 30$ m/s

Problem 4/29

▶**4/30** The 25-Mg flatcar supports a 7.5-Mg vehicle on a 5° ramp built on the flatcar. If the vehicle is released from rest with the flatcar also at rest, determine the velocity v of the flatcar when the vehicle has rolled $s = 12$ m down the ramp just before hitting the stop at B. Neglect all friction and treat the vehicle and the flatcar as particles.

Ans. $v = 1.186$ m/s

Problem 4/30

4/6 STEADY MASS FLOW

The momentum relation developed in Art. 4/4 for a general system of mass provides us with a direct means of analyzing the action of mass flow where a change of momentum occurs. The dynamics of mass flow is of great importance in the description of fluid machinery of all types including turbines, pumps, nozzles, air-breathing jet engines, and rockets. The treatment of mass flow in this article is not intended to take the place of a study of fluid mechanics, but merely to present the basic principles and equations of momentum that find important usage in fluid mechanics and in the general flow of mass whether the form be liquid, gaseous, or granular.

One of the most important cases of mass flow occurs during steady-flow conditions where the rate at which mass enters a given volume equals the rate at which mass leaves the same volume. The volume in question may be enclosed by a rigid container, fixed or moving, such as the nozzle of a jet aircraft or rocket, the space between blades in a gas turbine, the volume within the casing of a centrifugal pump, or the volume within the bend of a pipe through which a fluid is flowing at a steady rate. The design of such fluid machines is dependent on the analysis of the forces and moments that are developed through the corresponding momentum changes of the flowing mass.

Consider a rigid container, shown in section in Fig. 4/5a, into which mass flows in a steady stream at the rate m' through the entrance section of area A_1. Mass leaves the container through the exit section of area A_2 at the same rate, so that there is no accumulation or depletion of the total mass within the container during the period of observation. The velocity of the entering stream is $\mathbf{v}_1$ normal to A_1 and that of the leaving stream is $\mathbf{v}_2$ normal to A_2. If ρ_1 and ρ_2 are the respective densities of the two streams, the continuity of flow requires that

$$\rho_1 A_1 v_1 = \rho_2 A_2 v_2 = m' \qquad (4/17)$$

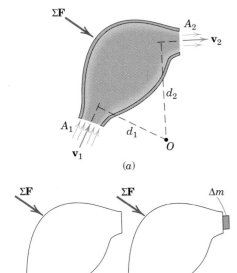

(a)

(b)

Figure 4/5

The forces which act are described by isolating either the mass of fluid within the container or the entire container and the fluid within it. We would use the first approach if the forces between the container and the fluid were to be described, and we would adopt the second approach when the forces external to the container are desired. The latter situation is our primary interest, in which case, the *system isolated* consists of the fixed structure of the container and the fluid within it at a particular instant of time. This isolation is described by a free-body diagram of the mass within a closed space volume defined by the exterior surface of the container and the entrance and exit surfaces. We must account for all forces applied *externally* to this system, and in Fig. 4/5a the vector sum of this external force system is denoted by $\Sigma\mathbf{F}$. Included in $\Sigma\mathbf{F}$ are the forces exerted on the container at points of its attachment to other structures, including attachments at A_1 and A_2, if present; the forces acting on the fluid within the container at A_1 and A_2 due to any static pressure that may exist in the fluid at these positions; and the weight of the fluid and structure if appreciable. The resultant $\Sigma\mathbf{F}$ of all of these external forces must equal $\dot{\mathbf{G}}$, the time rate of change of the linear momentum of the isolated system, by Eq. 4/6 which was developed in Art. 4/4 for any system of constant mass, rigid or nonrigid.

The expression for $\dot{\mathbf{G}}$ may be obtained by an incremental analysis. Figure 4/5b illustrates the system at time t when the system mass is that of the container, the mass within it, and an increment Δm about to enter during time Δt. At time $t + \Delta t$ the same total mass is that of the container, the mass within it, and an equal increment Δm that leaves the container in time Δt. The linear momentum of the container and mass within it between the two sections A_1 and A_2 remains unchanged during Δt so that the change in momentum of the system in time Δt is

$$\Delta\mathbf{G} = (\Delta m)\mathbf{v}_2 - (\Delta m)\mathbf{v}_1 = \Delta m(\mathbf{v}_2 - \mathbf{v}_1)$$

Division by Δt and passage to the limit yield $\dot{\mathbf{G}} = m'\,\Delta\mathbf{v}$, where

$$m' = \lim_{\Delta t \to 0}\left(\frac{\Delta m}{\Delta t}\right) = \frac{dm}{dt}$$

Thus, by Eq. 4/6

$$\boxed{\Sigma\mathbf{F} = m'\Delta\mathbf{v}} \tag{4/18}$$

Equation 4/18 establishes the relation between the resultant force on a steady-flow system and the corresponding mass flow rate and vector velocity increment.*

Alternatively, we may note that the time rate of change of linear momentum is the vector difference between the rate at which linear momentum leaves the system and the rate at which linear momentum enters the system. Thus, we may write $\dot{\mathbf{G}} = m'\mathbf{v}_2 - m'\mathbf{v}_1 = m'\Delta\mathbf{v}$, which agrees with the foregoing result.

We can now see one of the powerful applications of our general force-momentum equation that we derived for any mass system. Our system here includes a body which is rigid (the structural container for the mass stream) and particles which are in motion (the

---

*We must be careful not to interpret dm/dt as the time derivative of the mass of the isolated system. That derivative is zero since the system mass is constant for a steady-flow process. To help avoid confusion, the symbol m' rather than dm/dt is used to represent the steady mass flow rate.

flow of mass). By defining the boundary of the system, the mass within which is constant for steady flow conditions, we are able to utilize the generality of Eq. 4/6. However, we must be very careful to account for *all* external forces acting on the system, and they become clear if our free-body diagram is correct.

A similar formulation is obtained for the case of angular momentum in steady-flow systems. The resultant moment of all external forces about some fixed point O on or off the system, Fig. 4/5a, equals the time rate of change of angular momentum of the system about O. This fact was established in Eq. 4/7 which, for the case of steady flow in a single plane, becomes

$$\Sigma M_O = m'(v_2 d_2 - v_1 d_1) \qquad (4/19)$$

When the velocities of the incoming and outgoing flows are not in the same plane, the equation may be written in vector form as

$$\boxed{\Sigma \mathbf{M}_O = m'(\mathbf{d}_2 \times \mathbf{v}_2 - \mathbf{d}_1 \times \mathbf{v}_1)} \qquad (4/19a)$$

where $\mathbf{d}_1$ and $\mathbf{d}_2$ are the position vectors to the centers of A_1 and A_2 from the fixed reference O. In both relations, the mass center G may be used alternatively as a moment center by virtue of Eq. 4/9.

Equations 4/18 and 4/19 are very simple relations which find important use in describing relatively complex fluid actions. It should be noted that these equations relate *external* forces to the resultant changes in momentum and are independent of the flow path and momentum changes *internal* to the system.

The foregoing analysis may also be applied to systems which move with constant velocity by noting that the basic relations $\Sigma \mathbf{F} = \dot{\mathbf{G}}$ and $\Sigma \mathbf{M}_O = \dot{\mathbf{H}}_O$ or $\Sigma \mathbf{M}_G = \dot{\mathbf{H}}_G$ apply to systems moving with constant velocity as discussed in Arts. 3/12 and 4/4. The only restriction is that the mass within the system remain constant with respect to time.

Three examples of the analysis of steady mass flow are given in the following sample problems, which illustrate the application of the principles embodied in Eqs. 4/18 and 4/19.

Sample Problem 4/5

The smooth vane shown diverts the open stream of fluid of cross-sectional area A, mass density ρ, and velocity v. (a) Determine the force components R and F required to hold the vane in a fixed position. (b) Find the forces when the vane is given a constant velocity u less than v and in the direction of v.

Fixed vane

Solution. **Part (a).** The free-body diagram of the vane together with the fluid portion undergoing the momentum change is shown. The momentum equation may be applied to the isolated system for the change in motion in both the x- and y-directions. With the vane stationary, the magnitude of the exit velocity v' equals that of the entering velocity v with fluid friction neglected. The changes in the velocity components are then

① $$\Delta v_x = v' \cos \theta - v = -v(1 - \cos \theta)$$

and

$$\Delta v_y = v' \sin \theta - 0 = v \sin \theta$$

The mass rate of flow is $m' = \rho A v$, and substitution into Eq. 4/18 gives

$$[\Sigma F_x = m'\Delta v_x] \qquad -F = \rho A v[-v(1 - \cos \theta)]$$

$$F = \rho A v^2(1 - \cos \theta) \qquad Ans.$$

$$[\Sigma F_y = m'\Delta v_y] \qquad R = \rho A v[v \sin \theta]$$

$$R = \rho A v^2 \sin \theta \qquad Ans.$$

Part (b). In the case of the moving vane, the final velocity v' of the fluid upon exit is the vector sum of the velocity u of the vane plus the velocity of the fluid relative to the vane $v - u$. This combination is shown in the velocity diagram to the right of the figure for the exit conditions. The x-component of v' is the sum of the components of its two parts, so $v'_x = (v - u) \cos \theta + u$. The change in x-velocity of the stream is

$$\Delta v_x = (v - u) \cos \theta + (u - v) = -(v - u)(1 - \cos \theta)$$

The y-component of v' is $(v - u) \sin \theta$, so that the change in the y-velocity of the stream is $\Delta v_y = (v - u) \sin \theta$.

The mass rate of flow m' is the mass undergoing momentum change per unit of time. This rate is the mass flowing over the vane per unit time and *not* the rate of issuance from the nozzle. Thus,

$$m' = \rho A(v - u)$$

The impulse-momentum principle of Eq. 4/18 applied in the positive coordinate direction gives

② $$[\Sigma F_x = m'\Delta v_x] \qquad -F = \rho A(v - u)[-(v - u)(1 - \cos \theta)]$$

$$F = \rho A(v - u)^2(1 - \cos \theta) \qquad Ans.$$

$$[\Sigma F_y = m'\Delta v_y] \qquad R = \rho A(v - u)^2 \sin \theta \qquad Ans.$$

Moving vane

① Be careful with algebraic signs when using Eq. 4/18. The change in v_x is the final value minus the initial value measured in the positive x-direction. Also we must be careful to write $-F$ for ΣF_x.

② Observe that for given values of u and v, the angle for maximum force F is $\theta = 180°$.

Sample Problem 4/6

For the moving vane of Sample Problem 4/5, determine the optimum speed u of the vane for the generation of maximum power by the action of the fluid on the vane.

Solution. The force R shown with the figure for Sample Problem 4/5 is normal to the velocity of the vane so does no work. The work done by the force F shown is negative, but the power developed by the force (reaction to F) exerted by the fluid on the moving vane is

$$[P = Fu] \qquad P = \rho A(v - u)^2 u(1 - \cos \theta)$$

The velocity of the vane for maximum power for the one blade in the stream is specified by

$$\left[\frac{dP}{du} = 0\right] \qquad \rho A(1 - \cos \theta)(v^2 - 4uv + 3u^2) = 0$$

① $$(v - 3u)(v - u) = 0 \qquad u = \frac{v}{3} \qquad \textit{Ans.}$$

The second solution $u = v$ gives a minimum condition of zero power. An angle $\theta = 180°$ completely reverses the flow and clearly produces both maximum force and maximum power for any value of u.

① The result here applies to a single vane only. In the case of multiple vanes, such as the blades on a turbine disk, the rate at which fluid issues from the nozzles is the same rate at which fluid is undergoing momentum change. Thus, $m' = \rho A v$ rather than $\rho A(v - u)$. With this change, the optimum value of u turns out to be $u = v/2$.

Sample Problem 4/7

The offset nozzle has a discharge area A at B and an inlet area A_0 at C. A liquid enters the nozzle at a static gage pressure p through the fixed pipe and issues from the nozzle with a velocity v in the direction shown. If the constant density of the liquid is ρ, write expressions for the tension T, shear Q, and bending moment M in the pipe at C.

Solution. The free-body diagram of the nozzle and the fluid within it shows the tension T, shear Q, and bending moment M acting on the flange of the nozzle where it attaches to the fixed pipe. The force pA_0 on the fluid within the nozzle due to the static pressure is an additional external force.

Continuity of flow with constant density requires that

$$Av = A_0 v_0$$

where v_0 is the velocity of the fluid at the entrance to the nozzle. The momentum principle of Eq. 4/18 applied to the system in the two coordinate directions gives

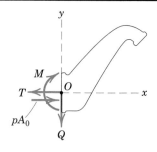

$$[\Sigma F_x = m' \Delta v_x] \qquad pA_0 - T = \rho Av(v \cos \theta - v_0)$$

① $$T = pA_0 + \rho Av^2 \left(\frac{A}{A_0} - \cos \theta\right) \qquad \textit{Ans.}$$

$$[\Sigma F_y = m' \Delta v_y] \qquad -Q = \rho Av(-v \sin \theta - 0)$$

$$Q = \rho Av^2 \sin \theta \qquad \textit{Ans.}$$

The moment principle of Eq. 4/19 applied in the clockwise sense gives

② $$[\Sigma M_O = m'(v_2 d_2 - v_1 d_1)] \qquad M = \rho Av(va \cos \theta + vb \sin \theta - 0)$$

$$M = \rho Av^2(a \cos \theta + b \sin \theta) \qquad \textit{Ans.}$$

① Again, be careful to observe the correct algebraic signs of the terms on both sides of Eqs. 4/18 and 4/19.

② The forces and moment acting on the pipe are equal and opposite to those shown acting on the nozzle.

Sample Problem 4/8

An air-breathing jet aircraft of total mass m flying with a constant speed v consumes air at the mass rate m_a' and exhausts burned gas at the mass rate m_g' with a velocity u relative to the aircraft. Fuel is consumed at the constant rate m_f'. The total aerodynamic forces acting on the aircraft are the lift L, normal to the direction of flight, and the drag D, opposite to the direction of flight. Any force due to the static pressure across the inlet and exhaust surfaces is assumed to be included in D. Write the equation for the motion of the aircraft and identify the thrust T.

Solution. The free-body diagram of the aircraft together with the air, fuel, and exhaust gas within it is given and shows only the weight, lift, and drag forces as defined. We attach axes x-y to the aircraft and apply our momentum equation relative to the moving system.

The fuel will be treated as a steady stream entering the aircraft with no velocity relative to the system and leaving with a relative velocity u in the exhaust stream. We now apply Eq. 4/18 relative to the reference axes and treat the air and fuel flows separately. For the air flow, the change in velocity in the x-direction relative to the moving system is

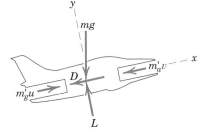

$$\Delta v_a = -u - (-v) = -(u - v)$$

and for the fuel flow the x-change in velocity relative to x-y is

$$\Delta v_f = -u - (0) = -u$$

Thus, we have

$$[\Sigma F_x = m' \Delta v_x] \qquad -mg \sin \theta - D = -m_a'(u - v) - m_f'u$$
$$= -m_g'u + m_a'v$$

where the substitution $m_g' = m_a' + m_f'$ has been made. Changing signs gives

$$m_g'u - m_a'v = mg \sin \theta + D$$

which is the equation of motion of the system.

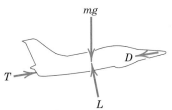

If we modify the boundaries of our system to expose the interior surfaces on which the air and gas act, we will have the simulated model shown, where the air exerts a force $m_a'v$ on the interior of the turbine and the exhaust gas reacts against the interior surfaces with the force $m_g'u$.

The commonly used model is shown in the final diagram, where the net effect of air and exhaust momentum changes is replaced by a simulated thrust

$$T = m_g'u - m_a'v \qquad \qquad \textit{Ans.}$$

applied to the aircraft from a presumed external source.

Inasmuch as m_f' is generally only 2 percent or less of m_a', we can use the approximation $m_g' \cong m_a'$ and express the thrust as

$$T \cong m_g'(u - v) \qquad \qquad \textit{Ans.}$$

We have analyzed the case of constant velocity. Although our Newtonian principles do not generally hold relative to accelerating axes, it can be shown that we may use the $F = ma$ equation for the simulated model and write $T - mg \sin \theta - D = m\dot{v}$ with virtually no error.

① Note that the boundary of the system cuts across the air stream at the entrance to the air scoop and across the exhaust stream at the nozzle.

② We are permitted to use moving axes which translate with constant velocity. See Arts. 3/14 and 4/2.

③ Riding with the aircraft, we observe the air entering our system with a velocity $-v$ measured in the plus x-direction and leaving the system with an x-velocity of $-u$. The final value minus the initial one gives the expression cited, namely, $-u - (-v) = -(u - v)$.

④ We now see that the "thrust" is, in reality, not a force external to the entire airplane shown in the first figure but can be modeled as an external force.

PROBLEMS

Introductory problems

4/31 A large rocket ready for vertical launch has a total mass of $2.7(10^3)$ Mg. At launch, fuel is burned at the rate of 13 Mg/s with an exhaust velocity of 2400 m/s. Determine the initial acceleration a of the rocket. Assume atmospheric pressure across the exit plane of the exhaust nozzle.

Ans. $a = 1.746$ m/s^2

4/32 The experimental race car is propelled by a rocket motor and is designed to reach a maximum speed $v = 500$ km/h under the thrust T of its motor. Prior wind-tunnel tests disclose that the wind resistance at this speed is 1000 N. If the rocket motor is burning fuel at the rate of 1.6 kg/s, determine the velocity u of the exhaust gases relative to the car.

Problem 4/32

4/33 When only the air of a sand-blasting gun is turned on, the force of the air on a flat surface normal to the stream and close to the nozzle is 20 N. With the nozzle in the same position, the force increases to 30 N when sand is admitted to the stream. If sand is being consumed at the rate of 4.5 kg/min, calculate the velocity v of the sand particles as they strike the surface.

Ans. $v = 133.3$ m/s

4/34 In an operational design test of its equipment, the fire truck's water cannon is delivering fresh water through its 50-mm-diameter nozzle at the rate of 5.30 m^3/min at the 20° angle. Calculate the total friction force F exerted by the pavement on the tires of the truck, which remains in a fixed position with its brakes locked.

Problem 4/34

4/35 The jet water ski has reached its maximum velocity of 70 km/h when operating in salt water. The water intake is in the horizontal tunnel in the bottom of the hull, so the water enters the intake at the velocity of 70 km/h relative to the ski. The motorized pump discharges water from the horizontal exhaust nozzle of 50-mm diameter at the rate of 0.082 m^3/s. Calculate the resistance R of the water to the hull at the operating speed.

Ans. $R = 1885$ N

Problem 4/35

4/36 The fire tug discharges a stream of salt water (density 1030 kg/m³) with a nozzle velocity of 40 m/s at the rate of 0.080 m³/s. Calculate the propeller thrust T which must be developed by the tug to maintain a fixed position while pumping.

Problem 4/36

4/37 A jet-engine noise suppressor consists of a movable duct that is secured directly behind the jet exhaust by cable A and deflects the blast directly upward. During a ground test, the engine sucks in air at the rate of 43 kg/s and burns fuel at the rate of 0.8 kg/s. The exhaust velocity is 720 m/s. Determine the tension T in the cable.

Ans. $T = 32.6$ kN

Problem 4/37

Representative problems

4/38 The rocket shown is designed to test the operation of a new guidance system. When it has reached a certain altitude beyond the effective influence of the earth's atmosphere, its mass has decreased to 2.80 Mg, and its trajectory is 30° from the vertical. Rocket fuel is being consumed at the rate of 120 kg/s with an exhaust velocity of 640 m/s relative to the nozzle. Gravitational acceleration is 9.34 m/s² at its altitude. Calculate the n- and t-components of the acceleration of the rocket.

Problem 4/38

4/39 A jet of fresh water under pressure issues from the 20-mm-diameter fixed nozzle with a velocity $v = 40$ m/s and is diverted into the two equal streams. Neglect any energy loss in the streams and compute the force F required to hold the vane in place.

Ans. $F = 938$ N

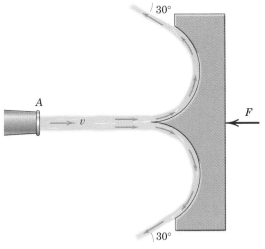

Problem 4/39

4/40 Salt water is being discharged into the atmosphere from the two 30° outlets at the total rate of 30 m³/min. Each of the discharge nozzles has a flow diameter of 100 mm, and the inside diameter of the pipe at the connecting section A is 250 mm. The pressure of the water at section A-A is 550 kPa. If each of the six bolts at the flange A-A is tightened to a tension of 10 kN, calculate the average pressure p on the flange gasket, which has an area of 24.0×10^3 mm². The pipe above the flange and the water within it have a mass of 60 kg.

Problem 4/40

4/41 The 250-g ball is supported by the vertical stream of fresh water which issues from the 12-mm-diameter nozzle with a velocity of 10 m/s. Calculate the height h of the ball above the nozzle. Assume that the stream remains intact and there is no energy lost in the jet stream.

Ans. $h = 4.86$ m

Problem 4/41

4/42 A jet of fluid with cross-sectional area A and mass density ρ issues from the nozzle with a velocity v and impinges on the inclined trough shown in section. Some of the fluid is diverted in each of the two directions. If the trough is smooth, the velocity of both diverted streams remains v, and the only force which can be exerted on the trough is normal to the bottom surface. Hence, the trough will be held in position by forces whose resultant is F normal to the trough. By writing impulse-momentum equations for the directions along and normal to the trough, determine the force F required to support the trough. Also find the volume rate of flow Q_1 and Q_2 for the two streams.

Problem 4/42

4/43 A jet-engine thrust reverser to reduce aircraft speed of 200 km/h after landing employs folding vanes which deflect the exhaust gases in the direction indicated. If the engine is consuming 50 kg of air per second and 0.65 kg of fuel per second, calculate the braking thrust as a fraction n of the engine thrust without the deflector vanes. The exhaust gases have a velocity of 650 m/s relative to the nozzle.

Ans. $n = 0.638$

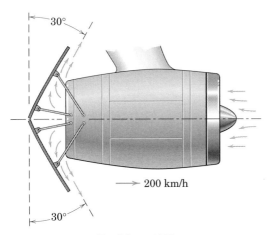

Problem 4/43

4/44 The axial-flow fan C pumps air through the duct of circular cross section and exhausts it with a velocity v at B. The air densities at A and B are ρ_A and ρ_B, respectively, and the corresponding pressures are p_A and p_B. The fixed deflecting blades at D restore axial flow to the air after it passes through the propeller blades C. Write an expression for the resultant horizontal force R exerted on the fan unit by the flange and bolts at A.

Problem 4/44

4/45 One of the most advanced methods for cutting metal plates uses a high-velocity water jet which carries an abrasive garnet powder. The jet issues from the 0.25-mm-diameter nozzle at A and follows the path shown through the thickness t of the plate. As the plate is slowly moved to the right, the jet makes a narrow precision slot in the plate. The water-abrasive mixture is used at the low rate of $2(10^{-3})$ m³/min and has a density of 1100 kg/m³. Water issues from the bottom of the plate with a velocity which is 60 percent of the impinging nozzle velocity. Calculate the horizontal force F required to hold the plate against the jet.

Ans. $F = 23.0$ N

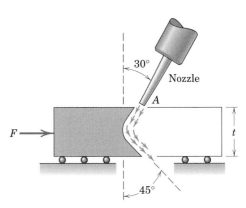

Problem 4/45

4/46 In the static test of a jet engine and exhaust nozzle assembly, air is sucked into the engine at the rate of 30 kg/s and fuel is burned at the rate of 1.6 kg/s. The flow area, static pressure, and axial-flow velocity for the three sections shown are as follows:

	Sec. A	Sec. B	Sec. C
Flow area, m^2	0.15	0.16	0.06
Static pressure, kPa	−14	140	14
Axial-flow velocity, m/s	120	315	600

Determine the tension T in the diagonal member of the supporting test stand and calculate the force F exerted on the nozzle flange at B by the bolts and gasket to hold the nozzle to the engine housing.

Problem 4/46

4/47 The diverter section of pipe between A and B is designed to allow the parallel pipes to clear an obstruction. The flange of the diverter is secured at C by a heavy bolt. The pipe carries fresh water at the steady rate of 20 m^3/min under a static pressure of 900 kPa entering the diverter. The inside diameter of the pipe at A and at B is 100 mm. The tensions in the pipe at A and B are balanced by the pressure in the pipe acting over the flow area. There is no shear or bending of the pipes at A or B. Calculate the moment M supported by the bolt at C.

Ans. $M = 2830$ N·m

Problem 4/47

4/48 The sump pump has a net mass of 310 kg and pumps fresh water against a 6-m head at the rate of 0.125 m^3/s. Determine the vertical force R between the supporting base and the pump flange at A during operation. The mass of water in the pump may be taken as the equivalent of a 200-mm-diameter column 6 m in height.

Problem 4/48

4/49 The experimental ground-effect machine has a total mass of 2.2 Mg. It hovers close to the ground
by pumping air at atmospheric pressure through
the circular intake duct at B and discharging it
horizontally under the periphery of the skirt C.
For an intake velocity v of 45 m/s, calculate the
average air pressure p under the 6-m-diameter
machine at ground level. The density of the air is
1.206 kg/m^3.

Ans. p = 1.035 kPa

Problem 4/49

4/50 The ducted fan unit of mass m is supported in the
vertical position on its flange at A. The unit draws
in air with a density ρ and a velocity u through
section A and discharges it through section B with
a velocity v. Both inlet and outlet pressures are
atmospheric. Write an expression for the force R
applied to the flange of the fan unit by the supporting slab.

Problem 4/50

4/51 The fire hose is looped around through 180° from
where it is attached to the fire hydrant. Water
pressure in the hydrant is 800 kPa and the nozzle
velocity is 24 m/s. The inside diameter of the hose
at the hydrant is 80 mm, and the flow rate is 3
m^3/min. Calculate the force F required to hold the
nozzle of the flexible hose in place and find the
tension T in the hose at the hydrant. Neglect any
friction between the hose and its horizontal supporting surface.

Ans. F = 1200 N, T = 4.52 kN

Problem 4/51

4/52 A rotary snow plow mounted on a large truck eats
its way through a snow drift on a level road at a
constant speed of 20 km/h. The plow discharges
60 Mg of snow per minute from its 45° chute with
a velocity of 12 m/s relative to the plow. Calculate
the tractive force P on the tires in the direction of
motion necessary to move the plow and find the
corresponding lateral force R between the tires
and the road.

Problem 4/52

4/53 A vertical force P acting on the 1.2-kg cylindrical valve A, shown in section, serves to limit the flow of fresh water from the top of the vertical pipe B of 80-mm inside diameter. Water is fed through the bottom inlet of the pipe. Calculate the force P required to maintain the valve in the position shown under a flow rate of 2.3 m^3/min and a static pressure of 80 kPa in the water at section C. (*Suggestion:* Choose the valve and shaded portion of water above section C as the combined free body.)

Ans. $P = 671$ N

Problem 4/53

4/54 The industrial blower sucks in air through the axial opening A with a velocity v_1 and discharges it at atmospheric pressure and temperature through the 150-mm-diameter duct B with a velocity v_2. The blower handles 16 m^3 of air per minute with the motor and fan running at 3450 rev/min. If the motor requires 0.32 kW of power under no load (both ducts closed), calculate the power P consumed while air is being pumped.

Problem 4/54

4/55 The military jet aircraft has a total mass of 10 Mg and is poised for takeoff with brakes set while the engine is revved up to maximum power. At this condition, air with a density of 1.206 kg/m^3 is sucked into the intake ducts at the rate of 48 kg/s with a static pressure of -2.0 kPa (gage) across the duct entrance. The total cross-sectional area of both intake ducts (one on each side) is 1.160 m^2. The air–fuel ratio is 18, and the exhaust velocity u is 940 m/s with zero back pressure (gage) across the exhaust nozzle. Compute the initial acceleration a of the aircraft upon release of the brakes.

Ans. $a = 4.83$ m/s^2

Problem 4/55

4/56 The helicopter shown has a mass m and hovers in position by imparting downward momentum to a column of air defined by the slipstream boundary shown. Find the downward velocity v given to the air by the rotor at a section in the stream below the rotor, where the pressure is atmospheric and the stream radius is r. Also find the power P required of the engine. Neglect the rotational energy of the air, any temperature rise due to air friction, and any change in air density ρ.

Problem 4/56

4/57 The VTOL (vertical takeoff and landing) military aircraft is capable of rising vertically under the action of its jet exhaust, which can be "vectored" from $\theta \cong 0$ for takeoff and hovering to $\theta = 90°$ for forward flight. The loaded aircraft has a mass of 8600 kg. At full takeoff power, its turbo-fan engine consumes air at the rate of 90 kg/s and has an air–fuel ratio of 18. Exhaust-gas velocity is 1020 m/s with essentially atmospheric pressure across the exhaust nozzles. Air with a density of 1.206 kg/m³ is sucked into the intake scoops at a pressure of -2 kPa (gage) over the total inlet area of 1.10 m². Determine the angle θ for vertical takeoff and the corresponding vertical acceleration a_y of the aircraft.

Ans. $\theta = 2.31°, a_y = 1.448$ m/s²

Problem 4/57

4/58 The sprinkler is made to rotate at the constant angular velocity ω and distributes water at the volume rate Q. Each of the four nozzles has an exit area A. Write an expression for the torque M on the shaft of the sprinkler necessary to maintain the given motion. For a given pressure and, hence, flow rate Q, at what speed ω_0 will the sprinkler operate with no applied torque? Let ρ be the density of the water.

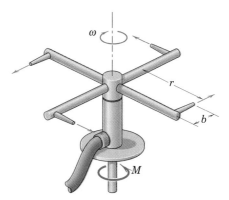

Problem 4/58

4/59 A high-speed jet of air issues from the 40-mm-diameter nozzle A with a velocity v of 240 m/s and impinges on the vane OB, shown in its edge view. The vane and its right-angle extension have negligible mass compared with the attached 6-kg cylinder and are freely pivoted about a horizontal axis through O. Calculate the angle θ assumed by the vane with the horizontal. The air density under the prevailing conditions is 1.206 kg/m³.

Ans. $\theta = 38.2°$

Problem 4/59

▶4/60 The centrifugal pump handles 20 m³ of fresh water per minute with inlet and outlet velocities of 18 m/s. The impeller is turned clockwise through the shaft at O by a motor which delivers 40 kW at a pump speed of 900 rev/min. With the pump filled but not turning, the vertical reactions at C and D are each 250 N. Calculate the forces exerted by the foundation on the pump at C and D while the pump is running. The tensions in the connecting pipes at A and B are exactly balanced by the respective forces due to the static pressure in the water. (*Suggestion:* Isolate the entire pump and water within it between sections A and B and apply the momentum principle to the entire system.)

Ans. C = 4340 N up, D = 3840 N down

▶4/61 An axial section of the suction nozzle A for a bulk wheat unloader is shown here. The outer pipe is secured to the inner pipe by several longitudinal webs which do not restrict the flow of air. A vacuum of 230 mm of mercury (p = -30.7 kPa gage) is maintained in the inner pipe, and the pressure across the bottom of the outer pipe is atmospheric (p = 0). Air at 1.206 kg/m³ is drawn in through the space between the pipes at a rate of 16 Mg/h at atmospheric pressure and draws with it 135 Mg of wheat per hour up the pipe at a velocity of 40 m/s. If the nozzle unit below section A-A has a mass of 30 kg, calculate the compression C in the connection at A-A.

Ans. C = 265 N

Problem 4/60

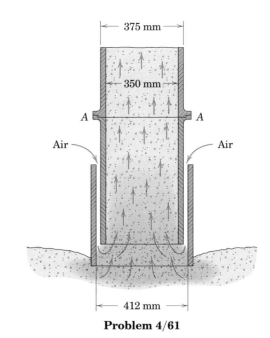

Problem 4/61

▶**4/62** The valve, which is screwed into the fixed pipe at section *A-A*, is designed to discharge fresh water at the rate of 1.30 m³/min into the atmosphere in the *x-y* plane as shown. Water pressure at *A-A* is 1000 kPa gage. The flow area at *A-A* has a diameter of 50 mm, and the diameter of the discharge area at *B* is 25 mm. Neglect the weight of the valve and water within it and compute the shear *V*, tension *F*, torsion *T*, and bending moment *M* at section *A-A*.

> *Ans.* $V = 733$ N, $F = 1588$ N
> $T = 36.6$ N·m, $M = 54.8$ N·m

Problem 4/62

▶**4/63** A test vehicle designed for impact studies has a mass $m = 1.4$ Mg and is accelerated from rest by the impingement of a high-velocity water jet upon its curved deflector attached to the rear of the vehicle. The jet of fresh water is produced by the air-operated piston and issues from the 140-mm-diameter nozzle with a velocity $v = 150$ m/s. Frictional resistance of the vehicle, treated as a particle, amounts to 10 percent of its weight. Determine the velocity u of the vehicle 3 seconds after release from rest. (*Hint:* Adapt the results of Sample Problem 4/5.)

> *Ans.* $u = 131.0$ m/s

Problem 4/63

▶**4/64** In the figure is shown an impulse-turbine wheel for a hydroelectric power plant which is to operate with a static head of water of 300 m at each of its six nozzles and is to rotate at the speed of 270 rev/min. Each wheel and generator unit is to develop an output power of 22 000 kW. The efficiency of the generator may be taken to be 0.90, and an efficiency of 0.85 for the conversion of the kinetic energy of the water jets to energy delivered by the turbine may be expected. The mean peripheral speed of such a wheel for greatest efficiency will be about 0.47 times the jet velocity. If each of the buckets is to have the shape shown, determine the necessary jet diameter *d* and wheel diameter *D*. Assume that the water acts on the bucket which is at the tangent point of each jet stream.

> *Ans.* $d = 165.3$ mm, $D = 2.55$ m

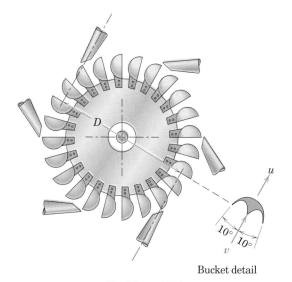

Problem 4/64

4/7 VARIABLE MASS

In Art. 4/4 we extended the equations for the motion of a particle to include a system of particles. This extension led to the very general expressions $\Sigma \mathbf{F} = \dot{\mathbf{G}}$, $\Sigma \mathbf{M}_O = \dot{\mathbf{H}}_O$, and $\Sigma \mathbf{M}_G = \dot{\mathbf{H}}_G$, which are Eqs. 4/6, 4/7, and 4/9, respectively. In their derivation, the summations were taken over a fixed collection of particles, so that the mass of the system to be analyzed was constant. In Art. 4/6 these momentum principles were extended in Eqs. 4/18 and 4/19 to describe the action of forces on a system defined by a geometric volume through which passes a steady flow of mass. The amount of mass within this volume was, therefore, constant with respect to time and we were, thus, permitted to use Eqs. 4/6, 4/7, and 4/9. When the mass within the boundary of a system under consideration is not constant, the foregoing relationships are no longer valid.*

We will now develop the equation for the linear motion of a system whose mass varies with time. Consider first a body that gains mass by overtaking and swallowing a stream of matter, Fig. 4/6a. The mass of the body and its velocity at any instant are m and v, respectively. The stream of matter is assumed to be moving in the same direction as m with a constant velocity v_0 less than v. By virtue of Eq. 4/18, the force exerted by m on the particles of the stream to accelerate them from a velocity v_0 to a greater velocity v is $R = m'(v - v_0) = \dot{m}u$, where the time rate of increase of m is $m' = \dot{m}$ and where u is the magnitude of the relative velocity with which the particles approach m. In addition to R, all other forces acting

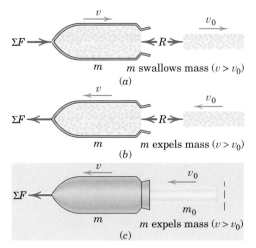

m swallows mass $(v > v_0)$
(a)

m expels mass $(v > v_0)$
(b)

m expels mass $(v > v_0)$
(c)

Figure 4/6

*In relativistic mechanics the mass is found to be a function of velocity, and its time derivative has a meaning different from that in Newtonian mechanics.

on m in the direction of its motion are denoted by ΣF. The equation of motion of m from Newton's second law is, therefore, $\Sigma F - R = m\dot{v}$ or

$$\boxed{\Sigma F = m\dot{v} + \dot{m}u} \qquad\qquad \textbf{(4/20)}$$

Similarly, if the body loses mass by expelling it rearward so that its velocity v_0 is less than v, Fig. 4/6b, the force R required to decelerate the particles from a velocity v to a lesser velocity v_0 is $R = m'(-v_0 - [-v]) = m'(v - v_0)$. But $m' = -\dot{m}$ since m is decreasing. Also the relative velocity with which the particles leave m is $u = v - v_0$. Thus, the force R becomes $R = -\dot{m}u$. If ΣF denotes the resultant of all other forces acting on m in the direction of its motion, Newton's second law requires $\Sigma F + R = m\dot{v}$ or

$$\Sigma F = m\dot{v} + \dot{m}u$$

which is the same relationship as in the case where m is gaining mass. We may use Eq. 4/20, therefore, as the equation of motion of m, whether it is gaining or losing mass.

A frequent error in the use of the force-momentum equation is to express the partial force sum ΣF as

$$\Sigma F = \frac{d}{dt}(mv) = m\dot{v} + \dot{m}v$$

From this expansion we see that the direct differentiation of the linear momentum gives the correct force ΣF *only* when the body picks up mass initially at rest or when it expels mass which is left with zero absolute velocity. In both instances, $v_0 = 0$ and $u = v$.

We may also obtain Eq. 4/20 by a direct differentiation of the momentum from the basic relation $\Sigma F = \dot{G}$, provided a proper system of constant total mass is chosen. To illustrate this approach, we take the case where m is losing mass and use Fig. 4/6c which shows the system of m and an arbitrary portion m_0 of the stream of ejected mass. The mass of this system is $m + m_0$ and is constant. The ejected stream of mass is assumed to move undisturbed once separated from m, and the only force external to the entire system is ΣF which is applied directly to m as before. The reaction $R = -\dot{m}u$ is internal to the system and is not disclosed as an external force on the system. With constant total mass, the momentum principle $\Sigma F = \dot{G}$ is applicable and we have

$$\Sigma F = \frac{d}{dt}(mv + m_0 v_0) = m\dot{v} + \dot{m}v + \dot{m}_0 v_0 + m_0 \dot{v}_0$$

Clearly, $\dot{m}_0 = -\dot{m}$, and the velocity of the ejected mass with respect to m is $u = v - v_0$. Also $\dot{v}_0 = 0$ since m_0 moves undisturbed with no acceleration once free of m. Thus, the relation becomes

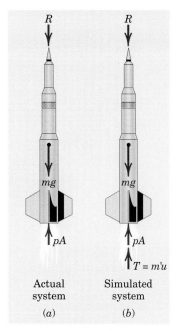

Actual
system

Simulated
system

(a)

(b)

Figure 4/7

$$\Sigma F = m\dot{v} + \dot{m}u$$

which is identical to the result of the previous formulation, Eq. 4/20.

The case of m losing mass is clearly descriptive of rocket propulsion. Figure 4/7a shows a vertically ascending rocket, the system for which is the mass within the volume defined by the exterior surface of the rocket and the exit plane across the nozzle. External to this system, the free-body diagram discloses the instantaneous values of gravitational attraction mg, aerodynamic resistance R, and the force pA due to the average static pressure p across the nozzle exit plane of area A. Also the rate of mass flow is $m' = -\dot{m}$. Thus, we may write the equation of motion of the rocket, $\Sigma F = m\dot{v} + \dot{m}u$, as $pA - mg - R = m\dot{v} + \dot{m}u$, or

$$m'u + pA - mg - R = m\dot{v} \qquad (4/21)$$

Equation 4/21 is of the form "$\Sigma F = ma$" where the first term in "ΣF" is the thrust $T = m'u$. Thus, the rocket may be simulated as a body to which an external thrust T is applied, Fig. 4/7b, and the problem may then be analyzed like any other $F = ma$ problem, except that m is a function of time.

It may be observed that, during the initial stages of motion when the magnitude of the velocity v of the rocket is less than the relative exhaust velocity u, the absolute velocity v_0 of the exhaust gases will be directed rearward. On the other hand, when the rocket reaches a velocity v whose magnitude is greater than u, the absolute velocity v_0 of the exhaust gases will be directed forward. For a given mass rate of flow, the rocket thrust T depends only on the relative exhaust velocity u and not on the magnitude or on the direction of the absolute velocity v_0 of the exhaust gases.

In the foregoing treatment of bodies whose mass changes with time, we have assumed that all elements of the mass m of the body were moving with the same velocity v at any instant of time and that the particles of mass that were added to or expelled from the body underwent an abrupt transition of velocity upon entering or leaving the body. Thus, this velocity change has been modeled as a mathematical discontinuity. In reality, this change in velocity cannot be discontinuous even though the transition may be rapid. In the case of a rocket, for example, the velocity change occurs continuously in the space between the combustion zone and the exit plane of the exhaust nozzle. A more general analysis* of variable-mass dynamics removes this restriction of discontinuous velocity change and introduces a slight correction to Eq. 4/20.

---

*For a development of the equations which describe the general motion of a time-dependent system of mass, see Art. 53 of the first author's *Dynamics, 2nd Edition SI Version*, 1975, John Wiley & Sons.

Sample Problem 4/9

The end of a chain of length L and mass ρ per unit length which is piled on a platform is lifted vertically with a constant velocity v by a variable force P. Find P as a function of the height x of the end above the platform. Also find the energy lost during the lifting of the chain.

Solution I (variable-mass approach). Equation 4/20 will be used and applied to the moving part of the chain of length x which is gaining mass. The force summation ΣF includes all forces acting on the moving part except the force exerted by the particles that are being attached. From the diagram we have

$$\Sigma F_x = P - \rho g x$$

The velocity is constant so that $\dot{v} = 0$. The rate of increase of mass is $\dot{m} = \rho v$, and the relative velocity with which the attaching particles approach the moving part is $u = v - 0 = v$. Thus, Eq. 4/20 becomes

① $\quad [\Sigma F = m\dot{v} + \dot{m}u] \quad P - \rho g x = 0 + \rho v(v) \quad P = \rho(gx + v^2) \quad$ *Ans.*

We now see that the force P consists of the two parts, $\rho g x$, which is the weight of the moving part of the chain, and ρv^2, which is the added force required to change the momentum of the links on the platform from a condition at rest to a velocity v.

Solution II (constant-mass approach). The principle of impulse and momentum for a system of particles expressed by Eq. 4/6 will be applied to the entire chain considered as the system of constant mass. The free-body diagram of the system shows the unknown force P, the total weight of all links $\rho g L$, and the force $\rho g(L - x)$ exerted by the platform on those links that are at rest on it. The momentum of the system at any position is $G_x = \rho x v$ and the momentum equation gives

② $\quad \left[\Sigma F_x = \dfrac{dG_x}{dt}\right] \quad P + \rho g(L - x) - \rho g L = \dfrac{d}{dt}(\rho x v) \quad P = \rho(gx + v^2)$

Ans.

Again the force P is seen to be equal to the weight of the portion of the chain which is off the platform plus the added term which accounts for the time rate of increase of momentum of the chain.

Energy loss. Each link on the platform acquires its velocity abruptly through an impact with the link above it which lifts it off the platform. The succession of impacts gives rise to an energy loss ΔE (negative work $-\Delta E$) so that the work-energy equation becomes $U'_{1\text{-}2} = \int P \, dx - \Delta E = \Delta T + \Delta V_g$, where

$$\int P \, dx = \int_0^L (\rho g x + \rho v^2) \, dx = \tfrac{1}{2}\rho g L^2 + \rho v^2 L$$

$$\Delta T = \tfrac{1}{2}\rho L v^2 \qquad \Delta V_g = \rho g L \frac{L}{2} = \tfrac{1}{2}\rho g L^2$$

Substituting into the work-energy equation gives

$$\tfrac{1}{2}\rho g L^2 + \rho v^2 L - \Delta E = \tfrac{1}{2}\rho L v^2 + \tfrac{1}{2}\rho g L^2 \qquad \Delta E = \tfrac{1}{2}\rho L v^2 \quad \textit{Ans.}$$

---

① The model of Fig. 4/6a shows the mass being added to the leading end of the moving part. With the chain the mass is added to the trailing end, but the effect is the same.

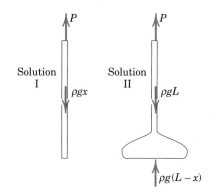

Solution I $\quad \rho g x$ $\qquad$ Solution II $\quad \rho g L$ $\qquad \rho g(L - x)$

② We must be very careful not to use $\Sigma F = \dot{G}$ for a system whose mass is changing. Thus, we have taken the total chain as the system since its mass is constant.

③ Note that $U'_{1\text{-}2}$ includes work done by internal nonelastic forces, such as the link-to-link impact forces, where this work is converted into heat and acoustical energy loss ΔE.

Sample Problem 4/10

Replace the open-link chain of Sample Problem 4/9 by a flexible but inextensible rope or bicycle-type chain of length L and mass ρ per unit length. Determine the force P required to elevate the end of the rope with a constant velocity v and determine the corresponding reaction R between the coil and the platform.

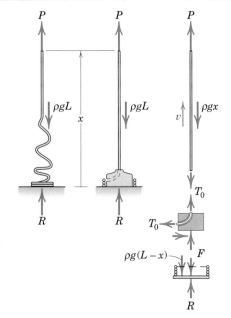

Solution. The free-body diagram of the coil and moving portion of the rope is shown in the left-hand figure. Because of some resistance to bending and some lateral motion, the transition from rest to vertical velocity v will occur over an appreciable segment of the rope. Nevertheless, assume first that all moving elements have the same velocity so that Eq. 4/6 for the system gives

$$\left[\Sigma F_x = \frac{dG_x}{dt} \right] \qquad P + R - \rho g L = \frac{d}{dt}(\rho x v) \qquad P + R = \rho v^2 + \rho g L$$

We assume further that all elements of the coil of rope are at rest on the platform and transmit no force to the platform other than their weight, so that $R = \rho g(L - x)$. Substitution into the foregoing relation gives

$$P + \rho g(L - x) = \rho v^2 + \rho g L \qquad \text{or} \qquad P = \rho v^2 + \rho g x$$

which is the same result as that for the chain in Sample Problem 4/9.

The total work done on the rope by P becomes

$$U'_{1\text{-}2} = \int P\, dx = \int_0^x (\rho v^2 + \rho g x)\, dx = \rho v^2 x + \tfrac{1}{2}\rho g x^2$$

Substitution into the work-energy equation gives

$$[U'_{1\text{-}2} = \Delta T + \Delta V_g] \qquad \rho v^2 x + \tfrac{1}{2}\rho g x^2 = \Delta T + \rho g x\, \frac{x}{2} \qquad \Delta T = \rho x v^2$$

which is twice the kinetic energy $\tfrac{1}{2}\rho x v^2$ of vertical motion. Thus, an equal amount of kinetic energy is unaccounted for. This conclusion largely negates our assumption of one-dimensional x-motion.

In order to produce a one-dimensional model that retains the inextensibility property assigned to the rope, it is necessary to impose a physical constraint at the base to guide the rope into vertical motion and at the same time preserve a smooth transition from rest to upward velocity v without energy loss. Such a guide is included in the free-body diagram of the entire rope in the middle figure and is represented schematically in the middle free-body diagram of the right-hand figure.

For a conservative system, the work-energy equation gives

$$[dU' = dT + dV_g] \qquad P\, dx = d(\tfrac{1}{2}\rho x v^2) + d\left(\rho g x\, \frac{x}{2} \right)$$

$$P = \tfrac{1}{2}\rho v^2 + \rho g x$$

Substitution into the impulse-momentum equation $\Sigma F_x = \dot{G}_x$ gives

$$\tfrac{1}{2}\rho v^2 + \rho g x + R - \rho g L = \rho v^2 \qquad R = \tfrac{1}{2}\rho v^2 + \rho g(L - x)$$

Although this force, which exceeds the weight by $\tfrac{1}{2}\rho v^2$, is unrealistic experimentally, it would be present in the idealized model.

Equilibrium of the vertical section requires

$$T_0 = P - \rho g x = \tfrac{1}{2}\rho v^2 + \rho g x - \rho g x = \tfrac{1}{2}\rho v^2$$

Because it requires a force of ρv^2 to change the momentum of the rope elements, the restraining guide must supply the balance $F = \tfrac{1}{2}\rho v^2$ which, in turn, is transmitted to the platform.

① Perfect flexibility would not permit any resistance to bending.

② Remember that v is constant and equals $\dot{x}$. Also note that this same relation applies to the chain of Sample Problem 4/9.

③ This added term of unaccounted-for kinetic energy exactly equals the energy lost by the chain during the impact of its links.

④ This restraining guide may be visualized as a canister of negligible mass rotating within the coil with an angular velocity v/r and connected to the platform through its shaft. As it turns, it feeds the rope from a rest position to an upward velocity v, as indicated in the accompanying figure.

$\omega = v/r$

⑤ Note that the mass center of the section of length x is a distance $x/2$ above the base.

Sample Problem 4/11

A rocket of initial total mass m_0 is fired vertically up from the north pole and accelerates until the fuel, which burns at a constant rate, is exhausted. The relative nozzle velocity of the exhaust gas has a constant value u, and the nozzle exhausts at atmospheric pressure throughout the flight. If the residual mass of the rocket structure and machinery is m_b when burnout occurs, determine the expression for the maximum velocity reached by the rocket. Neglect atmospheric resistance and the variation of gravity with altitude.

Solution I (F = ma solution). We adopt the approach illustrated with Fig. 4/7b and treat the thrust as an external force on the rocket. With the neglect of the back pressure p across the nozzle and the atmospheric resistance R, Eq. 4/21 or Newton's second law gives

$$T - mg = m\dot{v}$$

But the thrust is $T = m'u = -\dot{m}u$ so that the equation of motion becomes

$$-\dot{m}u - mg = m\dot{v}$$

Multiplication by dt, division by m, and rearrangement give

$$dv = -u\frac{dm}{m} - g\,dt$$

which is now in a form which can be integrated. The velocity v corresponding to the time t is given by the integration

$$\int_0^v dv = -u\int_{m_0}^m \frac{dm}{m} - g\int_0^t dt$$

or

$$v = u\ln\frac{m_0}{m} - gt$$

Since the fuel is burned at the constant rate $m' = -\dot{m}$, the mass at any time t is $m = m_0 + \dot{m}t$. If we let m_b stand for the mass of the rocket when burnout occurs, then the time at burnout becomes $t_b = (m_b - m_0)/\dot{m} = (m_0 - m_b)/(-\dot{m})$. This time gives the condition for maximum velocity, which is

$$v_{max} = u\ln\frac{m_0}{m_b} + \frac{g}{\dot{m}}(m_0 - m_b) \qquad\qquad Ans.$$

The quantity $\dot{m}$ is a negative number since the mass decreases with time.

Solution II (variable-mass solution). If we use Eq. 4/20, then $\Sigma F = -mg$ and the equation becomes

$$[\Sigma F = m\dot{v} + \dot{m}u] \qquad\qquad -mg = m\dot{v} + \dot{m}u$$

But $\dot{m}u = -m'u = -T$ so that the equation of motion becomes

$$T - mg = m\dot{v}$$

which is the same as formulated with *Solution I*.

① The neglect of atmospheric resistance is not a bad assumption for a first approximation inasmuch as the velocity of the ascending rocket is smallest in the dense part of the atmosphere and greatest in the rarefied region. Also for an altitude of 320 km, the acceleration due to gravity is 91 percent of the value at the surface of the earth.

② Vertical launch from the north pole is taken only to eliminate any complication due to the earth's rotation in figuring the absolute trajectory of the rocket.

PROBLEMS

Introductory problems

4/65 At the instant of vertical launch the rocket expels
exhaust at the rate of 220 kg/s with an exhaust
velocity of 820 m/s. If the initial vertical acceleration is 6.80 m/s^2, calculate the total mass of the
rocket and fuel at launch.

Ans. m = 10.86 Mg

Problem 4/65

4/66 The space shuttle, together with its central fuel
tank and two booster rockets, has a total mass of
2.04(10^6) kg at lift-off. Each of the two booster
rockets produces a thrust of 11.80(10^6) N, and
each of the three main engines of the shuttle produces a thrust of 2.00(10^6) N. The specific impulse
(ratio of exhaust velocity to gravitational acceleration) for each of the three main engines of the
shuttle is 455 s. Calculate the initial vertical acceleration a of the assembly with all five engines
operating and find the rate at which fuel is being
consumed by each of the shuttle's three engines.

Problem 4/66

4/67 A tank truck for washing down streets has a total mass of 10 Mg when its tank is full. With the spray turned on, 40 kg of water per second issue from the nozzle with a velocity of 20 m/s relative to the truck at the 30° angle shown. If the truck is to accelerate at the rate of 0.6 m/s^2 when starting on a level road, determine the required tractive force P between the tires and the road (*a*) when the spray is turned on and (*b*) when the spray is turned off.

Ans. (*a*) $P = 5.31$ kN, (*b*) $P = 6.00$ kN

Problem 4/67

4/68 A small rocket of initial mass m_0 is fired vertically up near the surface of the earth (g constant), and the mass rate of exhaust m' and the relative exhaust velocity u are constant. Determine the velocity v as a function of the time t of flight if the air resistance is neglected and if the mass of the rocket case and machinery is negligible compared with the mass of the fuel carried.

4/69 The magnetometer boom for a spacecraft consists of a large number of triangular-shaped units that spring into their deployed configuration upon release from the canister in which they were folded and packed prior to release. Write an expression for the force F which the base of the canister must exert on the boom during its deployment in terms of the increasing length x and its time derivatives. The mass of the boom per unit of deployed length is ρ. Treat the supporting base on the spacecraft as a fixed platform and assume that the deployment takes place outside of any gravitational field. Neglect the dimension b compared with x.

Ans. $F = \rho(x\ddot{x} + \dot{x}^2)$

Problem 4/69

4/70 A 2-stage rocket is fired vertically up and is above the atmosphere when the first stage burns out and the second stage separates and ignites. The second stage carries 1200 kg of fuel and has an empty mass of 200 kg. Upon ignition the second stage burns fuel at the rate of 5.2 kg/s and has a constant exhaust velocity of 3000 m/s relative to its nozzle. Determine the acceleration of the second stage 60 seconds after ignition and find the maximum acceleration and the time t after ignition at which it occurs. Neglect the variation of g and take it to be 8.70 m/s^2 for its range of altitude averaging about 400 km.

Representative problems

4/71 At a bulk loading station, gravel leaves the hopper at the rate of 100 kg/s with a velocity of 3 m/s in the direction shown and is deposited on the moving flat-bed truck. The tractive force between the driving wheels and the road is 1.7 kN, which overcomes the 900 N of frictional road resistance. Determine the acceleration a of the truck 4 seconds after the hopper is opened over the truck bed, at which instant the truck has a forward speed of 2.5 km/h. The mass of the empty truck is 5.4 Mg.

Ans. $a = 0.1518$ m/s^2

Problem 4/71

4/72 Fresh water issues from the two 30-mm-diameter holes in the bucket with a velocity of 2.5 m/s in the directions shown. Calculate the force P required to give the bucket an upward acceleration of 0.5 m/s^2 from rest if it contains 20 kg of water at that time. The empty bucket has a mass of 0.6 kg.

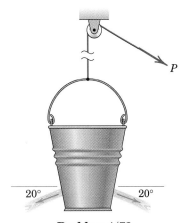

Problem 4/72

4/73 The railroad coal car has an empty mass of 25 Mg and carries a total load of 90 Mg of coal. The bins are equipped with bottom doors which permit discharging coal through an opening between the rails. If the car dumps coal at the rate of 10 Mg/s in a downward direction relative to the car, and if frictional resistance to motion is 20 N per megagram of total remaining mass, determine the coupler force P required to give the car an acceleration of 0.045 m/s^2 in the direction of P at the instant when half the coal has been dumped.

Ans. $P = 4.55$ kN

Problem 4/73

4/74 A coil of heavy flexible cable with a total length of 100 m and a mass of 1.2 kg/m is to be laid along a straight horizontal line. The end is secured to a post at A, and the cable peels off the coil and emerges through the horizontal opening in the cart as shown. The cart and drum together have a mass of 40 kg. If the cart is moving to the right with a velocity of 2 m/s when 30 m of cable remain in the drum and the tension in the rope at the post is 2.4 N, determine the force P required to give the cart and drum an acceleration of 0.3 m/s^2. Neglect all friction.

Problem 4/74

4/75 By lowering a scoop as it skims the surface of a body of water, the aircraft (nicknamed the "Super Scooper") is able to ingest 4.5 m^3 of fresh water during a 12-second run. The plane then flies to a fire area and makes a massive water drop with the ability to repeat the procedure as many times as are necessary. The plane approaches its run with a velocity of 280 km/h and an initial mass of 16.4 Mg. As the scoop enters the water, the pilot advances the throttle to provide an additional 300 hp (223.8 kW) needed to prevent undue deceleration. Determine the initial deceleration when the scooping action starts. (Neglect the difference between the average and the initial rates of water intake.)

Ans. $a = -1.603$ m/s^2 (deceleration)

Problem 4/75

4/76 The end of a pile of loose-link chain of mass ρ per unit length is being pulled horizontally along the surface by a constant force P. If the coefficient of kinetic friction between the chain and the surface is μ_k, determine the acceleration a of the chain in terms of x and $\dot{x}$.

Problem 4/76

4/77 The upper end of the open-link chain of length L and mass ρ per unit length is lowered at a constant speed v by the force P. Determine the reading R of the platform scale in terms of x.

Ans. $R = \rho g x + \rho v^2$

Problem 4/77

4/78 The upper end of the open-link chain of length L and mass ρ per unit length is released from rest with the lower end just touching the platform of the scale. Determine the expression for the force F read on the scale as a function of the distance x through which the upper end has fallen. (*Comment:* The chain acquires a free-fall velocity of $\sqrt{2gx}$ because the links on the scale exert no force on those above, which are still falling freely. Work the problem in two ways: first, by evaluating the time rate of change of momentum for the entire chain and, second, by considering the force F to be composed of the weight of the links at rest on the scale plus the force necessary to divert an equivalent stream of fluid.)

Problem 4/78

4/79 A small rocket-propelled vehicle has an initial mass of 60 kg, including 10 kg of fuel. Fuel is burned at the constant rate of 1 kg/s with an exhaust velocity relative to the nozzle of 120 m/s. Upon ignition the vehicle is released from rest on the 10° incline. Calculate the maximum velocity v reached by the vehicle. Neglect all friction.

Ans. $v = 4.84$ m/s

10°

Problem 4/79

4/80 The chain of length L and mass ρ per unit length is released from rest on the smooth horizontal surface with a negligibly small overhang x to initiate motion. Determine (*a*) the acceleration a as a function of x, (*b*) the tension T in the chain at the smooth corner as a function of x, and (*c*) the velocity v of the last link A as it reaches the corner.

Problem 4/80

4/81 Determine the force P required to give the open-link chain of total length L a constant velocity $v = \dot{y}$. The chain has a mass ρ per unit length. Also, by applying the impulse-momentum equation to the left-hand portion of the system, verify that the force R supporting the pile of chain equals the weight of the pile. Neglect the small size and mass of the pulley and any friction in the pulley.

Ans. $P = \rho v^2 + \rho g(h - y)$

Problem 4/81

4/82 A coal car with an empty mass of 25 Mg is moving freely with a speed of 1.2 m/s under a hopper which opens and releases coal into the moving car at the constant rate of 4 Mg per second. Determine the distance x moved by the car during the time that 32 Mg of coal are deposited in the car. Neglect any frictional resistance to rolling along the horizontal track.

Problem 4/82

4/83 The cart carries a pile of open-link chain of mass ρ per unit length. The chain passes freely through the hole in the cart and is brought to rest, link by link, by the tension T in the portion of the chain resting on the ground and secured at its end A. The cart and the chain on it move under the action of the constant force P and have a velocity v_0 and mass m_0 when $x = 0$. Determine expressions for the acceleration a and velocity v of the cart in terms of x if all friction is neglected. Also find T. Observe that the transition link 2 is decelerated from the velocity v to velocity zero by the tension T transmitted by the last horizontal link 1. Also note that link 2 exerts no force on the following link 3 during the transition. Explain why the $\dot{m}u$ term is absent if Eq. 4/20 is applied to this problem.

$$Ans. \quad a = \frac{P}{m_0 - \rho x}$$

$$v = \sqrt{v_0{}^2 + \frac{2P}{\rho} \ln \frac{m_0}{m_0 - \rho x}}$$

$$T = \rho v^2$$

1   2   3
Transition link 2

Problem 4/83

4/84 The open-link chain of length L and mass ρ per unit length is released from rest in the position shown, where the bottom link is almost touching the platform and the horizontal section is supported on a smooth surface. Friction at the corner guide is negligible. Determine (*a*) the velocity v_1 of end A as it reaches the corner and (*b*) its velocity v_2 as it strikes the platform. (*c*) Also specify the total loss Q of energy.

Problem 4/84

4/85 In the figure is shown a system used to arrest the motion of an airplane landing on a field of restricted length. The plane of mass m rolling freely with a velocity v_0 engages a hook which pulls the ends of two heavy chains, each of length L and mass ρ per unit length, in the manner shown. A conservative calculation of the effectiveness of the device neglects the retardation of chain friction on the ground and any other resistance to the motion of the airplane. With these assumptions, compute the velocity v of the airplane at the instant when the last link of each chain is put in motion. Also determine the relation between the displacement x and the time t after contact with the chain. Assume each link of the chain acquires its velocity v suddenly upon contact with the moving links.

$$\text{Ans. } v = \frac{v_0}{1 + 2\rho L/m}$$

$$x = \frac{m}{\rho}\left[\sqrt{1 + \frac{2v_0 t\rho}{m}} - 1\right]$$

▶4/86 A metal canister of mass m_0 contains a coil of flexible cord of mass ρ per unit length. The end of the cord is secured to a support at the top of an elevator shaft, and the canister is released from rest. A thumb screw on the neck of the canister adjusts a spring clamp which applies a constant friction force $F/2$ on each side of the cord as it passes through the neck. Determine an expression for the acceleration a of the canister and the force P at the support in terms of x and the velocity v. The total length of the cord is L.

$$\text{Ans. } a = g + \frac{\rho v^2/2 - F}{m_0 + \rho(L - x)}$$

$$P = \rho g x + \rho v^2/2 + F$$

Problem 4/86

Problem 4/85

▶4/87 The chain of mass ρ per unit length passes over the small freely-turning pulley and is released from rest with only a small imbalance h to initiate motion. Determine the acceleration a and velocity v of the chain and the force R supported by the hook at A, all in terms of h as it varies from essentially zero to H. Neglect the weight of the pulley and its supporting frame and the weight of the small amount of chain in contact with the pulley. (*Hint:* The force R does not equal two times the equal tensions T in the chain tangent to the pulley.)

$$\text{Ans. } a = (h/H)g, \; v = h\sqrt{g/H}$$
$$R = 2\rho g[H - (2h^2/H)]$$

Problem 4/87

▶4/88 The free end of the open-link chain of total length L and mass ρ per unit length is released from rest at $x = 0$. Determine the force R on the fixed end and the tension T_1 in the chain at the lower end of the nonmoving part in terms of x. Also find the total loss Q of energy when $x = L$.

$$\text{Ans. } R = \tfrac{1}{2}\rho g(L + 3x), \; T_1 = \rho g x, \; Q = \tfrac{1}{4}\rho g L^2$$

Problem 4/88

▶4/89 Replace the chain of Prob. 4/88 by a flexible rope or bicycle chain of mass ρ per unit length and total length L. The free end is released from rest at $x = 0$ and falls under the influence of gravity. Determine the acceleration a of the free end, the force R at the fixed end, and the tension T_1 in the rope at the loop, all in terms of x. (Note that a is greater than g. What happens to the energy of the system when $x = L$?)

$$\text{Ans. } a = g\left[1 + \frac{x(L - x/2)}{(L - x)^2}\right]$$
$$R = \tfrac{1}{2}\rho g\left[(L + x) + \frac{x(L - x/2)}{L - x}\right]$$
$$T_1 = \tfrac{1}{2}\rho g\,\frac{x(L - x/2)}{L - x}$$

▶4/90 The free end of the flexible and inextensible rope of mass ρ per unit length and total length L is given a constant upward velocity v. Write expressions for P, the force R supporting the fixed end, and the tension T_1 in the rope at the loop in terms of x. (For the loop of negligible size, the tension is the same on both sides.)

$$Ans.\ T_1 = \tfrac{1}{4}\rho v^2$$
$$P = \tfrac{1}{2}\rho(\tfrac{1}{2}v^2 + gx)$$
$$R = \tfrac{1}{4}\rho v^2 + \rho g(L - x/2)$$

Problem 4/90

▶4/91 One end of the pile of chain falls through a hole in its support and pulls the remaining links after it in a steady flow. If the links which are initially at rest acquire the velocity of the chain suddenly and without frictional resistance or interference from the support or from adjacent links, find the velocity v of the chain as a function of x if $v = 0$ when $x = 0$. Also find the acceleration a of the falling chain and the energy Q lost from the system as the last link leaves the platform. (*Hint:* Apply Eq. 4/20 and treat the product xv as the variable when solving the differential equation. Also note at the appropriate step that $dx = v\,dt$.) The total length of the chain is L, and its mass per unit length is ρ.

$$Ans.\ v = \sqrt{\frac{2gx}{3}}, \ a = \frac{g}{3}, \ Q = \frac{\rho g L^2}{6}$$

Problem 4/91

PART II

DYNAMICS OF RIGID BODIES

Rigid-body kinematics describes the relations between the linear and angular motions of bodies that retain their shape. The design of gears, cams, connecting links, and many other moving machine parts are kinematic problems. The design of the control mechanism shown is a good example of rigid-body kinematics where the relations between input and output motions require precise analysis.

PLANE KINEMATICS OF RIGID BODIES

5

5/1 INTRODUCTION

In Chapter 2 on particle kinematics, we developed the relationships governing the displacement, velocity, and acceleration of points as they moved along straight or curved paths. In rigid-body kinematics we use these same relationships but must also account for the rotational motion of the body as well. Thus, rigid-body kinematics involves both linear and angular quantities.

The description of the motion of rigid bodies is useful in two important ways. First, it is frequently necessary to generate, transmit, or control certain desired motions by the use of cams, gears, and linkages of various types. Here an analysis of the displacement, velocity, and acceleration of the motion is necessary to determine the design geometry of the mechanical parts. Furthermore, as a result of the motion generated, forces are frequently developed that must be accounted for in the design of the parts. Second, it is often necessary to determine the motion of a rigid body resulting from the forces applied to it. Calculation of the motion of a rocket under the influence of its jet thrust and gravitational attraction is an example of such a problem. In both types of problems, we need to have command of the principles of rigid-body kinematics. In this chapter we will cover the kinematics of motion which may be analyzed as taking place in a single plane. In Chapter 7 we will present a brief introduction to the kinematics of motion which requires three spatial coordinates.

A rigid body was defined in the previous chapter as a system of particles for which the distances between the particles remain unchanged. Thus, if each particle of such a body is located by a position vector from reference axes attached to and rotating with the body, there will be no change in any position vector as measured from these axes. This formulation is, of course, an ideal one since all solid materials change shape to some extent when forces are applied to them. Nevertheless, if the movements associated with the changes in shape are very small compared with the overall movements of the body as a whole, then the ideal concept of rigidity is quite acceptable. As mentioned in Chapter 1, the displacements due to the flutter of an aircraft wing, for instance, are of no consequence in the description of the flight path of aircraft as a whole, for which the rigid-body assumption is clearly in order. On the other hand, if the problem is one of describing, as a function of time, the internal wing stress due to wing flutter, then the relative motions of portions of the wing become of prime importance and cannot be neglected. In this case, the wing may not be considered a rigid body. In the present chapter and in the two that follow, essentially all of the material is based on the assumption of rigidity.

A rigid body executes plane motion when all parts of the body move in parallel planes. For convenience, we generally consider the *plane of motion* to be the plane that contains the mass center, and we treat the body as a thin slab with motion confined to the plane of the slab. This idealization fits a very large category of rigid-body motions encountered in engineering.

The plane motion of a rigid body may be divided into several categories, as represented in Fig. 5/1.

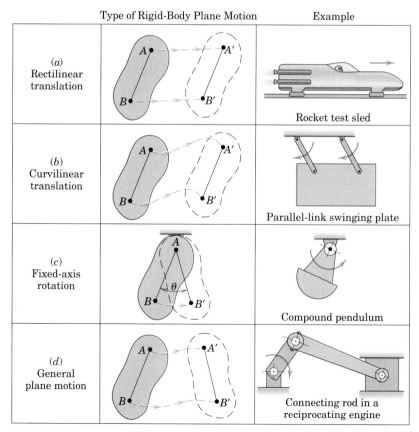

Type of Rigid-Body Plane Motion		Example
(a) Rectilinear translation		Rocket test sled
(b) Curvilinear translation		Parallel-link swinging plate
(c) Fixed-axis rotation		Compound pendulum
(d) General plane motion		Connecting rod in a reciprocating engine

Figure 5/1

Translation is defined as any motion in which every line in the body remains parallel to its original position at all times. In translation there is *no rotation of any line in the body*. In *rectilinear translation*, part (*a*) of Fig. 5/1, all points in the body move in parallel straight lines. In *curvilinear translation*, part (*b*), all points move on congruent curves. We note that in each of the two cases of translation, the motion of the body is completely specified by the motion of any point in the body, since all points have the same motion. Thus, our earlier study of the motion of a point (particle) in Chapter 2 enables us to describe completely the translation of a rigid body.

Rotation about a fixed axis, part (*c*) of Fig. 5/1, is the angular motion about the axis. It follows that all particles move in circular paths about the axis of rotation, and all lines in the body which are perpendicular to the axis of rotation (including those that do not pass through the axis) rotate through the same angle in the same time. Again, our discussion in Chapter 2 on the circular motion of a point enables us to describe the motion of a rotating rigid body, which is treated in the next article.

General plane motion of a rigid body, part (*d*) of Fig. 5/1, is a combination of translation and rotation. We will utilize the principles of relative motion covered in Art. 2/8 in the description of general plane motion.

Note that in each of the examples cited, the actual paths of all particles in the body are projected onto the single plane of motion as represented in each figure.

Determination of the plane motion of rigid bodies is accomplished either by directly calculating the absolute displacements and their time derivatives from the geometry involved or by utilizing the principles of relative motion. Each method is important and useful and will be taken up in turn in the articles which follow.

5/2 ROTATION

The rotation of a rigid body is described by its angular motion. Figure 5/2 shows a rigid body which is rotating as it undergoes plane motion in the plane of the figure. The angular positions of any two lines 1 and 2 attached to the body are specified by θ_1 and θ_2 measured from any convenient fixed reference direction. Because the angle β is invariant, the relation $\theta_2 = \theta_1 + \beta$ upon differentiation with respect to time gives $\dot{\theta}_2 = \dot{\theta}_1$ and $\ddot{\theta}_2 = \ddot{\theta}_1$ or, during a finite interval, $\Delta\theta_2 = \Delta\theta_1$. Thus, *all lines on a rigid body in its plane of motion have the same angular displacement, the same angular velocity, and the same angular acceleration*. It should be noted that the angular motion of a line depends only on its angular displacement with respect to any arbitrary fixed reference and on the time derivatives of the displacement. Angular motion does not

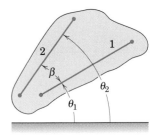

Figure 5/2

require the presence of a fixed axis, normal to the plane of motion, about which the line and the body rotate.

(a) Angular motion relations. The angular velocity ω and angular acceleration α of a rigid body in plane rotation are, respectively, the first and second time derivatives of the angular position coordinate θ of any line in the plane of motion of the body. These definitions give

$$\omega = \frac{d\theta}{dt} = \dot{\theta}$$

$$\alpha = \frac{d\omega}{dt} = \dot{\omega} \quad \text{or} \quad \alpha = \frac{d^2\theta}{dt^2} = \ddot{\theta} \qquad (5/1)$$

$$\omega \, d\omega = \alpha \, d\theta \quad \text{or} \quad \dot{\theta} \, d\dot{\theta} = \ddot{\theta} \, d\theta$$

The third relation is obtained by eliminating dt from the first two. In each of these relations, the positive direction for ω and α, clockwise or counterclockwise, is the same as that chosen for θ. Equations 5/1 should be recognized as analogous to the defining equations for the rectilinear motion of a particle, expressed by Eqs. 2/1, 2/2, and 2/3. In fact, all relations that were described for rectilinear motion in Art. 2/2 apply to the case of rotation in a plane if the linear quantities s, v, and a are replaced by their respective equivalent angular quantities θ, ω, and α. As we proceed further with rigid-body dynamics, we will find that the analogies between the relationships for linear and angular motion are almost complete throughout kinematics and kinetics. These relations are important to recognize, as they help to demonstrate the symmetry and unity found throughout mechanics.

For rotation with *constant* angular acceleration, the integrals of Eqs. 5/1 become

$$\omega = \omega_0 + \alpha t$$

$$\omega^2 = \omega_0{}^2 + 2\alpha(\theta - \theta_0)$$

$$\theta = \theta_0 + \omega_0 t + \tfrac{1}{2}\alpha t^2$$

Here θ_0 and ω_0 are the values of the angular position coordinate and angular velocity, respectively, at $t = 0$, and t is the duration of the interval of motion considered. The student should be able to carry out these integrations easily, as they are completely analogous to the corresponding equations for rectilinear motion with constant acceleration covered in Art. 2/2.

The graphical relationships described for s, v, a, and t in Figs. 2/3 and 2/4 may be used for θ, ω, and α merely by the substitution of corresponding symbols. The student should sketch these graphical relations for plane rotation. The mathematical procedures for obtaining velocity and displacement from acceleration described for rectilinear motion may be applied to rotation by merely replacing the linear quantities by their corresponding angular quantities.

(b) Rotation about a fixed axis. When a body rotates about a fixed axis, all points other than those on the axis move in concentric circles about the fixed axis. Thus, for the rigid body in Fig. 5/3 rotating about a fixed axis normal to the plane of the figure through O, any point such as A moves in a circle of radius r. From our previous discussion in Art. 2/5, we are already familiar with the relationships between the linear motion of A and the angular motion of

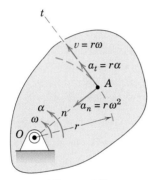

Figure 5/3

the line normal to its path, which is also the angular motion of the rigid body. With the notation $\omega = \dot{\theta}$ and $\alpha = \dot{\omega} = \ddot{\theta}$ for the angular velocity and angular acceleration, respectively, of the body we have Eqs. 2/11, rewritten as

$$
\begin{aligned}
v &= r\omega \\
a_n &= r\omega^2 = v^2/r = v\omega \\
a_t &= r\alpha
\end{aligned}
\qquad (5/2)
$$

These quantities may be expressed alternatively using the cross-product relationship of vector notation. The vector formulation is especially important in three-dimensional analysis of motion. The angular velocity of the rotating body may be expressed by the vector $\boldsymbol{\omega}$ normal to the plane of rotation and having a sense governed by the right-hand rule, as shown in Fig. 5/4a. From the definition of the vector cross product, we see that the vector $\mathbf{v}$ is ob-

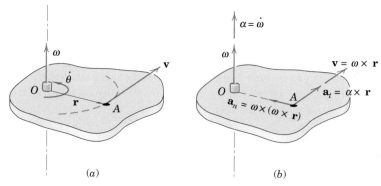

Figure 5/4

tained by crossing $\boldsymbol{\omega}$ into $\mathbf{r}$. This cross product gives the correct magnitude and direction for $\mathbf{v}$ and we write

$$\mathbf{v} = \dot{\mathbf{r}} = \boldsymbol{\omega} \times \mathbf{r}$$

The order of the vectors to be crossed must be retained. The reverse order gives $\mathbf{r} \times \boldsymbol{\omega} = -\mathbf{v}$.

The acceleration of A is obtained by differentiating the cross-product expression for $\mathbf{v}$, which gives

$$\mathbf{a} = \dot{\mathbf{v}} = \boldsymbol{\omega} \times \dot{\mathbf{r}} + \dot{\boldsymbol{\omega}} \times \mathbf{r}$$
$$= \boldsymbol{\omega} \times (\boldsymbol{\omega} \times \mathbf{r}) + \dot{\boldsymbol{\omega}} \times \mathbf{r}$$
$$= \boldsymbol{\omega} \times \mathbf{v} + \boldsymbol{\alpha} \times \mathbf{r}$$

Here $\boldsymbol{\alpha} = \dot{\boldsymbol{\omega}}$ stands for the angular acceleration of the body. Thus, the vector equivalents to Eqs. 5/2 are

$$\boxed{\begin{aligned} \mathbf{v} &= \boldsymbol{\omega} \times \mathbf{r} \\ \mathbf{a}_n &= \boldsymbol{\omega} \times (\boldsymbol{\omega} \times \mathbf{r}) \\ \mathbf{a}_t &= \boldsymbol{\alpha} \times \mathbf{r} \end{aligned}} \qquad (5/3)$$

and are shown in Fig. 5/4b.

For three-dimensional motion of a rigid body, the angular-velocity vector $\boldsymbol{\omega}$ may change direction as well as magnitude, and in this case, the angular acceleration, which is the time derivative of angular velocity, $\boldsymbol{\alpha} = \dot{\boldsymbol{\omega}}$, will no longer be in the same direction as $\boldsymbol{\omega}$.

Sample Problem 5/1

A flywheel rotating freely at 1800 rev/min clockwise is subjected to a variable counterclockwise torque which is first applied at time $t = 0$. The torque produces a counterclockwise angular acceleration $\alpha = 4t$ rad/s^2, where t is the time in seconds during which the torque is applied. Determine (a) the time required for the flywheel to reduce its clockwise angular speed to 900 rev/min, (b) the time required for the flywheel to reverse its direction of rotation, and (c) the total number of revolutions, clockwise plus counterclockwise, turned by the flywheel during the first 14 seconds of torque application.

Solution. The counterclockwise direction will be taken arbitrarily as positive.

(a) Since α is a known function of the time, we may integrate it to obtain angular velocity. With the initial angular velocity of ① $-1800(2\pi)/60 = -60\pi$ rad/s, we have

$$[d\omega = \alpha \, dt] \qquad \int_{-60\pi}^{\omega} d\omega = \int_{0}^{t} 4t \, dt \qquad \omega = -60\pi + 2t^2$$

Substituting the clockwise angular speed of 900 rev/min or $\omega = -900(2\pi)/60 = -30\pi$ rad/s gives

$$-30\pi = -60\pi + 2t^2 \qquad t^2 = 15\pi \qquad t = 6.86 \text{ s} \qquad Ans.$$

(b) The flywheel changes direction when its angular velocity is momentarily zero. Thus,

$$0 = -60\pi + 2t^2 \qquad t^2 = 30\pi \qquad t = 9.71 \text{ s} \qquad Ans.$$

(c) The total number of revolutions through which the flywheel turns during 14 seconds is the number of clockwise turns N_1 during the first 9.71 seconds, plus the number of counterclockwise turns N_2 during the remainder of the interval. Integrating the expression for ω in terms of t gives us the angular displacement in radians. Thus, for the first interval

$$[d\theta = \omega \, dt] \qquad \int_{0}^{\theta_1} d\theta = \int_{0}^{9.71} (-60\pi + 2t^2) \, dt$$

$$\theta_1 = [-60\pi t + \tfrac{2}{3}t^3]_{0}^{9.71} = -1220 \text{ rad}$$

② or $N_1 = 1220/2\pi = 194.2$ revolutions clockwise.

For the second interval

$$\int_{0}^{\theta_2} d\theta = \int_{9.71}^{14} (-60\pi + 2t^2) \, dt$$

③ $$\theta_2 = [-60\pi t + \tfrac{2}{3}t^3]_{9.71}^{14} = 410 \text{ rad}$$

or $N_2 = 410/2\pi = 65.3$ revolutions counterclockwise. Thus, the total number of revolutions turned during the 14 seconds is

$$N = N_1 + N_2 = 194.2 + 65.3 = 259 \text{ rev} \qquad Ans.$$

We have plotted ω versus t and we see that θ_1 is represented by the negative area and θ_2 by the positive area. If we had integrated over the entire interval in one step, we would have obtained $|\theta_2| - |\theta_1|$.

① We must be very careful to be consistent with our algebraic signs. The lower limit is the negative (clockwise) value of the initial angular velocity. Also we must convert revolutions to radians since α is in radian units.

② Again note that the minus sign signifies clockwise in this problem.

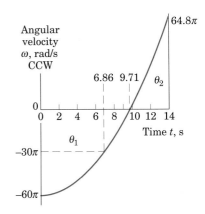

③ We could have converted the original expression for α into the units of rev/s^2, in which case our integrals would have come out directly in revolutions.

Sample Problem 5/2

The pinion A of the hoist motor drives gear B, which is attached to the hoisting drum. The load L is lifted from its rest position and acquires an upward velocity of 2 m/s in a vertical rise of 0.8 m with constant acceleration. As the load passes this position, compute (*a*) the acceleration of point C on the cable in contact with the drum and (*b*) the angular velocity and angular acceleration of the pinion A.

Solution. (*a*) If the cable does not slip on the drum, the vertical velocity and acceleration of the load L are, of necessity, the same as the tangential velocity v and tangential acceleration a_t of point C. For the rectilinear motion of L with constant acceleration, the *n*- and *t*-components of the acceleration of C become

$$[v^2 = 2as] \qquad a = a_t = v^2/2s = 2^2/[2(0.8)] = 2.5 \text{ m/s}^2$$

① $[a_n = v^2/r] \qquad a_n = 2^2/0.400 = 10 \text{ m/s}^2$

$$[a = \sqrt{a_n^2 + a_t^2}] \qquad a_C = \sqrt{(10)^2 + (2.5)^2} = 10.31 \text{ m/s}^2 \quad Ans.$$

① Recognize that a point on the cable changes the direction of its velocity after it contacts the drum and acquires a normal component of acceleration.

(*b*) The angular motion of gear A is determined from the angular motion of gear B by the velocity v_1 and tangential acceleration a_1 of their common point of contact. First, the angular motion of gear B is determined from the motion of point C on the attached drum. Thus,

$$[v = r\omega] \qquad \omega_B = v/r = 2/0.400 = 5 \text{ rad/s}$$

$$[a_t = r\alpha] \qquad \alpha_B = a_t/r = 2.5/0.400 = 6.25 \text{ rad/s}^2$$

Then from $v_1 = r_A\omega_A = r_B\omega_B$ and $a_1 = r_A\alpha_A = r_B\alpha_B$, we have

$$\omega_A = \frac{r_B}{r_A}\omega_B = \frac{0.300}{0.100}5 = 15 \text{ rad/s CW} \qquad Ans.$$

$$\alpha_A = \frac{r_B}{r_A}\alpha_B = \frac{0.300}{0.100}6.25 = 18.75 \text{ rad/s}^2 \text{ CW} \qquad Ans.$$

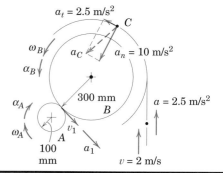

Sample Problem 5/3

The right-angled bar rotates clockwise with an angular velocity which is decreasing at the rate of 4 rad/s². Write the vector expressions for the velocity and acceleration of point A when $\omega = 2$ rad/s.

Solution. Using the right-hand rule gives

$$\omega = -2\mathbf{k} \text{ rad/s} \qquad \text{and} \qquad \alpha = +4\mathbf{k} \text{ rad/s}^2$$

The velocity and acceleration of A become

$$[\mathbf{v} = \boldsymbol{\omega} \times \mathbf{r}] \qquad \mathbf{v} = -2\mathbf{k} \times (0.4\mathbf{i} + 0.3\mathbf{j}) = 0.6\mathbf{i} - 0.8\mathbf{j} \text{ m/s} \qquad Ans.$$

$$[\mathbf{a}_n = \boldsymbol{\omega} \times (\boldsymbol{\omega} \times \mathbf{r})] \quad \mathbf{a}_n = -2\mathbf{k} \times (0.6\mathbf{i} - 0.8\mathbf{j}) = -1.6\mathbf{i} - 1.2\mathbf{j} \text{ m/s}^2$$

$$[\mathbf{a}_t = \boldsymbol{\alpha} \times \mathbf{r}] \qquad \mathbf{a}_t = 4\mathbf{k} \times (0.4\mathbf{i} + 0.3\mathbf{j}) = -1.2\mathbf{i} + 1.6\mathbf{j} \text{ m/s}^2$$

$$[\mathbf{a} = \mathbf{a}_n + \mathbf{a}_t] \qquad \mathbf{a} = -2.8\mathbf{i} + 0.4\mathbf{j} \text{ m/s}^2 \qquad Ans.$$

The magnitudes of $\mathbf{v}$ and $\mathbf{a}$ are

$$v = \sqrt{0.6^2 + 0.8^2} = 1.0 \text{ m/s} \quad \text{and} \quad a = \sqrt{2.8^2 + 0.4^2} = 2.83 \text{ m/s}^2$$

PROBLEMS

Introductory problems

5/1 The square plate rotates about the fixed pivot O. At the instant represented, its angular velocity is $\omega = 6$ rad/s and its angular acceleration is $\alpha = 4$ rad/s^2, in the directions shown in the figure. Determine the velocity and acceleration of (*a*) point A and (*b*) point B.

$$Ans.\ (a)\ \mathbf{v}_A = 270\mathbf{i}\ \text{mm/s}$$
$$\mathbf{a}_A = -180\mathbf{i} - 1620\mathbf{j}\ \text{mm/s}^2$$
$$(b)\ \mathbf{v}_B = 270\mathbf{i} + 180\mathbf{j}\ \text{mm/s}$$
$$\mathbf{a}_B = 900\mathbf{i} - 1740\mathbf{j}\ \text{mm/s}^2$$

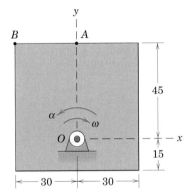

Dimensions in millimeters

Problem 5/1

5/2 The T-shaped body rotates about a horizontal axis through point O. At the instant represented, its angular velocity is $\omega = 3$ rad/s and its angular acceleration is $\alpha = 14$ rad/s^2 in the directions indicated. Determine the velocity and acceleration of (*a*) point A and (*b*) point B. Express your results in terms of components along the n- and t-axes shown.

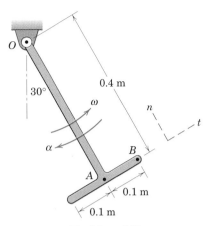

Problem 5/2

5/3 The flywheel has a diameter of 600 mm and rotates with increasing speed about its z-axis shaft. When point P on the rim crosses the y-axis with $\theta = 90°$, it has an acceleration given by $\mathbf{a} = -1.8\mathbf{i} - 4.8\mathbf{j}$ m/s^2. For this instant, determine the angular velocity ω and the angular acceleration α of the flywheel.

$$Ans.\ \alpha = 6\ \text{rad/s}^2,\ \omega = 4\ \text{rad/s}$$

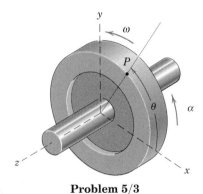

Problem 5/3

5/4 If the acceleration of point P on the rim of the flywheel of Prob. 5/3 is $\mathbf{a} = -3.02\mathbf{i} - 1.624\mathbf{j}$ m/s^2 when $\theta = 60°$, determine the angular velocity ω and angular acceleration α of the 600-mm-diameter flywheel for this position.

5/5 The angular velocity of a gear is controlled according to $\omega = 12 - 3t^2$ where ω, in radians per second, is positive in the clockwise sense and where t is the time in seconds. Find the net angular displacement $\Delta\theta$ from the time $t = 0$ to $t = 3$ s. Also find the total number of revolutions N through which the gear turns during the 3 seconds.

Ans. $\Delta\theta = 9$ rad, $N = 3.66$ rev

5/6 The angular acceleration of a body which is rotating about a fixed axis is given by $\alpha = -k\omega^2$, where the constant $k = 0.1$ (no units). Determine the angular displacement and time elapsed when the angular velocity has been reduced to one-third its initial value $\omega_0 = 12$ rad/s.

5/7 The rigid link moves from position ABC to position $A'B'C'$ while end A moves 100 mm to the left with a constant velocity of 25 mm/s. Determine the average angular velocity ω_{av} of the arm BC during this interval. Assume counterclockwise motion.

Ans. $\omega_{av} = 0.393$ rad/s

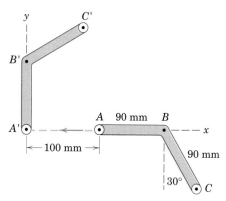

Problem 5/7

5/8 The angular position of a radial line in a rotating disk is given by the clockwise angle $\theta = 2t^3 - 3t^2 + 4$, where θ is in radians and t is in seconds. Calculate the angular displacement $\Delta\theta$ of the disk during the interval in which its angular acceleration increases from 42 rad/s^2 to 66 rad/s^2.

5/9 The right-angle bar rotates about the z-axis through O with an angular acceleration $\alpha = 3$ rad/s^2 in the direction shown. Determine the ve-

locity and acceleration of point P when the angular velocity reaches the value $\omega = 2$ rad/s.

Ans. $\mathbf{v}_P = -0.4\mathbf{i} + \mathbf{j}$ m/s
$\mathbf{a}_P = -1.4\mathbf{i} - 2.3\mathbf{j}$ m/s^2

Problem 5/9

5/10 The angular position of a radial line on a rotating disk is given by $\theta = (-1 + 1.5t)e^{-0.5t}$, where θ is in radians and t is in seconds. Plot the angular position, angular velocity, and angular acceleration versus time for the first 20 seconds of motion. Determine the time at which the acceleration is zero.

Representative problems

5/11 The mass center G of the car has a velocity of 60 km/h at position A and 1.52 s later at B has a velocity of 80 km/h. The radius of curvature of the road at B is 60 m. Calculate the angular velocity ω of the car at B and the average angular velocity ω_{av} of the car between A and B.

Ans. $\omega = 0.373$ rad/s, $\omega_{av} = 0.344$ rad/s

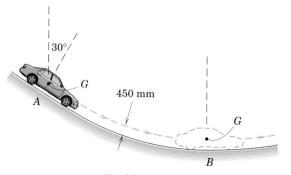

Problem 5/11

5/12 Magnetic tape is being fed over and around the light pulleys mounted in a computer. If the speed v of the tape is constant and if the magnitude of the acceleration of point A on the tape is 4/3 times that of point B, calculate the radius r of the smaller pulley.

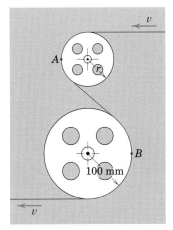

Problem 5/12

5/13 The solid cylinder rotates about its z-axis. At the instant represented, point P on the rim has a velocity whose x-component is -1.28 m/s, and $\theta = 20°$. Determine the angular velocity $\boldsymbol{\omega}$ of line AB on the face of the cylinder. Does the element line BC have an angular velocity?

Ans. $\boldsymbol{\omega} = 24.9\mathbf{k}$ rad/s, No

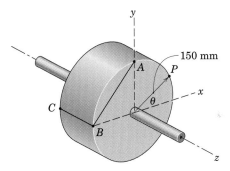

Problem 5/13

5/14 The rotating arm starts from rest and acquires a rotational speed $N = 600$ rev/min in 2 seconds with constant angular acceleration. Find the time t after starting before the acceleration vector of end P makes an angle of $45°$ with the arm OP.

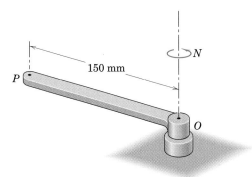

Problem 5/14

5/15 The circular disk rotates with a constant angular velocity $\omega = 40$ rad/s about its axis, which is inclined in the y-z plane at the angle $\theta = \tan^{-1}\frac{3}{4}$. Determine the vector expressions for the velocity and acceleration of point P, whose position vector at the instant shown is $\mathbf{r} = 375\mathbf{i} + 400\mathbf{j} - 300\mathbf{k}$ mm. (Check the magnitudes of your results from the scalar values $v = r\omega$ and $a_n = r\omega^2$.)

Ans. $\mathbf{v} = -20\mathbf{i} + 12\mathbf{j} - 9\mathbf{k}$ m/s
$\mathbf{a} = -600\mathbf{i} - 640\mathbf{j} + 480\mathbf{k}$ m/s^2

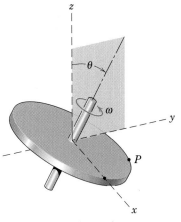

Problem 5/15

5/16 The two V-belt pulleys form an integral unit and rotate about the fixed axis at O. At a certain instant, point A on the belt of the smaller pulley has a velocity $v_A = 1.5$ m/s, and point B on the belt of the larger pulley has an acceleration $a_B = 45$ m/s^2 as shown. For this instant determine the magnitude of the acceleration $\mathbf{a}_C$ of point C and sketch the vector in your solution.

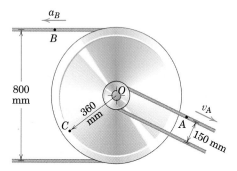

Problem 5/16

5/17 The two attached pulleys are driven by the belt with increasing speed. When the belt reaches a speed $v = 0.6$ m/s, the total acceleration of point P is 8 m/s^2. For this instant determine the angular acceleration α of the pulleys and the acceleration of point B on the belt.

Ans. $\alpha = 17.44$ rad/s^2, $a_B = 1.744$ m/s^2

Problem 5/17

5/18 The flywheel is rotating with an angular velocity ω_0 at time $t = 0$ when a torque is applied to increase its angular velocity. If the torque is controlled so that the angle θ between the total acceleration of point A on the rim and the radial line to A remains constant, determine the angular velocity ω and the angular acceleration α as functions of the time t.

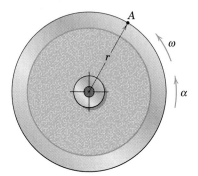

Problem 5/18

5/19 The rotation of an element in a certain mechanism is controlled so that the rate of change of its angular velocity ω with respect to its angular displacement θ is a constant k. If the angular velocity of the element is ω_0 at the start when $\theta = 0$ and $t = 0$, derive the expressions for θ, ω, and the angular acceleration α as functions of the time t.

Ans. $\theta = \dfrac{\omega_0}{k}(e^{kt} - 1)$, $\omega = \omega_0 e^{kt}$, $\alpha = \omega_0 k e^{kt}$

5/20 The design characteristics of a gear-reduction unit are under review. Gear B is rotating clockwise with a speed of 300 rev/min when a torque is applied to gear A at time $t = 2$ s to give gear A a counterclockwise acceleration α which varies with time for a duration of 4 seconds as shown. Determine the speed N of gear B when $t = 6$ s.

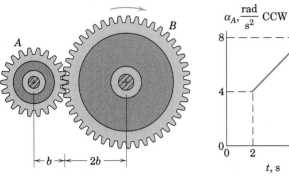

Problem 5/20

5/21 The circular disk rotates about its z-axis with an angular velocity in the direction shown. At a certain instant the magnitude of the velocity of point A is 3 m/s and is decreasing at the rate of 7.2 m/s^2. Write the vector expressions for the angular acceleration $\boldsymbol{\alpha}$ of the disk and the total acceleration of point B at this instant.

$$Ans. \quad \boldsymbol{\alpha} = -36\mathbf{k} \text{ rad/s}^2$$
$$\mathbf{a}_B = 5.4\mathbf{i} - 33.8\mathbf{j} \text{ m/s}^2$$

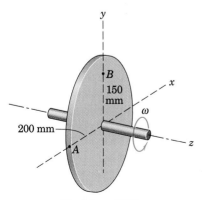

Problem 5/21

▶**5/22** A V-belt speed-reduction drive is shown where pulley A drives the two integral pulleys B which in turn drive pulley C. If A starts from rest at time $t = 0$ and is given a constant angular acceleration α_1, derive expressions for the angular velocity of C and the magnitude of the acceleration of a point P on the belt, both at time t.

$$Ans. \quad \omega_C = \left(\frac{r_1}{r_2}\right)^2 \alpha_1 t$$
$$a_P = \frac{r_1^2}{r_2} \alpha_1 \sqrt{1 + \left(\frac{r_1}{r_2}\right)^4 \alpha_1^2 t^4}$$

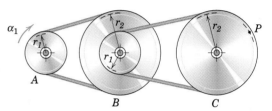

Problem 5/22

5/3   *ABSOLUTE MOTION*

We now develop the approach of absolute-motion analysis to describe the plane kinematics of rigid bodies. In this approach, we make use of the geometric relations which define the configuration of the body involved and then proceed to take the time derivatives of the defining geometric relations to obtain velocities and accelerations. In Art. 2/9 of Chapter 2 on particle kinematics, we introduced the application of absolute-motion analysis for the constrained motion of connected particles. For the pulley configurations treated, the relevant velocities and accelerations were determined by successive differentiation of the lengths of the connecting cables. In this earlier treatment, the geometric relations were quite simple, and no angular quantities had to be considered. Now that we will be dealing with rigid-body motion, however, we find that our defining geometric relations include both linear and angular measurements and, therefore, the time derivatives of these quantities will involve both linear and angular velocities and linear and angular accelerations.

In absolute-motion analysis, it is essential that we be totally consistent with the mathematics of the description. For example, if the angular position of a moving line in the plane of motion is specified by its counterclockwise angle θ measured from some convenient fixed reference axis, then the positive sense for both angular velocity $\dot{\theta}$ and angular acceleration $\ddot{\theta}$ will also be counterclockwise. A negative sign for either quantity will, of course, indicate a clockwise angular motion. The defining relations for linear motion, Eqs. 2/1, 2/2, and 2/3, and the relations involving angular motion, Eqs. 5/1 and 5/2 or 5/3, will find repeated use in the motion analysis and should be mastered thoroughly.

The absolute-motion approach to rigid-body kinematics is quite straightforward, provided the configuration lends itself to a geometric description which is not overly complex. If the geometric configuration is awkward or complex, analysis by the principles of relative motion may be preferable. Relative-motion analysis is treated in this chapter beginning with Art. 5/4. The choice between absolute- and relative-motion analyses is best made after experience has been gained with both approaches.

The three sample problems which follow illustrate the application of absolute-motion analysis to three commonly encountered situations, and they should be studied thoroughly. The kinematics of a rolling wheel, treated in Sample Problem 5/4, is especially basic and will find repeated use in the problem work because the rolling wheel in various forms is such a common element in mechanical systems.

Sample Problem 5/4

A wheel of radius r rolls on a flat surface without slipping. Determine the angular motion of the wheel in terms of the linear motion of its center O. Also determine the acceleration of a point on the rim of the wheel as the point comes into contact with the surface on which the wheel rolls.

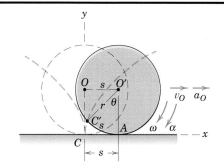

Solution.   The figure shows the wheel rolling to the right from the dotted to the full position without slipping. The linear displacement of the center O is s, which is also the arc length $C'A$ along the rim on which the wheel rolls. The radial line CO rotates to the new position $C'O'$ through the angle θ, where θ is measured from the vertical direction. If the wheel does not slip, the arc $C'A$ must equal the distance s. Thus, the displacement relationship and its two time derivatives give

$$s = r\theta$$

① $$v_O = r\omega \qquad\qquad \textit{Ans.}$$

$$a_O = r\alpha$$

where $v_O = \dot{s}$, $a_O = \dot{v}_O = \ddot{s}$, $\omega = \dot{\theta}$, and $\alpha = \dot{\omega} = \ddot{\theta}$. The angle θ, of course, must be in radians. The acceleration a_O will be directed in the sense opposite to that of v_O if the wheel is slowing down. In this event, the angular acceleration α will have the opposite sense to ω.

The origin of fixed coordinates is taken arbitrarily but conveniently at the point of contact between C on the rim of the wheel and the ground. When point C has moved along its cycloidal path to C', its new coordinates and their time derivatives become

$$x = s - r\sin\theta = r(\theta - \sin\theta) \qquad y = r - r\cos\theta = r(1 - \cos\theta)$$

$$\dot{x} = r\dot{\theta}(1 - \cos\theta) = v_O(1 - \cos\theta) \quad \dot{y} = r\dot{\theta}\sin\theta = v_O\sin\theta$$

$$\ddot{x} = \dot{v}_O(1 - \cos\theta) + v_O\dot{\theta}\sin\theta \qquad \ddot{y} = \dot{v}_O\sin\theta + v_O\dot{\theta}\cos\theta$$

$$\quad = a_O(1 - \cos\theta) + r\omega^2\sin\theta \qquad = a_O\sin\theta + r\omega^2\cos\theta$$

For the desired instant of contact, $\theta = 0$ and

② $$\ddot{x} = 0 \quad \text{and} \quad \ddot{y} = r\omega^2 \qquad\qquad \textit{Ans.}$$

Thus, the acceleration of the point C on the rim at the instant of contact with the ground depends only on r and ω and is directed toward the center of the wheel. If desired, the velocity and acceleration of C at any position θ may be obtained by writing the expressions $\mathbf{v} = \dot{x}\mathbf{i} + \dot{y}\mathbf{j}$ and $\mathbf{a} = \ddot{x}\mathbf{i} + \ddot{y}\mathbf{j}$.

Application of the kinematic relationships for a wheel which rolls without slipping should be recognized for various configurations of rolling wheels such as those illustrated on the right. If a wheel slips as it rolls, the foregoing relations are no longer valid.

① These three relations are not entirely unfamiliar at this point, and their application to the rolling wheel should be mastered thoroughly.

② Clearly, when $\theta = 0$, the point of contact has zero velocity so that $\dot{x} = \dot{y} = 0$. The acceleration of the contact point on the wheel will also be obtained by the principles of relative motion in Art. 5/6.

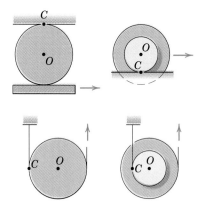

Sample Problem 5/5

The load L is being hoisted by the pulley and cable arrangement shown. Each cable is wrapped securely around its respective pulley so it does not slip. The two pulleys to which L is attached are fastened together to form a single rigid body. Calculate the velocity and acceleration of the load L and the corresponding angular velocity ω and angular acceleration α of the double pulley under the following conditions:

Case (a)   Pulley 1: $\omega_1 = \dot{\omega}_1 = 0$ (pulley at rest)

Pulley 2: $\omega_2 = 2$ rad/s, $\alpha_2 = \dot{\omega}_2 = -3$ rad/s^2

Case (b)   Pulley 1: $\omega_1 = 1$ rad/s, $\alpha_1 = \dot{\omega}_1 = 4$ rad/s^2

Pulley 2: $\omega_2 = 2$ rad/s, $\alpha_2 = \dot{\omega}_2 = -2$ rad/s^2

$r_1 = 100$ mm $r_2 = 100$ mm

100 mm   200 mm

Solution.   The tangential displacement, velocity, and acceleration of a point on the rim of pulley 1 or 2 equal the corresponding vertical motions of point A or B since the cables are assumed to be inextensible.

Case (a).   With A momentarily at rest, line AB rotates to AB' through the angle $d\theta$ during time dt. From the diagram we see that the displacements and their time derivatives give

$$ds_B = \overline{AB}\, d\theta \qquad v_B = \overline{AB}\omega \qquad (a_B)_t = \overline{AB}\alpha$$

$$ds_O = \overline{AO}\, d\theta \qquad v_O = \overline{AO}\omega \qquad a_O = \overline{AO}\alpha$$

With $v_D = r_2\omega_2 = 0.1(2) = 0.2$ m/s, $a_D = r_2\alpha_2 = 0.1(-3) = -0.3$ m/s^2, we have for the angular motion of the double pulley

$$\omega = v_B/\overline{AB} = v_D/\overline{AB} = 0.2/0.3 = 2/3 \text{ rad/s (CCW)} \qquad Ans.$$

$$\alpha = (a_B)_t/\overline{AB} = a_D/\overline{AB} = -0.3/0.3 = -1 \text{ rad/s}^2 \text{ (CW)} \qquad Ans.$$

The corresponding motion of O and the load L is

$$v_O = \overline{AO}\omega = 0.1(2/3) = 0.0667 \text{ m/s} \qquad Ans.$$

$$a_O = \overline{AO}\alpha = 0.1(-1) = -0.1 \text{ m/s}^2 \qquad Ans.$$

Case (b).   With point C, and hence point A, in motion, line AB moves to $A'B'$ during time dt. From the diagram for this case, we see that the displacements and their time derivatives give

$$ds_B - ds_A = \overline{AB}\, d\theta \qquad v_B - v_A = \overline{AB}\omega \qquad (a_B)_t - (a_A)_t = \overline{AB}\alpha$$

$$ds_O - ds_A = \overline{AO}\, d\theta \qquad v_O - v_A = \overline{AO}\omega \qquad a_O - (a_A)_t = \overline{AO}\alpha$$

With $v_C = r_1\omega_1 = 0.1(1) = 0.1$ m/s, $v_D = r_2\omega_2 = 0.1(2) = 0.2$ m/s, $a_C = r_1\alpha_1 = 0.1(4) = 0.4$ m/s^2 $a_D = r_2\alpha_2 = 0.1(-2) = -0.2$ m/s^2 we have for the angular motion of the double pulley

$$\omega = \frac{v_B - v_A}{\overline{AB}} = \frac{v_D - v_C}{\overline{AB}} = \frac{0.2 - 0.1}{0.3} = 1/3 \text{ rad/s (CCW)} \qquad Ans.$$

$$\alpha = \frac{(a_B)_t - (a_A)_t}{\overline{AB}} = \frac{a_D - a_C}{\overline{AB}} = \frac{-0.2 - 0.4}{0.3} = -2 \text{ rad/s}^2 \text{ (CW)} \qquad Ans.$$

The corresponding motion of O and the load L is

$$v_O = v_A + \overline{AO}\omega = v_C + \overline{AO}\omega = 0.1 + 0.1(1/3) = 0.1333 \text{ m/s} \qquad Ans.$$

$$a_O = (a_A)_t + \overline{AO}\alpha = a_C + \overline{AO}\alpha = 0.4 + 0.1(-2) = 0.2 \text{ m/s}^2 \qquad Ans.$$

① Recognize that the inner pulley is a wheel rolling along the fixed line of the left-hand cable. Thus, the expressions of Sample Problem 5/4 hold.

Case (a)

② Since B moves along a curved path, in addition to its tangential component of acceleration $(a_B)_t$, it will also have a normal component of acceleration toward O which does not affect the angular acceleration of the pulley.

③ The diagrams show these quantities and the simplicity of their linear relationships. The visual picture of the motion of O and B as AB rotates through the angle $d\theta$ should clarify the analysis.

Case (b)

④ Again, as in case (a), the differential rotation of line AB as seen from the figure establishes the relation between the angular velocity of the pulley and the linear velocities of points A, O, and B. The negative sign for $(a_B)_t = a_D$ produces the acceleration diagram shown but does not destroy the linearity of the relationships.

Sample Problem 5/6

Motion of the equilateral triangular plate ABC in its plane is controlled by the hydraulic cylinder D. If the piston rod in the cylinder is moving upward at the constant rate of 0.3 m/s during an interval of its motion, calculate for the instant when $\theta = 30°$ the velocity and acceleration of the center of the roller B in the horizontal guide and the angular velocity and angular acceleration of edge CB.

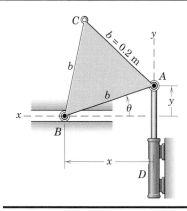

Solution. With the x-y coordinates chosen as shown, the given motion of A is $v_A = \dot{y} = 0.3$ m/s and $a_A = \ddot{y} = 0$. The accompanying motion of B is given by x and its time derivatives, which may be obtained from $x^2 + y^2 = b^2$. Differentiating gives

$$x\dot{x} + y\dot{y} = 0 \qquad \dot{x} = -\frac{y}{x}\dot{y}$$

$$x\ddot{x} + \dot{x}^2 + y\ddot{y} + \dot{y}^2 = 0 \qquad \ddot{x} = -\frac{\dot{x}^2 + \dot{y}^2}{x} - \frac{y}{x}\ddot{y}$$

① Observe that it is simpler to differentiate a product than a quotient. Thus, differentiate $x\dot{x} + y\dot{y} = 0$ rather than $\dot{x} = -y\dot{y}/x$.

With $y = b \sin\theta$, $x = b \cos\theta$, and $\ddot{y} = 0$, the expressions become

$$v_B = \dot{x} = -v_A \tan\theta$$

$$a_B = \ddot{x} = -\frac{v_A^2}{b}\sec^3\theta$$

Substituting the numerical values $v_A = 0.3$ m/s and $\theta = 30°$ gives

$$v_B = -0.3\left(\frac{1}{\sqrt{3}}\right) = -0.1732 \text{ m/s} \qquad \textit{Ans.}$$

$$a_B = -\frac{(0.3)^2(2/\sqrt{3})^3}{0.2} = -0.693 \text{ m/s}^2 \qquad \textit{Ans.}$$

The negative signs indicate that the velocity and acceleration of B are both to the right since x and its derivatives are positive to the left.

The angular motion of CB is the same as that of every line on the plate, including AB. Differentiating $y = b\sin\theta$ gives

$$\dot{y} = b\dot{\theta}\cos\theta \qquad \omega = \dot{\theta} = \frac{v_A}{b}\sec\theta$$

The angular acceleration is

$$\alpha = \dot{\omega} = \frac{v_A}{b}\dot{\theta}\sec\theta\tan\theta = \frac{v_A^2}{b^2}\sec^2\theta\tan\theta$$

Substitution of the numerical values gives

$$\omega = \frac{0.3}{0.2}\frac{2}{\sqrt{3}} = 1.732 \text{ rad/s} \qquad \textit{Ans.}$$

$$\alpha = \frac{(0.3)^2}{(0.2)^2}\left(\frac{2}{\sqrt{3}}\right)^2\frac{1}{\sqrt{3}} = 1.732 \text{ rad/s}^2 \qquad \textit{Ans.}$$

Both ω and α are counterclockwise since their signs are positive in the sense of the positive measurement of θ.

PROBLEMS

Introductory problems

5/23 Slider *A* moves in the horizontal slot with a constant speed *v* for a short interval of motion. Determine the angular velocity ω of bar *AB* in terms of the displacement x_A.

$$Ans. \quad \omega = \frac{\sqrt{3}v}{2L\sqrt{1 - \frac{3}{4}\frac{x_A^2}{L^2}}}$$

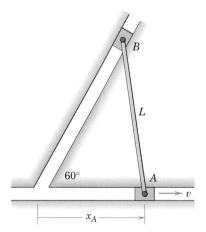

Problem 5/23

5/24 Point *A* is given a constant acceleration *a* to the right starting from rest with *x* essentially zero. Determine the angular velocity ω of link *AB* in terms of *x* and *a*.

Problem 5/24

5/25 The wheel of radius *r* rolls without slipping, and its center *O* has a constant velocity v_O to the right. Determine expressions for the velocity *v* and acceleration *a* of point *A* on the rim by differentiating its *x*- and *y*-coordinates. Represent your result graphically as vectors on your sketch and show that *v* is the vector sum of two v_O vectors.

$$Ans. \quad v = v_O\sqrt{2(1 + \sin\theta)}, \quad a = \frac{v_O^2}{r} \text{ toward } O$$

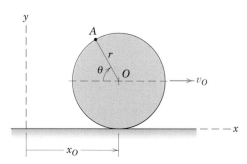

Problem 5/25

5/26 The telephone-cable reel rolls without slipping on the horizontal surface. If point *A* on the cable has a velocity $v_A = 0.8$ m/s to the right, compute the velocity of the center *O* and the angular velocity ω of the reel. (Be careful not to make the mistake of assuming that the reel rolls to the left.)

Problem 5/26

5/27 In a mechanism designed to convert linear motion to angular motion, the hydraulic cylinder gives pin *A* a constant downward velocity *v* for a short interval of motion. Determine the angular acceleration α of the slotted links in terms of θ.

$$Ans. \quad \alpha = -\frac{v^2}{b^2} \sin 2\theta \cos^2 \theta$$

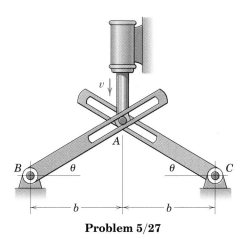

Problem 5/27

5/28 Slider *A* moves with constant speed *v* on the straight guide for a short interval, while slider *B* moves on the circular guide whose center is at *O*. Determine the angular velocity ω of link *AB* as a function of the displacement *s* of slider *A*.

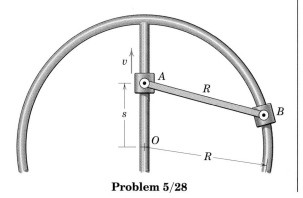

Problem 5/28

5/29 The small vehicle rides on rails and is driven by the 400-mm-diameter friction wheel turned by an electric motor. Determine the speed *v* of the vehicle if the friction-drive wheel is rotating at a speed of 300 rev/min and if no slipping occurs.

$$Ans. \quad v = 6.28 \text{ m/s}$$

Problem 5/29

5/30 The elements of the electric-motor drive for a diesel-electric locomotive are shown. The quill gear is integral with the 1000-mm-diameter wheel and is driven by the pinions of the two electric motors mounted to the wheel carriage. (Gear teeth are not shown.) For a locomotive speed of 90 km/h, what is the rotational speed *N* of the motor pinions?

Problem 5/30

5/31 The linear actuator is designed for rapid horizontal velocity v of jaw C for a slow change in the distance between A and B. If the hydraulic cylinder decreases this distance at the rate u, determine the horizontal velocity of jaw C in terms of the angle θ.

$$\text{Ans. } v = \frac{7}{2}u \cot \frac{\theta}{2}$$

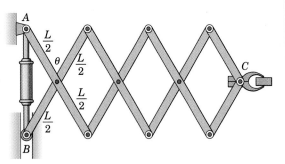

Problem 5/31

5/32 The elements of a wheel-and-disk mechanical integrator are shown in the figure. The integrator wheel A turns about its fixed shaft and is driven by friction from disk B with no slipping occurring tangent to its rim. The distance y is a variable and can be controlled at will. Show that the angular displacement of the integrator wheel is given by $z = (1/b) \int y \, dx$, where x is the angular displacement of the disk B.

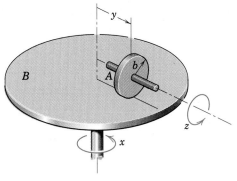

Problem 5/32

Representative problems

5/33 Calculate the angular velocity ω of the slender bar AB as a function of the distance x and the constant angular velocity ω_0 of the drum.

$$\text{Ans. } \omega = \frac{rh\omega_0}{x^2 + h^2}$$

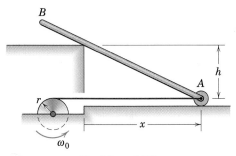

Problem 5/33

5/34 Rotation of the lever OA is controlled by the motion of the contacting circular disk whose center is given a horizontal velocity v. Determine the expression for the angular velocity ω of the lever OA in terms of x.

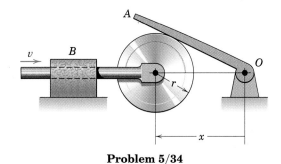

Problem 5/34

5/35 Boom OA is being elevated by the rope-and-pulley arrangement shown. If point B on the rope is given a constant velocity $v_B = 3.2$ m/s, determine the angular velocity ω and angular acceleration α of the boom for $\theta = 30°$.

     Ans. $\omega = 0.1342$ rad/s, $\alpha = -0.001916$ rad/s^2

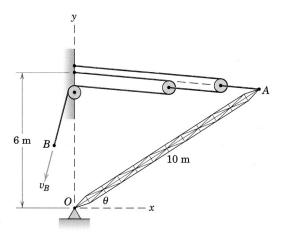

Problem 5/35

5/36 The telescoping link is hinged at O, and its end A is given a constant upward velocity of 200 mm/s by the piston rod of the fixed hydraulic cylinder B. Calculate the angular velocity $\dot{\theta}$ and the angular acceleration $\ddot{\theta}$ of link OA for the instant when $y = 600$ mm.

Problem 5/36

5/37 Derive an expression for the upward velocity v of the car hoist in terms of θ. The piston rod of the hydraulic cylinder is extending at the rate $\dot{s}$.

$$Ans.\ \ v = \frac{2\dot{s}\sqrt{b^2 + L^2 - 2bL \cos \theta}}{L \tan \theta}$$

Problem 5/37

5/38 Determine the acceleration of the shaft B for $\theta = 60°$ if the crank OA has an angular acceleration $\ddot{\theta} = 8$ rad/s^2 and an angular velocity $\dot{\theta} = 4$ rad/s at this position. The spring maintains contact between the roller and the surface of the plunger.

Problem 5/38

5/39 A roadway speed bump is being installed on a level road to remind motorists of the existing speed limit. If the driver of the car experiences at G a vertical acceleration of as much as g, up or down, he is expected to realize that his speed is bordering on being excessive. For the speed bump with the cosine contour shown, derive an expression for the height h of the bump which will produce a vertical component of acceleration at G of g at a car speed v. Compute h if $b = 1$ m and $v = 20$ km/h. Neglect the effects of suspension-spring flexing and finite wheel diameter.

$$Ans. \ h = 4g\left(\frac{b}{\pi v}\right)^2, \ h = 128.8 \text{ mm}$$

Problem 5/39

5/40 Motion of the wheel as it rolls up the fixed rack on its geared hub is controlled through the peripheral cable by the driving wheel D, which turns counterclockwise at the constant rate $\omega_0 = 4$ rad/s for a short interval of motion. By examining the geometry of a small (differential) rotation of line $AOCB$ as it pivots momentarily about the contact point C, determine the angular velocity ω of the wheel and the velocities of point A and the center O. Also find the acceleration of point C.

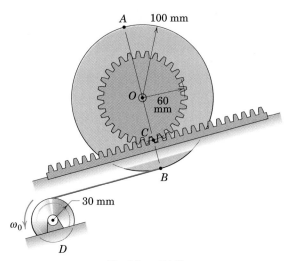

Problem 5/40

5/41 The flywheel turns clockwise with a constant speed of 600 rev/min. The connecting link AB slides through the pivoted collar at C. Calculate the angular velocity ω of AB for the instant when $\theta = 60°$.

$$Ans. \ \omega = 17.95 \text{ rad/s CW}$$

Problem 5/41

5/42 The rod OB slides through the collar pivoted to the rotating link at A. If CA has an angular velocity $\omega = 3$ rad/s for an interval of motion, calculate the angular velocity of OB when $\theta = 45°$.

$\overline{CA} = 200$ mm

400 mm

Problem 5/42

5/43 A variable-speed belt drive consists of the two pulleys each of which is constructed of two cones which turn as a unit but are capable of being drawn together or separated so as to change the effective radius of the pulley. If the angular velocity ω_1 of pulley 1 is constant, determine the expression for the angular acceleration $\alpha_2 = \dot{\omega}_2$ of pulley 2 in terms of the rates of change $\dot{r}_1$ and $\dot{r}_2$ of the effective radii.

$$Ans.\quad \alpha_2 = \frac{\dot{r}_1 r_2 - r_1 \dot{r}_2}{r_2^2}\omega_1$$

Problem 5/43

5/44 The slotted arm pivots about O and maintains the relation between the motions of sliders A and B and their control rods. Each small pivoted block is pinned to its respective slider and is constrained to slide in its rotating slot. Show that the displacement x is proportional to the reciprocal of y. Then establish the relation between the velocities v_A and v_B. Also, if v_A is constant for a short interval of motion, determine the acceleration of B.

Problem 5/44

5/45 It is desired to design a system for controlling the rate of extension $\dot{x}$ of the fire-truck ladder during elevation of the ladder so that the bucket B will have a vertical motion only. Determine $\dot{x}$ in terms of the elongation rate $\dot{c}$ of the hydraulic cylinder for given values of θ and x.

$$Ans.\quad \dot{x} = \frac{L + x}{b}\frac{\tan\theta}{\cos\frac{1}{2}(\theta + \delta)}\dot{c}, \text{ where } \delta = \sin^{-1}(h/b)$$

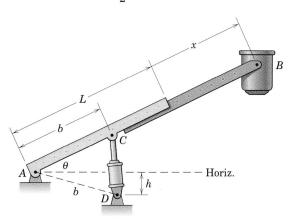

Problem 5/45

5/46 The mechanism is designed to produce small oscillations of the shaft A and attached fork through the rotation of the offset circular cam, which rotates about O with a constant angular velocity ω_0. Determine an expression for the angular velocity ω of the fork and its shaft A as a function of the cam angle θ.

Problem 5/46

5/47 The piston rod of the hydraulic cylinder gives point B a velocity v_B as shown. Determine the magnitude v_C of the velocity of end C of link ABC in terms of θ.

$$Ans.\ v_C = \frac{v_B}{2}\sqrt{8 + \sec^2\frac{\theta}{2}}$$

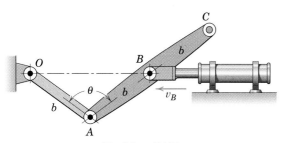

Problem 5/47

5/48 Show that the expressions $v = r\omega$ and $a_t = r\alpha$ hold for the motion of the center O of the wheel which rolls on the concave or convex circular arc, where ω and α are the absolute angular velocity and acceleration, respectively, of the wheel. (*Hint:* Follow the example of Sample Problem 5/4 and allow the wheel to roll a small distance. Be very careful to identify the correct *absolute* angle through which the wheel turns in each case in determining its angular velocity and angular acceleration.)

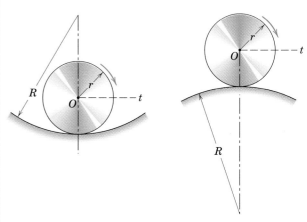

Problem 5/48

5/49 The two gears form an integral unit and roll on the fixed rack. The large gear has 48 teeth, and the worm turns with a speed of 120 rev/min. Find the velocity v_O of the center O of the gear.

$$Ans.\ v_O = 26.2\ \text{mm/s}$$

Problem 5/49

5/50 Angular oscillation of the slotted link is achieved by the crank *OA*, which rotates clockwise at the steady speed N = 120 rev/min. Determine an expression for the angular velocity $\dot{\beta}$ of the slotted link in terms of θ.

Problem 5/50

▶5/51 The Geneva wheel is a mechanism for producing intermittent rotation. Pin P in the integral unit of wheel A and locking plate B engages the radial slots in wheel C thus turning wheel C one-fourth of a revolution for each revolution of the pin. At the engagement position shown, θ = 45°. For a constant clockwise angular velocity ω_1 = 2 rad/s of wheel A, determine the corresponding counterclockwise angular velocity ω_2 of wheel C for θ = 20°. (Note that the motion during engagement is governed by the geometry of triangle O_1O_2P with changing θ.)

            Ans. ω_2 = 1.923 rad/s

Problem 5/51

▶5/52 The rod AB slides through the pivoted collar as end A moves along the slot. If A starts from rest at x = 0 and moves to the right with a constant acceleration of 0.1 m/s², calculate the angular acceleration α of AB at the instant when x = 150 mm.

            Ans. α = 0.1408 rad/s² CCW

Problem 5/52

▶5/53 The punch is operated by a simple harmonic oscillation of the pivoted sector given by $\theta = \theta_0 \sin 2\pi t$ where the amplitude is $\theta_0 = \pi/12$ rad (15°) and the time for one complete oscillation is 1 second. Determine the acceleration of the punch when (a) θ = 0 and (b) $\theta = \pi/12$.

            Ans. (a) a = 0.909 m/s² up
                      (b) a = 0.918 m/s² down

Problem 5/53

▶5/54 One of the most common mechanisms is the slider-crank. Express the angular velocity ω_{AB} and angular acceleration α_{AB} of the connecting rod AB in terms of the crank angle θ for a given constant crank speed ω_0. Take ω_{AB} and α_{AB} to be positive counterclockwise.

$$Ans. \quad \omega_{AB} = \frac{r\omega_0}{l} \frac{\cos\theta}{\sqrt{1 - \frac{r^2}{l^2}\sin^2\theta}}$$

$$\alpha_{AB} = \frac{r\omega_0^2}{l}\sin\theta \frac{\frac{r^2}{l^2} - 1}{\left(1 - \frac{r^2}{l^2}\sin^2\theta\right)^{3/2}}$$

Problem 5/54

5/4 *RELATIVE VELOCITY*

The second approach to rigid-body kinematics is to use the principles of relative motion. In Art. 2/8 we developed these principles for measurements relative to translating axes and applied the relative-velocity equation

$$\mathbf{v}_A = \mathbf{v}_B + \mathbf{v}_{A/B} \tag{2/20}$$

to the motions of two particles A and B.

We now choose two points on the *same* rigid body for our two particles. The consequence of this choice is that the motion of one point as seen by an observer translating with the other point must be circular since the radial distance to the observed point from the reference point does not change. This observation is the *key* to the successful understanding of a large majority of problems in the plane motion of rigid bodies. This concept is illustrated in Fig. 5/5*a*, which shows a rigid body moving in the plane of the figure from position AB to $A'B'$ during time Δt. This movement may be visualized as occurring in two parts. First, the body translates to the parallel position $A''B'$ with the displacement $\Delta \mathbf{r}_B$. Second, the body rotates about B' through the angle $\Delta \theta$. From the nonrotating reference axes x'-y' attached to the reference point B', it is seen that this remaining motion of the body is one of simple rotation about B', giving rise to the displacement $\Delta \mathbf{r}_{A/B}$ of A with respect to B. To the nonrotating observer attached to B, the body appears to undergo fixed-axis rotation about B with A executing circular motion as emphasized in Fig. 5/5*b*. Therefore, the relationships developed for circular motion in Arts. 2/5 and 5/2 and cited as Eqs. 2/11 and 5/2 (or 5/3) describe the relative portion of A's motion.

Point B was arbitrarily chosen as the reference point for attachment of our nonrotating reference axes x-y. Point A could have been used just as well, in which case we observe B to have circular motion about A considered fixed as shown in Fig. 5/5*c*. We see that

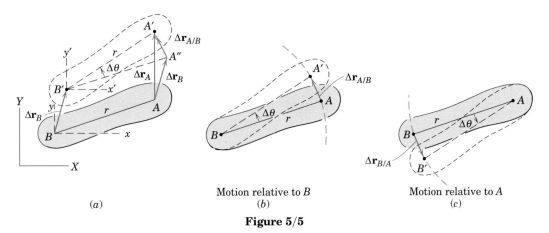

Motion relative to B
(b)

Motion relative to A
(c)

(a)

Figure 5/5

the sense of the rotation, counterclockwise in this example, is the same whether we choose A or B as the reference, and we see that $\Delta \mathbf{r}_{B/A} = -\Delta \mathbf{r}_{A/B}$.

With B as the reference point, the total displacement of A is seen from Fig. 5/5a to be

$$\Delta \mathbf{r}_A = \Delta \mathbf{r}_B + \Delta \mathbf{r}_{A/B}$$

where $\Delta \mathbf{r}_{A/B}$ has the magnitude $r\Delta\theta$ as $\Delta\theta$ approaches zero. We note that the *relative linear motion* $\Delta \mathbf{r}_{A/B}$ is accompanied by the *absolute angular motion* $\Delta\theta$, as seen from the translating axes x'-y'. Dividing the expression for $\Delta \mathbf{r}_A$ by the corresponding time interval Δt and passing to the limit give the relative-velocity equation

$$\boxed{\mathbf{v}_A = \mathbf{v}_B + \mathbf{v}_{A/B}} \tag{5/4}$$

This expression is the same as Eq. 2/20, with the one restriction that the distance r between A and B remains constant. The magnitude of the relative velocity is thus seen to be $v_{A/B} = \lim_{\Delta t \to 0} (|\Delta \mathbf{r}_{A/B}|/\Delta t) = \lim_{\Delta t \to 0} (r\Delta\theta/\Delta t)$ which, with $\omega = \dot{\theta}$, becomes

$$\boxed{v_{A/B} = r\omega} \tag{5/5}$$

With $\mathbf{r}$ representing the vector $\mathbf{r}_{A/B}$, from the first of Eqs. 5/3 we may write the relative velocity as the vector

$$\boxed{\mathbf{v}_{A/B} = \boldsymbol{\omega} \times \mathbf{r}} \tag{5/6}$$

where $\boldsymbol{\omega}$ is the angular-velocity vector normal to the plane of the motion in the sense determined by the right-hand rule. A critical observation seen from Figs. 5/5b and c is that the relative linear velocity is always perpendicular to the line joining the two points in question.

The application of Eq. 5/4 is clarified by visualizing the separate translation and rotation components of the equation. These components are emphasized in Fig. 5/6, which shows a rigid body

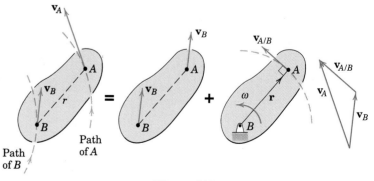

Figure 5/6

in plane motion. With B chosen as the reference point, the velocity of A is the vector sum of the translational portion $\mathbf{v}_B$, plus the rotational portion $\mathbf{v}_{A/B} = \boldsymbol{\omega} \times \mathbf{r}$, which has the magnitude $v_{A/B} = r\omega$, where $|\boldsymbol{\omega}| = \dot{\theta}$, the *absolute* angular velocity of AB. The fact that the *relative linear velocity* is *always perpendicular* to the line joining the two points in question is an important key to the solution of many problems. The student should draw the equivalent diagram where point A is used as the reference point rather than B.

A second use of Eq. 5/4 for relative-velocity problems in plane motion may be made for constrained sliding contact between two links in a mechanism. In this case, we choose points A and B as coincident points, one on each link, for the instant under consideration. In contrast to the previous example, in this case, the two points are on different bodies so they are not a fixed distance apart. This second use of the relative-velocity equation is illustrated in Sample Problem 5/10.

Solution of the relative-velocity equation may be carried out by scalar or vector algebra, or a graphical analysis may be employed. In any event, a sketch of the vector polygon which represents the vector equation should be made to reveal the physical relationships involved. From this sketch, scalar component equations may be written by projecting the vectors along convenient directions. Usually, a simultaneous solution may be avoided by a careful choice of the projections. Alternatively, each term in the relative-motion equation may be written in terms of its $\mathbf{i}$- and $\mathbf{j}$-components, from which two scalar equations result when the equality is applied, separately, to the coefficients of the $\mathbf{i}$- and $\mathbf{j}$-terms. Many problems lend themselves to a graphical solution, particularly when the given geometry results in an awkward mathematical expression. In this case, the known vectors are first constructed in their correct positions using a convenient scale. Next, the unknown vectors, which complete the polygon and satisfy the vector equation, are measured directly from the drawing. The choice of method to be used depends on the particular problem at hand, the accuracy required, and individual preference and experience. All three of the approaches are illustrated in the sample problems which follow.

Regardless of which method of solution we employ, we note that the single vector equation in two dimensions is equivalent to two scalar equations, so that two scalar unknowns at the most can be determined. The unknowns, for instance, might be the magnitude of one vector and the direction of another. A systematic identification of the knowns and unknowns should be made before a solution is attempted.

Sample Problem 5/7

The wheel of radius $r = 300$ mm rolls to the right without slipping and has a velocity $v_O = 3$ m/s of its center O. Calculate the velocity of point A on the wheel for the instant represented.

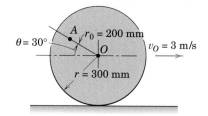

Solution I (scalar-geometric). The center O is chosen as the reference point for the relative-velocity equation since its motion is given. We therefore write

$$\mathbf{v}_A = \mathbf{v}_O + \mathbf{v}_{A/O}$$

where the relative-velocity term is observed from the translating axes x-y attached to O. The angular velocity of AO is the same as that of the wheel which, from Sample Problem 5/4, is $\omega = v_O/r = 3/0.3 = 10$ rad/s. Thus, from Eq. 5/5 we have

$$[v_{A/O} = r_0\dot{\theta}] \qquad v_{A/O} = 0.2(10) = 2 \text{ m/s}$$

① which is normal to AO as shown. The vector sum $\mathbf{v}_A$ is shown on the diagram and may be calculated from the law of cosines. Thus,

② $$v_A^2 = 3^2 + 2^2 + 2(3)(2)\cos 60° = 19 \text{ (m/s)}^2 \quad v_A = 4.36 \text{ m/s} \quad Ans.$$

The contact point C momentarily has zero velocity and can be used alternatively as the reference point, in which case, the relative-velocity equation becomes $\mathbf{v}_A = \mathbf{v}_C + \mathbf{v}_{A/C} = \mathbf{v}_{A/C}$ where

$$v_{A/C} = \overline{AC}\omega = \frac{\overline{AC}}{\overline{OC}} v_O = \frac{0.436}{0.300}(3) = 4.36 \text{ m/s} \quad v_A = v_{A/C} = 4.36 \text{ m/s}$$

③ The distance $\overline{AC} = 436$ mm is calculated separately. We see that $\mathbf{v}_A$ is normal to AC since A is momentarily rotating about point C.

Solution II (vector). We will now use Eq. 5/6 and write

$$\mathbf{v}_A = \mathbf{v}_O + \mathbf{v}_{A/O} = \mathbf{v}_O + \boldsymbol{\omega} \times \mathbf{r}_0$$

where

④ $$\boldsymbol{\omega} = -10\mathbf{k} \text{ rad/s}$$

$$\mathbf{r}_0 = 0.2(-\mathbf{i}\cos 30° + \mathbf{j}\sin 30°) = -0.1732\mathbf{i} + 0.1\mathbf{j} \text{ m}$$

$$\mathbf{v}_O = 3\mathbf{i} \text{ m/s}$$

We now solve the vector equation

$$\mathbf{v}_A = 3\mathbf{i} + \begin{vmatrix} \mathbf{i} & \mathbf{j} & \mathbf{k} \\ 0 & 0 & -10 \\ -0.1732 & 0.1 & 0 \end{vmatrix} = 3\mathbf{i} + 1.732\mathbf{j} + 1.0\mathbf{i}$$

$$= 4\mathbf{i} + 1.732\mathbf{j} \text{ m/s} \qquad\qquad Ans.$$

The magnitude $v_A = \sqrt{4^2 + (1.732)^2} = \sqrt{19} = 4.36$ m/s and direction agree with the previous solution.

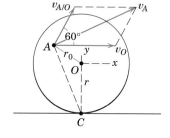

① Be sure to visualize $v_{A/O}$ as the velocity that A appears to have in its circular motion relative to O.

② The vectors may also be laid off to scale graphically and the magnitude and direction of v_A measured directly from the diagram.

③ The velocity of any point on the wheel is easily determined by using the contact point C as the reference point. The student should construct the velocity vectors for a number of points on the wheel for practice.

④ The vector $\boldsymbol{\omega}$ is directed into the paper by the right-hand rule, whereas the positive z-direction is out from the paper; hence, the minus sign.

Sample Problem 5/8

Crank CB oscillates about C through a limited arc, causing crank OA to oscillate about O. When the linkage passes the position shown with CB horizontal and OA vertical, the angular velocity of CB is 2 rad/s counterclockwise. For this instant, determine the angular velocities of OA and AB.

Solution I (vector). The relative-velocity equation $\mathbf{v}_A = \mathbf{v}_B + \mathbf{v}_{A/B}$ is rewritten as

$$\boldsymbol{\omega}_{OA} \times \mathbf{r}_A = \boldsymbol{\omega}_{CB} \times \mathbf{r}_B + \boldsymbol{\omega}_{AB} \times \mathbf{r}_{A/B}$$

① We are using here the first of Eqs. 5/3 and Eq 5/6.

where $\boldsymbol{\omega}_{OA} = \omega_{OA}\mathbf{k}$, $\boldsymbol{\omega}_{CB} = 2\mathbf{k}$ rad/s, $\boldsymbol{\omega}_{AB} = \omega_{AB}\mathbf{k}$,

$$\mathbf{r}_A = 100\mathbf{j} \text{ mm} \qquad \mathbf{r}_B = -75\mathbf{i} \text{ mm} \qquad \mathbf{r}_{A/B} = -175\mathbf{i} + 50\mathbf{j} \text{ mm}$$

Substitution gives

$$\omega_{OA}\mathbf{k} \times 100\mathbf{j} = 2\mathbf{k} \times (-75\mathbf{i}) + \omega_{AB}\mathbf{k} \times (-175\mathbf{i} + 50\mathbf{j})$$

$$-100\omega_{OA}\mathbf{i} = -150\mathbf{j} - 175\omega_{AB}\mathbf{j} - 50\omega_{AB}\mathbf{i}$$

Matching coefficients of the respective $\mathbf{i}$- and $\mathbf{j}$-terms gives

$$-100\omega_{OA} + 50\omega_{AB} = 0 \qquad 25(6 + 7\omega_{AB}) = 0$$

the solutions of which are

$$\omega_{AB} = -6/7 \text{ rad/s} \qquad \text{and} \qquad \omega_{OA} = -3/7 \text{ rad/s} \qquad Ans.$$

② The minus signs in the answers indicate that the vectors $\boldsymbol{\omega}_{AB}$ and $\boldsymbol{\omega}_{OA}$ are in the negative $\mathbf{k}$-direction. Hence, the angular velocities are clockwise.

Solution II (scalar-geometric). Solution by the scalar geometry of the vector triangle is particularly simple here since $\mathbf{v}_A$ and $\mathbf{v}_B$ are at right angles for this special position of the linkages. First, we compute v_B, which is

$$[v = r\omega] \qquad v_B = 0.075(2) = 0.150 \text{ m/s}$$

and represent it in its correct direction as shown. The vector $\mathbf{v}_{A/B}$ must be perpendicular to AB, and the angle θ between $\mathbf{v}_{A/B}$ and $\mathbf{v}_B$ is also the angle made by AB with the horizontal direction. This angle is given by

$$\tan\theta = \frac{100 - 50}{250 - 75} = \frac{2}{7}$$

The horizontal vector $\mathbf{v}_A$ completes the triangle for which we have

$$v_{A/B} = v_B/\cos\theta = 0.150/\cos\theta$$

$$v_A = v_B \tan\theta = 0.150(2/7) = 0.30/7 \text{ m/s}$$

③ Always make certain that the sequence of vectors in the vector polygon agrees with the equality of vectors specified by the vector equation.

The angular velocities become

$$[\omega = v/r] \qquad \omega_{AB} = v_{A/B}/\overline{AB} = \frac{0.150}{\cos\theta}\frac{\cos\theta}{0.250 - 0.075}$$

$$= 6/7 \text{ rad/s CW} \qquad Ans.$$

$$[\omega = v/r] \qquad \omega_{OA} = v_A/\overline{OA} = \frac{0.30}{7}\frac{1}{0.100} = 3/7 \text{ rad/s CW} \qquad Ans.$$

Sample Problem 5/9

The common configuration of a reciprocating engine is that of the slider-crank mechanism shown. If the crank OB has a clockwise rotational speed of 1500 rev/min, determine for the position where $\theta = 60°$ the velocity of the piston A, the velocity of point G on the connecting rod, and the angular velocity of the connecting rod.

Solution. The velocity of the crank pin B as a point on AB is easily found, so that B will be used as the reference point for determining the velocity of A. The relative-velocity equation may now be written

$$\mathbf{v}_A = \mathbf{v}_B + \mathbf{v}_{A/B}$$

The crank-pin velocity is

① $[v = r\omega]$ $\qquad v_B = 0.125 \dfrac{1500(2\pi)}{60} = 19.63$ m/s

① Remember always to convert ω to radians per unit time when using $v = r\omega$.

and is normal to OB. The direction of $\mathbf{v}_A$ is, of course, along the horizontal cylinder axis. The direction of $\mathbf{v}_{A/B}$ must be perpendicular to the line AB as explained in the present article and as indicated on the diagram, where the reference point B is shown as fixed. We obtain this direction by computing angle β from the law of sines, which gives

$$\frac{125}{\sin \beta} = \frac{350}{\sin 60°} \qquad \beta = \sin^{-1} 0.309 = 18.02°$$

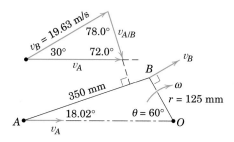

We now complete the sketch of the velocity triangle where the angle between $\mathbf{v}_{A/B}$ and $\mathbf{v}_A$ is $90° - 18.02° = 72.0°$ and the third angle is $180° - 30° - 72.0° = 78.0°$. Vectors $\mathbf{v}_A$ and $\mathbf{v}_{A/B}$ are shown with their proper sense such that the head-to-tail sum of $\mathbf{v}_B$ and $\mathbf{v}_{A/B}$ equals $\mathbf{v}_A$. The magnitudes of the unknowns are now calculated from the trigonometry of the vector triangle or are scaled from the diagram if a graphical solution is used. Solving for $\mathbf{v}_A$ and $\mathbf{v}_{A/B}$ by the law of sines gives

② $\qquad \dfrac{v_A}{\sin 78.0°} = \dfrac{19.63}{\sin 72.0°} \qquad v_A = 20.2$ m/s $\qquad$ *Ans.*

② A graphical solution to this problem is the quickest to achieve, although its accuracy is limited. Solution by vector algebra can, of course, be used but would involve somewhat more labor in this problem.

$$\frac{v_{A/B}}{\sin 30°} = \frac{19.63}{\sin 72.0°} \qquad v_{A/B} = 10.32 \text{ m/s}$$

The angular velocity of AB is counterclockwise, as revealed by the sense of $\mathbf{v}_{A/B}$, and is

$[\omega = v/r]$ $\qquad \omega_{AB} = \dfrac{v_{A/B}}{AB} = \dfrac{10.32}{0.350} = 29.5$ rad/s $\qquad$ *Ans.*

We now determine the velocity of G by writing

$$\mathbf{v}_G = \mathbf{v}_B + \mathbf{v}_{G/B}$$

where $\quad v_{G/B} = \overline{GB}\omega_{AB} = \dfrac{\overline{GB}}{\overline{AB}} v_{A/B} = \dfrac{100}{350}(10.32) = 2.95$ m/s.

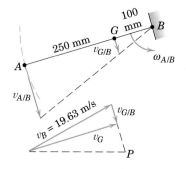

As seen from the diagram, $\mathbf{v}_{G/B}$ has the same direction as $\mathbf{v}_{A/B}$. The vector sum is shown on the last diagram. We can calculate v_G with some geometric labor or simply measure its magnitude and direction from the velocity diagram drawn to scale. For simplicity we adopt the latter procedure here and obtain

$$v_G = 19.24 \text{ m/s} \qquad \textit{Ans.}$$

As seen, the diagram may be superposed directly on the first velocity diagram.

Sample Problem 5/10

The power screw turns at a speed that gives the threaded collar C a velocity of 0.25 m/s vertically down. Determine the angular velocity of the slotted arm when $\theta = 30°$.

$\theta = 30°$

450 mm

Solution. The angular velocity of the arm can be found if the velocity of a point on the arm is known. We choose a point A on the arm coincident with the pin B of the collar for this purpose. If we use B as our reference point and write $\mathbf{v}_A = \mathbf{v}_B + \mathbf{v}_{A/B}$, we see from the diagram, which shows the arm and points A and B an instant before and an instant after coincidence, that $\mathbf{v}_{A/B}$ has a direction along the slot away from O.

① Physically, of course, this point does not exist, but we can imagine such a point in the middle of the slot and attached to the arm.

The magnitudes of $\mathbf{v}_A$ and $\mathbf{v}_{A/B}$ are the only unknowns in the vector equation, so that it may now be solved. We draw the known vector $\mathbf{v}_B$ and then obtain the intersection P of the known directions of $\mathbf{v}_{A/B}$ and $\mathbf{v}_A$. The solution gives

$$v_A = v_B \cos \theta = 0.25 \cos 30° = 0.217 \text{ m/s}$$

$$[\omega = v/r] \qquad \omega = v_A/\overline{OA} = \frac{0.217}{0.450/\cos 30°}$$

$$= 0.417 \text{ rad/s CCW} \qquad Ans.$$

② Always identify the knowns and unknowns before attempting the solution of a vector equation.

We note the difference between this problem of constrained sliding contact between two links and the three preceding sample problems of relative velocity, where no sliding contact occurred and where the points A and B were located on the same rigid body in each case.

PROBLEMS

Introductory problems

5/55 The speed of the center of the earth as it orbits the sun is $v = 107\ 257$ km/h, and the absolute angular velocity of the earth about its north-south spin axis is $\omega = 7.292(10^{-5})$ rad/s. Use the value $R = 6371$ km for the earth's radius and determine the velocities of points A, B, C, D, all of which are on the equator.

$$\begin{aligned}
Ans.\ \mathbf{v}_A &= -1672\mathbf{i} + 107\ 257\mathbf{j} \text{ km/h} \\
\mathbf{v}_B &= 105\ 585\mathbf{j} \text{ km/h} \\
\mathbf{v}_C &= 1672\mathbf{i} + 107\ 257\mathbf{j} \text{ km/h} \\
\mathbf{v}_D &= 108\ 929\mathbf{j} \text{ km/h}
\end{aligned}$$

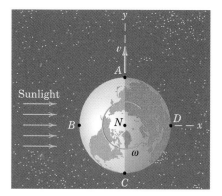

Problem 5/55

5/56 Vertex A of the equilateral triangular plate has a velocity $v_A = 0.8$ m/s to the right, and the counterclockwise angular velocity of the plate is $\omega = 5$ rad/s. Determine the velocity of vertex C for the instant shown.

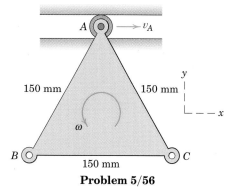

Problem 5/56

5/57 End A of the 600-mm link has a velocity of 5 m/s in the direction shown. At the same instant end B has a velocity whose magnitude is 6 m/s as indicated. Find the angular velocity ω of the link in two ways.

$$Ans.\ \omega = 11.09 \text{ rad/s CW}$$

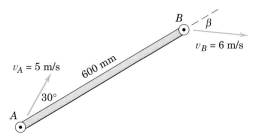

Problem 5/57

5/58 The magnitude of the absolute velocity of point A on the automobile tire is 12 m/s when A is in the position shown. What are the corresponding velocity v_O of the car and the angular velocity ω of the wheel? (The wheel rolls without slipping.)

Problem 5/58

5/59 For a short interval, collars A and B are sliding along the fixed vertical shaft with velocities $v_A =$ 2 m/s and $v_B = 3$ m/s in the directions shown. Determine the magnitude of the velocity of point C for the position $\theta = 60°$.

Ans. $v_C = 1.528$ m/s

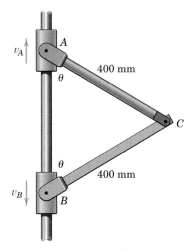

Problem 5/59

5/60 The two pulleys are riveted together to form a single rigid unit, and each of the two cables is securely wrapped around its respective pulley. If point A on the hoisting cable has a velocity $v = 0.9$ m/s, determine the magnitudes of the velocity of point O and the velocity of point B on the larger pulley for the position shown.

Problem 5/60

5/61 Determine the angular velocity of the telescoping link AB for the position shown where the driving links have the angular velocities indicated.

Ans. $\omega_{AB} = 0.96$ rad/s CCW

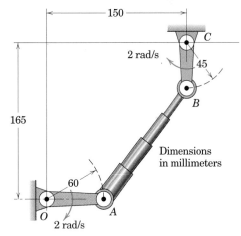

Problem 5/61

5/62 For the instant represented point B crosses the horizontal axis through point O with a downward velocity $v = 0.6$ m/s. Determine the corresponding value of the angular velocity ω_{OA} of link OA.

Problem 5/62

5/63 At the instant represented, the velocity of point A of the 1.2-m bar is 3 m/s to the right. Determine the speed v_B of point B and the angular velocity ω of the bar. The diameter of the small end wheels may be neglected.

Ans. $v_B = 4.38$ m/s, $\omega = 3.23$ rad/s CCW

Problem 5/63

Representative problems

5/64 The wheel of radius r rolls without slipping and has an angular velocity ω. Write an expression for the velocity of point A in terms of θ and show that the velocities of A and B are perpendicular to one another.

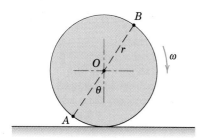

Problem 5/64

5/65 The rider of the bicycle shown pumps steadily to maintain a constant speed of 16 km/h against a slight head wind. Calculate the maximum and minimum magnitudes of the absolute velocity of the pedal A.

Ans. $(v_A)_{max} = 5.33$ m/s
$(v_A)_{min} = 3.56$ m/s

Problem 5/65

5/66 For an interval of its motion the piston rod of the hydraulic cylinder has a velocity $v_A = 1.2$ m/s as shown. At a certain instant $\theta = \beta = 60°$. For this instant determine the angular velocity ω_{BC} of link BC.

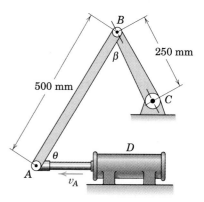

Problem 5/66

5/67 The rotation of the gear is controlled by the horizontal motion of end A of the rack AB. If the piston rod has a constant velocity $\dot{x} = 300$ mm/s during a short interval of motion, determine the angular velocity ω_0 of the gear and the angular velocity ω_{AB} of AB at the instant when $x = 800$ mm.

Ans. $\omega_{AB} = 0.0968$ rad/s CCW
$\omega_0 = 1.452$ rad/s CW

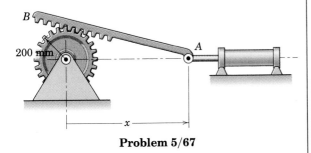

Problem 5/67

5/68 Determine the angular velocity of link BC for the instant indicated. In case (a), the center O of the disk is a fixed pivot, while in case (b), the disk rolls without slipping on the horizontal surface. In both cases, the disk has clockwise angular velocity ω. Neglect the small distance of pin A from the edge of the disk.

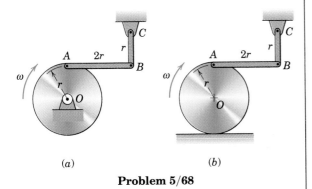

Problem 5/68

5/69 For the instant represented, crank OB has a clockwise angular velocity $\omega = 0.8$ rad/s and is passing the horizontal position. Determine the corresponding velocity of the guide roller A in the 20° slot and the velocity of point C midway between A and B.

Ans. $v_A = 0.226$ m/s, $v_C = 0.1747$ m/s

Problem 5/69

5/70 Rod CB slides through the pivoted collar attached to link OA. If CB has a clockwise angular velocity of 2 rad/s, determine the angular velocity ω_{OA} of link OA when $\theta = 60°$.

Problem 5/70

5/71 The unit at A consists of a high-torque geared motor which rotates link AB at the constant rate $\dot{\theta} = 0.5$ rad/s. Unit A is free to roll along the horizontal surface. Determine the velocity v_A of unit A when θ reaches $60°$.

Ans. $v_A = 305$ mm/s

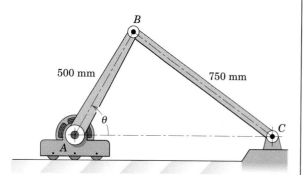

Problem 5/71

5/72 Each of the sliding bars A and B engages its respective rim of the two riveted wheels without slipping. Determine the magnitude of the velocity of point P for the position shown.

Problem 5/72

5/73 The linkage of Prob. 5/42 is repeated here. If CA has a clockwise angular velocity $\omega = 3$ rad/s for an interval of motion, determine the angular velocity of rod OB when $\theta = 30°$.

Ans. $\omega_{OB} = 1.430$ rad/s CCW

Problem 5/73

5/74 The elements of a simplified clam-shell bucket for a dredge are shown. The cable which opens and closes the bucket passes through the block at O. With O as a fixed point, determine the angular velocity ω of the bucket jaws when $\theta = 45°$ as they are closing. The upward velocity of the control cable is 0.5 m/s as it passes through the block.

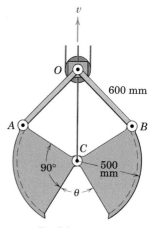

Problem 5/74

5/75 The elements of a switching device are shown. If the vertical control rod has a downward velocity v of 0.9 m/s when $\theta = 60°$ and if roller A is in continuous contact with the horizontal surface, determine the magnitude of the velocity of C for this instant.

$$Ans. \ v_C = 1.873 \text{ m/s}$$

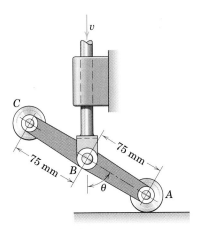

Problem 5/75

5/76 Horizontal motion of the piston rod of the hydraulic cylinder controls the rotation of link OB about O. For the instant represented, $v_A = 2$ m/s and OB is horizontal. Determine the angular velocity ω of OB for this instant.

Problem 5/76

5/77 The elements of the mechanism for deployment of a spacecraft magnetometer boom are shown. Determine the angular velocity of the boom when the driving link OB crosses the y-axis with an angular velocity $\omega_{OB} = 0.5$ rad/s if $\tan \theta = 4/3$ at this instant.

$$Ans. \ \boldsymbol{\omega}_{CA} = 0.429\mathbf{k} \text{ rad/s}$$

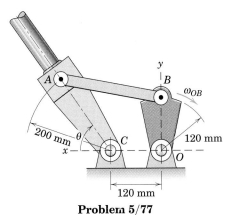

Problem 5/77

5/78 For the instant represented the rotating link D has an angular velocity $\omega = 2$ rad/s, and its slot is vertical. Also $\theta = 60°$ momentarily. Determine the velocity of end A of link AB for this instant.

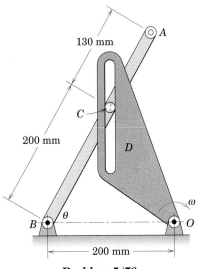

Problem 5/78

5/79 In the four-bar linkage shown control link OA has a counterclockwise angular velocity $\omega_0 = 10$ rad/s during a short interval of motion. When link CB passes the vertical position shown, point A has coordinates $x = -60$ mm and $y = 80$ mm. By means of vector algebra determine the angular velocity of AB and BC.

Ans. $\boldsymbol{\omega}_{AB} = 2.5\mathbf{k}$ rad/s, $\boldsymbol{\omega}_{BC} = 5.83\mathbf{k}$ rad/s

Problem 5/79

5/80 Determine the angular velocity ω of the telescoping link AB at the instant represented. The angular velocity of each of the driving links is shown.

Problem 5/80

5/81 The Geneva mechanism of Prob. 5/51 is shown again here. By relative-motion principles determine the angular velocity ω_2 of wheel C for $\theta = 20°$. Wheel A has a constant clockwise angular velocity $\omega_1 = 2$ rad/s.

Ans. $\omega_2 = 1.923$ rad/s

Problem 5/81

5/82 Motion of the threaded collars A and B is controlled by the rotation of their respective lead screws. If A has a velocity to the right of 75 mm/s and B has a velocity to the left of 50 mm/s when $x = 150$ mm, determine the angular velocity ω of ACD at this instant.

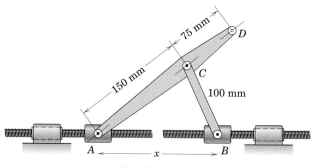

Problem 5/82

5/83 In the design of a produce-processing plant, roller trays of produce are to be oscillated under water spray by the action of the connecting link AB and crank OB. For the instant when $\theta = 15°$, the angular velocity of AB is 0.086 rad/s clockwise. Find the corresponding angular velocity $\dot\theta$ of the crank and the velocity v_A of the tray. Solve the relative-velocity equation by either vector algebra or vector geometry.

$$Ans.\ \omega_{OB} = 0.944\ \text{rad/s CCW}$$
$$v_A = 0.298\ \text{m/s to the left}$$

Problem 5/83

5/84 Slotted arm OB oscillates about the vertical by the action of the rotating crank CA of 125-mm length, where the end pin A engages the slot. For a crank speed N of 120 rev/min, determine the angular velocity ω_{OB} of the slotted arm for the position $\theta = 60°$. For this position also specify the velocity of A relative to the slot, the magnitude of which is the time rate of change of the distance $\overline{OA}$.

Problem 5/84

5/85 The mechanism is designed to convert from one rotation to another. Rotation of link BC is controlled by the rotation of the curved slotted arm OA which engages pin P. For the instant represented $\theta = 30°$ and the angle β between the tangent to the curve at P and the horizontal is 40°. If the angular velocity of OA is 3 rad/s for this position, determine the velocity of point C.

$$Ans.\ v_C = 4.33\ \text{m/s}$$

Problem 5/85

5/86 A mechanism for pushing small boxes from an assembly line onto a conveyor belt is shown with arm OD and crank CB in their vertical positions. The crank revolves clockwise at a constant rate of 1 revolution every 2 seconds. For the position shown, determine the speed at which the box is being shoved horizontally onto the conveyor belt.

Problem 5/86

▶5/87 At the instant represented, a = 150 mm and b =
125 mm, and the distance $a + b$ between A and C
is decreasing at the rate of 0.2 m/s. Determine the
common velocity v of points B and D for this instant.

Ans. v = 0.0536 m/s

▶5/88 The wheel rolls without slipping. For the instant
portrayed, when O is directly under point C, link
OA has a velocity v = 1.5 m/s to the right and
θ = 30°. Determine the angular velocity ω of the
slotted link.

Ans. ω = 18.22 rad/s CCW

Problem 5/87

Problem 5/88

5/5 INSTANTANEOUS CENTER OF ZERO VELOCITY

In the previous article, we determined the velocity of a point on a rigid body in plane motion by adding the relative velocity due to rotation about a convenient reference point to the velocity of the reference point. In the present article, we will solve the problem by choosing a unique reference point which momentarily has zero velocity. As far as velocities are concerned, the body may be considered to be in pure rotation about an axis, normal to the plane of motion, passing through this point. This axis is called the *instantaneous axis* of zero velocity, and the intersection of this axis with the plane of motion is known as the *instantaneous center* of zero velocity. This approach provides us with a valuable means for visualizing and analyzing velocities in plane motion.

The existence of the instantaneous center is easily shown. For the body in Fig. 5/7a, let us assume that the directions of the absolute velocities of any two points A and B on the body are known and are not parallel. If there is a point about which A has absolute circular motion at the instant considered, this point must lie on the normal to $\mathbf{v}_A$ through A. Similar reasoning applies to B, and the intersection C of these two perpendiculars fulfills the requirement for an absolute center of rotation *at the instant considered*. Point C is the instantaneous center of zero velocity and may lie on or off the body. If it lies off the body, it may be visualized as lying on the body extended. The instantaneous center is not a fixed point in the body nor a fixed point in the plane.

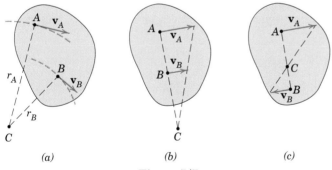

| (a) | (b) | (c) |

Figure 5/7

If the magnitude of the velocity of one of the points, say, v_A, is also known, the angular velocity ω of the body and the linear velocity of every point in the body are easily obtained. Thus, the angular velocity of the body, Fig. 5/7a, is

$$\omega = \frac{v_A}{r_A}$$

which, of course, is also the angular velocity of *every* line in the body.

Therefore, the velocity of B is $v_B = r_B\omega = (r_B/r_A)v_A$. Once the instantaneous center is located, the direction of the instantaneous velocity of every point in the body is readily found since it must be perpendicular to the radial line joining the point in question with C.

If the velocities of two points in a body having plane motion are parallel, Fig. 5/7b or 5/7c, the line joining the points is perpendicular to the direction of the velocities, and the instantaneous center C is located by direct proportion as shown. We can readily see from Fig. 5/7b that as the parallel velocities approach equality in magnitude, the instantaneous center C moves farther away from the body and approaches infinity in the limit as the body gives up its angular velocity and translates only.

As the body changes its position, the instantaneous center C also changes its position both in space and on the body. The locus of the instantaneous centers in space is known as the *space centrode*, and the locus of the positions of the instantaneous centers on the body is known as the *body centrode*. At the instant considered, the two curves are tangent at the position of point C. It may be shown that the body centrode curve rolls on the space centrode curve during the motion of the body, as indicated schematically in Fig. 5/8.

Whereas the instantaneous center of zero velocity as a point on the body is momentarily at rest, its acceleration generally is *not* zero. Thus, this point may *not* be used as an instantaneous center of zero acceleration in a manner analogous to its use for finding velocity. An instantaneous center of zero acceleration does exist for bodies in general plane motion, but its location and use represent a specialized topic in mechanism kinematics and will not be treated here.

Figure 5/8

Sample Problem 5/11

The wheel of Sample Problem 5/7, shown again here, rolls to the right without slipping, with its center O having a velocity $v_O = 3$ m/s. Locate the instantaneous center of zero velocity and use it to find the velocity of point A for the position indicated.

Solution. The point on the rim of the wheel in contact with the ground has no velocity if the wheel is not slipping; it is, therefore, the instantaneous center C of zero velocity. The angular velocity of the wheel becomes

$[\omega = v/r]$ $\qquad \omega = v_O/\overline{OC} = 3/0.300 = 10$ rad/s

The distance from A to C is

① $\overline{AC} = \sqrt{(0.300)^2 + (0.200)^2 - 2(0.300)(0.200) \cos 120°} = 0.436$ m

The velocity of A becomes

② $[v = r\omega]$ $\qquad v_A = \overline{AC}\omega = 0.436(10) = 4.36$ m/s $\qquad$ *Ans.*

The direction of $\mathbf{v}_A$ is perpendicular to AC as shown.

① Be sure to recognize that the cosine of $120°$ is itself negative.

② From the results of this problem, you should be able to visualize and sketch the velocities of all points on the wheel.

Sample Problem 5/12

Arm OB of the linkage has a clockwise angular velocity of 10 rad/s in the position shown where $\theta = 45°$. Determine the velocity of A, the velocity of D, and the angular velocity of link AB for the instant shown.

Solution. The directions of the velocities of A and B are tangent to their circular paths about the fixed centers O' and O as shown. The intersection of the two perpendiculars to the velocities from A and B ① locates the instantaneous center C for the link AB. The distances $\overline{AC}$, $\overline{BC}$, and $\overline{DC}$ shown on the diagram are computed or scaled from the drawing. The angular velocity of BC, considered a line on the body extended, is equal to the angular velocity of AC, DC, and AB and is

$[\omega = v/r]$ $\qquad \omega_{BC} = \dfrac{v_B}{\overline{BC}} = \dfrac{\overline{OB}\omega_{OB}}{\overline{BC}} = \dfrac{150\sqrt{2}(10)}{350\sqrt{2}}$

$\qquad\qquad\qquad = 4.29$ rad/s CCW $\qquad$ *Ans.*

Thus, the velocities of A and D are

$[v = r\omega]$ $\qquad v_A = 0.350(4.29) = 1.500$ m/s $\qquad$ *Ans.*

$\qquad\qquad v_D = 0.381(4.29) = 1.632$ m/s $\qquad$ *Ans.*

in the directions shown.

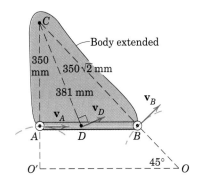

① For the instant depicted, we should visualize link AB and its body extended to be rotating as a single unit about point C.

PROBLEMS

Introductory problems

5/89 The figure for Prob. 5/76 is repeated here. Solve for the angular velocity of *OB* in that problem by the method of this article.

Ans. $\omega = 8.59$ rad/s

Problem 5/89

5/90 For the instant represented, corner *A* of the rectangular plate has a velocity $v_A = 2.8$ m/s and the plate has a clockwise angular velocity $\omega = 12$ rad/s. Determine the magnitude of the corresponding velocity of point *B*.

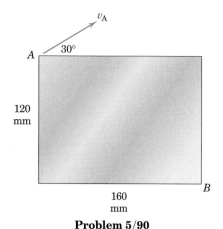

Problem 5/90

5/91 For the instant represented, when crank *OA* passes the horizontal position, determine the velocity of the center *G* of link *AB* by the method of this article.

Ans. $v_G = 277$ mm/s

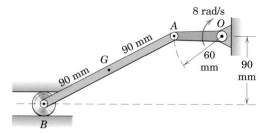

Problem 5/91

5/92 The bar *AB* has a counterclockwise angular velocity of 6 rad/s. Construct the velocity vectors for points *A* and *G* of the bar and specify their magnitudes if the instantaneous center of zero velocity for the bar is (*a*) at C_1, (*b*) at C_2, and (*c*) at C_3.

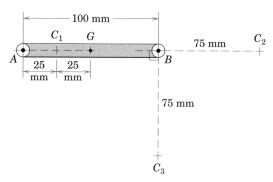

Problem 5/92

5/93 Motion of the bar is controlled by the constrained paths of *A* and *B*. If the angular velocity of the bar is 2 rad/s counterclockwise as the position $\theta = 45°$ is passed, determine the velocity v_A and the magnitude of the velocity of *P*.

Ans. $v_A = 0.707$ m/s, $v_P = 1.581$ m/s

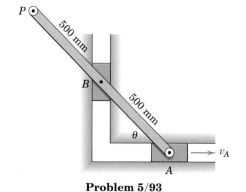

Problem 5/93

5/94 At a certain instant vertex B of the right-triangular plate has a velocity of 200 mm/s in the direction shown. If the instantaneous center of zero velocity for the plate is 40 mm from point B and if the angular velocity of the plate is clockwise, determine the velocity of point D.

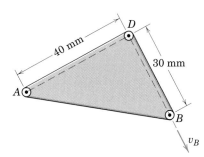

Problem 5/94

5/95 The linkage of Prob. 5/69 is repeated here. At the instant represented, crank OB has a clockwise angular velocity $\omega = 0.8$ rad/s and is passing the horizontal position. By the method of this article, determine the corresponding velocity of the guide roller A in the 20° slot and the velocity of point C midway between A and B.

Ans. $v_A = 226$ mm/s, $v_C = 174.7$ mm/s

Problem 5/95

5/96 The linkage is reproduced here to the scale indicated. For the position shown, link O_1A has a clockwise angular velocity of 4 rad/s. By direct measurements from the figure, determine the corresponding velocity of vertex D as accurately as you can.

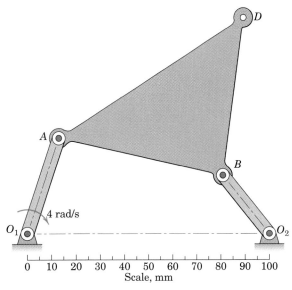

Problem 5/96

5/97 The blade of a rotary power mower turns counterclockwise at the angular speed of 1800 rev/min. If the body centrode is a circle of radius 0.75 mm, compute the velocity v_O of the mower.

Ans. $v_O = 0.1414$ m/s

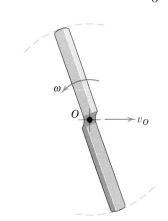

Problem 5/97

5/98 The shaft of the wheel unit rolls without slipping on the fixed horizontal surface, and point O has a velocity of 0.8 m/s to the right. By the method of this article, determine the velocities of points A, B, C, and D.

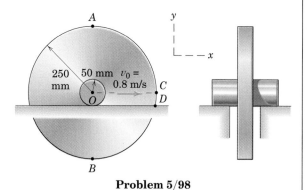

Problem 5/98

5/99 The rear driving wheel of a car has a diameter of 650 mm and has an angular speed N of 200 rev/min on an icy road. If the instantaneous center of zero velocity is 100 mm from the point of contact of the tire with the road, determine the velocity v of the car and the slipping velocity v_s of the tire on the ice.

$Ans.$ $v = 4.71$ m/s, $v_s = 2.09$ m/s

Problem 5/99

Representative problems

5/100 The mechanism of Prob. 5/23 is repeated here. At the instant when $x_A = 0.85L$, the velocity of the slider at A is $v = 2$ m/s to the right. Determine the corresponding velocity of slider B and the angular velocity ω of bar AB if $L = 0.8$ m.

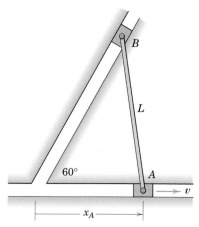

Problem 5/100

5/101 The gear D (teeth not shown) rotates clockwise about O with a constant angular velocity of 4 rad/s. The 90° sector AOB is mounted on an independent shaft at O, and each of the small gears at A and B meshes with gear D. If the sector has a counterclockwise angular velocity of 3 rad/s at the instant represented, determine the corresponding angular velocity ω of each of the small gears.

$Ans.$ $\omega = 31$ rad/s CCW

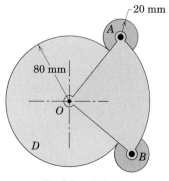

Problem 5/101

5/102 The hydraulic cylinder produces a limited horizontal motion of point A. If $v_A = 4$ m/s when $\theta = 45°$, determine the magnitude of the velocity of D and the angular velocity ω of ABD for this position.

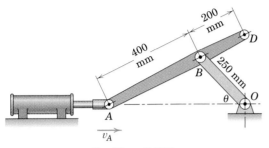

Problem 5/102

5/103 A device which tests the resistance to wear of two materials A and B is shown. If the link EO has a velocity of 1.2 m/s to the right when $\theta = 45°$, determine the rubbing velocity v_A.

         Ans. $v_A = 2.76$ m/s

Problem 5/103

5/104 The attached wheels roll without slipping on the plates A and B, which are moving in opposite directions as shown. If $v_A = 60$ mm/s to the right and $v_B = 200$ mm/s to the left, determine the velocities of the center O and the point P for the position shown.

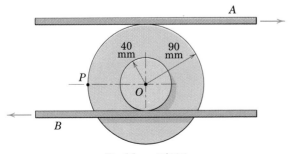

Problem 5/104

5/105 Rotation of the lever OA is controlled by the motion of the contacting circular disk whose center is given a horizontal velocity v. Determine the expression for the angular velocity ω of lever AO in terms of x. Also determine the angular velocity ω_w of the wheel if it does not slip on the lever.

$$Ans. \quad \omega = \frac{v}{x\sqrt{(x/r)^2 - 1}}, \quad \omega_w = \frac{v}{r}\sqrt{1 - (r/x)^2}$$

Problem 5/105

5/106 The rectangular body B is pivoted to the crank OA at A and is supported by the wheel at D. If OA has a counterclockwise angular velocity of 2 rad/s, determine the velocity of point E and the angular velocity of body B when the crank OA passes the vertical position shown.

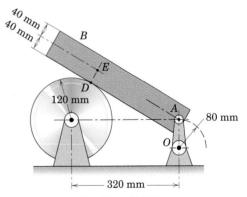

Problem 5/106

5/107 Horizontal oscillation of the spring-loaded plunger E is controlled by varying the air pressure in the horizontal pneumatic cylinder F. If the plunger has a velocity of 2 m/s to the right when $\theta = 30°$, determine the downward velocity v_D of roller D in the vertical guide and find the angular velocity ω of ABD for this position.

Ans. $v_D = 2.31$ m/s, $\omega = 13.33$ rad/s

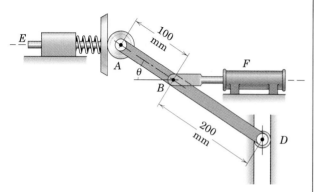

Problem 5/107

5/108 The small cylinder rolls on the surface of the large cylinder without slipping. By using the instantaneous center of zero velocity of the small cylinder, determine the velocity v_A of point A shown where θ is increasing at the rate $\dot{\theta}$.

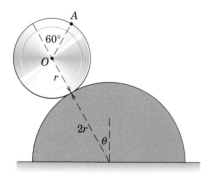

Problem 5/108

5/109 The flexible band F is attached at E to the rotating sector and leads over the guide pulley. Determine the angular velocities of AD and BD for the position shown if the band has a velocity of 4 m/s.

Ans. $\omega_{AD} = 12.5$ rad/s, $\omega_{BD} = 7.5$ rad/s

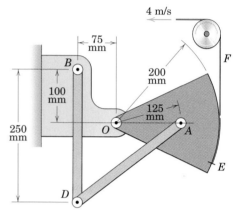

Problem 5/109

5/110 The mechanism of Prob. 5/47 is repeated here. By the method of this article determine the expression for the magnitude of the velocity of point C in terms of the velocity v_B of the piston rod and the angle θ.

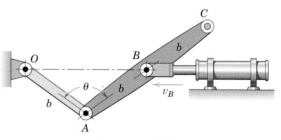

Problem 5/110

5/111 In the design of the mechanism shown, collar A is to slide along the fixed shaft as angle θ increases. When $\theta = 30°$, the control link at D is to have a downward component of velocity of 0.60 m/s. Determine the corresponding velocity of collar A by the method of this article.

Ans. $v_A = 0.509$ m/s

Problem 5/111

5/112 Motion of the roller A against its restraining spring is controlled by the downward motion of the plunger E. For an interval of motion the velocity of E is $v = 0.2$ m/s. Determine the velocity of A when θ becomes 90°.

Problem 5/112

5/113 In the design of this mechanism, upward motion of the plunger G controls the motion of a control rod attached at A. Point B of link AH is confined to move with the sliding collar on the fixed vertical shaft ED. If G has a velocity $v_G = 2$ m/s for a short interval, determine the velocity of A for the position $\theta = 45°$.

Ans. $v_A = 5.95$ m/s

Problem 5/113

▶5/114 Determine the angular velocity ω of the ram head *AE* of the rock crusher in the position for which $\theta = 60°$. The crank *OB* has an angular speed of 60 rev/min. When *B* is at the bottom of its circle, *D* and *E* are on a horizontal line through *F*, and lines *BD* and *AE* are vertical. The dimensions are $\overline{OB} = 100$ mm, $\overline{BD} = 750$ mm, and $\overline{AE} = \overline{ED} = \overline{DF} = 375$ mm. Carefully construct the configuration graphically, and use the method of this article.

Ans. $\omega = 1.11$ rad/s CW

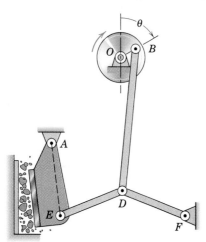

Problem 5/114

▶5/115 The shaft at *O* drives the arm *OA* at a clockwise speed of 90 rev/min about the fixed bearing at *O*. Use the method of the instantaneous center of zero velocity to determine the rotational speed of gear *B* (gear teeth not shown) if (*a*) ring gear *D* is fixed and (*b*) ring gear *D* rotates counterclockwise about *O* with a speed of 80 rev/min.

Ans. (*a*) $\omega_B = 360$ rev/min
(*b*) $\omega_B = 600$ rev/min

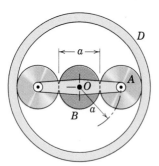

Problem 5/115

▶5/116 The large roller bearing rolls to the left on its outer race with a velocity of its center *O* of 0.9 m/s. At the same time the central shaft and inner race rotate counterclockwise with an angular speed of 240 rev/min. Determine the angular velocity ω of each of the rollers.

Ans. $\omega = 10.73$ rad/s CW

Problem 5/116

5/6 RELATIVE ACCELERATION

The relative-acceleration relationship may be obtained from the equation $\mathbf{v}_A = \mathbf{v}_B + \mathbf{v}_{A/B}$ for relative velocities using nonrotating reference axes for two points A and B in plane motion by differentiation with respect to time to get $\dot{\mathbf{v}}_A = \dot{\mathbf{v}}_B + \dot{\mathbf{v}}_{A/B}$ or

$$\mathbf{a}_A = \mathbf{a}_B + \mathbf{a}_{A/B} \tag{5/7}$$

In words, Eq. 5/7 states that the acceleration of point A equals the acceleration of point B plus (vectorially) the acceleration that A appears to have to a nonrotating observer moving with B. If points A and B are located on the same rigid body in the plane of motion, the distance r between them remains constant so that the observer moving with B perceives A to have circular motion about B, as we saw in Art. 5/4 with the relative-velocity relationship. With the relative motion being circular, it follows that the relative-acceleration term will have both a normal component directed from A toward B due to the change of direction of $\mathbf{v}_{A/B}$ and a tangential component perpendicular to AB due to the change in magnitude of $\mathbf{v}_{A/B}$. These acceleration components for circular motion, cited in Eqs. 5/2, were covered earlier in Art. 2/5 and should be thoroughly familiar by now. Thus, we may write

$$\mathbf{a}_A = \mathbf{a}_B + (\mathbf{a}_{A/B})_n + (\mathbf{a}_{A/B})_t \tag{5/8}$$

where the magnitudes of the relative-acceleration components are

$$(a_{A/B})_n = v_{A/B}^2/r = r\omega^2$$
$$(a_{A/B})_t = \dot{v}_{A/B} = r\alpha \tag{5/9}$$

or in vector notation the acceleration components become

$$(\mathbf{a}_{A/B})_n = \boldsymbol{\omega} \times (\boldsymbol{\omega} \times \mathbf{r})$$
$$(\mathbf{a}_{A/B})_t = \boldsymbol{\alpha} \times \mathbf{r} \tag{5/9a}$$

In these relationships, $\boldsymbol{\omega}$ is the angular velocity and $\boldsymbol{\alpha}$ is the angular acceleration of the body. The vector locating A from B is $\mathbf{r}$. It is important to observe that the *relative* acceleration terms depend on the respective *absolute* angular velocity and *absolute* angular acceleration.

The meaning of Eqs. 5/8 and 5/9 is illustrated in Fig. 5/9, which shows a rigid body in plane motion with points A and B moving along separate curved paths with absolute accelerations $\mathbf{a}_A$ and $\mathbf{a}_B$. Contrary to the case with velocities, the accelerations $\mathbf{a}_A$ and $\mathbf{a}_B$

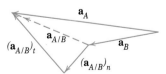

Figure 5/9

are, in general, not tangent to the paths described by A and B when these paths are curvilinear. The figure shows the acceleration of A to be composed of two parts: the acceleration of B and the acceleration of A with respect to B. A sketch showing the reference point as fixed is useful in disclosing the correct sense of each of the two components of the relative-acceleration term.

Alternatively, we may write the acceleration equation in the reverse order, which puts the nonrotating reference axes on A rather than B. This order gives

$$\mathbf{a}_B = \mathbf{a}_A + \mathbf{a}_{B/A}$$

Here $\mathbf{a}_{B/A}$ and its n- and t-components are the negatives of $\mathbf{a}_{A/B}$ and its n- and t-components. The student should make a sketch corresponding to Fig. 5/9 for this reversed sequence of terms.

As in the case of the relative-velocity equation, we can handle the solution to Eq. 5/8 in three different ways, namely, by scalar algebra and geometry, by vector algebra, or by graphical construction. The student will be well advised to become familiar with all three techniques. In any event, a sketch of the vector polygon representing the vector equation should be made with close attention paid to the head-to-tail combination of vectors which agrees with the equation. Known vectors should be added first, and the unknown vectors will become the closing legs of the vector polygon. It is vital that we visualize the vectors in their geometrical sense, as only then can the full significance of the acceleration equation be perceived. Before attempting a solution, we should identify the knowns and unknowns, keeping in mind that a solution to a vector equation in two dimensions can be carried out when the unknowns have been reduced to two scalar quantities. These quantities may be the magnitude or direction of any of the terms of the equation. We observe that when both points move on curved paths, there will, in general, be six scalar quantities to account for in Eq. 5/8.

Because the normal acceleration components depend on velocities, it is generally necessary to solve for the velocities before the acceleration calculations can be made. The reference point in the relative-acceleration equation is chosen as some point on the body in question whose acceleration is either known or can be easily found. Care must be exercised *not* to use the instantaneous center of zero velocity as the reference point unless its acceleration is known and accounted for. An instantaneous center of zero acceleration exists for a rigid body in general plane motion, but will not be discussed here since its use is somewhat specialized.

Sample Problem 5/13

The wheel of radius r rolls to the left without slipping and, at the instant considered, the center O has a velocity $\mathbf{v}_O$ and an acceleration $\mathbf{a}_O$ to the left. Determine the acceleration of points A and C on the wheel for the instant considered.

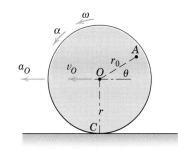

Solution. From our previous analysis of Sample Problem 5/4, we know that the angular velocity and angular acceleration of the wheel are

$$\omega = v_O/r \quad \text{and} \quad \alpha = a_O/r$$

The acceleration of A is written in terms of the given acceleration of O. Thus,

$$\mathbf{a}_A = \mathbf{a}_O + \mathbf{a}_{A/O} = \mathbf{a}_O + (\mathbf{a}_{A/O})_n + (\mathbf{a}_{A/O})_t$$

The relative-acceleration terms are viewed as though O were fixed, and for this relative circular motion they have the magnitudes

$$(a_{A/O})_n = r_0\omega^2 = r_0\left(\frac{v_O}{r}\right)^2$$

$$(a_{A/O})_t = r_0\alpha = r_0\left(\frac{a_O}{r}\right)$$

① and the directions shown.

Adding the vectors head-to-tail gives $\mathbf{a}_A$ as shown. In a numerical problem, we may obtain the combination algebraically or graphically. The algebraic expression for the magnitude of $\mathbf{a}_A$ is found from the square root of the sum of the squares of its components. If we use n- and t-directions, we have

②
$$\begin{aligned}
a_A &= \sqrt{(a_A)_n{}^2 + (a_A)_t{}^2} \\
&= \sqrt{[a_O\cos\theta + (a_{A/O})_n]^2 + [a_O\sin\theta + (a_{A/O})_t]^2} \\
&= \sqrt{(r\alpha\cos\theta + r_0\omega^2)^2 + (r\alpha\sin\theta + r_0\alpha)^2} \quad \text{Ans.}
\end{aligned}$$

The direction of $\mathbf{a}_A$ can be computed if desired.

The acceleration of the instantaneous center C of zero velocity, considered a point on the wheel, is obtained from the expression

$$\mathbf{a}_C = \mathbf{a}_O + \mathbf{a}_{C/O}$$

where the components of the relative-acceleration term are $(a_{C/O})_n = r\omega^2$ directed from C to O and $(a_{C/O})_t = r\alpha$ directed to the right on account of the counterclockwise angular acceleration of line CO about O. The terms are added together in the lower diagram, and it is seen that

③
$$a_C = r\omega^2 \quad \text{Ans.}$$

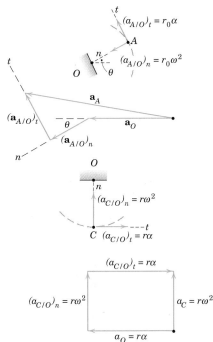

① The counterclockwise angular acceleration α of OA determines the positive direction of $(a_{A/O})_t$. The normal component $(a_{A/O})_n$ is, of course, directed toward the reference center O.

② If the wheel were rolling to the right with the same velocity v_O but still had an acceleration a_O to the left, note that the solution for a_A would be unchanged.

③ We note that the acceleration of the instantaneous center of zero velocity is independent of α and is directed toward the center of the wheel. This conclusion is a useful result to remember.

Sample Problem 5/14

The linkage of Sample Problem 5/8 is repeated here. Crank CB has a constant counterclockwise angular velocity of 2 rad/s in the position shown during a short interval of its motion. Determine the angular acceleration of links AB and OA for this position. Solve by using vector algebra.

Solution. We first solve for the velocities which were obtained in Sample Problem 5/8. They are

$$\omega_{AB} = -6/7 \text{ rad/s} \quad \text{and} \quad \omega_{OA} = -3/7 \text{ rad/s}$$

where the counterclockwise direction ($+\mathbf{k}$-direction) is taken as positive. The acceleration equation is

$$\mathbf{a}_A = \mathbf{a}_B + (\mathbf{a}_{A/B})_n + (\mathbf{a}_{A/B})_t$$

where, from Eqs. 5/3 and 5/9a, we may write

①
$$\mathbf{a}_A = \boldsymbol{\alpha}_{OA} \times \mathbf{r}_A + \boldsymbol{\omega}_{OA} \times (\boldsymbol{\omega}_{OA} \times \mathbf{r}_A)$$
$$= \alpha_{OA}\mathbf{k} \times 100\mathbf{j} + (-\tfrac{3}{7}\mathbf{k}) \times (-\tfrac{3}{7}\mathbf{k} \times 100\mathbf{j})$$
$$= -100\alpha_{OA}\mathbf{i} - 100(\tfrac{3}{7})^2\mathbf{j} \text{ mm/s}^2$$

$$\mathbf{a}_B = \boldsymbol{\alpha}_{CB} \times \mathbf{r}_B + \boldsymbol{\omega}_{CB} \times (\boldsymbol{\omega}_{CB} \times \mathbf{r}_B)$$
$$= \mathbf{0} + 2\mathbf{k} \times (2\mathbf{k} \times [-75\mathbf{i}])$$
$$= 300\mathbf{i} \text{ mm/s}^2$$

②
$$(\mathbf{a}_{A/B})_n = \boldsymbol{\omega}_{AB} \times (\boldsymbol{\omega}_{AB} \times \mathbf{r}_{A/B})$$
$$= -\tfrac{6}{7}\mathbf{k} \times [(-\tfrac{6}{7}\mathbf{k}) \times (-175\mathbf{i} + 50\mathbf{j})]$$
$$= (\tfrac{6}{7})^2(175\mathbf{i} - 50\mathbf{j}) \text{ mm/s}^2$$

$$(\mathbf{a}_{A/B})_t = \boldsymbol{\alpha}_{AB} \times \mathbf{r}_{A/B}$$
$$= \alpha_{AB}\mathbf{k} \times (-175\mathbf{i} + 50\mathbf{j})$$
$$= -50\alpha_{AB}\mathbf{i} - 175\alpha_{AB}\mathbf{j} \text{ mm/s}^2$$

We now substitute these results into the relative-acceleration equation and equate separately the coefficients of the $\mathbf{i}$-terms and the coefficients of the $\mathbf{j}$-terms to give

$$-100\alpha_{OA} = 429 - 50\alpha_{AB}$$
$$-18.37 = -36.7 - 175\alpha_{AB}$$

The solutions are

$$\alpha_{AB} = -0.1050 \text{ rad/s}^2 \quad \text{and} \quad \alpha_{OA} = -4.34 \text{ rad/s}^2 \quad Ans.$$

Since the unit vector $\mathbf{k}$ points out from the paper in the positive z-direction, we see that the angular accelerations of AB and OA are both clockwise (negative).

It is recommended that the student sketch each of the acceleration vectors in its proper geometric relationship according to the relative-acceleration equation to help clarify the meaning of the solution.

① Remember to preserve the order of the factors in the cross products.

② In expressing the term $\mathbf{a}_{A/B}$, be certain that $\mathbf{r}_{A/B}$ is written as the vector from B to A and not the reverse.

Sample Problem 5/15

The slider-crank mechanism of Sample Problem 5/9 is repeated here. The crank OB has a constant clockwise angular speed of 1500 rev/min. For the instant when the crank angle θ is 60°, determine the acceleration of the piston A and the angular acceleration of the connecting rod AB.

Solution. The acceleration of A may be expressed in terms of the acceleration of the crank pin B. Thus,

$$\mathbf{a}_A = \mathbf{a}_B + (\mathbf{a}_{A/B})_n + (\mathbf{a}_{A/B})_t$$

① Point B moves in a circle of 5-in. radius with a constant speed so that it has only a normal component of acceleration directed from B to O.

$$[a_n = r\omega^2] \qquad a_B = 0.125 \left(\frac{1500[2\pi]}{60} \right)^2 = 3080 \text{ m/s}^2$$

The relative-acceleration terms are visualized with A rotating in a circle relative to B, which is considered fixed, as shown. From Sample Problem 5/9, the angular velocity of AB for these same conditions is $\omega_{AB} = 29.5$ rad/s so that

② $[a_n = r\omega^2] \qquad (a_{A/B})_n = 0.350 \ (29.5)^2 = 305 \text{ m/s}^2$

directed from A to B. The tangential component $(\mathbf{a}_{A/B})_t$ is known in direction only since its magnitude depends on the unknown angular acceleration of AB. We also know the direction of $\mathbf{a}_A$ since the piston is confined to move along the horizontal axis of the cylinder. There are now only two unknowns left in the equation, namely, the magnitudes of $\mathbf{a}_A$ and $(\mathbf{a}_{A/B})_t$, so the solution may be carried out.

If we adopt an algebraic solution using the geometry of the acceleration polygon, we first compute the angle between AB and the horizontal. With the law of sines, this angle becomes 18.02°. Equating separately the horizontal components and the vertical components of the terms in the acceleration equation, as seen from the acceleration polygon, gives

$$a_A = 3080 \cos 60° + 305 \cos 18.02° - (a_{A/B})_t \sin 18.02°$$

$$0 = 3080 \sin 60° - 305 \sin 18.02° - (a_{A/B})_t \cos 18.02°$$

The solution to these equations gives the magnitudes

$$(a_{A/B})_t = 2710 \text{ m/s}^2 \quad \text{and} \quad a_A = 994 \text{ m/s}^2 \qquad \textit{Ans.}$$

With the sense of $(\mathbf{a}_{A/B})_t$ also determined from the diagram, the angular acceleration of AB is seen from the figure representing rotation relative to B to be

$$[\alpha = a_t/r] \qquad \alpha_{AB} = 2710/0.350 = 7740 \text{ rad/s}^2 \text{ clockwise} \qquad \textit{Ans.}$$

③ If we adopt a graphical solution, we begin with the known vectors $\mathbf{a}_B$ and $(\mathbf{a}_{A/B})_n$ and add them head-to-tail using a convenient scale. Next we construct the direction of $(\mathbf{a}_{A/B})_t$ through the head of the last vector. The solution of the equation is obtained by the intersection P of this last line with a horizontal line through the starting point representing the known direction of the vector sum $\mathbf{a}_A$. Scaling the magnitudes from the diagram gives values that agree with the calculated results

$$a_A = 994 \text{ m/s}^2 \quad \text{and} \quad (a_{A/B})_t = 2710 \text{ m/s}^2 \qquad \textit{Ans.}$$

① If the crank OB had an angular acceleration, $\mathbf{a}_B$ would also have a tangential component of acceleration.

② Alternatively, the relation $a_n = v^2/r$ may be used for calculating $(a_{A/B})_n$, provided the relative velocity $v_{A/B}$ is used for v. The equivalence is easily seen when it is recalled that $v_{A/B} = r\omega$.

③ Except where extreme accuracy is required, do not hesitate to use a graphical solution, as it is quick and reveals the physical relationships among the vectors. The known vectors, of course, may be added in any order as long as the governing equation is satisfied.

PROBLEMS

Introductory problems

5/117 The two rotor blades of 800-mm radius rotate counterclockwise with a constant angular velocity $\omega = \dot{\theta} = 2$ rad/s about the shaft at O mounted in the sliding block. The acceleration of the block is $a_O = 3$ m/s². Determine the magnitude of the acceleration of the tip A of the blade when (*a*) $\theta = 0$, (*b*) $\theta = 90°$, and (*c*) $\theta = 180°$. Does the velocity of O or the sense of ω enter into the calculation?

> *Ans.* (*a*) $a_A = 0.2$ m/s²
> (*b*) $a_A = 4.39$ m/s²
> (*c*) $a_A = 6.2$ m/s²

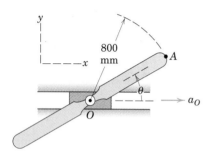

Problem 5/117

5/118 The 3.6-m steel beam is being hoisted from its horizontal position by the two cables attached at A and B. If the initial angular accelerations are $\alpha_1 = 0.2$ rad/s² and $\alpha_2 = 0.6$ rad/s², determine the initial values of (*a*) the angular acceleration of the beam, (*b*) the acceleration of point C, and (*c*) the distance d from A to the point on the center line of the beam which has zero acceleration.

Problem 5/118

5/119 A container for waste materials is dumped by the hydraulically-activated linkage shown. If the piston rod starts from rest in the position indicated and has an acceleration of 0.5 m/s² in the direction shown, compute the initial angular acceleration of the container.

> *Ans.* $\alpha_{AB} = 0.1768$ rad/s² CW

Problem 5/119

5/120 The center O of the wheel is mounted on the sliding block which has an acceleration $a_O = 8$ m/s² to the right. At the instant when $\theta = 45°$, $\dot{\theta} = 3$ rad/s and $\ddot{\theta} = -8$ rad/s². For this instant determine the magnitudes of the acceleration of points A and B.

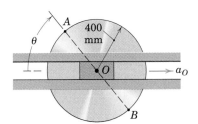

Problem 5/120

5/121 Determine the acceleration of point B on the equator of the earth, repeated here from Prob. 5/55. Use the data given with that problem and assume that the earth's orbital path is circular, consulting Table D/2 as necessary. Consider the center of the sun fixed.

Ans. $\mathbf{a}_B = 0.0279\mathbf{i}$ m/s²

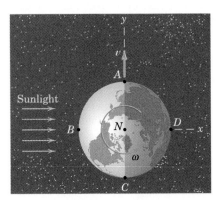

Problem 5/121

5/122 The center O of the disk has the velocity and acceleration shown in the figure. If the disk rolls without slipping on the horizontal surface, determine the velocity of A and the acceleration of B for the instant represented.

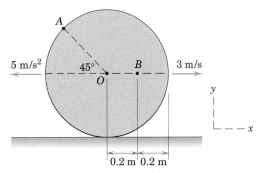

Problem 5/122

5/123 A car has a forward acceleration $a = 4$ m/s² without slipping its 600-mm-diameter wheels. Determine the velocity v of the car when a point P on the wheel in the position shown will have zero horizontal component of acceleration.

Ans. $v = 1.702$ m/s

Problem 5/123

5/124 The wheel-and-shaft unit of Prob. 5/98 is repeated here with the additional specification that the acceleration of the center O is 1.4 m/s^2 to the left as shown. If the shaft rolls on the fixed horizontal surface without slipping, determine the accelerations of points A and D.

Problem 5/124

Representative problems

5/125 The bar AB from Prob. 5/63 is repeated here. If the velocity of point A is 3 m/s to the right and is constant for an interval including the position shown, determine the tangential acceleration of point B along its path and the angular acceleration of the bar.

$Ans.$ $(a_B)_t = -23.9$ m/s^2, $\alpha = 36.2$ rad/s^2 CW

Problem 5/125

5/126 The punch of Prob. 5/53 is shown again here where the sector is given an oscillation specified by $\theta = \theta_0 \sin 2\pi t$. Calculate the acceleration of the punch when $\theta = 0$ if $\theta_0 = \pi/12$ rad.

Problem 5/126

5/127 The load L is lowered by the two pulleys which are fastened together and rotate as a single unit. For the instant represented, drum A has a counterclockwise angular velocity of 4 rad/s, which is decreasing by 4 rad/s each second. Simultaneously drum B has a clockwise angular velocity of 6 rad/s, which is increasing by 2 rad/s each second. Calculate the acceleration of points C and D and the load L.

$Ans.$ $\mathbf{a}_C = 0.9\mathbf{j} - 0.08\mathbf{i}$ m/s^2
$\mathbf{a}_D = 0.22\mathbf{j} - 0.6\mathbf{i}$ m/s^2
$\mathbf{a}_L = 0.3\mathbf{j}$ m/s^2

Problem 5/127

5/128 Calculate the angular acceleration of the plate in the position shown where control link AO has a constant angular velocity $\omega_{OA} = 4$ rad/s and $\theta = 60°$ for both links.

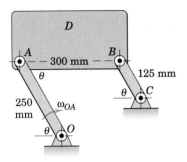

Problem 5/128

5/129 The sliding collar moves up and down the shaft, causing an oscillation of crank OB. If the velocity of A is not changing as it passes the null position where AB is horizontal and OB is vertical, determine the angular acceleration of OB in that position.

$$Ans. \ \ \alpha_{OB} = \frac{v_A{}^2}{rl} \ \text{CW}$$

Problem 5/129

5/130 Link OA has a constant counterclockwise angular velocity ω during a short interval of its motion. For the position shown determine the angular accelerations of AB and BC.

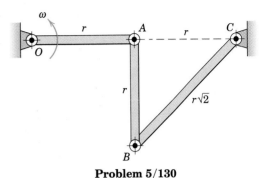

Problem 5/130

5/131 The configuration shown is designed to produce oscillation of the bin of small parts about C, as a part of a production process. Obtain an expression for the angular acceleration α of the bin for the position where the connecting link AB and the crank OA are in the vertical position with BC horizontal. The crank is rotating clockwise with a constant angular velocity ω for a short interval of motion.

$$Ans. \ \ \alpha_{BC} = \frac{r\omega^2}{L}\left(1 + \frac{r}{b}\right) \ \text{CW}$$

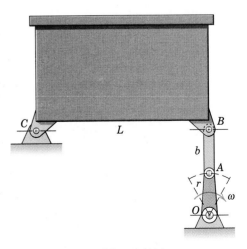

Problem 5/131

5/132 At the instant represented $\theta = 45°$, and the triangular plate ABC has a counterclockwise angular velocity of 20 rad/s and a clockwise angular acceleration of 100 rad/s². Determine the magnitudes of the corresponding velocity **v** and acceleration **a** of the piston rod of the hydraulic cylinder attached to C.

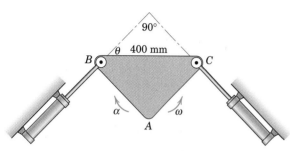

Problem 5/132

5/133 The linkage of Prob. 5/66 is shown again here. For the instant when $\theta = \beta = 60°$, the hydraulic cylinder gives A a velocity $v_A = 1.2$ m/s which is increasing by 0.9 m/s each second. For this instant determine the angular acceleration of link BC.

Ans. $\alpha_{BC} = 2.08$ rad/s² CCW

Problem 5/133

5/134 The triangular plate has a clockwise angular velocity of 3 rad/s and OA has zero angular acceleration for the instant represented. Determine the angular acceleration of plate ABD and link BC for this instant.

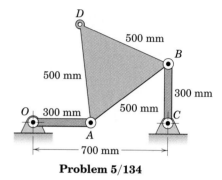

Problem 5/134

5/135 For the linkage shown, if $v_A = 500$ mm/s and is constant when the two links become perpendicular to one another, determine the angular acceleration of CB for this position.

Ans. $\alpha_{CB} = 5.76$ rad/s² CW

Problem 5/135

5/136 The deployment mechanism for the spacecraft magnetometer boom of Prob. 5/77 is shown again here. The driving link OB has a constant clockwise angular velocity ω_{OB} of 0.5 rad/s as it crosses the vertical position. Determine the angular acceleration α_{CA} of the boom for the position shown where $\tan \theta = 4/3$.

Problem 5/136

5/137 The linkage of Prob. 5/79 is shown again here. If OA has a constant counterclockwise angular velocity $\omega_0 = 10$ rad/s, calculate the angular acceleration of link AB for the position where the coordinates of A are $x = -60$ mm and $y = 80$ mm. Link BC is vertical for this position. Solve by vector algebra. (Use the results of Prob. 5/79 for the angular velocities of AB and BC which are $\omega_{BC} = 5.83\mathbf{k}$ rad/s and $\omega_{AB} = 2.5\mathbf{k}$ rad/s.)

Ans. $\alpha_{AB} = 10.42\mathbf{k}$ rad/s^2

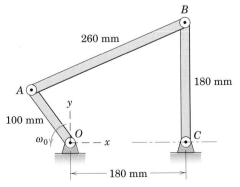

Problem 5/137

5/138 The revolving crank ED and connecting link CD cause the rigid frame ABO to oscillate about O. For the instant represented ED and CD are both perpendicular to FO, and the crank ED has an angular velocity of 0.4 rad/s and an angular acceleration of 0.06 rad/s^2, both counterclockwise. For this instant determine the acceleration of point A with respect to point B.

Problem 5/138

5/139 Plane motion of the triangular plate ABC is controlled by crank OA and link DB. For the instant represented, when OA and DB are vertical, OA has a clockwise angular velocity of 3 rad/s and a counterclockwise angular acceleration of 10 rad/s^2. Determine the angular acceleration of DB for this instant.

Ans. $\alpha_{DB} = 1.234$ rad/s^2 CCW

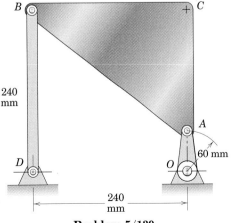

Problem 5/139

5/140 The elements of a power hacksaw are shown in the figure. The saw blade is mounted in a frame which slides along the horizontal guide. If the motor turns the flywheel at a constant counterclockwise speed of 60 rev/min, determine the acceleration of the blade for the position where $\theta = 90°$, and find the corresponding angular acceleration of the link AB.

Problem 5/140

5/141 For a short interval of motion link OA has a constant angular velocity $\omega = 4$ rad/s. Determine the angular acceleration α_{AB} of link AB for the instant when OA is parallel to the horizontal axis through B.

Ans. $\alpha_{AB} = 1.688$ rad/s^2 CCW

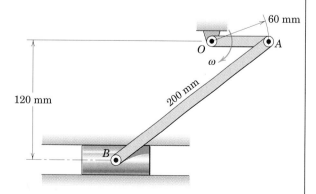

Problem 5/141

5/142 The simplified clam-shell bucket of Prob. 5/74 is shown again here. With the block at O considered fixed and with the constant velocity of the control cable at C equal to 0.5 m/s, determine the angular acceleration α of the right-hand bucket jaw when $\theta = 45°$ as the bucket jaws are closing.

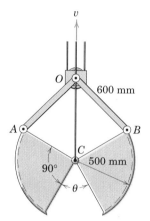

Problem 5/142

5/143 In the design of this linkage, motion of the square plate is controlled by the two pivoted links. Link OA has a constant angular velocity $\omega = 4$ rad/s during a short interval of motion. For the instant represented, $\theta = \tan^{-1} 4/3$ and AB is parallel to the x-axis. For this instant, determine the angular acceleration of both the plate and link CB.

Ans. $\alpha_{CB} = 16.08$ rad/s^2 CCW
$\alpha_{AB} = 3.81$ rad/s^2 CCW

Problem 5/143

5/144 Motion of link *ABC* is controlled by the horizontal movement of the piston rod of the hydraulic cylinder *D* and by the vertical guide for the pinned slider at *B*. For the instant when $\theta = 45°$, the piston rod is retracting at the constant rate $v_C = 180$ mm/s. Determine the acceleration of point *A* for this instant.

Problem 5/145

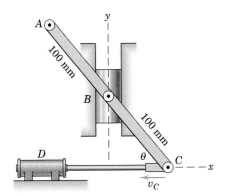

Problem 5/144

5/145 The mechanism of Prob. 5/109 is repeated here where the flexible band *F* attached to the sector at *E* is given a constant velocity of 4 m/s as shown. For the instant when *BD* is perpendicular to *OA*, determine the angular acceleration of *BD*.
Ans. $\alpha_{BD} = 46.9$ rad/s^2 CW

5/146 Elements of the switching device of Prob. 5/75 are shown again here. If the velocity *v* of the control rod is 0.9 m/s and is decreasing at the rate of 6 m/s^2 when $\theta = 60°$, determine the magnitude of the acceleration of *C*.

Problem 5/146

▶5/147 An oil pumping rig is shown in the figure. The flexible pump rod D is fastened to the sector at E and is always vertical as it enters the fitting below D. The link AB causes the beam BCE to oscillate as the weighted crank OA revolves. If OA has a constant clockwise speed of 1 rev every 3 s, determine the acceleration of the pump rod D when the beam and the crank OA are both in the horizontal position shown.

Ans. $a_D = 0.568$ m/s² down

▶5/148 An intermittent-drive mechanism for perforated tape F consists of the link DAB driven by the crank OB. The trace of the motion of the finger at D is shown by the dashed line. Determine the magnitude of the acceleration of D at the instant shown when both OB and CA are horizontal if OB has a constant clockwise rotational velocity of 120 rev/min.

Ans. $a_D = 1997$ mm/s²

Problem 5/147

Problem 5/148

5/7 *MOTION RELATIVE TO ROTATING AXES*

In our discussion of the relative motion of particles in Art. 2/8 and in our use of the relative-motion equations for the plane motion of rigid bodies in this present chapter, we have made all of our relative-velocity and relative-acceleration measurements from *nonrotating* reference axes. The solution of many problems in kinematics where motion is generated within or observed from a system that itself is rotating is greatly facilitated by the use of rotating reference axes. An example of such a motion would be the movement of a fluid particle along the curved vane of a centrifugal pump, where the path relative to the vanes of the impeller becomes an important design consideration.

We begin the description of motion using rotating axes by considering the plane motion of two particles A and B in the fixed X-Y plane, Fig. 5/10*a*. For the time being, we will consider A and B to be moving independently of one another for the sake of generality. The motion of A will now be observed from a moving reference frame x-y which has its origin attached to B and which rotates with an angular velocity $\omega = \dot{\theta}$. We may write this angular velocity as the vector $\boldsymbol{\omega} = \omega\mathbf{k} = \dot{\theta}\mathbf{k}$, where the vector is normal to the plane of motion and where its positive sense is in the positive z-direction (out from the paper), as established by the right-hand rule. The absolute position vector of A is given by

$$\mathbf{r}_A = \mathbf{r}_B + \mathbf{r} = \mathbf{r}_B + (x\mathbf{i} + y\mathbf{j}) \qquad (5/10)$$

where $\mathbf{i}$ and $\mathbf{j}$ are unit vectors attached to the x-y frame and $\mathbf{r} = x\mathbf{i} + y\mathbf{j}$ stands for $\mathbf{r}_{A/B}$, the position vector of A with respect to B. The velocity and acceleration equations require successive differentiation of the position-vector equation with respect to time. In contrast to the case of translating axes treated in Art. 2/8, the unit vectors $\mathbf{i}$ and $\mathbf{j}$ are now rotating with the x-y axes and, therefore, have time derivatives which must be evaluated. These derivatives may be seen from Fig. 5/10*b*, which shows the infinitesimal change in each unit vector during time dt as the reference axes rotate through an angle $d\theta = \omega \, dt$. The differential change in $\mathbf{i}$ is $d\mathbf{i}$, and it has the direction of $\mathbf{j}$ and a magnitude equal to the angle $d\theta$ times the length of the vector $\mathbf{i}$, which is unity. Thus, $d\mathbf{i} = d\theta\mathbf{j}$. Similarly, the unit vector $\mathbf{j}$ has an infinitesimal change $d\mathbf{j}$ that points in the negative x-direction, so that $d\mathbf{j} = -d\theta\mathbf{i}$. Dividing by dt and replacing $d\mathbf{i}/dt$ by $\dot{\mathbf{i}}$, $d\mathbf{j}/dt$ by $\dot{\mathbf{j}}$, and $d\theta/dt$ by $\dot{\theta} = \omega$ result in

$$\dot{\mathbf{i}} = \omega\mathbf{j} \qquad \text{and} \qquad \dot{\mathbf{j}} = -\omega\mathbf{i}$$

When the cross product is introduced, we see from Fig. 5/10*c* that $\boldsymbol{\omega} \times \mathbf{i} = \omega\mathbf{j}$ and $\boldsymbol{\omega} \times \mathbf{j} = -\omega\mathbf{i}$. Thus, the time derivatives of the unit vectors may be written as

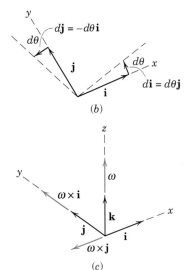

Figure 5/10

$$\boxed{\dot{\mathbf{i}} = \boldsymbol{\omega} \times \mathbf{i} \quad \text{and} \quad \dot{\mathbf{j}} = \boldsymbol{\omega} \times \mathbf{j}} \tag{5/11}$$

(a) Relative velocity. We shall now use the expressions of Eqs. 5/11 when we take the time derivative of the position-vector equation for A and B to obtain the relative-velocity relation. Differentiation of Eq. 5/10 gives

$$\dot{\mathbf{r}}_A = \dot{\mathbf{r}}_B + \frac{d}{dt}(x\mathbf{i} + y\mathbf{j})$$

$$= \dot{\mathbf{r}}_B + (x\dot{\mathbf{i}} + y\dot{\mathbf{j}}) + (\dot{x}\mathbf{i} + \dot{y}\mathbf{j})$$

But $x\dot{\mathbf{i}} + y\dot{\mathbf{j}} = \boldsymbol{\omega} \times x\mathbf{i} + \boldsymbol{\omega} \times y\mathbf{j} = \boldsymbol{\omega} \times (x\mathbf{i} + y\mathbf{j}) = \boldsymbol{\omega} \times \mathbf{r}$. Also, since the observer in x-y measures velocity components $\dot{x}$ and $\dot{y}$, we see that $\dot{x}\mathbf{i} + \dot{y}\mathbf{j} = \mathbf{v}_{\text{rel}}$, which is the velocity relative to the x-y frame of reference. Thus, the relative-velocity equation becomes

$$\boxed{\mathbf{v}_A = \mathbf{v}_B + \boldsymbol{\omega} \times \mathbf{r} + \mathbf{v}_{\text{rel}}} \tag{5/12}$$

Comparison of Eq. 5/12 with Eq. 2/20 for nonrotating reference axes shows that $\mathbf{v}_{A/B} = \boldsymbol{\omega} \times \mathbf{r} + \mathbf{v}_{\text{rel}}$, from which we conclude that the term $\boldsymbol{\omega} \times \mathbf{r}$ is the difference between the relative velocity as measured from nonrotating and rotating axes.

To illustrate further the meaning of the last two terms in Eq. 5/12, the motion of particle A relative to the rotating x-y plane is shown in Fig. 5/11 as taking place in a curved slot in a plate that represents the rotating x-y reference system. The velocity of A as measured relative to the plate, $\mathbf{v}_{\text{rel}}$, would be tangent to the path fixed in the x-y plate and would have a magnitude $\dot{s}$, where s is measured along the path. This relative velocity may also be viewed as the velocity $\mathbf{v}_{A/P}$ relative to a point P attached to the plate and coincident with A at the instant under consideration. The term $\boldsymbol{\omega} \times \mathbf{r}$ has a magnitude $r\dot{\theta}$ and a direction normal to $\mathbf{r}$ and is the velocity relative to B of point P as seen from nonrotating axes attached to B.

The following comparison will help to establish the equivalence of, and clarify the differences between, the relative-velocity equations written for rotating and nonrotating reference axes:

$$\mathbf{v}_A = \mathbf{v}_B + \boldsymbol{\omega} \times \mathbf{r} + \mathbf{v}_{\text{rel}}$$

$$\mathbf{v}_A = \mathbf{v}_B + \underbrace{\mathbf{v}_{P/B}}_{} + \mathbf{v}_{A/P}$$

$$\mathbf{v}_A = \underbrace{\mathbf{v}_P}_{} + \mathbf{v}_{A/P} \tag{5/12a}$$

$$\mathbf{v}_A = \mathbf{v}_B + \underbrace{\mathbf{v}_{A/B}}_{}$$

In the second equation, the term $\mathbf{v}_{P/B}$ is measured from a nonrotating position—otherwise, it would be zero. The term $\mathbf{v}_{A/P}$ is the

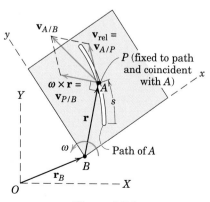

Figure 5/11

same as $\mathbf{v}_{rel}$ and is the velocity of A as measured in the x-y frame. In the third equation, $\mathbf{v}_P$ is the absolute velocity of P and represents the effect of the moving coordinate system, both translational and rotational. The fourth equation is the same as that developed for nonrotating axes, Eq. 2/20, and it is seen that $\mathbf{v}_{A/B} = \mathbf{v}_{P/B} + \mathbf{v}_{A/P} = \boldsymbol{\omega} \times \mathbf{r} + \mathbf{v}_{rel}$.

(b) Transformation of a time derivative. Equation 5/12 represents a transformation of the time derivative of the position vector between rotating and nonrotating axes. We may easily generalize this result to apply to the time derivative of any vector quantity $\mathbf{V} = V_x\mathbf{i} + V_y\mathbf{j}$. Accordingly, the total time derivative as taken in X-Y is

$$\left(\frac{d\mathbf{V}}{dt}\right)_{XY} = (\dot{V}_x\mathbf{i} + \dot{V}_y\mathbf{j}) + (V_x\dot{\mathbf{i}} + V_y\dot{\mathbf{j}})$$

The first two terms in the expression represent that part of the total derivative of $\mathbf{V}$ which is measured relative to the x-y reference system, and the second two terms represent that part of the derivative due to the rotation of the reference system. With the expressions for $\mathbf{i}$ and $\mathbf{j}$ from Eqs. 5/11, we may now write

$$\left(\frac{d\mathbf{V}}{dt}\right)_{XY} = \left(\frac{d\mathbf{V}}{dt}\right)_{xy} + \boldsymbol{\omega} \times \mathbf{V} \qquad\qquad (5/13)$$

Here $\boldsymbol{\omega} \times \mathbf{V}$ represents the difference between the time derivative of the vector as measured in a fixed reference and its time derivative as measured in the rotating reference. As we shall see in Art. 7/2 where three-dimensional motion is introduced, Eq. 5/13 is valid in three dimensions, as well as in two dimensions.

The physical significance of Eq. 5/13 is illustrated in Fig. 5/12, which shows the vector $\mathbf{V}$ at time t as observed both in the fixed axes X-Y and in the rotating axes x-y. Since we are dealing with the effects of rotation only, the vector may be shown through the origin of coordinates without loss of generality. During time dt, the vector swings to position $\mathbf{V}'$, and the observer in x-y measures the components dV due to its change in magnitude and $V\,d\beta$ due to its rotation $d\beta$ relative to x-y. To the rotating observer, then, the derivative $(d\mathbf{V}/dt)_{xy}$ which he or she measures has the components dV/dt and $V\,d\beta/dt = V\dot{\beta}$. The remaining part of the total time derivative not measured by the rotating observer has the magnitude $V\,d\theta/dt$ and, expressed as a vector, is $\boldsymbol{\omega} \times \mathbf{V}$. Thus, we see physically from the diagram that

$$(\dot{\mathbf{V}})_{XY} = (\dot{\mathbf{V}})_{xy} + \boldsymbol{\omega} \times \mathbf{V}$$

which is Eq. 5/13.

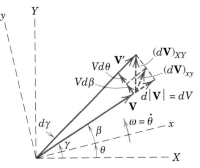

Figure 5/12

(c) Relative acceleration.   The relative-acceleration equation may be obtained by differentiating the relative-velocity relation, Eq. 5/12. Thus,

$$\mathbf{a}_A = \mathbf{a}_B + \dot{\boldsymbol{\omega}} \times \mathbf{r} + \boldsymbol{\omega} \times \dot{\mathbf{r}} + \dot{\mathbf{v}}_{\text{rel}}$$

From the derivation of Eq. 5/12 we saw that

$$\dot{\mathbf{r}} = \frac{d}{dt}(x\mathbf{i} + y\mathbf{j}) = (\dot{x}\mathbf{i} + \dot{y}\mathbf{j}) + (x\dot{\mathbf{i}} + y\dot{\mathbf{j}})$$

$$= \boldsymbol{\omega} \times \mathbf{r} + \mathbf{v}_{\text{rel}}$$

Therefore, the third term on the right side of the acceleration equation becomes

$$\boldsymbol{\omega} \times \dot{\mathbf{r}} = \boldsymbol{\omega} \times (\boldsymbol{\omega} \times \mathbf{r} + \mathbf{v}_{\text{rel}}) = \boldsymbol{\omega} \times (\boldsymbol{\omega} \times \mathbf{r}) + \boldsymbol{\omega} \times \mathbf{v}_{\text{rel}}$$

With the aid of Eqs. 5/11, the last term on the right side of the equation for $\mathbf{a}_A$ becomes

$$\dot{\mathbf{v}}_{\text{rel}} = \frac{d}{dt}(\dot{x}\mathbf{i} + \dot{y}\mathbf{j}) = (\ddot{x}\mathbf{i} + \ddot{y}\mathbf{j}) + (\dot{x}\dot{\mathbf{i}} + \dot{y}\dot{\mathbf{j}})$$

$$= \boldsymbol{\omega} \times (\dot{x}\mathbf{i} + \dot{y}\mathbf{j}) + (\ddot{x}\mathbf{i} + \ddot{y}\mathbf{j})$$

$$= \boldsymbol{\omega} \times \mathbf{v}_{\text{rel}} + \mathbf{a}_{\text{rel}}$$

Substitution into the expression for $\mathbf{a}_A$ and collection of terms give

$$\boxed{\mathbf{a}_A = \mathbf{a}_B + \dot{\boldsymbol{\omega}} \times \mathbf{r} + \boldsymbol{\omega} \times (\boldsymbol{\omega} \times \mathbf{r}) + 2\boldsymbol{\omega} \times \mathbf{v}_{\text{rel}} + \mathbf{a}_{\text{rel}}} \quad (5/14)$$

Equation 5/14 is the general vector expression for the absolute acceleration of a particle A in terms of its acceleration $\mathbf{a}_{\text{rel}}$ measured relative to a moving coordinate system which rotates with an angular velocity $\boldsymbol{\omega}$ and an angular acceleration $\dot{\boldsymbol{\omega}}$. The terms $\dot{\boldsymbol{\omega}} \times \mathbf{r}$ and $\boldsymbol{\omega} \times (\boldsymbol{\omega} \times \mathbf{r})$ are shown in Fig. 5/13. They represent, respectively, the normal and tangential components of the acceleration $\mathbf{a}_{P/B}$ of the coincident point P in its circular motion with respect to B. This motion would be observed from a set of nonrotating axes moving with B. The magnitude of $\dot{\boldsymbol{\omega}} \times \mathbf{r}$ is $r\ddot{\theta}$ and its direction is tangent to the circle. The magnitude of $\boldsymbol{\omega} \times (\boldsymbol{\omega} \times \mathbf{r})$ is $r\omega^2$ and its direction is from P to B along the normal to the circle. The acceleration of A relative to the plate along the path, $\mathbf{a}_{\text{rel}}$, may be expressed in rectangular, normal and tangential, or polar coordinates in the rotating system. Generally, n- and t-components are most frequently used, and these components are depicted in Fig. 5/13. The tangential component would have the magnitude $(a_{\text{rel}})_t = \ddot{s}$, where s is the distance measured along the path to A. The normal component would have the magnitude v_{rel}^2/ρ, where ρ is the radius of curvature of the path as measured in x-y. The sense of this vector would always be toward the center of curvature.

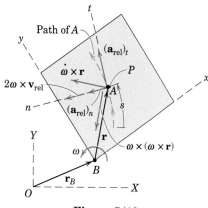

Figure 5/13

The term $2\boldsymbol{\omega} \times \mathbf{v}_{\text{rel}}$, shown in Fig. 5/13, is known as the *Coriolis* acceleration* and represents the difference between the acceleration of A relative to P as measured from nonrotating axes and from rotating axes. The direction is always normal to the term $\mathbf{v}_{\text{rel}}$, and the sense is established by the right-hand rule for the cross product.

The Coriolis acceleration is difficult to visualize because it is composed of two separate physical effects. To help with this visualization, we will consider the simplest possible motion in which this term appears. In Fig. 5/14a we have a rotating disk with a radial slot in which a small particle A is confined to slide. Let the disk turn with a constant angular velocity $\omega = \dot{\theta}$ and let the particle move along the slot with a constant speed $v_{\text{rel}} = \dot{x}$ relative to the slot. The velocity of A has the components $\dot{x}$ due to motion along the slot and $x\omega$ due to the rotation of the slot. The changes in these two velocity components due to the rotation of the disk are shown in the b-part of the figure for the interval dt, during which the x-y axes rotate with the disk through the angle $d\theta$ to x'-y'. The velocity increment due to the change in direction of $\mathbf{v}_{\text{rel}}$ is $\dot{x}\, d\theta$ and that due to the change in magnitude of $x\omega$ is $\omega\, dx$, both being in the y-direction normal to the slot. Dividing each increment by dt and adding give the sum $\omega\dot{x} + \dot{x}\omega = 2\dot{x}\omega$, which is the magnitude of the Coriolis acceleration $2\boldsymbol{\omega} \times \mathbf{v}_{\text{rel}}$. Dividing the remaining velocity increment $x\omega\, d\theta$ due to the change in direction of $x\omega$ by dt gives $x\omega\dot{\theta}$ or $x\omega^2$, which is the acceleration of a point P fixed to the slot and momentarily coincident with the particle A. We now see how Eq. 5/14 fits these results. With the origin B in that equation taken at the fixed center O, $\mathbf{a}_B = \mathbf{0}$. With constant angular velocity, $\dot{\boldsymbol{\omega}} \times \mathbf{r} = \mathbf{0}$. With $\mathbf{v}_{\text{rel}}$ constant in magnitude and no curvature to the slot, $\mathbf{a}_{\text{rel}} = \mathbf{0}$. We are left with

$$\mathbf{a}_A = \boldsymbol{\omega} \times (\boldsymbol{\omega} \times \mathbf{r}) + 2\boldsymbol{\omega} \times \mathbf{v}_{\text{rel}}$$

Replacing $\mathbf{r}$ by $x\mathbf{i}$, $\boldsymbol{\omega}$ by $\omega\mathbf{k}$, and $\mathbf{v}_{\text{rel}}$ by $\dot{x}\mathbf{i}$ gives

$$\mathbf{a}_A = -x\omega^2\mathbf{i} + 2\dot{x}\omega\mathbf{j}$$

which checks our analysis from Fig. 5/14. We also note that this same result is contained in our polar-coordinate analysis of plane curvilinear motion in Eq. 2/14 when we let $\ddot{r} = 0$ and $\ddot{\theta} = 0$ and replace r by x and $\dot{\theta}$ by ω. If the slot in the disk of Fig. 5/14 had been curved, we would have had a normal component of acceleration relative to the slot so that $\mathbf{a}_{\text{rel}}$ would not be zero.

The following comparison will help to establish the equivalence of, and clarify the differences between, the relative-acceleration equations written for rotating and nonrotating reference axes:

*Named after the French military engineer G. Coriolis (1792–1843), who was the first to call attention to this term.

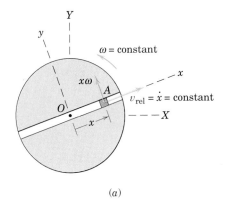

$\omega = $ constant

$v_{\text{rel}} = \dot{x} = $ constant

(a)

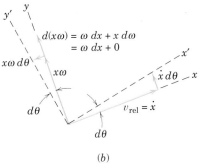

$d(x\omega) = \omega\, dx + x\, d\omega$
$= \omega\, dx + 0$

$v_{\text{rel}} = \dot{x}$

(b)

Figure 5/14

$$\mathbf{a}_A = \mathbf{a}_B + \underbrace{\dot{\boldsymbol{\omega}} \times \mathbf{r} + \boldsymbol{\omega} \times (\boldsymbol{\omega} \times \mathbf{r})} + \underbrace{2\boldsymbol{\omega} \times \mathbf{v}_{\text{rel}} + \mathbf{a}_{\text{rel}}}$$

$$\mathbf{a}_A = \mathbf{a}_B + \underbrace{\mathbf{a}_{P/B}} + \mathbf{a}_{A/P} \qquad (5/14a)$$

$$\mathbf{a}_A = \underbrace{\mathbf{a}_P} + \underbrace{\mathbf{a}_{A/P}}$$

$$\mathbf{a}_A = \mathbf{a}_B + \underbrace{\mathbf{a}_{A/B}}$$

The equivalence of $\mathbf{a}_{P/B}$ and $\dot{\boldsymbol{\omega}} \times \mathbf{r} + \boldsymbol{\omega} \times (\boldsymbol{\omega} \times \mathbf{r})$, as shown in the second equation, has already been described. From the third equation where $\mathbf{a}_B + \mathbf{a}_{P/B}$ has been combined to give $\mathbf{a}_P$, it is seen that the relative-acceleration term $\mathbf{a}_{A/P}$, unlike the corresponding relative-velocity term, is not equal to the relative acceleration $\mathbf{a}_{\text{rel}}$ measured from the rotating x-y frame of reference. The Coriolis term is, therefore, the difference between the acceleration $\mathbf{a}_{A/P}$ of A relative to P as measured in a nonrotating system and the acceleration $\mathbf{a}_{\text{rel}}$ of A relative to P as measured in a rotating system. From the fourth equation, it is seen that the acceleration $\mathbf{a}_{A/B}$ of A with respect to B as measured in a nonrotating system, Eq. 2/21, is a combination of the last four terms in the first equation for the rotating system.

The results expressed by Eq. 5/14 may be visualized somewhat more simply by writing the acceleration of A in terms of the acceleration of the coincident point P. Since the acceleration of P is $\mathbf{a}_P = \mathbf{a}_B + \dot{\boldsymbol{\omega}} \times \mathbf{r} + \boldsymbol{\omega} \times (\boldsymbol{\omega} \times \mathbf{r})$, we may rewrite Eq. 5/14 as

$$\boxed{\mathbf{a}_A = \mathbf{a}_P + 2\boldsymbol{\omega} \times \mathbf{v}_{\text{rel}} + \mathbf{a}_{\text{rel}}} \qquad (5/14b)$$

When the equation is written in this form, point P may not be picked at random because it is the one point attached to the rotating reference frame coincident with A at the instant of analysis. Again, reference to Fig. 5/13 should be made to clarify the meaning of each of the terms in Eq. 5/14 and its equivalent, Eq. 5/14b.

In summary, once we have chosen our rotating reference system, we must recognize the following quantities:

$\mathbf{v}_B$ = absolute velocity of the origin B of the rotating axes

$\mathbf{a}_B$ = absolute acceleration of the origin B of the rotating axes

$\mathbf{r}$ = position vector of the coincident point P measured from B

$\boldsymbol{\omega}$ = angular velocity of the rotating axes

$\dot{\boldsymbol{\omega}}$ = angular acceleration of the rotating axes

$\mathbf{v}_{\text{rel}}$ = velocity of A measured relative to the rotating axes

$\mathbf{a}_{\text{rel}}$ = acceleration of A measured relative to the rotating axes

Also, it is well to be reminded that our vector analysis depends on the consistent use of a right-handed set of coordinate axes. Finally, note is made of the fact that Eqs. 5/12 and 5/14, developed here for plane motion, hold equally well for space motion, and this extension of generality will be covered in Art. 7/6.

Sample Problem 5/16

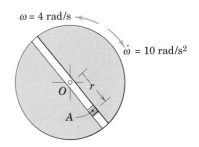

At the instant represented, the disk with the radial slot is rotating about O with a counterclockwise angular velocity of 4 rad/s which is decreasing at the rate of 10 rad/s². The motion of slider A is separately controlled, and at this instant, $r = 150$ mm, $\dot{r} = 125$ mm/s, and $\ddot{r} = 2025$ mm/s². Determine the absolute velocity and acceleration of A for this position.

Solution. We have motion relative to a rotating path, so that a rotating coordinate system with origin at O is indicated. We attach x-y axes to the disk and use the unit vectors $\mathbf{i}$ and $\mathbf{j}$.

Velocity. With the origin at O, the term $\mathbf{v}_B$ of Eq. 5/12 disappears and we have

① $$\mathbf{v}_A = \boldsymbol{\omega} \times \mathbf{r} + \mathbf{v}_{\text{rel}}$$

The angular velocity as a vector is $\boldsymbol{\omega} = 4\mathbf{k}$ rad/s, where $\mathbf{k}$ is the unit
② vector normal to the x-y plane in the $+z$-direction. Our relative-velocity equation becomes

$$\mathbf{v}_A = 4\mathbf{k} \times 0.150\mathbf{i} + 0.125\mathbf{i} = 0.600\mathbf{j} + 0.125\mathbf{i} \text{ m/s} \qquad Ans.$$

in the direction indicated and has the magnitude

$$v_A = \sqrt{(0.600)^2 + (0.125)^2} = 0.613 \text{ m/s} \qquad Ans.$$

Acceleration. Equation 5/14 written for zero acceleration of the origin of the rotating coordinate system is

$$\mathbf{a}_A = \boldsymbol{\omega} \times (\boldsymbol{\omega} \times \mathbf{r}) + \dot{\boldsymbol{\omega}} \times \mathbf{r} + 2\boldsymbol{\omega} \times \mathbf{v}_{\text{rel}} + \mathbf{a}_{\text{rel}}$$

The terms become

③ $$\boldsymbol{\omega} \times (\boldsymbol{\omega} \times \mathbf{r}) = 4\mathbf{k} \times (4\mathbf{k} \times 0.150\mathbf{i}) = 4\mathbf{k} \times 0.6\mathbf{j} = -2.4\mathbf{i} \text{ m/s}^2$$

$$\dot{\boldsymbol{\omega}} \times \mathbf{r} = -10\mathbf{k} \times 0.150\mathbf{i} = -1.5\mathbf{j} \text{ m/s}^2$$

$$2\boldsymbol{\omega} \times \mathbf{v}_{\text{rel}} = 2(4\mathbf{k}) \times 0.125\mathbf{i} = 1.0\mathbf{j} \text{ m/s}^2$$

$$\mathbf{a}_{\text{rel}} = 2.025\mathbf{i} \text{ m/s}^2$$

The total acceleration is, therefore,

$$\mathbf{a}_A = (2.025 - 2.4)\mathbf{i} + (1.0 - 1.5)\mathbf{j} = -0.375\mathbf{i} - 0.5\mathbf{j} \text{ m/s}^2 \qquad Ans.$$

in the direction indicated and has the magnitude

$$a_A = \sqrt{(0.375)^2 + (0.5)^2} = 0.625 \text{ m/s}^2 \qquad Ans.$$

Vector notation is certainly not essential to the solution of this problem. The student should be able to work out the steps with scalar notation just as easily. The correct direction of the Coriolis-acceleration term may always be found by the direction in which the head of the $\mathbf{v}_{\text{rel}}$ vector would move if rotated about its tail in the sense of $\boldsymbol{\omega}$ as shown.

① This equation is the same as $\mathbf{v}_A = \mathbf{v}_P + \mathbf{v}_{A/P}$, where P is a point attached to the disk coincident with A at this instant.

② Note that the x-y-z axes chosen constitute a right-handed system.

③ Be sure to recognize that $\boldsymbol{\omega} \times (\boldsymbol{\omega} \times \mathbf{r})$ and $\dot{\boldsymbol{\omega}} \times \mathbf{r}$ represent the normal and tangential components of acceleration of a point P on the disk coincident with A. This description becomes that of Eq. 5/14b.

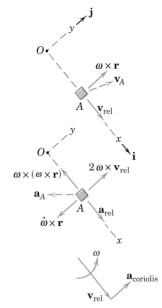

Sample Problem   5/17

The pin A of the hinged link AC is confined to move in the rotating slot of link OD. The angular velocity of OD is $\omega = 2$ rad/s clockwise and is constant for the interval of motion concerned. For the position where $\theta = 45°$ with AC horizontal, determine the velocity of pin A and the velocity of A relative to the rotating slot in OD.

Solution.   Motion of a point (pin A) along a rotating path (the slot) suggests the use of rotating coordinate axes x-y attached to arm OD. With the origin at the fixed point O, the term $\mathbf{v}_B$ of Eq. 5/12 vanishes, and we have $\mathbf{v}_A = \boldsymbol{\omega} \times \mathbf{r} + \mathbf{v}_{\text{rel}}$.

The velocity of A in its circular motion about C is

$$\mathbf{v}_A = \boldsymbol{\omega}_{CA} \times \mathbf{r}_{CA} = \omega_{CA}\mathbf{k} \times (225/\sqrt{2})(-\mathbf{i} - \mathbf{j}) = (225/\sqrt{2})\omega_{CA}(\mathbf{i} - \mathbf{j})$$

① where the angular velocity ω_{CA} is arbitrarily assigned in a clockwise sense in the positive z-direction ($+\mathbf{k}$).

The angular velocity $\boldsymbol{\omega}$ of the rotating axes is that of the arm OD and, by the right-hand rule, is $\boldsymbol{\omega} = \omega\mathbf{k} = 2\mathbf{k}$ rad/s. The vector from the origin to the point P on OD coincident with A is $\mathbf{r} = \overline{OP}\mathbf{i} = \sqrt{(450 - 225)^2 + (225)^2}\,\mathbf{i} = 225\sqrt{2}\mathbf{i}$ mm. Thus,

$$\boldsymbol{\omega} \times \mathbf{r} = 2\mathbf{k} \times 225\sqrt{2}\mathbf{i} = 450\sqrt{2}\mathbf{j} \text{ mm/s}$$

Finally, the relative-velocity term $\mathbf{v}_{\text{rel}}$ is the velocity measured by an observer attached to the rotating reference frame and is $\mathbf{v}_{\text{rel}} = \dot{x}\mathbf{i}$. Substitution into the relative-velocity equation gives

$$(225/\sqrt{2})\omega_{CA}(\mathbf{i} - \mathbf{j}) = 450\sqrt{2}\mathbf{j} + \dot{x}\mathbf{i}$$

Equating separately the coefficients of the $\mathbf{i}$ and $\mathbf{j}$ terms yields

$$(225/\sqrt{2})\omega_{CA} = \dot{x} \quad \text{and} \quad -(225/\sqrt{2})\omega_{CA} = 450\sqrt{2}$$

giving

$$\omega_{CA} = -4 \text{ rad/s} \quad \text{and} \quad \dot{x} = v_{\text{rel}} = -450\sqrt{2} \text{ mm/s} \quad \textit{Ans.}$$

With a negative value for ω_{CA}, the actual angular velocity of CA is counterclockwise, so the velocity of A is up with a magnitude of

② $$v_A = 225(4) = 900 \text{ mm/s} \quad \textit{Ans.}$$

Geometric clarification of the terms is helpful and is easily shown. Using the equivalence between the third and the first of Eqs. 5/12a with $\mathbf{v}_B = \mathbf{0}$ enables us to write $\mathbf{v}_A = \mathbf{v}_P + \mathbf{v}_{A/P}$, where P is the point on the rotating arm OD coincident with A. Clearly, $v_P = \overline{OP}\omega = 225\sqrt{2}(2) = 450\sqrt{2}$ mm/s and its direction is normal to OD. The relative velocity $\mathbf{v}_{A/P}$, which is the same as $\mathbf{v}_{\text{rel}}$, is seen from the figure to be along the slot toward O. This conclusion becomes clear when it is observed that A is approaching P along the slot from below before coincidence and is receding from P upward along the slot following coincidence. The velocity of A is tangent to its circular arc about C. The vector equation may now be satisfied since there are only two remaining scalar unknowns, namely, the magnitude of $\mathbf{v}_{A/P}$ and the magnitude of $\mathbf{v}_A$. For the 45° position, the figure requires $v_{A/P} = 450\sqrt{2}$ mm/s and $v_A = 900$ mm/s, each in its direction shown. The angular velocity of AC is

$$[\omega = v/r] \quad \omega_{AC} = v_A/\overline{AC} = 900/225 = 4 \text{ rad/s counterclockwise}$$

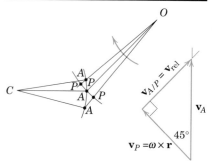

① It is clear enough physically that CA will have a counterclockwise angular velocity for the conditions specified, so we anticipate a negative value for ω_{CA}.

② Solution of the problem is not restricted to the reference axes used. Alternatively, the origin of the x-y axes, still attached to CD, could be chosen at the coincident point P on CD. This choice would merely replace the $\boldsymbol{\omega} \times \mathbf{r}$ term by its equal, $\mathbf{v}_P$. As a further selection, all vector quantities could be expressed in terms of X-Y components using unit vectors $\mathbf{I}$ and $\mathbf{J}$.

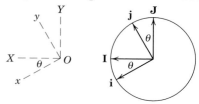

A direct conversion between the two reference systems is obtained from the geometry of the unit circle and gives

$$\mathbf{i} = \mathbf{I} \cos \theta - \mathbf{J} \sin \theta$$

and

$$\mathbf{j} = \mathbf{I} \sin \theta + \mathbf{J} \cos \theta$$

Sample Problem 5/18

For the conditions of Sample Problem 5/17, determine the angular acceleration of AC and the acceleration of A relative to the rotating slot in arm OD.

Solution. We attach the rotating coordinate system x-y to arm OD and use Eq. 5/14. With the origin at the fixed point O, the term $\mathbf{a}_B$ becomes zero so that

$$\mathbf{a}_A = \dot{\boldsymbol{\omega}} \times \mathbf{r} + \boldsymbol{\omega} \times (\boldsymbol{\omega} \times \mathbf{r}) + 2\boldsymbol{\omega} \times \mathbf{v}_{rel} + \mathbf{a}_{rel}$$

From the solution to Sample Problem 5/17, we make use of the values $\boldsymbol{\omega} = 2\mathbf{k}$ rad/s, $\boldsymbol{\omega}_{CA} = -4\mathbf{k}$ rad/s, $\mathbf{v}_{rel} = -450\sqrt{2}\mathbf{i}$ mm/s and write

$$\mathbf{a}_A = \dot{\boldsymbol{\omega}}_{CA} \times \mathbf{r}_{CA} + \boldsymbol{\omega}_{CA} \times (\boldsymbol{\omega}_{CA} \times \mathbf{r}_{CA})$$

$$= \dot{\omega}_{CA}\mathbf{k} \times \frac{225}{\sqrt{2}}(-\mathbf{i} - \mathbf{j}) - 4\mathbf{k} \times \left(-4\mathbf{k} \times \frac{225}{\sqrt{2}}[-\mathbf{i} - \mathbf{j}]\right)$$

$$\dot{\boldsymbol{\omega}} \times \mathbf{r} = \mathbf{0} \text{ since } \boldsymbol{\omega} = \text{constant}$$

$$\boldsymbol{\omega} \times (\boldsymbol{\omega} \times \mathbf{r}) = 2\mathbf{k} \times (2\mathbf{k} \times 225\sqrt{2}\mathbf{i}) = -900\sqrt{2}\mathbf{i} \text{ mm/s}^2$$

$$2\boldsymbol{\omega} \times \mathbf{v}_{rel} = 2(2\mathbf{k}) \times (-450\sqrt{2}\mathbf{i}) = -1800\sqrt{2}\mathbf{j} \text{ mm/s}^2$$

$$\mathbf{a}_{rel} = \ddot{x}\mathbf{i}$$

① Substitution into the relative-acceleration equation yields

$$\frac{1}{\sqrt{2}}(225\dot{\omega}_{CA} + 3600)\mathbf{i} + \frac{1}{\sqrt{2}}(-225\dot{\omega}_{CA} + 3600)\mathbf{j}$$

$$= -900\sqrt{2}\mathbf{i} - 1800\sqrt{2}\mathbf{j} + \ddot{x}\mathbf{i}$$

Equating separately the $\mathbf{i}$ and $\mathbf{j}$ terms gives

$$(225\dot{\omega}_{CA} + 3600)/\sqrt{2} = -900\sqrt{2} + \ddot{x}$$

and $\quad (-225\dot{\omega}_{CA} + 3600)/\sqrt{2} = -1800\sqrt{2}$

Solving for the two unknowns gives

$$\dot{\omega}_{CA} = 32 \text{ rad/s}^2 \quad \text{and} \quad \ddot{x} = a_{rel} = 8910 \text{ mm/s}^2 \quad Ans.$$

If desired, the acceleration of A may also be written as

$$\mathbf{a}_A = (225/\sqrt{2})(32)(\mathbf{i} - \mathbf{j}) + (3600/\sqrt{2})(\mathbf{i} + \mathbf{j}) = 7637\mathbf{i} - 2550\mathbf{j} \text{ mm/s}^2$$

We make use here of the geometric representation of the relative-acceleration equation to further clarify the problem. The geometric approach may be used as an alternative solution. Again, we introduce point P on OD coincident with A. The equivalent scalar terms are

$$(a_A)_t = |\dot{\boldsymbol{\omega}}_{CA} \times \mathbf{r}_{CA}| = r\dot{\omega}_{CA} = r\alpha_{CA} \text{ normal to } CA, \text{ sense unknown}$$

$$(a_A)_n = |\boldsymbol{\omega}_{CA} \times (\boldsymbol{\omega}_{CA} \times \mathbf{r}_{CA})| = r\omega_{CA}^2 \text{ from } A \text{ to } C$$

$$(a_P)_n = |\boldsymbol{\omega} \times (\boldsymbol{\omega} \times \mathbf{r})| = \overline{OP}\omega^2 \text{ from } P \text{ to } O$$

$$(a_P)_t = |\dot{\boldsymbol{\omega}} \times \mathbf{r}| = r\dot{\omega} = 0 \text{ since } \boldsymbol{\omega} = \text{constant}$$

$$|2\boldsymbol{\omega} \times \mathbf{v}_{rel}| = 2\omega v_{rel} \text{ directed as shown}$$

$$a_{rel} = \ddot{x} \text{ along } OD, \text{ sense unknown}$$

We start with the known vectors and add them head-to-tail for each side of the equation beginning at R and ending at S, where the intersection ② of the known directions of $(\mathbf{a}_A)_t$ and $\mathbf{a}_{rel}$ establishes the solution. Closure of the polygon determines the sense of each of the two unknown vectors, ② and their magnitudes are easily calculated from the figure geometry.

① If the slot had been curved with a radius of curvature ρ, the term $\mathbf{a}_{rel}$ would have had a component v_{rel}^2/ρ normal to the slot and directed toward the center of curvature in addition to its component along the slot.

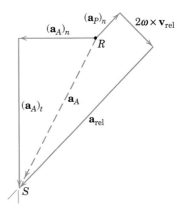

② It is always possible to avoid a simultaneous solution by projecting the vectors onto the perpendicular to one of the unknowns.

PROBLEMS

Introductory problems

5/149 An airplane flies north at a fixed altitude with a constant speed of 800 km/h. Determine the Coriolis component of its acceleration when a latitude of 30° north is reached. Consider the center of the earth fixed and refer to Table D/2 for the angular velocity of the earth.

Ans. $a_{\text{cor}} = 0.01620 \text{ m/s}^2$ west

5/150 The disk rotates about a fixed axis through O with angular velocity $\omega = 5$ rad/s and angular acceleration $\alpha = 3$ rad/s² at the instant represented, in the directions shown. The slider A moves in the straight slot. Determine the absolute velocity and acceleration of A for the same instant, when $y = 250$ mm, $\dot{y} = -600$ mm/s, and $\ddot{y} = 750$ mm/s².

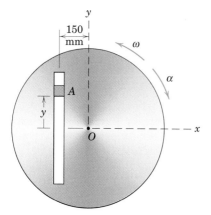

Problem 5/150

5/151 The disk rotates about a fixed axis through O with angular velocity $\omega = 5$ rad/s and angular acceleration $\alpha = 3$ rad/s² in the directions shown at a certain instant. The small sphere A moves in the circular slot, and at the same instant, $\beta = 30°$, $\dot{\beta} = 2$ rad/s, and $\ddot{\beta} = -4$ rad/s². Determine the absolute velocity and acceleration of A at this instant.

Ans. $\mathbf{v}_A = 1.312\mathbf{i} + 2.27\mathbf{j}$ m/s
$\mathbf{a}_A = 14.60\mathbf{i} - 11.46\mathbf{j}$ m/s²

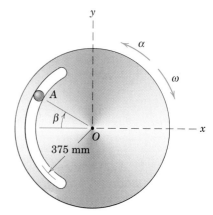

Problem 5/151

5/152 The slotted wheel rolls to the right without slipping, with a constant speed $v = 0.6$ m/s of its center O. Simultaneously, motion of the sliding block A is controlled by a mechanism not shown so that $\dot{x} = 450$ mm/s with $\ddot{x} = 0$. Determine the acceleration of A for the instant when $x = 150$ mm and $\theta = 30°$.

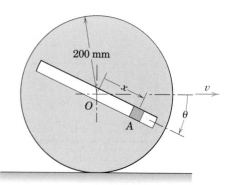

Problem 5/152

5/153 The disk rolls without slipping on the horizontal surface, and at the instant represented, the center O has the velocity and acceleration shown in the figure. For this instant, the particle A has the indicated speed u and time-rate-of-change of speed $\dot{u}$, both relative to the disk. Determine the absolute velocity and acceleration of particle A.

Ans. $\mathbf{v}_A = -3.4\mathbf{i}$ m/s, $\mathbf{a}_A = 2\mathbf{i} - 0.667\mathbf{j}$ m/s^2

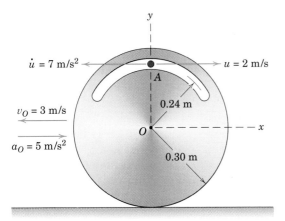

Problem 5/153

5/154 A stationary pole A is viewed by an observer P who is sitting on a small merry-go-round which rotates about a fixed vertical axis at B with a constant angular velocity Ω as shown. Determine the apparent velocity of A as seen by the observer P. Does this velocity depend on the location of the observer on the merry-go-round?

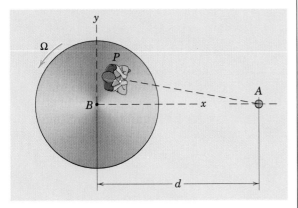

Problem 5/154

5/155 Car B is rounding the curve with a constant speed of 54 km/h, and car A is approaching car B in the intersection with a constant speed of 72 km/h. Determine the velocity which car A appears to have to an observer riding in and turning with car B. The x-y axes are attached to car B. Is this apparent velocity the negative of the velocity that B appears to have to a nonrotating observer in car A? The distance separating the two cars at the instant depicted is 40 m.

Ans. $\mathbf{v}_{\text{rel}} = 20\mathbf{i} - 9\mathbf{j}$ m/s, No

Problem 5/155

5/156 Ship A is proceeding north at a constant speed of 12 knots, and ship B is moving at 10 knots while turning to port (left) at the constant rate of 10 deg/min. When the ships are separated by a distance of 2 nautical miles in the relative positions shown, with B momentarily heading west and A crossing its bow, the navigator of B measures the apparent velocity of A. Find the magnitude of this velocity and specify the clockwise angle β which it makes with the north direction.

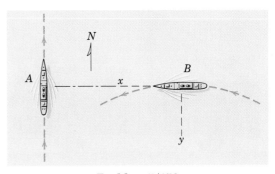

Problem 5/156

5/157 A vehicle A travels west at high speed on a perfectly straight road B which is tangent to the earth's surface at the equator. The road has no curvature whatsoever in the vertical plane. Determine the necessary speed v_{rel} of the vehicle relative to the road which will give rise to zero acceleration of the vehicle in the vertical direction. Assume that the center of the earth has no acceleration.

Ans. $v_{rel} = 837$ km/h

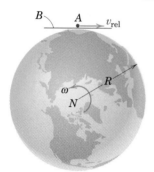

Problem 5/157

Representative problems

5/158 The fire truck is moving forward at a speed of 60 km/h and is decelerating at the rate of 3 m/s². Simultaneously, the ladder is being raised and extended. At the instant considered the angle θ is 30° and is increasing at the constant rate of 10 deg/s. Also at this instant the extension b of the ladder is 1.5 m, with $\dot{b} = 0.6$ m/s and $\ddot{b} = -0.3$ m/s². For this instant determine the acceleration of the end A of the ladder (a) with respect to the truck and (b) with respect to the ground.

Problem 5/158

5/159 The air transport B is flying with a constant speed of 800 km/h in a horizontal arc of 15-km radius. When B reaches the position shown, aircraft A, flying southwest at a constant speed of 600 km/h, crosses the radial line from B to the center of curvature C of its path. Write the vector expression, using the x-y axes attached to B, for the velocity of A as measured by an observer in and turning with B.

Ans. $\mathbf{v}_{rel} = -117.9\mathbf{i} - 222\mathbf{j}$ m/s

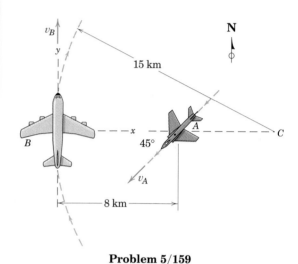

Problem 5/159

5/160 For the conditions of Prob. 5/159, obtain the vector expression for the acceleration which aircraft A appears to have to an observer in and turning with aircraft B, to which axes x-y are attached. Use the results cited in Prob. 5/159 for $\mathbf{v}_{rel}$.

5/161 For an alternative solution to Prob. 5/158 assign r-θ coordinates with origin at B as shown. Then make use of the polar-coordinate relations for the acceleration of A relative to B. The r- and θ- components of the absolute acceleration should coincide with the components along and normal to the ladder which would be found in Prob. 5/158.

Ans. $\mathbf{a}_A = -3.13\mathbf{e}_r + 1.709\mathbf{e}_\theta \text{ m/s}^2$

Problem 5/161

5/162 The slotted disk sector rotates with a constant counterclockwise angular velocity $\omega = 3$ rad/s. Simultaneously the slotted arm OC oscillates about the line OB (fixed to the disk) so that θ changes at the constant rate of 2 rad/s except at the extremities of the oscillation during reversal of direction. Determine the total acceleration of the pin A when $\theta = 30°$ and $\dot\theta$ is positive (clockwise).

Problem 5/162

5/163 For the instant represented, link CB is rotating counterclockwise at a constant rate $N = 4$ rad/s, and its pin A causes a clockwise rotation of the slotted member ODE. Determine the angular velocity ω and angular acceleration α of ODE for this instant.

Ans. $\omega = 4$ rad/s CW, $\alpha = 64.0$ rad/s² CCW

Problem 5/163

5/164 Cars A and B are both traveling on the curved intersecting roads with equal constant speeds of 48 km/h. For the positions shown obtain the vector expressions for the velocity and acceleration which A appears to have to an observer in B who rotates with the car. The x-y axes are attached to car B.

Problem 5/164

5/165 Two satellites are in circular equatorial orbits of different altitudes. Satellite A is in a geosynchronous orbit (one with the same period as the earth's rotation so that it "hovers" over the same spot on the equator). Satellite B has an orbit of radius $r_B = 30\ 000$ km. Calculate the velocity which A appears to have to an observer fixed in B when the elevation angle θ is (a) 0° and (b) 90°. The x-y axes are attached to B whose antenna always points toward the center of the earth ($-y$-direction). Consult Art. 3/13 and Appendix D for necessary orbital information.

$$Ans.\ (a)\ \mathbf{v}_{rel} = 5250\mathbf{i} - 5190\mathbf{j}\ \text{km/h}$$
$$(b)\ \mathbf{v}_{rel} = 7380\mathbf{i}\ \text{km/h}$$

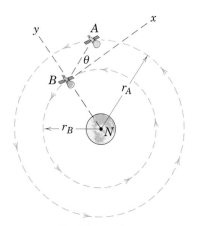

Problem 5/165

5/166 A radar station B situated at the equator observes a satellite A in a circular equatorial orbit of 200-km altitude and moving from west to east. For the instant when the satellite is 30° above the horizon, determine the difference between the velocity of the satellite relative to the radar station, as measured from a nonrotating frame of reference, and the velocity as measured relative to the reference frame of the radar station.

Problem 5/166

5/167 The spring-mounted collar oscillates on the shaft according to $x = 0.04 \sin \pi t$, where x is in meters and t is in seconds. Simultaneously the frame rotates about the bearing at O with an angular velocity $\omega = 2 \sin (\pi t/2)$ rad/s. Determine the acceleration of the center C of the collar (a) when $t = 3$ s and (b) when $t = 1/2$ s.

$$Ans.\ (a)\ \mathbf{a}_C = -0.297\mathbf{j}\ \text{m/s}^2$$
$$(b)\ \mathbf{a}_C = -0.919\mathbf{i} - 0.311\mathbf{j}\ \text{m/s}^2$$

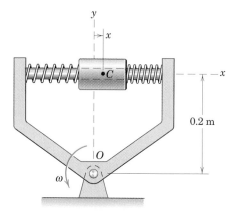

Problem 5/167

5/168 The figure shows the vanes of a centrifugal-pump impeller which turns with a constant clockwise speed of 200 rev/min. The fluid particles are observed to have an absolute velocity whose component in the r-direction is 3 m/s at discharge from the vane. Furthermore, the magnitude of the velocity of the particles measured relative to the vane is increasing at the rate of 24 m/s² just before they leave the vane. Determine the magnitude of the total acceleration of a fluid particle an instant before it leaves the impeller. The radius of curvature ρ of the vane at its end is 200 mm.

Problem 5/168

5/169 A test chamber A is used to study motion sickness. It is capable of oscillation about a horizontal axis through O according to $\theta = \theta_0 \sin 2\pi f_1 t$, and at the same time the chamber has a linear motion $y = y_0 \sin 2\pi f_2 t$ relative to the frame. For a certain series of tests, the amplitudes are set at $\theta_0 = \pi/4$ radians and $y_0 = 150$ mm, while the corresponding frequencies are $f_1 = \frac{1}{4}$ cycle/s and $f_2 = \frac{1}{2}$ cycle/s. Determine the vector expression for the acceleration of point C in the chamber at the instant when $t = 2$ s.

$$\text{Ans. } \mathbf{a}_C = \frac{0.3\pi^3}{8}(\mathbf{i} - \pi\mathbf{j}) \text{ m/s}^2$$

Problem 5/169

5/170 Link OA has a constant clockwise angular velocity of 3 rad/s for a brief interval of its rotation. Determine the angular acceleration α_{BC} of BC for the instant when $\theta = 60°$. First use a rotating-frame analysis, and then verify your result with an absolute-motion approach.

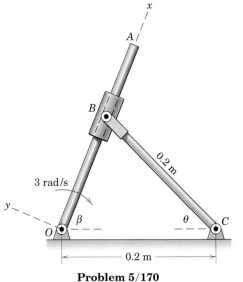

Problem 5/170

5/171 The pin A in the bell crank AOD is guided by the flanges of the collar B, which slides with a constant velocity v_B of 0.9 m/s along the fixed shaft for an interval of motion. For the position $\theta = 30°$ determine the acceleration of the plunger CE, whose upper end is positioned by the radial slot in the bell crank.

$$\text{Ans. } a_C = 24.9 \text{ m/s}^2 \text{ up}$$

Problem 5/171

5/172 The crank OA revolves clockwise with a constant angular velocity of 10 rad/s within a limited arc of its motion. For the position $\theta = 30°$ determine the angular velocity of the slotted link CB and the acceleration of A as measured relative to the slot in CB.

Problem 5/172

5/173 The Geneva wheel of Prob. 5/51 is shown again here. Determine the angular acceleration α_2 of wheel C for the instant when $\theta = 20°$. Wheel A has a constant clockwise angular velocity of 2 rad/s.

Ans. $\alpha_2 = 16.53$ rad/s^2 CCW

Problem 5/173

▶5/174 The mechanism shown is a device to produce high torque in the shaft at O. The gear unit, pivoted at C, turns the right-handed screw at a constant speed $N = 100$ rev/min in the direction shown and advances the threaded collar at A along the screw toward C. Determine the time rate of change $\dot{\omega}_{AO}$ of the angular velocity of AO as it passes the vertical position shown. The screw has 3 single threads per centimeter of length.

Ans. $\dot{\omega}_{AO} = -4.41(10^{-4})$ rad/s^2

Problem 5/174

▶5/175 Determine the angular acceleration of link EC in the position shown, where $\omega = \dot{\beta} = 2$ rad/s and $\ddot{\beta} = 6$ rad/s^2 when $\theta = \beta = 60°$. Pin A is fixed to link EC. The circular slot in link DO has a radius of curvature of 150 mm. In the position shown, the tangent to the slot at the point of contact is parallel to AO.

Ans. $\alpha_{EC} = 12$ rad/s^2 CCW

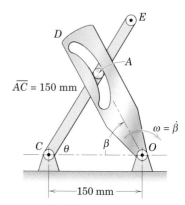

Problem 5/175

▶5/176 The space shuttle A is in an equatorial circular orbit of 240-km altitude and is moving from west to east. Determine the velocity and acceleration which it appears to have to an observer B fixed to and rotating with the earth at the equator as the shuttle passes overhead. Use $R = 6378$ km for the radius of the earth. Also use Fig. 1/1 for the appropriate value of g and carry out your calculations to 4-figure accuracy.

Ans. $\mathbf{v}_{rel} = -26\,220\mathbf{i}$ km/h
$\mathbf{a}_{rel} = -8.018\mathbf{j}$ m/s^2
(using $g = 9.814$ m/s^2)

Problem 5/176

5/8 PROBLEM FORMULATION AND REVIEW

In Chapter 5 we have applied our knowledge of basic kinematics, which we developed in Chapter 2, to the plane motion of rigid bodies. We approached the problem essentially in two ways. First, we wrote an equation which described the general geometric configuration of a given problem in terms of knowns and unknowns. Then we differentiated the equation with respect to time to obtain velocities and accelerations, both linear and angular. We called this approach an absolute-motion analysis, and it is satisfactory when the geometric expression and its time derivatives can be easily written.

In our second approach, we applied the principles of relative motion to rigid bodies and found that this approach enabled us to solve many problems that would be too awkward to handle by mathematical differentiation. The relative-velocity equation, the instantaneous center of zero velocity, and the relative-acceleration equation all require that we visualize clearly and analyze correctly the case of circular motion of one point around another point, as viewed from nonrotating axes. Recognition of the relationships for circular motion is certainly an essential key without which little progress can be made.

The relative-velocity and relative-acceleration relationships are vector equations which we may solve in any one of three ways: by a scalar-geometric analysis of the vector polygon, by vector algebra, or by a graphical construction of the vector polygon. Each method has its advantages, and the student is urged to gain experience with all three. If one uses a vector-algebra approach, he or she should sketch the geometry of the vectors in any event, because the ability to visualize physical-geometrical relationships of force and motion is probably the most important experience of all in the study of mechanics.

In Chapter 6 we shall study the kinetics of rigid bodies in plane motion. Here we find that the ability to solve for the linear and angular accelerations of rigid bodies is absolutely necessary in order to apply the force and moment equations which relate the applied forces to the resulting motions. Thus, the material of Chapter 5 is essential to that in Chapter 6.

Finally, in Chapter 5 we introduced rotating coordinate systems that enable us to solve problems where the motion is observed relative to a rotating frame of reference. Whenever a point moves along a path that itself is turning, analysis by rotating axes is indicated if a relative-motion approach is used. In deriving Eq. 5/12 for relative velocity and Eq. 5/14 for relative acceleration, where the relative terms are measured from a rotating reference system, it was necessary for us to account for the time derivatives of the unit vectors **i** and **j**. Equations 5/12 and 5/14 also apply to spatial motion, as will be shown in Chapter 7.

REVIEW PROBLEMS

5/177 The oscillating produce tray of Prob. 5/83 is shown again here. If the crank OB has a constant counterclockwise angular velocity of 0.944 rad/s, determine the angular velocity of AB when $\theta = 20°$.

Ans. $\omega_{AB} = 0.1133$ rad/s CW

Problem 5/177

5/178 Magnetic tape is fed over and around the light pulleys mounted in a computer frame. If the speed v of the tape is constant and if the ratio of the magnitudes of the acceleration of points A and B is 2/3, determine the radius r of the larger pulley.

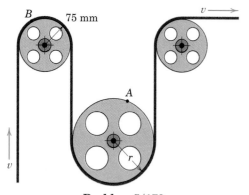

Problem 5/178

5/179 The helicopter is flying in the horizontal x-direction with a velocity $v = 200$ km/h, and the plane of rotation of the 9-m-diameter rotor is tilted 10° from the horizontal x-y plane. The rotor blades rotate with an angular velocity $\Omega = 800$ rev/min. For the instant represented write the vector expressions for the absolute velocities of rotor tip A and rotor tip B.

Ans. $\mathbf{v}_A = 427\mathbf{i} - 65.5\mathbf{k}$ m/s
$\mathbf{v}_B = 55.6\mathbf{i} + 377\mathbf{j}$ m/s

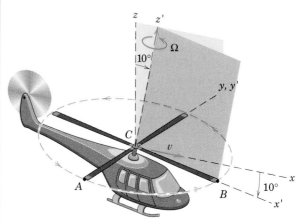

Problem 5/179

5/180 If end A of the rigid link AB has a velocity v_A in the direction shown, show that it is impossible for end B to have a parallel velocity v_B of magnitude less than or greater than v_A.

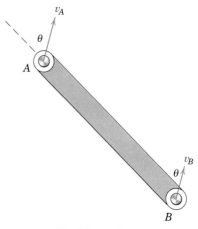

Problem 5/180

5/181 The load L is being elevated by the downward velocities of ends A and B of the cable. Determine the acceleration of point P on the top of the sheave for the instant when $v_A = 0.6$ m/s, $\dot{v}_A = 0.15$ m/s^2, $v_B = 0.9$ m/s, and $\dot{v}_B = -0.15$ m/s^2.

Ans. $a_P = 3/16$ m/s^2

Problem 5/181

5/182 The large power-cable reel is rolled up the incline by the vehicle as shown. The vehicle starts from rest with $x = 0$ for the reel and accelerates at the constant rate of 0.6 m/s^2. For the instant when $x = 1.8$ m, calculate the acceleration of point P on the reel in the position shown.

Problem 5/182

5/183 At the instant represented $x = 50$ mm and $\dot{s} = 1.6$ m/s. Determine the corresponding velocity of point B.

Ans. $v_B = 1.029$ m/s

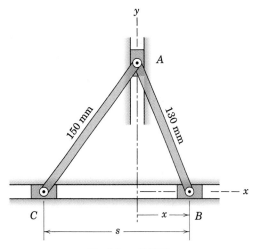

Problem 5/183

5/184 Tape is being transferred from reel A to reel B. If reel B rotates with a constant angular velocity ω_0, determine an expression for the angular acceleration $\alpha = \dot{\omega}$ of reel A at any instant when the tape radii of A and B are r and r_0, respectively. The thickness of the tape is b. Neglect the very small angular motion of the tape between the reels.

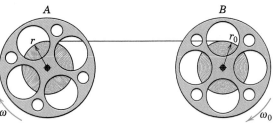

Problem 5/184

5/185 The tilting device maintains a sloshing water bath for washing vegetable produce. If the crank *OA* oscillates about the vertical and has a clockwise angular velocity of 4π rad/s when *OA* is vertical, determine the angular velocity of the basket in the position shown where $\theta = 30°$.

Ans. $\omega_{DB} = 3.24$ rad/s CW

Problem 5/185

5/186 Determine the angular acceleration of the basket of the vegetable washer of Prob. 5/185 for the position where *OA* is vertical. In this position *OA* has an angular velocity of 4π rad/s and no angular acceleration.

5/187 Each of the sliding blocks has a constant velocity 0.6 m/s, *A* to the right and *C* to the left. Determine the magnitude of the velocity of pin *B* at the instant when $x = 120$ mm.

Ans. $v_B = 0.512$ m/s

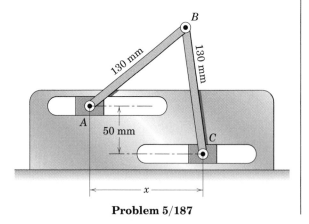

Problem 5/187

5/188 The isosceles triangular plate is guided by the two vertex rollers *A* and *B* which are confined to move in the perpendicular slots. The control rod gives *A* a constant velocity v_A to the left for an interval of motion. Determine the value of θ for which the horizontal component of the velocity of *C* is zero.

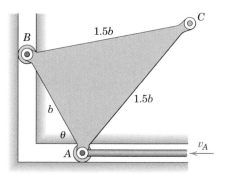

Problem 5/188

5/189 The figure illustrates a commonly used quick-return mechanism which produces a slow cutting stroke of the tool (attached to *D*) and a rapid return stroke. If the driving crank *OA* is turning at the constant rate $\dot{\theta} = 3$ rad/s, determine the magnitude of the velocity of point *B* for the instant when $\theta = 30°$.

Ans. $v_B = 288$ mm/s

Problem 5/189

5/190 The wheel rolls on the circular surface without slipping. In the bottom position, it has an angular velocity ω and an angular acceleration α, both clockwise. For this position, obtain expressions for the acceleration of point C on the wheel in contact with the path and for the acceleration of point A.

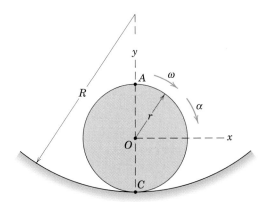

Problem 5/190

5/191 For the position shown where $\theta = 30°$, point A on the sliding collar has a constant velocity $v = 0.3$ m/s with corresponding lengthening of the hydraulic cylinder AC. For this same position BD is horizontal and DE is vertical. Determine the angular acceleration α_{DE} of DE at this instant.

Ans. $\alpha_{DE} = 2.45$ rad/s² CCW

Problem 5/191

▶5/192 Near the end of its takeoff roll, the airplane is "rotating" (nose pitching up) just prior to lift-off. The velocity and acceleration of the aircraft, expressed in terms of the motion of the wheel assembly C, are v_C and a_C, both directed horizontally forward. The pitch angle is θ, and the pitch rate $\omega = \dot{\theta}$ is increasing at the rate $\alpha = \dot{\omega}$. If a person A is walking forward in the center aisle with a velocity and an acceleration v_{rel} and a_{rel}, both measured forward relative to the cabin, derive expressions for the velocity and acceleration of A as observed by a ground-fixed observer.

Ans. $\mathbf{v}_A = (v_C \cos\theta - \omega h + \dot{L})\mathbf{i} +$
$(\omega L - v_C \sin\theta)\mathbf{j}$
$\mathbf{a}_A = (a_C \cos\theta - \alpha h - L\omega^2 + \ddot{L})\mathbf{i} +$
$(-a_C \sin\theta - h\omega^2 + L\alpha + 2\omega\dot{L})\mathbf{j}$

Problem 5/192

▶5/193 The crank OB revolves clockwise at the constant rate ω_0 of 5 rad/s. For the instant when $\theta = 90°$ determine the angular acceleration α of the rod BD, which slides through the pivoted collar at C.

Ans. $\alpha = 6.25$ rad/s² CW

Problem 5/193

*Computer-oriented problems

*5/194 For the Geneva wheel of Prob. 5/51, shown again here, write the expression for the angular velocity ω_2 of the slotted wheel C during engagement of pin P and plot ω_2 for the range $-45° \le \theta \le 45°$. The driving wheel A has a constant angular velocity $\omega_1 = 2$ rad/s.

$$200/\sqrt{2} \text{ mm} \qquad 200/\sqrt{2} \text{ mm}$$

Problem 5/194

*5/195 The mechanism of Prob. 5/84 is repeated here. For a constant speed $N = 120$ rev/min of crank CA, determine and plot the angular velocity $\dot\beta$ of arm OB as a function of θ through 360°, where β is the angle between OC and OB. Find θ for zero angular velocity of OB.

$$\text{Ans. } \dot\beta = 2\pi \frac{5\cos\theta - 1}{13 - 5\cos\theta}, \theta = 78.5°, 282°$$

625 mm

125 mm

Problem 5/195

*5/196 A constant torque M exceeds the moment about O due to the force F on the plunger, and an angular acceleration $\ddot\theta = 100(1 - \cos\theta)$ rad/s² results. If the crank OA is released from rest at B, where $\theta = 30°$, and strikes the stop at C, where $\theta = 150°$, plot the angular velocity $\dot\theta$ as a function of θ and find the time t for the crank to rotate from $\theta = 90°$ to $\theta = 150°$.

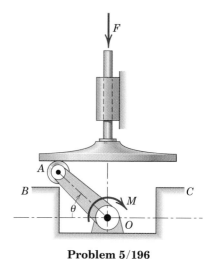

Problem 5/196

*5/197 The crank rotates clockwise at the content rate $\dot\theta = 3$ rad/s. The connecting link AB passes through the pivoted collar at C. Determine the maximum velocity v at which AB passes through the collar and the corresponding value of θ. Plot $\overline{l}$ vs. θ for $0 \le \theta \le 180°$, where l is the distance $\overline{AC}$.

$$\text{Ans. } v = 240 \text{ mm/s at } \theta = 70.5°$$

80 mm

240 mm

Problem 5/197

*5/198 For the slider-crank configuration shown, derive the expression for the velocity v_A of the piston (taken positive to the right) as a function of θ. Substitute the numerical data of Sample Problem 5/15 and calculate v_A as a function of θ for $0 \leq \theta \leq 180°$. Plot v_A versus θ and find its maximum magnitude and the corresponding value of θ. (By symmetry anticipate the results for $180° \leq \theta \leq 360°$.)

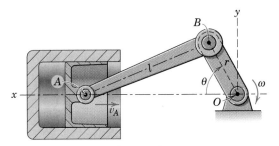

Problem 5/198

*5/199 For the slider-crank of Prob. 5/198 derive the expression for the acceleration a_A of the piston (taken positive to the right) as a function of θ for $\omega = \dot{\theta} = $ constant. Substitute the numerical data of Sample Problem 5/15 and calculate a_A as a function of θ for $0 \leq \theta \leq 180°$. Plot a_A versus θ and find the value of θ for which $a_A = 0$. (By symmetry anticipate the results for $180° \leq \theta \leq 360°$.)

Ans. $a_A = r\omega^2 \left[\cos \theta + \dfrac{r}{l} \dfrac{1 - 2 \sin^2 \theta + \dfrac{r^2}{l^2} \sin^4 \theta}{(1 - \dfrac{r^2}{l^2} \sin^2 \theta)^{3/2}} \right]$

$a_A = 0$ at $\theta = 72.3°$

*5/200 Bar OA rotates about the fixed pivot O with constant angular velocity $\dot{\beta} = 0.8$ rad/s. Pin A is fixed to bar OA and is engaged in the slot of member BD, which rotates about a fixed axis through point B. Determine and plot over the range $0 \leq \beta \leq 360°$ the angular velocity and angular acceleration of BD and the velocity and acceleration of pin A relative to member BD.

Problem 5/200

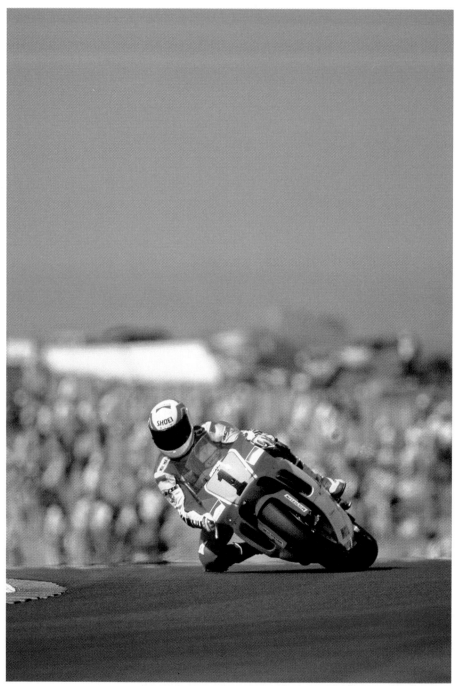

The forces that cause or accompany the motion of a rigid body, which may translate and rotate simultaneously, must be calculated to establish design criteria. The motorcycle may be simulated as a rigid body where the forward and lateral forces exerted by the road on the tires and the weight of the motorcycle give it its acceleration in rounding the curve.

PLANE KINETICS OF RIGID BODIES

6

6/1 INTRODUCTION

The kinetics of rigid bodies treats the relationships between the forces which act on the bodies from sources external to their boundaries and the corresponding translational and rotational motions of the bodies. In Chapter 5 we developed the kinematic relationships for the plane motion of rigid bodies, and we will use these relationships extensively in this present chapter, where the effects of forces on the two-dimensional motion of rigid bodies are examined.

For our purpose in Chapter 6, a body which can be approximated as a thin slab with its motion confined to the plane of the slab will be treated under the category of plane motion. The plane of motion will contain the mass center, and all forces which act on the body will be projected onto the plane of motion. A body which has appreciable dimensions normal to the plane of motion but is symmetrical about that plane of motion through the mass center may be treated as having plane motion. These idealizations clearly fit a very large category of rigid-body motions.

In Chapter 3 we found that two force equations of motion were required to define the plane motion of a particle whose motion has two linear components. For the plane motion of a rigid body, an additional equation is needed to specify the state of rotation of the body. Thus, two force equations and one moment equation or their equivalent are required to determine the state of rigid-body plane motion.

The relationships which form the basis for most of the analysis of rigid-body motion were developed in Chapter 4 for a general system of particles. Frequent reference will be made to these equations as they are further developed in Chapter 6 and applied specifically to the plane motion of rigid bodies. It is advisable to refer to the developments of Chapter 4 frequently as Chapter 6 is studied. Also, it is inadvisable to proceed further until a firm grasp of the calculation of velocities and accelerations as developed in Chapter 5 for rigid-body plane motion is well in hand. Without the ability to determine accelerations correctly from the principles of kinematics, it is frequently useless to attempt application of the force and moment principles of motion. Consequently, it is essential to master the necessary kinematics, including the calculation of relative accelerations, before proceeding.

Basic to the approach to kinetics is the isolation of the body or system to be analyzed. This isolation was illustrated and used in Chapter 3 for particle kinetics and will be employed consistently in the present chapter. For problems involving the instantaneous relationships among force, mass, and acceleration or momentum, the body or system should be explicitly defined by isolating it with its *free-body diagram*. When the principles of work and energy are employed, an *active-force diagram* which shows only those external forces which do work on the system may be used in lieu of the free-body diagram. *No solution of a problem should be attempted without first defining the complete external boundary of the body or system and identifying all external forces which act on it.*

Chapter 6 is organized into the same three sections in which we treated the kinetics of particles in Chapter 3. Section A relates the forces and moments to the instantaneous linear and angular accelerations. Section B treats the solution of problems by the method of work and energy. Section C covers the methods of impulse and momentum. Virtually all of the basic concepts and approaches covered in these three sections were treated in Chapter 3 on particle kinetics. This repetition will result in accelerated progress in Chapter 6, provided the kinematics of rigid-body plane motion is well in hand. In each of the three sections, the three types of motion, namely, translation, fixed-axis rotation, and general plane motion, will be treated.

In the kinetics of rigid bodies which have angular motion, it is necessary to introduce a property of the body which accounts for the radial distribution of its mass with respect to a particular axis of rotation normal to the plane of motion. This property is known as the *mass moment of inertia* of the body, and it is essential that we be able to calculate this property in order to solve rotational problems. Familiarity with the calculation of mass moments of inertia is assumed of the reader at this point. Appendix B treats this topic for those who need instruction or review.

SECTION A. FORCE, MASS, AND ACCELERATION

6/2 GENERAL EQUATIONS OF MOTION

In Arts. 4/2 and 4/4 we derived the force and moment vector equations of motion for a perfectly general system of mass. We now apply these results by starting, first, with a general rigid body in three dimensions. The force equation, Eq. 4/1,

$$\Sigma \mathbf{F} = m\bar{\mathbf{a}} \qquad [4/1]$$

tells us that the resultant $\Sigma\mathbf{F}$ of the external forces acting on the body equals the mass m of the body times the acceleration $\bar{\mathbf{a}}$ of its mass center G. The moment equation taken about the mass center, Eq. 4/9,

$$\Sigma\mathbf{M}_G = \dot{\mathbf{H}}_G \qquad [4/9]$$

tells us that the resultant moment about the mass center of the external forces on the body equals the time rate of change of the angular momentum of the body about the mass center.

We recall from our study of statics that a general system of forces acting on a rigid body may be replaced by a resultant force applied at a chosen point and a corresponding couple. By replacing the external forces by their equivalent force-couple system with the resultant force acting through the mass center, we may then visualize the action of the forces and the corresponding dynamic response of the body with the aid of Fig. 6/1. The a-part of the figure represents the free-body diagram of the body. The b-part of the figure shows the equivalent force-couple system with the resultant force applied through G. The c-part of the figure represents the

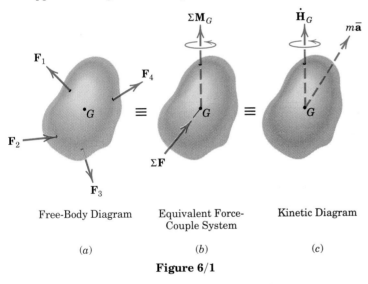

Free-Body Diagram   Equivalent Force-   Kinetic Diagram
                 Couple System

        (a) (b) (c)

Figure 6/1

resulting dynamic effects as specified by Eqs. 4/1 and 4/9 and is termed here the *kinetic diagram*. The equivalence between the free-body diagram and the kinetic diagram enables us to clearly visualize and easily remember the separate translational and rotational effects of the forces applied to a rigid body. We shall express this equivalence mathematically as we apply these results to the treatment of rigid-body plane motion which follows.

Plane-motion equations. We now apply the foregoing relationships to the case of plane motion. Figure 6/2 represents a rigid body moving with plane motion in the x-y plane. The mass center

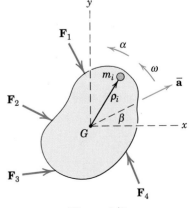

Figure 6/2

G has an acceleration $\bar{\mathbf{a}}$, and the body has an angular velocity $\boldsymbol{\omega} = \omega\mathbf{k}$ and an angular acceleration $\boldsymbol{\alpha} = \alpha\mathbf{k}$, both taken in the positive z-direction. Because the z-direction of both $\boldsymbol{\omega}$ and $\boldsymbol{\alpha}$ remains perpendicular to the plane of motion, we may use scalar notation ω and $\alpha = \dot{\omega}$ to represent the angular velocity and angular acceleration.

The angular momentum about the mass center for the general system was expressed in Eq. 4/8a as $\mathbf{H}_G = \Sigma\boldsymbol{\rho}_i \times m_i\dot{\boldsymbol{\rho}}_i$ where $\boldsymbol{\rho}_i$ is the position vector relative to G of the representative particle of mass m_i. For our rigid body, the velocity of m_i relative to G is $\dot{\boldsymbol{\rho}}_i = \boldsymbol{\omega} \times \boldsymbol{\rho}_i$, which has a magnitude $\rho_i\omega$ and lies in the plane of motion normal to $\boldsymbol{\rho}_i$. The product $\boldsymbol{\rho}_i \times \dot{\boldsymbol{\rho}}_i$ is then a vector normal to the x-y plane in the sense of $\boldsymbol{\omega}$, and its magnitude is $\rho_i^2\omega$. Thus, the magnitude of $\mathbf{H}_G$ becomes $H_G = \Sigma\rho_i^2 m_i\omega = \omega\Sigma\rho_i^2 m_i$. The summation, which may also be written as $\int \rho^2\,dm$, is defined as the *mass moment of inertia* $\bar{I}$ of the body about the z-axis through G. (See Appendix B for a discussion of the calculation of mass moments of inertia.) We may now write

$$H_G = \bar{I}\omega$$

where $\bar{I}$ is a constant property of the body and is a measure of the rotational inertia or resistance to change in rotational velocity due to the radial distribution of mass around the z-axis through G. With this substitution, our moment equation, Eq. 4/9, becomes

$$\Sigma M_G = \dot{H}_G = \bar{I}\dot{\omega} = \bar{I}\alpha$$

where $\alpha = \dot{\omega}$ is the angular acceleration of the body.

The moment equation and the vector form of the generalized Newton's second law of motion, Eq. 4/1, are now written

$$\boxed{\begin{aligned} \Sigma\mathbf{F} &= m\bar{\mathbf{a}} \\[4pt] \Sigma M_G &= \bar{I}\alpha \end{aligned}}$$

(6/1)

Equations 6/1 are the general equations of motion for a rigid body in plane motion. In applying Eqs. 6/1, we express the vector force equation in terms of its two scalar components using x-y, n-t, or r-θ coordinates according to the description that is most convenient for the problem at hand.

As an alternative approach, it is instructive to derive the moment equation by referring directly to the forces which act on the representative particle of mass m_i, as shown in Fig. 6/3. The acceleration of m_i equals the vector sum of $\bar{a}$ and the relative terms $\rho_i\omega^2$ and $\rho_i\alpha$, where the mass center G is used as the reference point. It follows that the resultant of all forces on m_i has the components $m_i\bar{a}$, $m_i\rho_i\omega^2$, and $m_i\rho_i\alpha$ in the directions shown. The sum of the moments of these force components about G in the sense of α becomes

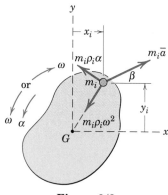

Figure 6/3

$$M_{G_i} = m_i\rho_i^2\alpha + (m_i\bar{a}\sin\beta)x_i - (m_i\bar{a}\cos\beta)y_i$$

Similar moment expressions exist for all particles in the body, and the sum of these moments about G for the resultant forces acting on all particles may be written as

$$\Sigma M_G = \Sigma m_i\rho_i^2\alpha + \bar{a}\sin\beta\,\Sigma m_ix_i - \bar{a}\cos\beta\,\Sigma m_iy_i$$

But the origin of coordinates is taken at the mass center, so that $\Sigma m_ix_i = m\bar{x} = 0$ and $\Sigma m_iy_i = m\bar{y} = 0$. Thus, the moment sum becomes

$$\Sigma M_G = \Sigma m_i\rho_i^2\alpha = \bar{I}\alpha$$

as before. The contribution to ΣM_G of the forces internal to the body is, of course, zero since they occur in pairs of equal and opposite forces of action and reaction between interacting particles. Thus, ΣM_G, as before, represents the sum of moments about the mass center G of only the external forces acting on the body, as disclosed by the free-body diagram. We note that the force component $m_i\rho_i\omega^2$ has no moment about G and conclude, therefore, that the angular velocity ω has no influence on the moment equation about the mass center G.

The results embodied in our basic equations of motion for a rigid body in plane motion, Eqs. 6/1, are represented diagrammatically in Fig. 6/4, which is the two-dimensional counterpart of the *a*- and *c*-parts of Fig. 6/1 for a general three-dimensional body. The free-body diagram discloses the forces and moments appearing on

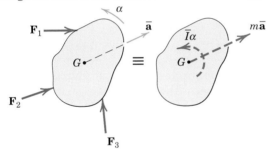

Free-Body Diagram   Kinetic Diagram

Figure 6/4

the left-hand side of our equations of motion. The kinetic diagram discloses the resulting dynamic response in terms of the translational term $m\bar{\mathbf{a}}$ and the rotational term $\bar{I}\alpha$ which appear on the right-hand side of Eqs. 6/1. As previously mentioned, the translational term $m\bar{\mathbf{a}}$ will be expressed by its *x-y*, *n-t*, or *r-θ* components once the appropriate inertial reference system is designated. The equivalence depicted in Fig. 6/4 is basic to our understanding of the kinetics of plane motion and will be employed frequently in the solution of problems. Representation of the resultants $m\bar{\mathbf{a}}$ and $\bar{I}\alpha$ will help ensure that the force and moment sums determined from the free-body diagram are equated to their proper resultants.

Alternative moment equation. In Art. 4/4 of Chapter 4 on systems of particles, we developed a general moment equation about an arbitrary point P, Eq. 4/11, which is

$$\Sigma \mathbf{M}_P = \dot{\mathbf{H}}_G + \bar{\boldsymbol{\rho}} \times m\bar{\mathbf{a}} \qquad [4/11]$$

where $\bar{\boldsymbol{\rho}}$ is the vector from P to the mass center G and $\bar{\mathbf{a}}$ is the mass-center acceleration. As we have shown earlier in this article, for a rigid body in plane motion $\dot{\mathbf{H}}_G$ becomes $\bar{I}\alpha$. Also, the cross product $\bar{\boldsymbol{\rho}} \times m\bar{\mathbf{a}}$ is simply the moment of magnitude $m\bar{a}d$ of $m\bar{\mathbf{a}}$ about P. Therefore, for the two-dimensional body illustrated in Fig. 6/5 with its free-body diagram and kinetic diagram, we may rewrite Eq. 4/11 simply as

$$\boxed{\Sigma M_P = \bar{I}\alpha + m\bar{a}d} \qquad (6/2)$$

Free-Body Diagram   Kinetic Diagram

Figure 6/5

Clearly, all three terms are positive in the counterclockwise sense for the example shown, and the choice of P eliminates reference to $\mathbf{F}_1$ and $\mathbf{F}_3$. If we had wished to eliminate reference to $\mathbf{F}_2$ and $\mathbf{F}_3$, for example, by choosing their intersection as the reference point, then P would lie on the opposite side of the $m\bar{\mathbf{a}}$ vector, and the clockwise moment of $m\bar{\mathbf{a}}$ about P would be a negative term in the equation. Equation 6/2 is easily remembered as it is merely an expression of the familiar principle of moments, where the sum of the moments about P equals the combined moment about P of their sum, expressed by the resultant couple $\Sigma M_G = \bar{I}\alpha$ and the resultant force $\Sigma \mathbf{F} = m\bar{\mathbf{a}}$.

In Art. 4/4 we also developed an alternative moment equation about P, Eq. 4/13, which is

$$\Sigma \mathbf{M}_P = (\dot{\mathbf{H}}_P)_{\text{rel}} + \bar{\boldsymbol{\rho}} \times m\mathbf{a}_P \qquad [4/13]$$

For rigid-body plane motion, if P is chosen as a point *fixed* to the body, then in scalar form $(\dot{\mathbf{H}}_P)_{\text{rel}}$ becomes $I_P\alpha$, where I_P is the mass moment of inertia about an axis through P and α is the angular acceleration of the body. So we may write the equation as

$$\boxed{\Sigma \mathbf{M}_P = I_P \alpha + \bar{\rho} \times m\mathbf{a}_P} \qquad (6/3)$$

where the acceleration of P is $\mathbf{a}_P$ and the position vector from P to G is $\bar{\rho}$. When $\bar{\rho} = \mathbf{0}$, point P becomes the mass center G, and Eq. 6/3 reduces to the scalar form $\Sigma M_G = \bar{I}\alpha$, previously derived. When point P becomes a point O fixed in an inertial reference system and attached to the body (or body extended), then $\mathbf{a}_P \equiv \mathbf{0}$, and Eq. 6/3 in scalar form reduces to

$$\boxed{\Sigma M_O = I_O \alpha} \qquad (6/4)$$

Equation 6/4 then applies to the rotation of a rigid body about a nonaccelerating point O fixed to the body and is the two-dimensional simplification of Eq. 4/7.

Unconstrained and constrained motion. The motion of a rigid body may be unconstrained or constrained. The rocket moving in a vertical plane, Fig. 6/6a, is an example of unconstrained motion as there are no physical confinements to its motion. The two components $\bar{a}_x$ and $\bar{a}_y$ of the mass-center acceleration and the angular acceleration α may be determined independently of one another by direct application of Eqs. 6/1. The bar in Fig. 6/6b, on the other hand, represents a constrained motion, where the vertical and horizontal guides for the ends of the bar impose a kinematic relationship between the acceleration components of the mass center and the angular acceleration of the bar. Thus, it is necessary to determine this kinematic relationship from the principles established in Chapter 5 and to combine it with the force and moment equations of motion before a solution can be carried out. In general, dynamics problems which involve physical constraints to motion require a kinematic analysis relating linear to angular acceleration before the force and moment equations of motion can be solved. It is for this

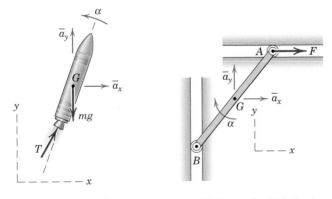

(a) Unconstrained Motion   (b) Constrained Motion

Figure 6/6

reason that an understanding of the principles and methods of Chapter 5 is so vital to the work of Chapter 6.

System of interconnected bodies. Upon occasion, in problems dealing with two or more connected rigid bodies whose motions are related kinematically, it is convenient to analyze the bodies as an entire system. Figure 6/7 illustrates two rigid bodies hinged at *A* and subjected to the external forces shown. The forces in the connection at *A* are internal to the system and are not disclosed.

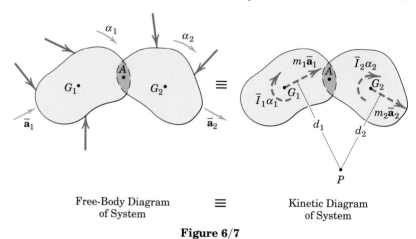

Free-Body Diagram   ≡   Kinetic Diagram
of System   of System

Figure 6/7

The resultant of all external forces must equal the vector sum of the two resultants $m_1\bar{\mathbf{a}}_1$ and $m_2\bar{\mathbf{a}}_2$, and the sum of the moments about some arbitrary point such as *P* of all external forces must equal the moment of the resultants, $\bar{I}_1\alpha_1 + \bar{I}_2\alpha_2 + m_1\bar{a}_1d_1 + m_2\bar{a}_2d_2$. Thus, we may state

$$\Sigma\mathbf{F} = \Sigma m\bar{\mathbf{a}}$$

$$\Sigma M_P = \Sigma\bar{I}\alpha + \Sigma m\bar{a}d$$

(6/5)

where the summations on the right-hand side of the equations represent as many terms as there are separate bodies. If there are more than three remaining unknowns in a system, however, the three independent scalar equations of motion, when applied to the system, are not sufficient to solve the problem. In this case, more advanced methods such as virtual work (Art. 6/7) or Lagrange's equations (not discussed in this book*) could be employed or else the

*When an interconnected system has more than one degree of freedom, that is, requires more than one coordinate to specify completely the configuration of the system, the more advanced equations of Lagrange are generally used. See the first author's *Dynamics, 2nd Edition SI Version*, 1975, John Wiley & Sons, for a treatment of Lagrange's equations.

system dismembered and each part analyzed separately with the resulting equations solved simultaneously.

Analysis procedure. In the solution of force-mass-acceleration problems for the plane motion of rigid bodies, the following steps should be taken after the conditions and requirements of the problem are clearly in mind.

(1) Kinematics. First, identify the class of motion and then solve for any needed linear and angular accelerations which can be determined solely from given kinematic information. In the case of constrained plane motion, it is usually necessary to establish the relation between the linear acceleration of the mass center and the angular acceleration of the body by first solving the appropriate relative-velocity and relative-acceleration equations. Again, it is emphasized that success in working force-mass-acceleration problems in this chapter is contingent on the ability to describe the necessary kinematics, so that frequent review of Chapter 5 is recommended.

(2) Diagrams. Always draw the complete free-body diagram of the body to be analyzed. Assign a convenient inertial coordinate system and label all known and unknown quantities. The kinetic diagram should also be constructed so as to clarify the equivalence between the applied forces and the resulting dynamic response.

(3) Equations of motion. Apply the three equations of motion from Eqs. 6/1, being consistent with the algebraic signs in relation to the choice of reference axes. Equation 6/2 or 6/3 may be employed as an alternative to one of Eqs. 6/1. Combine these relations with the results from any needed kinematic analysis. Count the number of unknowns and be certain that there are an equal number of independent equations available. For a solvable rigid-body problem in plane motion, there can be no more than the five scalar unknowns which can be determined from the three scalar equations of motion, obtained from Eqs. 6/1, and the two scalar component relations that come from the relative-acceleration equation.

The sequence of steps to be followed in any given problem will depend on the conditions of the problem. To organize a solution in a logical sequence of steps, the student should recognize the dependency of a required quantity on intermediate quantities and then trace the dependencies back to the given information. This establishes the reverse order of the steps to be followed in carrying out the actual solution.

The foregoing developments will now be applied to the three cases of motion in a plane, namely, translation, fixed-axis rotation, and general plane motion in the three articles which follow.

6/3 TRANSLATION

Rigid-body translation in plane motion was described in Art. 5/1 and illustrated in Figs. 5/1*a* and 5/1*b*, where we saw that every line in a translating body remains parallel to its original position at all times. In rectilinear translation all points move in straight lines, whereas in curvilinear translation all points move on congruent curved paths. In either case, there is no angular motion of the translating body, so that both ω and α are zero. Therefore, from the moment relation of Eqs. 6/1, we see that all reference to the moment of inertia is eliminated for a translating body.

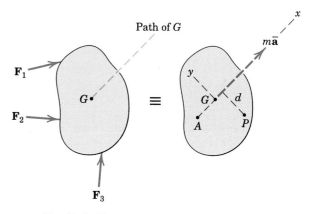

Free-Body Diagram   Kinetic Diagram

(*a*) Rectilinear Translation
($\alpha = 0$, $\omega = 0$)

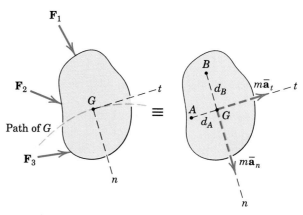

Free-Body Diagram   Kinetic Diagram

(*b*) Curvilinear Translation
($\alpha = 0$, $\omega = 0$)

Figure 6/8

For a translating body, then, our general equations for plane motion, Eqs. 6/1, may be written

$$\Sigma \mathbf{F} = m\bar{\mathbf{a}}$$

$$\Sigma M_G = \bar{I}\alpha = 0$$

(6/6)

For rectilinear translation, illustrated in Fig. 6/8a, if the x-axis is chosen in the direction of the acceleration, then the two scalar force equations become $\Sigma F_x = m\bar{a}_x$ and $\Sigma F_y = m\bar{a}_y = 0$. For curvilinear translation, Fig. 6/8b, if we use n-t coordinates, the two scalar force equations become $\Sigma F_n = m\bar{a}_n$ and $\Sigma F_t = m\bar{a}_t$. In both cases, $\Sigma M_G = 0$.

We may also employ the alternative moment equation, Eq. 6/2, with the aid of the kinetic diagram. For rectilinear translation we see that $\Sigma M_P = m\bar{a}d$ and $\Sigma M_A = 0$. For curvilinear translation the kinetic diagram permits us to write $\Sigma M_A = m\bar{a}_n d_A$ in the clockwise sense and $\Sigma M_B = m\bar{a}_t d_B$ in the counterclockwise sense. Thus, we have complete freedom to choose a convenient moment center.

Sample Problem 6/1

The 1500-kg pickup truck reaches a speed of 50 km/h from rest in a distance of 60 m up the 10-percent incline with constant acceleration. Calculate the normal force under each pair of wheels and the friction force under the rear driving wheels. The effective coefficient of friction between the tires and the road is known to be at least 0.8.

600 mm

1500 mm   1500 mm

10

Solution. We will assume that the mass of the wheels is negligible compared with the total mass of the truck. The truck may now be simulated by a single rigid body in rectilinear translation with an acceleration of

$[v^2 = 2as]$ $\qquad \bar{a} = \dfrac{(50/3.6)^2}{2(60)} = 1.608 \text{ m/s}^2$

The free-body diagram of the complete truck shows the normal forces N_1 and N_2, the friction force F in the direction to oppose the slipping of the driving wheels, and the weight W represented by its two components. With $\theta = \tan^{-1} 1/10 = 5.71°$, these two components are $W \cos \theta = 1500(9.81) \cos 5.71° = 14.64(10^3)$ N and $W \sin \theta = 1500(9.81) \sin 5.71° = 1464$ N. The kinetic diagram shows the resultant, which passes through the mass center and is in the direction of its acceleration. Its magnitude is

$$m\bar{a} = 1500(1.608) = 2410 \text{ N}$$

Applying the three equations of motion, Eqs. 6/1, for the three unknowns gives

$[\Sigma F_x = m\bar{a}_x]$ $\qquad F - 1464 = 2410 \qquad F = 3880 \text{ N} \qquad$ *Ans.*

$[\Sigma F_y = m\bar{a}_y = 0]$ $\qquad N_1 + N_2 - 14.64(10^3) = 0 \qquad$ *(a)*

$[\Sigma M_G = \bar{I}\alpha = 0]$ $\qquad 1.5N_1 + 3880(0.6) - N_2(1.5) = 0 \qquad$ *(b)*

Solving *(a)* and *(b)* simultaneously gives

$$N_1 = 6550 \text{ N} \qquad N_2 = 8100 \text{ N} \qquad \text{*Ans.*}$$

In order to support a friction force of 3880 N, a coefficient of friction of at least $F/N_2 = 3880/8100 = 0.48$ is required. Since our coefficient of friction is at least 0.8, the surfaces are rough enough to support the calculated value of F so that our result is correct.

Alternative solution. From the kinetic diagram we see that N_1 and N_2 can be obtained independently of one another by writing separate moment equations about A and B.

$[\Sigma M_A = m\bar{a}d]$ $\quad 3N_2 - 1.5(14.64)(10^3) - 0.6(1464) = 2410(0.6)$

$$N_2 = 8100 \text{ N} \qquad \text{*Ans.*}$$

$[\Sigma M_B = m\bar{a}d]$ $\quad 14.64(10^3)(1.5) - 1464(0.6) - 3N_1 = 2410(0.6)$

$$N_1 = 6550 \text{ N} \qquad \text{*Ans.*}$$

① Without this assumption, we would be obliged to account for the relatively small additional forces which produce moments to give the wheels their angular acceleration.

② Recall that 3.6 km/h is 1 m/s.

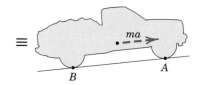

③ We must be careful not to use the friction equation $F = \mu N$ here since we do not have a case of slipping or impending slipping. If the given coefficient of friction were less than 0.48, the friction force would be μN_2, and the car would be unable to attain the acceleration of 1.608 m/s². In this case, the unknowns would be N_1, N_2, and a.

④ The left-hand side of the equation is evaluated from the free-body diagram, and the right-hand side from the kinetic diagram. The positive sense for the moment sum is arbitrary but must be the same for both sides of the equation. In this problem, we have taken the clockwise sense as positive for the moment of the resultant force about B.

Sample Problem 6/2

The vertical bar AB has a mass of 150 kg with center of mass G midway between the ends. The bar is elevated from rest at $\theta = 0$ by means of the parallel links of negligible mass, with a constant couple $M = 5$ kN·m applied to the lower link at C. Determine the angular acceleration α of the links as a function of θ and find the force B in the link DB at the instant when $\theta = 30°$.

Solution. The motion of the bar is seen to be curvilinear translation since the bar itself does not rotate during the motion. With the circular motion of the mass center G, we choose n- and t-coordinates as the most convenient description. With negligible mass of the links, the tangential component A_t of the force at A is obtained from the free-body diagram of AC, where $\Sigma M_C \cong 0$ and $A_t = M/\overline{AC} = 5/1.5 = 3.33$ kN. ① The force at B is along the link. All applied forces are shown on the free-body diagram of the bar, and the kinetic diagram is also indicated, where the $m\overline{a}$ resultant is shown in terms of its two components. ②

The sequence of solution is established by noting that A_n and B depend on the n-summation of forces and, hence, on $m\overline{r}\omega^2$ at $\theta = 30°$. The value of ω depends on the variation of $\alpha = \ddot{\theta}$ with θ. This dependency is established from a force summation in the t-direction for a general value of θ, where $\overline{a}_t = (\overline{a}_t)_A = \overline{AC}\alpha$. Thus, we begin with

$$[\Sigma F_t = m\overline{a}_t] \qquad 3.33 - 0.15(9.81) \cos \theta = 0.15(1.5\alpha)$$

$$\alpha = 14.81 - 6.54 \cos \theta \text{ rad/s}^2 \qquad Ans.$$

With α a known function of θ, the angular velocity ω of the links is obtained from

$$[\omega\, d\omega = \alpha\, d\theta] \qquad \int_0^\omega \omega\, d\omega = \int_0^\theta (14.81 - 6.54 \cos \theta)\, d\theta$$

$$\omega^2 = 29.60 - 13.08 \sin \theta$$

Substitution for $\theta = 30°$ gives

$$(\omega^2)_{30°} = 8.97 \text{ (rad/s)}^2 \qquad \alpha_{30°} = 9.15 \text{ rad/s}^2$$

and

$$m\overline{r}\omega^2 = 0.15(1.5)(8.97) = 2.02 \text{ kN}$$

$$m\overline{r}\alpha = 0.15(1.5)(9.15) = 2.06 \text{ kN}$$

The force B may be obtained by a moment summation about A, which eliminates A_n and A_t and the weight. Or a moment summation may be taken about the intersection of A_n and the line of action of $m\overline{r}\alpha$, which eliminates A_n and $m\overline{r}\alpha$. Using A as a moment center gives

$$[\Sigma M_A = m\overline{a}d] \qquad 1.8 \cos 30° B = 2.02(1.2) \cos 30° + 2.06(0.6)$$

$$B = 2.14 \text{ kN} \qquad Ans.$$

The component A_n could be obtained from a force summation in the n-direction or from a moment summation about G or about the intersection of B and the line of action of $m\overline{r}\alpha$.

① Generally speaking, the best choice of reference axes is to make them coincide with the directions in which the components of the mass-center acceleration are expressed. Examine the consequences of choosing horizontal and vertical axes in this problem.

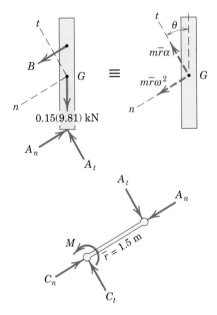

② The force and moment equations for a body of negligible mass become the same as the equations of equilibrium. Link BD, therefore, acts as a two-force member in equilibrium.

PROBLEMS

Introductory problems

6/1 What acceleration a of the collar along the horizontal guide will result in a steady-state 15° deflection of the pendulum from the vertical? The slender rod of length l and the particle each have mass m. Friction at the pivot P is negligible.

Ans. $a = 0.268g$

Problem 6/1

6/2 The rear-wheel-drive lawn mower, when placed into gear while at rest, is observed to momentarily spin its rear tires as it accelerates. If the coefficients of friction between the rear tires and the ground are $\mu_s = 0.7$ and $\mu_k = 0.5$, determine the forward acceleration a of the mower. The mass of the mower and attached bag is 50 kg with center of mass at G. Assume that the operator does not push on the handle, so that $P = 0$.

Problem 6/2

6/3 The bicyclist applies the brakes as he descends the 10° incline. What deceleration a would cause the dangerous condition of tipping about the front wheel A? The combined center of mass of the rider and bicycle is at G.

Ans. $a = 0.510g$

Problem 6/3

6/4 Determine the minimum speed v and the corresponding angle θ in order that the motorcycle may ride on the vertical wall of a cylindrical track. The effective coefficient of friction between the tires and the wall is 0.70. (Note that the forces and acceleration lie in the plane of the figure, so the problem may be treated as one of translatory plane motion.)

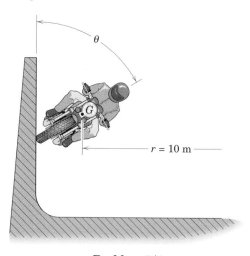

Problem 6/4

6/5 The 1650-kg car has its mass center at G. Calculate the normal forces N_A and N_B between the road and the front and rear pairs of wheels under conditions of maximum acceleration. The mass of the wheels is small compared with the total mass of the car. The coefficient of static friction between the road and the rear driving wheels is 0.8.

Ans. $N_A = 6.85$ kN, $N_B = 9.34$ kN

Problem 6/5

6/6 A drill press with mass center at G is transported on the flat bed of a truck. If the driver neglects to secure the drill press to the truck, determine the maximum acceleration a which the truck may have before the drill press slips or tips about its base at A. The coefficient of static friction between the truck bed and the drill press is 0.60.

Problem 6/6

6/7 A passenger car of an overhead monorail system is driven by one of its two small wheels A or B. Select the one for which the car can be given the greater acceleration without slipping the driving wheel and compute the maximum acceleration if the effective coefficient of friction is limited to 0.25 between the wheels and the rail. Neglect the small mass of the wheels.

Ans. B, $a = 1.344$ m/s^2

Problem 6/7

6/8 The device shown consists of a vertical frame A to which are freely pivoted a geared sector at O and a balanced gear and attached pointer at C. Under a steady horizontal acceleration a to the right, the sector undergoes a clockwise angular displacement, thus causing the pointer to register a steady counterclockwise angle θ from the zero-acceleration position at $\theta = 0$. Determine the acceleration corresponding to an angle θ.

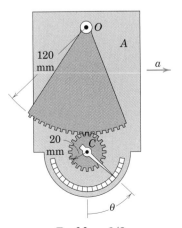

Problem 6/8

Representative problems

6/9 Arm *AB* of a classifying accelerometer has a mass of 0.1 kg with mass center at *G* and is pivoted freely to the frame *F* at *A*. The torsional spring at *A* is set to preload the arm with an applied clockwise moment of 225 N · mm. Determine the downward acceleration *a* of the frame at which the contacts at *B* will separate and break the electrical circuit.

Ans. a = 72.3 m/s^2

Problem 6/9

6/10 The arm *OA* of the classifying accelerometer has a mass of 0.25 kg with center of mass at *G*. The adjusting screw and spring are preset to a force of 12 N at *B*. At what acceleration *a* would the electrical contacts at *A* be on the verge of opening? Motion is in the vertical plane of the figure.

Problem 6/10

6/11 The uniform 100-kg log is supported by the two cables and used as a battering ram. If the log is released from rest in the position shown, calculate the initial tension induced in each cable immediately after release and the corresponding angular acceleration α of the cables.

Ans. T$_A$ = 212 N, *T$_B$* = 637 N
α = 2.45 rad/s^2

Problem 6/11

6/12 Determine the value of the force *P* which would cause the cabinet to begin to tip. What coefficient μ$_s$ of static friction is necessary to ensure that tipping occurs without slipping?

Problem 6/12

6/13 Solid homogeneous cylinders 400 mm high and 250 mm in diameter are supported by a flat conveyor belt which moves horizontally. If the speed of the belt increases according to $v = 1.2 + 0.9t^2$ m/s, where t is the time in seconds measured from the instant the increase begins, calculate the value of t for which the cylinders begin to tip over. Cleats on the belt prevent the cylinders from slipping.

Ans. $t = 3.41$ s

Problem 6/13

6/14 The design of an assembly-line conveyor system is under review. The conveyor belt A, moving at 0.5 m/s, transfers uniform steel cylinders to the chute B, where they slide with increasing speed. Each cylinder has a diameter $d = 80$ mm and a height $h = 180$ mm. If the coefficient of kinetic friction between the cylinders and the chute is $\mu_k = 0.15$ and the chute angle is $\theta = 15°$, find the acceleration a of the cylinders along the chute and prove that they will not tip over.

Problem 6/14

6/15 A cleated conveyor belt transports solid homogeneous cylinders up a 15° incline. The diameter of the cylinder is half their height. Determine the maximum acceleration which the belt may have without tipping the cylinders as it starts.

Ans. $a = 0.224g$

Problem 6/15

6/16 The cart B moves to the right with acceleration $a = 2g$. If the steady-state angular deflection of the uniform slender rod of mass $3m$ is observed to be 20°, determine the value of the torsional spring constant K. The spring, which exerts a moment of magnitude $M = K\theta$ on the rod, is undeformed when the rod is vertical. The values of m and l are 0.5 kg and 0.6 m, respectively. Treat the small end sphere of mass m as a particle.

Problem 6/16

6/17 The loaded trailer has a mass of 900 kg with center of mass at G and is attached at A to a rear-bumper hitch. If the car and trailer reach a velocity of 60 km/h on a level road in a distance of 30 m from rest with constant acceleration, compute the vertical component of the force supported by the hitch at A. Neglect the small friction force exerted on the relatively light wheels.

Ans. $A_y = 1389$ N

Problem 6/17

6/18 The uniform 40-kg plank is supported in the pickup truck at its end A and at B where it rests on the smooth top of the cab. Calculate the contact force at B if the truck starts forward with an acceleration $a = 4$ m/s^2.

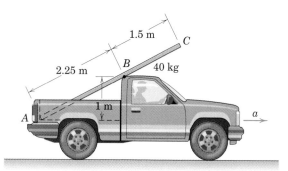

Problem 6/18

6/19 The coefficient of static friction at both ends of the uniform bar is 0.40. Determine the maximum horizontal acceleration a which the truck may have without causing the bar to slip. (*Suggestion:* The problem can be solved by using only one equation, a moment equation. The location of the moment center may be determined graphically.)

Ans. $a = 2.38$ m/s^2

Problem 6/19

6/20 The two uniform identical bars are freely hinged at the lower ends and are supported at the upper ends by small rollers of negligible mass which roll on a horizontal rail. Determine the steady-state angle θ assumed by the bars when they are accelerating under the action of a constant force F. Also find the vertical forces on the rollers at A and B.

Problem 6/20

6/21 Determine the magnitude P and direction θ of the force required to impart a rearward acceleration $a = 1.5 \text{ m/s}^2$ to the loaded wheelbarrow with no rotation from the position shown. The combined mass of the wheelbarrow and its load is 190 kg with center of mass at G. Compare the normal force at B under acceleration with that for static equilibrium in the position shown. Neglect the friction and mass of the wheel.

Ans. $P = 439 \text{ N}, \theta = 49.6°$
$B = 1530 \text{ N}, B_{\text{st}} = 1553 \text{ N}$

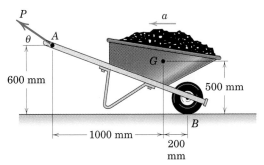

Problem 6/21

6/22 A jet transport with a landing speed of 200 km/h reduces its speed to 60 km/h with a negative thrust R from its jet thrust reversers in a distance of 425 m along the runway with constant deceleration. The total mass of the aircraft is 140 Mg with mass center at G. Compute the reaction N under the nose wheel B toward the end of the braking interval and prior to the application of mechanical braking. At the lower speed aerodynamic forces on the aircraft are small and may be neglected.

Problem 6/22

6/23 Design tests of the landing sequence for the lunar excursion module are conducted using the pendulum model suspended by the parallel wires A and B. If the model has a mass of 10 kg with mass center at G, and if $\dot{\theta} = 2 \text{ rad/s}$ when $\theta = 60°$, calculate the tension in each of the wires for this instant.

Ans. $T_A = 147.9 \text{ N}, T_B = 21.1 \text{ N}$

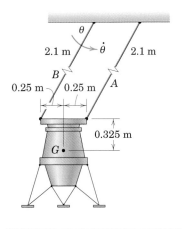

Problem 6/23

6/24 The parallelogram linkage shown moves in the vertical plane with the uniform 8-kg bar EF attached to the plate at E by a pin which is welded both to the plate and to the bar. A torque (not shown) is applied to link AB through its lower pin to drive the links in a clockwise direction. When θ reaches 60°, the links have an angular acceleration and an angular velocity of 6 rad/s^2 and 3 rad/s, respectively. For this instant calculate the magnitudes of the force F and torque M supported by the pin at E.

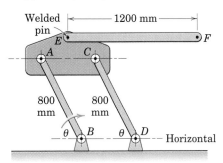

Problem 6/24

6/25 The figure shows the Saturn V mobile launch platform A together with the umbilical tower B, unfueled rocket C, and crawler-transporter D which carries the system to the launch site. The approximate dimensions of the structure and locations of the mass centers G are given. The approximate masses are $m_A = 3$ Gg, $m_B = 3.3$ Gg, $m_C = 0.23$ Gg, and $m_D = 3$ Gg. The minimum stopping distance from the top speed of 1.5 km/h is 0.1 m. Compute the vertical component of the reaction under the front crawler unit F during the period of maximum deceleration.

Ans. $F = 59.5$ MN

Problem 6/25

6/26 The loaded pickup truck, which has a mass of 1800 kg with mass center at G_1, is hauling the 900-kg trailer with mass center at G_2. While going down a 10-percent grade, the driver applies his brakes and slows down from 96 km/h to 48 km/h in a distance of 110 m. For this interval, compute the x- and y-components of the force exerted on the trailer hitch at D by the truck. Also find the corresponding normal force under each pair of wheels at B and C. Neglect the rotational effect of the wheels.

Problem 6/26

6/27 The van seen from the rear is traveling at a speed v around a turn of mean radius r banked inward at an angle θ. The effective coefficient of friction between the tires and the road is μ. Determine (a) the proper bank angle for a given v to eliminate any tendency to slip or tip, and (b) the maximum speed v before the van tips or slips for a given θ. Note that the forces and the acceleration lie in the plane of the figure so that the problem may be treated as one of plane motion even though the velocity is normal to this plane.

Ans. (a) $\theta = \tan^{-1} \dfrac{v^2}{gr}$

(b) Slips first if $v^2 = gr \dfrac{\mu + \tan \theta}{1 - \mu \tan \theta}$, $\mu < \dfrac{b}{2h}$

Tips first if $v^2 = gr \dfrac{b/2h + \tan \theta}{1 - \dfrac{b}{2h} \tan \theta}$, $\mu > \dfrac{b}{2h}$

Problem 6/27

6/28 The parallelogram linkage is used to transfer crates from platform A to platform B and is hydraulically operated. The oil pressure in the cylinder is programmed to provide a smooth transition of motion from $\theta = 0$ to $\theta = \theta_0 = \pi/3$ rad given by $\theta = \dfrac{\pi}{6}\left(1 - \cos\dfrac{\pi t}{2}\right)$ where t is in seconds. Determine the force at D on the pin (a) just after the start of the motion with θ and t essentially zero and (b) when $t = 1$ s. The crate and platform have a combined mass of 200 kg with mass center at G. The mass of each link is small and may be neglected.

Problem 6/28

6/29 The homogeneous 20-kg rectangular plate is supported in the vertical plane by the light parallel links shown. If a couple $M = 110$ N·m is applied to the end of link AB with the system initially at rest, calculate the force supported by the pin at C as the plate lifts off its support with $\theta = 30°$.

$Ans.$ $C = 218$ N

Problem 6/29

6/30 The tandem unit A of the road grader has a mass of 3000 kg and is freely pivoted to the motive unit B at O, which is also the mass center of A. The mass of unit B alone is 10 Mg, including wheels C, with mass center at G. Find the minimum distance s in which the grader can stop when traveling on a level road at 40 km/h (blade retracted and engine disengaged) so that the rear pair of wheels of the tandem unit A will not lift off the ground. Brakes are on the tandem wheels only. Treat each of the two units as a rigid body.

Problem 6/30

6/31 The two wheels of the vehicle are connected by a 20-kg link AB with center of mass at G. The link is pinned to the wheel at B, and the pin at A fits into a smooth horizontal slot in the link. If the vehicle has a constant speed of 4 m/s, determine the magnitude of the force supported by the pin at B for the position $\theta = 30°$.

Ans. $B = 188.3$ N

▶ **6/32** The uniform 200-kg bar AB is raised in the vertical plane by the application of a constant couple $M = 3$ kN·m applied to the link at C. The mass of the links is small and may be neglected. If the bar starts from rest at $\theta = 0$, determine the magnitude of the force supported by the pin at A as the position $\theta = 60°$ is passed.

Ans. $A = 2.02$ kN

Problem 6/31

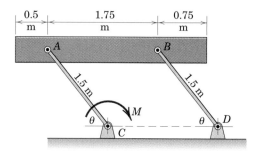

Problem 6/32

6/4 *FIXED-AXIS ROTATION*

Rotation of a rigid body about a fixed axis O was described in Art. 5/2 and illustrated in Fig. 5/1c. For this motion, we saw that all points in the body describe circles about the rotation axis, and all lines of the body in the plane of motion have the same angular velocity ω and angular acceleration α. The acceleration components of the mass center for circular motion are most easily expressed in n-t coordinates, so we have $a_n = \bar{r}\omega^2$ and $a_t = \bar{r}\alpha$, as shown in Fig. 6/9a for rotation of the rigid body about the fixed axis through O. The b-part of the figure represents the free-body diagram, and the equivalent kinetic diagram in the c-part of the figure shows the force resultant $m\bar{\mathbf{a}}$ in terms of its n- and t-components and the resultant couple $\bar{I}\alpha$. Our general equations for plane motion, Eqs. 6/1, are directly applicable and are repeated here.

Fixed-Axis Rotation

(a)

$$\Sigma\mathbf{F} = m\bar{\mathbf{a}}$$
$$\Sigma M_G = \bar{I}\alpha$$

[6/1]

Thus, the two scalar components of the force equation become $\Sigma F_n = m\bar{r}\omega^2$ and $\Sigma F_t = m\bar{r}\alpha$. In applying the moment equation about G, it is essential to account for the moment of the force applied to the body at O, so this force must not be omitted from the free-body diagram.

For fixed-axis rotation, it is generally useful to apply a moment equation directly about the rotation axis O. We derived this equation previously as Eq. 6/4, which is repeated here.

$$\Sigma M_O = I_O\alpha$$

[6/4]

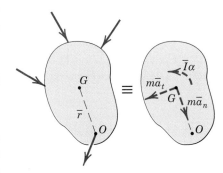

Free-Body Diagram   Kinetic Diagram
        (b)   (c)

Figure 6/9

From the kinetic diagram in Fig. 6/9c, we may obtain Eq. 6/4 very easily by evaluating the moment of the resultants about O, which becomes $\Sigma M_O = \bar{I}\alpha + m\bar{a}_t\bar{r}$. Substitution of the transfer-of-axis relation for mass moments of inertia, $I_O = \bar{I} + m\bar{r}^2$, gives $\Sigma M_O = (I_O - m\bar{r}^2)\alpha + m\bar{r}^2\alpha = I_O\alpha$.

For the common case of rotation of a rigid body about a fixed axis through its mass center G, clearly, $\bar{\mathbf{a}} = \mathbf{0}$, and therefore $\Sigma\mathbf{F} = \mathbf{0}$. The resultant of the applied forces then is the couple $\bar{I}\alpha$.

We may combine the resultant force $m\bar{a}_t$ and resultant couple $\bar{I}\alpha$ by moving $m\bar{a}_t$ to a parallel position through point Q on line OG, Fig. 6/10, located by $m\bar{r}\alpha q = \bar{I}\alpha + m\bar{r}\alpha(\bar{r})$. Using the transfer-of-axis theorem and $I_O = k_O^2 m$ gives $q = k_O^2/\bar{r}$. Point Q is called the *center of percussion* and has the unique property that the resultant of all forces applied to the body must pass through it. It follows that the sum of the moments of all forces about the center of percussion is always zero, $\Sigma M_Q = 0$.

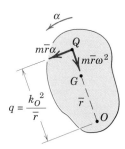

Figure 6/10

Sample Problem 6/3

The 300-kg concrete block is elevated by the hoisting mechanism shown, where the cables are securely wrapped around the respective drums. The drums, which are fastened together and turn as a single unit about their mass center at O, have a combined mass of 150 kg and a radius of gyration about O of 450 mm. If a constant tension P of 1.8 kN is maintained by the power unit at A, determine the vertical acceleration of the block and the resultant force on the bearing at O.

600 mm 300 mm
$P = 1.8$ kN
$m = 150$ kg
$k_O = 450$ mm
45°
A
300 kg

Solution I. The free-body and kinetic diagrams of the drums and concrete block are drawn showing all forces which act, including the ① components O_x and O_y of the bearing reaction. The resultant of the force system on the drums for centroidal rotation is the couple $\bar{I}\alpha = I_O\alpha$, where

② $[I = k^2m]$ $\bar{I} = I_O = (0.450)^2 150 = 30.4$ kg·m²

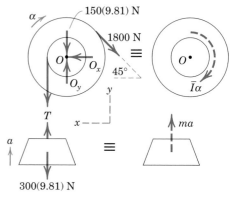

150(9.81) N
1800 N
45°
$\bar{I}\alpha$

T

300(9.81) N

ma

Taking moments about the mass center O for the pulley in the sense of the angular acceleration α gives

$[\Sigma M_G = \bar{I}\alpha]$ $\quad 1800(0.600) - T(0.300) = 30.4\alpha$ $\quad$ (a)

The acceleration of the block is described by

$[\Sigma F_y = ma_y]$ $\quad T - 300(9.81) = 300a$ $\quad$ (b)

From $a_t = r\alpha$, we have $a = 0.300\alpha$. With this substitution, Eqs. (a) and (b) are combined to give

$\quad T = 3250$ N $\quad \alpha = 3.44$ rad/s² $\quad a = 1.031$ m/s² $\quad$ Ans.

The bearing reaction is computed from its components. Since $\bar{a} = 0$, we use the equilibrium equations

$[\Sigma F_x = 0]$ $\quad O_x - 1800 \cos 45° = 0$ $\quad O_x = 1273$ N
$[\Sigma F_y = 0]$ $\quad O_y - 150(9.81) - 3250 - 1800 \sin 45° = 0$ $\quad O_y = 6000$ N
$\quad O = \sqrt{(1273)^2 + (6000)^2} = 6130$ N $\quad$ Ans.

Solution II. We may use a more condensed approach by drawing the free-body diagram of the entire system, thus eliminating reference to T, which becomes internal to the new system. From the kinetic diagram for the system, we see that the moment sum about O must equal the resultant couple $\bar{I}\alpha$ for the drums, plus the moment of the resultant ma for the block. Thus, from the principle of Eq. 6/5 we have

$[\Sigma M_O = \bar{I}\alpha + m\bar{a}d]$

$\quad 1800(0.600) - 300(9.81)(0.300) = 30.4\alpha + 300a(0.300)$

With $a = 0.300\alpha$, the solution gives, as before, $a = 1.031$ m/s².

We may equate the force sums on the entire system to the sums of the resultants. Thus,

$[\Sigma F_y = \Sigma m\bar{a}_y]$

$\quad O_y - 150(9.81) - 300(9.81) - 1800 \sin 45° = 150(0) + 300(1.031)$

$\quad O_y = 6000$ N

$[\Sigma F_x = \Sigma m\bar{a}_x]$ $\quad O_x - 1800 \cos 45° = 0$ $\quad O_x = 1273$ N

① Be alert to the fact that the tension T is not 300(9.81) N. If it were, the block would not accelerate.

② Do not overlook the need to express k_o in meters when using g in m/s².

150(9.81) N
1800 N
45°
$\bar{I}\alpha$
ma
300(9.81) N

Sample Problem 6/4

The pendulum has a mass of 7.5 kg with center of mass at G and has a radius of gyration about the pivot O of 295 mm. If the pendulum is released from rest at $\theta = 0$, determine the total force supported by the bearing at the instant when $\theta = 60°$. Friction in the bearing is negligible.

Solution. The free-body diagram of the pendulum in a general ① position is shown along with the corresponding kinetic diagram, where ① the components of the resultant force have been drawn through G.

The normal component O_n is found from a force equation in the n-direction, which involves the normal acceleration $\bar{r}\omega^2$. Since the angular velocity ω of the pendulum is found from the integral of the angular acceleration and since O_t depends on the tangential acceleration $\bar{r}\alpha$, it follows that α should be obtained first. To this end with $I_O = k_O^2 m$, the moment equation about O gives

② $[\Sigma M_O = I_O \alpha]$ $7.5(9.81)(0.25) \cos \theta = (0.295)^2(7.5)\alpha$

$$\alpha = 28.2 \cos \theta \ \text{rad/s}^2$$

and for $\theta = 60°$

$[\omega \, d\omega = \alpha \, d\theta]$ $\displaystyle\int_0^\omega \omega \, d\omega = \int_0^{\pi/3} 28.2 \cos \theta \, d\theta$

$$\omega^2 = 48.8 \ (\text{rad/s})^2$$

The remaining two equations of motion applied to the 60° position yield

$[\Sigma F_n = m\bar{r}\omega^2]$ $O_n - 7.5(9.81) \sin 60° = 7.5(0.25)(48.8)$

$$O_n = 155.2 \ \text{N}$$

$[\Sigma F_t = m\bar{r}\alpha]$ $-O_t + 7.5(9.81) \cos 60° = 7.5(0.25)(28.2) \cos 60°$

$$O_t = 10.37 \ \text{N}$$

$$O = \sqrt{(155.2)^2 + (10.37)^2} = 155.6 \ \text{N} \qquad Ans.$$

The proper sense for O_t may be observed at the outset by applying the moment equation $\Sigma M_G = \bar{I}\alpha$, where the moment about G due to O_t must be clockwise to agree with α. The force O_t may also be obtained initially by a moment equation about the center of percussion Q, shown in the lower figure, which avoids the necessity of computing α. First, we must obtain the distance q, which is

$[q = k_O^2/\bar{r}]$ $\displaystyle q = \frac{(0.295)^2}{0.250} = 0.348 \ \text{m}$

$[\Sigma M_Q = 0]$ $O_t(0.348) - 7.5(9.81)(\cos 60°)(0.348 - 0.250) = 0$

$$O_t = 10.37 \ \text{N} \qquad Ans.$$

① The acceleration components of G are, of course, $\bar{a}_n = \bar{r}\omega^2$ and $\bar{a}_t = \bar{r}\alpha$.

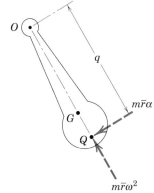

② Review the theory again and satisfy yourself that $\Sigma M_O = I_O \alpha = \bar{I}\alpha + m\bar{r}^2\alpha = m\bar{r}\alpha q$.

③ Note especially here that the force summations are taken in the positive direction of the acceleration components of the mass center G.

PROBLEMS

Introductory problems

6/33 The 20-kg uniform steel plate is freely hinged about the *z*-axis as shown. Calculate the force supported by each of the bearings at *A* and *B* an instant after the plate is released from rest in the horizontal *y-z* plane.

$$Ans. \ F_A = F_B = 24.5 \ N$$

Problem 6/33

6/34 Explain why a baseball bat should be swung so that the impact of the ball with the bat does not occur opposite the mass center G of the bat as shown. Where should impact occur?

Problem 6/34

6/35 The automotive dynamometer is able to simulate road conditions for an acceleration of 0.6*g* for the loaded pickup truck with a total mass of 2.8 Mg. Calculate the required moment of inertia of the dynamometer drum about its center *O* assuming that the drum turns freely during the acceleration phase of the test.

$$Ans. \ I_O = 2.27 \ Mg \cdot m^2$$

Problem 6/35

6/36 The square frame is composed of four equal lengths of uniform slender rod, and the ball attachment at *O* is suspended in a socket (not shown). Beginning from the position shown, the assembly is rotated 45° about axis *A-A* and released. Determine the initial angular acceleration of the frame. Repeat for a 45° rotation about axis *B-B*. Neglect the small mass, offset, and friction of the ball.

Problem 6/36

6/37 The uniform 8-kg slender bar is hinged about a horizontal axis through O and released from rest in the horizontal position. Determine the distance b from the mass center to O which will result in an initial angular acceleration of 16 rad/s², and find the force R on the bar at O just after release.

Ans. $b = 53.6$ mm, $R = 71.6$ N

G

$\leftarrow b \rightarrow$ O

$\leftarrow$ 300 mm $\rightarrow$ $\leftarrow$ 300 mm $\rightarrow$

Problem 6/37

6/38 The half ring of mass m and radius r is welded to a small horizontal shaft mounted in a bearing as shown. Neglect the mass of the shaft and determine the angular acceleration of the ring when a torque M is applied to the shaft.

M

r

Problem 6/38

6/39 An *air table* is used to study the elastic motion of flexible spacecraft models. Pressurized air escaping from numerous small holes in the horizontal surface provides a supporting air cushion which largely eliminates friction. The model shown consists of a cylindrical hub of radius r and four appendages of length l and small thickness t. The hub and the four appendages all have the same depth d and are constructed of the same material of density ρ. Assume that the spacecraft is rigid and determine the moment M which must be applied to the hub to spin the model from rest to an angular velocity ω in a time period of τ seconds. (Note that for a spacecraft with highly flexible appendages, the moment must be judiciously applied to the rigid hub to avoid undesirable large elastic deflections of the appendages.)

Ans. $M = \dfrac{\omega \rho d}{\tau}\left[\dfrac{1}{2}\pi r^4 + 4lt\left(\dfrac{1}{3}l^2 + rl + r^2\right)\right]$

Problem 6/39

6/40 The 750-mm slender bar has a mass of 9 kg and is mounted on a vertical shaft at O. If a torque $M = 10$ N·m is applied to the bar through its shaft, calculate the horizontal force R on the bearing as the bar starts to rotate.

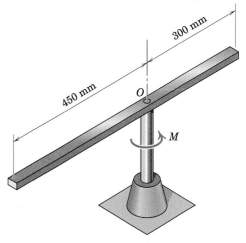

300 mm

450 mm

O

M

Problem 6/40

6/41 The bar A of mass m is formed into a 90° circular arc of radius r and is attached to the hub by the light rods. The curved bar oscillates about the vertical axis under the action of a torsional spring B. At the instant under consideration, the angular velocity is ω and the angular acceleration is α. Write expressions for the moment M exerted by the spring on the hub and the horizontal force R exerted by the shaft on the hub.

$$Ans. \ M = mr^2\alpha, \ R = \frac{2\sqrt{2}mr}{\pi}\sqrt{\alpha^2 + \omega^4}$$

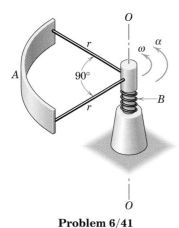

Problem 6/41

Representative problems

6/42 The uniform 20-kg bar is released from rest in the horizontal position shown and strikes the fixed corner B at the center of percussion of the bar. Determine the t-component of the force exerted by the bearing O on the bar just prior to impact, during impact, and just after impact.

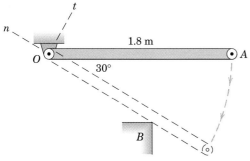

Problem 6/42

6/43 A gimbal pedestal supports a payload in the space shuttle and deploys it when the doors of the cargo bay are opened in orbit. The payload is modeled as a homogeneous rectangular block with a mass of 6000 kg. The torque on the gimbal axis O-O is 30 N·m supplied by a d-c brushless motor. With the shuttle orbiting in a "weightless" condition, determine the time t required to bring the payload from its stowed position at $\theta = 0$ to its deployed position at $\theta = 90°$ if the torque is applied for the first 45° of travel and then reversed for the remaining 45° to bring the payload to a stop ($\dot{\theta} = 0$).

$$Ans. \ t = 78.6 \text{ s}$$

Problem 6/43

6/44 A device for impact testing consists of a 34-kg pendulum with mass center at G and with radius of gyration about O of 620 mm. The distance b for the pendulum is selected so that the force on the bearing at O has the least possible value during impact with the specimen at the bottom of the swing. Determine b and calculate the magnitude of the total force R on the bearing O an instant after release from rest at $\theta = 60°$.

Problem 6/44

6/45 The solid cylindrical rotor B has a mass of 43 kg and is mounted on its central axis C-C. The frame A rotates about the fixed vertical axis O-O under the applied torque $M = 30$ N·m. The rotor may be unlocked from the frame by withdrawing the locking pin P. Calculate the angular acceleration α of the frame A if the locking pin is (a) in place and (b) withdrawn. Neglect all friction and the mass of the frame.

Ans. (a) $\alpha = 8.46$ rad/s^2
(b) $\alpha = 11.16$ rad/s^2

Problem 6/45

6/46 The mass of gear A is 20 kg and its centroidal radius of gyration is 150 mm. The mass of gear B is 10 kg and its centroidal radius of gyration is 100 mm. Calculate the angular acceleration of gear B when a torque of 12 N·m is applied to the shaft of gear A. Neglect friction.

$r_A = 240$ mm
$r_B = 180$ mm

Problem 6/46

6/47 A reel of flexible power cable is mounted on the dolly, which is fixed in position. There are 60 m of cable having a mass of 0.65 kg per meter of length wound on the reel at a radius of 375 mm. The empty spool has a mass of 28 kg and has a radius of gyration about its axis of 300 mm. A tension T of 90 N is required to overcome frictional resistance to turning. Calculate the angular acceleration α of the reel if a tension of 180 N is applied to the free end of the cable.

Ans. $\alpha = 4.22$ rad/s^2

Problem 6/47

6/48 A vibration test is run to check the design adequacy of bearings A and B. The unbalanced rotor and attached shaft have a combined mass of 2.8 kg. To locate the mass center, a torque of 0.660 N·m is applied to the shaft to hold it in equilibrium in a position rotated 90° from that shown. A constant torque $M = 1.5$ N·m is then applied to the shaft, which reaches a speed of 1200 rev/min in 18 revolutions starting from rest. (During each revolution the angular acceleration varies, but its average value is the same as for constant acceleration.) Determine (a) the radius of gyration k of the rotor and shaft about the rotation axis, (b) the force F which each bearing exerts on the shaft immediately after M is applied, and (c) the force R exerted by each bearing when the speed of 1200 rev/min is reached and M is removed. Neglect any frictional resistance and the bearing forces due to static equilibrium.

Problem 6/48

6/49 For Prob. 6/42, determine the n-component of the force exerted by the bearing at O an instant prior to impact of the bar with the corner B.

Ans. $O_n = 245$ N

6/50 Disk B has a mass of 22 kg and has a centroidal radius of gyration of 200 mm. The power unit C consists of a motor M and a disk A, which is driven at a constant angular speed of 1600

rev/min. The coefficients of static and kinetic friction between the two disks are $\mu_s = 0.8$ and $\mu_k = 0.6$, respectively. Disk B is initially stationary when contact with disk A is established by application of the constant force $P = 14$ N. Determine the angular acceleration α of B and the time t required for B to reach its steady-state speed.

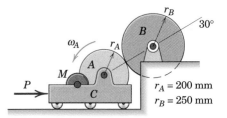

$r_A = 200$ mm
$r_B = 250$ mm

Problem 6/50

6/51 The robotic device consists of the stationary pedestal OA, arm AB pivoted at A, and arm BC pivoted at B. The rotation axes are normal to the plane of the figure. Estimate (a) the moment M_A applied to arm AB required to rotate it about joint A at 4 rad/sec² counterclockwise from the position shown with joint B locked and (b) the moment M_B applied to arm BC required to rotate it about joint B at the same rate with joint A locked. The mass of arm AB is 25 kg and that of BC is 4 kg, with the stationary portion of joint A excluded entirely and the mass of joint B divided equally between the two arms. Assume that the centers of mass G_1 and G_2 are in the geometric centers of the arms and model the arms as slender rods.

Ans. (a) $M_A = 109.8$ N·m
(b) $M_B = 5.51$ N·m

Problem 6/51

6/52 The uniform semicircular bar of mass m and radius r is hinged freely about a horizontal axis through A. If the bar is released from rest in the position shown, where AB is horizontal, determine the initial angular acceleration α of the bar and the expression for the force exerted on the bar by the pin at A. (Note carefully that the initial tangential acceleration of the mass center is not vertical.)

Problem 6/52

6/53 The 12-kg cylinder supported by the bearing brackets at A and B has a moment of inertia about the vertical z_0-axis through its mass center G equal to 0.080 kg·m². The disk and brackets have a moment of inertia about the vertical z-axis of rotation equal to 0.60 kg·m². If a torque $M = 16$ N·m is applied to the disk through its shaft with the disk initially at rest, calculate the horizontal x-components of force supported by the bearings at A and B.

Ans. $A = 22.1$ N, $B = 11.03$ N

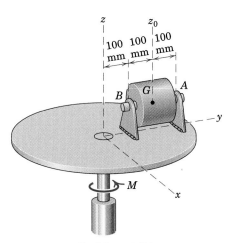

Problem 6/53

6/54 The narrow ring of mass m is free to rotate in the vertical plane about O. If the ring is released from rest at $\theta = 0$, determine expressions for the n- and t-components of the force at O in terms of θ.

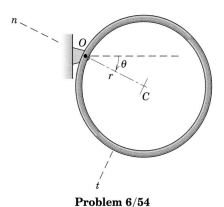

Problem 6/54

6/55 The uniform slender bar AB has a mass of 8 kg and swings in a vertical plane about the pivot at A. If $\dot{\theta} = 2$ rad/s when $\theta = 30°$, compute the force supported by the pin at A at that instant.

Ans. $A = 56.3$ N

Problem 6/55

6/56 The solid homogeneous cylinder has a mass of 100 kg and is mounted on a right-angled shaft which turns freely about the horizontal axis O-O. If the cylinder is released from rest with its own axis in the horizontal plane, calculate the initial angular acceleration of the assembly and the resultant force F exerted by the bearing A on the shaft. The mass of the shaft may be neglected.

6/58 The semicircular ring of mass m and radius r is welded to the vertical shaft, which permits the ring to rotate in the horizontal plane about axis O-O. If a torque M is applied to the ring through its shaft, determine the expression for the resulting angular acceleration α of the ring and the force F acting in the horizontal plane on the ring at O as the ring starts from rest.

Problem 6/56

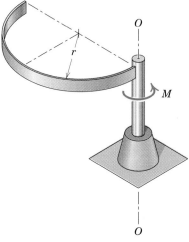

Problem 6/58

6/57 The right-angle plate is formed from a flat plate having a mass ρ per unit area and is welded to the horizontal shaft mounted in the bearing at O. If the shaft is free to rotate, determine the initial angular acceleration α of the plate when it is released from rest with the upper surface in the horizontal plane. Also determine the y- and z-components of the resultant force on the shaft at O.

$$Ans. \quad \alpha = \frac{3g}{10b}, \ O_y = \frac{9}{20}\rho bcg, \ O_z = \frac{37}{20}\rho bcg$$

Problem 6/57

6/59 Each of the uniform slender rods of mass m is welded at its end tangent to the rim of the circular disk. The disk rotates in a horizontal plane about a fixed vertical axis through its center O. Determine expressions for the bending moment M, the tension T, and the shear V transmitted by the weld to the rod if the disk has (a) a constant angular velocity ω about O and (b) a start-up counterclockwise angular acceleration α about O. Does the sense (CW or CCW) of ω make any difference? Analyze the forces in the horizontal plane only.

Ans. (a) $M = mrl\omega^2/2$, $V = mr\omega^2$, $T = ml\omega^2/2$
(b) $M = -ml^2\alpha/3$, $V = -ml\alpha/2$, $T = mr\alpha$

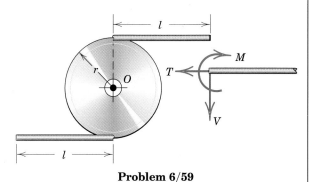

Problem 6/59

6/60 The uniform steel I-beam has a mass of 300 kg and is supported in the vertical plane as shown. Calculate the force R supported by the pin at O for the condition immediately after the support at B is suddenly removed. The mass of the bracket on the left end is small and may be neglected. Also treat the beam as a uniform slender bar.

Problem 6/60

6/61 The 24-kg uniform slender bar AB is mounted on end rollers of negligible mass and rotates about the fixed point O as it follows the circular path in the vertical plane. The bar is released from a position which gives it an angular velocity $\omega = 2$ rad/s as it passes the position $\theta = 45°$. Calculate the forces F_A and F_B exerted by the guide on the rollers for this instant.

Ans. $F_A = 108.3$ N, $F_B = 141.6$ N

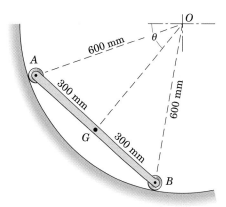

Problem 6/61

6/62 The uniform semicircular ring of mass $m = 2.5$ kg and mean radius $r = 200$ mm is mounted on spokes of negligible mass and pivoted about a horizontal axis through O. If the ring is released from rest in the position $\theta = 30°$, determine the force R supported by the bearing O just after release.

Problem 6/62

6/63 The 0.4-kg link B with center of mass 55 mm from O-O has a radius of gyration about O-O of 69 mm. The link is welded to the steel tube and is free to rotate about the fixed horizontal shaft at O-O. The mass of the tube is 0.92 kg. If the tube is released from rest with the link in the horizontal position, calculate the initial angular acceleration α of the assembly and the corresponding reaction O exerted by the shaft on the link.

Ans. $\alpha = 63.6$ rad/s^2, $O = 2.41$ N

62.5 mm

125 mm

50 mm

50 mm

B

Problem 6/63

6/64 A flexible cable 60 meters long with a mass of 0.160 kg per meter of length is wound around the reel. With $y = 0$, the weight of the 4-kg cylinder is required to start turning the reel to overcome friction in its bearings. Determine the downward acceleration a in meters per second squared of the cylinder as a function of y in meters. The empty reel has a mass of 16 kg with a radius of gyration about its bearing of 200 mm.

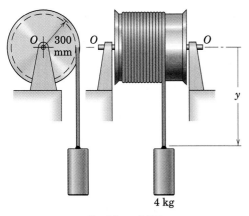

300 mm

y

4 kg

Problem 6/64

6/65 The figure shows a roll-off truck ramp for discharging loaded containers. The loaded 120-Mg container may be treated as a homogeneous solid rectangular block with mass center at G. If the supporting wheel A is restrained from movement, calculate the force F_B exerted by the ramp on the supporting wheel B when the truck starts from rest with a forward acceleration of 3 m/s^2. Neglect friction at B.

Ans. $F_B = 310$ kN

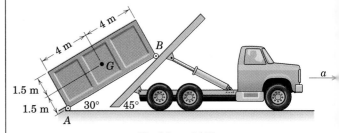

4 m

4 m

B

G

1.5 m

1.5 m

30° 45°

A

a

Problem 6/65

6/66 Prior to deployment of its two instrument arms AB, the spacecraft shown in the upper view is spinning at the constant rate of 1 revolution per second. Each instrument arm, shown in the lower view, has a mass of 10 kg with mass center at G. Calculate the tension T in the deployment cable prior to its release. Also find the magnitude of the force on the pin at A. Neglect any acceleration of the center O of the spacecraft.

B A

O ω

A B

O ω

400 mm

100 mm

T

A 500 mm G B

Problem 6/66

6/67 The curved bar of mass m is hinged to the rotating disk at O and bears against one of the smooth pins A and B which are fastened to the disk. If the disk rotates about its vertical axis C, determine the force exerted on the bar by the hinge at O and the reaction A or B on the bar (*a*) if the disk has a constant angular velocity ω and (*b*) as the disk starts from rest with a counterclockwise angular acceleration α.

Ans. (*a*) $O = \dfrac{2mr\omega^2}{\pi}, A = \dfrac{2mr\omega^2}{\pi}$

     (*b*) $O = mr\alpha \sqrt{1 + \dfrac{4}{\pi^2}}, B = mr\alpha \left(1 - \dfrac{2}{\pi}\right)$

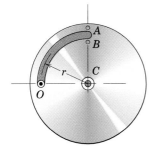

Problem 6/67

6/68 A horizontal shaft of radius r_0 rotates about its fixed axis through O with a constant angular acceleration α_0. A ring of mass m with radius r and negligible thickness rests on the shaft and turns with it. If friction is sufficient to prevent slipping, determine the steady-state angle θ assumed by the normal to the surfaces of contact at A. Specify the minimum coefficient of friction $\mu_{\min}$ to prevent slipping for a given value of α_0.

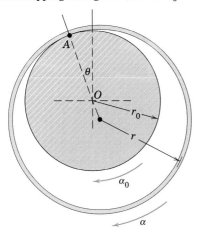

Problem 6/68

6/69 The split ring of mass m and radius r is rotating in a horizontal plane with an angular velocity ω about the vertical axis through O attached to one end of the ring. Determine the bending moment M and shear force V in the ring at A. Consider forces in the plane of the ring only.

Ans. $M = \dfrac{2}{\pi}mr^2\omega^2, V = \dfrac{1}{2}mr\omega^2$

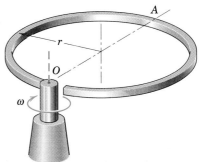

Problem 6/69

6/70 In 1993 the space shuttle captured the orbiting Hubble telescope and secured it to the cargo bay for repairs in the configuration simulated in the figure. During the repair period it was important to limit any angular acceleration of the shuttle so as not to induce excessive forces in the latches which secured the telescope to the shuttle. To illustrate the dynamics involved, replace the telescope by a 3000-kg homogeneous flat plate secured in the simplified manner shown in the exploded view. Calculate the maximum counterclockwise angular acceleration α which the shuttle may have so as not to exceed a tension T_A of 2 kN in the link at A. For this condition also calculate the corresponding force F_B supported by the pin at B. Make your calculation with respect to a nonrotating reference system moving with the shuttle in its orbit.

Problem 6/70

▶ **6/71** For the beam described in Prob. 6/60, determine the maximum angular velocity ω reached by the beam as it rotates in the vertical plane about the bearing at O. Also determine the corresponding force R supported by the pin at O for this condition. Again, treat the beam as a slender 300-kg bar and neglect the mass of the supporting bracket.

Ans. $\omega = 2.03$ rad/s, $R = 5.66$ kN

▶ **6/72** The uniform rectangular block is released from rest with θ essentially zero and pivots in the vertical plane about the center A of its lower face on the fixed corner. (*a*) If the block is observed to slip when $\theta = 30°$, find the coefficient of static friction between the block and the corner. (*b*) If the bottom face of the block is notched so that it cannot slip, find the angle θ at which contact between the block and the corner ceases

Ans. (*a*) $\mu_s = 0.229$
(*b*) $\theta = 54.6°$

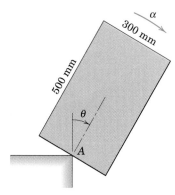

Problem 6/72

6/5 *GENERAL PLANE MOTION*

The dynamics of general plane motion of a rigid body combines translation and rotation. In Art. 6/2 we represented such a body in Fig. 6/4 with its free-body diagram and its kinetic diagram, which discloses the dynamic resultants of the applied forces. Figure 6/4 and Eqs. 6/1, which apply to general plane motion, are repeated here for convenient reference.

$$\Sigma \mathbf{F} = m\bar{\mathbf{a}}$$

$$\Sigma M_G = \bar{I}\alpha \qquad \text{[6/1]}$$

As previously, the two scalar components of the force equation are expressed in whatever coordinate system most readily describes the acceleration of the mass center. Direct application of these equations expresses the equivalence between the externally applied forces, as disclosed by the free-body diagram, and their force and moment resultants, as represented by the kinetic diagram.

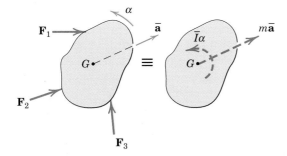

Free-Body Diagram   Kinetic Diagram

Figure 6/4

In Art. 6/2 we also showed, with the aid of Fig. 6/5, the application of the alternative relation for moments about any point P, Eq. 6/2. This figure and this equation are also repeated here for easy reference.

$$\Sigma M_P = \bar{I}\alpha + m\bar{a}d \qquad \text{[6/2]}$$

In some instances, it may be more convenient to use the alternative moment relation of Eq. 6/3 when moments are taken about a point P whose acceleration is known. It is also noted that the equation

Free-Body Diagram   Kinetic Diagram

Figure 6/5

for moments about a nonaccelerating point O on the body, Eq. 6/4, constitutes still another alternative moment relation and at times may be used to advantage.

In working a problem in general plane motion, we first observe whether the motion is unconstrained or constrained, as illustrated in the examples of Fig. 6/6. If the motion is constrained, we must account for the kinematic relationship between the linear and the angular accelerations and incorporate it into our force and moment equations of motion. If the motion is unconstrained, the accelerations can be determined independently of one another by direct application of the three motion equations, Eqs. 6/1.

In order for a rigid-body problem to be solvable, the number of unknowns cannot exceed the number of independent equations available to describe them, and a check on the sufficiency of the relationships should always be made. At the most, for plane motion we have three scalar equations of motion and two scalar components of the vector relative-acceleration equation for constrained motion. Thus, we can handle as many as five unknowns.

Strong emphasis should be placed on the importance of a clear choice of the body to be isolated and the representation of this isolation by a correct free-body diagram. Only after this vital step has been completed can we properly evaluate the equivalence between the external forces and their resultants. Of equal importance in the analysis of plane motion is a clear understanding of the kinematics involved. Very often, the difficulties experienced at this point have to do with kinematics, and a thorough review of the relative-acceleration relations for plane motion will be most helpful. In formulating the solution to a problem, we recognize that the directions of certain forces or accelerations may not be known at the outset, so that it may be necessary to make initial assumptions whose validity will be proved or disproved when the solution is carried out. It is essential, however, that all assumptions made be consistent with

the principle of action and reaction and with any kinematic requirements, which are also called conditions of constraint. Thus, if
a wheel is rolling on a horizontal surface, its center is constrained
to move on a horizontal line. Furthermore, if the unknown linear
acceleration a of the center of the wheel is assumed positive to the
right, then the unknown angular acceleration α would be positive
in a clockwise sense in order that $a = +r\alpha$, if we assume the wheel
does not slip. Also, we note that, for a wheel which rolls without
slipping, the static friction force between the wheel and its supporting surface is generally *less* than its maximum value, so that
$F \neq \mu_s N$. But if the wheel slips as it rolls, $a \neq r\alpha$, and a kinetic
friction force is generated which is given by $F = \mu_k N$. It may be
necessary to test the validity of either assumption, slipping or no
slipping, in a given problem. The difference between the static and
kinetic coefficients of friction, μ_s and μ_k, is sometimes ignored, in
which case, μ is used for either or both coefficients.

Sample Problem 6/5

A metal hoop with a radius $r = 150$ mm is released from rest on the 20° incline. If the coefficients of static and kinetic friction are $\mu_s = 0.15$ and $\mu_k = 0.12$, determine the angular acceleration α of the hoop and the time t for the hoop to move a distance of 3 m down the incline.

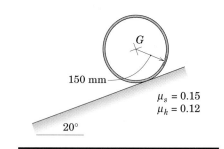

150 mm

$\mu_s = 0.15$
$\mu_k = 0.12$

20°

Solution. The free-body diagram shows the unspecified weight mg, the normal force N, and the friction force F acting on the hoop at the contact point C with the incline. The kinetic diagram shows the resultant force $m\bar{a}$ through G in the direction of its acceleration and the couple $\bar{I}\alpha$. The counterclockwise angular acceleration requires a counterclockwise moment about G, so F must be up the incline.

Assume that the hoop rolls without slipping, so that $\bar{a} = r\alpha$. Application of the components of Eqs. 6/1 with x- and y-axes assigned gives

$r = 150$ mm

$[\Sigma F_x = m\bar{a}_x]$ $mg \sin 20° - F = m\bar{a}$

$[\Sigma F_y = m\bar{a}_y = 0]$ $N - mg \cos 20° = 0$

① $[\Sigma M_G = \bar{I}\alpha]$ $Fr = mr^2\alpha$

Elimination of F between the first and third equations and substitution of the kinematic assumption $\bar{a} = r\alpha$ give

② $$\bar{a} = \frac{g}{2} \sin 20° = \frac{9.81}{2}(0.342) = 1.678 \text{ m/s}^2$$

Alternatively, with our assumption of $\bar{a} = r\alpha$ for pure rolling, a moment sum about C by Eq. 6/2 gives $\bar{a}$ directly. Thus,

$[\Sigma M_C = \bar{I}\alpha + m\bar{a}d]$ $mgr \sin 20° = mr^2 \dfrac{\bar{a}}{r} + m\bar{a}r$ $\bar{a} = \dfrac{g}{2} \sin 20°$

To check our assumption of no slipping, we calculate F and N and compare F with its limiting value. From the above equations,

$$F = mg \sin 20° - m\frac{g}{2} \sin 20° = 0.1710mg$$

$$N = mg \cos 20° = 0.940mg$$

But the maximum possible friction force is

$[F_{\max} = \mu_s N]$ $F_{\max} = 0.15(0.940mg) = 0.1410mg$

Because our calculated value of $0.1710mg$ exceeds the limiting value of $0.1410mg$, we conclude that our assumption of pure rolling was wrong. Therefore, the hoop slips as it rolls and $\bar{a} \neq r\alpha$. The friction force then becomes the kinetic value

$[F = \mu_k N]$ $F = 0.12(0.940mg) = 0.1128mg$

The motion equations now give

$[\Sigma F_x = m\bar{a}_x]$ $mg \sin 20° - 0.1128mg = m\bar{a}$

$\bar{a} = 0.229(9.81) = 2.25 \text{ m/s}^2$

③ $[\Sigma M_G = \bar{I}\alpha]$ $0.1128mg(r) = mr^2\alpha$

$$\alpha = \frac{0.1128(9.81)}{0.150} = 7.37 \text{ rad/s}^2 \qquad \textit{Ans.}$$

The time required for the center G of the hoop to move 3 m from rest with constant acceleration is

$[x = \frac{1}{2}a t^2]$ $t = \sqrt{\dfrac{2x}{\bar{a}}} = \sqrt{\dfrac{2(3)}{2.25}} = 1.633 \text{ s} \qquad \textit{Ans.}$

① Because all of the mass of a hoop is a distance r from its center G, its moment of inertia about G must be mr^2.

② Note that $\bar{a}$ is independent of both m and r.

③ Note that α is independent of m but dependent on r.

Sample Problem 6/6

The drum A is given a constant angular acceleration α_0 of 3 rad/s^2 and causes the 70-kg spool B to roll on the horizontal surface by means of the connecting cable, which wraps around the inner hub of the spool. The radius of gyration $\bar{k}$ of the spool about its mass center G is 250 mm, and the coefficient of static friction between the spool and the horizontal surface is 0.25. Determine the tension T in the cable and the friction force F exerted by the horizontal surface on the spool.

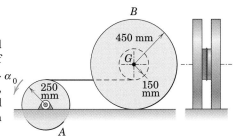

Solution. The free-body diagram and the kinetic diagram of the spool are drawn as shown. The correct direction of the friction force may be assigned in this problem by observing from both diagrams that with counterclockwise angular acceleration, a moment sum about point G (and also about point D) must be counterclockwise. A point on the connecting cable has an acceleration $a_t = r\alpha = 0.25(3) = 0.75$ m/s^2, which is also the horizontal component of the acceleration of point D on the spool. It will be assumed initially that the spool rolls without slipping, in which case it has a counterclockwise angular acceleration $\alpha = (a_D)_x/\overline{DC} = 0.75/0.30 = 2.5$ rad/s^2. The acceleration of the mass center G is, therefore, $\bar{a} = r\alpha = 0.45(2.5) = 1.125$ m/s^2.

With the kinematics determined, we now apply the three equations of motion, Eqs. 6/1.

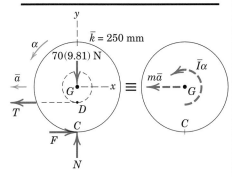

$$[\Sigma F_x = m\bar{a}_x] \qquad F - T = 70(-1.125) \qquad (a)$$

$$[\Sigma F_y = m\bar{a}_y] \qquad N - 70(9.81) = 0 \qquad N = 687 \text{ N}$$

② $$[\Sigma M_G = \bar{I}\alpha] \qquad F(0.450) - T(0.150) = 70(0.250)^2(2.5) \qquad (b)$$

Solving (a) and (b) simultaneously gives

$$F = 75.8 \text{ N} \quad \text{and} \quad T = 154.6 \text{ N} \qquad Ans.$$

To establish the validity of our assumption of no slipping, we see that the surfaces are capable of supporting a maximum friction force $F_{\max} = \mu_s N = 0.25(687) = 171.7$ N. Since only 75.8 N of friction force is required, we conclude that our assumption of rolling without slipping is valid.

If the coefficient of static friction had been 0.1, for example, then the friction force would have been limited to $0.1(687) = 68.7$ N, which is less than 75.8 N, and the spool would slip. In this event, the kinematic relation $\bar{a} = r\alpha$ would no longer hold. With $(a_D)_x$ known, the angular acceleration would be $\alpha = [\bar{a} - (a_D)_x]/\overline{GD}$. Using this relation along with $F = \mu_k N = 68.7$ N, we would then re-solve the three equations of motion for the unknowns T, $\bar{a}$, and α.

Alternatively, with point C as a moment center in the case of pure rolling, we may use Eq. 6/2 and obtain T directly. Thus,

$$[\Sigma M_C = \bar{I}\alpha + m\bar{a}r] \qquad 0.3T = 70(0.25)^2(2.5) + 70(1.125)(0.45)$$

$$T = 154.6 \text{ N} \qquad Ans.$$

where the previous kinematic results for no slipping have been incorporated. We could also write a moment equation about point D to obtain F directly.

① The relation between $\bar{a}$ and α is the kinematical requirement that accompanies the constraint to the motion imposed by the assumption that the spool rolls without slipping.

② Be careful not to make the mistake of using $\frac{1}{2}mr^2$ for $\bar{I}$ of the spool, which is not a uniform circular disk.

③ Our principles of relative acceleration are a necessity here. Hence, the relation $(a_{G/D})_t = \overline{GD}\alpha$ should be recognized.

④ The flexibility in the choice of moment centers provided by the kinetic diagram can usually be employed to simplify the analysis.

Sample Problem 6/7

The slender 30-kg bar AB moves in the vertical plane, with its ends constrained to follow the smooth horizontal and vertical guides. If the 150-N force is applied at A with the bar initially at rest in the position for which $\theta = 30°$, calculate the resulting angular acceleration of the bar and the forces on the small end rollers at A and B.

Solution. The bar undergoes constrained motion, so that we must establish the relationship between the mass-center acceleration and the angular acceleration. The relative-acceleration equation $\mathbf{a}_A = \mathbf{a}_B + \mathbf{a}_{A/B}$

① must be solved first, and then the equation $\overline{\mathbf{a}} = \mathbf{a}_G = \mathbf{a}_B + \mathbf{a}_{G/B}$ is next solved to obtain expressions relating $\overline{a}$ and α. With α assigned in its clockwise physical sense, the acceleration polygons which represent these equations are shown, and their solution gives

$$\overline{a}_x = \overline{a} \cos 30° = 0.6\alpha \cos 30° = 0.520\alpha \text{ m/s}^2$$

$$\overline{a}_y = \overline{a} \sin 30° = 0.6\alpha \sin 30° = 0.3\alpha \text{ m/s}^2$$

Next we construct the free-body diagram and the kinetic diagram as shown. With $\overline{a}_x$ and $\overline{a}_y$ now known in terms of α, the remaining unknowns are α and the forces A and B. We now apply Eqs. 6/1, which give

② $[\Sigma M_G = \overline{I}\alpha]$

$$150(0.6 \cos 30°) - A(0.6 \sin 30°) + B(0.6 \cos 30°) = \frac{1}{12} 30(1.2^2)\alpha$$

$[\Sigma F_x = m\overline{a}_x]$ $\quad 150 - B = 30(0.520\alpha)$

$[\Sigma F_y = m\overline{a}_y]$ $\quad A - 30(9.81) = 30(0.3\alpha)$

Solving the three equations simultaneously gives us the results

$$A = 337 \text{ N} \quad B = 76.8 \text{ N} \quad \alpha = 4.69 \text{ rad/s}^2 \quad \textit{Ans.}$$

As an alternative solution, we can use Eq. 6/2 with point C as the moment center and avoid the necessity of solving three equations simultaneously. This choice eliminates reference to forces A and B and gives α directly. Thus,

③ $[\Sigma M_C = \overline{I}\alpha + \Sigma m\overline{a}d]$

$$150(1.2 \cos 30°) - 30(9.81)(0.6 \sin 30°) = \frac{1}{12} 30(1.2)^2\alpha$$

$$+ 30(0.520\alpha)(0.6 \cos 30°) + 30(0.3\alpha)(0.6 \sin 30°)$$

$$67.6 = 14.40\alpha \quad \alpha = 4.69 \text{ rad/s}^2 \quad \textit{Ans.}$$

With α determined, we can now apply the force equations independently and get

$[\Sigma F_y = m\overline{a}_y]$ $\quad A - 30(9.81) = 30(0.3)(4.69)$ $\quad A = 337 \text{ N} \quad \textit{Ans.}$

$[\Sigma F_x = m\overline{a}_x]$ $\quad 150 - B = 30(0.520)(4.69)$ $\quad B = 76.8 \text{ N} \quad \textit{Ans.}$

① If the application of the relative-acceleration equations is not perfectly clear at this point, then review Art. 5/6. Note that the relative normal acceleration term is absent since there is no angular velocity of the bar.

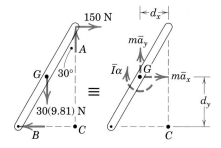

② Recall that the moment of inertia of a slender rod about its center is $\frac{1}{12}ml^2$.

③ From the kinetic diagram, $\Sigma m\overline{a}d = m\overline{a}_x d_y + m\overline{a}_y d_x$. Since both terms of the sum are clockwise, in the same sense as $\overline{I}\alpha$, they are positive.

Sample Problem 6/8

A car door is inadvertently left slightly open when the brakes are applied to give the car a constant rearward acceleration a. Derive expressions for the angular velocity of the door as it swings pass the 90° position and the components of the hinge reactions for any value of θ. The mass of the door is m, its mass center is a distance $\bar{r}$ from the hinge axis O, and the radius of gyration about O is k_O.

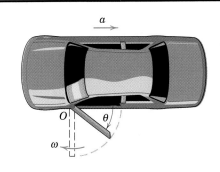

Solution. Because the angular velocity ω increases with θ, we need to find how the angular acceleration α varies with θ so that we may integrate it over the interval to obtain ω. We obtain α from a moment equation about O. First, we draw the free-body diagram of the door in the horizontal plane for a general position θ. The only forces in this plane are the components of the hinge reaction shown here in the x- and y-directions. On the kinetic diagram, in addition to the resultant couple $\bar{I}\alpha$ shown in the sense of α, we represent the resultant force $m\bar{a}$ in terms of its components by using an equation of relative acceleration with respect to O. This equation becomes the kinematic equation of constraint and is

$$\bar{\mathbf{a}} = \mathbf{a}_G = \mathbf{a}_O + (\mathbf{a}_{G/O})_n + (\mathbf{a}_{G/O})_t$$

The magnitudes of the $m\bar{a}$ components are then

$$ma_O = ma \qquad m(a_{G/O})_n = m\bar{r}\omega^2 \qquad m(a_{G/O})_t = m\bar{r}\alpha$$

where $\omega = \dot{\theta}$ and $\alpha = \ddot{\theta}$.

For a given angle θ, the three unknowns are α, O_x, and O_y. We can eliminate O_x and O_y by a moment equation about O, which gives

$$[\Sigma M_O = \bar{I}\alpha + \Sigma m\bar{a}d] \qquad 0 = m(k_O{}^2 - \bar{r}^2)\alpha + m\bar{r}\alpha(\bar{r}) - ma(\bar{r}\sin\theta)$$

Solving for α gives

$$\alpha = \frac{a\bar{r}}{k_O{}^2}\sin\theta$$

Now we integrate α first to a general position and get

$$[\omega\,d\omega = \alpha\,d\theta] \qquad \int_0^\omega \omega\,d\omega = \int_0^\theta \frac{a\bar{r}}{k_O{}^2}\sin\theta\,d\theta$$

$$\omega^2 = \frac{2a\bar{r}}{k_O{}^2}(1 - \cos\theta)$$

For $\theta = \pi/2$, $\qquad \omega = \dfrac{1}{k_O}\sqrt{2a\bar{r}}$ *Ans.*

To find O_x and O_y for any given value of θ, the force equations give

$$[\Sigma F_x = m\bar{a}_x] \qquad O_x = ma - m\bar{r}\omega^2\cos\theta - m\bar{r}\alpha\sin\theta$$

$$= m\left[a - \frac{2a\bar{r}^2}{k_O{}^2}(1 - \cos\theta)\cos\theta - \frac{a\bar{r}^2}{k_O{}^2}\sin^2\theta\right]$$

$$= ma\left[1 - \frac{\bar{r}^2}{k_O{}^2}(1 + 2\cos\theta - 3\cos^2\theta)\right] \quad \textit{Ans.}$$

$$[\Sigma F_y = m\bar{a}_y] \qquad O_y = m\bar{r}\alpha\cos\theta - m\bar{r}\omega^2\sin\theta$$

$$= m\bar{r}\frac{a\bar{r}}{k_O{}^2}\sin\theta\cos\theta - m\bar{r}\frac{2a\bar{r}}{k_O{}^2}(1 - \cos\theta)\sin\theta$$

$$= \frac{ma\bar{r}^2}{k_O{}^2}(3\cos\theta - 2)\sin\theta \qquad \textit{Ans.}$$

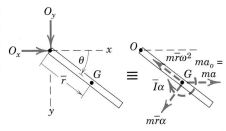

① Point O is chosen because it is the only point on the door whose acceleration is known.

② Be careful to place $m\bar{r}\alpha$ in the sense of positive α with respect to rotation about O.

③ The free-body diagram shows that there is zero moment about O. We use the transfer-of-axis theorem here and substitute $k_O{}^2 = \bar{k}^2 + \bar{r}^2$. If this relation is not totally familiar, review Art. B/1 in Appendix B.

④ We may also use Eq. 6/3 with O as a moment center

$$\Sigma M_O = \bar{I}_O\alpha + \bar{\rho} \times m\mathbf{a}_O$$

where the scalar values of the terms are $\bar{I}_O\alpha = mk_O{}^2\alpha$ and $\bar{\rho} \times m\mathbf{a}_O$ becomes $-\bar{r}ma\sin\theta$.

⑤ The kinetic diagram shows clearly the terms that make up $m\bar{a}_x$ and $m\bar{a}_y$.

PROBLEMS

Introductory problems

6/73 The uniform square steel plate has a mass of 6 kg and is resting on a smooth horizontal surface in the *x-y* plane. If a horizontal force $P = 120$ N is applied to one corner in the direction shown, determine the magnitude of the initial acceleration of corner *A*.

Ans. $a_A = 63.2$ m/s^2

Problem 6/73

6/74 The 30-kg solid circular disk is initially at rest on the horizontal surface when a 12-N force *P*, constant in magnitude and direction, is applied to the cord wrapped securely around its periphery. Friction between the disk and the surface is negligible. Calculate the angular velocity ω of the disk after the 12-N force has been applied for 2 seconds and find the linear velocity v of the center of the disk after it has moved 1.2 meters from rest.

Problem 6/74

6/75 The uniform 50-kg pole *ABC* is balanced in the vertical position when a 500-N horizontal force is suddenly applied at *B*. If the coefficient of kinetic friction between the pole and the ground is 0.3, determine the initial acceleration of point *A*.

Ans. $a_A = 5.89$ m/s^2 to the right

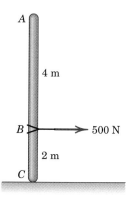

Problem 6/75

6/76 The spacecraft is spinning with a constant angular velocity ω about the *z*-axis at the same time that its mass center *O* is traveling with a velocity v_O in the *y*-direction. If a tangential hydrogen-peroxide jet is fired when the craft is in the position shown, determine the expression for the absolute acceleration of point *A* on the spacecraft rim at the instant the jet force is *F*. The radius of gyration of the craft about the *z*-axis is *k*, and its mass is *m*.

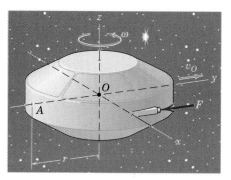

Problem 6/76

6/77 A long cable of length L and mass ρ per unit length is wrapped around the periphery of a spool of negligible mass. One end of the cable is fixed, and the spool is released from rest in the position shown. Find the initial acceleration a of the center of the spool.   *Ans.* $a = g/2$

Problem 6/77

6/78 Above the earth's atmosphere at an altitude of 400 km where the acceleration due to gravity is 8.69 m/s^2, a certain rocket has a total remaining mass of 300 kg and is directed 30° from the vertical. If the thrust T from the rocket motor is 4 kN and if the rocket nozzle is tilted through an angle of 1° as shown, calculate the angular acceleration α of the rocket and the x- and y-components of the acceleration of its mass center G. The rocket has a centroidal radius of gyration of 1.5 m.

Problem 6/78

6/79 What should be the radius r_0 of the circular groove in order that there be no friction force acting between the wheel and the horizontal surface regardless of the magnitude of the force P applied to the cord? The centroidal radius of gyration of the wheel is $\bar{k}$.

Ans. $r_0 = \bar{k}^2/r$

Problem 6/79

6/80 The 10-kg wheel with a radius of gyration of 180 mm about its center O is released from rest on the 60° incline and slips as it rolls. If the kinetic coefficient of friction is $\mu_k = 0.30$, calculate the acceleration a_O of the center O of the wheel and its angular acceleration α.

Problem 6/80

6/81 How large would the coefficient of static friction μ_s have to be in order that the wheel of Prob. 6/80 not slip as it rolls?

Ans. $(\mu_s)_{min} = 0.775$

6/82 The uniform bar of mass m and length L is moving horizontally with a velocity v on its light end rollers. Determine the force under roller B an instant after it passes point C and prior to mechanical interference with the path. At what velocity v will the force under roller B reach zero?

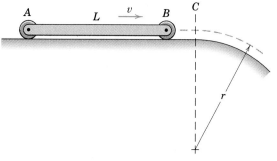

Problem 6/82

6/83 The figure for Prob. 5/118 is repeated here. The 3.6-m steel beam has a mass of 125 kg and is hoisted from rest where the tension in each of the cables is 613 N. If the hoisting drums are given initial angular accelerations $\alpha_1 = 4$ rad/s^2 and $\alpha_2 = 6$ rad/s^2, calculate the corresponding tensions T_A and T_B in the cables. The beam may be treated as a slender bar.

Ans. $T_A = 637$ N, $T_B = 707$ N

Problem 6/83

Representative problems

6/84 The hoop of radius r and negligible mass has four equally spaced spokes, each of mass m and length r. If the hoop is released on the incline and rolls without slipping, determine the acceleration a of the center of the hoop and the friction force F acting on the hoop. For a given angle θ, what is the minimum value μ_s of the static coefficient of friction for which no slipping will occur?

Problem 6/84

6/85 The mass center G of the 10-kg wheel is off center by 10 mm. If G is in the position shown as the wheel rolls without slipping through the bottom of the circular path of 2-m radius with an angular velocity ω of 10 rad/s, compute the force P exerted by the path on the wheel. (Be careful to use the correct mass-center acceleration.)

Ans. $P = 100.3$ N

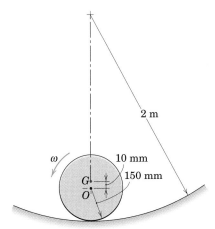

Problem 6/85

6/86 Determine the angular acceleration of each of the two wheels as they roll without slipping down the inclines. For wheel A investigate the case where the mass of the rim and spokes is negligible and the mass of the bar is concentrated along its centerline. For wheel B assume that the thickness of the rim is negligible compared with its radius so that all of the mass is concentrated in the rim. Also specify the minimum coefficient of static friction μ_s required to prevent each wheel from slipping.

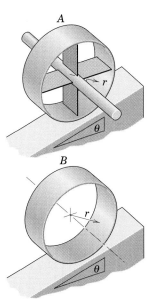

Problem 6/86

6/87 The uniform slender bar rests on a smooth horizontal surface when a force F is applied normal to the bar at point A. Point A is observed to have an initial acceleration a_A of 20 m/s^2, and the bar has a corresponding angular acceleration α of 18 rad/s^2. Determine the distance b.

Ans. $b = 0.553$ m

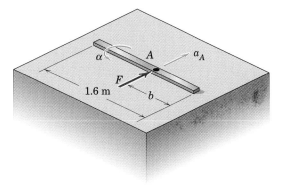

Problem 6/87

6/88 Each of the two solid disk wheels has a mass of 10 kg, and the inner solid cylinder has a mass of 8 kg. The disk wheels and the inner disk are mounted on the small central shaft O-O and can rotate independently of each other. Friction in the shaft bearings is negligible, whereas friction between the incline and the large disk wheels is sufficient to prevent slippage of the wheels. Determine the acceleration of the center O after the assembly is released on the 10° incline. The cord wrapped securely around the inner cylinder is fastened to point A.

Problem 6/88

6/89 The circular disk of 200-mm radius has a mass of 25 kg with centroidal radius of gyration $\bar{k} = 175$ mm and has a concentric circular groove of 75-mm radius cut into it. A steady force T is applied at an angle θ to a cord wrapped around the groove as shown. If $T = 30$ N, $\theta = 0$, $\mu_s = 0.10$, and $\mu_k = 0.08$, determine the angular acceleration α of the disk, the acceleration a of its mass center G, and the friction force F which the surface exerts on the disk.

$$Ans. \ \alpha = 2.12 \ \text{rad/s}^2 \ \text{CW}, \ a = 0.425 \ \text{m/s}^2$$
$$F = 19.38 \ \text{N}$$

Problem 6/89

6/90 The truck, initially at rest with a solid cylindrical roll of paper in the position shown, moves forward with a constant acceleration a. Find the distance s which the truck goes before the paper rolls off the edge of its horizontal bed. Friction is sufficient to prevent slipping.

Problem 6/90

6/91 The fairing which covers the spacecraft package in the nose of the booster rocket is jettisoned when the rocket is in space where gravitational attraction is negligible. A mechanical actuator moves the two halves slowly from the closed position I to position II at which point the fairings are released to rotate freely about their hinges at O under the influence of a constant acceleration a of the rocket. When position III is reached, the hinge at O is released and the fairings drift away from the rocket. Determine the angular velocity ω of the fairings at the 90° position. The mass of each fairing is m with center of mass at G and radius of gyration k_O about O.

$$Ans. \ \omega = \frac{1}{k_O}\sqrt{2a\bar{r}}$$

Problem 6/91

6/92 A uniform slender rod of length l and mass m is secured to a circular hoop of radius l as shown. The mass of the hoop is negligible. If the rod and hoop are released from rest on a horizontal surface in the position illustrated, determine the initial values of the friction force F and normal force N under the hoop if friction is sufficient to prevent slipping.

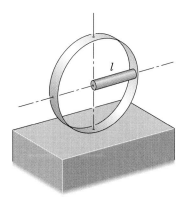

Problem 6/92

6/93 The uniform semicylindrical shell of mass m and radius r is released from rest in the position shown with its lower edge resting on the horizontal surface. Determine the minimum coefficient of static friction μ_s which is necessary to prevent any initial slipping of the shell.

Ans. $\mu_s = 0.399$

Problem 6/93

6/94 A uniform flat plate of mass m in the form of a quarter-circular sector of radius r is secured to the hoop of negligible mass in the position shown. If the hoop and sector are released from rest in this position on a horizontal surface and roll without slipping, determine the initial friction force F.

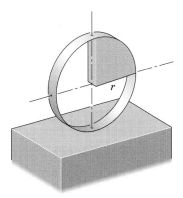

Problem 6/94

6/95 End A of the uniform 5-kg bar is pinned freely to the collar, which has an acceleration $a = 4 \text{ m/s}^2$ along the fixed horizontal shaft. If the bar has a clockwise angular velocity $\omega = 2$ rad/s as it swings past the vertical, determine the components of the force on the bar at A for this instant.

Ans. $A_x = 5$ N, $A_y = 57.1$ N

Problem 6/95

6/96 Small ball-bearing rollers mounted on the ends of the slender bar of mass m and length l constrain the motion of the bar in the horizontal x-y slots. If a couple M is applied to the bar initially at rest at $\theta = 45°$, determine the forces exerted on the rollers at A and B as the bar starts to move.

Problem 6/96

6/97 The slender rod of mass m and length l is released from rest in the vertical position with the small roller at end A resting on the incline. Determine the initial acceleration of A.

$$\text{Ans. } a_A = \frac{g \sin \theta}{1 - \frac{3}{4} \cos^2 \theta}$$

Problem 6/97

6/98 The yo-yo has a mass m and a radius of gyration k about its center O. The cord has a maximum length $y = L$ and is wound around the small inner hub of radius r with its end secured to a point on the hub. If the yo-yo is released from the position $y = 0$ with a downward velocity v_O of its center O, determine the tension T in the cord and the acceleration a of its center during its downward and upward motions. Also find the maximum downward velocity v of its center.

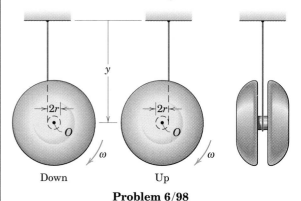

Down   Up

Problem 6/98

6/99 The figure shows the edge view of a uniform concrete slab with a mass of 12 Mg. The slab is being hoisted slowly by the winch D with cable attached to the dolly. At the position $\theta = 60°$, the distance x from the fixed ground position to the dolly is equal to the length $L = 4$ m of the slab. If the hoisting cable should break at this position, determine the initial acceleration a_A of the small dolly, whose mass is negligible, and the initial tension T in the fixed cable. End A of the slab will not slip on the dolly.

$$\text{Ans. } a_A = 8.50 \text{ m/s}^2, \ T = 29.4 \text{ kN}$$

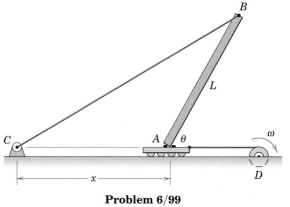

Problem 6/99

6/100 Determine the maximum horizontal force P that may be applied to the cart of mass M for which the wheel will not slip as it begins to roll on the cart. The wheel has mass m, rolling radius r, and radius of gyration $\bar{k}$. The coefficients of static and kinetic friction between the wheel and the cart are μ_s and μ_k, respectively.

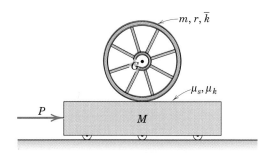

Problem 6/100

6/101 In an investigation of whiplash resulting from rear-end collisions, sudden rotation of the head is modeled by a homogeneous solid sphere of mass m and radius r pivoted about a tangent axis (at the neck). If this axis at O is given a constant acceleration a with the head initially at rest, determine expressions for the initial angular acceleration α of the head and its angular velocity ω as a function of the angle θ of rotation. Assume that the neck is relaxed so that no moment is applied to the head at O.

$$Ans. \quad \alpha = \frac{5a}{7r}, \quad \omega = \sqrt{\frac{10}{7r}}\sqrt{g(1 - \cos\theta) + a\sin\theta}$$

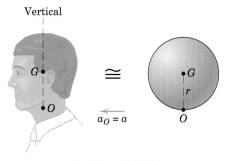

Problem 6/101

6/102 The hydraulic cylinder BC of the dump truck is broken and is disconnected. The driver (who has passed a course in dynamics) decides to calculate the minimum acceleration a of the truck required to tilt the dump about its pivot at A. He then proceeds to calculate the initial angular acceleration α of the dump if the truck is given an acceleration of $1.2a$. What are his correct answers for a and α and would he be able to carry out the experiment? The dump container may be modeled as a homogeneous and solid rectangular block with mass center at G.

Problem 6/102

6/103 The uniform 3.6-m pole is hinged to the truck bed and released from the vertical position as the truck starts from rest with an acceleration of 0.9 m/s^2. If the acceleration remains constant during the motion of the pole, calculate the angular velocity ω of the pole as it reaches the horizontal position.

$$Ans. \quad \omega = 2.99 \text{ rad/s}$$

Problem 6/103

6/104 The truck carries a 1500-mm-diameter spool of cable with a mass of 0.75 kg per meter of length. There are 150 turns on the full spool. The empty spool has a mass of 140 kg with radius of gyration of 530 mm. The truck alone has a mass of 2030 kg with mass center at G. If the truck starts from rest with an initial acceleration of $0.2g$, determine (a) the tension T in the cable where it attaches to the wall and (b) the normal reaction under each pair of wheels. Neglect the rotational inertia of the truck wheels.

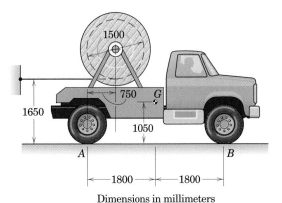

Dimensions in millimeters

Problem 6/104

6/105 In a study of head injury against the instrument panel of a car during sudden or crash stops where lap belts without shoulder straps are used, the segmented human model shown in the figure is analyzed. The hip joint O is assumed to remain fixed relative to the car, and the torso above the hip is treated as a rigid body of mass m freely pivoted at O. The center of mass of the torso is at G with the initial position of OG taken as vertical. The radius of gyration of the torso about O is k_O. If the car is brought to a sudden stop with a constant deceleration a, determine the velocity v relative to the car with which the model's head strikes the instrument panel. Substitute the values $m = 50$ kg, $\bar{r} = 450$ mm, $r = 800$ mm, $k_O = 550$ mm, $\theta = 45°$, and $a = 10g$ and compute v.

Ans. v = 11.73 m/s

Problem 6/105

6/106 The unbalanced 20-kg wheel with mass center at G has a radius of gyration about G of 202 mm. The wheel rolls down the 20° incline without slipping. In the position shown the wheel has an angular velocity of 3 rad/s. Calculate the friction force F acting on the wheel at this position.

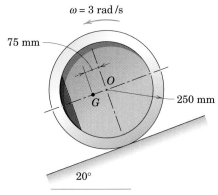

Problem 6/106

6/107 The 0.6-kg connecting rod AB of a certain internal-combustion engine has a mass center at G and has a radius of gyration about G of 28 mm. The piston and piston pin A have a combined mass of 0.82 kg. The engine is running at a constant speed of 3000 rev/min, so that the angular velocity of the crank is $3000(2\pi)/60 = 100\pi$ rad/s. Neglect the weights of the components and the force exerted by the gas in the cylinder compared with the dynamic forces generated and calculate the magnitude of the force on the piston pin A for the crank angle $\theta = 90°$. (*Suggestion:* Use the alternative moment relation, Eq. 6/3, with B as the moment center.)

Ans. $A = 1522$ N

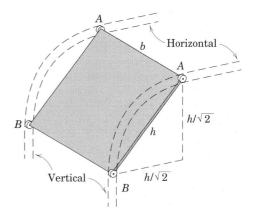

Problem 6/108

6/109 The T-shaped body of mass m is composed of two identical slender bars welded together. If the body is released from rest in the vertical plane in the position shown, determine the initial acceleration of point A. Neglect the small mass and friction of the roller.

Ans. $a_A = \dfrac{14}{109} g$ down the slot

Problem 6/107

Problem 6/109

6/108 The overhead garage door is a homogeneous rectangular panel of mass m and is guided by its corner rollers that run in the tracks shown (dashed). If the door is released from rest in the position shown, determine the force exerted on the door by each of the rollers at A and B. Neglect any friction.

6/110 The uniform square plate of mass m is suspended in the vertical plane by its corner rollers A and B as it moves horizontally to the right with a velocity v. Determine the force that each roller exerts on the plate an instant after corner B has entered the circular portion of its guide.

Problem 6/110

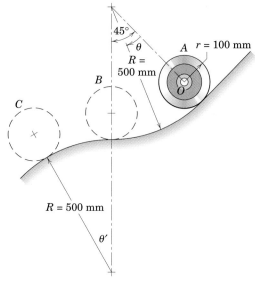

Problem 6/111

▶**6/111** The 5-kg wheel with a radius of gyration of 80 mm about its center O is released from rest at position A and rolls down the circular paths without slipping. Determine the normal force N exerted by the path on the wheel (*a*) just before the wheel reaches position B and (*b*) just after it passes B. Then find the angle θ' at which the wheel loses contact with the path. (*Caution:* Note that the angular velocity of the wheel for the initial concave portion of the path is $\omega = v_O/r = \dfrac{R-r}{r}(-\dot{\theta})$ and for the convex portion is $\omega = v_O/r = \dfrac{R+r}{r}\dot{\theta}'$.)

Ans. (*a*) $N = 66.6$ N, (*b*) $N = 37.4$ N
    (*c*) $\theta' = 48.9°$

▶**6/112** The Ferris wheel at an amusement park has an even number n of gondolas, each freely pivoted at its point of support on the wheel periphery. Each gondola has a loaded mass m, a radius of gyration k about its point of support A, and a mass center a distance h from A. The wheel structure has a moment of inertia I_O about its bearing at O. Determine an expression for the tangential force F which must be transmitted to the wheel periphery at C in order to give the wheel an initial angular acceleration α starting from rest. *Suggestion:* Analyze the gondolas in pairs A and B. Be careful not to assume that the initial angular acceleration of the gondolas is the same as that of the wheel. (*Note:* An American engineer named George Washington Gale Ferris, Jr., created a giant amusement-wheel ride for the World's Columbian Exposition in Chicago in 1893. The wheel was 76 m in diameter with 36 gondolas, each of which carried up to 60 passengers. Fully loaded, the wheel and gondolas had a mass of 1100 Mg. The ride was powered by a 746-kW steam engine.)

$$\textit{Ans. } F = \left\{ mRn \left(1 - \frac{h^2}{2k^2} \right) + \frac{I_O}{R} \right\} \alpha$$

Problem 6/112

SECTION B. WORK AND ENERGY

6/6 WORK-ENERGY RELATIONS

In our study of the kinetics of particles in Arts. 3/6 and 3/7, we developed and applied the principles of work and energy to the motion of a particle and to selected cases of connected particles. We found that these principles were especially useful in describing motion resulting from the cumulative effect of forces acting through distances. Furthermore, when the forces were conservative, we were able to determine velocity changes by analyzing the energy conditions at the beginning and end of the motion interval. For finite displacements, the work-energy method eliminates the necessity for determining the acceleration and integrating it over the interval to obtain the velocity change. These same advantages are realized when we extend the work-energy principles to describe rigid-body motion. Before carrying out this extension we strongly recommend that the definitions and concepts of work, kinetic energy, gravitational and elastic potential energy, conservative forces, and power treated in Arts. 3/6 and 3/7 be carefully reviewed. We shall assume familiarity with these quantities as we apply them to rigid-body problems. In Arts. 4/3 and 4/4 on the kinetics of systems of particles, we extended the principles of Arts. 3/6 and 3/7 to encompass any general system of mass particles, which includes rigid bodies, and this discussion should also be reviewed.

(a) Work of forces and couples. The work done by a force **F** has been treated in detail in Art. 3/6 and is given by

$$U = \int \mathbf{F} \cdot d\mathbf{r} \qquad \text{or} \qquad U = \int (F \cos \alpha) \, ds$$

where $d\mathbf{r}$ is the infinitesimal vector displacement of the point of application of **F**, as shown in Fig. 3/2a. In the equivalent scalar form of the integral, α is the angle between **F** and the direction of the displacement, and ds is the magnitude of the vector displacement $d\mathbf{r}$. We frequently have occasion to evaluate the work done by a couple M which acts on a rigid body during its motion. Figure 6/11 shows a couple $M = Fb$ acting on a rigid body which moves in the plane of the couple. During time dt the body rotates through an angle $d\theta$, and line AB moves to $A'B'$. We may consider this motion in two parts, first a translation to $A'B''$ and then a rotation $d\theta$ about A'. We see immediately that during the translation the work done by one of the forces cancels that done by the other force, so that the net work done is $dU = F(b \, d\theta) = M \, d\theta$ due to the rotational part of the motion. If the couple acts in the sense opposite to the rotation, the work done is negative. During a finite rotation, the work done

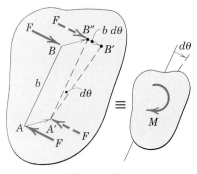

Figure 6/11

by a couple M whose plane is parallel to the plane of motion is, therefore,

$$U = \int M \, d\theta$$

(b) Kinetic energy. We now use the familiar expression for the kinetic energy of a particle to develop expressions for the kinetic energy of a rigid body for each of the three classes of rigid-body plane motion illustrated in Fig. 6/12.

(Translation) The translating rigid body of Fig. 6/12a has a mass m and all of its particles have a common velocity v. The kinetic energy of any particle of mass m_i of the body is $T_i = \frac{1}{2}m_i v^2$, so for the entire body $T = \Sigma\frac{1}{2}m_i v^2 = \frac{1}{2}v^2\Sigma m_i$ or

$$\boxed{T = \tfrac{1}{2}mv^2} \qquad (6/7)$$

This expression holds for both rectilinear and curvilinear translation.

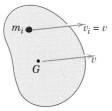

(a) Translation

(Fixed-axis rotation) The rigid body in Fig. 6/12b rotates with an angular velocity ω about the fixed axis through O. The kinetic energy of a representative particle of mass m_i is $T_i = \frac{1}{2}m_i(r_i\omega)^2$. Thus, for the entire body $T = \frac{1}{2}\omega^2\Sigma m_i r_i^2$. But the moment of inertia of the body about O is $I_O = \Sigma m_i r_i^2$, so

$$\boxed{T = \tfrac{1}{2}I_O\omega^2} \qquad (6/8)$$

(b) Fixed-Axis Rotation

The similarity in form of the kinetic energy expressions for translation and rotation should be noted. The reader should verify that the dimensions of the two expressions are identical.

(General plane motion) The rigid body in Fig. 6/12c executes plane motion where, at the instant considered, the velocity of its mass center G is $\bar{v}$ and its angular velocity is ω. The velocity v_i of a representative particle of mass m_i may be expressed in terms of the mass-center velocity $\bar{v}$ and the velocity $\rho_i\omega$ relative to the mass center as shown. With the aid of the law of cosines, we write the kinetic energy of the body as the sum ΣT_i of the kinetic energies of all its particles. Thus,

$$T = \Sigma\tfrac{1}{2}m_i v_i^2 = \Sigma\tfrac{1}{2}m_i(\bar{v}^2 + \rho_i^2\omega^2 + 2\bar{v}\rho_i\omega \cos\theta)$$

Since ω and $\bar{v}$ are common to all terms in the third summation, we may factor them out. Thus, the third term in the expression for T becomes

$$\omega\bar{v}\Sigma m_i\rho_i \cos\theta = \omega\bar{v}\Sigma m_i y_i = 0$$

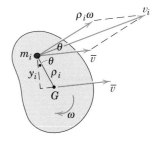

(c) General Plane Motion

Figure 6/12

since $\Sigma m_i y_i = m\bar{y} = 0$. The kinetic energy of the body is then $T = \frac{1}{2}\bar{v}^2\Sigma m_i + \frac{1}{2}\omega^2\Sigma m_i \rho_i^2$ or

$$T = \tfrac{1}{2}m\bar{v}^2 + \tfrac{1}{2}\bar{I}\omega^2 \qquad (6/9)$$

where $\bar{I}$ is the moment of inertia of the body about its mass center. This expression for kinetic energy clearly shows the separate contributions to the total kinetic energy resulting from the translational velocity $\bar{v}$ of the mass center and the rotational velocity ω about the mass center.

The kinetic energy of plane motion may also be expressed in terms of the rotational velocity about the instantaneous center C of zero velocity. Since C momentarily has zero velocity, the proof leading to Eq. 6/8 for the fixed point O holds equally well for point C, so that, alternatively, we may write the kinetic energy of a rigid body in plane motion as

$$T = \tfrac{1}{2}I_C\omega^2 \qquad (6/10)$$

In Art. 4/3 we derived Eq. 4/4 for the kinetic energy of any system of mass. We now see that this expression becomes equivalent to Eq. 6/9 when the mass system is rigid. For a rigid body, the quantity $\dot{\rho}_i$ in Eq. 4/4 is the velocity of the representative particle relative to the mass center and is the vector $\boldsymbol{\omega} \times \boldsymbol{\rho}_i$, which has the magnitude $\rho_i\omega$. The summation term in Eq. 4/4 becomes $\Sigma\frac{1}{2}m_i(\rho_i\omega)^2 = \frac{1}{2}\omega^2\Sigma m_i\rho_i^2 = \frac{1}{2}\bar{I}\omega^2$, which brings Eq. 4/4 into agreement with Eq. 6/9.

(c) Potential energy. Gravitational potential energy V_g and elastic potential energy V_e were covered in detail in Art. 3/7. Recall that the symbol U' (rather than U) is used to denote the work done by all forces except the weight and elastic forces, which are accounted for in the potential-energy terms.

Work-energy equation. The work-energy relation, Eq. 3/17, was introduced in Art. 3/7 for particle motion and was generalized in Art. 4/3 to include the motion of a general system of particles. This equation

$$U'_{1\text{-}2} = \Delta T + \Delta V_g + \Delta V_e \qquad [4/3]$$

applies to any mechanical system. For application to the motion of a single rigid body, the term ΔV_e disappears, and the total work $U'_{1\text{-}2}$ done on the body by the external forces (other than gravity forces) equals the corresponding change ΔT in the kinetic energy of the body plus the change ΔV_g in its potential energy of position in the gravitational field. Alternatively, the equation may be written

$U_{1-2} = \Delta T$, provided that the work of gravitational forces is included in the expression for U.

When applied to an interconnected system of rigid bodies, Eq. 4/3 will include the change ΔV_e in the stored elastic energy in the connections. The term U'_{1-2} will include the work of all forces external to the system (other than gravitational forces), as well as the negative work of internal friction forces if any. The term ΔT is the sum of the changes in kinetic energy of all moving parts during the interval of motion in question, and ΔV_g is the sum of the changes in gravitational potential energy for the various members.

The work-energy relationship may be written alternatively in the form of Eq. 3/17a where the terms correspond to the natural sequence of events.

When the work-energy principle is applied to a single rigid body, either a *free-body diagram* or an *active-force diagram* should be used. In the case of an interconnected system of rigid bodies, an active-force diagram of the entire system should be drawn in order to isolate the system and disclose all forces that do work on the system. Diagrams should also be drawn to disclose the initial and final positions of the system for the given interval of motion.

The work-energy equation provides a direct relationship between the forces that do work and the corresponding changes in the motion of a mechanical system. However, if there is appreciable internal mechanical friction, then the system must be dismembered in order to disclose the kinetic-friction forces and account for the negative work that they do. When the system is dismembered, one of the primary advantages of the work-energy approach is automatically lost. The work-energy method is most useful for analyzing conservative systems of interconnected bodies, where energy loss due to the negative work of friction forces is negligible.

(d) Power. The concept of power was discussed in Art. 3/6 on work-energy for particle motion. It is recalled that power is the time rate at which work is performed. For a force $\mathbf{F}$ acting on a rigid body in plane motion, the power developed by that force at a given instant was given by Eq. 3/12 and is the rate at which it is doing work or

$$P = \frac{dU}{dt} = \frac{\mathbf{F} \cdot d\mathbf{r}}{dt} = \mathbf{F} \cdot \mathbf{v}$$

where $d\mathbf{r}$ and $\mathbf{v}$ are, respectively, the differential displacement and the velocity of the point of application of the force. Similarly, for a couple M acting on the body, the power developed by the couple at a given instant is the rate at which it is doing work or

$$P = \frac{dU}{dt} = \frac{M \, d\theta}{dt} = M\omega$$

where $d\theta$ and ω are, respectively, the differential angular displacement and the angular velocity of the body. If the senses of M and ω are the same, the power is positive and energy is supplied to the body. Conversely, if M and ω have opposite senses, the power is negative and energy is removed from the body. If the force **F** and the couple M act simultaneously, the total instantaneous power is

$$P = \mathbf{F} \cdot \mathbf{v} + M\omega$$

We may also express power by evaluating the rate at which the total mechanical energy of a rigid body or a system of rigid bodies is changing. The work-energy relation, Eq. 4/3, for an infinitesimal displacement is

$$dU' = dT + dV_g + dV_e$$

where dU' is the work of the active forces and couples applied to the body or to the system of bodies. Excluded from dU' are the works of gravitational forces and spring forces, which are accounted for in the terms dV_g and dV_e. Dividing by dt gives the total power of the active forces and couples as

$$P = \frac{dU'}{dt} = \dot{T} + \dot{V}_g + \dot{V}_e = \frac{d}{dt}(T + V)$$

Thus, we see that the power developed by the active forces and couples equals the rate of change of the total mechanical energy of the body or system of bodies. We note from Eq. 6/9 that, for a given body, the first term may be written

$$\dot{T} = \frac{dT}{dt} = \frac{d}{dt}\left(\frac{1}{2}m\overline{\mathbf{v}}\cdot\overline{\mathbf{v}} + \frac{1}{2}\overline{I}\omega^2\right)$$

$$= \frac{1}{2}m(\overline{\mathbf{a}}\cdot\overline{\mathbf{v}} + \overline{\mathbf{v}}\cdot\overline{\mathbf{a}}) + \overline{I}\omega\dot{\omega}$$

$$= m\overline{\mathbf{a}}\cdot\overline{\mathbf{v}} + \overline{I}\alpha(\omega) = \mathbf{R}\cdot\overline{\mathbf{v}} + \overline{M}\omega$$

where **R** is the resultant of *all* forces acting on the body and $\overline{M}$ is the resultant moment about the mass center G of *all* forces. The dot product accounts for the case of curvilinear motion of the mass center, where $\overline{\mathbf{a}}$ and $\overline{\mathbf{v}}$ are not in the same direction.

Sample Problem 6/9

The wheel rolls up the incline on its hubs without slipping and is pulled by the 100-N force applied to the cord wrapped around its outer rim. If the wheel starts from rest, compute its angular velocity ω after its center has moved a distance of 3 m up the incline. The wheel has a mass of 40 kg with center of mass at O and has a centroidal radius of gyration of 150 mm. Determine the power input from the 100-N force at the end of the 3-m motion interval.

Solution. Of the four forces shown on the free-body diagram of the wheel, only the 100-N pull and the weight of $40(9.81) = 392$ N do work. The friction force does no work as long as the wheel does not slip. By use of the concept of the instantaneous center C of zero velocity, we see that a point A on the cord to which the 100-N force is applied has a velocity $v_A = [(200 + 100)/100]v$. Hence, point A on the cord moves a distance of $(200 + 100)/100 = 3$ times as far as the center O. Thus, with the effect of the weight included in the U-term, the work done on the wheel becomes

$$U_{1\text{-}2} = 100 \, \frac{200 + 100}{100} \, (3) - (392 \sin 15°)(3) = 595 \text{ J}$$

The wheel has general plane motion, so that the change in its kinetic energy is

$$[T = \tfrac{1}{2}m\bar{v}^2 + \tfrac{1}{2}\bar{I}\omega^2] \quad \Delta T = \left[\frac{1}{2}40(0.10\omega)^2 + \frac{1}{2}40(0.15)^2\omega^2\right] - 0$$
$$= 0.650\omega^2$$

The work-energy equation gives

$$[U_{1\text{-}2} = \Delta T] \qquad 595 = 0.650\omega^2 \qquad \omega = 30.3 \text{ rad/s}$$

Alternatively, the kinetic energy of the wheel may be written

$$[T = \tfrac{1}{2}I_C\omega^2] \qquad T = \frac{1}{2} \, 40[(0.15)^2 + (0.10)^2]\omega^2 = 0.650\omega^2$$

The power input from the 100-N force when $\omega = 30.3$ rad/s is

$$[P = \mathbf{F} \cdot \mathbf{v}] \qquad P_{100} = 100(0.3)(30.3) = 908 \text{ W} \qquad \qquad \textit{Ans.}$$

① Since the velocity of the instantaneous center C on the wheel is zero, it follows that the rate at which the friction force does work is continuously zero. Hence, F does no work as long as the wheel does not slip. If the wheel were rolling on a moving platform, however, the friction force would do work, even though the wheel were not slipping.

② Note that the component of the weight down the plane does negative work.

③ Be careful to use the correct radius in the expression $v = r\omega$ for the velocity of the center of the wheel.

④ Recall that $I_C = \bar{I} + m\overline{OC}^2$, where $\bar{I} = I_O = mk_O^2$.

⑤ The velocity here is that of the application point of the 100-N force.

Sample Problem 6/10

The 1200-mm slender bar has a mass of 20 kg with mass center at B and is released from rest in the position for which θ is essentially zero. Point B is confined to move in the smooth vertical guide, while end A moves in the smooth horizontal guide and compresses the spring as the bar falls. Determine (a) the angular velocity of the bar as the position $\theta = 30°$ is passed and (b) the velocity with which B strikes the horizontal surface if the stiffness of the spring is 5 kN/m.

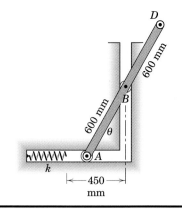

Solution. With the friction and mass of the small rollers at A and B neglected, the system may be treated as being conservative.

Part (a). For the first interval of motion from $\theta = 0$ to $\theta = 30°$, the spring is not engaged, so that there is no V_e term in the energy equation. If we adopt the alternative of treating the work of the weight in the V_g term, then there are no other forces that do work, and $U'_{1\text{-}2} = 0$.

Since we have a constrained plane motion, there is a kinematic relation between the velocity v_B of the center of mass and the angular velocity ω of the bar. This relation is easily obtained by using the instantaneous center C of zero velocity and noting that $v_B = \overline{CB}\omega$. Thus, the kinetic energy of the bar in the 30° position becomes

$$[T = \tfrac{1}{2}m\bar{v}^2 + \tfrac{1}{2}\bar{I}\omega^2]$$

$$T = \frac{1}{2}\,20(0.300\omega)^2 + \frac{1}{2}\left(\frac{1}{12}\,20[1.2]^2\right)\omega^2 = 2.10\omega^2$$

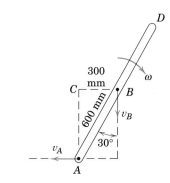

The change in gravitational potential energy is the weight times the change in height of the mass center and equals

$$[\Delta V_g = W\Delta h] \quad \Delta V_g = 20(9.81)(0.600\cos 30° - 0.600) = -15.77 \text{ J}$$

We now substitute into the energy equation and get

$$[U'_{1\text{-}2} = \Delta T + \Delta V_g] \quad 0 = 2.10\omega^2 - 15.77 \quad \omega = 2.74 \text{ rad/s} \quad \textit{Ans.}$$

Part (b). For the entire interval of motion, we include the spring as a part of the system, where

$$[V_e = \tfrac{1}{2}kx^2] \quad \Delta V_e = \frac{1}{2}\,(5000)(0.600 - 0.450)^2 - 0 = 56.3 \text{ J}$$

In the final horizontal position, point A has no velocity, so that the bar is, in effect, rotating about A. Hence, its kinetic energy is

$$[T = \tfrac{1}{2}I_A\omega^2] \quad T = \frac{1}{2}\left(\frac{1}{3}\,20[1.2]^2\right)\left(\frac{v_B}{0.6}\right)^2 = 13.33v_B^2$$

The change in gravitational potential energy is

$$[\Delta V_g = W\Delta h] \quad \Delta V_g = 20(9.81)(-0.600) = -117.7 \text{ J}$$

Substituting into the energy equation gives

$$[U'_{1\text{-}2} = \Delta T + \Delta V_g + \Delta V_e] \quad 0 = (13.33v_B^2 - 0) - 117.7 + 56.3$$

$$v_B = 2.15 \text{ m/s} \quad \textit{Ans.}$$

Alternatively, if the bar alone constitutes the system, the active-force diagram shows the weight, which does positive work, and the spring force kx, which does negative work. We would then write

$$[U_{1\text{-}2} = \Delta T] \quad 117.7 - 56.3 = 13.33v_B^2$$

which is identical with the previous result.

① We recognize that the forces acting on the bar at A and B are normal to the respective directions of motion and, hence, do no work.

② Note that we have kept all quantities in base units.

(Alternative Active-force Diagram)

Sample Problem 6/11

In the mechanism shown, each of the two wheels has a mass of 30 kg and a centroidal radius of gyration of 100 mm. Each link OB has a mass of 10 kg and may be treated as a slender bar. The 7-kg collar at B slides on the fixed vertical shaft with negligible friction. The spring has a stiffness $k = 30$ kN/m and is contacted by the bottom of the collar when the links reach the horizontal position. If the collar is released from rest at the position $\theta = 45°$ and if friction is sufficient to prevent the wheels from slipping, determine (a) the velocity v_B of the collar as it first strikes the spring and (b) the maximum deformation x of the spring.

Solution. The mechanism executes plane motion and is conservative with the neglect of kinetic friction losses. The datum for zero gravitational potential energy V_g is conveniently taken through O as shown.

(a) For the interval from $\theta = 45°$ to $\theta = 0$, we note that ΔT_{wheels} is zero since each wheel starts from rest and momentarily comes to rest at $\theta = 0$. Also, at the lower position each link is merely rotating about its point O so that

$$\Delta T = [2(\tfrac{1}{2}I_O\omega^2) - 0]_{\text{links}} + [\tfrac{1}{2}mv^2 - 0]_{\text{collar}}$$

$$= \frac{1}{3}\,10(0.375)^2\left(\frac{v_B}{0.375}\right)^2 + \frac{1}{2}\,7v_B^2 = 6.83v_B^2$$

The collar at B drops a distance $0.375/\sqrt{2} = 0.265$ m so that

$$\Delta V = \Delta V_g = 0 - 2(10)(9.81)\frac{0.265}{2} - 7(9.81)(0.265) = -44.2 \text{ J}$$

① Also, $U'_{1\text{-}2} = 0$. Hence,

$$[U'_{1\text{-}2} = \Delta T + \Delta V] \quad 0 = 6.83v_B^2 - 44.2 \qquad v_B = 2.54 \text{ m/s} \qquad Ans.$$

(b) At the condition of maximum deformation x of the spring, all parts are momentarily at rest, which makes $\Delta T = 0$. Thus,

$$[U'_{1\text{-}2} = \Delta T + \Delta V_g + \Delta V_e]$$

$$0 = 0 - 2(10)(9.81)\left(\frac{0.265}{2} + \frac{x}{2}\right) - 7(9.81)(0.265 + x) + \tfrac{1}{2}(30)(10^3)x^2$$

Solution for the positive value of x gives

$$x = 60.1 \text{ mm} \qquad Ans.$$

It should be noted that the results of parts (a) and (b) involve a very simple net energy change despite the fact that the mechanism has undergone a fairly complex sequence of motions. Solution of this and similar problems by other than a work-energy approach is not an inviting prospect.

① With the work of the weight of the collar B included in the ΔV_g term, there are no other forces external to the system which do work. The friction force acting under each wheel does no work since the wheel does not slip, and, of course, the normal force does no work here. Hence, $U'_{1\text{-}2} = 0$.

PROBLEMS

(In the following problems neglect any energy loss due to kinetic friction unless otherwise indicated.)

Introductory problems

6/113 The log is suspended by the two parallel 5-m cables and used as a battering ram. At what angle θ should the log be released from rest in order to strike the object to be smashed with a velocity of 4 m/s?

Ans. $\theta = 33.2°$

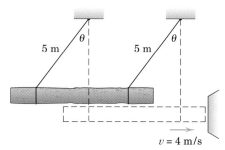

Problem 6/113

6/114 The uniform rectangular plate has a mass of 300 kg and is supported in the vertical plane by the two parallel links of negligible mass and by the cable AC. If the cable suddenly breaks, determine the angular velocity ω of the links an instant before the plate strikes the horizontal surface E. Also find the force in member DC at the same instant.

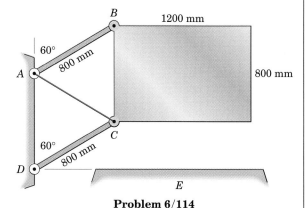

Problem 6/114

6/115 The homogeneous rectangular crate has a mass of 120 kg and is supported in the horizontal position by the cable at A and the corner hinge at O. If the cable at A is suddenly released, calculate the angular velocity ω of the crate just before it strikes the 30° incline. Does the weight of the crate influence the results, other quantities unchanged?

Ans. $\omega = 3.07$ rad/s

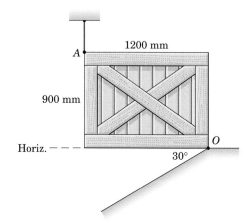

Problem 6/115

6/116 The square frame is made from four slender rods, each of mass m and length b. The frame is rotating in its plane with an angular velocity ω. Determine the linear velocity v of the center C which will make the kinetic energy of translation equal to the kinetic energy of rotation.

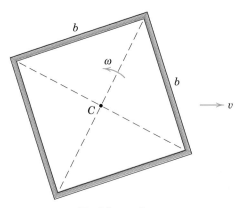

Problem 6/116

6/117 The two uniform rectangular panels, each with a mass of 30 kg, are hinged about their common edge AB and released from rest on the horizontal surface in the position shown. If hinge AB strikes the horizontal surface with a velocity of 3.6 m/s, calculate the energy loss Q due to friction prior to impact.

Ans. $Q = 23.3$ J

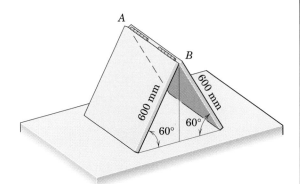

Problem 6/117

6/118 The 15-kg wheel is released from rest and rolls on its hubs without slipping. Calculate the velocity v of the center O of the wheel after it has moved a distance $x = 3$ m down the incline. The radius of gyration of the wheel about O is 125 mm.

Problem 6/118

6/119 A steady 22-N force is applied normal to the handle of the hand-operated grinder. The gear inside the housing with its shaft and attached handle have a combined mass of 1.8 kg and a radius of gyration about their axis of 72 mm. The grinding wheel with its attached shaft and pinion (inside housing) have a combined mass of 0.55 kg and a radius of gyration of 54 mm. If the gear ratio between gear and pinion is 4:1, calculate the speed N of the grinding wheel after 6 complete revolutions of the handle starting from rest.

Ans. $N = 3220$ rev/min

Problem 6/119

Representative problems

6/120 A constant force F is applied in the vertical direction to the symmetrical linkage starting from
the rest position shown. Determine the angular
velocity ω which the links acquire as they reach
the position $\theta = 0$. Each link has a mass m_0. The
wheel is a solid circular disk of mass m and rolls
on the horizontal surface without slipping.

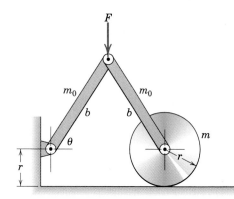

Problem 6/120

6/121 The figure shows the cross section AB of a garage
door which is a rectangular 2.5-m by 5-m panel
of uniform thickness with a mass of 200 kg. The
door is supported by the struts of negligible mass
and hinged at O. Two spring-and-cable assemblies, one on each side of the door, control the
movement. When the door is in the horizontal
open position, each spring is unextended. If the
door is given a slight imbalance from the open
position and allowed to fall, determine the value
of the spring constant k for each spring which will
limit the angular velocity of the door to 1.5 rad/s
when edge B strikes the floor.

                              Ans. $k = 1.270$ kN/m

Problem 6/121

6/122 The 5.5-kg lever OA with 250-mm radius of gyration about O is initially at rest in the vertical
position ($\theta = 90°$) where the attached spring of
stiffness $k = 525$ N/m is unstretched. Calculate
the constant moment M applied to the lever
through its shaft at O which will give the lever
an angular velocity $\omega = 4$ rad/s as the lever
reaches the horizontal position $\theta = 0$.

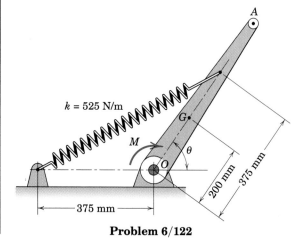

Problem 6/122

6/123 The 50-kg flywheel has a radius of gyration $\bar{k} = 0.4$ m about its shaft axis and is subjected to the torque $M = 2(1 - e^{-0.1\theta})$ N·m, where θ is in radians. If the flywheel is at rest when $\theta = 0$, determine its angular velocity after 5 revolutions.

Ans. $\omega = 3.31$ rad/s

Problem 6/123

6/124 The drum of 375-mm radius and its shaft have a mass of 41 kg and a radius of gyration of 300 mm about the axis of rotation. A total of 18 m of flexible steel cable with a mass of 3.08 kg per meter of length is wrapped around the drum with one end secured to the surface of the drum. The free end of the cable has an initial overhang $x = 0.6$ m as the drum is released from rest. Determine the angular velocity ω of the drum for the instant when $x = 6$ m. Assume that the center of mass of the portion of cable remaining on the drum lies on the shaft axis when $x = 6$ m. Neglect friction.

375 mm

x

Problem 6/124

6/125 The two identical links, each of length b and mass m, may be treated as uniform slender bars. If they are released from rest in the position shown with end A constrained by the smooth vertical guide, determine the velocity v with which A reaches O with θ essentially zero.

Ans. $v = \sqrt{6gb \sin{(\theta/2)}}$

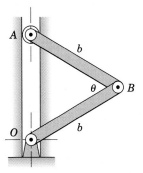

Problem 6/125

6/126 For the pivoted slender rod of length l, determine the distance x for which the angular velocity will be a maximum as the bar passes the vertical position after being released in the horizontal position shown. State the corresponding angular velocity.

Problem 6/126

6/127 Specify the unstretched length l_0 of the spring of stiffness $k = 1400$ N/m which will result in a velocity of 0.25 m/s for the contact at A if the toggle is given a slight nudge from its null position at $\theta = 0$. The toggle has a mass of 1.5 kg and a radius of gyration about O of 55 mm. Motion occurs in the horizontal plane.

Ans. $l_0 = 90.0$ mm

Problem 6/127

6/128 For the assembly shown, arm OA has a mass of 0.8 kg and a radius of gyration about O of 140 mm. Gear B has a mass of 0.9 kg and may be treated as a solid circular disk. Gear C is fixed in the vertical plane and cannot rotate. If a constant moment $M = 4$ N·m is applied to arm OA, initially at rest in the horizontal position shown, calculate the velocity v of point A as it reaches the top at A'.

Problem 6/128

6/129 The center of the 100-kg wheel with centroidal radius of gyration of 100 mm has a velocity of 0.6 m/s down the incline in the position shown. Calculate the normal reaction N under the wheel as it rolls past position A. Assume that no slipping occurs.

Ans. $N = 1696$ N

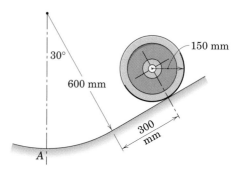

Problem 6/129

6/130 The electric motor shown is delivering 4 kW at 1725 rev/min to a pump which it drives. Calculate the angle δ through which the motor deflects under load if the stiffness of each of its four spring mounts is 15 kN/m. In what direction does the motor shaft turn?

Problem 6/130

6/131 A lid-support mechanism is being designed for a storage chest to limit the angular velocity of the 4.5-kg uniform lid to 1.5 rad/s for $\theta = 0$ when it is released from rest with θ essentially equal to 90°. Two identical mechanisms are included as indicated on the pictorial sketch. Specify the necessary stiffness k of each of the two springs, which are compressed 50 mm upon closure. Neglect the weight of the links and any friction in the sliding collars C. Also, the thickness of the lid is small compared with its other dimensions.

Ans. $k = 4.25$ kN/m

Problem 6/132

6/133 Under active development is the storage of energy in high-speed rotating disks where friction is effectively eliminated by encasing the rotor in an evacuated enclosure and by using magnetic bearings. For a 10-kg rotor with a radius of gyration of 90 mm rotating initially at 80 000 rev/min, calculate the power P which can be extracted from the rotor by applying a constant 2.10-N·m retarding torque (*a*) when the torque is first applied and (*b*) at the instant when the torque has been applied for 120 seconds.

Ans. (*a*) $P = 17.59$ kW, (*b*) $P = 11.06$ kW

6/134 The body shown is constructed of uniform slender rod and consists of a ring of radius r attached to a straight rod section of length $2r$. The body pivots freely about a ball-and-socket joint at O. If the body is at rest in the vertical position shown and is given a slight nudge, compute its angular velocity ω after a 90° rotation about (*a*) axis *A-A* and (*b*) axis *B-B*.

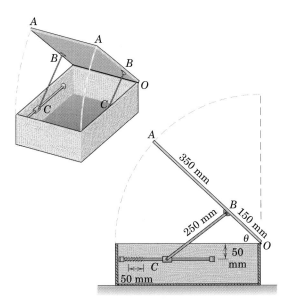

Problem 6/131

6/132 The sheave of 400-mm radius has a mass of 50 kg and a radius of gyration of 300 mm. The sheave and its 100-kg load are suspended by the cable and the spring, which has a stiffness of 1.5 kN/m. If the system is released from rest with the spring initially stretched 100 mm, determine the velocity of O after it has dropped 50 mm.

Problem 6/134

6/135 Motive power for the experimental 10-Mg bus comes from the energy stored in a rotating flywheel which it carries. The flywheel has a mass of 1500 kg and a radius of gyration of 500 mm and is brought up to a maximum speed of 4000 rev/min. If the bus starts from rest and acquires a speed of 72 km/h at the top of a hill 20 m above the starting position, compute the reduced speed N of the flywheel. Assume that 10 percent of the energy taken from the flywheel is lost. Neglect the rotational energy of the wheels of the bus.

Ans. N = 3720 rev/min

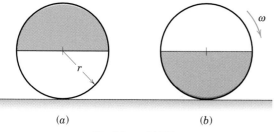

Problem 6/135

6/136 The semicircular disk of mass $m = 2$ kg is mounted in the light hoop of radius $r = 150$ mm and released from rest in position (a). Determine the angular velocity ω of the hoop and the normal force N under the hoop as it passes position (b) after rotating through 180°. The hoop rolls without slipping.

    (a)   (b)

Problem 6/136

6/137 The figure shows the cross section of a 100-kg garage door which is a uniform rectangular panel 2.4 by 2.4 m. The door carries two spring assemblies, one on each side of the door, like the one shown. Each spring has a stiffness of 700 N/m and is unstretched when the door is in the open position shown. If the door is released from rest in this position, calculate the velocity of the edge at A as it strikes the garage floor.

Ans. v = 3.04 m/s

Problem 6/137

6/138 The small vehicle is designed for high-speed travel over the snow. The endless tread for each side of the vehicle has a mass ρ per unit length and is driven by the front wheels. Determine that portion M of the constant front-axle torque required to give both vehicle treads their motion corresponding to a vehicle velocity v achieved with constant acceleration in a distance s from rest on level terrain.

Problem 6/138

6/139 The 10-kg double wheel with radius of gyration of 125 mm about O is connected to the spring of stiffness $k = 600$ N/m by a cord which is wrapped securely around the inner hub. If the wheel is released from rest on the incline with the spring stretched 225 mm, calculate the maximum velocity v of its center O during the ensuing motion. The wheel rolls without slipping.

Ans. $v_{\max} = 1.325$ m/s

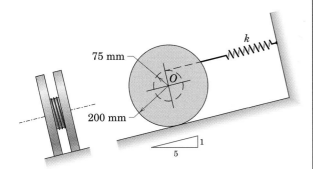

Problem 6/139

6/140 Each of the two hinges at A and B of the uniform lid of mass m of a child's toy chest contains a torsion spring which exerts a resisting moment $M = K\theta$ on the lid as it is being closed. (a) Specify the torsional stiffness K of each spring which will result in zero angular velocity of the lid as it reaches the horizontal closed position ($\theta = \pi/2$) when released from rest at $\theta = 0$. (b) What would be the angular acceleration α of the lid when it is released from rest in the closed position? Would these hinges be a practical solution?

Problem 6/140

6/141 The 30-kg uniform circular disk is at rest in position A when a constant counterclockwise couple $M = 40$ N·m is applied to it to roll the disk up the circular surface. If the disk rolls without slipping, compute the velocity v of its center O as the top position B is reached. (*Caution*: Be careful to establish the correct angle through which the disk rotates.)

Ans. $v = 2.10$ m/s

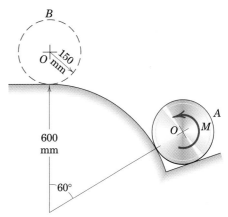

Problem 6/141

6/142 The figure shows the cross section of a uniform 100-kg ventilator door hinged about its upper horizontal edge at O. The door is controlled by the spring-loaded cable which passes over the small pulley at A. The spring has a stiffness of 200 N per meter of stretch and is undeformed when $\theta = 0$. If the door is released from rest in the horizontal position, determine the maximum angular velocity ω reached by the door and the corresponding angle θ.

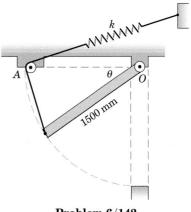

Problem 6/142

6/143 Each of the two links has a mass of 2 kg and a centroidal radius of gyration of 60 mm. The slider at B has a mass of 3 kg and moves freely in the vertical guide. The spring has a stiffness of 6 kN/m. If a constant torque $M = 20$ N·m is applied to link OA through its shaft at O starting from the rest position at $\theta = 45°$, determine the angular velocity ω of OA when $\theta = 0$.

Ans. $\omega = 9.51$ rad/s

Problem 6/143

6/144 A small experimental vehicle has a total mass m of 500 kg including wheels and driver. Each of the four wheels has a mass of 40 kg and a centroidal radius of gyration of 400 mm. Total frictional resistance R to motion is 400 N and is measured by towing the vehicle at a constant speed on a level road with engine disengaged. Determine the power output of the engine for a speed of 72 km/h up the 10-percent grade (*a*) with zero acceleration and (*b*) with an acceleration of 3 m/s². (*Hint:* Power equals the time rate of increase of the total energy of the vehicle plus the rate at which frictional work is overcome.)

Problem 6/144

6/145 The uniform 15-kg semicircular disk is supported in the equilibrium position shown by the two cables which are wrapped around its attached hubs and lead to the identical springs. Each spring has a stiffness $k = 2.6$ kN/m. If the disk is rotated 90° so that its mass center is in the lowest possible position and then released from rest, calculate the angular velocity ω of the disk as it passes the equilibrium position. Neglect the mass of the hubs and shaft.

Ans. $\omega = 7.11$ rad/s

Problem 6/145

6/146 The uniform slender rod of length l is released from rest in the dashed vertical position. With what speed v_A does end A strike the 30° incline? Neglect the small mass and friction of the end rollers.

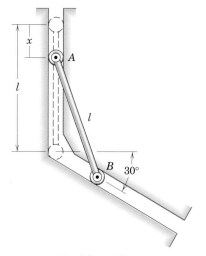

Problem 6/146

6/147 The 45-kg uniform circular disk with its attached 9-kg slender bar is released from rest in the position shown and rolls without slipping on the horizontal surface. Calculate the velocity v_O of the center O when the mass center of the bar is directly below the center O of the disk.

Ans. $v_O = 0.954$ m/s

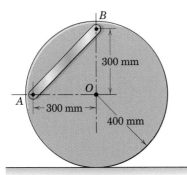

Problem 6/147

6/148 The open square frame is constructed of four identical slender rods each of length b. If the frame is released from rest in the position shown, determine the speed of corner A (a) after A has dropped a distance b and (b) after A has dropped a distance $2b$. The small wheels roll without friction in the slots of the vertical surface.

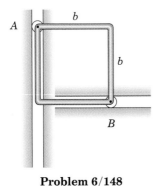

Problem 6/148

6/149 A solid roll of wrapping paper with an initial radius r_0 is released from rest on the incline and allowed to unroll with the free end clamped at the top. Determine the velocity of the roll in terms of the distance x that it has moved down the slope. The total length of the paper in the roll is L. Can you reconcile the difference between the initial energy of the roll and the final energy of the paper after all motion has ceased?

Ans. $v = 2\sqrt{\dfrac{gx}{3}\dfrac{L - x/2}{L - x}}\sin\theta$

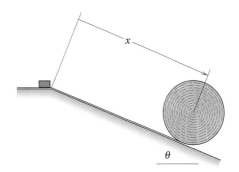

Problem 6/149

6/150 Motion of the 600-mm slender bar of mass 4 kg is controlled by the constrained movement of its small rollers A and B of negligible mass and friction. The bar starts from rest in the horizontal position with $\theta = 0$ and moves in the vertical plane under the action of the constant force $P = 50$ N applied normal to the bar at end C. Calculate the velocity v with which roller A strikes the wall of the vertical guide at $\theta = 90°$.

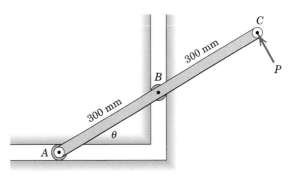

Problem 6/150

▶ **6/151** The 7-kg wheel with radius of gyration of 100 mm about its center O is constrained by the band which is secured at A and wraps around the wheel and ends at point B. The band has a mass of 1.2 kg per meter of length. If the wheel is released from rest in the position shown with a steady 18-N force applied to the band at B, calculate the velocity of the center O of the wheel when it has rolled 250 mm down the incline.

Ans. $v = 1.231$ m/s

▶ **6/152** The uniform bar ABC has a mass m and starts from rest with $\theta = 180°$ where A, B, C, and O are collinear. If the applied force P is constant in magnitude, determine the angular velocity ω of the bar as B reaches O with $\theta = 0$. The mass of the roller at B and the mass of the strut OC are negligible. (*Hint:* Replace P by a force P at B and a couple.)

Ans. $\omega = \sqrt{\dfrac{6P\pi}{13mb}}$

Problem 6/152

Problem 6/151

6/7 ACCELERATION FROM WORK-ENERGY; VIRTUAL WORK

In addition to the determination of velocities resulting from the action of forces acting over finite displacements, we may use the work-energy equation to establish the instantaneous accelerations of the members of a system of interconnected bodies as a result of the active forces applied. We may also modify the equation to establish the configuration of such a system when it undergoes a constant acceleration.

For an infinitesimal interval of motion, Eq. 4/3 becomes

$$dU' = dT + dV$$

The term dU' represents the total work done by all active non-potential forces acting on the system under consideration during the infinitesimal displacement of the system. The work of potential forces is included in the dV-term. If we use the subscript i to denote a representative body of the interconnected system, the differential change in kinetic energy T for the entire system becomes

$$dT = d(\Sigma\tfrac{1}{2}m_i\bar{v}_i^2 + \Sigma\tfrac{1}{2}\bar{I}_i\omega_i^2) = \Sigma m_i\bar{v}_i \, d\bar{v}_i + \Sigma\bar{I}_i\omega_i \, d\omega_i$$

where $d\bar{v}_i$ and $d\omega_i$ are the respective changes in the magnitudes of the velocities and where the summation is taken over all bodies of the system. But for each body, $m_i\bar{v}_i \, d\bar{v}_i = m_i\bar{\mathbf{a}}_i\cdot d\bar{\mathbf{s}}_i$ and $\bar{I}_i\omega_i \, d\omega_i = \bar{I}_i\alpha_i \, d\theta_i$, where $d\bar{\mathbf{s}}_i$ represents the infinitesimal linear displacement of the center of mass and where $d\theta_i$ represents the infinitesimal angular displacement of the body in the plane of motion. We note that $\bar{\mathbf{a}}_i\cdot d\bar{\mathbf{s}}_i$ is identical to $(\bar{a}_i)_t \, d\bar{s}_i$, where $(\bar{a}_i)_t$ is the component of $\bar{\mathbf{a}}_i$ along the tangent to the curve described by the mass center of the body in question. Also α_i represents $\ddot{\theta}_i$, the angular acceleration of the representative body. Consequently, for the entire system

$$dT = \Sigma m_i\bar{\mathbf{a}}_i\cdot d\bar{\mathbf{s}}_i + \Sigma\bar{I}_i\alpha_i \, d\theta_i$$

This change may also be written as

$$dT = \Sigma\mathbf{R}_i\cdot d\bar{\mathbf{s}}_i + \Sigma\mathbf{M}_{G_i}\cdot d\boldsymbol{\theta}_i$$

where $\mathbf{R}_i$ and $\mathbf{M}_{G_i}$ are the resultant force and resultant couple acting on body i and where $d\boldsymbol{\theta}_i = d\theta_i\mathbf{k}$. These last two equations merely show us that the differential change in kinetic energy equals the differential work done on the system by the resultant forces and resultant couples acting on all the bodies of the system.

The term dV represents the differential change in the total gravitational potential energy V_g and the total elastic potential energy V_e and has the form

$$dV = d(\Sigma m_igh_i + \Sigma\tfrac{1}{2}k_jx_j^2) = \Sigma m_ig \, dh_i + \Sigma k_jx_j \, dx_j$$

where h_i represents the vertical distance of the center of mass of the representative body of mass m_i above any convenient datum plane and where x_j stands for the deformation, tensile or compressive, of a representative elastic member of the system (spring) whose stiffness is k_j.

The complete expression for dU' may now be written as

$$dU' = \Sigma m_i \bar{\mathbf{a}}_i \cdot d\bar{\mathbf{s}}_i + \Sigma \bar{I}_i \alpha_i \, d\theta_i + \Sigma m_i g \, dh_i + \Sigma k_j x_j \, dx_j \quad \text{(6/11)}$$

In the application of Eq. 6/11 to a system of one degree of freedom, the terms $m_i \bar{\mathbf{a}}_i \cdot d\bar{\mathbf{s}}_i$ and $\bar{I}_i \alpha_i \, d\theta_i$ will be positive if the accelerations are in the same direction as the respective displacements and negative if they are in the opposite direction. Equation 6/11 has the advantage of relating the accelerations to the active forces directly, which eliminates the need for dismembering the system and then eliminating the internal forces and reactive forces by simultaneous solution of the force-mass-acceleration equations for each member.

Virtual work. In Eq. 6/11 the differential motions are differential changes in the real or actual displacements which occur. For a mechanical system which assumes a steady-state configuration during constant acceleration, we often find it convenient to introduce the concept of *virtual work*. The concepts of virtual work and virtual displacement have been introduced and used to establish equilibrium configurations for static systems of interconnected bodies (see Chapter 7 of *Vol. 1 Statics*). A virtual displacement is any assumed and arbitrary displacement, linear or angular, away from the natural or actual position. For a system of connected bodies, the virtual displacements must be consistent with the constraints of the system. For example, when one end of a link is hinged about a fixed pivot, the virtual displacement of the other end must be normal to the line joining the two ends. Such requirements for displacements consistent with the constraints are purely kinematic and provide what are known as the *equations of constraint*. If a set of virtual displacements satisfying the equations of constraint and therefore consistent with the constraints is given to a mechanical system, the proper relationship between the coordinates which specify the configuration of the system will be established by applying the work-energy relationship of Eq. 6/11, expressed in terms of virtual changes. Thus,

$$\delta U' = \Sigma m_i \bar{\mathbf{a}}_i \cdot \delta \bar{\mathbf{s}}_i + \Sigma \bar{I}_i \alpha_i \, \delta\theta_i + \Sigma m_i g \, \delta h_i + \Sigma k_j x_j \, \delta x_j \quad \text{(6/11a)}$$

It is customary to use the differential symbol d to refer to differential changes in the *real* displacements, whereas the symbol δ is used to signify differential changes which are assumed or *virtual* changes.

80 N

80 mm

A   B

Sample Problem 6/12

The movable rack A has a mass of 3 kg, and rack B is fixed. The gear has a mass of 2 kg and a radius of gyration of 60 mm. In the position shown, the spring, which has a stiffness of 1.2 kN/m, is stretched a distance of 40 mm. For the instant represented, determine the acceleration a of rack A under the action of the 80-N force. The plane of the figure is vertical.

Solution. The given figure represents the active-force diagram for the entire system, which is conservative.

During an infinitesimal upward displacement dx of rack A, the work dU' done on the system is $80\ dx$, where x is in meters, and this work equals the sum of the corresponding changes in the total energy of the system. These changes, which appear in Eq. 6/11, are as follows:

$$[dT = \Sigma m_i \bar{\mathbf{a}}_i \cdot d\bar{\mathbf{s}}_i + \Sigma \bar{I}_i \alpha_i\, d\theta_i]$$

$$dT_{\text{rack}} = 3a\ dx$$

$$dT_{\text{gear}} = 2\,\frac{a}{2}\,\frac{dx}{2} + 2(0.06)^2\,\frac{a/2}{0.08}\,\frac{dx/2}{0.08} = 0.781a\ dx$$

The change in potential energies of the system from Eq. 6/11, becomes

$$[dV = \Sigma m_i g\, dh_i + \Sigma k_j x_j\, dx_j]$$

$$dV_{\text{rack}} = 3g\ dx = 3(9.81)\ dx = 29.4\ dx$$

$$dV_{\text{gear}} = 2g(dx/2) = g\ dx = 9.81\ dx$$

$$dV_{\text{spring}} = k_j x_j\, dx_j = 1200(0.04)\ dx/2 = 24\ dx$$

Substitution into Eq. 6/11 gives us

$$80\ dx = 3a\ dx + 0.781a\ dx + 29.4\ dx + 9.81\ dx + 24\ dx$$

Canceling dx and solving for a give

$$a = 16.76/3.78 = 4.43\ \text{m/s}^2 \qquad\qquad Ans.$$

We see that using the work-energy method for an infinitesimal displacement has given us the direct relation between the applied force and the resulting acceleration. It was unnecessary to dismember the system, draw two free-body diagrams, apply $\Sigma F = m\bar{a}$ twice, apply $\Sigma M_G = \bar{I}\alpha$ and $F = kx$, eliminate unwanted terms, and finally solve for a.

① Note that none of the remaining forces external to the system do any work. The work done by the weight and by the spring is accounted for in the potential energy terms.

② Note that $\bar{a}_i$ for the gear is its mass-center acceleration, which is half that for the rack A. Also its displacement is $dx/2$. For the rolling gear, the angular acceleration from $a = r\alpha$ becomes $\alpha_i = (a/2)/0.08$, and the angular displacement from $ds = r\,d\theta$ becomes $d\theta_i = (dx/2)/0.08$.

③ Note here that the displacement of the spring is one-half that of the rack. Hence, $x_i = x/2$.

Sample Problem 6/13

A constant force P is applied to end A of the two identical and uniform links and causes them to move to the right in their vertical plane with a horizontal acceleration a. Determine the steady-state angle θ made by the bars with one another.

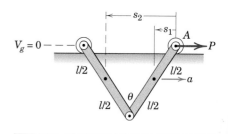

Solution. The figure constitutes the active-force diagram for the system. To find the steady-state configuration, consider a virtual displacement of each bar from the natural position assumed during the acceleration. Measurement of the displacement with respect to end A eliminates any work done by force P during the virtual displacement. Thus,

$$\delta U' = 0$$

The virtual change in kinetic energy from Eq. 6/11a is

$$\delta T = \Sigma m \bar{\mathbf{a}} \cdot \delta \bar{\mathbf{s}} = ma(-\delta s_1) + ma(-\delta s_2)$$

$$= -ma \left[\delta\left(\frac{l}{2} \sin \frac{\theta}{2}\right) + \delta\left(\frac{3l}{2} \sin \frac{\theta}{2}\right) \right]$$

$$= -ma \left(l \cos \frac{\theta}{2} \, \delta\theta \right)$$

We choose the horizontal line through A as the datum for zero potential energy. Thus, the potential energy of the links is

$$V_g = 2mg\left(-\frac{l}{2} \cos \frac{\theta}{2} \right)$$

and the virtual change in potential energy becomes

$$\delta V_g = \delta\left(-2mg \, \frac{l}{2} \cos \frac{\theta}{2} \right) = \frac{mgl}{2} \sin \frac{\theta}{2} \, \delta\theta$$

Substitution into the work-energy equation for virtual changes, Eq. 6/11a, gives

$$[\delta U' = \delta T + \delta V_g] \qquad 0 = -mal \cos \frac{\theta}{2} \, \delta\theta + \frac{mgl}{2} \sin \frac{\theta}{2} \, \delta\theta$$

from which

$$\theta = 2 \tan^{-1} \frac{2a}{g} \qquad\qquad Ans.$$

Again, in this problem we see that the work-energy approach obviated the necessity for dismembering the system, drawing separate free-body diagrams, applying motion equations, eliminating unwanted terms, and solving for θ.

① Note that we use the symbol δ to refer to an assumed or virtual differential change rather than the symbol d, which refers to an infinitesimal change in the real displacement.

② In evaluating δT, we are finding the variation of T resulting from a virtual change in displacement. Therefore, we must use the expression $m\mathbf{a} \cdot \delta\mathbf{s}$ and not the expression $\delta(\frac{1}{2}mv^2)$ which represents the variation of T resulting from a variation in v.

③ We have chosen to use the angle θ to describe the configuration of the links, although we could have used the distance between the two ends of the links just as well.

PROBLEMS

Introductory problems

6/153 The two uniform slender bars are hinged at O and supported on the horizontal surface by their end rollers of negligible mass. If the bars are released from rest in the position shown, determine their initial angular acceleration α as they collapse in the vertical plane. (*Suggestion*: Make use of the instantaneous center of zero velocity in writing the expression for dT.)

$$Ans. \quad \alpha = \frac{3g \cos \theta}{2b}$$

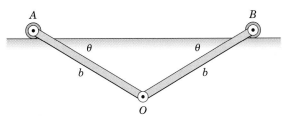

Problem 6/153

6/154 The position of the horizontal platform of mass m_0 is controlled by the parallel slender links of mass m and $2m$. Determine the initial angular acceleration α of the links as they start from their supported position shown under the action of a force P applied normal to AB at its end.

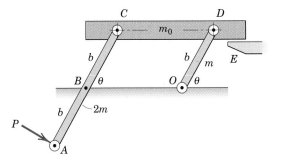

Problem 6/154

6/155 The uniform slender bar of mass m is shown in its equilibrium position in the vertical plane before the couple M is applied to the end of the bar. Determine the initial angular acceleration α of the bar upon application of M. The mass of each guide roller is negligible.

$$Ans. \quad \alpha = \frac{M}{mb^2(\cos^2 \theta + \frac{1}{3})}$$

Problem 6/155

6/156 The mechanism shown moves in the vertical plane. The vertical bar AB has a mass of 4.5 kg, and each of the two links has a mass of 2.7 kg with mass center at G and with a radius of gyration of 250 mm about its bearing (O or C). The spring has a stiffness of 220 N/m and an unstretched length of 450 mm. If the support at D is suddenly withdrawn, determine the initial angular acceleration α of the links.

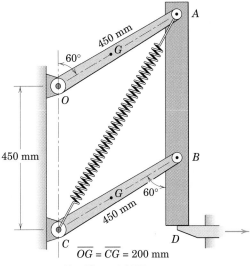

$\overline{OG} = \overline{CG} = 200$ mm

Problem 6/156

Representative problems

6/157 The load of mass m is given an upward acceleration a from its supported rest position by the application of the forces P. Neglect the mass of the links compared with m and determine the initial acceleration a.

$$\text{Ans.} \quad a = \frac{2P}{5m} \tan \frac{\theta}{2} - g$$

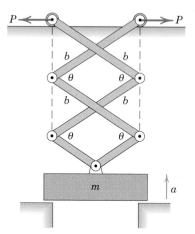

Problem 6/157

6/158 The cargo box of the food-delivery truck for aircraft servicing has a loaded mass m and is elevated by the application of a couple M on the lower end of the link which is hinged to the truck frame. The horizontal slots allow the linkage to unfold as the cargo box is elevated. Determine the upward acceleration of the box in terms of h for a given value of M. Neglect the mass of the links.

Problem 6/158

6/159 The box and load of the dump truck have a mass m with mass center at G and a moment of inertia I_A about the pivot at A. Determine the angular acceleration α of the box when it is started from rest in the position shown under the application of the couple M to link CD. The figure $ABDC$ is a parallelogram. Neglect the mass of the links.

$$\text{Ans.} \quad \alpha = [M - mg(b \cos \theta - a \sin \theta)]/I_A$$

Problem 6/159

6/160 Each of the uniform bars OA and OB has a mass of 2 kg and is freely hinged at O to the vertical shaft, which is given an upward acceleration $a = g/2$. The links which connect the light collar C to the bars have negligible mass, and the collar slides freely on the shaft. The spring has a stiffness $k = 130$ N/m and is uncompressed for the position equivalent to $\theta = 0$. Calculate the angle θ assumed by the bars under conditions of steady acceleration.

Dimensions in millimeters

Problem 6/160

6/167 The aerial tower shown is designed to elevate a workman in a vertical direction. An internal mechanism at B maintains the angle between AB and BC at twice the angle θ between BC and the ground. If the combined mass of the man and the cab is 200 kg and if all other masses are neglected, determine the torque M applied to BC at C and the torque M_B in the joint at B required to give the cab an initial vertical acceleration of 1.2 m/s² when it is started from rest in the position $\theta = 30°$.   *Ans.* $M_B = 11.44$ kN·m
$$M = 0$$

Problem 6/167

6/168 The uniform arm OA has a mass of 4 kg, and the gear D has a mass of 5 kg with a radius of gyration about its center of 64 mm. The large gear B is fixed and cannot rotate. If the arm and small gear are released from rest in the position shown in the vertical plane, calculate the initial angular acceleration α of OA.

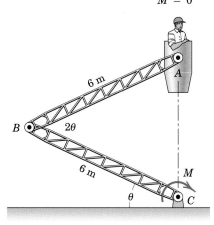

Problem 6/168

6/169 The vehicle is used to transport supplies to and from the bottom of the 25-percent grade. Each pair of wheels, one at A and the other at B, has a mass of 140 kg with a radius of gyration of 150 mm. The drum C has a mass of 40 kg and a radius of gyration of 100 mm. The total mass of the vehicle is 520 kg. The vehicle is released from rest with a restraining force T of 500 N in the control cable which passes around the drum and is secured at D. Determine the initial acceleration a of the vehicle. The wheels roll without slipping.   *Ans.* $a = 0.341$ m/s²

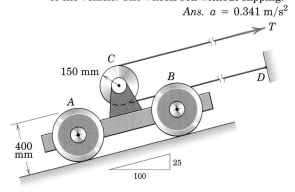

Problem 6/169

▶ **6/170** The chain of total length πr with mass ρ per unit length is supported in the vertical plane on the fixed circular surface with its upper end initially secured at point A. If the fastening at A is suddenly released, determine the initial acceleration a of the chain. Assume that the surface is sufficiently smooth so that friction may be neglected.

$$Ans. \ a = \left(\frac{1}{\pi} + \frac{1}{2}\right)g$$

Problem 6/170

SECTION C. IMPULSE AND MOMENTUM

6/8 IMPULSE-MOMENTUM EQUATIONS

The principles of impulse and momentum were developed and used in Articles 3/9 and 3/10 for the description of particle motion. In this treatment, we observed that these principles were of particular importance when the applied forces were expressible as functions of the time and when interactions between particles occurred during short periods of time, such as with impact. Similar advantages result when the impulse-momentum principles are applied to the motion of rigid bodies.

In Art. 4/2 the impulse-momentum principles were extended to cover any defined system of mass particles without restriction as to the connections between the particles of the system. These extended relations all apply to the motion of a rigid body, which is merely a special case of a general system of mass. We will now apply these equations directly to rigid-body motion in two dimensions.

(a) Linear momentum. In Art. 4/4 we defined the linear momentum of a mass system as the vector sum of the linear momenta of all of its particles and wrote $\mathbf{G} = \Sigma m_i \mathbf{v}_i$. With $\mathbf{r}_i$ representing the position vector to m_i, we have $\mathbf{v}_i = \dot{\mathbf{r}}_i$ and $\mathbf{G} = \Sigma m_i \dot{\mathbf{r}}_i$ which, for a system whose total mass is constant, may be written as $\mathbf{G} = d(\Sigma m_i \mathbf{r}_i)/dt$. When we substitute the principle of moments $m\bar{\mathbf{r}} = \Sigma m_i \mathbf{r}_i$ to locate the mass center, the momentum becomes $\mathbf{G} = d(m\bar{\mathbf{r}})/dt = m\dot{\bar{\mathbf{r}}}$, where $\dot{\bar{\mathbf{r}}}$ is the velocity $\bar{\mathbf{v}}$ of the mass center. Therefore, as before, we find that the linear momentum of any mass system, rigid or nonrigid, is

$$\boxed{\mathbf{G} = m\bar{\mathbf{v}}} \qquad [4/5]$$

In the derivation of Eq. 4/5, we note that it was unnecessary to employ the kinematic condition for a rigid body, Fig. 6/13, which is $\mathbf{v}_i = \bar{\mathbf{v}} + \boldsymbol{\omega} \times \boldsymbol{\rho}_i$. In that case, we obtain the same result by writing $\mathbf{G} = \Sigma m_i(\bar{\mathbf{v}} + \boldsymbol{\omega} \times \boldsymbol{\rho}_i)$. The first sum is $\bar{\mathbf{v}}\Sigma m_i = m\bar{\mathbf{v}}$, and the second sum becomes $\boldsymbol{\omega} \times \Sigma m_i \boldsymbol{\rho}_i = \boldsymbol{\omega} \times m\bar{\boldsymbol{\rho}} = \mathbf{0}$ since $\boldsymbol{\rho}_i$ is measured from the mass center, making $\bar{\boldsymbol{\rho}}$ zero.

Next in Art. 4/4 we rewrote Newton's generalized second law as Eq. 4/6. This equation and its integrated form are

$$\boxed{\Sigma\mathbf{F} = \dot{\mathbf{G}}} \quad \text{and} \quad \boxed{\int_{t_1}^{t_2} \Sigma\mathbf{F}\, dt = \mathbf{G}_2 - \mathbf{G}_1} \qquad (6/12)$$

Equation 6/12 may be written in its scalar component form, which, for plane motion in the x-y plane, gives

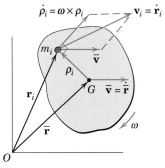

Figure 6/13

$$\boxed{\begin{aligned} \Sigma F_x &= \dot{G}_x \\ \Sigma F_y &= \dot{G}_y \end{aligned}} \quad \text{and} \quad \boxed{\begin{aligned} \int_{t_1}^{t_2} \Sigma F_x \, dt &= G_{x_2} - G_{x_1} \\ \int_{t_1}^{t_2} \Sigma F_y \, dt &= G_{y_2} - G_{y_1} \end{aligned}} \qquad \textbf{(6/12a)}$$

In words, the first of Eqs. 6/12 and 6/12a states that the resultant force equals the time rate of change of momentum. The integrated form of Eqs. 6/12 and 6/12a states that the linear impulse on the body during the interval $t_2 - t_1$ equals the corresponding change in linear momentum. The impulse-momentum relationships may be alternatively written in the form corresponding to Eq. 3/23a, where the order of the terms corresponds to the natural sequence of events. As in the force-mass-acceleration formulation, the force summations in Eqs. 6/12 and 6/12a must include *all* forces acting externally on the body considered. We emphasize, therefore, that in the use of the impulse-momentum equations, it is essential to construct the complete free-body diagram so as to disclose all forces which appear in the force summation. In contrast to the method of work and energy, all forces exert impulses, whether they do work or not.

(b) Angular momentum.   Angular momentum is defined as the moment of linear momentum. In Art. 4/4 we expressed the angular momentum about the mass center of any prescribed system of mass as $\mathbf{H}_G = \Sigma \boldsymbol{\rho}_i \times m_i \mathbf{v}_i$, which is merely the vector sum of the moments about G of the linear momenta of all particles. We showed in Art. 4/4 that this vector sum could also be written as $\mathbf{H}_G = \Sigma \boldsymbol{\rho}_i \times m_i \dot{\boldsymbol{\rho}}_i$, where $\dot{\boldsymbol{\rho}}_i$ is the velocity of m_i with respect to G. Although we have simplified this expression in Art. 6/2 in the course of deriving the moment equation of motion, we will pursue this same expression again for sake of emphasis by using the rigid body in plane motion represented in Fig. 6/13. The relative velocity becomes $\dot{\boldsymbol{\rho}}_i = \boldsymbol{\omega} \times \boldsymbol{\rho}_i$, where the angular velocity of the body is $\boldsymbol{\omega} = \omega \mathbf{k}$. The unit vector $\mathbf{k}$ is directed into the paper for the sense of $\boldsymbol{\omega}$ shown. Since $\boldsymbol{\rho}_i$, $\dot{\boldsymbol{\rho}}_i$, and $\boldsymbol{\omega}$ are at right angles to one another, the magnitude of $\dot{\boldsymbol{\rho}}_i$ is $\rho_i \omega$, and the magnitude of $\boldsymbol{\rho}_i \times m_i \dot{\boldsymbol{\rho}}_i$ is $\rho_i^2 \omega m_i$. Thus, we may write $\mathbf{H}_G = \Sigma \rho_i^2 m_i \omega \mathbf{k} = \bar{I} \omega \mathbf{k}$, where $\bar{I} = \Sigma m_i \rho_i^2$ is the mass moment of inertia of the body about its mass center. Because the angular-momentum vector is always normal to the plane of motion, vector notation is generally unnecessary, and we may write the angular momentum about the mass center as the scalar

$$\boxed{H_G = \bar{I} \omega} \qquad \textbf{(6/13)}$$

This angular momentum appears in the moment-angular-momentum relation, Eq. 4/9, which in scalar notation for plane motion, along with its integrated form, is

$$\Sigma M_G = \dot{H}_G \qquad \text{and} \qquad \int_{t_1}^{t_2} \Sigma M_G \, dt = H_{G_2} - H_{G_1} \qquad \textbf{(6/14)}$$

In words, the first of Eqs. 6/14 states that the sum of the moments about the mass center of *all* forces acting on the body equals the time rate of change of angular momentum about the mass center. The integrated form of Eq. 6/14 states that the angular impulse about the mass center of all forces acting on the body during the interval $t_2 - t_1$ equals the corresponding change in the angular momentum about G. Again, as with the linear case, the angular-impulse and angular momentum relationship may be alternatively written in the form equivalent to Eq. 3/29a. The sense for positive rotation must be clearly established, and the algebraic signs of ΣM_G, H_{G_2}, and H_{G_1} must be consistent with this choice. Again, a free-body diagram is essential.

With the moments about G of the linear momenta of all particles accounted for by $H_G = \bar{I}\omega$, it follows that we may represent the linear momentum $\mathbf{G} = m\bar{\mathbf{v}}$ as a vector through the mass center G, as shown in Fig. 6/14a. Thus, $\mathbf{G}$ and $\mathbf{H}_G$ have vector properties analogous to those of the resultant force and couple.

With the establishment of the linear and angular momentum resultants in Fig. 6/14a, which represents the momentum diagram, the angular momentum H_O about any point O is easily written as

$$H_O = \bar{I}\omega + m\bar{v}d \qquad \textbf{(6/15)}$$

This expression holds at any particular instant of time about O, which may be a fixed or moving point on or off the body.

When a body rotates about a fixed point O on the body or body extended, as shown in Fig. 6/14b, the relations $\bar{v} = \bar{r}\omega$ and $d = \bar{r}$ may be substituted into the expression for H_O, giving $H_O = (\bar{I}\omega + m\bar{r}^2\omega)$. But $\bar{I} + m\bar{r}^2 = I_O$ so that

$$H_O = I_O\omega \qquad \textbf{(6/16)}$$

In Art. 4/2 we derived Eq. 4/7, which is the moment-angular-momentum equation about a fixed point O. This equation, written in scalar notation for plane motion along with its integrated form, is

$$\Sigma M_O = I_O\dot{\omega} \qquad \text{and} \qquad \int_{t_1}^{t_2} \Sigma M_O \, dt = I_O(\omega_2 - \omega_1) \qquad \textbf{(6/17)}$$

(a)

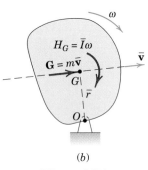

(b)

Figure 6/14

In addition to drawing a complete free-body diagram in order that the force and moment summations may be correctly evaluated when we apply the linear and angular impulse-momentum equations for rigid-body motion, we find it useful to draw the momentum diagram, which indicates the resultant linear-momentum vector and the angular-momentum couple. We caution the reader here not to add linear momentum and angular momentum for the same reason that force and moment cannot be added directly.

(c) Interconnected rigid bodies. The equations of impulse and momentum may also be used for a system of interconnected rigid bodies since the momentum principles are applicable to any general system of constant mass. In Fig. 6/15 are shown the combined free-body diagram and momentum diagram for two interconnected bodies. Equations 4/6 and 4/7, which are $\Sigma \mathbf{F} = \dot{\mathbf{G}}$ and $\Sigma \mathbf{M}_O = \dot{\mathbf{H}}_O$ where O is a fixed reference point, may be written for each member of the system and added. The sums would be

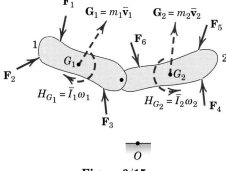

Figure 6/15

$$\Sigma \mathbf{F} = \dot{\mathbf{G}}_1 + \dot{\mathbf{G}}_2 + \cdots \tag{6/18}$$

$$\Sigma \mathbf{M}_O = \dot{\mathbf{H}}_{O_1} + \dot{\mathbf{H}}_{O_2} + \cdots$$

In integrated form for a finite time interval, these expressions are

$$\int_{t_1}^{t_2} \Sigma \mathbf{F}\, dt = (\Delta \mathbf{G})_{\text{system}} \qquad \int_{t_1}^{t_2} \Sigma \mathbf{M}_O\, dt = (\Delta \mathbf{H}_O)_{\text{system}} \tag{6/19}$$

We note that the equal and opposite actions and reactions in the connections are internal to the system and cancel one another so they will not be involved in the force and moment summations. Also, point O is one fixed reference point for the entire system.

(d) Conservation of momentum. In Part *(b)* of Art. 4/5, the principles of conservation of momentum for a general mass system were expressed by Eqs. 4/15 and 4/16. These principles are applicable to either a single rigid body or to a system of interconnected rigid bodies. Thus, if $\Sigma \mathbf{F} = \mathbf{0}$ for a given interval of time, then

$$\boxed{\Delta \mathbf{G} = \mathbf{0}} \qquad \qquad \textbf{[4/15]}$$

which says that the linear momentum vector undergoes no change in the absence of a resultant linear impulse. For the system of interconnected rigid bodies, there may be linear momentum changes of individual parts of the system during the interval, but there will be no resultant momentum change for the system if there is no resultant linear impulse.

Similarly, if the resultant moment about a given fixed point O or about the mass center is zero during a particular interval of time for a single rigid body or for a system of interconnected rigid bodies, then

$$\boxed{\Delta \mathbf{H}_O = \mathbf{0}} \quad \text{or} \quad \boxed{\Delta \mathbf{H}_G = \mathbf{0}} \qquad \text{[4/16]}$$

which says that the angular momentum either about the fixed point or about the mass center undergoes no change in the absence of a corresponding resultant angular impulse. Again, in the case of the interconnected system, there may be angular-momentum changes of individual components during the interval, but there will be no resultant angular-momentum change for the system if there is no resultant angular impulse about the fixed point or the mass center. Either of Eqs. 4/16 may hold without the other. In the case of an interconnected system, the use of the center of mass for the system is generally inconvenient. As was illustrated previously in Articles 3/9 and 3/10 in the chapter on particle motion, the use of momentum principles greatly facilitates the analysis of situations where forces and couples act for very short periods of time.

(e) Impact of rigid bodies. Impact phenomena involve a fairly complex interrelationship of energy and momentum transfer, energy dissipation, elastic and plastic deformation, relative impact velocity, and body geometry. In Art. 3/12 we treated the impact of bodies modeled as particles and considered only the case of central impact, where the contact forces of impact passed through the mass centers of the bodies, as would always happen with colliding smooth spheres, for example. To relate the conditions after impact to those before impact required the introduction of the so-called coefficient of restitution e or impact coefficient, which compares the relative separation velocity with the relative approach velocity measured along the direction of the contact forces. Although in the classical theory of impact, e was considered a constant for given materials, more modern investigations show that e is highly dependent on geometry and impact velocity as well as on materials. At best, even for spheres and rods under direct central and longitudinal impact, the coefficient of restitution is a complex and variable factor of limited use.

Any attempt to extend this simplified theory of impact utilizing a coefficient of restitution for the noncentral impact of rigid bodies of varying shape is a gross oversimplification that has little practical value. For this reason, we shall not include such an exercise in this book, even though such a theory is easily developed and appears in certain references. We can and do, however, make full use of the principles of conservation of linear and angular momentum when they are applicable in discussing the impact and other interactions of rigid bodies.

Sample Problem 6/14

The force P, which is applied to the cable wrapped around the central hub of the symmetrical wheel, is increased slowly according to $P = 6.5t$, where P is in newtons and t is the time in seconds after P is first applied. Determine the angular velocity ω of the wheel 10 s after P is applied if the wheel is rolling to the left with a velocity of its center of 0.9 m/s at time $t = 0$. The wheel, which has a mass of 60 kg and a radius of gyration about its center of 250 mm, rolls without slipping.

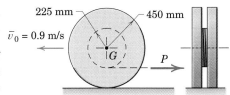

Solution. The free-body diagram of the wheel for any position within the interval is shown. Also indicated are the initial linear and angular momenta at time $t = 0$ and the final linear and angular momenta at time $t = 10$ s. The correct direction of the friction force F is that to oppose the slipping which would occur without friction.

Application of the linear impulse-momentum equation and the angular impulse-momentum equation over the *entire* interval gives

$$\left[\int_{t_1}^{t_2} \Sigma F_x \, dt = G_{x_2} - G_{x_1}\right] \quad \int_0^{10} (6.5t - F) \, dt = 60[0.450\omega - (-0.9)]$$

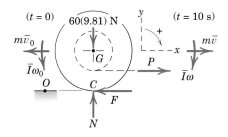

$$\left[\int_{t_1}^{t_2} \Sigma M_G \, dt = H_{G_2} - H_{G_1}\right]$$

$$\int_0^{10} [0.450F - 0.225(6.5t)] \, dt = 60(0.250)^2\left[\omega - \left(-\frac{0.9}{0.450}\right)\right]$$

Since the force F is variable, it must remain under the integral sign. We eliminate F between the two equations by multiplying the second one by 0.450 and adding to the first one. Integrating and solving for ω give

$$\omega = 2.60 \text{ rad/s clockwise} \qquad \textit{Ans.}$$

Alternative Solution. We could avoid the necessity of a simultaneous solution by applying the second of Eqs. 6/17 about a fixed point O on the horizontal surface. The moments of the 60(9.81)-N weight and the equal and opposite force N cancel one another, and F is eliminated since its moment about O is zero. Thus, the angular momentum about O becomes $H_O = \bar{I}\omega + m\bar{v}r = m\bar{k}^2\omega + mr^2\omega = m(\bar{k}^2 + r^2)\omega$, where $\bar{k}$ is the centroidal radius of gyration and r is the 0.450-m rolling radius. Thus, we see that $H_O = H_C$ since $\bar{k}^2 + r^2 = k_C^2$ and $H_C = I_C\omega = mk_C^2\omega$. Equation 6/17 now gives

$$\left[\int_{t_1}^{t_2} \Sigma M_O \, dt = H_{O_2} - H_{O_1}\right]$$

$$\int_0^{10} 6.5t(0.450 - 0.225) \, dt = 60[(0.250)^2 + (0.450)^2]\left[\omega - \left(-\frac{0.9}{0.450}\right)\right]$$

Solution of this one equation is equivalent to the simultaneous solution of the two previous equations.

① Also, we note the clockwise imbalance of moments about C, which causes a clockwise angular acceleration as the wheel rolls without slipping. Since the moment sum about G must also be in the clockwise sense of α, the friction force must act to the left to provide it.

② Note carefully the signs of the momentum terms. The final linear velocity is assumed in the positive x-direction, so G_{x_2} is positive. The initial linear velocity is negative, so G_{x_1} is negative. When we subtract the negative term, we get the double minus sign.

③ Since the wheel rolls without slipping, a positive x-velocity requires a clockwise angular velocity, and vice versa. Again, we subtract a minus quantity.

Sample Problem 6/15

The sheave E of the hoisting rig shown has a mass of 30 kg and a centroidal radius of gyration of 250 mm. The 40-kg load D which is carried by the sheave has an initial downward velocity $v_0 = 1.2$ m/s at the instant when a clockwise torque is applied to the hoisting drum A to maintain essentially a constant force $F = 380$ N in the cable at B. Compute the angular velocity ω of the sheave 5 s after the torque is applied to the drum and find the tension T in the cable at O during the interval. Neglect all friction.

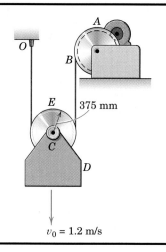

$v_0 = 1.2$ m/s

Solution. The load and the sheave taken together constitute the system, and its free-body diagram is shown. The tension T in the cable at O and the final angular velocity ω of the sheave are the two unknowns. We eliminate T initially by applying the moment-angular-momentum equation about the fixed point O, taking counterclockwise as positive.

$$\left[\int_{t_1}^{t_2} \Sigma M_O \, dt = H_{O_2} - H_{O_1} \right]$$

$$\int_{t_1}^{t_2} \Sigma M_O \, dt = \int_0^5 [380(0.750) - (30 + 40)(9.81)(0.375)] \, dt$$

$$= 137.4 \text{ N·m·s}$$

①
$$(H_{O_2} - H_{O_1})_D = mv_2 d - mv_1 d = md(v_2 - v_1)$$

$$= 40(0.375)[v - (-1.2)] = 15(0.375\omega + 1.2)$$

$$= 5.63\omega + 18 \text{ N·m·s}$$

$$(H_{O_2} - H_{O_1})_E = \bar{I}(\omega_2 - \omega_1) + md(\bar{v}_2 - \bar{v}_1)$$

$$= 30(0.250)^2[\omega - (-1.2/0.375)]$$

$$+ 30(0.375)[0.375\omega - (-1.2)]$$

$$= 6.09\omega + 19.50 \text{ N·m·s}$$

Substituting into the momentum equation gives

$$137.4 = 5.63\omega + 18 + 6.09\omega + 19.50$$

$$\omega = 8.53 \text{ rad/s counterclockwise} \qquad \textit{Ans.}$$

The force-linear-momentum equation is now applied to the system to determine T. With the positive direction up, we have

$$\left[\int_{t_1}^{t_2} \Sigma F \, dt = G_2 - G_1 \right]$$

$$\int_0^5 [T + 380 - 70(9.81)] \, dt = 70[0.375(8.53) - (-1.2)]$$

$$5T = 1841 \qquad T = 368 \text{ N} \qquad \textit{Ans.}$$

If we had taken our moment equation around the center C of the sheave instead of point O, it would contain both unknowns T and ω, and we would be obliged to solve it simultaneously with the foregoing force equation, which would also contain the same two unknowns.

① Watch for the double minus sign in subtracting a negative quantity. Also, the units of angular momentum, which are those of angular impulse, may be written as kg·m^2/s.

Sample Problem 6/16

The uniform rectangular block of dimensions shown is sliding to the left on the horizontal surface with a velocity v when it strikes the small step in the surface. Assume negligible rebound at the step and compute the minimum value of v which will permit the block to pivot about the edge of the step and just reach the standing position A with no velocity. Compute the percentage energy loss $\Delta E/E$ for $b = c$.

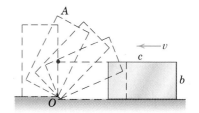

Solution. It will be assumed that the edge of the step O acts as a latch on the corner of the block, so that the block pivots about O. Furthermore, the height of the step is assumed to be negligible compared with the dimensions of the block. During impact the only force which exerts a moment about O is the weight mg, but the angular impulse due to the weight is extremely small since the time of impact is negligible. Thus, we may assume that the angular momentum about O is conserved.

The initial angular momentum of the block about O just before impact is the moment of its linear momentum and is $H_O = mv(b/2)$. The velocity of the center of mass G immediately after impact is $\bar{v}$, and the angular velocity is $\omega = \bar{v}/\bar{r}$. The angular momentum about O just after impact when the block is starting its rotation about O is

③ $[H_O = I_O\omega]$ $\quad H_O = \left\{ \dfrac{1}{12} m(b^2 + c^2) + m\left[\left(\dfrac{c}{2}\right)^2 + \left(\dfrac{b}{2}\right)^2 \right] \right\} \omega$

$$= \dfrac{m}{3} (b^2 + c^2)\omega$$

Conservation of angular momentum gives

$[\Delta H_O = 0]$ $\quad \dfrac{m}{3} (b^2 + c^2)\omega = mv\dfrac{b}{2}$ $\quad \omega = \dfrac{3vb}{2(b^2 + c^2)}$

This angular velocity will be sufficient to raise the block just past position A if the kinetic energy of rotation equals the increase in potential energy. Thus,

$[\Delta T + \Delta V_g = 0]$ $\quad \dfrac{1}{2} I_O\omega^2 - mg\left[\sqrt{\left(\dfrac{b}{2}\right)^2 + \left(\dfrac{c}{2}\right)^2} - \dfrac{b}{2} \right] = 0$

$$\dfrac{1}{2}\dfrac{m}{3} (b^2 + c^2)\left[\dfrac{3vb}{2(b^2 + c^2)} \right]^2 - \dfrac{mg}{2} (\sqrt{b^2 + c^2} - b) = 0$$

$$v = 2\sqrt{\dfrac{g}{3}\left(1 + \dfrac{c^2}{b^2}\right)(\sqrt{b^2 + c^2} - b)} \qquad \textit{Ans.}$$

The percentage loss of energy is

$$\dfrac{\Delta E}{E} = \dfrac{\frac{1}{2}mv^2 - \frac{1}{2}I_O\omega^2}{\frac{1}{2}mv^2} = 1 - \dfrac{k_O{}^2\omega^2}{v^2} = 1 - \left(\dfrac{b^2 + c^2}{3}\right)\left[\dfrac{3b}{2(b^2 + c^2)} \right]^2$$

$$= 1 - \dfrac{3}{4\left(1 + \dfrac{c^2}{b^2}\right)}, \qquad \Delta E/E = 62.5\% \qquad \text{for } b = c \qquad \textit{Ans.}$$

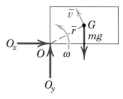

① If the corner of the block struck a spring instead of the rigid step, then the time of the interaction during compression of the spring could become appreciable, and the angular impulse about the fixed point at the end of the spring due to the moment of the weight would have to be accounted for.

② Notice the abrupt change in direction and magnitude of the velocity of G during the impact.

③ Be sure to use the transfer theorem $I_O = \bar{I} + m\bar{r}^2$ correctly here.

PROBLEMS

Introductory problems

6/171 The constant tensions of 200 N and 160 N are applied to the hoisting cable as shown. If the velocity v of the load is 2 m/s down and the angular velocity ω of the pulley is 8 rad/s counterclockwise at time $t = 0$, determine v and ω after the cable tensions have been applied for 5 s. Note the independence of the results.

Ans. $v = 0.379$ m/s up
$\omega = 56.0$ rad/s CW

Problem 6/171

6/172 The center O of the wheel has a velocity $v_1 = 2$ m/s up the 10-percent incline at time $t = 0$. Find the velocity v_2 of the wheel when $t = 6$ s. The wheel has a radius of gyration of 90 mm and rolls without slipping.

Problem 6/172

6/173 The uniform slender rod is at rest on a smooth horizontal surface when struck at point A by a bullet travelling normal to the rod. Calculate the distance x from the end of the bar to point C, the instantaneous center of zero velocity, about which the rod begins to rotate during the impact.

Ans. $x = 100$ mm

Problem 6/173

6/174 The 28-g bullet has a horizontal velocity of 500 m/s as it strikes the 25-kg compound pendulum, which has a radius of gyration $k_O = 925$ mm. If the distance $h = 1075$ mm, calculate the angular velocity ω of the pendulum with its embedded bullet immediately after the impact.

Problem 6/174

6/175 The 10-kg wheel has a diameter of 400 mm and a radius of gyration about its axis of 180 mm. The wheel carries a 4-kg shaft of small diameter through its hub as shown. If the wheel has an angular velocity $\omega = 4$ rad/s as it rolls without slipping on the horizontal surface, calculate the distance h to the axis A–A, parallel to the shaft axis, about which the angular momentum of the combined body is zero.

Ans. h = 316 mm

Problem 6/175

6/176 The cable drum has a mass of 800 kg with radius of gyration of 480 mm about its center O and is mounted in bearings on the 1200-kg carriage. The carriage is initially moving to the left with a speed of 1.5 m/s and the drum is rotating counterclockwise with an angular velocity of 3 rad/s when a constant horizontal tension $T = 400$ N is applied to the cable at time $t = 0$. Determine the velocity v of the carriage and the angular velocity ω of the drum when $t = 10$ s. Neglect the mass of the carriage wheels.

Problem 6/176

6/177 The center O of the 2-kg wheel, with radius of gyration of 60 mm about O, has a velocity $v_O = 0.3$ m/s down the 15° incline when a force $P = 10$ N is applied to the cord wrapped around its inner hub. If the wheel rolls without slipping, calculate the velocity v of the center O when P has been applied for 5 seconds.

Ans. v = 1.575 m/s

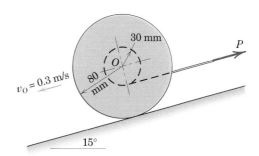

Problem 6/177

6/178 Determine the angular momentum of the earth about the center of the sun. Assume a homogeneous earth and a circular earth orbit of radius $149.6(10^6)$ km; consult Table D/2 for other needed information. Comment on the relative contributions of the terms $\bar{I}\omega$ and $m\bar{v}d$.

Problem 6/178

Representative problems

6/179 The homogeneous circular cylinder of mass m and radius R carries a slender rod of mass $m/2$ attached to it as shown. If the cylinder rolls on the surface without slipping with a velocity v_O of its center O, determine the angular momenta H_G and H_O of the system about its center of mass G and about O for the instant shown.

$Ans.$ $H_G = \frac{11}{16}mRv_O$, $H_O = \frac{37}{32}mRv_O$

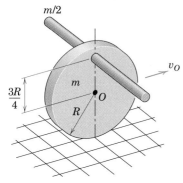

Problem 6/179

6/180 The uniform circular disk of 200-mm radius has a mass of 25 kg and is mounted on the rotating bar OA in three different ways. In each case the bar rotates about its vertical shaft at O with a clockwise angular velocity $\omega = 4$ rad/s. In case (a) the disk is welded to the bar. In case (b) the disk, which is pinned freely at A, moves with curvilinear translation and therefore has no rigid-body rotation. In case (c) the relative angle between the disk and the bar is increasing at the rate $\dot{\theta} = 8$ rad/s. Calculate the angular momentum of the disk about point O for each case.

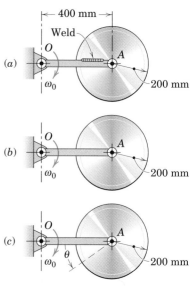

Problem 6/180

6/181 The large rotor has a mass of 60 kg and a radius of gyration about its vertical axis of 200 mm. The small rotor is a solid circular disk with a mass of 8 kg and is initially rotating with an angular velocity $\omega_1 = 80$ rad/s with the large rotor at rest. A spring-loaded pin P which rotates with the large rotor is released and bears against the periphery of the small disk, bringing it to a stop relative to the large rotor. Neglect any bearing friction and calculate the final angular velocity of the assembly.

$Ans.$ $\omega = 1.221$ rad/s

Problem 6/181

6/182 Just after leaving the platform, the diver's fully-extended 80-kg body has a rotational speed of 0.3 rev/s about an axis normal to the plane of the trajectory. Estimate the angular velocity N later in the dive when the diver has assumed the tuck position. Make reasonable assumptions concerning the mass moment of inertia of the body in each configuration.

Problem 6/182

6/183 The slender bar of mass m and length b is pivoted at its lower end at O in the manner shown in the separate detail of the support O. The bar is released from rest in the vertical position 1. When the middle of the bar strikes the pivot at A in position 2, it becomes latched to the pivot, and simultaneously the connection at O becomes disengaged. Determine the angular velocity ω_3 of the bar just after it engages the pivot at A in position 3.

Ans. $\omega_3 = \sqrt{3g/b}$

Problem 6/183

6/184 Two small variable-thrust jets are actuated to keep the spacecraft angular velocity about the z-axis constant at $\omega_0 = 1.25$ rad/s as the two telescopic booms are extended from $r_1 = 1.2$ m to $r_2 = 4.5$ m at a constant rate over a two-minute period. Determine the necessary thrust T for each jet as a function of time where $t = 0$ is the time when the telescoping action is begun. The small 10-kg experiment modules at the ends of the booms may be treated as particles, and the mass of the rigid booms is negligible.

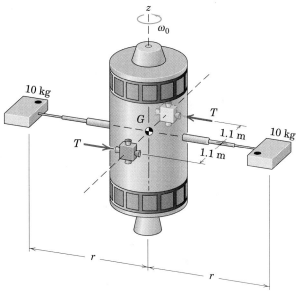

Problem 6/184

6/185   The 3.6-kg slotted circular disk has a radius of gyration about its center O of 150 mm and initially is rotating freely about a fixed vertical axis through O with a speed $N_1 = 600$ rev/min. The 0.9-kg uniform slender bar A is initially at rest relative to the disk in the centered slot position as shown. A slight disturbance causes the bar to slide to the end of the slot where it comes to rest relative to the disk. Calculate the new angular speed N_2 of the disk, assuming the absence of friction in the shaft bearing at O. Does the presence of any friction in the slot affect the final result?

Ans. $N_2 = 569$ rev/min

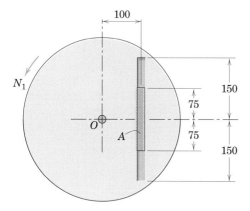

Dimensions in millimeters

Problem 6/185

6/186   A 55-kg dynamics instructor is demonstrating the principles of angular momentum to her class. She stands on a freely rotating platform with her body aligned with the vertical platform axis. With the platform not rotating, she holds a modified bicycle wheel so that its axis is vertical. She then turns the wheel axis to a horizontal orientation without changing the 600-mm distance from the centerline of her body to the wheel center, and her students observe a platform rotation rate of 30 rev/min. If the rim-weighted wheel has a mass of 10 kg and a centroidal radius of gyration $\bar{k} = 300$ mm, and is spinning at a fairly constant rate of 250 rev/min, estimate the mass moment of inertia I of the instructor (in the posture shown) about the vertical platform axis.

Problem 6/186

6/187   Each of the two 300-mm uniform rods A has a mass of 1.5 kg and is hinged at its end to the rotating base B. The 4-kg base has a radius of gyration of 40 mm and is initially rotating freely about its vertical axis with a speed of 300 rev/min and with the rods latched in the vertical positions. If the latches are released and the rods assume the horizontal positions, calculate the new rotational speed N of the assembly.

Ans. $N = 32.0$ rev/min

Problem 6/187

6/188 The 74-kg ice skater with arms extended horizontally spins about a vertical axis with a rotational speed of 1 rev/s. Estimate his rotational speed N if he fully retracts his arms, bringing his hands very close to the centerline of his body. As a reasonable approximation, model the extended arms as uniform slender rods, each of which is 680 mm long with a mass of 7 kg. Model the torso as a solid 60-kg cylinder 330 mm in diameter. Treat the man with arms retracted as a solid 74-kg cylinder of 330-mm diameter. Neglect friction at the skate-ice interface.

Problem 6/188

6/189 The uniform 78-kg concrete block falls from rest in the horizontal position shown, strikes the fixed corner A, and pivots around it with no rebound. Calculate the angular velocity ω of the block immediately after it hits the corner and the percentage loss n of energy due to the impact.

Ans. $\omega = 1.605$ rad/s, $n = 91.7\%$

Problem 6/189

6/190 A man of mass m stands at point A on the horizontal turntable, initially at rest. The turntable is free to rotate about O with negligible friction and has a moment of inertia I about O. If the man starts to walk clockwise in a circle of radius r, determine the angle θ through which the turntable rotates until the man reaches his starting point in the new position A'. If the man walks with a speed v_r relative to the turntable, what is the corresponding angular velocity ω of the turntable?

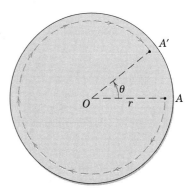

Problem 6/190

6/191 The uniform circular disk D has a mass of 4 kg and is free to turn about the bearing axis C-C. The 2.5-kg arm B is fastened to the vertical shaft O-O. The arm may be approximated as a slender bar 300 mm long, and the moment of inertia of the vertical shaft O-O may be neglected. The 3-kg rod A is fastened securely to the arm. If the initial angular velocity of D is $\omega_0 = 7$ rad/s in the direction shown and the arm B is at rest, determine the angular velocity ω_B of the arm after a torque $M = 1.5$ N·m has been applied to the shaft for 4 s.

Ans. $\omega_B = 34.0$ rad/s

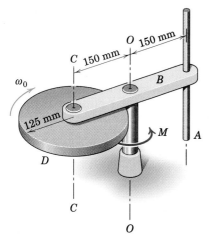

Problem 6/191

6/192 The elements of a spacecraft with axial mass symmetry and a reaction-wheel control system are shown in the figure. When the motor exerts a torque on the reaction wheel, an equal and opposite torque is exerted on the spacecraft, thereby changing its angular momentum in the z-direction. If all system elements start from rest and the motor exerts a constant torque M for a time period t, determine the final angular velocity of (a) the spacecraft and (b) the wheel relative to the spacecraft. The mass moment of inertia about the z-axis of the entire spacecraft, including the wheel, is I and that of the wheel alone is I_w. The spin axis of the wheel is coincident with the z-axis of symmetry of the spacecraft.

Problem 6/192

6/193 The solid circular cylinder of radius r is at rest on the flat belt when a force P is applied to the belt. If P is sufficient to cause slipping between the belt and the cylinder at all times, determine the time t required for the cylinder to reach the dashed position. Also determine the angular velocity ω of the cylinder in this same position. The coefficient of friction between the cylinder and the belt is μ_k.

Ans. $t = \sqrt{\dfrac{2s}{\mu_k g}}$, $\omega = \dfrac{2}{r}\sqrt{2\mu_k gs}$

Problem 6/193

6/194 The 17.5-Mg lunar landing module with center of mass at G has a radius of gyration of 1.8 m about G. The module is designed to contact the lunar surface with a vertical free-fall velocity of 8 km/h. If one of the four legs hits the lunar surface on a small incline and suffers no rebound, compute the angular velocity ω of the module immediately after impact as it pivots about the contact point. The 9-m dimension is the distance across the diagonal of the square formed by the four feet as corners.

Problem 6/194

6/195 The slotted circular disk whose mass is 6 kg has a radius of gyration about O of 175 mm. The disk carries the four steel balls, each of mass 0.15 kg and located as shown, and rotates freely about a vertical axis through O with an angular speed of 120 rev/min. Each of the small balls is held in place by a latching device not shown. If the balls are released while the disk is rotating and come to rest relative to the disk at the outer ends of the slots, compute the new angular velocity ω of the disk. Also find the magnitude $|\Delta E|$ of the energy loss due to the impact of the balls with the ends of the slots. Neglect the diameter of the balls and discuss this approximation.

Ans. $\omega = 109.6$ rev/min, $|\Delta E| = 1.298$ J

Problem 6/195

6/196 The body of the spacecraft has a mass of 160 kg and has a radius of gyration about its z-axis of 0.45 m. Each of the two solar panels may be treated as a uniform flat plate of 8-kg mass. If the spacecraft is rotating about its z-axis at the angular rate of 1.0 rad/s with $\theta = 0$, determine the angular rate ω after the panels are rotated to the position $\theta = \pi/2$ by an internal mechanism. Neglect the small momentum change of the body about the y-axis.

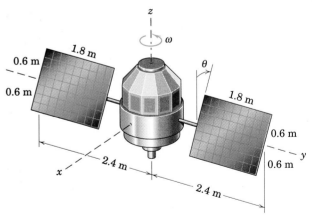

Problem 6/196

6/197 The preliminary design of a unit for automatically reducing the speed of a freely rotating assembly is shown. Initially the unit is rotating freely about a vertical axis through O at a speed of 600 rev/min with the arms secured in the positions shown by AB. When the arms are released, they swing outward and become latched in the dashed positions shown. The disk has a mass of 30 kg with a radius of gyration of 90 mm about O. Each arm has a length of 160 mm and a mass of 0.84 kg and may be treated as a uniform slender rod. Determine the new speed N of rotation and calculate the loss ΔQ of energy of the system. Would the results be affected by either the direction of rotation or the sequence of release of the rods?

Ans. $N = 504$ rev/min, $\Delta Q = 98.1$ J

Problem 6/197

6/198 The homogeneous sphere of mass m and radius r is projected along the incline of angle θ with an initial speed v_0 and no angular velocity ($\omega_0 = 0$). If the coefficient of kinetic friction is μ_k, determine the time duration t of the period of slipping. In addition, state the velocity v of the mass center G and the angular velocity ω at the end of the period of slipping.

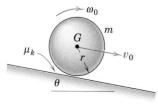

Problem 6/198

6/199 In the rotating assembly shown, arm OA and the attached motor housing B have a combined mass of 4.5 kg and have a radius of gyration about the z-axis of 175 mm. The motor armature and attached 125-mm-radius disk have a combined mass of 7 kg and a radius of gyration of 100 mm about their own axis. The entire assembly is free to rotate about the z-axis. If the motor is turned on with OA initially at rest, determine the angular speed N of OA when the motor has reached a speed of 300 rev/min *relative* to arm OA.

Ans. $N = 37.4$ rev/min

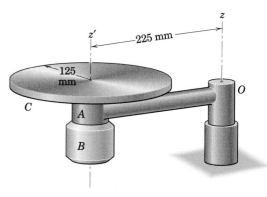

Problem 6/199

6/200 Determine the minimum velocity v which the wheel may have and just roll over the obstruction. The centroidal radius of gyration of the wheel is k, and it is assumed that the wheel does not slip.

Problem 6/200

6/201 The three bars are free to rotate about the fixed horizontal x-axis at O. Each of the bars A has a mass of 20 kg, a radius of gyration about the x-axis of 300 mm, and a mass center at G_1. The 8-kg bar B, with a radius of gyration about the same axis of 220 mm and a mass center at G_2, is released from rest in the horizontal (y-axis) position and becomes attached to bars A by a latch at C at the bottom of its swing. Calculate the angle θ through which the three bars rotate as a unit and find the loss ΔQ of energy due to the impact.

Ans. $\theta = 8.97°, \Delta Q = 12.75$ J

Problem 6/201

6/202 The small gear is made to rotate in a horizontal plane about the large stationary gear by means of the torque M applied to the arm OA. The small gear has a mass of 3 kg and may be treated as a circular disk. The 2-kg arm OA has a radius of gyration about the fixed bearing at O of 150 mm. Determine the constant torque M required to give the arm OA an absolute angular velocity of 20 rad/s in 3 s, starting from rest. Neglect friction.

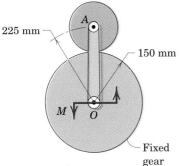

Problem 6/202

6/203 A frozen-juice can rests on the horizontal rack of a freezer door as shown. With what maximum angular velocity Ω can the door be "slammed" shut against its seal and not dislodge the can? Assume that the can rolls without slipping on the corner of the rack, and neglect the dimension d compared with the 500-mm distance.

Ans. $\Omega = 1.135$ rad/s

Problem 6/203

6/204 The uniform slender bar of mass m and length l has no angular velocity as end A strikes the ground against the stop with no rebound. If $\alpha = 15°$, what is the minimum magnitude of the initial velocity $\mathbf{v}_1$ for which the bar will rotate about A to the vertical position?

Problem 6/204

▶6/205 The 30-kg wheel has a radius of gyration about its center of 75 mm and is rotating clockwise at the rate of 300 rev/min when it is released onto the incline with no velocity of its center O. While the wheel is slipping, it is observed that the center O remains in a fixed position. Determine the coefficient of kinetic friction μ_k and the time t during which slipping occurs. Also determine the velocity v of the center 4 seconds after the wheel has stopped slipping.

Ans. $\mu_k = 0.1763$, $t = 1.037$ s, $v = 4.36$ m/s

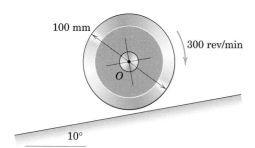

Problem 6/205

▶6/206 The preliminary design of the gear assembly calls for the arm AOB to reach a speed of 600 rev/min in 5 seconds from rest under the action of the steady moment M applied to the arm through its shaft at O. The outer ring gear is fixed. Each of the two arms AOB (one on each side of the gears) has a mass of 2.1 kg with a radius of gyration about O of 75 mm. Each of the gears C and D has a mass of 1.2 kg with a radius of gyration of 50 mm about its center. Calculate M. (If H_O is the angular momentum of the entire system about O, is it correct to write $M = \dot{H}_O$?

Ans. $M = 0.711$ N·m, No

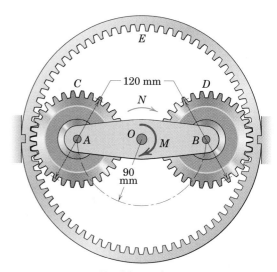

Problem 6/206

6/9   *PROBLEM FORMULATION AND REVIEW*

In Chapter 6 we have made use of essentially all the elements of dynamics studied so far. We have found that a knowledge of kinematics, using both absolute- and relative-motion analysis, is an essential part of the solution to problems in rigid-body kinetics. Our approach in Chapter 6 paralleled Chapter 3, where we developed the kinetics of particles using force-mass-acceleration, work-energy, and impulse-momentum methods.

The following outline will help to summarize important considerations in the solution of rigid-body problems in plane motion:

(a) *Identification of body or system.*   It is essential to make an unambiguous decision as to which body or system of bodies is to be analyzed and then isolate the selected body or system by drawing its free-body diagram, kinetic diagram, or active-force diagram, whichever is appropriate.

(b) *Type of motion.*   Next identify the category of motion as rectilinear translation, curvilinear translation, fixed-axis rotation, or general plane motion. Always see that the kinematics of the problem is properly described before attempting to solve the kinetic equations.

(c) *Coordinate system.*   Choose an appropriate coordinate system. The geometry of the particular kinematics involved is usually the deciding factor. Designate the positive sense for moment and force summations and be consistent with the choice.

(d) *Principle and method.*   If the instantaneous relationship between the applied forces and the acceleration is desired, then the equivalence between the forces and their $m\bar{\mathbf{a}}$ and $\bar{I}\alpha$ resultants, as disclosed by the free-body and kinetic diagrams, will indicate the most direct approach to a solution.

When motion occurs over an interval of displacement, the work-energy approach is indicated, and we relate initial to final velocities without calculating the acceleration. We have seen the advantage of this approach for interconnected mechanical systems with negligible internal friction.

If the interval of motion is specified in terms of time rather than displacement, the impulse-momentum approach is indicated. When the angular motion of a rigid body is suddenly changed, we use the principle of conservation of angular momentum.

(e) *Assumptions and approximations.*   By now we should have acquired a feel for the practical significance of certain assumptions and approximations, such as treating a rod as an ideal slender bar and neglecting friction when it is minimal. These and other idealizations are important to the process of obtaining solutions to real problems.

REVIEW PROBLEMS

6/207 A person who walks through the revolving door exerts a 90-N horizontal force on one of the four door panels. If each panel is modeled by a 60-kg uniform rectangular plate which is 1.2 m in length as viewed from above, determine the angular acceleration of the door unit. Neglect friction.

Ans. $\alpha = 0.604$ rad/s^2

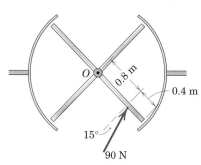

Problem 6/207

6/208 The bar is suspended by a light cord from a fixed point O and is released from rest in the position shown. With the aid of a free-body diagram, prove that the bar will not remain in line with the cord as it swings.

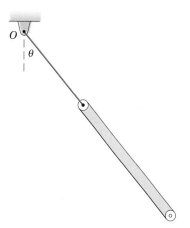

Problem 6/208

6/209 The laminate roller consists of the 4-kg uniform bar with two rollers of negligible mass which apply force to the laminates being glued to the wood filler. Without calculating the acceleration of the bar, determine the force exerted by each roller on the laminate when the horizontal 50-N force is applied to the end of the bar.

Ans. $A = 143.4$ N, $B = 104.2$ N

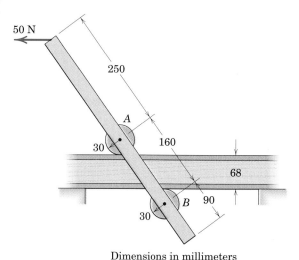

Dimensions in millimeters

Problem 6/209

6/210 The dump truck carries 5 m^3 of dirt with a density of 1600 kg/m^3, and the elevating mechanism rotates the dump about the pivot A at a constant angular rate of 4 deg/s. The mass center of the dump and load is at G. Determine the maximum power P required during the tilting of the load.

Problem 6/210

6/211 The forklift truck with center of mass at G_1 has a mass of 1600 kg including the vertical mast. The fork and load have a combined mass of 900 kg with center of mass at G_2. The roller guide at B is capable of supporting horizontal force only, whereas the connection at C, in addition to supporting horizontal force, also transmits the vertical elevating force. If the fork is given an upward acceleration which is sufficient to reduce the force under the rear wheels at A to zero, calculate the corresponding reaction at B.

Ans. $B = 10.46$ kN

Problem 6/211

6/212 A space telescope is shown in the figure. One of the reaction wheels of its attitude-control system is spinning as shown at 10 rad/s, and at this speed the friction in the wheel bearing causes an internal moment of 10^{-6} N·m. Both the wheel speed and the friction moment may be considered constant over a time span of several hours. If the mass moment of inertia of the entire spacecraft about the x-axis is $150(10^3)$kg·m², determine how much time passes before the line of sight of the initially stationary spacecraft drifts by 1 arcsecond, which is 1/3600 degree. All other elements are fixed relative to the spacecraft, and no torquing of the reaction wheel shown is performed to correct the attitude drift. Neglect external torques.

Problem 6/212

6/213 The uniform slender bar has a mass of 30 kg and is released from rest in the near-vertical position shown, where the spring of stiffness 150 N/m is unstretched. Calculate the velocity with which end A strikes the horizontal surface.

Ans. $v_A = 3.70$ m/s

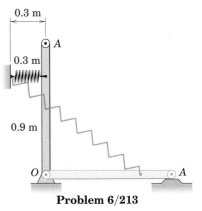

Problem 6/213

6/214 A cylindrical shell of 400-mm diameter and mass m is rotating about its central horizontal axis with an angular velocity $\omega_0 = 30$ rad/s when it is released onto a horizontal surface with no velocity of its center ($v_0 = 0$). If slipping between the shell and the surface occurs for 1.5 s, calculate the coefficient of kinetic friction μ_k and the maximum velocity v reached by the center of the shell.

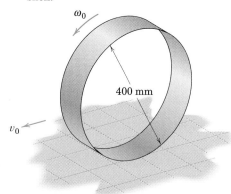

Problem 6/214

6/215 The uniform rectangular block with the given dimensions is dropped from rest from the position shown. Corner A strikes the ledge at B and becomes latched to it. Determine the angular velocity ω of the block immediately after it becomes attached to B. Also find the percentage n of energy loss during the corner attachment for the case $b = c$.

$$Ans. \ \omega = \frac{3b\sqrt{2gh}}{2(b^2 + c^2)}, \ n = 62.5\%$$

Problem 6/215

6/216 Each of the solid square blocks is allowed to fall by rotating clockwise from the rest positions shown. The support at O in case (a) is a hinge and in case (b) is a small roller. Determine the angular velocity ω of each block as edge OC becomes horizontal just before striking the supporting surface.

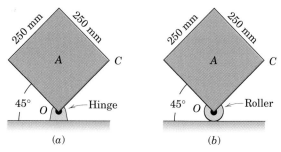

Problem 6/216

6/217 The 3-kg pendulum with mass center at G is pivoted at A to the fixed support CA. It has a radius of gyration of 425 mm about $O-O$ and swings through an amplitude $\theta = 60°$. For the instant when the pendulum is in the extreme position, calculate the moments M_x, M_y, and M_z applied by the base support to the column at C.

$$Ans. \ M_x = -5.64 \ \text{N·m}$$
$$M_y = 1.976 \ \text{N·m}$$
$$M_z = -2.26 \ \text{N·m}$$

Problem 6/217

6/218 The link OA and pivoted circular disk are released from rest in the position shown and swing in the vertical plane about the fixed bearing at O. The 6-kg link OA has a radius of gyration about O of 375 mm. The disk has a mass of 8 kg. The two bearings are assumed to be frictionless. Find the force F_O exerted at O on the link (a) just after release and (b) as OA swings through the vertical position OA'.

Problem 6/218

6/219 In an acrobatic stunt, man A of mass m_A drops from a raised platform onto the end of the light but strong beam with a velocity v_0. The boy of mass m_B is propelled upward with a velocity v_B. For a given ratio $n = m_B/m_A$ determine b in terms of L to maximize the upward velocity of the boy. Assume that both man and boy act as rigid bodies.

$$Ans. \ b = \frac{L}{1 + \sqrt{n}}$$

Problem 6/219

6/220 Wheel A has a mass of 50 kg with a 250-mm radius of gyration about its center O and is held initially at rest on the inclined 25-kg slab B. The wheel is released when a force $P = 180$ N is applied to the slab. Calculate the acceleration a_B of the slab, the acceleration a_O of the center of the wheel, and the minimum value $(\mu_s)_{min}$ of the coefficient of static friction for which no slipping between the wheel and the slab will occur.

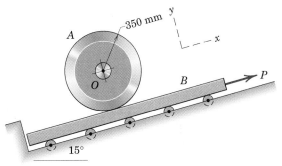

Problem 6/220

6/221 The car with standard rear-wheel drive has a mass of 1600 kg with center of mass at G. The effective coefficient of friction between the tires and the road is 0.80. Treat the car and wheels as a single rigid body by neglecting the rotational inertia of the wheels and calculate the maximum acceleration a which the car is capable of reaching. Then calculate the torque M applied to each wheel by its axle. Each rear wheel has a mass of 32 kg, a diameter of 620 mm, and a radius of gyration of 210 mm.

$$Ans. \ a = 4.62 \ m/s^2, M = 1166 \ N \cdot m$$

Problem 6/221

6/222 The mechanical flyball governor operates with a vertical shaft O-O. As the shaft speed N is increased, the rotational radius of the two 1.5-kg balls tends to increase, and the 9-kg mass A is lifted up by the collar B. Determine the steady-state value of β for a rotational speed of 150 rev/min. Neglect the mass of the arms and collar.

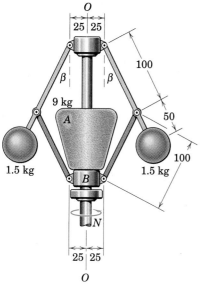

Dimensions in millimeters

Problem 6/222

6/223 The small block of mass m slides in the smooth radial slot of the disk, which turns freely in its bearing. If the block is displaced slightly from the center position when the angular velocity of the disk is ω_0, determine its radial velocity v_r as a function of the radial distance r. The mass moment of inertia of the disk about its axis of rotation is I_O.

Ans. $v_r = r\omega_0 \sqrt{\dfrac{I_0}{I_0 + mr^2}}$

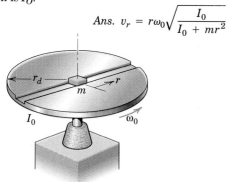

Problem 6/223

▶**6/224** The split ring of radius r is rotating about a vertical axis through its center O with a constant angular velocity ω. Use a differential element of the ring and derive expressions for the shear force N and rim tension T in the ring in terms of the angle θ. Determine the bending moment M_C at point C by using one-half of the ring as a free body. The mass of the ring per unit length of rim is ρ.

Ans. $N = \rho r^2 \omega^2 \sin \theta$
$T = \rho r^2 \omega^2 (1 + \cos \theta)$
$M_C = 2\rho r^3 \omega^2$

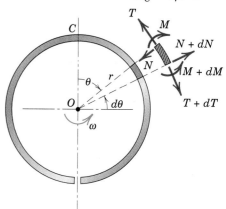

Problem 6/224

▶**6/225** Before it hits the ground a falling chimney, such as the one shown, will usually crack at the point where the bending moment is greatest. Show that the position of maximum moment occurs at the center of percussion relative to the upper end for a slender chimney of constant cross section. Neglect any restraining moment at the base.

Problem 6/225

▶ **6/226** The mass center G of the 70-kg person shown has a maximum velocity of 4 m/s at the bottom of the swing. For the dimensions given, calculate the bending moment M supported by the lumbar vertebrae of the spine for this position. The mid-lumbar vertebrae and the mass center G of the body are essentially coincident. The mass center of the lower portion of the body is at G_1 for the position shown. Assume that the masses of the upper and lower parts of the body are equal. Neglect any forces supported by soft tissue of the body on the section of the torso at G. Also neglect the mass of the swing and the mass of the person's arms.

Ans. $M = 58.4$ N·m

$$\overline{OG} = \overline{OG_1} = 2300 \text{ mm}$$
$$\overline{1\text{-}2} = \overline{2\text{-}3} = 575 \text{ mm}$$
$$\overline{1\text{-}3} = 500 \text{ mm}$$
$$\overline{1\text{-}G} = 455 \text{ mm}$$

Problem 6/226

▶ **6/227** The two slender bars, each having a mass of 4 kg, are hinged at B and pivoted at C. If a horizontal impulse $\int F dt = 14$ N·s is applied to the end A of the lower bar during an interval of 0.1 s during which the bars are still essentially in their vertical rest positions, compute the angular velocity ω_2 of the upper bar immediately after the impulse.

Ans. $\omega_2 = 2.50$ rad/s

Problem 6/227

Computer-oriented problems

*6/228 The uniform 1.2-m slender bar with light end rollers is released from rest in the vertical plane with θ essentially zero. Determine and graph the velocity of A as a function of θ and find the maximum velocity of A and the corresponding angle θ.

Problem 6/228

*6/229 The 8-kg pendulum with mass center at G and a radius of gyration of 235 mm about O is released from rest at the horizontal position $\theta = 0$. Plot the n-component of force supported by the bearing at O from $\theta = 0$ to $\theta = 90°$. Indicate the maximum magnitude of the t-component of force at the bearing O.

Ans. $(O_t)_{max} = 9.70$ N ($-t$ direction) at $\theta = 0$

$\overline{OG} = 220$ mm

Problem 6/229

*6/230 The 5-kg spring-loaded plunger is designed to oscillate in the vertical direction under the action of the two springs, each of which has a stiffness $k = 1050$ N/m and is unstretched when $x = 0$. If the plunger is released from rest in the position $x = 225$ mm, plot its velocity v in terms of x and find v_{max} and the corresponding value of x. Assume negligible friction in the guide bearing.

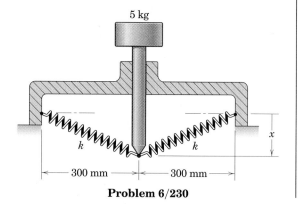

5 kg

k k

x

← 300 mm → ← 300 mm →

Problem 6/230

*6/231 The system of Prob. 6/16 is repeated here. The cart B moves to the right with acceleration $a = 2g$. If $m = 0.5$ kg, $l = 0.6$ m, and $K = 75$ N·m/rad, determine the steady-state angular deflection θ of the uniform slender rod of mass $3m$. Treat the small end sphere of mass m as a particle. The spring, which exerts a moment of magnitude $M = K\theta$ on the rod, is undeformed when the rod is vertical.

Ans. $\theta = 12.17°$

Problem 6/231

*6/232 The steel I-beam is to be transported by the overhead trolley to which it is hinged at O. If the trolley starts from rest with $\theta = \dot{\theta} = 0$ and is given a constant horizontal acceleration $a = 2$ m/s^2, find the maximum values of $\dot{\theta}$ and θ. The magnitude of the initial swing would constitute a shop safety consideration.

Problem 6/232

*6/233 The uniform power pole of mass m and length L is hoisted into a vertical position with its lower end supported by a fixed pivot at O. The guy wires supporting the pole are accidentally released, and the pole falls to the ground. Plot the x- and y-components of the force exerted on the pole at O in terms of θ from 0 to 90°. Can you explain why O_y increases again after going to zero?

$$\text{Ans. } O_x = \frac{3mg}{4} \sin\theta(3\cos\theta - 2)$$

$$O_y = \frac{mg}{4}(3\cos\theta - 1)^2$$

Problem 6/233

Problem 6/234

*6/235 The uniform 100-kg beam AB is hanging initially at rest with $\theta = 0$ when the constant force $P = 300$ N is applied to the cable. Determine (a) the maximum angular velocity reached by the beam with the corresponding angle θ and (b) the maximum angle θ_{max} reached by the beam.

$$\text{Ans. } \omega_{max} = 0.680 \text{ rad/s at } \theta = 22.4°$$
$$\theta_{max} = 45.9°$$

*6/234 For a train traveling at 160 km/h around a horizontal curve of radius 1.9 km, calculate the elevation angle β of the track so that passengers will feel only a force normal to their seats and the rails will exert no side thrust against the wheels, as indicated in part (a) of the figure. An experimental train rounds this same curve at a speed of 260 km/h with cars which are automatically tilted an angle θ with respect to the rails, as shown in part (b) of the figure. This angle reduces the side thrust F felt by the passengers. Determine the tilt angle θ required to limit F to 30 percent of the side thrust they would feel if both θ and β were zero.

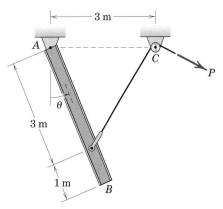

Problem 6/235

*6/236 The 4-kg bar OA with mass center G is hinged in the vertical plane to the moving carriage at O. The attached spring is unstretched when $\theta = 0$. Determine the steady-state angle θ assumed by the bar during a steady horizontal acceleration $a = 4$ m/s^2 of the carriage. (*Hint:* There are two values of θ, one stable and the other unstable.)

Problem 6/236

*6/237 The 18-m telephone pole of essentially uniform diameter is being hoisted into the vertical position by two cables attached at B as shown. The end O rests on a fixed support and cannot slip. When the pole is nearly vertical, the fitting at B suddenly breaks, releasing both cables. When the angle θ reaches 10°, the speed of the upper end A of the pole is 1.35 m/s. From this point, calculate the time t which the workman would have to get out of the way before the pole hits the ground. With what speed v_A does end A hit the ground?

Ans. $t = 2.83$ s, $v_A = 22.9$ m/s

Problem 6/237

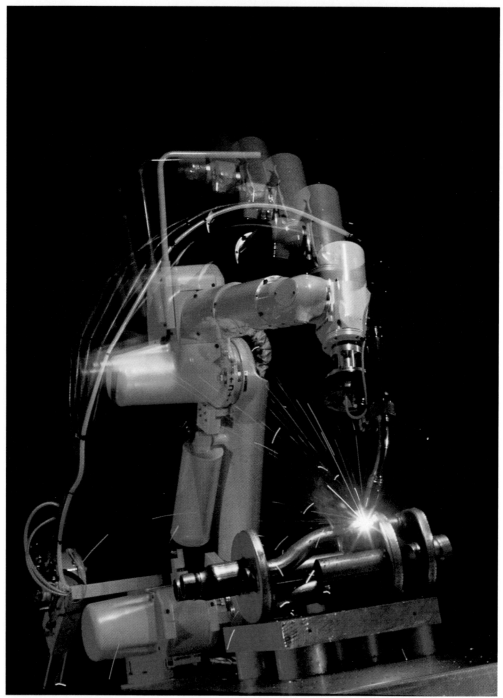

There is an increasing number of rigid bodies whose motions are three-dimensional. Adding a third dimension considerably complicates the mathematics. The multiple-exposure photo of a computer-controlled robotic welder is a good example of the application of three-dimensional dynamics in modern engineering where precise positioning in space is essential.

INTRODUCTION TO THREE-DIMENSIONAL DYNAMICS OF RIGID BODIES

7

7/1 INTRODUCTION

Although a large percentage of dynamics problems in engineering lend themselves to solution by means of the principles of plane motion, modern developments have focused increasing attention on problems which call for the analysis of motion in three dimensions. Inclusion of the third dimension adds considerable complexity to the kinematic and kinetic relationships. Not only does the added dimension introduce a third component to vectors which represent force, linear velocity, linear acceleration, and linear momentum, but the introduction of the third dimension adds the possibility of two additional components for vectors representing angular quantities including moments of forces, angular velocity, angular acceleration, and angular momentum. It is in three-dimensional motion that the full power of vector analysis is utilized.

A good background in the dynamics of plane motion is extremely useful in the study of three-dimensional dynamics, as the approach to problems and many of the terms are the same as or analogous to those in two dimensions. If the study of three-dimensional dynamics is undertaken without the benefit of prior study of plane-motion dynamics, more time will be required to master the principles and to become familiar with the approach to problems.

The treatment presented in Chapter 7 is not intended as a complete development of the three-dimensional motion of rigid bodies but merely as a basic introduction to the subject. This introduction should, however, be sufficient to solve many of the more common problems in three-dimensional motion and also to lay the foundation for more advanced study. We shall proceed as we did for particle motion and for rigid-body plane motion by first examining the necessary kinematics and then proceeding to the kinetics.

SECTION A. KINEMATICS

7/2 *TRANSLATION*

Figure 7/1 shows a rigid body translating in three-dimensional space. Any two points in the body, such as A and B, will move along parallel straight lines if the motion is one of *rectilinear translation* or will move along congruent curves if the motion is one of *curvilinear translation*. In either case, every line in the body, such as AB, remains parallel to its original position. The position vectors and their first and second time derivatives are

$$\mathbf{r}_A = \mathbf{r}_B + \mathbf{r}_{A/B} \qquad \mathbf{v}_A = \mathbf{v}_B \qquad \mathbf{a}_A = \mathbf{a}_B$$

where $\mathbf{r}_{A/B}$ remains constant and therefore has no time derivative. Thus, all points in the body have the same velocity and the same acceleration. The kinematics of translation presents no special difficulty, and further elaboration is unnecessary.

Figure 7/1

7/3 *FIXED-AXIS ROTATION*

Consider now the *rotation* of a rigid body about a fixed axis n-n in space with an angular velocity $\boldsymbol{\omega}$, as shown in Fig. 7/2. The angular velocity is a vector in the direction of the rotation axis with a sense established by the familiar right-hand rule. For fixed-axis rotation, $\boldsymbol{\omega}$ does not change its direction since it lies along the axis. We choose the origin O of the fixed coordinate system on the rotation axis for convenience. Any point such as A which is not on the axis moves in a circular arc in a plane normal to the axis and has a velocity

$$\boxed{\mathbf{v} = \boldsymbol{\omega} \times \mathbf{r}} \tag{7/1}$$

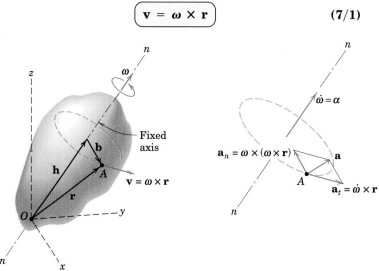

Figure 7/2

which may be seen by replacing **r** by **h** + **b** and noting that $\boldsymbol{\omega} \times \mathbf{h}$ = **0**. The acceleration of A is given by the time derivative of Eq. 7/1. Thus,

$$\mathbf{a} = \dot{\boldsymbol{\omega}} \times \mathbf{r} + \boldsymbol{\omega} \times (\boldsymbol{\omega} \times \mathbf{r}) \qquad (7/2)$$

where $\dot{\mathbf{r}}$ has been replaced by its equal, $\mathbf{v} = \boldsymbol{\omega} \times \mathbf{r}$. The normal and tangential components of **a** for the circular motion have the familiar magnitudes $a_n = |\boldsymbol{\omega} \times (\boldsymbol{\omega} \times \mathbf{r})| = b\omega^2$ and $a_t = |\dot{\boldsymbol{\omega}} \times \mathbf{r}| = b\alpha$, where $\alpha = \dot{\omega}$. Inasmuch as both **v** and **a** are perpendicular to $\boldsymbol{\omega}$ and $\dot{\boldsymbol{\omega}}$, it follows that $\mathbf{v} \cdot \boldsymbol{\omega} = 0$, $\mathbf{v} \cdot \dot{\boldsymbol{\omega}} = 0$, $\mathbf{a} \cdot \boldsymbol{\omega} = 0$, and $\mathbf{a} \cdot \dot{\boldsymbol{\omega}} = 0$ for fixed-axis rotation.

7/4   PARALLEL-PLANE MOTION

When all points in a rigid body move in planes which are parallel to a fixed plane P, Fig. 7/3, we have a general form of plane motion. The reference plane is customarily taken through the mass center G and is referred to as the plane of motion. Since each point in the body, such as A', has a motion identical with the motion of the corresponding point, such as point A in plane P, it follows that the kinematics of plane motion covered in Chapter 5 provides a complete description of the motion when applied to the reference plane.

7/5   ROTATION ABOUT A FIXED POINT

When a body rotates about a fixed point, the angular-velocity vector no longer remains fixed in direction, and this change calls for a more general concept of rotation.

We must first examine the conditions under which rotation vectors obey the parallelogram law of addition and may, therefore, be treated as proper vectors. Consider a solid sphere, Fig. 7/4, which is cut from a rigid body confined to rotate about the fixed point O.

Figure 7/3

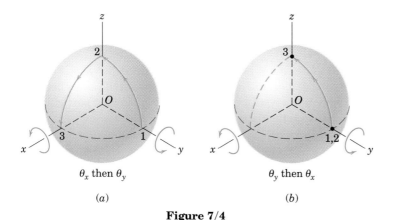

θ_x then θ_y θ_y then θ_x

(a)   (b)

Figure 7/4

The *x-y-z* axes here are taken fixed in space and do not rotate with the body. In the *a*-part of the figure, two successive 90° rotations of the sphere about, first, the *x*-axis and, second, the *y*-axis result in the motion of a point which is initially on the *y*-axis in position 1, to positions 2 and 3, successively. On the other hand, if the order of the rotations is reversed, the point suffers no motion during the *y*-rotation but moves to point 3 during the 90° rotation about the *x*-axis. Thus, the two cases do not yield the same final position, and it is evident from this one special example that finite rotations do not generally obey the parallelogram law of vector addition and are not commutative. Thus, finite rotations may *not* be treated as proper vectors.

Infinitesimal rotations, however, do obey the parallelogram law of vector addition. This fact is shown in Fig. 7/5, which represents the combined effect of two infinitesimal rotations $d\boldsymbol{\theta}_1$ and $d\boldsymbol{\theta}_2$ of a rigid body about the respective axes through the fixed point O. As a result of $d\boldsymbol{\theta}_1$, point A has a displacement $d\boldsymbol{\theta}_1 \times \mathbf{r}$, and likewise $d\boldsymbol{\theta}_2$ causes a displacement $d\boldsymbol{\theta}_2 \times \mathbf{r}$ of point A. Either order of ad-

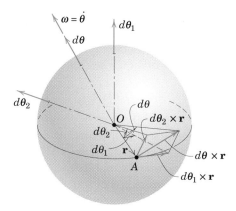

Figure 7/5

dition of these infinitesimal displacements clearly produces the same resultant displacement, which is $d\boldsymbol{\theta}_1 \times \mathbf{r} + d\boldsymbol{\theta}_2 \times \mathbf{r} = (d\boldsymbol{\theta}_1 + d\boldsymbol{\theta}_2) \times \mathbf{r}$. Hence, the two rotations are equivalent to the single rotation $d\boldsymbol{\theta} = d\boldsymbol{\theta}_1 + d\boldsymbol{\theta}_2$. It follows that the angular velocities $\boldsymbol{\omega}_1 = \dot{\boldsymbol{\theta}}_1$ and $\boldsymbol{\omega}_2 = \dot{\boldsymbol{\theta}}_2$ may be added vectorially to give $\boldsymbol{\omega} = \dot{\boldsymbol{\theta}} = \boldsymbol{\omega}_1 + \boldsymbol{\omega}_2$. We conclude, therefore, that at any instant of time a body with one fixed point is rotating instantaneously about a particular axis passing through the fixed point.

To aid in visualizing the concept of the instantaneous axis of rotation, we will cite a specific example. Figure 7/6 represents a solid cylindrical rotor made of clear plastic containing many black particles embedded in the plastic. The rotor is spinning about its shaft axis at the steady rate ω_1, and its shaft, in turn, is rotating about the fixed vertical axis at the steady rate ω_2, with rotations in

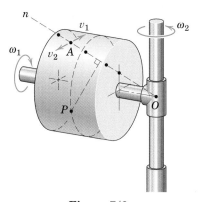

Figure 7/6

the directions indicated. If the rotor is photographed at a certain instant during its motion, the resulting picture would show one line of black dots sharply defined, indicating that, momentarily, their velocity was zero. This line of points with no velocity establishes the instantaneous position of the axis of rotation O-n. Any dot on this line, such as A, would have equal and opposite velocity components, v_1 due to ω_1 and v_2 due to ω_2. All other dots, such as the one at P, would appear blurred, and their movements would show as short streaks in the form of small circular arcs in planes normal to the axis O-n. Thus, all particles of the body, except those on line O-n, are momentarily rotating in circular arcs about the instantaneous axis of rotation. If a succession of photographs were taken, we would observe that the rotation axis would be defined by a new series of sharply-defined dots and that the axis would change position both in space and relative to the body. For rotation of a rigid body about a fixed point, then, it is seen that the rotation axis is, in general, not a line fixed in the body.

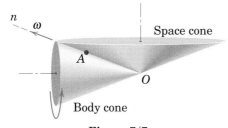

Figure 7/7

(a) Body and space cones. Relative to the plastic cylinder of Fig. 7/6, the instantaneous axis of rotation O-A-n generates a right-circular cone about the cylinder axis called the *body cone*. As the two rotations continue and the cylinder swings around the vertical axis, the instantaneous axis of rotation also generates a right-circular cone about the vertical axis called the *space cone*. These cones are shown in Fig. 7/7 for this particular example. We see that the body cone rolls on the space cone and that the angular velocity $\boldsymbol{\omega}$ of the body is a vector which lies along the common element of the two cones. For a more general case where the rotations are not steady, the space and body cones are not right-circular cones, Fig. 7/8, but the body cone still rolls on the space cone.

(b) Angular acceleration. The angular acceleration $\boldsymbol{\alpha}$ of a rigid body in three-dimensional motion is the time derivative of its angular velocity, $\boldsymbol{\alpha} = \dot{\boldsymbol{\omega}}$. In contrast to the case of rotation in a single plane where the scalar α measures only the change in magnitude of the angular velocity, in three-dimensional motion the vector $\boldsymbol{\alpha}$ reflects the change in direction of $\boldsymbol{\omega}$ as well as its change in magnitude. Thus in Fig. 7/8 where the tip of the angular velocity

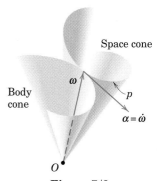

Figure 7/8

vector $\boldsymbol{\omega}$ follows the space curve p and changes in both magnitude and direction, the angular acceleration $\boldsymbol{\alpha}$ becomes a vector tangent to this curve in the direction of the change in $\boldsymbol{\omega}$.

When the magnitude of $\boldsymbol{\omega}$ remains constant, the angular acceleration $\boldsymbol{\alpha}$ is normal to $\boldsymbol{\omega}$. For this case, if we let $\boldsymbol{\Omega}$ stand for the angular velocity with which the vector $\boldsymbol{\omega}$ itself rotates (*precesses*) as it forms the space cone, the angular acceleration may be written

$$\boldsymbol{\alpha} = \boldsymbol{\Omega} \times \boldsymbol{\omega} \qquad (7/3)$$

This relation is easily seen from Fig. 7/9 where the vectors $\boldsymbol{\alpha}$, $\boldsymbol{\omega}$, and $\boldsymbol{\Omega}$ in the lower figure bear exactly the same relationship to each other as do the vectors $\mathbf{v}$, $\mathbf{r}$, and $\boldsymbol{\omega}$ in the upper figure for relating the velocity of a point A on a rigid body to its position vector from O and the angular velocity of the body.

If we use Fig. 7/2 to represent a rigid body rotating about a fixed point O with the instantaneous axis of rotation n-n, we see that the velocity $\mathbf{v}$ and acceleration $\mathbf{a} = \dot{\mathbf{v}}$ of any point A in the body are given by the same expressions as apply to the case in which the axis is fixed, namely,

$$\mathbf{v} = \boldsymbol{\omega} \times \mathbf{r} \qquad [7/1]$$

$$\mathbf{a} = \dot{\boldsymbol{\omega}} \times \mathbf{r} + \boldsymbol{\omega} \times (\boldsymbol{\omega} \times \mathbf{r}) \qquad [7/2]$$

The one difference between the case of rotation about a fixed axis and rotation about a fixed point lies in the fact that for rotation about a fixed point, the angular acceleration $\boldsymbol{\alpha} = \dot{\boldsymbol{\omega}}$ will have a component normal to $\boldsymbol{\omega}$ due to the change in direction of $\boldsymbol{\omega}$, as well as a component in the direction of $\boldsymbol{\omega}$ to reflect any change in the magnitude of $\boldsymbol{\omega}$. Although any point on the rotation axis n-n momentarily will have zero velocity, it will *not* have zero acceleration as long as $\boldsymbol{\omega}$ is changing its direction. On the other hand, for rotation about a fixed axis, $\boldsymbol{\alpha} = \dot{\boldsymbol{\omega}}$ has only the one component along the fixed axis to reflect the change in the magnitude of $\boldsymbol{\omega}$. Furthermore, points that lie on the fixed rotation axis clearly have no velocity or acceleration.

Although the discussion in this article has been developed for the case of rotation about a fixed point, we observe that rotation is a function *solely* of angular change, so that the expressions for $\boldsymbol{\omega}$ and $\boldsymbol{\alpha}$ do not depend on the fixity of the point around which rotation occurs. Thus, rotation may take place independently of the linear motion of the rotation point. This conclusion is the three-dimensional counterpart of the concept of rotation of a rigid body in plane motion described in Art. 5/2 and used throughout Chapters 5 and 6.

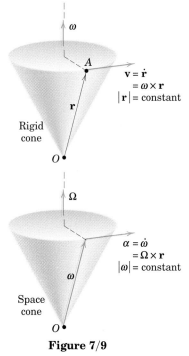

Figure 7/9

Sample Problem 7/1

The 0.8-m arm OA for a remote-control mechanism is pivoted about the horizontal x-axis of the clevis, and the entire assembly rotates about the z-axis with a constant speed $N = 60$ rev/min. Simultaneously, the arm is being raised at the constant rate $\dot\beta = 4$ rad/s. For the position where $\beta = 30°$, determine (a) the angular velocity of OA, (b) the angular acceleration of OA, (c) the velocity of point A, and (d) the acceleration of point A. If, in addition to the motion described, the vertical shaft and point O had a linear motion, say, in the z-direction, would that motion change the angular velocity or angular acceleration of OA?

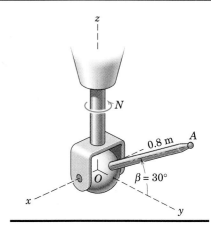

Solution. (a) Since the arm OA is rotating about both the x- and the z-axes, it has the components $\omega_x = \dot\beta = 4$ rad/s and $\omega_z = 2\pi N/60 = 2\pi(60)/60 = 6.28$ rad/s. The angular velocity is

$$\boldsymbol{\omega} = \boldsymbol{\omega}_x + \boldsymbol{\omega}_z = 4\mathbf{i} + 6.28\mathbf{k} \text{ rad/s} \qquad Ans.$$

(b) The angular acceleration of OA is

$$\boldsymbol{\alpha} = \dot{\boldsymbol{\omega}} = \dot{\boldsymbol{\omega}}_x + \dot{\boldsymbol{\omega}}_z$$

Since $\boldsymbol{\omega}_z$ is not changing in magnitude or direction, $\dot{\boldsymbol{\omega}}_z = \mathbf{0}$. But $\boldsymbol{\omega}_x$ is changing direction and thus has a derivative which, from Eq. 7/3, is

$$\dot{\boldsymbol{\omega}}_x = \boldsymbol{\omega}_z \times \boldsymbol{\omega}_x = 6.28\mathbf{k} \times 4\mathbf{i} = 25.1\mathbf{j} \text{ rad/s}^2$$

① Therefore,

$$\boldsymbol{\alpha} = 25.1\mathbf{j} + \mathbf{0} = 25.1\mathbf{j} \text{ rad/s}^2 \qquad Ans.$$

(c) With the position vector of A given by $\mathbf{r} = 0.693\mathbf{j} + 0.4\mathbf{k}$ m, the velocity of A from Eq. 7/1 becomes

$$\mathbf{v} = \boldsymbol{\omega} \times \mathbf{r} = \begin{vmatrix} \mathbf{i} & \mathbf{j} & \mathbf{k} \\ 4 & 0 & 6.28 \\ 0 & 0.693 & 0.4 \end{vmatrix} = -4.35\mathbf{i} - 1.60\mathbf{j} + 2.77\mathbf{k} \text{ m/s} \qquad Ans.$$

(d) The acceleration of A from Eq. 7/2 is

$$\mathbf{a} = \dot{\boldsymbol{\omega}} \times \mathbf{r} + \boldsymbol{\omega} \times (\boldsymbol{\omega} \times \mathbf{r})$$

$$= \boldsymbol{\alpha} \times \mathbf{r} + \boldsymbol{\omega} \times \mathbf{v}$$

$$= \begin{vmatrix} \mathbf{i} & \mathbf{j} & \mathbf{k} \\ 0 & 25.1 & 0 \\ 0 & 0.693 & 0.4 \end{vmatrix} + \begin{vmatrix} \mathbf{i} & \mathbf{j} & \mathbf{k} \\ 4 & 0 & 6.28 \\ -4.35 & -1.60 & 2.77 \end{vmatrix}$$

$$= (10.05\mathbf{i}) + (10.05\mathbf{i} - 38.4\mathbf{j} - 6.40\mathbf{k})$$

②
$$= 20.1\mathbf{i} - 38.4\mathbf{j} - 6.40\mathbf{k} \text{ m/s}^2 \qquad Ans.$$

The angular motion of OA depends only on the angular changes N and $\dot\beta$, so any linear motion of O does not affect $\boldsymbol{\omega}$ and $\boldsymbol{\alpha}$.

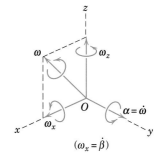

① Alternatively, consider axes x-y-z to be attached to the vertical shaft and clevis so that they rotate. The derivative of $\boldsymbol{\omega}_x$ becomes $\dot{\boldsymbol{\omega}}_x = 4\dot{\mathbf{i}}$. But from Eq. 5/11, we have $\dot{\mathbf{i}} = \boldsymbol{\omega}_z \times \mathbf{i} = 6.28\mathbf{k} \times \mathbf{i} = 6.28\mathbf{j}$. Thus, $\boldsymbol{\alpha} = \dot{\boldsymbol{\omega}}_x = 4(6.28)\mathbf{j} = 25.1\mathbf{j} \text{ rad/s}^2$ as before.

② To compare methods, it is suggested that these results for $\mathbf{v}$ and $\mathbf{a}$ be obtained by applying Eqs. 2/18 and 2/19 for particle motion in spherical coordinates, changing symbols as necessary.

Sample Problem 7/2

The electric motor with an attached disk is running at a constant low speed of 120 rev/min in the direction shown. Its housing and mounting base are initially at rest. The entire assembly is next set in rotation about the vertical Z-axis at the constant rate $N = 60$ rev/min with a fixed angle γ of 30°. Determine (a) the angular velocity and angular acceleration of the disk, (b) the space and body cones, and (c) the velocity and acceleration of point A at the top of the disk for the instant shown.

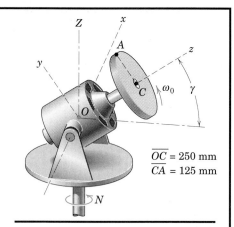

$\overline{OC} = 250$ mm
$\overline{CA} = 125$ mm

Solution. The axes x-y-z with unit vectors $\mathbf{i}, \mathbf{j}, \mathbf{k}$ are attached to the motor frame, with the z-axis coinciding with the rotor axis and the x-axis coinciding with the horizontal axis through O about which the motor tilts. The Z-axis is vertical and carries the unit vector $\mathbf{K} = \mathbf{j} \cos \gamma + \mathbf{k} \sin \gamma$.

(a) The rotor and disk have two components of angular velocity: $\omega_0 = 120(2\pi)/60 = 4\pi$ rad/s about the z-axis and $\Omega = 60(2\pi)/60 = 2\pi$ rad/s about the Z-axis. Thus, the angular velocity becomes

① $\omega = \omega_0 + \Omega = \omega_0\mathbf{k} + \Omega\mathbf{K}$

$= \omega_0\mathbf{k} + \Omega(\mathbf{j} \cos \gamma + \mathbf{k} \sin \gamma) = (\Omega \cos \gamma)\mathbf{j} + (\omega_0 + \Omega \sin \theta)\mathbf{k}$

$= (2\pi \cos 30°)\mathbf{j} + (4\pi + 2\pi \sin 30°)\mathbf{k} = \pi(\sqrt{3}\mathbf{j} + 5.0\mathbf{k})$ rad/s

Ans.

The angular acceleration of the disk from Eq. 7/3 is

② $\alpha = \dot{\omega} = \Omega \times \omega$

$= \Omega(\mathbf{j} \cos \gamma + \mathbf{k} \sin \gamma) \times [(\Omega \cos \gamma)\mathbf{j} + (\omega_0 + \Omega \sin \gamma)\mathbf{k}]$

$= \Omega(\omega_0 \cos \gamma + \Omega \sin \gamma \cos \gamma)\mathbf{i} - (\Omega^2 \sin \gamma \cos \gamma)\mathbf{i}$

③ $= (\Omega\omega_0 \cos \gamma)\mathbf{i} = \mathbf{i}(2\pi)(4\pi) \cos 30° = 68.4\mathbf{i}$ rad/s^2 *Ans.*

(b) The angular velocity vector ω is the common element of the space and body cones which may now be constructed as shown.

(c) The position vector of point A for the instant considered is

$$\mathbf{r} = 0.125\mathbf{j} + 0.250\mathbf{k} \text{ m}$$

From Eq. 7/1 the velocity of A is

$$\mathbf{v} = \omega \times \mathbf{r} = \begin{vmatrix} \mathbf{i} & \mathbf{j} & \mathbf{k} \\ 0 & \sqrt{3}\pi & 5\pi \\ 0 & 0.125 & 0.250 \end{vmatrix} = -0.1920\pi\mathbf{i} \text{ m/s} \quad \textit{Ans.}$$

From Eq. 7/2 the acceleration of point A is

$\mathbf{a} = \dot{\omega} \times \mathbf{r} + \omega \times (\omega \times \mathbf{r}) = \alpha \times \mathbf{r} + \omega \times \mathbf{v}$

$= 68.4\mathbf{i} \times (0.125\mathbf{j} + 0.250\mathbf{k}) + \pi(\sqrt{3}\mathbf{j} + 5\mathbf{k}) \times (-0.1920\pi\mathbf{i})$

$= -26.6\mathbf{j} + 11.83\mathbf{k} \text{ m/s}^2$ *Ans.*

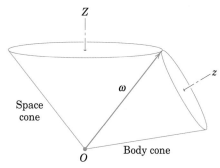

Space cone

Body cone

① Note that $\omega_0 + \Omega = \omega = \omega_y + \omega_z$ as shown on the vector diagram.

② Remember that Eq. 7/3 gives the complete expression for α only for steady precession where $|\omega|$ is constant, which applies to this problem.

③ Since the magnitude of ω is constant, α must be tangent to the base circle of the space cone, which puts it in the plus x-direction in agreement with our calculated conclusion.

PROBLEMS

Introductory problems

7/1 The mechanism to control the deployment of a spacecraft solar panel from position A to position B is to be designed. Determine the vector expression for the single angle θ of rotation of the panel which can achieve the required change of position. The side facing the plus x-direction in position A must face the plus z-direction in position B. (*Suggestion*: Determine the intersection of two planes, one containing all possible axes of rotation to accomplish the movement for edge O-1, and the other containing all such axes for edge O-2.)

$$\text{Ans. } \theta = \frac{\pi}{\sqrt{2}}(\mathbf{i} + \mathbf{k})$$

Problem 7/1

7/2 Consider the rectangular block with fixed axes x-y-z. Rotate the block about the x-axis through 90° and then from this new position rotate it about the y-axis through 90° and sketch the final position of the block. Repeat the rotations but in the reverse order and compare the two final positions. If each rotation is represented by a vector, what conclusion can be drawn concerning vector addition of finite rotations?

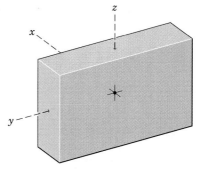

Problem 7/2

7/3 The solid cylinder is rotating about the fixed axis OA with a constant speed $N = 600$ rev/min in the direction shown. If the x- and y-components of the velocity of point P are 4 m/s and -2 m/s, determine its z-component of velocity and the radial distance R from P to the rotation axis. Also find the magnitude of the acceleration of P.

$$\text{Ans. } v_z = -2 \text{ m/s}, R = 78.0 \text{ mm}$$
$$a_P = 308 \text{ m/s}^2$$

Problem 7/3

7/4 The disk rotates with a spin velocity of 15 rad/s about its horizontal z-axis first in the direction (a) and second in the direction (b). The assembly rotates with the velocity $N = 10$ rad/s about the vertical axis. Construct the space and body cones for each case.

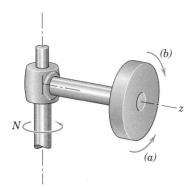

Problem 7/4

7/5 The rotor and shaft are mounted in a clevis which can rotate about the z-axis with an angular velocity Ω. With $\Omega = 0$ and θ constant, the rotor has an angular velocity $\boldsymbol{\omega}_0 = -4\mathbf{j} - 3\mathbf{k}$ rad/s. Find the velocity $\mathbf{v}_A$ of point A on the rim if its position vector at this instant is $\mathbf{r} = 0.5\mathbf{i} + 1.2\mathbf{j} + 1.1\mathbf{k}$ m. What is the rim speed v_B of any point B?

Ans. $\mathbf{v}_A = -0.8\mathbf{i} - 1.5\mathbf{j} + 2\mathbf{k}$ m/s
$v_B = 2.62$ m/s

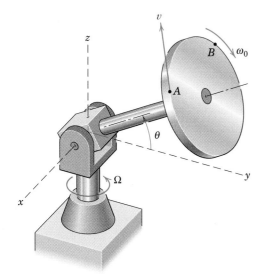

Problem 7/5

7/6 The rod is hinged about the axis O-O of the clevis, which is attached to the end of the vertical shaft. The shaft rotates with a constant angular velocity ω_0 as shown. If θ is decreasing at the constant rate $-\dot{\theta} = p$, write expressions for the angular velocity $\boldsymbol{\omega}$ and angular acceleration $\boldsymbol{\alpha}$ of the rod.

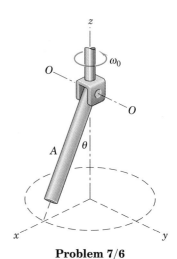

Problem 7/6

7/7 The rotor B spins about its inclined axis OA at the speed $N_1 = 200$ rev/min, where $\beta = 30°$. Simultaneously, the assembly rotates about the vertical z-axis at the rate N_2. If the total angular velocity of the rotor has a magnitude of 40 rad/s, determine N_2.

Ans. $N_2 = 440$ rev/min

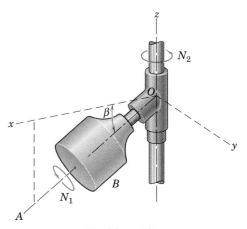

Problem 7/7

7/8 A slender rod bent into the shape shown rotates about the fixed line CD at a constant angular rate ω. Determine the velocity and acceleration of point A.

Problem 7/8

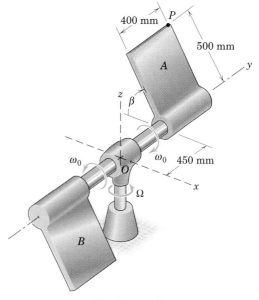

Problem 7/9

7/9 The panel assembly and attached x-y-z axes rotate with a constant angular velocity $\Omega = 0.6$ rad/s about the vertical z-axis. Simultaneously, the panels rotate about the y-axis as shown with a constant rate $\omega_0 = 2$ rad/s. Determine the angular acceleration α of panel A and find the acceleration of point P for the instant when $\beta = 90°$.

$$Ans. \ \alpha = -1.2\mathbf{i} \ \text{rad/s}^2$$
$$\mathbf{a}_P = 0.894\mathbf{j} - 2\mathbf{k} \ \text{m/s}^2$$

7/10 The circular disk is rotating about the fixed axis OC with a constant angular velocity $\omega = 20$ rad/s. At a certain instant, point A on the rim passes a point whose x-y-z coordinates are 40, 118, and 51 mm, respectively. Calculate the magnitudes of the velocity $\mathbf{v}$ and acceleration $\mathbf{a}$ of point A. Also find the radius R of the disk.

Problem 7/10

Representative problems

7/11 The bevel-gear unit shown in section is designed so that gear C is fixed and does not rotate. The base fixture at O, along with its shaft OB, makes 4 complete revolutions every second about the vertical z-axis in the direction indicated, causing bevel gear D to rotate. Draw the space and body cones for the motion and calculate the magnitude ω of the angular velocity of gear D, where $\beta = 14.86°$ for a tooth ratio of 39/20. Also find the component of angular velocity of D about its own axis O-n.

Ans. $\omega = 69.1$ rad/s, $\omega_n = 66.8$ rad/s

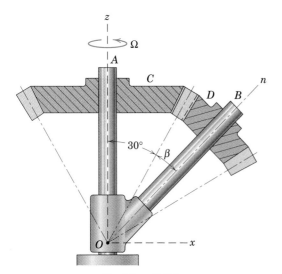

Problem 7/11

7/12 Use the results cited for Prob. 7/11 and determine the vector expression for the angular acceleration $\boldsymbol{\alpha}$ of bevel gear D.

7/13 The spool A rotates about its axis with an angular velocity of 20 rad/s, first in the sense of ω_a and second in the sense of ω_b. Simultaneously, the assembly rotates about the vertical axis with an angular velocity $\omega_1 = 10$ rad/s. Determine the magnitude ω of the total angular velocity of the spool and construct the body and space cones for the spool for each case.

Ans. $\omega = 26.5$ rad/s, 17.32 rad/s

Problem 7/13

7/14 The electric fan has a constant speed of 1720 rev/min in the direction indicated with its axis oriented as shown. If the x- and y-components of the velocity of point A on the blade tip are 15 m/s and -20 m/s, respectively, determine the magnitude v of the velocity of the blade tip and the diameter of the fan blades.

Problem 7/14

7/15 In manipulating the dumbbell, the jaws of the robotic device have an angular velocity $\omega_p = 2$ rad/s about the axis OG with γ fixed at 60°. The entire assembly rotates about the vertical Z-axis at the constant rate $\Omega = 0.8$ rad/s. Determine the angular velocity $\boldsymbol{\omega}$ and angular acceleration $\boldsymbol{\alpha}$ of the dumbbell. Express the results in terms of the given orientation of axes x-y-z, where the y-axis is parallel to the Y-axis.

$$Ans. \quad \boldsymbol{\omega} = -0.4\mathbf{i} + 2.69\mathbf{k} \text{ rad/s}$$
$$\boldsymbol{\alpha} = 0.8\mathbf{j} \text{ rad/s}^2$$

7/16 The robot shown has five degrees of rotational freedom. The x-y-z axes are attached to the base ring, which rotates about the z-axis at the rate ω_1. The arm O_1O_2 rotates about the x-axis at the rate $\omega_2 = \dot{\theta}$. The control arm O_2A rotates about axis O_1-O_2 at the rate ω_3 and about a perpendicular axis through O_2 which is momentarily parallel to the x-axis at the rate $\omega_4 = \dot{\beta}$. Finally, the jaws rotate about axis O_2-A at the rate ω_5. The magnitudes of all angular rates are constant. For the configuration shown, determine the magnitude ω of the total angular velocity of the jaws for $\theta = 60°$ and $\beta = 45°$ if $\omega_1 = 2$ rad/s, $\dot{\theta} = 1.5$ rad/s, and $\omega_3 = \omega_4 = \omega_5 = 0$. Also express the angular acceleration $\boldsymbol{\alpha}$ of arm O_1O_2 as a vector.

Problem 7/15

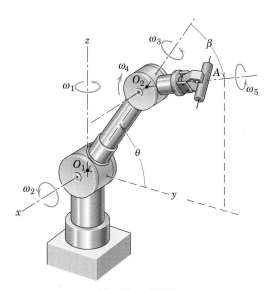

Problem 7/16

7/17 For the robot of Prob. 7/16, determine the angular velocity $\boldsymbol{\omega}$ and angular acceleration $\boldsymbol{\alpha}$ of the jaws A if $\theta = 60°$ and $\beta = 30°$, both constant, and if $\omega_1 = 2$ rad/s, $\omega_2 = \omega_3 = \omega_4 = 0$, and $\omega_5 = 0.8$ rad/s, all constant.

$$Ans.\ \boldsymbol{\omega} = 0.693\mathbf{j} + 2.40\mathbf{k} \text{ rad/s}$$
$$\boldsymbol{\alpha} = -1.386\mathbf{i} \text{ rad/s}^2$$

7/18 The circular disk B of radius r rolls without slipping in a circle of radius b on the fixed disk C. Write expressions for the angular velocity $\boldsymbol{\omega}$ and the angular acceleration $\boldsymbol{\alpha}$ of disk B if its axle rotates about the vertical z-axis at the constant angular velocity ω_0.

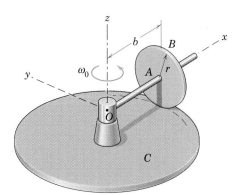

Problem 7/18

7/19 If disk C in Prob. 7/18 rotates clockwise with a constant angular velocity Ω about the z-axis when viewed from above and the angular rate ω_0 of OA remains unchanged, determine expressions for the angular velocity $\boldsymbol{\omega}$ and angular acceleration $\boldsymbol{\alpha}$ of disk B.

$$Ans.\ \boldsymbol{\omega} = -\frac{b}{r}(\omega_0 + \Omega)\mathbf{i} + \omega_0\mathbf{k}$$
$$\boldsymbol{\alpha} = -\frac{b}{r}\omega_0(\omega_0 + \Omega)\mathbf{j}$$

7/20 The circular disk of 120-mm radius rotates about the z-axis at the constant rate $\omega_z = 20$ rad/s, and the entire assembly rotates about the fixed x-axis at the constant rate $\omega_x = 10$ rad/s. Calculate the magnitudes of the velocity $\mathbf{v}$ and acceleration $\mathbf{a}$ of point B for the instant when $\theta = 30°$.

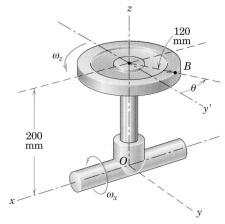

Problem 7/20

7/21 The design of the rotating arm OA of a control mechanism requires that it rotate about the vertical Z-axis at the constant rate $\Omega = \dot{\beta} = \pi$ rad/s. Simultaneously, OA oscillates according to $\theta = \theta_0 \sin 4\Omega t$, where $\theta_0 = \pi/6$ radians and t is in seconds measured from the time when $\beta = 0$. Determine the angular velocity $\boldsymbol{\omega}$ and the angular acceleration $\boldsymbol{\alpha}$ of OA for the instant (a) when $t = 1/2$ s and (b) when $t = 1/8$ s. The x-y reference axes rotate in the X-Y plane with the angular velocity Ω.

$$Ans.\ (a)\ \boldsymbol{\omega} = \pi\left(-\frac{2\pi}{3}\mathbf{j} + \mathbf{k}\right),\ \boldsymbol{\alpha} = \frac{2\pi^3}{3}\mathbf{i}$$

$$(b)\ \boldsymbol{\omega} = \pi\mathbf{k},\ \boldsymbol{\alpha} = \frac{8\pi^3}{3}\mathbf{j}$$

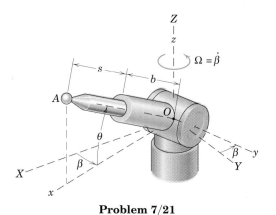

Problem 7/21

7/22 For the rotating and oscillating control arm OA of Prob. 7/21, determine the velocity $\mathbf{v}$ and acceleration $\mathbf{a}$ of the ball tip A for the condition when $t = 1/2$ s. Distance $b = 120$ mm and $s = 100$ mm, and $\theta = \theta_0 \sin 4\Omega t$ as defined in Prob. 7/21 with $\Omega = \pi$ rad/s and $\theta_0 = \pi/6$ rad.

7/23 If the angular velocity $\boldsymbol{\omega}_0 = -4\mathbf{j} - 3\mathbf{k}$ rad/s of the rotor in Prob. 7/5 is constant in magnitude, determine the angular acceleration $\boldsymbol{\alpha}$ of the rotor for (a) $\Omega = 0$ and $\dot{\theta} = 2$ rad/s (both constant) and (b) $\theta = \tan^{-1}\left(\frac{3}{4}\right)$ and $\Omega = 2$ rad/s (both constant). Find the magnitude of the acceleration of point A in each case, where A has the position vector $\mathbf{r} = 0.5\mathbf{i} + 1.2\mathbf{j} + 1.1\mathbf{k}$ m at the instant represented.

Ans. (a) $\boldsymbol{\alpha} = 6\mathbf{j} - 8\mathbf{k}$ rad/s^2, $a_A = 21.2$ m/s^2
(b) $\boldsymbol{\alpha} = 8\mathbf{i}$ rad/s^2, $a_A = 10.67$ m/s^2

7/24 The vertical shaft and attached clevis rotate about the z-axis at the constant rate $\Omega = 4$ rad/s. Simultaneously, the shaft B revolves about its axis OA at the constant rate $\omega_0 = 3$ rad/s, and the angle γ is decreasing at the constant rate of $\pi/4$ rad/s. Determine the angular velocity $\boldsymbol{\omega}$ and the magnitude of the angular acceleration $\boldsymbol{\alpha}$ of shaft B when $\gamma = 30°$. The x-y-z axes are attached to the clevis and rotate with it.

Problem 7/24

▶7/25 The right-circular cone A rolls on the fixed right-circular cone B at a constant rate and makes one complete trip around B every 4 s. Compute the magnitude of the angular acceleration $\boldsymbol{\alpha}$ of cone A during its motion.

Ans. $\alpha = 6.32$ rad/s^2

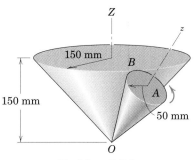

Problem 7/25

▶7/26 The end of link A is welded to the yoke which is pivoted about the z-axis to the collar C. The collar may rotate about the x-axis of the fixed shaft. Link A and its yoke can rotate about both the x- and z-axes but not about the y-axis. Regardless of the motion of the other end of link A, show that the angular velocity $\boldsymbol{\omega}$ of link A and its yoke must satisfy the relation $\boldsymbol{\omega} \cdot \mathbf{h} \times (\mathbf{r} \times \mathbf{h}) = 0$. Vectors $\mathbf{r}$ and $\mathbf{h}$ are any vectors directed, respectively, along the link and along the fixed shaft. The axis of the link A is normal to the yoke axis (z-direction) and, hence, lies in the x-y plane.

Problem 7/26

▶7/27 The solid right-circular cone of base radius r and height h rolls on a flat surface without slipping. The center B of the circular base moves in a circular path around the z-axis with a constant speed v. Determine the angular velocity $\boldsymbol{\omega}$ and the angular acceleration $\boldsymbol{\alpha}$ of the solid cone.

$$Ans. \ \boldsymbol{\omega} = v\sqrt{\frac{1}{r^2} + \frac{1}{h^2}}\,\mathbf{i}$$

$$\boldsymbol{\alpha} = -\frac{v^2}{h^2}\left(\frac{r}{h} + \frac{h}{r}\right)\mathbf{j}$$

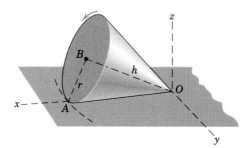

Problem 7/27

▶7/28 The pendulum oscillates about the x-axis according to $\theta = \dfrac{\pi}{6}\sin 3\pi t$ radians, where t is the time in seconds. Simultaneously, the shaft OA revolves about the vertical z-axis at the constant rate $\omega_z = 2\pi$ rad/s. Determine the velocity $\mathbf{v}$ and acceleration $\mathbf{a}$ of the center B of the pendulum as well as its angular acceleration $\boldsymbol{\alpha}$ for the instant when $t = 0$.

$$Ans. \ \mathbf{v} = -0.359\mathbf{j} \text{ m/s}$$
$$\mathbf{a} = 8.45\mathbf{i} + 4.87\mathbf{k} \text{ m/s}^2$$
$$\boldsymbol{\alpha} = -31.0\mathbf{j} \text{ rad/s}^2$$

Problem 7/28

7/6   *GENERAL MOTION*

The kinematic analysis of a rigid body which has general three-dimensional motion is best accomplished with the aid of our principles of relative motion. These principles have been applied to problems in plane motion and will now be extended to space motion. We will make use of both translating reference axes and rotating reference axes.

(a) Translating reference axes.   In Fig. 7/10 is shown a rigid body which has an angular velocity $\boldsymbol{\omega}$. We choose any convenient point B as the origin of a translating reference system x-y-z. The velocity $\mathbf{v}$ and acceleration $\mathbf{a}$ of any other point A in the body are given by the relative-velocity and relative-acceleration expressions

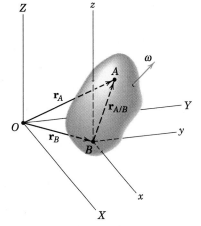

$$\mathbf{v}_A = \mathbf{v}_B + \mathbf{v}_{A/B} \qquad [5/4]$$

$$\mathbf{a}_A = \mathbf{a}_B + \mathbf{a}_{A/B} \qquad [5/7]$$

which were developed in Arts. 5/4 and 5/6 for the plane motion of rigid bodies and which hold equally well for any plane in space in which the three vectors for each of the equations must lie.

In applying these relations to rigid-body motion in space, we note from Fig. 7/10 that the distance $\overline{AB}$ remains constant. Thus, from an observer's position on x-y-z, the body appears to rotate about the point B and point A appears to lie on a spherical surface with B as the center. Consequently, we may view the general motion as a translation of the body with the motion of B plus a rotation of the body about B.

Figure 7/10

The relative-motion terms represent the effect of the rotation about B and are identical to the velocity and acceleration expressions discussed in the previous article for rotation of a rigid body about a fixed point. Therefore, the relative-velocity and relative-acceleration equations may be written

$$\boxed{\begin{aligned} \mathbf{v}_A &= \mathbf{v}_B + \boldsymbol{\omega} \times \mathbf{r}_{A/B} \\ \mathbf{a}_A &= \mathbf{a}_B + \dot{\boldsymbol{\omega}} \times \mathbf{r}_{A/B} + \boldsymbol{\omega} \times (\boldsymbol{\omega} \times \mathbf{r}_{A/B}) \end{aligned}} \qquad \textbf{(7/4)}$$

where $\boldsymbol{\omega}$ and $\dot{\boldsymbol{\omega}}$ are the instantaneous angular velocity and angular acceleration, respectively, of the body.

The selection of the reference point B is quite arbitrary in theory. In practice, point B is chosen for convenience as some point in the body whose motion is known in whole or in part. If point A is chosen as the reference point, the relative-motion equations become

$$\mathbf{v}_B = \mathbf{v}_A + \boldsymbol{\omega} \times \mathbf{r}_{B/A}$$

$$\mathbf{a}_B = \mathbf{a}_A + \dot{\boldsymbol{\omega}} \times \mathbf{r}_{B/A} + \boldsymbol{\omega} \times (\boldsymbol{\omega} \times \mathbf{r}_{B/A})$$

where $\mathbf{r}_{B/A} = -\mathbf{r}_{A/B}$. It should be clear that $\boldsymbol{\omega}$ and, hence, $\dot{\boldsymbol{\omega}}$ are the same vectors for either formulation since the absolute angular motion of the body is independent of the choice of reference point. When we come to the kinetic equations for general motion, we will see that the mass center of a body is frequently the most convenient reference point to choose.

If points A and B in Fig. 7/10 represent the ends of a rigid control link in a spatial mechanism where the end connections act as ball-and-socket joints (as in Sample Problem 7/3), it is necessary to impose certain kinematical requirements. Clearly, any rotation of the link about its own axis AB does not affect the action of the link. Thus, the angular velocity $\boldsymbol{\omega}_n$ whose vector is normal to the link describes its action. It is necessary, therefore, that $\boldsymbol{\omega}_n$ and $\mathbf{r}_{A/B}$ be at right angles, and this condition is satisfied if $\boldsymbol{\omega}_n \cdot \mathbf{r}_{A/B} = 0$. Similarly, it is only the component $\boldsymbol{\alpha}_n{}^*$ of the angular acceleration of the link normal to AB that affects its action, so that $\boldsymbol{\alpha}_n \cdot \mathbf{r}_{A/B} = 0$ must also hold.

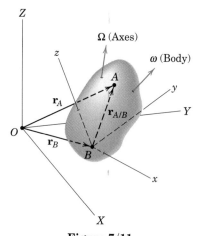

Figure 7/11

(b) Rotating reference axes. A more general formulation of the motion of a rigid body in space calls for the use of reference axes which rotate as well as translate. The description of Fig. 7/10 is modified in Fig. 7/11 to show reference axes whose origin is attached to the reference point B as before, but which rotate with an absolute angular velocity $\boldsymbol{\Omega}$ which may be different from the absolute angular velocity $\boldsymbol{\omega}$ of the body.

We now make use of Eqs. 5/11, 5/12, 5/13, and 5/14 developed in Art. 5/7 for describing the plane motion of a rigid body with the use of rotating axes. The extension of these relations from two to three dimensions is easily accomplished by merely including the z-component of the vectors, and this step is left to the student to carry out. Replacing $\boldsymbol{\omega}$ in these equations by the angular velocity $\boldsymbol{\Omega}$ of our rotating x-y-z axes gives us

$$\dot{\mathbf{i}} = \boldsymbol{\Omega} \times \mathbf{i} \qquad \dot{\mathbf{j}} = \boldsymbol{\Omega} \times \mathbf{j} \qquad \dot{\mathbf{k}} = \boldsymbol{\Omega} \times \mathbf{k} \qquad (7/5)$$

for the time derivatives of the rotating unit vectors attached to x-y-z. The expressions for the velocity and acceleration of point A become

$$\mathbf{v}_A = \mathbf{v}_B + \boldsymbol{\Omega} \times \mathbf{r}_{A/B} + \mathbf{v}_{\text{rel}}$$

$$\mathbf{a}_A = \mathbf{a}_B + \dot{\boldsymbol{\Omega}} \times \mathbf{r}_{A/B} + \boldsymbol{\Omega} \times (\boldsymbol{\Omega} \times \mathbf{r}_{A/B}) + 2\boldsymbol{\Omega} \times \mathbf{v}_{\text{rel}} + \mathbf{a}_{\text{rel}}$$

$$(7/6)$$

---

*It may be shown that $\boldsymbol{\alpha}_n = \dot{\boldsymbol{\omega}}_n$ if the angular velocity of the link about its own axis is not changing. See the first author's *Dynamics, 2nd Edition SI Version*, 1975, John Wiley & Sons, Art. 37.

where $\mathbf{v}_{\mathrm{rel}} = \dot{x}\mathbf{i} + \dot{y}\mathbf{j} + \dot{z}\mathbf{k}$ and $\mathbf{a}_{\mathrm{rel}} = \ddot{x}\mathbf{i} + \ddot{y}\mathbf{j} + \ddot{z}\mathbf{k}$ are, respectively, the velocity and acceleration of point A measured relative to x-y-z by an observer attached to x-y-z. Again, we note that $\boldsymbol{\Omega}$ is the angular velocity of the axes and may be different from the angular velocity $\boldsymbol{\omega}$ of the body. Also we note that $\mathbf{r}_{A/B}$ remains constant in magnitude for a rigid body, but it will change direction with respect to x-y-z when the angular velocity $\boldsymbol{\Omega}$ of the axes is different from the angular velocity $\boldsymbol{\omega}$ of the body. We observe further that, if x-y-z are rigidly attached to the body, $\boldsymbol{\Omega} = \boldsymbol{\omega}$ and $\mathbf{v}_{\mathrm{rel}}$ and $\mathbf{a}_{\mathrm{rel}}$ are both zero, which makes the equations identical to Eqs. 7/4.

In Art. 5/7 we also developed the relationship (Eq. 5/13) between the time derivative of a vector $\mathbf{V}$ as measured in the fixed X-Y system and the time derivative of $\mathbf{V}$ as measured relative to the rotating x-y system. For our three-dimensional case, this relation becomes

$$\left(\frac{d\mathbf{V}}{dt}\right)_{XYZ} = \left(\frac{d\mathbf{V}}{dt}\right)_{xyz} + \boldsymbol{\Omega} \times \mathbf{V} \qquad (7/7)$$

When we apply this transformation to the relative-position vector $\mathbf{r}_{A/B} = \mathbf{r}_A - \mathbf{r}_B$ for our rigid body of Fig. 7/11, we get

$$\left(\frac{d\mathbf{r}_A}{dt}\right)_{XYZ} = \left(\frac{d\mathbf{r}_B}{dt}\right)_{XYZ} + \left(\frac{d\mathbf{r}_{A/B}}{dt}\right)_{xyz} + \boldsymbol{\Omega} \times \mathbf{r}_{A/B}$$

or

$$\mathbf{v}_A = \mathbf{v}_B + \mathbf{v}_{\mathrm{rel}} + \boldsymbol{\Omega} \times \mathbf{r}_{A/B}$$

which gives us the first of Eqs. 7/6.

Equations 7/6 are particularly useful when the reference axes are attached to a moving body within which relative motion occurs.

Equation 7/7 may be recast as the vector operator

$$\left(\frac{d[\;\;]}{dt}\right)_{XYZ} = \left(\frac{d[\;\;]}{dt}\right)_{xyz} + \boldsymbol{\Omega} \times [\;\;] \qquad (7/7a)$$

where $[\;\;]$ stands for any vector $\mathbf{V}$ expressible both in X-Y-Z and in x-y-z. If we apply the operator to itself, we obtain the second time derivative, which becomes

$$\left(\frac{d^2[\;\;]}{dt^2}\right)_{XYZ} = \left(\frac{d^2[\;\;]}{dt^2}\right)_{xyz} + \dot{\boldsymbol{\Omega}} \times [\;\;] + \boldsymbol{\Omega} \times (\boldsymbol{\Omega} \times [\;\;])$$

$$+ 2\boldsymbol{\Omega} \times \left(\frac{d[\;\;]}{dt}\right)_{xyz} \qquad (7/7b)$$

This exercise is left to the student, and the form of Eq. 7/7b is observed to be the same as that of the second of Eqs. 7/6 expressed for $\mathbf{a}_{A/B} = \mathbf{a}_A - \mathbf{a}_B$.

Sample Problem 7/3

Crank CB rotates about the horizontal axis with an angular velocity $\omega_1 = 6$ rad/s which is constant for a short interval of motion which includes the position shown. The link AB has a ball-and-socket fitting on each end and connects crank DA with CB. For the instant shown, determine the angular velocity ω_2 of crank DA and the angular velocity ω_n of link AB.

Solution. The relative-velocity relation, Eq. 7/4, will be solved ① first using translating reference axes attached to B. The equation is

$$\mathbf{v}_A = \mathbf{v}_B + \boldsymbol{\omega}_n \times \mathbf{r}_{A/B}$$

② where $\boldsymbol{\omega}_n$ is the angular velocity of link AB taken normal to AB. The velocities of A and B are

$$[v = r\omega] \qquad \mathbf{v}_A = 50\omega_2\mathbf{j} \qquad \mathbf{v}_B = 100(6)\mathbf{i} = 600\mathbf{i} \text{ mm/s}$$

Also $\mathbf{r}_{A/B} = 50\mathbf{i} + 100\mathbf{j} + 100\mathbf{k}$ mm. Substitution into the velocity relation gives

$$50\omega_2\mathbf{j} = 600\mathbf{i} + \begin{vmatrix} \mathbf{i} & \mathbf{j} & \mathbf{k} \\ \omega_{nx} & \omega_{ny} & \omega_{nz} \\ 50 & 100 & 100 \end{vmatrix}$$

Expanding the determinant and equating the coefficients of the $\mathbf{i}, \mathbf{j}, \mathbf{k}$ terms give

$$-6 = \qquad\quad + \omega_{ny} - \omega_{nz}$$
$$\omega_2 = -2\omega_{nx} \qquad\quad + \omega_{nz}$$
$$0 = \quad 2\omega_{nx} - \omega_{ny}$$

These equations may be solved for ω_2, which becomes

$$\omega_2 = 6 \text{ rad/s} \qquad\qquad Ans.$$

As they stand, the three equations incorporate the fact that $\boldsymbol{\omega}_n$ is normal to $\mathbf{v}_{A/B}$, but they cannot be solved until the requirement that $\boldsymbol{\omega}_n$ be normal to $\mathbf{r}_{A/B}$ is included. Thus,

$$[\boldsymbol{\omega}_n \cdot \mathbf{r}_{A/B} = 0] \qquad 50\omega_{nx} + 100\omega_{ny} + 100\omega_{nz} = 0$$

Combination with two of the three previous equations yields the solutions

$$\omega_{nx} = -\tfrac{4}{3} \text{ rad/s} \qquad \omega_{ny} = -\tfrac{8}{3} \text{ rad/s} \qquad \omega_{nz} = \tfrac{10}{3} \text{ rad/s}$$

Thus,

$$\boldsymbol{\omega}_n = \tfrac{2}{3}(-2\mathbf{i} - 4\mathbf{j} + 5\mathbf{k}) \text{ rad/s}$$

with

$$\omega_n = \tfrac{2}{3}\sqrt{2^2 + 4^2 + 5^2} = 2\sqrt{5} \text{ rad/s} \qquad\qquad Ans.$$

① We select B as the reference point since its motion can easily be determined from the given angular velocity ω_1 of CB.

② The angular velocity $\boldsymbol{\omega}$ of AB is taken as a vector $\boldsymbol{\omega}_n$ normal to AB since any rotation of the link about its own axis AB has no influence on the behavior of the linkage.

③ The relative-velocity equation may be written as $\mathbf{v}_A - \mathbf{v}_B = \mathbf{v}_{A/B} = \boldsymbol{\omega}_n \times \mathbf{r}_{A/B}$, which requires that $\mathbf{v}_{A/B}$ be perpendicular to both $\boldsymbol{\omega}_n$ and $\mathbf{r}_{A/B}$. This equation alone does not incorporate the additional requirement that $\boldsymbol{\omega}_n$ be perpendicular to $\mathbf{r}_{A/B}$. Thus, we must also satisfy $\boldsymbol{\omega}_n \cdot \mathbf{r}_{A/B} = 0$.

Sample Problem 7/4

Determine the angular acceleration $\dot{\omega}_2$ of crank AD in Sample Problem 7/3 for the conditions cited. Also find the angular acceleration $\dot{\omega}_n$ of link AB.

Solution. The accelerations of the links may be found from the second of Eqs. 7/4, which may be written

$$\mathbf{a}_A = \mathbf{a}_B + \dot{\boldsymbol{\omega}}_n \times \mathbf{r}_{A/B} + \boldsymbol{\omega}_n \times (\boldsymbol{\omega}_n \times \mathbf{r}_{A/B})$$

where $\boldsymbol{\omega}_n$, as in Sample Problem 7/3, is the angular velocity of AB taken normal to AB. The angular acceleration of AB is written as $\dot{\boldsymbol{\omega}}_n$.

① In terms of their normal and tangential components, the accelerations of A and B are

$$\mathbf{a}_A = 50\omega_2{}^2\mathbf{i} + 50\dot{\omega}_2\mathbf{j} = 1800\mathbf{i} + 50\dot{\omega}_2\mathbf{j} \text{ mm/s}^2$$

$$\mathbf{a}_B = 100\omega_1{}^2\mathbf{k} + (0)\mathbf{i} = 3600\mathbf{k} \text{ mm/s}^2$$

Also

$$\boldsymbol{\omega}_n \times (\boldsymbol{\omega}_n \times \mathbf{r}_{A/B}) = -\omega_n{}^2\mathbf{r}_{A/B} = -20(50\mathbf{i} + 100\mathbf{j} + 100\mathbf{k}) \text{ mm/s}^2$$

$$\dot{\boldsymbol{\omega}}_n \times \mathbf{r}_{A/B} = (100\dot{\omega}_{ny} - 100\dot{\omega}_{nz})\mathbf{i}$$
$$+ (50\dot{\omega}_{nz} - 100\dot{\omega}_{nx})\mathbf{j} + (100\dot{\omega}_{nx} - 50\dot{\omega}_{ny})\mathbf{k}$$

Substitution into the relative-acceleration equation and equating respective coefficients of $\mathbf{i}$, $\mathbf{j}$, $\mathbf{k}$ give

$$28 = \dot{\omega}_{ny} - \dot{\omega}_{nz}$$
$$\dot{\omega}_2 + 40 = -2\dot{\omega}_{nx} + \dot{\omega}_{nz}$$
$$-32 = 2\dot{\omega}_{nx} - \dot{\omega}_{ny}$$

Solution of these equations for $\dot{\omega}_2$ gives

$$\dot{\omega}_2 = -36 \text{ rad/s}^2 \qquad Ans.$$

② The vector $\dot{\boldsymbol{\omega}}_n$ is normal to $\mathbf{r}_{A/B}$ but is not normal to $\mathbf{v}_{A/B}$, as was the case with $\boldsymbol{\omega}_n$.

$$[\dot{\boldsymbol{\omega}}_n \cdot \mathbf{r}_{A/B} = 0] \qquad 2\dot{\omega}_{nx} + 4\dot{\omega}_{ny} + 4\dot{\omega}_{nz} = 0$$

which, when combined with the preceding relations for these same quantities, gives

$$\dot{\omega}_{nx} = -8 \text{ rad/s}^2 \qquad \dot{\omega}_{ny} = 16 \text{ rad/s}^2 \qquad \dot{\omega}_{nz} = -12 \text{ rad/s}^2$$

Thus,

$$\dot{\boldsymbol{\omega}}_n = 4(-2\mathbf{i} + 4\mathbf{j} - 3\mathbf{k}) \text{ rad/s}^2 \qquad Ans.$$

and

$$|\dot{\boldsymbol{\omega}}_n| = 4\sqrt{2^2 + 4^2 + 3^2} = 4\sqrt{29} \text{ rad/s}^2 \qquad Ans.$$

① If the link AB had an angular velocity component along AB, then a change in both magnitude and direction of this component could occur which would contribute to the actual angular acceleration of the link as a rigid body. However, since any rotation about its own axis AB has no influence on the motion of the cranks at C and D, we shall concern ourselves only with $\dot{\boldsymbol{\omega}}_n$.

② The component of $\dot{\boldsymbol{\omega}}_n$ which is not normal to $\mathbf{v}_{A/B}$ gives rise to the change in direction of $\mathbf{v}_{A/B}$.

Sample Problem 7/5

The motor housing and its bracket rotate about the Z-axis at the constant rate $\Omega = 3$ rad/s. The motor shaft and disk have a constant angular velocity of spin $p = 8$ rad/s with respect to the motor housing in the direction shown. If γ is constant at 30°, determine the velocity and acceleration of point A at the top of the disk and the angular acceleration $\boldsymbol{\alpha}$ of the disk.

① **Solution.** The rotating reference axes x-y-z are attached to the motor housing, and the rotating base for the motor has the momentary orientation shown with respect to the fixed axes X-Y-Z. We will use both X-Y-Z components with unit vectors $\mathbf{I}$, $\mathbf{J}$, $\mathbf{K}$ and x-y-z components with unit vectors $\mathbf{i}$, $\mathbf{j}$, $\mathbf{k}$. The angular velocity of the x-y-z axes becomes $\boldsymbol{\Omega} = \Omega\mathbf{K} = 3\mathbf{K}$ rad/s.

① This choice for the reference axes provides a simple description for the motion of the disk relative to these axes.

Velocity. The velocity of A is given by the first of Eqs. 7/6

$$\mathbf{v}_A = \mathbf{v}_B + \boldsymbol{\Omega} \times \mathbf{r}_{A/B} + \mathbf{v}_{\text{rel}}$$

where

$$\mathbf{v}_B = \boldsymbol{\Omega} \times \mathbf{r}_B = 3\mathbf{K} \times 0.350\mathbf{J} = -1.05\mathbf{I} = -1.05\mathbf{i} \text{ m/s}$$

②
$$\boldsymbol{\Omega} \times \mathbf{r}_{A/B} = 3\mathbf{K} \times (0.300\mathbf{j} + 0.120\mathbf{k})$$
$$= (-0.9 \cos 30°)\mathbf{i} + (0.36 \sin 30°)\mathbf{i} = -0.599\mathbf{i} \text{ m/s}$$

$$\mathbf{v}_{\text{rel}} = \mathbf{p} \times \mathbf{r}_{A/B} = 8\mathbf{j} \times (0.300\mathbf{j} + 0.120\mathbf{k}) = 0.960\mathbf{i} \text{ m/s}$$

Thus,

$$\mathbf{v}_A = -1.05\mathbf{i} - 0.599\mathbf{i} + 0.960\mathbf{i} = -0.689\mathbf{i} \text{ m/s} \qquad \textit{Ans.}$$

Acceleration. The acceleration of A is given by the second of Eqs. 7/6

$$\mathbf{a}_A = \mathbf{a}_B + \dot{\boldsymbol{\Omega}} \times \mathbf{r}_{A/B} + \boldsymbol{\Omega} \times (\boldsymbol{\Omega} \times \mathbf{r}_{A/B}) + 2\boldsymbol{\Omega} \times \mathbf{v}_{\text{rel}} + \mathbf{a}_{\text{rel}}$$

where

$$\mathbf{a}_B = \boldsymbol{\Omega} \times (\boldsymbol{\Omega} \times \mathbf{r}_B) = 3\mathbf{K} \times (3\mathbf{K} \times 0.350\mathbf{J}) = -3.15\mathbf{J}$$
$$= 3.15(-\mathbf{j} \cos 30° + \mathbf{k} \sin 30°) = -2.73\mathbf{j} + 1.575\mathbf{k} \text{ m/s}^2$$

$$\dot{\boldsymbol{\Omega}} = 0$$

②
$$\boldsymbol{\Omega} \times (\boldsymbol{\Omega} \times \mathbf{r}_{A/B}) = 3\mathbf{K} \times [3\mathbf{K} \times (0.300\mathbf{j} + 0.120\mathbf{k})]$$
$$= 3\mathbf{K} \times (-0.599\mathbf{i}) = -1.557\mathbf{j} + 0.899\mathbf{k} \text{ m/s}^2$$

$$2\boldsymbol{\Omega} \times \mathbf{v}_{\text{rel}} = 2(3\mathbf{K}) \times 0.960\mathbf{i} = 5.76\mathbf{J}$$
$$= 5.76(\mathbf{j} \cos 30° - \mathbf{k} \sin 30°) = 4.99\mathbf{j} - 2.88\mathbf{k} \text{ m/s}^2$$

$$\mathbf{a}_{\text{rel}} = \mathbf{p} \times (\mathbf{p} \times \mathbf{r}_{A/B}) = 8\mathbf{j} \times [8\mathbf{j} \times (0.300\mathbf{j} + 0.120\mathbf{k})]$$
$$= -7.68\mathbf{k} \text{ m/s}^2$$

② Note that $\mathbf{K} \times \mathbf{i} = \mathbf{J} = \mathbf{j} \cos \gamma - \mathbf{k} \sin \gamma$, $\mathbf{K} \times \mathbf{j} = -\mathbf{i} \cos \gamma$, and $\mathbf{K} \times \mathbf{k} = \mathbf{i} \sin \gamma$.

Substitution into the expression for $\mathbf{a}_A$ and collecting terms give us

$$\mathbf{a}_A = 0.703\mathbf{j} - 8.09\mathbf{k} \text{ m/s}^2$$

and
$$a_A = \sqrt{(0.703)^2 + (8.09)^2} = 8.12 \text{ m/s}^2 \qquad \textit{Ans.}$$

Angular acceleration. Since the precession is steady, we may use Eq. 7/3 to give us

$$\boldsymbol{\alpha} = \dot{\boldsymbol{\omega}} = \boldsymbol{\Omega} \times \boldsymbol{\omega} = 3\mathbf{K} \times (3\mathbf{K} + 8\mathbf{j})$$
$$= 0 + (-24 \cos 30°)\mathbf{i} = -20.8\mathbf{i} \text{ rad/s}^2 \qquad \textit{Ans.}$$

PROBLEMS

Introductory problems

7/29 The wheels of the jet plane are spinning at their angular rate corresponding to a takeoff speed of 150 km/h. The retracting mechanism operates with θ increasing at the rate of 30° per second. Calculate the angular acceleration $\boldsymbol{\alpha}$ of the wheels for these conditions.

Ans. $\boldsymbol{\alpha} = 77.9\mathbf{i}$ rad/s²

Problem 7/29

7/30 The collar at O and attached shaft OC rotate about the fixed x_0-axis at the constant rate $\Omega = 4$ rad/s. Simultaneously, the circular disk rotates about OC at the constant rate $p = 10$ rad/s. Determine the magnitude of the total angular velocity $\boldsymbol{\omega}$ of the disk and find its angular acceleration $\boldsymbol{\alpha}$.

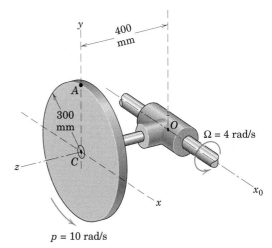

Problem 7/30

7/31 An unmanned radar-radio controlled aircraft with tilt-rotor propulsion is being designed for reconnaissance purposes. Vertical rise begins with $\theta = 0$ and is followed by horizontal flight as θ approaches 90°. If the rotors turn at a constant speed N of 360 rev/min, determine the angular acceleration $\boldsymbol{\alpha}$ of rotor A for $\theta = 30°$ if $\dot{\theta}$ is constant at 0.2 rad/s.

Ans. $\boldsymbol{\alpha} = -1.2\pi(\sqrt{3}\mathbf{i} + \mathbf{k})$ rad/s²

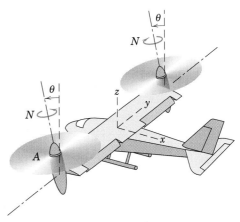

Problem 7/31

7/32 The square plate is rotating about its central z-axis at the rate $\dot{\phi}$. The x-y axes lie in the plane of the plate with the x-axis remaining parallel to the X-axis of the fixed X-Y-Z reference frame. The center line OB of the shaft, and, hence, the x-y-z axes, rotate about the fixed X-axis at the constant rate $\dot{\theta}$. Determine expressions for the angular acceleration of the plate as measured relative to the rotating x-y-z axes and as measured in the fixed X-Y-Z axes. Solve in two different ways. Express your results using unit vectors $\mathbf{i}$, $\mathbf{j}$, and $\mathbf{k}$.

Problem 7/32

7/33 The design of an internal bevel gear and ring gear assembly is shown in section. The ring gear D is fixed and does not rotate. The collar E, which carries the shaft of the bevel gear C, rotates about the central axis OA at the steady rate $N = 5$ revolutions per second. Thus, the center line OB of the bevel gear generates a cone of half-vertex angle γ. For a tooth ratio of 33/12, $\beta = 14.90°$. Determine the angular velocity $\boldsymbol{\omega}$ and angular acceleration $\boldsymbol{\alpha}$ of gear C. (The motion geometry is shown in the auxiliary sketch where bevel gear C is represented by the shaded circular disk which rolls on the ring gear with its axis OB generating the cone shown in orange dashes. The space trajectory of the tooth contact point P is the fixed ring-gear circle D.)

> *Ans.* $\boldsymbol{\omega} = 43.3(\mathbf{i} + \mathbf{k})$ rad/s
> $\boldsymbol{\alpha} = -1361\mathbf{j}$ rad/s^2

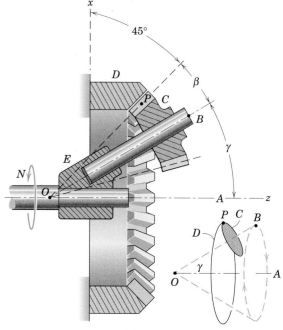

Problem 7/33

7/34 The collars at the ends of the telescoping link AB slide along the fixed shafts shown. During an interval of motion, $v_A = 125$ mm/s and $v_B = 50$ mm/s. Determine the vector expression for the angular velocity $\boldsymbol{\omega}_n$ of the center line of the link for the position where $y_A = 100$ mm and $y_B = 50$ mm.

Problem 7/34

Representative problems

7/35 For the rotating square plate of Prob. 7/32, determine the velocity and acceleration of the corner A for the position where $\phi = 0$ and $\theta = 30°$ if $\dot{\phi} \neq 0$ and $\ddot{\theta} = 0$. The distance $\overline{OB}$ is R.

$$Ans. \; \mathbf{v}_A = \left(\frac{b}{\sqrt{2}} \dot{\phi} - R\dot{\theta} \right) \mathbf{j}$$

$$\mathbf{a}_A = (-R\dot{\theta} + b\sqrt{2}\dot{\phi}) \dot{\theta}\mathbf{k} - \frac{b}{\sqrt{2}} \dot{\phi}^2 \mathbf{i}$$

7/36 The small motor M is pivoted about the x-axis through O and gives its shaft OA a constant speed p rad/s in the direction shown relative to its housing. The entire unit is then set into rotation about the vertical Z-axis at the constant angular velocity Ω rad/s. Simultaneously, the motor pivots about the x-axis at the constant rate $\dot{\beta}$ for an interval of motion. Determine the angular acceleration $\boldsymbol{\alpha}$ of the shaft OA in terms of β. Express your result in terms of the unit vectors for the rotating x-y-z axes.

7/37 The flight simulator is mounted on six hydraulic actuators connected in pairs to their attachment points on the underside of the simulator. By programming the actions of the activators a variety of flight conditions can be simulated with translational and rotational displacements through a limited range of motion. Axes x-y-z are attached to the simulator with origin B at the center of the volume. For the instant represented, B has a velocity and an acceleration in the horizontal y-direction of 0.96 m/s and 1.2 m/s², respectively. Simultaneously, the angular velocities and their time rates of change are $\omega_x = 1.4$ rad/s, $\dot{\omega}_x = 2$ rad/s², $\omega_y = 1.2$ rad/s, $\dot{\omega}_y = 3$ rad/s², $\omega_z = \dot{\omega}_z = 0$. For this instant determine the magnitudes of the velocity and acceleration of point A.

$$Ans. \; v_A = 2.04 \text{ m/s}, \; a_A = 6.23 \text{ m/s}^2$$

Problem 7/37

Problem 7/36

7/38 The spacecraft is revolving about its z-axis, which has a fixed space orientation, at the constant rate $p = \frac{1}{10}$ rad/s. Simultaneously its solar panels are unfolding at the rate $\dot{\beta}$ which is programmed to produce the variation with β shown in the graph. Determine the angular acceleration $\boldsymbol{\alpha}$ of panel A an instant (a) before and an instant (b) after it reaches the position $\beta = 18°$.

Problem 7/38

7/39 The collar and clevis A are given a constant upward velocity of 0.2 m/s for an interval of motion and cause the ball end of the bar to slide in the radial slot in the rotating disk. Determine the angular acceleration of the bar when the bar passes the position for which $z = 75$ mm. The disk turns at the constant rate of 2 rad/s.

$$Ans. \ \boldsymbol{\alpha} = -3\mathbf{i} - 4\mathbf{j} \text{ rad/s}^2$$

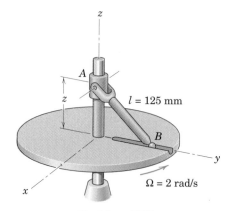

Problem 7/39

7/40 The circular disk is spinning about its own axis (y-axis) at the constant rate $p = 10\pi$ rad/s. Simultaneously the frame is rotating about the Z-axis at the constant rate $\Omega = 4\pi$ rad/s. Calculate the angular acceleration $\boldsymbol{\alpha}$ of the disk and the acceleration of point A at the top of the disk. Axes x-y-z are attached to the frame, which has the momentary orientation shown with respect to the fixed axes X-Y-Z.

Problem 7/40

7/41 For the conditions described in Prob. 7/36, determine the velocity $\mathbf{v}$ and acceleration $\mathbf{a}$ of the center A of the ball tool in terms of β.

Ans. $\mathbf{v}_A = -\Omega(R + b\sin\beta)\mathbf{i} + (b\dot{\beta}\cos\beta)\mathbf{j}$
$- (b\dot{\beta}\sin\beta)\mathbf{k}$

$\mathbf{a}_A = -(2b\Omega\dot{\beta}\cos\beta)\mathbf{i} - [\Omega^2(R + b\sin\beta)$
$+ b\dot{\beta}^2\sin\beta]\mathbf{j} - (b\dot{\beta}^2\cos\beta)\mathbf{k}$

7/42 The circular disk of 100-mm radius rotates about its z-axis at the constant speed $p = 240$ rev/min, and arm OCB rotates about the Y-axis at the constant speed $N = 30$ rev/min. Determine the velocity $\mathbf{v}$ and acceleration $\mathbf{a}$ of point A on the disk as it passes the position shown. Use reference axes x-y-z attached to the arm OCB.

Problem 7/42

7/43 Link AB is secured to the rotating arm OA and to the slider at B by a ball-and-socket joint at each end. Arm OA is confined to rotate about the fixed vertical shaft, and the slider is confined to move along the fixed rectangular bar. If the velocity of the slider is $v_B = 0.5$ m/s when $s = 50$ mm with OA in the position shown, calculate the corresponding angular velocity $\boldsymbol{\omega}_n$ of link AB. Use the nonrotating axes x-y-z attached to B.

Ans. $\boldsymbol{\omega}_n = \frac{5}{9}(4\mathbf{i} - 2\mathbf{j} + 5\mathbf{k})$ rad/s

Problem 7/43

7/44 The test chamber of a flight simulator consists of a drum that spins about the a-a axis at a constant angular rate p relative to the cylindrical housing A. The housing, in turn, is mounted on transverse horizontal bearings and rotates about the axis b-b at the constant rate $\dot{\theta}$. The entire assembly is made to rotate about the fixed vertical z-axis at the constant rate Ω. Write the expression for the angular acceleration $\boldsymbol{\alpha}$ of the spinning drum during the compounded motion for a given value of θ.

Problem 7/44

7/45 The circular disk of radius r is mounted on its shaft which is pivoted at O so that it may rotate about the vertical z_0-axis. If the disk rolls at constant speed without slipping and makes one complete turn around the circle of radius R in time τ, determine the expression for the absolute angular velocity $\boldsymbol{\omega}$ of the disk. Use axes x-y-z which rotate around the z_0-axis. (*Hint:* The absolute angular velocity of the disk equals the angular velocity of the axes plus (vectorially) the angular velocity relative to the axes as seen by holding x-y-z fixed and rotating the circular disk of radius R at the rate $2\pi/\tau$.)

Ans. $\boldsymbol{\omega} = \dfrac{2\pi}{\tau}\left[\left(-\dfrac{R}{r} + \dfrac{r}{R}\right)\mathbf{j} + \dfrac{\sqrt{R^2 - r^2}}{R}\mathbf{k}\right]$

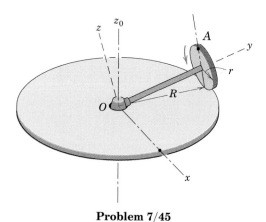

Problem 7/45

7/46 Determine the angular acceleration $\boldsymbol{\alpha}$ for the rolling circular disk of Prob. 7/45. Use the results cited in the answer for that problem.

7/47 The center O of the spacecraft is moving through space with a constant velocity. During the period of motion prior to stabilization, the spacecraft has a constant rotational rate $\Omega = \frac{1}{2}$ rad/s about its z-axis. The x-y-z axes are attached to the body of the craft, and the solar panels rotate about the y-axis at the constant rate $\dot{\theta} = \frac{1}{4}$ rad/s with respect to the spacecraft. If $\boldsymbol{\omega}$ is the absolute angular velocity of the solar panels, determine $\dot{\boldsymbol{\omega}}$. Also find the acceleration of point A when $\theta = 30°$.

Ans. $\dot{\boldsymbol{\omega}} = \frac{1}{8}\mathbf{i}$ rad/s^2
$\mathbf{a}_A = 0.0938\mathbf{i} - 0.730\mathbf{j} - 0.0325\mathbf{k}$ m/s^2

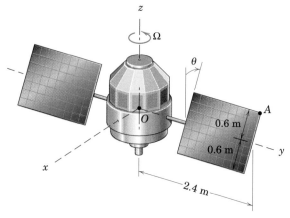

Problem 7/47

7/48 The thin circular disk of mass m and radius r is rotating about its z-axis with a constant angular velocity p, and the yoke in which it is mounted rotates about the X-axis through OB with a constant angular velocity ω_1. Simultaneously, the entire assembly rotates about the fixed Y-axis through O with a constant angular velocity ω_2. Determine the velocity $\mathbf{v}$ and acceleration $\mathbf{a}$ of point A on the rim of the disk as it passes the position shown where the x-y plane of the disk coincides with the X-Y plane. The x-y-z axes are attached to the yoke.

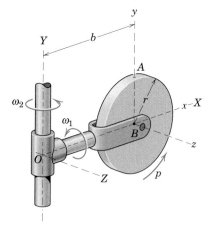

Problem 7/48

7/49 A simulator for perfecting the docking procedure for spacecraft consists of the frame A which is mounted on four air-bearing pads so that it can translate and rotate freely on the horizontal surface. Mounted in the frame is a drum B which can rotate about the horizontal axis of the frame A. The coordinate axes x-y-z are attached to the drum B, and the z-axis of the drum makes an angle β with the horizontal. Inside the drum is the simulated command module C which can rotate within the drum about its z-axis at a rate p. For a certain test run, frame A is rotating on the horizontal surface in the direction shown with a constant angular velocity of 0.2 rad/s. Simultaneously, the drum B is rotating about the x-axis at the constant rate $\dot{\beta} = 0.15$ rad/s, and the module C is turning inside the drum at the constant rate $p = 0.9$ rad/s in the direction indicated. For these conditions, determine the angular velocity $\boldsymbol{\omega}$ and the angular acceleration $\boldsymbol{\alpha}$ of the simulator C as it passes the position $\beta = 0$.

    Ans. $\boldsymbol{\omega}_{\beta=0} = 0.15\mathbf{i} + 0.2\mathbf{j} + 0.9\mathbf{k}$ rad/s

          $\boldsymbol{\alpha}_{\beta=0} = 0.18\mathbf{i} - 0.135\mathbf{j} - 0.030\mathbf{k}$ rad/s^2

Problem 7/49

7/50 Determine the velocity $\mathbf{v}$ of point A on the disk of Prob. 7/45 for the position shown.

7/51 Determine the acceleration $\mathbf{a}$ of point A on the disk of Prob. 7/45 for the position shown.

$$Ans. \ \mathbf{a}_A = \left(\frac{2\pi}{\tau}\right)^2 \left[\sqrt{R^2 - r^2} \left(\frac{2r^2}{R^2} - 3\right)\mathbf{j} \right.$$
$$\left. + \left(3r - \frac{R^2}{r} - \frac{2r^3}{R^2}\right)\mathbf{k} \right]$$

▶**7/52** For a short interval of motion, collar A moves along its fixed shaft with a velocity $v_A = 2$ m/s in the Y-direction. Collar B, in turn, slides along its fixed vertical shaft. Link AB is 700 mm in length and can turn within the clevis at A to allow for the angular change between the clevises. For the instant when A passes the position where $y = 200$ mm, determine the velocity of collar B using nonrotating axes attached to B and find the component $\boldsymbol{\omega}_n$, normal to AB, of the angular velocity of the link. Also solve for $\mathbf{v}_B$ by differentiating the appropriate relation $x^2 + y^2 + z^2 = l^2$.

    Ans. $\mathbf{v}_B = -\frac{2}{3}\mathbf{k}$ m/s

        $\boldsymbol{\omega}_n = \frac{10}{49}(\frac{40}{3}\mathbf{i} - 2\mathbf{j} + 6\mathbf{k})$ rad/s

Problem 7/52

▶7/53 The motor turns the disk at the constant speed $p = 30$ rad/s. The motor is also swiveling about the horizontal axis BO (y-axis) at the constant speed $\dot{\theta} = 2$ rad/s. Simultaneously, the entire assembly is rotating about the vertical axis C-C at the constant rate $q = 8$ rad/s. For the instant when $\theta = 30°$, determine the angular acceleration α of the disk and the acceleration $\mathbf{a}$ of point A at the bottom of the disk. Axes x-y-z are attached to the motor housing, and plane O-x_0-y is horizontal.

   Ans. $\mathbf{a} = -52.1\mathbf{i} - 9.23\mathbf{j} + 120.2\mathbf{k}$ m/s^2
   $\alpha = 13.86\mathbf{i} + 208\mathbf{j} + 52\mathbf{k}$ rad/s^2

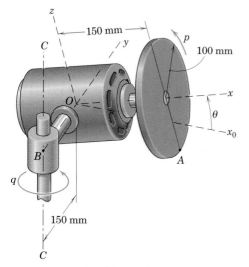

Problem 7/53

▶7/54 The gyro rotor shown is spinning at the constant rate of 100 rev/min relative to the x-y-z axes in the direction indicated. If the angle γ between the gimbal ring and the horizontal X-Y plane is made to increase at the constant rate of 4 rad/s and if the unit is forced to precess about the vertical at the constant rate $N = 20$ rev/min, calculate the magnitude of the angular acceleration α of the rotor when $\gamma = 30°$. Solve by using Eq. 7/7 applied to the angular velocity of the rotor.

   Ans. $\alpha = 42.8$ rad/s^2

Problem 7/54

SECTION B. KINETICS

7/7 ANGULAR MOMENTUM

The force equation for a mass system, rigid or nonrigid, Eq. 4/1 or 4/6, is the generalization of Newton's second law for the motion of a particle and should require no further explanation. The moment equation for three-dimensional motion, however, is not nearly as simple as the third of Eqs. 6/1 for plane motion since the change of angular momentum has a number of additional components that are absent in plane motion.

We consider now a rigid body moving with any general motion in space, Fig. 7/12a. Axes x-y-z are *attached* to the body with origin at the mass center G. Thus, the angular velocity ω of the body becomes the angular velocity of the x-y-z axes as observed from the fixed reference axes X-Y-Z. The absolute angular momentum $\mathbf{H}_G$ of the body about its mass center G is the sum of the moments about G of the linear momenta of all elements of the body and was expressed in part (b) of Art. 4/4 as $\mathbf{H}_G = \Sigma(\boldsymbol{\rho}_i \times m_i\mathbf{v}_i)$, where $\mathbf{v}_i$ is the absolute velocity of the mass element m_i. But for the rigid body, $\mathbf{v}_i = \bar{\mathbf{v}} + \boldsymbol{\omega} \times \boldsymbol{\rho}_i$, where $\boldsymbol{\omega} \times \boldsymbol{\rho}_i$ is the relative velocity of m_i with respect to G as seen from nonrotating axes. Thus, we may write

$$\mathbf{H}_G = -\bar{\mathbf{v}} \times \Sigma m_i\boldsymbol{\rho}_i + \Sigma[\boldsymbol{\rho}_i \times m_i(\boldsymbol{\omega} \times \boldsymbol{\rho}_i)]$$

where we have factored out $\bar{\mathbf{v}}$ from the first summation terms by reversing the order of the cross product and changing the sign. With origin at the mass center G, the first term in $\mathbf{H}_G$ is zero since $\Sigma m_i\boldsymbol{\rho}_i = m\bar{\boldsymbol{\rho}} = \mathbf{0}$. The second term with the substitution of dm for m_i and $\boldsymbol{\rho}$ for $\boldsymbol{\rho}_i$ gives

$$\mathbf{H}_G = \int [\boldsymbol{\rho} \times (\boldsymbol{\omega} \times \boldsymbol{\rho})]\, dm \qquad (7/8)$$

Before expanding the integrand of Eq. 7/8, we consider also the case of a rigid body rotating about a fixed point O, Fig. 7/12b. The x-y-z axes are attached to the body, and both body and axes have an angular velocity $\boldsymbol{\omega}$. The angular momentum about O was expressed in part (b) of Art. 4/4 and is $\mathbf{H}_O = \Sigma(\mathbf{r}_i \times m_i\mathbf{v}_i)$, where, for the rigid body, $\mathbf{v}_i = \boldsymbol{\omega} \times \mathbf{r}_i$. Thus, with the substitution of dm for m_i and $\mathbf{r}$ for $\mathbf{r}_i$, the angular momentum is

$$\mathbf{H}_O = \int [\mathbf{r} \times (\boldsymbol{\omega} \times \mathbf{r})]\, dm \qquad (7/9)$$

We observe now that for the two cases of Figs. 7/12a and 7/12b, the position vectors $\boldsymbol{\rho}_i$ and $\mathbf{r}_i$ are given by the same expression $x\mathbf{i} + y\mathbf{j} + z\mathbf{k}$. Thus, Eqs. 7/8 and 7/9 are identical in form, and the symbol $\mathbf{H}$ will be used here for either case. We now carry out the expansion of the integrand in the two expressions for angular mo-

(a)

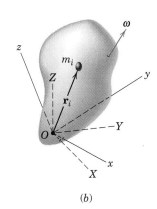

(b)

Figure 7/12

mentum, with recognition of the fact that the components of $\boldsymbol{\omega}$ are invariant with respect to the integrals over the body and, hence, become constant multipliers of the integrals. The cross-product expansion applied to the triple vector product upon collection of terms gives

$$
\begin{aligned}
d\mathbf{H} = \ \mathbf{i}[(y^2 + z^2)\omega_x & & -xy\omega_y & & -xz\omega_z] \ dm \\
+\mathbf{j}[& -yx\omega_x + (z^2 + x^2)\omega_y & & -yz\omega_z] \ dm \\
+\mathbf{k}[& -zx\omega_x & -zy\omega_y + (x^2 + y^2)\omega_z] \ dm
\end{aligned}
$$

Now let

$$
\boxed{
\begin{aligned}
I_{xx} &= \int (y^2 + z^2) \, dm & I_{xy} &= \int xy \, dm \\[2mm]
I_{yy} &= \int (z^2 + x^2) \, dm & I_{xz} &= \int xz \, dm \\[2mm]
I_{zz} &= \int (x^2 + y^2) \, dm & I_{yz} &= \int yz \, dm
\end{aligned}
}
\qquad \textbf{(7/10)}
$$

The quantities I_{xx}, I_{yy}, I_{zz} are known as the *moments of inertia* of the body about the respective axes, and I_{xy}, I_{xz}, I_{yz} are known as the *products of inertia* with respect to the coordinate axes. These quantities describe the manner in which the mass of a rigid body is distributed with respect to the chosen axes. The calculation of moments and products of inertia is explained fully in Appendix B. The double subscripts for the moments and products of inertia preserve a symmetry of notation which has special meaning in their description by tensor notation.[*] It is observed that $I_{xy} = I_{yx}, I_{xz} = I_{zx}$, and $I_{yz} = I_{zy}$. With the substitutions of Eqs. 7/10, the expression for $\mathbf{H}$ becomes

$$
\boxed{
\begin{aligned}
\mathbf{H} = \ & (\ I_{xx}\omega_x - I_{xy}\omega_y - I_{xz}\omega_z)\mathbf{i} \\
& +(-I_{yx}\omega_x + I_{yy}\omega_y - I_{yz}\omega_z)\mathbf{j} \\
& +(-I_{zx}\omega_x - I_{zy}\omega_y + I_{zz}\omega_z)\mathbf{k}
\end{aligned}
}
\qquad \textbf{(7/11)}
$$

and the components of $\mathbf{H}$ are clearly

$$
\boxed{
\begin{aligned}
H_x &= \ I_{xx}\omega_x - I_{xy}\omega_y - I_{xz}\omega_z \\
H_y &= -I_{yx}\omega_x + I_{yy}\omega_y - I_{yz}\omega_z \\
H_z &= -I_{zx}\omega_x - I_{zy}\omega_y + I_{zz}\omega_z
\end{aligned}
}
\qquad \textbf{(7/12)}
$$

[*]See, for example, the first author's *Dynamics, 2nd Edition SI Version*, 1975, John Wiley & Sons, Art. 41.

Equation 7/11 is the general expression for the angular momentum either about the mass center G or about a fixed point O for a rigid body rotating with an instantaneous angular velocity $\boldsymbol{\omega}$.

Attention is called to the fact that in each of the two cases represented, the reference axes x-y-z are *attached* to the rigid body. This attachment makes the moment-of-inertia integrals and the product-of-inertia integrals of Eqs. 7/10 invariant with time. If the x-y-z axes were to rotate with respect to an irregular body, then these inertia integrals would be functions of the time, which would introduce an undesirable complexity into the angular-momentum relations. An important exception occurs when a rigid body is spinning about an axis of symmetry, in which case, the inertia integrals are not affected by the angular position of the body about its spin axis. Thus, it is frequently convenient to permit a body with axial symmetry to rotate relative to the reference system about one of the coordinate axes. In addition to the momentum components due to the angular velocity $\boldsymbol{\Omega}$ of the reference axes, then, an added angular-momentum component along the spin axis due to the relative spin about the axis would have to be accounted for.

The array of moments and products of inertia

$$\begin{bmatrix} I_{xx} & -I_{xy} & -I_{xz} \\ -I_{yz} & I_{yy} & -I_{yz} \\ -I_{zx} & -I_{zy} & I_{zz} \end{bmatrix}$$

which appear in Eq. 7/12 is known as the *inertia matrix* or *inertia tensor*. As we change the orientation of the axes relative to the body, the moments and products of inertia will also change in value. It can be shown* that there is one unique orientation of axes x-y-z for a given origin for which the products of inertia vanish and the moments of inertia I_{xx}, I_{yy}, I_{zz} take on stationary values. For this orientation, the inertia matrix takes the form

$$\begin{bmatrix} I_{xx} & 0 & 0 \\ 0 & I_{yy} & 0 \\ 0 & 0 & I_{zz} \end{bmatrix}$$

and is said to be diagonalized. The axes x-y-z for which the products of inertia vanish are called the *principal axes of inertia*, and I_{xx}, I_{yy}, and I_{zz} are called the *principal moments of inertia*. The principal moments of inertia for a given origin represent the maximum, the minimum, and an intermediate value of the moments of inertia.

---

*See, for example, the first author's *Dynamics, 2nd Edition SI Version*, 1975, John Wiley & Sons, Art. 41.

If the coordinate axes coincide with the principal axes of inertia, Eq. 7/11 for the angular momentum about the mass center or about a fixed point becomes

$$\mathbf{H} = I_{xx}\omega_x\mathbf{i} + I_{yy}\omega_y\mathbf{j} + I_{zz}\omega_z\mathbf{k} \qquad (7/13)$$

It is always possible to locate the principal axes of inertia for a general three-dimensional rigid body. Thus, we can express its angular momentum by Eq. 7/13, although it may not always be convenient to do so for geometric reasons. Except when the body rotates about one of the principal axes of inertia or when $I_{xx} = I_{yy} = I_{zz}$, the vectors $\mathbf{H}$ and $\boldsymbol{\omega}$ have different directions.

The momentum properties of a rigid body may be represented by the resultant linear-momentum vector $\mathbf{G} = m\bar{\mathbf{v}}$ through the mass center and the resultant angular-momentum vector $\mathbf{H}_G$ about the mass center, as shown in Fig. 7/13. Although $\mathbf{H}_G$ has the properties of a free vector, we represent it through G for convenience. These vectors have properties analogous to those of a force and a couple. Thus, the angular momentum about any point P equals the free vector $\mathbf{H}_G$ plus the moment of the linear-momentum vector $\mathbf{G}$ about P. Therefore, we may write

$$\mathbf{H}_P = \mathbf{H}_G + \bar{\mathbf{r}} \times \mathbf{G} \qquad (7/14)$$

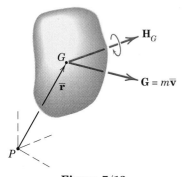

Figure 7/13

This relation, which was derived previously in Chapter 4 as Eq. 4/10, also applies to a fixed point O on the body or body extended, where O merely replaces P. Equation 7/14 constitutes a transfer theorem for angular momentum.

7/8 KINETIC ENERGY

In Art. 4/3 of the chapter on the dynamics of systems of particles, we developed the expression for the kinetic energy T of any general system of mass, rigid or nonrigid, and obtained the result

$$T = \tfrac{1}{2}m\bar{v}^2 + \Sigma\tfrac{1}{2}m_i|\dot{\boldsymbol{\rho}}_i|^2 \qquad [4/4]$$

where $\bar{v}$ is the velocity of the mass center and $\boldsymbol{\rho}_i$ is the position vector of a representative element of mass m_i with respect to the mass center. We identified the first term as the kinetic energy due to the translation of the system and the second term as the kinetic energy associated with the motion relative to the mass center. The translational term may be written alternatively as

$$\tfrac{1}{2}m\bar{v}^2 = \tfrac{1}{2}m\dot{\bar{\mathbf{r}}}\cdot\dot{\bar{\mathbf{r}}} = \tfrac{1}{2}\bar{\mathbf{v}}\cdot\mathbf{G}$$

where $\dot{\bar{\mathbf{r}}}$ is the velocity $\bar{\mathbf{v}}$ of the mass center and $\mathbf{G}$ is the linear momentum of the body.

For a rigid body, the relative term becomes the kinetic energy due to rotation about the mass center. Since $\dot{\boldsymbol{\rho}}_i$ is the velocity of the representative particle with respect to the mass center, then for the rigid body we may write it as $\dot{\boldsymbol{\rho}}_i = \boldsymbol{\omega} \times \boldsymbol{\rho}_i$, where $\boldsymbol{\omega}$ is the angular velocity of the body. With this substitution, the relative term in the kinetic energy expression becomes

$$\Sigma \tfrac{1}{2} m_i |\dot{\boldsymbol{\rho}}_i|^2 = \Sigma \tfrac{1}{2} m_i (\boldsymbol{\omega} \times \boldsymbol{\rho}_i) \cdot (\boldsymbol{\omega} \times \boldsymbol{\rho}_i)$$

If we use the fact that the dot and the cross may be interchanged in the triple scalar product, that is, $\mathbf{P} \times \mathbf{Q} \cdot \mathbf{R} = \mathbf{P} \cdot \mathbf{Q} \times \mathbf{R}$, we may write

$$(\boldsymbol{\omega} \times \boldsymbol{\rho}_i) \cdot (\boldsymbol{\omega} \times \boldsymbol{\rho}_i) = \boldsymbol{\omega} \cdot \boldsymbol{\rho}_i \times (\boldsymbol{\omega} \times \boldsymbol{\rho}_i)$$

Since $\boldsymbol{\omega}$ is the same factor in all terms of the summation, it may be factored out to give

$$\Sigma \tfrac{1}{2} m_i |\dot{\boldsymbol{\rho}}_i|^2 = \tfrac{1}{2} \boldsymbol{\omega} \cdot \Sigma \boldsymbol{\rho}_i \times m_i (\boldsymbol{\omega} \times \boldsymbol{\rho}_i) = \tfrac{1}{2} \boldsymbol{\omega} \cdot \mathbf{H}_G$$

where $\mathbf{H}_G$ is the same as the integral expressed by Eq. 7/8. Thus, the general expression for the kinetic energy of a rigid body moving with mass-center velocity $\bar{\mathbf{v}}$ and angular velocity $\boldsymbol{\omega}$ is

$$\boxed{T = \tfrac{1}{2} \bar{\mathbf{v}} \cdot \mathbf{G} + \tfrac{1}{2} \boldsymbol{\omega} \cdot \mathbf{H}_G} \qquad (7/15)$$

Expansion of this vector equation by substitution of the expression for $\mathbf{H}_G$ written from Eq. 7/11 yields

$$T = \tfrac{1}{2} m \bar{v}^2 + \tfrac{1}{2} (\bar{I}_{xx} \omega_x{}^2 + \bar{I}_{yy} \omega_y{}^2 + \bar{I}_{zz} \omega_z{}^2)$$
$$- (\bar{I}_{xy} \omega_x \omega_y + \bar{I}_{xz} \omega_x \omega_z + \bar{I}_{yz} \omega_y \omega_z) \qquad (7/16)$$

If the axes coincide with the principal axes of inertia, the kinetic energy is merely

$$T = \tfrac{1}{2} m \bar{v}^2 + \tfrac{1}{2} (\bar{I}_{xx} \omega_x{}^2 + \bar{I}_{yy} \omega_y{}^2 + \bar{I}_{zz} \omega_z{}^2) \qquad (7/17)$$

When a rigid body is pivoted about a fixed point O or when there is a point O in the body that momentarily has zero velocity, the kinetic energy is $T = \Sigma \tfrac{1}{2} m_i \dot{\mathbf{r}}_i \cdot \dot{\mathbf{r}}_i$. This expression reduces to

$$\boxed{T = \tfrac{1}{2} \boldsymbol{\omega} \cdot \mathbf{H}_O} \qquad (7/18)$$

where $\mathbf{H}_O$ is the angular momentum about O, as may be seen by replacing $\boldsymbol{\rho}_i$ in the previous derivation by $\mathbf{r}_i$, the position vector from O. Equations 7/15 and 7/18 are the three-dimensional counterparts of Eqs. 6/9 and 6/8 for plane motion.

Sample Problem 7/6

The bent plate has a mass of 70 kg per square meter of surface area and revolves about the z-axis at the rate $\omega = 30$ rad/s. Determine (a) the angular momentum $\mathbf{H}$ of the plate about point O and (b) the kinetic energy T of the plate. Neglect the mass of the hub and the thickness of the plate compared with its surface dimensions.

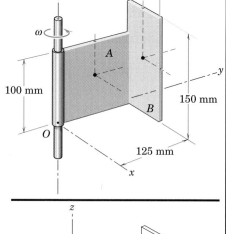

Solution. The moments and products of inertia are written with the aid of Eqs. B/3 and B/9 in Appendix B by transfer from the parallel centroidal axes for each part. First, the masses of the parts are $m_A = (0.100)(0.125)(70) = 0.875$ kg, $m_B = (0.075)(0.150)(70) = 0.788$ kg.

Part A

$$[I_{xx} = \bar{I}_{xx} + md^2] \quad I_{xx} = \frac{0.875}{12}[(0.100)^2 + (0.125)^2]$$

$$+ 0.875[(0.050)^2 + (0.0625)^2] = 0.007\ 47 \text{ kg·m}^2$$

$$[I_{yy} = \tfrac{1}{3}ml^2] \quad I_{yy} = \frac{0.875}{3}(0.100)^2 = 0.002\ 92 \text{ kg·m}^2$$

$$[I_{zz} = \tfrac{1}{3}ml^2] \quad I_{zz} = \frac{0.875}{3}(0.125)^2 = 0.004\ 56 \text{ kg·m}^2$$

$$\left[I_{xy} = \int xy\ dm, \quad I_{xz} = \int xz\ dm\right] \quad I_{xy} = 0 \quad I_{xz} = 0$$

$$[I_{yz} = \bar{I}_{yz} + md_y d_z] \quad I_{yz} = 0 + 0.875(0.0625)(0.050) = 0.002\ 73 \text{ kg·m}^2$$

Part B

$$[I_{xx} = \bar{I}_{xx} + md^2] \quad I_{xx} = \frac{0.788}{12}(0.150)^2 + 0.788[(0.125)^2 + (0.075)^2]$$

$$= 0.018\ 21 \text{ kg·m}^2$$

$$[I_{yy} = \bar{I}_{yy} + md^2] \quad I_{yy} = \frac{0.788}{12}[(0.075)^2 + (0.150)^2]$$

$$+ 0.788[(0.0375)^2 + (0.075)^2] = 0.007\ 38 \text{ kg·m}^2$$

$$[I_{zz} = \bar{I}_{zz} + md^2] \quad I_{zz} = \frac{0.788}{12}(0.075)^2 + 0.788[(0.125)^2 + (0.0375)^2]$$

$$= 0.013\ 78 \text{ kg·m}^2$$

$$[I_{xy} = \bar{I}_{xy} + md_x d_y] \quad I_{xy} = 0 + 0.788(0.0375)(0.125) = 0.003\ 69 \text{ kg·m}^2$$

$$[I_{xz} = \bar{I}_{xz} + md_x d_z] \quad I_{xz} = 0 + 0.788(0.0375)(0.075) = 0.002\ 21 \text{ kg·m}^2$$

$$[I_{yz} = \bar{I}_{yz} + md_y d_z] \quad I_{yz} = 0 + 0.788(0.125)(0.075) = 0.007\ 38 \text{ kg·m}^2$$

The sum of the respective inertia terms gives for the two plates together

$$I_{xx} = 0.0257 \text{ kg·m}^2 \qquad I_{xy} = 0.003\ 69 \text{ kg·m}^2$$
$$I_{yy} = 0.010\ 30 \text{ kg·m}^2 \qquad I_{xz} = 0.002\ 21 \text{ kg·m}^2$$
$$I_{zz} = 0.018\ 34 \text{ kg·m}^2 \qquad I_{yz} = 0.010\ 12 \text{ kg·m}^2$$

(a) The angular momentum of the body is given by Eq. 7/11, where $\omega_z = 30$ rad/s and ω_x and ω_y are zero. Thus,

$$\mathbf{H}_O = 30(-0.002\ 21\mathbf{i} - 0.010\ 12\mathbf{j} + 0.018\ 34\mathbf{k}) \text{ N·m·s} \quad Ans.$$

(b) The kinetic energy from Eq. 7/18 becomes

$$T = \tfrac{1}{2}\boldsymbol{\omega} \cdot \mathbf{H}_O = \tfrac{1}{2}(30\mathbf{k}) \cdot 30(-0.002\ 21\mathbf{i} - 0.010\ 12\mathbf{j} + 0.018\ 34\mathbf{k})$$

$$= 8.25 \text{ J}$$

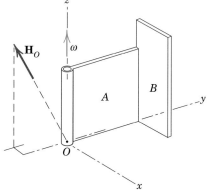

① The parallel-axis theorems for transferring moments and products of inertia from centroidal axes to parallel axes are explained in Appendix B and are most useful relations.

② Recall that the units of angular momentum may also be written in the base units as kg·m²/s.

PROBLEMS

Introductory problems

7/55 The bent rod has a mass ρ per unit length and rotates about the z-axis with an angular velocity ω. Determine the angular momentum $\mathbf{H}_O$ of the rod about the fixed origin O of the axes which are attached to the rod. Also find the kinetic energy T of the rod.

Ans. $\mathbf{H}_O = \rho b^3(-\frac{1}{2}\mathbf{i} - \frac{3}{2}\mathbf{j} + \frac{8}{3}\mathbf{k})\omega$, $T = \frac{4}{3}\rho b^3\omega^2$

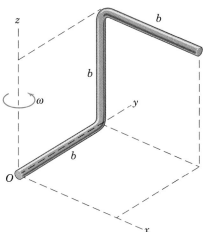

Problem 7/55

7/56 The slender shaft carries two offset particles each of mass m and rotates about the z-axis with an angular velocity ω as indicated. Write an expression for the angular momentum $\mathbf{H}$ for the system about the origin O of the x-y-z axes for the position shown. Write the kinetic energy T of the system by inspection and verify your result by applying Eq. 7/18.

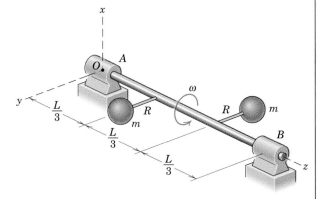

Problem 7/56

7/57 The aircraft landing gear viewed from the front is being retracted immediately after takeoff, and the wheel is spinning at the rate corresponding to the takeoff speed of 200 km/h. The 45-kg wheel has a radius of gyration about its z-axis of 370 mm. Neglect the thickness of the wheel and calculate the angular momentum of the wheel about G and about A for the position where θ is increasing at the rate of 30° per second.

Ans. $\mathbf{H}_G = -1.613\mathbf{j} - 744\mathbf{k}$ kg·m²/s
$\mathbf{H}_A = -2.70\mathbf{j} - 744\mathbf{k}$ kg·m²/s

Problem 7/57

7/58 Use the results of Prob. 7/55 and determine the angular momentum $\mathbf{H}_G$ of the bent rod of that problem about its mass center G using the given reference axes.

7/59 The rectangular steel plate of mass 12 kg is welded to the shaft with its plane tilted 15° from the plane (*x-y*) normal to the shaft axis. The shaft and plate are rotating about the fixed *z*-axis at the rate $N = 300$ rev/min. Determine the angular momentum $\mathbf{H}_O$ of the plate about the given axes and find its kinetic energy T.

Ans. $\mathbf{H}_O = -0.707\mathbf{j} + 4.45\mathbf{k}$ kg·m²/s

$T = 69.9$ J

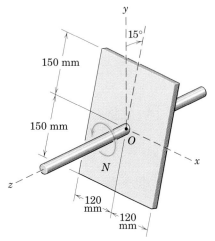

Problem 7/59

Representative problems

7/60 The solid half-circular cylinder of mass *m* revolves about the *z*-axis with an angular velocity ω as shown. Determine its angular momentum $\mathbf{H}$ with respect to the *x-y-z* axes.

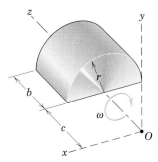

Problem 7/60

7/61 The elements of a reaction-wheel attitude-control system for a spacecraft are shown in the figure. Point *G* is the center of mass for the system of the spacecraft and wheels, and *x, y, z* are principal axes for the system. Each wheel has a mass *m* and a moment of inertia *I* about its own axis and spins with a relative angular velocity *p* in the direction indicated. The center of each wheel, which may be treated as a thin disk, is a distance *b* from *G*. If the spacecraft has angular velocity components Ω_x, Ω_y, and Ω_z, determine the angular momentum $\mathbf{H}_G$ of the three wheels as a unit.

Ans. $\mathbf{H}_G = Ip(\mathbf{i} + \mathbf{j} + \mathbf{k}) + 2(I + mb^2)\mathbf{\Omega}$

where $\mathbf{\Omega} = \Omega_x\mathbf{i} + \Omega_y\mathbf{j} + \Omega_z\mathbf{k}$

Problem 7/61

7/62 The solid circular cylinder of mass *m*, radius *r*, and length *b* revolves about its geometric axis at an angular rate *p* rad/s. Simultaneously, the bracket and attached shaft revolve about the *x*-axis at the rate ω rad/s. Write the expression for the angular momentum $\mathbf{H}_O$ of the cylinder about *O* with reference axes as shown.

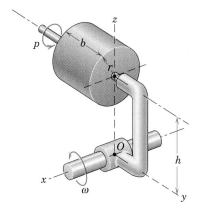

Problem 7/62

7/64 Each of the slender rods of length l and mass m is welded to the circular disk which rotates about the vertical z-axis with an angular velocity ω. Each rod makes an angle β with the vertical and lies in a plane parallel to the y-z plane. Determine an expression for the angular momentum $\mathbf{H}_O$ of the two rods about the origin O of the axes.

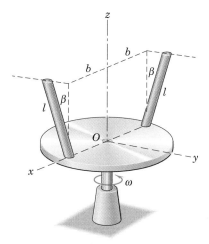

Problem 7/64

7/63 The slender steel rod AB has a mass of 2.8 kg and is secured to the rotating shaft by the rod OG and its fittings at O and G. The angle β remains constant at 30°, and the entire rigid assembly rotates about the z-axis at the steady rate $N = 600$ rev/min. Calculate the angular momentum $\mathbf{H}_O$ of AB and its kinetic energy T.

$$Ans. \ \mathbf{H}_O = 3.11\mathbf{j} + 33.5\mathbf{k} \text{ kg·m}^2/\text{s}$$
$$T = 1054 \text{ J}$$

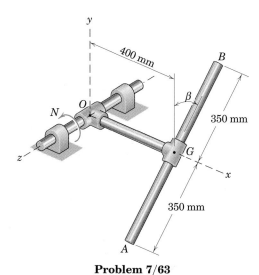

Problem 7/63

7/65 The space capsule shown has a mass m with mass center G. Its radius of gyration about its z-axis of rotational symmetry is k and that about either the x- or y-axis is k'. In space, the capsule spins within its x-y-z reference frame at the rate $p = \dot{\phi}$. Simultaneously, a point C on the z-axis moves in a circle about the z_0-axis with a frequency f (rotations per unit time). The z_0-axis has a constant direction in space. Determine the angular momentum $\mathbf{H}_G$ of the capsule relative to the axes designated. Note that the x-axis always lies in the z-z_0 plane and that the y-axis is therefore normal to z_0.

$$\text{Ans. } \mathbf{H}_G = -(2\pi m f k'^2 \sin\theta)\mathbf{i}$$
$$+ mk^2(2\pi f \cos\theta + p)\mathbf{k}$$

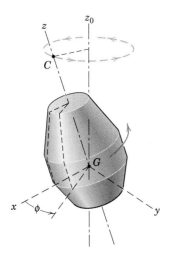

Problem 7/65

7/66 The semicircular ring of mass m and radius r is rotating about the vertical z-axis with an angular velocity ω. The plane of the ring makes an angle β with respect to the horizontal x-y plane. Determine the angular momentum $\mathbf{H}_O$ of the ring about O and find its kinetic energy T.

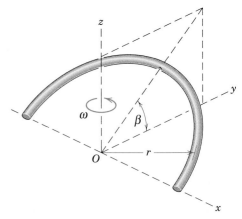

Problem 7/66

7/67 The uniform slender rod AB of mass m and length L is secured to the z-axis shaft and collars as shown, and the entire rigid assembly rotates about the z-axis with an angular velocity ω. Determine the angular momentum $\mathbf{H}_O$ of AB with respect to O.

$$\text{Ans. } \mathbf{H}_O = \frac{mLr}{6}\sin\theta\sqrt{1 - \frac{4r^2\sin^2\theta}{L^2}}\,\omega\mathbf{i}$$
$$+ mr^2\left(1 - \tfrac{2}{3}\sin^2\theta\right)\omega\mathbf{k}$$

Problem 7/67

7/68 The two uniform slender bars, each of mass m and length L, are welded together at right angles and rotate as a rigid unit about the z-axis with an angular velocity ω. Determine the momentum $\mathbf{H}_O$ of the unit about O and its kinetic energy T.

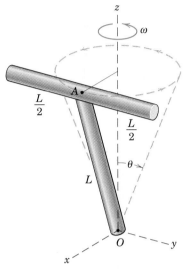

Problem 7/68

7/69 The 100-mm-radius wheel has a mass of 3 kg and turns about its y'-axis with an angular velocity $p = 40\pi$ rad/s in the direction shown. Simultaneously the fork rotates about its x-axis shaft with an angular velocity $\omega = 10\pi$ rad/s as indicated. Calculate the angular momentum of the wheel about its center O'. Also compute the kinetic energy of the wheel.

Ans. $\mathbf{H}_{O'} = 0.236(\mathbf{i} + 8\mathbf{j})$ kg·m²/s
$T = 215$ J

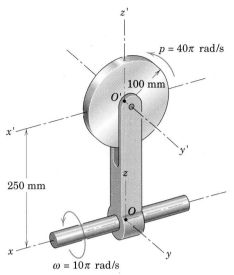

$p = 40\pi$ rad/s

100 mm

250 mm

$\omega = 10\pi$ rad/s

Problem 7/69

7/70 The solid circular disk of mass $m = 2$ kg and radius $r = 100$ mm rolls in a circle of radius $b = 200$ mm on the horizontal plane without slipping. If the center line OC of the axle of the wheel rotates about the z-axis with an angular velocity $\omega = 4\pi$ rad/s, determine the expression for the angular momentum of the disk with respect to the fixed point O. Also compute the kinetic energy of the wheel.

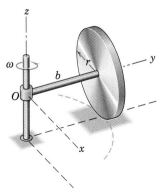

Problem 7/70

7/71 The uniform circular disk of 100-mm radius and small thickness has a mass of 3.6 kg and is spinning about its y'-axis at the rate $N = 300$ rev/min with its plane of rotation tilted at a constant angle $\beta = 20°$ from the vertical x-z plane. Simultaneously, the assembly rotates about the fixed z-axis at the rate $p = 60$ rev/min. Calculate the angular momentum $\mathbf{H}_O$ of the disk alone about the origin O of the x-y-z coordinates. Also calculate the kinetic energy T of the disk.

$$Ans. \ \mathbf{H}_O = 0.550\mathbf{j} + 1.670\mathbf{k} \ \text{kg·m}^2/\text{s}$$
$$T = 22.3 \text{ J}$$

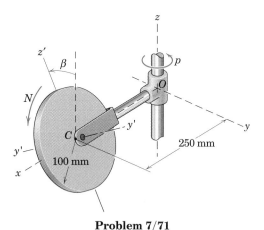

Problem 7/71

7/72 Rework Prob. 7/71 if β, instead of being constant at 20°, is increasing at the steady rate of 120 rev/min. Find the angular momentum $\mathbf{H}_O$ of the disk for the instant when $\beta = 20°$. Also compute the kinetic energy T of the disk. Is T dependent on β?

7/73 In a test of the solar panels for a spacecraft, the model shown is rotated about the vertical axis at the angular rate ω. If the mass per unit area of panel is ρ, write the expression for the angular momentum $\mathbf{H}_O$ of the assembly about the axes

shown in terms of θ. Also determine the maximum, minimum, and intermediate values of the moment of inertia about the axes through O.

$$Ans. \ \mathbf{H}_O = \left(\frac{m}{6}b^2\omega \sin 2\theta\right)\mathbf{i}$$
$$+ m\omega\left(\frac{c^2}{3} + \frac{b^2}{3}\cos^2\theta + a^2 + ac\right)\mathbf{k}$$
$$I_{max} = m\left(\frac{c^2 + b^2}{3} + a^2 + ac\right)$$
$$I_{inter} = m(\tfrac{1}{3}c^2 + a^2 + ac)$$
$$I_{min} = \tfrac{1}{3}mb^2$$

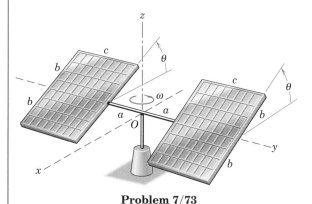

Problem 7/73

▶**7/74** The half-cylindrical shell of mass m, radius r, and length b revolves about one edge along the z-axis with a constant rate ω as shown. Determine the angular momentum $\mathbf{H}$ of the shell with respect to the x-y-z axes.

$$Ans. \ \mathbf{H} = mr\omega\left(\frac{b}{2}\mathbf{i} + \frac{b}{\pi}\mathbf{j} + 2r\mathbf{k}\right)$$

Problem 7/74

7/9 MOMENTUM AND ENERGY EQUATIONS OF MOTION

With the description of angular momentum, inertial properties, and kinetic energy of a rigid body established in the previous two articles, we are ready to apply the general momentum and energy equations of motion.

(a) Momentum equations.   In Art. 4/4 of Chapter 4, we established the general linear- and angular-momentum equations for a system of constant mass. These equations are

$$\Sigma \mathbf{F} = \dot{\mathbf{G}} \qquad\qquad\qquad \textbf{[4/6]}$$

$$\Sigma \mathbf{M} = \dot{\mathbf{H}} \qquad\qquad \textbf{[4/7] or [4/9]}$$

The general moment relation, Eq. 4/7 or 4/9, is expressed here by the single equation $\Sigma \mathbf{M} = \dot{\mathbf{H}}$, where the terms are taken either about a fixed point O or about the mass center G. In the derivation of the moment principle, the derivative of $\mathbf{H}$ was taken with respect to an absolute coordinate system. When $\mathbf{H}$ is expressed in terms of components measured relative to a moving coordinate system x-y-z which has an angular velocity $\mathbf{\Omega}$, then by Eq. 7/7 the moment relation becomes

$$\Sigma \mathbf{M} = \left(\frac{d\mathbf{H}}{dt}\right)_{xyz} + \mathbf{\Omega} \times \mathbf{H}$$

$$= (\dot{H}_x \mathbf{i} + \dot{H}_y \mathbf{j} + \dot{H}_z \mathbf{k}) + \mathbf{\Omega} \times \mathbf{H}$$

The terms in parentheses represent that part of $\dot{\mathbf{H}}$ due to the change in magnitude of the components of $\mathbf{H}$, and the cross-product term represents that part due to the changes in direction of the components of $\mathbf{H}$. Expansion of the cross product and rearrangement of terms give

$$\Sigma \mathbf{M} = (\dot{H}_x - H_y \Omega_z + H_z \Omega_y)\mathbf{i}$$
$$+ (\dot{H}_y - H_z \Omega_x + H_x \Omega_z)\mathbf{j} \qquad\qquad \textbf{(7/19)}$$
$$+ (\dot{H}_z - H_x \Omega_y + H_y \Omega_x)\mathbf{k}$$

Equation 7/19 is the most general form of the moment equation about a fixed point O or about the mass center G. The Ω's are the angular velocity components of rotation of the reference axes, and the H-components in the case of a rigid body are as defined in Eq. 7/12, where the ω's are the components of the angular velocity of the body. We now apply Eq. 7/19 to a rigid body where the coordinate axes are *attached* to the body. Under these conditions, when expressed in the x-y-z coordinates, the *moments and products of inertia are invariant with time,* and $\mathbf{\Omega} = \boldsymbol{\omega}$. Thus for axes attached to the body, the three scalar components of Eq. 7/19 become

$$\Sigma M_x = \dot{H}_x - H_y\omega_z + H_z\omega_y$$

$$\Sigma M_y = \dot{H}_y - H_z\omega_x + H_x\omega_z \qquad \textbf{(7/20)}$$

$$\Sigma M_z = \dot{H}_z - H_x\omega_y + H_y\omega_x$$

Equations 7/20 are the general moment equations for rigid-body motion with axes *attached to the body*. They hold with respect to axes through a fixed point O or through the mass center G.

In Art. 7/7 it was mentioned that, in general, for any origin fixed to a rigid body, there are three principal axes of inertia with respect to which the products of inertia vanish. If the reference axes coincide with the principal axes of inertia with origin at the mass center G or at a point O fixed to the body and fixed in space, the factors I_{xy}, I_{yz}, I_{xz} will be zero, and Eqs. 7/20 become

$$\Sigma M_x = I_{xx}\dot{\omega}_x - (I_{yy} - I_{zz})\omega_y\omega_z$$

$$\Sigma M_y = I_{yy}\dot{\omega}_y - (I_{zz} - I_{xx})\omega_z\omega_x \qquad \textbf{(7/21)}$$

$$\Sigma M_z = I_{zz}\dot{\omega}_z - (I_{xx} - I_{yy})\omega_x\omega_y$$

These relations, known as *Euler's equations**, are extremely useful in the study of rigid-body motion.

(b) Energy equations. The resultant of all external forces acting on a rigid body may be replaced by the resultant force $\Sigma\mathbf{F}$ acting through the mass center and a resultant couple $\Sigma\mathbf{M}_G$ acting about the mass center. Work is done by the resultant force and the resultant couple at the respective rates $\Sigma\mathbf{F}\cdot\overline{\mathbf{v}}$ and $\Sigma\mathbf{M}_G\cdot\boldsymbol{\omega}$, where $\overline{\mathbf{v}}$ is the linear velocity of the mass center and $\boldsymbol{\omega}$ is the angular velocity of the body. Integration over the time from condition 1 to condition 2 gives the total work done during the time interval. Equating the works done to the respective changes in kinetic energy as expressed in Eq. 7/15 gives

$$\int_{t_1}^{t_2} \Sigma\mathbf{F}\cdot\overline{\mathbf{v}}\, dt = \tfrac{1}{2}\overline{\mathbf{v}}\cdot\mathbf{G}\Big|_1^2 \qquad \int_{t_1}^{t_2} \Sigma\mathbf{M}_G\cdot\boldsymbol{\omega}\, dt = \tfrac{1}{2}\boldsymbol{\omega}\cdot\mathbf{H}_G\Big|_1^2 \qquad (7/22)$$

These equations express the change in translational kinetic energy and the change in rotational kinetic energy, respectively, for the interval during which $\Sigma\mathbf{F}$ or $\Sigma\mathbf{M}_G$ acts, and the sum of the two expressions equals ΔT.

The work-energy relationship, developed in Chapter 4 for a general system of particles and given by

---

*Named after Leonhard Euler (1707–1783), a Swiss mathematician.

$$U'_{1\text{-}2} = \Delta T + \Delta V_e + \Delta V_g \qquad [4/3]$$

was used in Chapter 6 for rigid bodies in plane motion. The equation is equally applicable to rigid-body motion in three dimensions. As we have seen previously, the work-energy approach is of great advantage when analyzing end-point conditions of motion (initial and final). Here the work $U'_{1\text{-}2}$ done during the interval by all active forces external to the body or system is equated to the sum of the corresponding changes in kinetic energy ΔT, elastic potential energy ΔV_e, and gravitational potential energy ΔV_g. The potential-energy changes are determined in the usual way, as described previously in Art. 3/7.

We shall limit our application of the equations developed in this article to two problems of special interest, parallel-plane motion and gyroscopic motion, discussed in the next two articles.

7/10 *PARALLEL-PLANE MOTION*

When all particles of a rigid body move in planes which are parallel to a fixed plane, the body has a general form of plane motion, as described in Art. 7/4 and pictured in Fig. 7/3. Every line in such a body which is normal to the fixed plane remains parallel to itself at all times. We take the mass center G as the origin of coordinates x-y-z which are attached to the body, with the x-y plane coinciding with the plane of motion P. The components of the angular velocity of both body and attached axes become $\omega_x = \omega_y = 0$, $\omega_z \neq 0$. For this case, the angular-momentum components from Eq. 7/12 become

$$H_x = -I_{xz}\omega_z \qquad H_y = -I_{yz}\omega_z \qquad H_z = I_{zz}\omega_z$$

and the moment relations of Eqs. 7/20 reduce to

$$\boxed{\begin{aligned} \Sigma M_x &= -I_{xz}\dot{\omega}_z + I_{yz}\omega_z{}^2 \\ \Sigma M_y &= -I_{yz}\dot{\omega}_z - I_{xz}\omega_z{}^2 \\ \Sigma M_z &= I_{zz}\dot{\omega}_z \end{aligned}} \qquad (7/23)$$

It is seen that the third moment equation is equivalent to the second of Eqs. 6/1, where the z-axis passes through the mass center, or to Eq. 6/4 if the z-axis passes through a fixed point O.

Equations 7/23 hold for an origin of coordinates at the mass center, as shown in Fig. 7/3, or for any origin on a fixed axis of rotation. The three independent force equations of motion which also apply to parallel-plane motion are clearly

$$\Sigma F_x = m\bar{a}_x \qquad \Sigma F_y = m\bar{a}_y \qquad \Sigma F_z = 0$$

Equations 7/23 find special use in describing the effect of dynamic imbalance in rotating shafts and in rolling bodies.

Sample Problem 7/7

The two circular disks, each of mass m_1, are connected by the curved bar bent into quarter-circular arcs and welded to the disks. The bar has a mass m_2. The total mass of the assembly is $m = 2m_1 + m_2$. If the disks roll without slipping on a horizontal plane with a constant velocity v of the disk centers, determine the value of the friction force under each disk at the instant represented when the plane of the curved bar is horizontal.

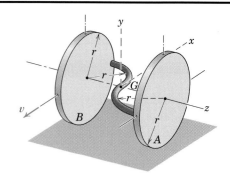

Solution.   The motion is identified as parallel-plane motion since the planes of motion of all parts of the system are parallel. The free-body diagram shows the normal forces and friction forces at A and B and the total weight mg acting through the mass center G, which we take as the origin of coordinates which rotate with the body.

We now apply Eqs. 7/23, where $I_{yz} = 0$ and $\dot{\omega}_z = 0$. The moment equation about the y-axis requires determination of I_{xz}. From the diagram showing the geometry of the curved rod and with ρ standing for the mass of the rod per unit length, we have

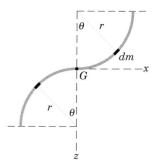

① $$\left[I_{xz} = \int xz \, dm \right] \qquad I_{xz} = \int_0^{\pi/2} (r \sin \theta)(-r + r \cos \theta)\rho r \, d\theta$$

$$+ \int_0^{\pi/2} (-r \sin \theta)(r - r \cos \theta)\rho r \, d\theta$$

Evaluating the integrals gives

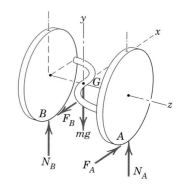

$$I_{xz} = -\rho r^3/2 - \rho r^3/2 = -\rho r^3 = -\frac{m_2 r^2}{\pi}$$

The second of Eqs. 7/23 with $\omega_z = v/r$ and $\dot{\omega}_z = 0$ gives

$$[\Sigma M_y = -I_{xz}\omega_z^2] \qquad F_A r + F_B r = -\left(-\frac{m_2 r^2}{\pi} \right) \frac{v^2}{r^2}$$

$$F_A + F_B = \frac{m_2 v^2}{\pi r}$$

But with $\bar{v} = v$ constant, $\bar{a}_x = 0$ so that

$$[\Sigma F_x = 0] \qquad F_A - F_B = 0 \qquad F_A = F_B$$

Thus,

$$F_A = F_B = \frac{m_2 v^2}{2\pi r} \qquad\qquad Ans.$$

We also note for the given position that with $I_{yz} = 0$ and $\dot{\omega}_z = 0$, the moment equation about the x-axis gives

② $$[\Sigma M_x = 0] \qquad -N_A r + N_B r = 0 \qquad N_A = N_B = mg/2$$

① We must be very careful to observe the correct sign for each of the coordinates of the mass element dm that make up the product xz.

② When the plane of the curved bar is not horizontal, the normal forces under the disks are no longer equal.

PROBLEMS

Introductory problems

7/75 The slender shaft of Prob. 7/56, shown again here, carries two offset particles, each of mass m, and rotates with the constant angular rate ω as indicated. Determine the x- and y-components of the bearing reactions at A and B due to the dynamic imbalance of the shaft for the position shown.

$$\text{Ans. } A_x = 0, A_y = -mR\omega^2/3$$
$$B_x = 0, B_y = mR\omega^2/3$$

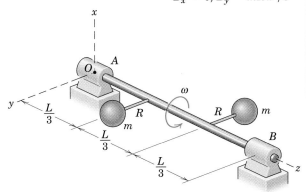

Problem 7/75

7/76 The uniform slender bar of length l and mass m is welded to the shaft, which rotates in bearings A and B with a constant angular velocity ω. Determine the expression for the force supported by the bearing at B as a function of θ. Consider only the force due to the dynamic imbalance and assume that the bearings can support radial forces only.

Problem 7/76

7/77 The paint stirrer shown in the figure is made from a rod of length $7b$ and mass ρ per unit length. Before immersion in the paint, the stirrer is rotating freely at a constant high angular velocity ω about its z-axis. Determine the bending moment M in the rod at the base O of the chuck.

$$\text{Ans. } M = M_x = -2\rho b^3 \omega^2$$

Problem 7/77

7/78 The dynamic imbalance of a certain crankshaft is approximated by the physical model shown, where the shaft carries three small 0.6-kg spheres attached by rods of negligible mass. If the shaft rotates at the constant speed of 1200 rev/min, calculate the forces R_A and R_B acting on the bearings. Neglect the gravitational forces.

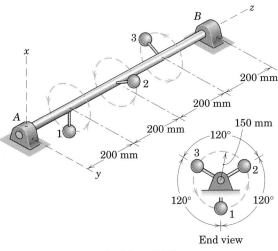

Problem 7/78

7/79 Each of the two right-angled bent rods has a masss of 1.2 kg and is parallel to the horizontal x-y plane. The rods are welded to the vertical shaft which rotates about the z-axis with a constant angular speed $N = 1200$ rev/min. Calculate the bending moment M in the shaft at its base O.

Ans. $M = 337$ N·m

Problem 7/79

Representative problems

7/80 Determine the bending moment M at the tangency point A in the semicircular rod of radius r and mass m as it rotates about the tangent axis with a constant and large angular velocity ω. Neglect the moment mgr produced by the weight of the rod.

Problem 7/80

7/81 The large satellite-tracking antenna has a moment of inertia I about its z-axis of symmetry and a moment of inertia I_O about each of the x- and y-axes. Determine the angular acceleration α of the antenna about the vertical Z-axis caused by a torque M applied about Z by the drive mechanism for a given orientation θ.

Ans. $\alpha = \dfrac{M}{I_O \cos^2 \theta + I \sin^2 \theta}$

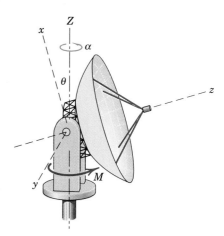

Problem 7/81

7/82 The plate has a mass of 3 kg and is welded to the fixed vertical shaft that rotates at the constant speed of 20π rad/s. Compute the moment **M** applied *to* the shaft *by* the plate due to dynamic imbalance.

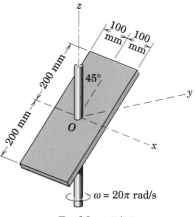

Problem 7/82

7/83 Each of the two semicircular disks has a mass of 1.20 kg and is welded to the shaft supported in bearings A and B as shown. Calculate the forces applied to the shaft by the bearings for a constant angular speed $N = 1200$ rev/min. Neglect the forces of static equilibrium.

Ans. $\mathbf{F}_A = 1608\mathbf{i}$ N, $\mathbf{F}_B = -1608\mathbf{i}$ N

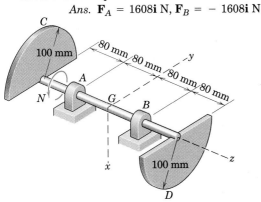

Problem 7/83

7/84 Solve Prob. 7/83 for the case where the assembly starts from rest with an initial angular acceleration $\alpha = 900$ rad/s² as a result of a starting torque (couple) M applied to the shaft in the same sense as N. Neglect the moment of inertia of the shaft about its z-axis and calculate M.

7/85 The uniform slender bar of mass ρ per unit length is freely pivoted about the y-axis at the clevis which rotates about the fixed vertical z-axis with a constant angular velocity ω. Determine the steady-state angle θ assumed by the bar. Length b is greater than length c.

$$\text{Ans. } \theta = \sin^{-1}\left(\frac{3g}{2\omega^2}\frac{b^2-c^2}{b^3+c^3}\right)$$

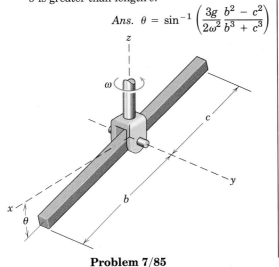

Problem 7/85

7/86 The uniform square flaps, each of mass m, are freely hinged at A and B to the square plate and attached shaft that rotate about the vertical z-axis with a constant angular velocity ω. Determine the angular velocity ω required to maintain a specified positive angle θ.

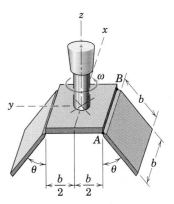

Problem 7/86

7/87 The circular disk of mass m and radius r is mounted on the vertical shaft with a small angle α between its plane and the plane of rotation of the shaft. Determine the expression for the bending moment $\mathbf{M}$ acting *on* the shaft due to the wobble of the disk at a shaft speed of ω rad/s.

Ans. $\mathbf{M} = (\frac{1}{8}mr^2\omega^2 \sin 2\alpha)\mathbf{j}$

Problem 7/87

7/88 The uniform square plate of mass m is welded at O to the end of the shaft which rotates about the vertical z-axis with a constant angular velocity ω. Determine the moment applied to the plate by the weld due only to the rotation.

Problem 7/88

7/89 For the plate of mass m in Prob. 7/88, determine the y- and z-components of the moment applied to the plate by the weld at O necessary to give the plate an angular acceleration $\alpha = \dot{\omega}$ starting from rest. Neglect the moment due to the weight.

Ans. $M_y = -\dfrac{1}{6}mb^2\alpha \sin 2\beta$

$$M_z = \frac{1}{12}mb^2\alpha(1 + 4\sin^2\beta)$$

7/90 Each of the quarter-circular plates has a mass of 2 kg and is secured to the vertical shaft mounted in the fixed bearing at O. Calculate the magnitude M of the bending moment in the shaft at O for a constant rotational speed $N = 300$ rev/min. Treat the plates as exact quarter-circular shapes.

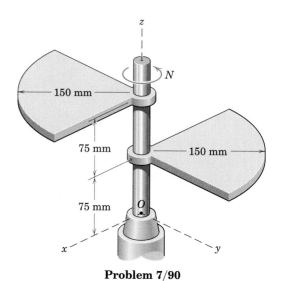

Problem 7/90

7/91 Calculate the bending moment M in the shaft at O for the rotating assembly of Prob. 7/90 as it starts from rest with an initial angular acceleration of 200 rad/s^2.

Ans. $M = 2.70$ N·m

7/92 The thin circular disk of mass m and radius R is hinged about its horizontal tangent axis to the end of a shaft rotating about its vertical axis with an angular velocity ω. Determine the steady-state angle β assumed by the plane of the disk with the vertical axis. Observe any limitation on ω to ensure that $\beta > 0$.

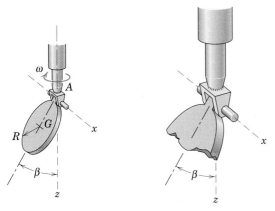

Detail of hinge at A

Problem 7/92

7/93 The partial ring has a mass ρ per unit length of rim and is welded to the shaft at A. Determine the bending moment M in the ring at A due to the rotation of the ring for an angular velocity ω.

$$Ans. \ M = \tfrac{1}{2}\rho r^3 \omega^2$$

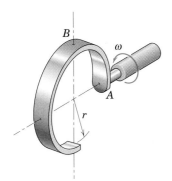

Problem 7/93

7/94 For the ring section of Prob. 7/93, determine the torsional moment T in the ring at B (moment about an axis tangent to the rim) as the ring starts from rest under the action of a torque M transmitted by the shaft to the ring at A. The mass per unit length of rim is ρ.

7/95 The uniform slender rod of length l is welded to the bracket at A on the underside of the disk B. The disk rotates about a vertical axis with a constant angular velocity ω. Determine the value of ω that will result in a zero moment supported by the weld at A for the position $\theta = 60°$ with $b = l/4$.

$$Ans. \ \omega = 2\sqrt{\frac{\sqrt{3}\,g}{l}}$$

Problem 7/95

7/96 The half-cylindrical shell of radius r, length $2b$, and mass m revolves about the vertical z-axis with a constant angular velocity ω as indicated. Determine the magnitude M of the bending moment in the shaft at A due to both the weight and the rotational motion of the shell.

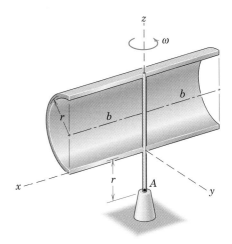

Problem 7/96

7/11   GYROSCOPIC MOTION: STEADY PRECESSION

One of the most interesting of all problems in dynamics is that of gyroscopic motion. This motion occurs whenever the axis about which a body is spinning is itself rotating about another axis. Although the complete description of this motion involves considerable complexity, the most common and useful examples of gyroscopic motion occur when the axis of a rotor spinning at constant speed turns (precesses) about another axis at a steady rate. Our discussion in this article will focus on this special case.

The gyroscope has important engineering applications. With a mounting in gimbal rings (see Fig. 7/19b), the gyro is free from external moments, and its axis will retain a fixed direction in space regardless of the rotation of the structure to which it is attached. In this way, the gyro is used for inertial guidance systems and other directional control devices. With the addition of a pendulous mass to the inner gimbal ring, the earth's rotation causes the gyro to precess so that the spin axis will always point north, and this action forms the basis of the gyro compass. The gyroscope has also found important use as a stabilizing device. The controlled precession of a large gyro mounted in a ship is used to produce a gyroscopic moment to counteract the rolling of a ship at sea. The gyroscopic effect is also an extremely important consideration in the design of bearings for the shafts of rotors which are subjected to forced precessions.

We shall first describe gyroscopic action with a simple physical approach which relies on our previous experience with the vector changes encountered in plane dynamics. This approach will help us gain a direct physical insight into gyroscopic action. Next, we will make use of the general momentum relation, Eq. 7/19, for a more complete description.

(a) Simplified approach. Figure 7/14 shows a symmetrical rotor spinning about the z-axis with a large angular velocity $\mathbf{p}$, known as the *spin velocity*. If we apply two forces F to the rotor axle to form a couple $\mathbf{M}$ whose vector is directed along the x-axis, we will find that the rotor shaft will rotate in the x-z plane about the y-axis in the sense indicated with a relatively slow angular velocity $\Omega = \dot{\psi}$ known as the *precession velocity*. Thus, we identify the spin axis ($\mathbf{p}$), the torque axis ($\mathbf{M}$), and the precession axis ($\mathbf{\Omega}$), where the usual right-hand rule identifies the sense of the rotation vectors. The rotor shaft does *not* turn about the x-axis in the sense of $\mathbf{M}$, as it would if the rotor were not spinning. To aid understanding of this phenomenon, a direct analogy may be made between the rotation vectors and the familiar vectors which describe the curvilinear motion of a particle.

Figure 7/15a shows a particle of mass m moving in the x-z plane with constant speed $|\mathbf{v}| = v$. The application of a force $\mathbf{F}$ normal to

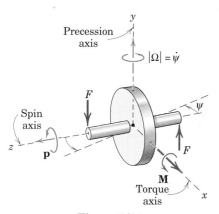

Figure 7/14

its linear momentum $\mathbf{G} = m\mathbf{v}$ causes a change $d\mathbf{G} = d(m\mathbf{v})$ in its momentum. We see that $d\mathbf{G}$, and hence $d\mathbf{v}$, is a vector in the direction of the normal force $\mathbf{F}$ according to Newton's second law $\mathbf{F} = \dot{\mathbf{G}}$, which may be written as $\mathbf{F}\, dt = d\mathbf{G}$. From Fig. 7/15$b$ we see that, in the limit, $\tan d\theta = d\theta = F\, dt/mv$ or $F = mv\dot{\theta}$. In vector notation with $\boldsymbol{\omega} = \dot{\theta}\mathbf{j}$, the force becomes

$$\mathbf{F} = m\boldsymbol{\omega} \times \mathbf{v}$$

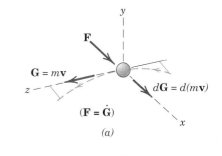
(a)

which is the vector equivalent of our familiar scalar relation $F_n = ma_n$ for the normal force on the particle as treated extensively in Chapter 3.

With these relations in mind, we now turn to our problem of rotation. Recall now the analogous equation $\mathbf{M} = \dot{\mathbf{H}}$ which we developed for any prescribed mass system, rigid or nonrigid, referred to its mass center (Eq. 4/9) or to a fixed point O (Eq. 4/7). We now apply this relation to our symmetrical rotor, as shown in Fig. 7/15c. For a high rate of spin $\mathbf{p}$ and a low precession rate Ω about the y-axis, the angular momentum is represented by the vector $\mathbf{H} = I\mathbf{p}$, where $I = I_{zz}$ is the moment of inertia of the rotor about the spin axis. Initially, we are neglecting the small component of angular momentum about the y-axis which accompanies the slow precession. The application of the couple $\mathbf{M}$ normal to $\mathbf{H}$ causes a change $d\mathbf{H} = d(I\mathbf{p})$ in the angular momentum. We see that $d\mathbf{H}$, and hence $d\mathbf{p}$, is a vector in the direction of the couple $\mathbf{M}$ since $\mathbf{M} = \dot{\mathbf{H}}$, which may also be written $\mathbf{M}\, dt = d\mathbf{H}$. Just as the change in the linear-momentum vector of the particle is in the direction of the applied force, so is the change in the angular-momentum vector of the gyro in the direction of the couple. Thus, we see that the vectors $\mathbf{M}$, $\mathbf{H}$, and $d\mathbf{H}$ are analogous to the vectors $\mathbf{F}$, $\mathbf{G}$, and $d\mathbf{G}$. With this insight, it is no longer strange to see the rotation vector undergo a change which is in the direction of $\mathbf{M}$, thereby causing the axis of the rotor to precess about the y-axis.

(b)

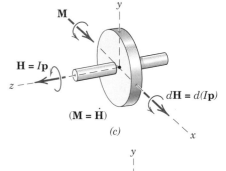
(c)

In Fig. 7/15d we see that during time dt the angular-momentum vector $I\mathbf{p}$ has swung through the angle $d\psi$, so that in the limit with $\tan d\psi = d\psi$, we have

$$d\psi = \frac{M\, dt}{Ip} \quad \text{or} \quad M = I\frac{d\psi}{dt}p$$

Substituting $\Omega = d\psi/dt$ for the magnitude of the precession velocity gives us

$$\boxed{M = I\Omega p} \tag{7/24}$$

We note that $\mathbf{M}$, Ω, and $\mathbf{p}$ as vectors are mutually perpendicular, and that their vector relationship may be represented by writing the equation in the cross-product form

(d)

Figure 7/15

Figure 7/16

Figure 7/17

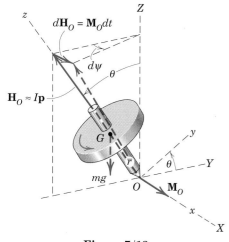

Figure 7/18

$$\boxed{\mathbf{M} = I\mathbf{\Omega} \times \mathbf{p}} \tag{7/24a}$$

which is completely analogous to the foregoing relation $\mathbf{F} = m\boldsymbol{\omega} \times \mathbf{v}$ for the curvilinear motion of a particle as developed from Figs. 7/15a and b. Equations 7/24 and 7/24a apply to moments taken about the mass center or about a fixed point on the axis of rotation.

The correct spatial relationship among the three vectors may be remembered from the fact that $d\mathbf{H}$, and hence $d\mathbf{p}$, is in the direction of $\mathbf{M}$, which establishes the correct sense for the precession $\mathbf{\Omega}$. Thus, the spin vector $\mathbf{p}$ always tends to rotate toward the torque vector $\mathbf{M}$. Figure 7/16 represents three orientations of the three vectors which are consistent with their correct order. Unless we establish this order correctly in a given problem, we are likely to arrive at a conclusion directly opposite to the correct one. It is emphasized that Eq. 7/24, like $\mathbf{F} = m\mathbf{a}$ and $M = I\alpha$, is an equation of motion, so that the couple $\mathbf{M}$ represents the couple due to *all* forces acting *on* the rotor, as disclosed by a correct *free-body diagram of the rotor*. Also it is pointed out that, when a rotor is forced to precess, as occurs with the turbine in a ship which is executing a turn, the motion will generate a *gyroscopic couple* $\mathbf{M}$ which obeys Eq. 7/24a in both magnitude and sense.

In the foregoing discussion of gyroscopic motion, it was assumed that the spin was large and the precession was small. Although we can see from Eq. 7/24 that for given values of I and M the precession $\mathbf{\Omega}$ must be small if p is large, let us now examine the influence of $\mathbf{\Omega}$ on the momentum relations. Again, we restrict our attention to steady precession, where $\mathbf{\Omega}$ has a constant magnitude. Figure 7/17 shows our same rotor again. Because it has a moment of inertia about the y-axis and an angular velocity of precession about this axis, there will be an additional component of angular momentum about the y-axis. Thus, we have the two components $H_z = Ip$ and $H_y = I_0\Omega$, where I_0 stands for I_{yy} and, again, I stands for I_{zz}. The total angular momentum is $\mathbf{H}$ as shown. The change in $\mathbf{H}$ remains $d\mathbf{H} = \mathbf{M}\,dt$ as previously, and the precession during time dt is the angle $d\psi = M\,dt/H_z = M\,dt/(Ip)$ as before. Thus, Eq. 7/24 is still valid and for steady precession is an exact description of the motion as long as the spin axis is perpendicular to the axis around which precession occurs.

Consider now the steady precession of a symmetrical top, Fig. 7/18, spinning about its axis with a high angular velocity p and supported at its point O. Here the spin axis makes an angle θ with the vertical Z-axis around which precession occurs. Again, we will neglect the small angular-momentum component due to the precession and consider H equal to $I\mathbf{p}$, the angular momentum about the axis of the top associated with the spin only. The moment about O is due to the weight and is $mg\bar{r}\sin\theta$, where $\bar{r}$ is the distance from

O to the mass center G. From the diagram, we see that the angular-momentum vector $\mathbf{H}_O$ has a change $d\mathbf{H}_O = \mathbf{M}_O\,dt$ in the direction of $\mathbf{M}_O$ during time dt and that θ is unchanged. The increment in precessional angle around the Z-axis is

$$d\psi = \frac{M_O\,dt}{Ip\,\sin\,\theta}$$

Substituting the values $M_O = mg\bar{r}\sin\theta$ and $\Omega = d\psi/dt$ gives

$$mg\bar{r}\,\sin\,\theta = I\Omega p\,\sin\,\theta \qquad \text{or} \qquad mg\bar{r} = I\Omega p$$

which is independent of θ. Introducing the radius of gyration so that $I = mk^2$ and solving for the precessional velocity give

$$\boxed{\Omega = \frac{g\bar{r}}{k^2 p}} \qquad\qquad (7/25)$$

Unlike Eq. 7/24, which is an exact description for the rotor of Fig. 7/17 with precession confined to the x-z plane, Eq. 7/25 is an approximation based on the assumption that the angular momentum associated with Ω is negligible compared with that associated with p. We shall see the amount of the error in part (b) of this article. On the basis of our analysis, the top will have a steady precession at the constant angle θ only if it is set in motion with a value of Ω which satisfies Eq. 7/25. When these conditions are not met, the precession becomes unsteady, and θ may oscillate with an amplitude which increases as the spin velocity decreases. The corresponding rise and fall of the rotation axis is known as *nutation*.

(b) More complete analysis. We will now make direct use of Eq. 7/19, which is the general angular-momentum equation for a rigid body, by applying it to a body spinning about its axis of rotational symmetry. This equation is valid for rotation about a fixed point or for rotation about the mass center. A spinning top, the rotor of a gyroscope, and a space capsule are examples of bodies whose motions can be described by the equations for rotation about a point. The general moment equations for this class of problems are fairly complex, and their complete solutions involve the use of elliptic integrals and somewhat lengthy computations. A large fraction of engineering problems where the motion is one of rotation about a point involve the steady precession of bodies of revolution which are spinning about their axes of symmetry. These conditions introduce simplification which greatly facilitates solution of the equations.

Consider a body with axial symmetry, Fig. 7/19a, rotating about a fixed point O on its axis, which is taken to be the z-direction. With O as origin, the x- and y-axes automatically become principal

(a)

(b)

(c)

Figure 7/19

axes of inertia along with the z-axis. This same description may be used for the rotation of a similar symmetrical body about its center of mass G, which is taken as the origin of coordinates as shown with the gimbaled gyroscope rotor of Fig. 7/19b. Again, the x- and y-axes are principal axes of inertia for point G. The same description may also be used to represent the rotation about the mass center of an axially symmetric body in space, such as the space capsule in Fig. 7/19c. In each case, we note that, regardless of the rotation of the axes or of the body relative to the axes (spin about the z-axis), the moments of inertia about the x- and y-axes remain constant with time. The principal moments of inertia are again designated $I_{zz} = I$ and $I_{xx} = I_{yy} = I_0$. The products of inertia are, of course, zero.

Before applying Eq. 7/19, we introduce a set of coordinates which provide a natural description for our problem. These coordinates are shown in Fig. 7/20 for the example of rotation about a fixed point O. The axes X-Y-Z are fixed in space, and plane A contains the X-Y axes and the fixed point O on the rotor axis. Plane B

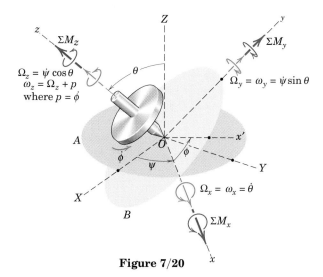

Figure 7/20

contains point O and is always normal to the rotor axis. Angle θ measures the inclination of the rotor axis from the vertical Z-axis and is also a measure of the angle between planes A and B. The intersection of the two planes is the x-axis, which is located by the angle ψ from the X-axis. The y-axis lies in plane B, and the z-axis coincides with the rotor axis. The angles θ and ψ completely specify the position of the rotor axis. The angular displacement of the rotor with respect to axes x-y-z is specified by the angle ϕ measured from the x-axis to the x'-axis, which is attached to the rotor. The spin velocity becomes $p = \dot{\phi}$.

The components of the angular velocity $\boldsymbol{\omega}$ of the rotor and the angular velocity $\boldsymbol{\Omega}$ of the axes x-y-z from Fig. 7/20 become

$$\Omega_x = \dot{\theta} \qquad\qquad \omega_x = \dot{\theta}$$
$$\Omega_y = \dot{\psi} \sin \theta \qquad \omega_y = \dot{\psi} \sin \theta$$
$$\Omega_z = \dot{\psi} \cos \theta \qquad \omega_z = \dot{\psi} \cos \theta + p$$

It is important to note that the axes and the body have identical x- and y-components of angular velocity, but that the z-components differ by the relative angular velocity p.

The angular-momentum components from Eq. 7/12 become

$$H_x = I_{xx}\omega_x = I_0\dot{\theta}$$
$$H_y = I_{yy}\omega_y = I_0\dot{\psi} \sin \theta$$
$$H_z = I_{zz}\omega_z = I(\dot{\psi} \cos \theta + p)$$

Substitution of the angular-velocity and angular-momentum components into Eq. 7/19 yields

$$\boxed{\begin{aligned}
\Sigma M_x &= I_0(\ddot{\theta} - \dot{\psi}^2 \sin \theta \cos \theta) + I\dot{\psi}(\dot{\psi} \cos \theta + p) \sin \theta \\
\Sigma M_y &= I_0(\ddot{\psi} \sin \theta + 2\dot{\psi}\dot{\theta} \cos \theta) - I\dot{\theta}(\dot{\psi} \cos \theta + p) \\
\Sigma M_z &= I\frac{d}{dt}(\dot{\psi} \cos \theta + p)
\end{aligned}}
\qquad \textbf{(7/26)}$$

Equations 7/26 are the general equations of rotation of a symmetrical body about either a fixed point O or the mass center G. In a given problem, the solution to the equations will depend on the moment sums applied to the body about the three coordinate axes. We will confine our use of these equations to two particular cases of rotation about a point which are described in the following sections.

(c) Steady-state precession. We now examine the conditions under which the rotor precesses at a steady rate $\dot{\psi}$ at a constant angle θ and with constant spin velocity p. Thus,

$$\dot{\psi} = \text{constant}, \quad \ddot{\psi} = 0$$
$$\theta = \text{constant}, \quad \dot{\theta} = \ddot{\theta} = 0$$
$$p = \text{constant}, \quad \dot{p} = 0$$

and Eqs. 7/26 become

$$\boxed{\begin{aligned}
\Sigma M_x &= \dot{\psi} \sin \theta[I(\dot{\psi} \cos \theta + p) - I_0\dot{\psi} \cos \theta] \\
\Sigma M_y &= 0 \\
\Sigma M_z &= 0
\end{aligned}}
\qquad \textbf{(7/27)}$$

From these results, we see that the required moment acting on the rotor about O (or about G) must be in the x-direction since the y- and z-components are zero. Furthermore, with the constant values of θ, $\dot{\psi}$, and p, the moment is constant in magnitude. It is also important to observe that the moment axis is perpendicular to the plane defined by the precession axis (Z-axis) and the spin axis (z-axis).

We may also obtain Eqs. 7/27 by observing that the components of $\mathbf{H}$ remain constant as observed in x-y-z so that $(\dot{\mathbf{H}})_{xyz} = \mathbf{0}$. Since in general $\Sigma \mathbf{M} = (\dot{\mathbf{H}})_{xyz} + \mathbf{\Omega} \times \mathbf{H}$, we have for the case of steady precession

$$\boxed{\Sigma \mathbf{M} = \mathbf{\Omega} \times \mathbf{H}} \tag{7/28}$$

which reduces to Eqs. 7/27 upon substitution of the values of $\mathbf{\Omega}$ and $\mathbf{H}$.

By far the most common engineering examples of gyroscopic motion occur when precession takes place about an axis which is normal to the rotor axis, as in Fig. 7/14. Thus with the substitution $\theta = \pi/2$, $\omega_z = p$, $\dot{\psi} = \Omega$, and $\Sigma M_x = M$, we have from Eqs. 7/27

$$M = I\Omega p \tag{7/24}$$

which we derived initially in this article from a direct analysis of this special case.

Now let us examine the steady precession of the rotor (symmetrical top) of Fig. 7/20 for any constant value of θ other than $\pi/2$. The moment ΣM_x about the x-axis is due to the weight of the rotor and is $mg\bar{r} \sin \theta$. Substitution into Eqs. 7/27 and rearrangement of terms give us

$$mg\bar{r} = I\dot{\psi}p - (I_0 - I)\dot{\psi}^2 \cos \theta$$

We see that $\dot{\psi}$ is small when p is large, so that the second term on the right-hand side of the equation becomes very small compared with $I\dot{\psi}p$. If we neglect this smaller term, we have $\dot{\psi} = mg\bar{r}/(Ip)$ which, upon use of the previous substitutions $\Omega = \dot{\psi}$ and $mk^2 = I$, becomes

$$\Omega = \frac{g\bar{r}}{k^2 p} \tag{7/25}$$

We derived this same relation earlier by assuming that the angular momentum was entirely along the spin axis.

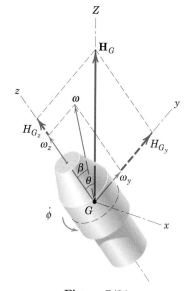

Figure 7/21

(d) Steady precession with zero moment. Consider now the motion of a symmetrical rotor with no external moment about its mass center. Such motion is encountered with spacecraft and projectiles which both spin and precess during flight. Figure 7/21 represents such a body. Here the Z-axis, which has a fixed direction in space, is chosen to coincide with the direction of the angular momentum $\mathbf{H}_G$, which is constant since $\Sigma\mathbf{M}_G = \mathbf{0}$. The x-y-z axes are attached in the manner described in Fig. 7/20. From Fig. 7/21 the three components of momentum are $H_{G_x} = 0$, $H_{G_y} = H_G \sin\theta$, $H_{G_z} = H_G \cos\theta$. From the defining relations, Eqs. 7/12, with the notation of this article, these components are also given by $H_{G_x} = I_0\omega_x$, $H_{G_y} = I_0\omega_y$, $H_{G_z} = I\omega_z$. Thus, $\omega_x = \Omega_x = 0$ so that θ is constant. This result means that the motion is one of steady precession about the constant $\mathbf{H}_G$ vector. With no x-component, the angular velocity $\boldsymbol{\omega}$ of the rotor lies in the y-z plane along with the Z-axis and makes an angle β with the z-axis. The relationship between β and θ is obtained from $\tan\theta = H_{G_y}/H_{G_z} = I_0\omega_y/(I\omega_z)$, which is

$$\tan\theta = \frac{I_0}{I}\tan\beta \qquad (7/29)$$

Thus, the angular velocity $\boldsymbol{\omega}$ makes a constant angle β with the spin axis.

The rate of precession is easily obtained from Eq. 7/27 with $M = 0$, which gives

$$\dot{\psi} = \frac{Ip}{(I_0 - I)\cos\theta} \qquad (7/30)$$

It is clear from this relation that the direction of the precession depends on the relative magnitudes of the two moments of inertia.

If $I_0 > I$, then $\beta < \theta$, as indicated in Fig. 7/22a, and the precession is said to be *direct*. Here the body cone rolls on the outside of the space cone.

If $I > I_0$, then $\theta < \beta$, as indicated in Fig. 7/22b, and the precession is said to be *retrograde*. In this instance, the space cone is internal to the body cone, and $\dot{\psi}$ and p have opposite signs.

If $I = I_0$, then $\theta = \beta$ from Eq. 7/29, and Fig. 7/22 shows that both angles must be zero to be equal. For this case, the body has no precession and merely rotates with an angular velocity $\mathbf{p}$. This condition occurs for a body with point symmetry, such as with a homogeneous sphere.

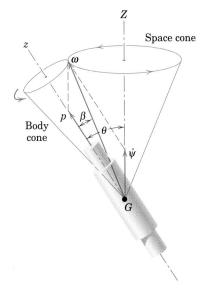

Direct precession $I_0 > I$

(a)

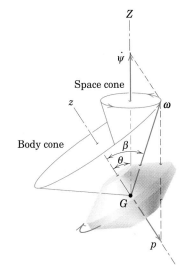

Retrograde precession $I_0 < I$

(b)

Figure 7/22

Sample Problem   7/8

The turbine rotor in a ship's power plant has a mass of 1000 kg, with center of mass at G and a radius of gyration of 200 mm. The rotor shaft is mounted in bearings A and B with its axis in the horizontal fore-and-aft direction and turns counterclockwise at a speed of 5000 rev/min when viewed from the stern. Determine the vertical components of the bearing reactions at A and B if the ship is making a turn to port (left) of 400-m radius at a speed of 25 knots (1 knot = 0.514 m/s). Does the bow of the ship tend to rise or fall because of the gyroscopic action?

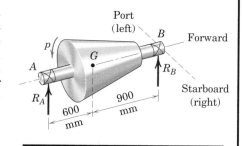

Solution.   The vertical component of the bearing reactions will equal the static reactions R_1 and R_2 due to the weight of the rotor, plus or minus the increment ΔR due to the gyroscopic effect. The moment principle from statics easily gives $R_1 = 5890$ N and $R_2 = 3920$ N. The given directions of the spin velocity $\mathbf{p}$ and the precession velocity $\mathbf{\Omega}$ are shown with the free-body diagram of the rotor. Since the spin axis always tends to rotate toward the torque axis, we see that the torque axis $\mathbf{M}$ points in the starboard direction as shown. The sense of the ΔR's is, therefore, up at B and down at A to produce the couple $\mathbf{M}$. Thus, the bearing reactions at A and B are

$$R_A = R_1 - \Delta R \quad \text{and} \quad R_B = R_2 + \Delta R$$

The precession velocity Ω is the speed of the ship divided by the radius of its turn.

$$[v = \rho\Omega] \qquad \Omega = \frac{25(0.514)}{400} = 0.0321 \text{ rad/s}$$

Equation 7/24 is now applied around the mass center G of the rotor to give

$$[M = I\Omega p] \qquad 1.500 \, \Delta R = 1000(0.200)^2(0.0321)\left[\frac{5000(2\pi)}{60}\right]$$

$$\Delta R = 449 \text{ N}$$

The required bearing reactions become

$$R_A = 5890 - 449 = 5440 \text{ N} \quad \text{and} \quad R_B = 3920 + 449 = 4370 \text{ N}$$
Ans.

We now observe that the forces just computed are those exerted *on* the rotor shaft *by* the structure of the ship. Consequently, from the principle of action and reaction, the equal and opposite forces are applied to the ship *by* the rotor shaft, as shown in the bottom sketch. Therefore, the effect of the gyroscopic couple is to generate the increments ΔR shown, and the bow will tend to fall and the stern to rise (but only slightly).

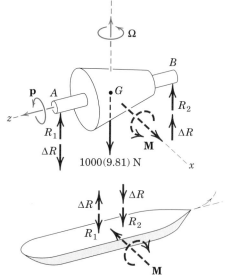

① If the ship is making a left turn, the rotation is counterclockwise when viewed from above, and the precession vector $\mathbf{\Omega}$ is up by the right-hand rule.

② After figuring the correct sense of $\mathbf{M}$ *on* the rotor, the common mistake is to apply it to the ship in the same sense, forgetting the action-and-reaction principle. Clearly, the results are then reversed. (Be certain not to make this mistake when operating a vertical gyro stabilizer in your yacht to counteract its roll!)

Sample Problem 7/9

A proposed space station is closely approximated by four uniform spherical shells, each of mass m and radius r. The mass of the connecting structure and internal equipment may be neglected as a first approximation. If the station is designed to rotate about its z-axis at the rate of one revolution every 4 seconds, determine (a) the number n of complete cycles of precession for each revolution about the z-axis if the plane of rotation deviates only slightly from a fixed orientation, and (b) find the period τ of precession if the spin axis z makes an angle of 20° with respect to the axis of fixed orientation about which precession occurs. Draw the space and body cones for this latter condition.

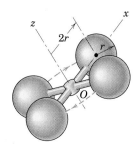

Solution. (a) The number of precession cycles or wobbles for each revolution of the station about the z-axis would be the ratio of the precessional velocity $\dot\psi$ to the spin velocity p, which, from Eq. 7/30, is

$$\frac{\dot\psi}{p} = \frac{I}{(I_0 - I)\cos\theta}$$

The moments of inertia are

$$I_{zz} = I = 4[\tfrac{2}{3}mr^2 + m(2r)^2] = \tfrac{56}{3}mr^2$$

$$I_{xx} = I_0 = 2(\tfrac{2}{3})mr^2 + 2[\tfrac{2}{3}mr^2 + m(2r)^2] = \tfrac{32}{3}mr^2$$

With θ very small, $\cos\theta \cong 1$, and the ratio of angular rates becomes

$$n = \frac{\dot\psi}{p} = \frac{\tfrac{56}{3}}{\tfrac{32}{3} - \tfrac{56}{3}} = -\frac{7}{3} \qquad Ans.$$

The minus sign indicates retrograde precession where, in the present case, $\dot\psi$ and p are essentially of opposite sense. Thus, the station will make seven wobbles for every three revolutions.

(b) For $\theta = 20°$ and $p = 2\pi/4$ rad/s, the period of precession or wobble is $\tau = 2\pi/|\dot\psi|$, so that from Eq. 7/30

$$\tau = \frac{2\pi}{2\pi/4}\left|\frac{I_0 - I}{I}\right|\cos\theta = 4(\tfrac{3}{7})\cos 20° = 1.611 \text{ s} \qquad Ans.$$

The precession is retrograde, and the body cone is external to the space cone as shown in the illustration where the body-cone angle, from Eq. 7/29, is

$$\tan\beta = \frac{I}{I_0}\tan\theta = \frac{56/3}{32/3}(0.364) = 0.637 \qquad \beta = 32.5°$$

① Our theory is based on the assumption that $I_{xx} = I_{yy}$ = the moment of inertia about any axis through G perpendicular to the z-axis. Such is the case here, and students should prove it to their own satisfaction.

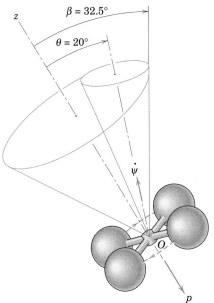

PROBLEMS

Introductory problems

7/97 A dynamics instructor demonstrates gyroscopic principles to his students. He suspends a rapidly spinning wheel with a string attached to one end of its horizontal axle. Describe the precession motion of the wheel.

Ans. Precession is CCW as viewed from above

Problem 7/97

7/98 The jet aircraft at the bottom of an inside vertical loop has a tendency, due to gyroscopic action of the engine rotor, to yaw to the right (as seen by the pilot and as indicated by the dashed orange wing-tip movements.) Determine the direction of rotation p_1 or p_2 of the engine rotor as depicted in the expanded view.

Problem 7/98

7/99 The student has volunteered to assist in a class-room demonstration involving a momentum wheel that is rapidly spinning with angular speed p as shown. The instructor has asked her to hold the axle of the wheel in the horizontal position shown and then attempt to tilt the axis upward in a vertical plane. What motion tendency of the wheel assembly will the student sense?

Ans. Clockwise rotation as seen from above

Problem 7/99

7/100 An experimental antipollution bus is powered by the kinetic energy stored in a large flywheel that spins at a high speed p in the direction indicated. As the bus encounters a short upward ramp, the front wheels rise, thus causing the flywheel to precess. What changes occur to the forces between the tires and the road during this sudden change?

Problem 7/100

7/101 The special-purpose fan is mounted as shown. The motor armature, shaft, and blades have a combined mass of 2.2 kg with radius of gyration of 60 mm. The axial position b of the 0.8-kg block A can be adjusted. With the fan turned off, the unit is balanced about the x-axis when b = 180 mm. The motor and fan operate at 1725 rev/min in the direction shown. Determine the value of b that will produce a steady precession of 0.2 rad/s about the positive y-axis.

Ans. b = 216 mm

Problem 7/102

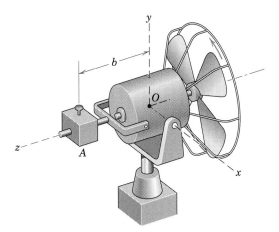

Problem 7/101

7/102 The two identical disks are rotating freely on the shaft, with angular velocities equal in magnitude and opposite in direction as shown. The shaft in turn is caused to rotate about the vertical axis in the sense indicated. Prove whether the shaft bends as in A or as in B because of gyroscopic action.

7/103 The 50-kg wheel is a solid circular disk which rolls on the horizontal plane in a circle of 600-mm radius. The wheel shaft is pivoted about the axis $O\text{-}O$ and is driven by the vertical shaft at the constant rate N = 48 rev/min about the Z-axis. Determine the normal force R between the wheel and the horizontal surface. Neglect the weight of the horizontal shaft.

Ans. R = 712 N

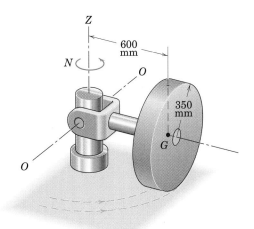

Problem 7/103

7/104 A top consists of a ring of mass $m = 0.52$ kg and mean radius $r = 60$ mm mounted on its central pointed shaft with spokes of negligible mass. The top is given a spin velocity of 10 000 rev/min and released on the horizontal surface with the point O remaining in a fixed position. The axis of the top is seen to make an angle of 15° with the vertical as it precesses. Determine the number N of precession cycles per minute. Also identify the direction of the precession and sketch the body and space cones.

Problem 7/104

7/105 The figure shows the side view of the wheel carriage (truck) of a railway passenger car where the vertical load is transmitted to the frame in which the journal wheel bearings are located. The lower view shows only one pair of wheels and their axle which rotates with the wheels. Each of the 825-mm-diameter wheels has a mass of 250 kg, and the 315-kg axle has a diameter of 125 mm. If the train is traveling at 130 km/h while rounding an 8° curve to the right (radius of curvature 218 m), calculate the change ΔR in the vertical force supported by each wheel due only to the gyroscopic action. As a close approximation, treat each wheel as a uniform circular disk and the axle as a uniform solid cylinder. Also assume both rails are in the same horizontal plane.

Ans. $\Delta R = 436$ N

Side view of carriage

View of wheels and axle
Problem 7/105

7/106 An airplane has just cleared the runway with a takeoff speed v. Each of its freely spinning wheels has a mass m, with a radius of gyration k about its axle. As seen from the front of the airplane, the wheel precesses at the angular rate Ω as the landing strut is folded into the wing about its pivot O. As a result of the gyroscopic action, the supporting member A exerts a torsional moment M on B to prevent the tubular member from rotating in the sleeve at B. Determine M and identify whether it is in the sense of M_1 or M_2.

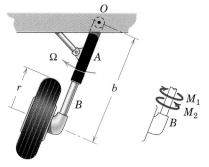

Problem 7/106

7/107 The 210-kg rotor of a turbojet aircraft engine has a radius of gyration of 220 mm and rotates counterclockwise at 18 000 rev/min when viewed from the front. If the aircraft is traveling at 1200 km/h and starts to execute an inside vertical loop of 3800-m radius, compute the gyroscopic moment M transmitted to the airframe. What correction to the controls does the pilot have to make in order to remain in the vertical plane?

Ans. $M = 1681$ N·m, Left rudder

Representative problems

7/108 A small air compressor for an aircraft cabin consists of the 3.50-kg turbine A that drives the 2.40-kg blower B at a speed of 20 000 rev/min. The shaft of the assembly is mounted transversely to the direction of flight and is viewed from the rear of the aircraft in the figure. The radii of gyration of A and B are 79.0 and 71.0 mm, respectively. Calculate the radial forces exerted on the shaft by the bearings at C and D if the aircraft executes a clockwise roll (rotation about the longitudinal flight axis) of 2 rad/s viewed from the rear of the aircraft. Neglect the small moments caused by the weights of the rotors. Draw a free-body diagram of the shaft as viewed from above and indicate the shape of its deflected center line.

Problem 7/108

7/109 The blades and hub of the helicopter rotor have a mass of 64 kg and have a radius of gyration of 3 m about the z-axis of rotation. With the rotor turning at 500 rev/min during a short interval following vertical liftoff, the helicopter tilts forward at the rate $\dot{\theta} = 10$ deg/s in order to acquire forward velocity. Determine the gyroscopic moment M transmitted to the body of the helicopter by its rotor and indicate whether the helicopter tends to deflect clockwise or counterclockwise, as viewed by a passenger facing forward.

Ans. $M = 5.26$ kN·m

Problem 7/109

7/110 The figure shows a gyro mounted with vertical axis and used to stabilize a hospital ship against rolling. The motor A turns the pinion that precesses the gyro by rotating the large precession gear B and attached rotor assembly about a horizontal transverse axis in the ship. The rotor turns inside the housing at a clockwise speed of 960 rev/min as viewed from the top and has a mass of 80 Mg with radius of gyration of 1.45 m. Calculate the moment exerted on the hull structure by the gyro if the motor turns the precession gear B at the rate of 0.320 rad/s. In which of the two directions, (a) or (b), should the motor turn in order to counteract a roll of the ship to port?

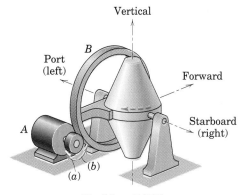

Problem 7/110

7/111 The primary structure of a proposed space station consists of five spherical shells connected by tubular spokes. The moment of inertia of the structure about its geometric axis A-A is twice as much as that about any axis through O normal to A-A. The station is designed to rotate about its geometric axis at the constant rate of 3 rev/min. If the spin axis A-A precesses about the Z-axis of fixed orientation and makes a very small angle with it, calculate the rate $\dot{\psi}$ at which the station wobbles. The mass center O has negligible acceleration.

Ans. $\dot{\psi} = -6$ rev/min, retrograde precession

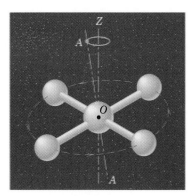

Problem 7/111

7/112 The conical shell is released in space with an angular velocity p about its central axis. If $h = 3r$ and the cone is observed to have a very slight wobble, determine the number n of complete wobbles for each revolution of the shell. Also indicate the ratio h/r for which no precession can occur.

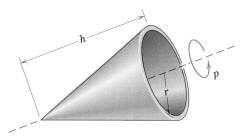

Problem 7/112

7/113 The electric motor has a total mass of 10 kg and is supported by the mounting brackets A and B attached to the rotating disk. The armature of the motor has a mass of 2.5 kg and a radius of gyration of 35 mm and turns counterclockwise at a speed of 1725 rev/min as viewed from A to B. The turntable revolves about its vertical axis at the constant rate of 48 rev/min in the direction shown. Determine the vertical components of the forces supported by the mounting brackets at A and B.

Ans. $R_A = 37.5$ N, $R_B = 60.6$ N

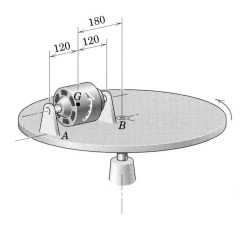

Dimensions in millimeters

Problem 7/113

7/114 The two solid cones with the same base and equal altitudes are spinning in space about their common axis at the rate p. For what ratio h/r will precession of their spin axis be impossible?

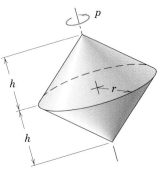

Problem 7/114

7/115 The solid cone of mass m, base radius r, and altitude h is spinning at a high rate p about its own axis and is released with its vertex O supported by a horizontal surface. Friction is sufficient to prevent the vertex from slipping in the x-y plane. Determine the direction of the precession Ω and the period τ of one complete rotation about the vertical z-axis.

Ans. $\Omega = \Omega\mathbf{k}, \ \tau = 4\pi r^2 p/(5gh)$

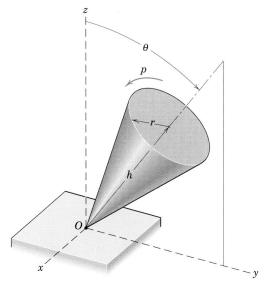

Problem 7/115

7/116 The tip of the cone in Prob. 7/115 is, in reality, somewhat rounded causing the point of contact P with the supporting surface to be slightly off the spin axis, as illustrated here. In the presence of kinetic friction, show why the angle θ will slowly decrease.

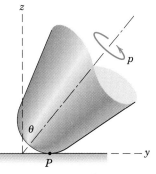

Problem 7/116

7/117 An experimental car is equipped with a gyro stabilizer to counteract completely the tendency of the car to tip when rounding a curve (no change in normal force between tires and road). The rotor of the gyro has a mass m_0 and a radius of gyration k, and is mounted in fixed bearings on a shaft that is parallel to the rear axle of the car. The center of mass of the car is a distance h above the road, and the car is rounding an unbanked level turn at a speed v. At what speed p should the rotor turn and in what direction to counteract completely the tendency of the car to overturn for either a right or a left turn? The combined mass of car and rotor is m.

Ans. $p = \dfrac{mvh}{m_0 k^2}$, opposite direction to car wheels

7/118 The cylindrical shell is rotating in space about its geometric axis. If the axis has a slight wobble, for what ratios of l/r will the motion be direct or retrograde precession?

Problem 7/118

7/119 A boy throws a thin circular disk (like a Frisbee) with a spin rate of 300 rev/min. The plane of the disk is seen to wobble through a total angle of 10°. Calculate the period τ of the wobble and indicate whether the precession is direct or retrograde.

Ans. $\tau = 0.0996$ s, retrograde

Problem 7/119

7/120 The uniform 650-mm rod has a mass of 3 kg and is welded centrally to the uniform 160-mm-radius circular disk which has a mass of 8 kg. The unit is given a spin velocity $p = 60$ rad/s in the direction shown. The axis of the rod is seen to wobble through a total angle of 30°. Calculate the angular velocity $\dot{\psi}$ of precession and determine whether it is $\dot{\psi}_1$ or $\dot{\psi}_2$.

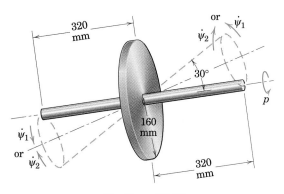

Problem 7/120

7/121 The open-ended thin-walled rectangular box of square cross section is rotating in space about its central longitudinal axis as shown. If the axis has a slight wobble, for what ratios l/b will the motion be direct or retrograde precession?

Ans. $l/b > \sqrt{2}$ direct, $l/b < \sqrt{2}$ retrograde

Problem 7/121

7/122 The figure shows a football in three common in-flight configurations. Case (*a*) is a perfectly thrown spiral pass with a spin rate of 120 rev/min. Case (*b*) is a wobbly spiral pass again with a spin rate of 120 rev/min about its own axis, but with the axis wobbling through a total angle of 20°. Case (*c*) is an end-over-end place kick with a rotational rate of 120 rev/min. For each case, specify the values of p, θ, β, and $\dot{\psi}$ as defined in this article. The moment of inertia about the ball's long axis is 0.3 of that about the transverse axis of symmetry.

Problem 7/122

7/123 The uniform slender bar of mass m and length l is centrally mounted on the shaft A-A, about which it rotates with a constant speed $\dot\phi = p$. Simultaneously, the yoke is forced to rotate about the x-axis with a constant speed ω_0. As a function of ϕ, determine the magnitude of the torque M required to maintain the constant speed ω_0. (*Hint:* Apply Eq. 7/19 to obtain the x-component of M.)

$$Ans.\ M = \tfrac{1}{12}ml^2p\omega_0 \sin 2\phi$$

7/125 Each of the three identical and equally spaced propeller blades has a moment of inertia I about the propeller z-axis. In addition to the angular velocity $p = \dot\phi$ of the propeller about the z-axis, the airplane is turning to the left at the angular rate Ω. Derive expressions for the x- and y-components of the bending moment M applied to the propeller shaft at the hub as functions of ϕ. Axes x-y rotate with the propeller.

$$Ans.\ M_x = 3I\Omega p \cos \phi,\ M_y = -3I\Omega p \sin \theta$$
$$M = 3I\Omega p$$

Problem 7/123

7/124 The 5-kg disk and hub A have a radius of gyration of 85 mm about the z_0-axis and spin at the rate $p = 1250$ rev/min. Simultaneously, the assembly rotates about the vertical z-axis at the rate $\Omega = 400$ rev/min. Calculate the gyroscopic moment **M** exerted *on* the shaft at C by the disk and the bending moment M_O in the shaft at O. Neglect the mass of the shaft but otherwise account for all forces acting on it.

Problem 7/125

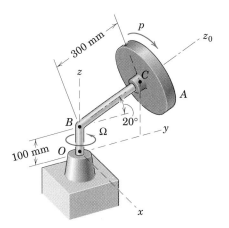

Problem 7/124

7/126 The housing of the electric motor is freely pivoted about the horizontal x-axis which passes through the mass center G of the rotor. If the motor is turning at the constant rate $\dot{\phi} = p$, determine the angular acceleration $\ddot{\psi}$ which will result from the application of the moment M about the vertical shaft if $\dot{\gamma} = \dot{\psi} = 0$. The mass of the frame and housing is considered negligible compared with the mass m of the rotor. The radius of gyration of the rotor about the z-axis is k_z and that about the x-axis is k_x.

Problem 7/126

▶7/127 The solid circular disk of mass m and small thickness is spinning freely on its shaft at the rate p. If the assembly is released in the vertical position at $\theta = 0$ with $\dot{\theta} = 0$, determine the horizontal components of the forces A and B exerted by the respective bearings on the horizontal shaft as the position $\theta = \pi/2$ is passed. Neglect the mass of the two shafts compared with m and neglect all friction. Solve by using the appropriate moment equations.

$$\text{Ans. } A_z = -\frac{m\dot{\theta}}{2}\left(\frac{r^2}{2b}p + l\dot{\theta}\right)$$

$$B_z = \frac{m\dot{\theta}}{2}\left(\frac{r^2}{2b}p - l\dot{\theta}\right)$$

$$\text{where } \dot{\theta} = 2\sqrt{\frac{2gl}{r^2 + 4l^2}}$$

Problem 7/127

▶7/128 The two solid homogeneous right circular cones, each of mass m, are fastened together at their vertices to form a rigid unit and are spinning about their axis of radial symmetry at the rate $p = 200$ rev/min. (a) Determine the ratio h/r for which the rotation axis will not precess. (b) Sketch the space and body cones for the cases where h/r is less than the critical ratio. (c) Sketch the space and body cones when $h = r$ and the precessional velocity is $\dot{\psi} = 18$ rad/s.

$$\text{Ans. (a) } h = \frac{r}{2}$$

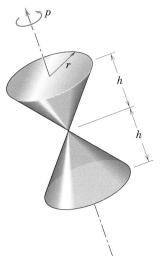

Problem 7/128

▶**7/129** The earth-scanning satellite is in a circular orbit of period τ. The angular velocity of the satellite about its y- or pitch-axis is $\omega = 2\pi/\tau$, and the angular rates about the x- and z-axes are zero. Thus, the x-axis of the satellite always points to the center of the earth. The satellite has a reaction-wheel attitude-control system consisting of the three wheels shown, each of which may be variably torqued by its individual motor. The angular rate Ω_z of the z-wheel relative to the satellite is Ω_0 at time $t = 0$, and the x- and y-wheels are at rest relative to the satellite at $t = 0$. Determine the axial torques M_x, M_y, and M_z that must be exerted by the motors on the shafts of their respective wheels in order that the angular velocity ω of the satellite will remain constant. The moment of inertia of each reaction wheel about its axis is I. The x and z reaction-wheel speeds are harmonic functions of the time with a period equal to that of the orbit. Plot the variations of the torques and the relative wheel speeds Ω_x, Ω_y, and Ω_z as functions of the time during one orbit period. (*Hint:* The torque to accelerate the x-wheel equals the reaction of the gyroscopic moment on the z-wheel, and vice versa.)

$$Ans. \ M_x = -I\omega\Omega_0 \cos \omega t$$
$$M_y = 0$$
$$M_z = -I\omega\Omega_0 \sin \omega t$$

▶**7/130** The elements of a gyrocompass are shown in the figure where the rotor, at north latitude γ on the surface of the earth, is mounted in a single gimbal ring which is free to rotate about the fixed vertical y-axis. The rotor axis is then able to rotate in the horizontal x-z plane, as measured by the angle β from the north direction. Assume that the gyro spins at the rate p and that it has mass moments of inertia about the spin axis and transverse axis through G of I and I_0, respectively. Show that the gyro axis oscillates about the north direction according to the equation $\ddot{\beta} + K^2\beta = 0$, where $K^2 = I\omega_0 p \cos \gamma/I_0$, and that the period of oscillation about the north direction for small values of β is $\tau = 2\pi\sqrt{I_0/(I\omega_0 p \cos \gamma)}$. (*Hint:* The components of the angular velocity $\boldsymbol{\Omega}$ of the axes in terms of the angular velocity ω_0 of the earth are

$$\Omega_x = -\omega_0 \cos \gamma \sin \beta$$
$$\Omega_y = \omega_0 \sin \gamma + \dot{\beta}$$
$$\Omega_z = \omega_0 \cos \gamma \cos \beta$$

which may be used in the y-component of the moment equations, Eqs. 7/19, to determine β as a function of time. Note that the square of the angular velocity ω_0 of the earth is small and may be neglected compared with the product $\omega_0 p$.)

Problem 7/129

Problem 7/130

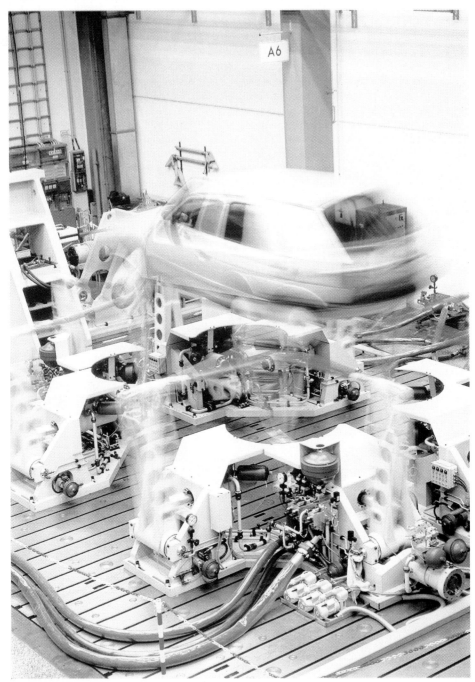

Vibration is an important subset of the discipline of dynamics. Bodies that are subjected to periodic disturbances can exhibit motion characterized by very large, even destructive, amplitudes. Shown here is an automobile undergoing shock and vibration testing by programmed hydraulic actuators. With such laboratory testing, a variety of road conditions and vehicle speeds can be simulated.

VIBRATION AND TIME RESPONSE

8

8/1 INTRODUCTION

An important and special class of problems in dynamics deals with the linear and angular motions of bodies which oscillate or respond to applied disturbances in the presence of restoring forces. A few examples of this class of dynamics problems are the response of an engineering structure to earthquakes, the vibration of an unbalanced rotating machine, the time response of the plucked string of a musical instrument, the wind-induced vibration of power lines, and the flutter of aircraft wings. In many cases, excessive vibration levels must be reduced to accommodate material limitations or human factors.

In the analysis of every engineering problem, the system under scrutiny must be represented by a physical model. It is often permissible to represent a *continuous* or *distributed-parameter system* (one in which the mass and spring elements are continuously spread over space) by a *discrete* or *lumped-parameter model* (one in which the mass and spring elements are separate and concentrated). Such a modeling scheme is especially desirable when some portions of a continuous system are relatively massive in comparison with other portions. For example, the physical model of a ship propeller shaft is often assumed to be a massless but twistable rod with a disk rigidly attached to each end—one disk representing the turbine and the other representing the propeller. As a second example, we observe that the mass of springs may often be neglected in comparison with that of attached bodies. It should be noted that not every system is reducible to a discrete model. For example, the transverse vibration of a diving board after the departure of the diver is a somewhat difficult problem of distributed-parameter vibration. In this chapter, we shall begin the study of discrete systems, limiting our discussion to those whose configurations may be described with one displacement variable. Such systems are said to possess *one degree of freedom*. For a more detailed study which includes the treatment of two or more degrees of freedom and continuous systems, the student should consult one of the many textbooks devoted solely to the subject of vibrations.

The remainder of Chapter 8 is divided into four sections: Article 8/2 treats the free vibration of particles and Art. 8/3 introduces the forced vibration of particles. Each of these two articles is subdivided into undamped- and damped-motion categories. In Art. 8/4 we discuss the vibration of rigid bodies. Finally, an energy approach to the solution of vibration problems is presented in Art. 8/5. The study of this chapter is greatly facilitated if one observes that the topic of vibrations is a direct application of the principles of kinetics as developed in Chapters 3 and 6. In particular, construction of a complete free-body diagram *drawn for an arbitrary positive value of the displacement variable*, followed by application of the appropriate governing equations of dynamics, will yield the equation of motion. From this equation of motion, which is a second-order ordinary differential equation, one may obtain all information of interest, such as the motion frequency, period, or the motion itself as a function of time.

8/2 FREE VIBRATION OF PARTICLES

When a spring-mounted body is disturbed from its equilibrium position, its ensuing motion in the absence of any imposed external forces is termed *free vibration*. In every actual case of free vibration, there exists some retarding or damping force which tends to diminish the motion. Common damping forces are those due to mechanical and fluid friction. In part (*a*) we consider the ideal case where the damping forces are small enough to be neglected. In part (*b*) we treat the case where the damping is appreciable and must be accounted for.

(a) Undamped free vibration. We begin by considering the horizontal vibration of the simple frictionless spring-mass system of Fig. 8/1*a*. Note that the variable x denotes the displacement of the mass from the equilibrium position, which, for this system, is also the position of zero spring deflection. Figure 8/1*b* shows a plot of the force F_s necessary to deflect the spring versus the corresponding spring deflection for various types of springs. Although nonlinear hard and soft springs are useful in some applications, we will restrict our attention to the linear spring. Such a spring exerts a restoring force $-kx$ on the mass—that is, when the mass is displaced to the right, the spring force is to the left, and vice versa. We must be careful to distinguish between the forces of magnitude F_s which must be applied at both ends of the massless spring to cause tension or compression and the force $F = -kx$ of equal magnitude which the spring exerts on the mass. The constant of proportionality k is known as the *spring constant, modulus,* or *stiffness* and has the units N/m.

The equation of motion for the body of Fig. 8/1*a* is obtained by first drawing its free-body diagram. Applying Newton's second law

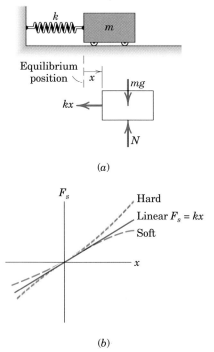

(*a*)

(*b*)

Figure 8/1

in the form $\Sigma F_x = m\ddot{x}$ gives

$$-kx = m\ddot{x} \quad \text{or} \quad m\ddot{x} + kx = 0 \qquad (8/1)$$

The oscillation of a mass subjected to a linear restoring force as described by this equation is called *simple harmonic motion* and is characterized by acceleration which is proportional to the displacement but of opposite sign. Equation 8/1 is normally written as

$$\boxed{\ddot{x} + \omega_n^2 x = 0} \qquad \textbf{(8/2)}$$

where

$$\boxed{\omega_n = \sqrt{k/m}} \qquad \textbf{(8/3)}$$

is a convenient substitution whose physical significance will be clarified shortly.

Because we anticipate an oscillatory motion, we look for a solution which gives x as a periodic function of time. Thus, a logical choice is

$$x = A \cos \omega_n t + B \sin \omega_n t \qquad (8/4)$$

or, alternatively,

$$x = C \sin (\omega_n t + \psi) \qquad (8/5)$$

Direct substitution of these expressions into Eq. 8/2 verifies that each expression is a valid solution to the equation of motion. The constants A and B, or C and ψ, are determined from knowledge of the initial displacement x_0 and initial velocity $\dot{x}_0$ of the mass. For example, if we work with the solution form of Eq. 8/4 and evaluate x and $\dot{x}$ at time $t = 0$, we obtain

$$x_0 = A \quad \text{and} \quad \dot{x}_0 = B\omega_n$$

Substitution of these values of A and B into Eq. 8/4 yields

$$x = x_0 \cos \omega_n t + \frac{\dot{x}_0}{\omega_n} \sin \omega_n t \qquad (8/6)$$

The constants C and ψ of Eq. 8/5 can be determined in terms of given initial conditions in a similar manner. Evaluation of Eq. 8/5 and its first time derivative at $t = 0$ gives

$$x_0 = C \sin \psi \quad \text{and} \quad \dot{x}_0 = C\omega_n \cos \psi$$

Solving for C and ψ yields

$$C = \sqrt{x_0^2 + (\dot{x}_0/\omega_n)^2} \qquad \psi = \tan^{-1}(x_0\omega_n/\dot{x}_0)$$

Substitution of these values into Eq. 8/5 gives

$$x = \sqrt{x_0^2 + (\dot{x}_0/\omega_n)^2} \sin [\omega_n t + \tan^{-1}(x_0\omega_n/\dot{x}_0)] \qquad (8/7)$$

Equations 8/6 and 8/7 represent two different mathematical expressions for the same time-dependent motion. We observe that $C = \sqrt{A^2 + B^2}$ and $\psi = \tan^{-1}(A/B)$.

The motion may be represented graphically, Fig. 8/2, where x is seen to be the projection onto a vertical axis of the rotating vector of length C. The vector rotates at the constant angular velocity $\omega_n = \sqrt{k/m}$, which is called the *natural circular frequency* and has the units radians per second. The number of complete cycles per

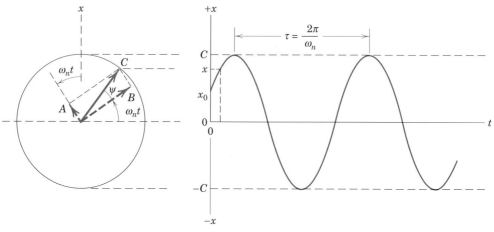

Figure 8/2

unit time is the *natural frequency* $f_n = \omega_n/2\pi$ and is expressed in hertz (1 hertz (Hz) = 1 cycle per second). The time required for one complete motion cycle (one rotation of the reference vector) is the *period* of the motion and is given by $\tau = 1/f_n = 2\pi/\omega_n$. We also see from the figure that x is the sum of the projections onto the vertical axis of two perpendicular vectors whose magnitudes are A and B and whose vector sum C is the *amplitude*. Vectors A, B, and C rotate together with the constant angular velocity ω_n. Thus, as we have already seen, $C = \sqrt{A^2 + B^2}$ and $\psi = \tan^{-1}(A/B)$.

As a further note on the free undamped vibration of particles, we see that, if the system of Fig. 8/1a is rotated 90° clockwise to obtain the system of Fig. 8/3 where the motion is vertical rather than horizontal, the equation of motion (and therefore all system properties) is unchanged if we continue to define x as the displacement from the equilibrium position. The equilibrium position now involves a nonzero spring deflection δ_{st}. From the free-body diagram of Fig. 8/3, Newton's second law gives

$$-k(\delta_{st} + x) + mg = m\ddot{x}$$

At the equilibrium position $x = 0$, the force sum must be zero, so that

$$-k\delta_{st} + mg = 0$$

Figure 8/3

Thus, we see that the pair of forces $-k\delta_{\text{st}}$ and mg on the left side of the motion equation cancel, giving

$$m\ddot{x} + kx = 0$$

which is identical to Eq. 8/1. The lesson here is that by defining the displacement variable to be zero at equilibrium rather than at the position of zero spring deflection, we may ignore the equal and opposite forces associated with equilibrium.*

(b) Damped free vibration. Every mechanical system possesses some inherent degree of friction, which acts as a consumer of mechanical energy. Precise mathematical models of the dissipative friction forces are usually complex. The dashpot or viscous damper is a device intentionally added to systems for the purpose of limiting or retarding vibration. It consists of a cylinder filled with a viscous fluid and a piston with holes or other passages by which the fluid can flow from one side of the piston to the other. Simple dashpots arranged as shown schematically in Fig. 8/4a exert a force F_d whose magnitude is proportional to the velocity of the mass, as depicted in Fig. 8/4b. The constant of proportionality c is known as the *viscous damping coefficient* and has units of $\text{N} \cdot \text{s/m}$. The direction of the damping force as applied to the mass is opposite to that of the velocity $\dot{x}$. Hence, the force on the mass is $-c\dot{x}$.

Complex dashpots with internal flow-rate-dependent one-way valves can produce different damping coefficients in extension and in compression; nonlinear characteristics are also possible. We shall restrict our attention to the simple linear dashpot.

The equation of motion for the body with damping is determined from the free-body diagram as shown in Fig. 8/4a. Newton's second law gives

$$-kx - c\dot{x} = m\ddot{x} \qquad \text{or} \qquad m\ddot{x} + c\dot{x} + kx = 0 \qquad (8/8)$$

In addition to the variable substitution $\omega_n = \sqrt{k/m}$, it is convenient, for reasons that will shortly become evident, to introduce the combination of constants

$$\zeta = c/(2m\omega_n)$$

The quantity ζ (zeta) is called the *viscous damping factor* or *damping ratio* and is a measure of the severity of the damping. The student should verify that ζ is nondimensional. Equation 8/8 may now be written as

$$\boxed{\ddot{x} + 2\zeta\omega_n\dot{x} + \omega_n^2 x = 0} \qquad (8/9)$$

(a)

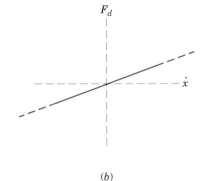

(b)

Figure 8/4

---

*For nonlinear systems, all forces, including the static forces associated with equilibrium, should be included in the analysis.

In order to solve the equation of motion, Eq. 8/9, we assume solutions of the form

$$x = Ae^{\lambda t}$$

Substitution into Eq. 8/9 yields the characteristic equation

$$\lambda^2 + 2\zeta\omega_n\lambda + \omega_n{}^2 = 0$$

whose roots are

$$\lambda_1 = \omega_n(-\zeta + \sqrt{\zeta^2 - 1}) \qquad \lambda_2 = \omega_n(-\zeta - \sqrt{\zeta^2 - 1})$$

By superposition, the general solution is

$$\begin{aligned} x &= A_1 e^{\lambda_1 t} + A_2 e^{\lambda_2 t} \\ &= A_1 e^{(-\zeta + \sqrt{\zeta^2-1})\omega_n t} + A_2 e^{(-\zeta - \sqrt{\zeta^2-1})\omega_n t} \end{aligned} \qquad (8/10)$$

Since $0 \leq \zeta \leq \infty$, the radicand $(\zeta^2 - 1)$ may be positive, negative, or even zero, giving rise to the following three categories of damped motion:

I. $\zeta > 1$ (*overdamped*). The roots λ_1 and λ_2 are distinct real negative numbers. The motion as given by Eq. 8/10 decays so that x approaches zero for large values of time t. There is no oscillation and therefore no period associated with the motion.

II. $\zeta = 1$ (*critically damped*). The roots λ_1 and λ_2 are equal real negative numbers ($\lambda_1 = \lambda_2 = -\omega_n$) and the solution to the differential equation for the special case of equal roots is given by

$$x = (A_1 + A_2 t)e^{-\omega_n t}$$

Again, the motion decays with x approaching zero for large time, and the motion is nonperiodic. A critically damped system, when excited with an initial velocity or displacement (or both), will approach equilibrium faster than will an overdamped system. Figure 8/5 depicts an actual

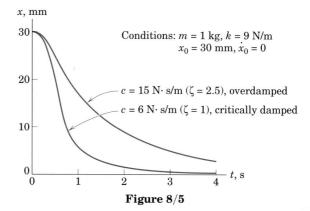

x, mm

Conditions: $m = 1$ kg, $k = 9$ N/m
$x_0 = 30$ mm, $\dot{x}_0 = 0$

$c = 15$ N·s/m ($\zeta = 2.5$), overdamped

$c = 6$ N·s/m ($\zeta = 1$), critically damped

t, s

Figure 8/5

response for both an overdamped and a critically damped system to an initial displacement x_0 with no initial velocity $\dot{x}_0$.

III. $\zeta < 1$ (*underdamped*). Noting that the radicand $(\zeta^2 - 1)$ is negative and recalling that $e^{(a+b)} = e^a e^b$ enable us to rewrite Eq. 8/10 as

$$x = \{A_1 e^{i\sqrt{1-\zeta^2}\omega_n t} + A_2 e^{-i\sqrt{1-\zeta^2}\omega_n t}\} e^{-\zeta \omega_n t}$$

where $i = \sqrt{-1}$. It is convenient to let a new variable ω_d represent the combination $\omega_n \sqrt{1 - \zeta^2}$. Thus,

$$x = \{A_1 e^{i\omega_d t} + A_2 e^{-i\omega_d t}\} e^{-\zeta \omega_n t}$$

Use of the Euler formula $e^{\pm ix} = \cos x \pm i \sin x$ allows the previous equation to be written as

$$\begin{aligned} x &= \{A_1(\cos \omega_d t + i \sin \omega_d t) \\ &\quad + A_2(\cos \omega_d t - i \sin \omega_d t)\} e^{-\zeta \omega_n t} \\ &= \{(A_1 + A_2) \cos \omega_d t + i(A_1 - A_2) \sin \omega_d t\} e^{-\zeta \omega_n t} \\ &= \{A_3 \cos \omega_d t + A_4 \sin \omega_d t\} e^{-\zeta \omega_n t} \end{aligned} \quad (8/11)$$

where $A_3 = (A_1 + A_2)$ and $A_4 = i(A_1 - A_2)$. We have shown with Eqs. 8/4 and 8/5 that the sum of two harmonics, such as those in the braces of Eq. 8/11, can be replaced by a single trigonometric function which involves a phase angle. Thus, Eq. 8/11 can be written as

$$x = \{C \sin (\omega_d t + \psi)\} e^{-\zeta \omega_n t}$$

or

$$x = Ce^{-\zeta \omega_n t} \sin (\omega_d t + \psi) \quad (8/12)$$

Equation 8/12 represents an exponentially decreasing harmonic function, as shown in Fig. 8/6 for specific numerical values. The frequency

$$\omega_d = \omega_n \sqrt{1 - \zeta^2}$$

is called the *damped natural frequency*. The *damped period* is given by $\tau_d = 2\pi/\omega_d = 2\pi/(\omega_n \sqrt{1 - \zeta^2})$.

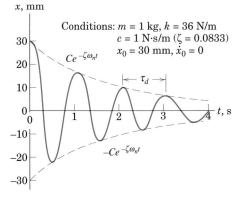

Conditions: $m = 1$ kg, $k = 36$ N/m
$c = 1$ N·s/m ($\zeta = 0.0833$)
$x_0 = 30$ mm, $\dot{x}_0 = 0$

Figure 8/6

It is important to note that the expressions developed for the constants C and ψ in terms of initial conditions for the case of no damping in part (*a*) are not valid for the case of damping in part (*b*). To find C and ψ if damping is present, one must begin anew, setting the general displacement expression of Eq. 8/12 and its first time derivative, both evaluated at time $t = 0$, equal to the initial displacement x_0 and initial velocity $\dot{x}_0$, respectively.

It is frequently desirable to experimentally determine the value of the damping ratio ζ for an underdamped system. The usual reason is that the value of the viscous damping coefficient c is not otherwise well known. The system is excited by initial conditions and a plot of the displacement x versus time t, such as that shown schematically in Fig. 8/7, is generated. Two successive amplitudes x_1 and x_2 are measured and their ratio

$$\frac{x_1}{x_2} = \frac{Ce^{-\zeta\omega_n t_1}}{Ce^{-\zeta\omega_n(t_1 + \tau_d)}} = e^{\zeta\omega_n \tau_d}$$

is formed. The *logarithmic decrement* δ is defined as

$$\delta = \ln\left(\frac{x_1}{x_2}\right) = \zeta\omega_n \tau_d = \zeta\omega_n \frac{2\pi}{\omega_n\sqrt{1 - \zeta^2}} = \frac{2\pi\zeta}{\sqrt{1 - \zeta^2}}$$

From this equation, we may solve for ζ and obtain

$$\zeta = \frac{\delta}{\sqrt{(2\pi)^2 + \delta^2}}$$

For a small damping ratio, $x_1 \cong x_2$ and $\delta \ll 1$, so that $\zeta \cong \delta/2\pi$. If x_1 and x_2 are so close in value that experimental distinction between them is impractical, the above analysis may be modified by using two observed amplitudes that are n cycles apart.

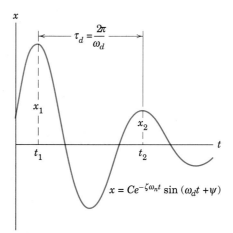

Figure 8/7

Sample Problem 8/1

A 10-kg body is suspended from a spring of constant $k = 2.5$ kN/m. At time $t = 0$, it has a downward velocity of 0.5 m/s as it passes through the position of static equilibrium. Determine

  (a) the static spring deflection δ_{st}
  (b) the natural frequency of the system in both rad/sec (ω_n) and cycles/sec (f_n)
  (c) the system period τ
  (d) the displacement x as a function of time, where x is measured from the position of static equilibrium
  (e) the maximum velocity v_{max} attained by the mass
  (f) the maximum acceleration a_{max} attained by the mass.

$k = 2.5$ kN/m

$m = 10$ kg

Solution. (a) From the spring relationship $F_s = kx$, we see that at equilibrium,

(1)
$$mg = k\delta_{st} \qquad \delta_{st} = \frac{mg}{k} = \frac{10(9.81)}{2500} = 0.0392 \text{ m or } 39.2 \text{ mm} \quad Ans.$$

(b)
$$\omega_n = \sqrt{\frac{k}{m}} = \sqrt{\frac{2500}{10}} = 15.81 \text{ rad/s} \qquad Ans.$$

$$f_n = (15.81)\left(\frac{1}{2\pi}\right) = 2.52 \text{ cycles/s (Hz)} \qquad Ans.$$

(c)
$$\tau = \frac{1}{f_n} = \frac{1}{2.52} = 0.397 \text{ s} \qquad Ans.$$

(2) (d) From Eq. 8/6:

$$x = x_0 \cos \omega_n t + \frac{\dot{x}_0}{\omega_n} \sin \omega_n t$$

$$= (0) \cos 15.81t + \frac{0.5}{15.81} \sin 15.81t$$

$$= 0.0316 \sin 15.81t \text{ m } \text{ or } 31.6 \sin 15.81t \text{ mm} \qquad Ans.$$

As an exercise, let us determine x from the alternative Eq. (8/7):

$$x = \sqrt{x_0{}^2 + (\dot{x}_0/\omega_n)^2} \, \sin\left[\omega_n t + \tan^{-1}\left(\frac{x_0\omega_n}{\dot{x}_0}\right)\right]$$

$$= \sqrt{0^2 + \left(\frac{0.5}{15.81}\right)^2} \, \sin\left\{15.81t + \tan^{-1}\left[\frac{(0)(15.81)}{0.5}\right]\right\}$$

$$= 0.0316 \sin 15.81t$$

(e) The velocity is $\dot{x} = 15.81(0.0316) \cos 15.81t = 0.5 \cos 15.81t$. Since the cosine function cannot be greater than 1 or less than -1, the maximum velocity v_{max} is 0.5 m/s, which, in this case, is the initial velocity. $\qquad Ans.$

(f) The acceleration is

$$\ddot{x} = -15.81(0.5) \sin 15.81t = -7.91 \sin 15.81t$$

The maximum acceleration a_{max} is 7.91 m/s^2. $\qquad Ans.$

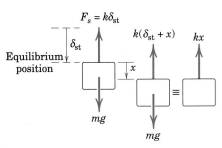

$F_s = k\delta_{st}$

$k(\delta_{st} + x)$ kx

Equilibrium position

δ_{st}

x

mg

mg

(1) The student should always exercise extreme caution in the matter of units. In the subject of vibrations, it is quite easy to commit errors due to mixing of feet and inches, cycles and radians, and other pairs that frequently enter the calculations.

(2) Recall that when we refer the motion to the position of static equilibrium, the equation of motion, and therefore its solution, for the present system is identical to that for the horizontally vibrating system.

Sample Problem 8/2

The 8-kg body is moved 0.2 m to the right of the equilibrium position and released from rest at time $t = 0$. Determine its displacement at time $t = 2$ s. The viscous damping coefficient c is 20 N·s/m, and the spring stiffness k is 32 N/m.

Solution. We must first determine whether the system is under-damped, critically damped, or overdamped. For that purpose, we compute the damping ratio ζ.

$$\omega_n = \sqrt{k/m} = \sqrt{32/8} = 2 \text{ rad/s} \qquad \zeta = \frac{c}{2m\omega_n} = \frac{20}{2(8)(2)} = 0.625$$

Since $\zeta < 1$, the system is underdamped. The damped natural frequency is $\omega_d = \omega_n\sqrt{1 - \zeta^2} = 2\sqrt{1 - (0.625)^2} = 1.561$ rad/s. The motion is given by Eq. 8/12 and is

$$x = Ce^{-\zeta\omega_n t} \sin(\omega_d t + \psi) = Ce^{-1.25t} \sin(1.561t + \psi)$$

The velocity is then

$$\dot{x} = -1.25Ce^{-1.25t} \sin(1.561t + \psi) + 1.561Ce^{-1.25t} \cos(1.561t + \psi)$$

Evaluating the displacement and velocity at time $t = 0$ gives

$$x_0 = C \sin \psi = 0.2 \qquad \dot{x}_0 = -1.25C \sin \psi + 1.561C \cos \psi = 0$$

Solving the two equations for C and ψ yields $C = 0.256$ m and $\psi = 0.896$ rad. Therefore, the displacement in meters is

$$x = 0.256e^{-1.25t} \sin(1.561t + 0.896)$$

① Evaluation for time $t = 2$ s gives $x_2 = -0.01616$ m. *Ans.*

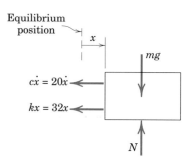

① We note that the exponential factor $e^{-1.25t}$ is 0.0821 at $t = 2$ s. Hence, $\zeta = 0.625$ represents severe damping, although the motion is still oscillatory.

Sample Problem 8/3

The two fixed counterrotating pulleys are driven at the same angular speed ω_0. A round bar is placed off center on the pulleys as shown. Determine the natural frequency of the resulting bar motion. The coefficient of kinetic friction between the bar and pulleys is μ_k.

Solution. The free-body diagram of the bar is constructed for an arbitrary displacement x from the central position as shown. The governing equations are

$[\Sigma F_x = m\ddot{x}] \qquad \mu_k N_A - \mu_k N_B = m\ddot{x}$

$[\Sigma F_y = 0] \qquad N_A + N_B - mg = 0$

① $[\Sigma M_A = 0] \qquad aN_B - \left(\dfrac{a}{2} + x\right)mg = 0$

Eliminating N_A and N_B from the first equation yields

② $$\ddot{x} + \frac{2\mu_k g}{a}x = 0$$

We recognize the form of this equation as that of Eq. 8/2, so that the natural frequency in radians per second is $\omega_n = \sqrt{2\mu_k g/a}$ and the natural frequency in cycles per second is

$$f_n = \frac{1}{2\pi}\sqrt{2\mu_k g/a} \qquad \text{Ans.}$$

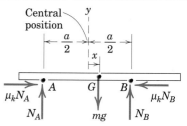

① Since the bar is slender and does not rotate, the use of a moment equilibrium equation is justified.

② We note that the angular speed ω_0 does not enter the equation of motion. The reason for this is our assumption that the kinetic friction force does not depend on the relative velocity at the contacting surface.

PROBLEMS

(Unless otherwise indicated, assume that all motion variables are referenced to the equilibrium position.)

Introductory problems—
undamped, free vibrations

8/1   When a 3-kg collar is placed upon the pan which is attached to the spring of unknown constant, the additional static deflection of the pan is observed to be 42 mm. Determine the spring constant k in N/m.

                   Ans. $k = 701$ N/m

Problem 8/1

8/2   Determine the natural frequency of the spring-mass system in both radians per second and cycles per second (Hz).

Problem 8/2

8/3   For the system of Prob. 8/2, determine the position x of the mass as a function of time if the mass is released from rest at time $t = 0$ from a position 50 mm to the left of the equilibrium position. Determine the maximum velocity and maximum acceleration of the mass over one cycle of motion.

$$Ans. \ x = -50 \cos 18t \text{ mm}$$
$$v_{max} = 0.9 \text{ m/s}$$
$$a_{max} = 16.2 \text{ m/s}^2$$

8/4   For the system of Prob. 8/2, determine the position x as a function of time if the mass is released at time $t = 0$ from a position 50 mm to the right of the equilibrium position with an initial velocity of 225 mm/s to the left. Determine the amplitude C and period τ of the motion.

8/5   For the spring-mass system shown, determine the static deflection δ_{st}, the system period τ, and the maximum velocity v_{max} which result if the cylinder is displaced 100 mm downward from its equilibrium position and released.

$$Ans. \ \delta_{st} = 0.200 \text{ m}, \ \tau = 0.898 \text{ s}$$
$$v_{max} = 0.7 \text{ m/s}$$

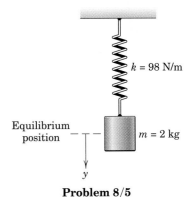

Problem 8/5

8/6   The cylinder of the system of Prob. 8/5 is displaced 100 mm downward from its equilibrium position and released at time $t = 0$. Determine the position y, velocity v, and acceleration a when $t = 3$ s. What is the maximum acceleration?

8/7 The vertical plunger has a mass of 2.5 kg and is supported by the two springs which are always in compression. Calculate the natural frequency f_n of vibration of the plunger if it is deflected from the equilibrium position and released from rest. Friction in the guide is negligible.

Ans. $f_n = 7.40$ Hz

$k_1 = 3.6$ kN/m

Fixed

$k_2 = 1.8$ kN/m

2.5 kg

Problem 8/7

8/8 In the equilibrium position, the 30-kg cylinder causes a static deflection of 50 mm in the coiled spring. If the cylinder is depressed an additional 25 mm and released from rest, calculate the resulting natural frequency f_n of vertical vibration of the cylinder in cycles per second (Hz).

30 kg

Problem 8/8

8/9 For the cylinder of Prob. 8/8, determine the vertical displacement x, measured positive down in millimeters from the equilibrium position, in terms of the time t in seconds measured from the instant of release from the position of 25 mm added deflection.

Ans. $x = 25 \cos 14.01t$ mm

Representative problems—
undamped, free vibrations

8/10 Calculate the natural frequency f_n of vertical oscillation of the spring-loaded cylinder when it is set into motion. Both springs are in tension at all times.

$k = 3000$ N/m

10 kg

$k = 3000$ N/m

Problem 8/10

8/11 The 0.1-kg slider oscillates in the fixed slot under the action of the three springs, each of stiffness $k = 90$ N/m. If the initial conditions at time $t = 0$ are $x_0 = 3$ mm and $\dot{x}_0 = 12$ mm/s, determine the position and velocity of the slider at time $t = 2$ s. What is the system period?

Ans. $x = -2.96$ mm, $\dot{x} = 27.0$ mm/s
$\tau = 0.1209$ s

x

k

m

k

k

Problem 8/11

8/12 An old car being moved by a magnetic crane pickup is dropped from a short distance above the ground. Neglect any damping effects of its worn-out shock absorbers and calculate the natural frequency f_n in cycles per second (Hz) of the vertical vibration which occurs after impact with the ground. Each of the four springs on the 1000-kg car has a constant of 17.5 kN/m. Because the center of mass is located midway between the axles and the car is level when dropped, there is no rotational motion. State any assumptions.

Electromagnet

Problem 8/12

8/13 During the design of the spring-support system for the 4000-kg weighing platform, it is decided that the frequency of free vertical vibration in the unloaded condition shall not exceed 3 cycles per second. (*a*) Determine the maximum acceptable spring constant k for each of the three identical springs. (*b*) For this spring constant, what would be the natural frequency f_n of vertical vibration of the platform loaded by the 40-Mg truck?

    Ans. (*a*) $k = 474$ kN/m, (*b*) $f_n = 0.905$ Hz

Problem 8/13

8/14 Replace the springs in each of the two cases shown by a single spring of stiffness k (equivalent spring stiffness) which will cause each mass to vibrate with its original frequency.

(*a*)   (*b*)

Problem 8/14

8/15 With the assumption of no slipping, determine the mass m of the block which must be placed on the top of the 6-kg cart in order that the system period be 0.75 s. What is the minimum coefficient of static friction μ_s for which the block will not slip relative to the cart if the cart is displaced 50 mm from the equilibrium position and released?

    Ans. $m = 2.55$ kg, $\mu_s = 0.358$

Problem 8/15

8/16 An energy-absorbing car bumper with its springs initially undeformed has an equivalent spring constant of 525 kN/m. If the 1200-kg car approaches a massive wall with a speed of 8 km/h, determine (*a*) the velocity *v* of the car as a function of time during contact with the wall, where $t = 0$ is the beginning of the impact, and (*b*) the maximum deflection x_{max} of the bumper.

Problem 8/16

8/17 A 55-kg woman stands in the center of an end-supported board and causes a midspan deflection of 22 mm. If she flexes her knees slightly in order to cause a vertical vibration, what is the frequency f_n of the motion? Assume elastic response of the board and neglect its relatively small mass.

Ans. $f_n = 3.36$ Hz

22 mm

Problem 8/17

8/18 Prove that the natural frequency f_n of oscillation for the mass *m* is independent of θ.

Problem 8/18

8/19 If both springs are unstretched when the mass is in the central position shown, determine the static deflection δ_{st} of the mass. What is the period of oscillatory motion about the position of static equilibrium?

$$Ans. \quad \delta_{st} = \frac{mg \sin \theta}{3k}, \quad \tau = 2\pi \sqrt{\frac{m}{3k}}$$

Problem 8/19

8/20 A small particle of mass *m* is attached to two highly tensioned wires as shown. Determine the system natural frequency ω_n for small vertical oscillations if the tension *T* in both wires is assumed to be constant. Is the calculation of the small static deflection of the particle necessary?

Problem 8/20

8/21 The cylindrical buoy floats in salt water (density 1030 kg/m³) and has a mass of 800 kg with a low center of mass to keep it stable in the upright position. Determine the frequency f_n of vertical oscillation of the buoy. Assume the water level remains undisturbed adjacent to the buoy.

Ans. $f_n = 0.301$ Hz

0.6 m

Problem 8/21

8/22 Shown in the figure is a model of a one-story building. The bar of mass m is supported by two light elastic upright columns whose upper and lower ends are fixed against rotation. For each column, if a force P and corresponding moment M were applied as shown in the right-hand part of the figure, the deflection δ would be given by $\delta = PL^3/12EI$, where L is the effective column length, E is Young's modulus, and I is the area moment of inertia of the column cross section with respect to its neutral axis. Determine the natural frequency of horizontal oscillation of the bar when the columns bend as shown in the figure.

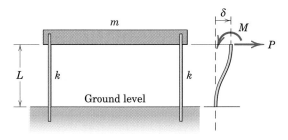

Problem 8/22

8/23 A 3-kg piece of putty is dropped 2 m onto the initially stationary 28-kg block, which is supported by four springs, each of which has a constant $k = 800$ N/m. Determine the displacement x as a function of time during the resulting vibration, where x is measured from the initial position of the block as shown.

Ans.
$$x = 9.20(1 - \cos 10.16t) + 59.7 \sin 10.16t \text{ mm}$$

Problem 8/23

8/24 Calculate the frequency f_n of vertical oscillation of the 25-kg block when it is set in motion. Each spring has a stiffness of 1200 N/m. Neglect the mass of the pulleys.

Problem 8/24

8/25 The weighing platform has a mass m and is connected to the spring of stiffness k by the system of levers shown. Derive the differential equation for small vertical oscillations of the platform and find the period τ. Designate y as the platform displacement from the equilibrium position and neglect the mass of the levers.

$$\text{Ans. } \ddot{y} + \frac{k}{m}\left(\frac{c}{b}\right)^4 y = 0, \ \tau = 2\pi \left(\frac{b}{c}\right)^2 \sqrt{\frac{m}{k}}$$

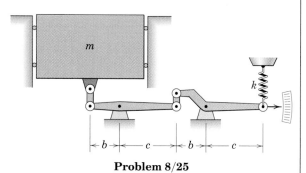

Problem 8/25

8/26 Calculate the natural frequency ω_n of the system shown in the figure. The mass and friction of the pulleys are negligible.

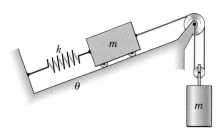

Problem 8/26

8/27 Calculate the natural circular frequency ω_n of the system shown in the figure. The mass and friction of the pulleys are negligible.

$$\text{Ans. } \omega_n = \sqrt{\frac{k}{5m}}$$

Problem 8/27

8/28 Derive the differential equation of motion for the system shown in terms of the variable x_1. The mass of the linkage is negligible. State the natural frequency ω_n' in rad/s for the case $k_1 = k_2 = k$ and $m_1 = m_2 = m$. Assume small oscillations throughout.

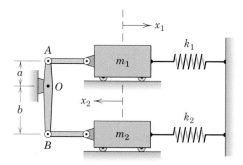

Problem 8/28

Introductory problems—
damped, free vibrations

8/29 Determine the value of the damping ratio ζ for the simple spring-mass-dashpot system shown.

$$\text{Ans. } \zeta = 0.460$$

Problem 8/29

8/30 Determine the value of the viscous damping co-efficient c for which the system shown is critically damped.

30 kN/m

35 kg

c

Problem 8/30

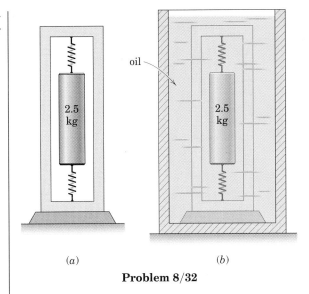

oil

2.5 kg

2.5 kg

(a)

(b)

Problem 8/32

8/31 The 3.6-kg body of Prob. 8/29 is released from rest a distance x_0 to the right of the equilibrium position. Determine the displacement x as a function of time t, where $t = 0$ is the time of release.

Ans. $x = x_0(\cos 10.72t + 0.518 \sin 10.72t)e^{-5.56t}$

8/32 The 2.5-kg spring-supported cylinder is set into free vertical vibration and is observed to have a period of 0.75 s in part (a) of the figure. The system is then completely immersed in an oil bath in part (b) of the figure, and the cylinder is displaced from its equilibrium position and released. Viscous damping ensues, and the ratio of two successive positive-displacement amplitudes is 4. Calculate the viscous damping ratio ζ, the viscous damping constant c, and the equivalent spring constant k.

8/33 A linear harmonic oscillator having a mass of 1.10 kg is set into motion with viscous damping. If the frequency is 10 Hz and if two successive amplitudes a full cycle apart are measured to be 4.65 mm and 4.30 mm as shown, compute the viscous damping coefficient c.

Ans. $c = 1.721$ N·s/m

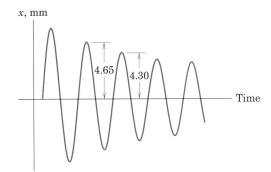

x, mm

4.65

4.30

Time

Problem 8/33

Representative problems—
damped, free vibrations

8/34 The figure represents the measured displacement-time relationship for a vibration with small damping where it is impractical to achieve accurate results by measuring the nearly equal amplitudes of two successive cycles. Modify the expression for the viscous damping factor ζ based on the measured amplitudes x_0 and x_N which are N cycles apart.

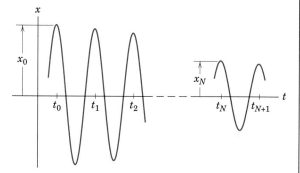

Problem 8/34

8/35 Determine the value of the viscous damping coefficient c for which the system has a damping ratio of (a) 0.5 and (b) 1.5.

Ans. (a) $c = 114.0$ N·s/m, (b) $c = 342$ N·s/m

Problem 8/35

8/36 Further design refinement for the weighing platform of Prob. 8/13 is shown here where two viscous dampers are to be added to limit the ratio of successive positive amplitudes of vertical vibration in the unloaded condition to 4. Determine the necessary viscous damping coefficient c for each of the dampers.

Problem 8/36

8/37 The system shown is released from rest from an initial position x_0. Determine the overshoot displacement x_1. Assume translational motion in the x-direction.

Ans. $x_1 = -0.1630x_0$

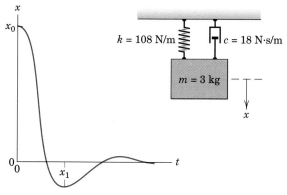

Problem 8/37

8/38 The mass of the system shown is released from rest at $x_0 = 150$ mm when $t = 0$. Determine the displacement x at $t = 0.5$ s if (a) $c = 200$ N·s/m and (b) $c = 300$ N·s/m.

Problem 8/38

8/39 The 2-kg mass is released from rest at a distance x_0 to the right of the equilibrium position. Determine the displacement x as a function of time.

Ans. $x = x_0[1.171e^{-2.67t} - 0.1708e^{-18.33t}]$

Problem 8/39

8/40 For the damped spring-mass system shown, determine the viscous damping coefficient for which critical damping will occur.

Problem 8/40

8/41 The cannon fires a 4.5-kg cannonball with an absolute velocity of 250 m/s at 20° to the horizontal. The combined mass of the cannon and its cart is 750 kg. If the recoil mechanism consists of the spring of constant $k = 27$ kN/m and the damper with viscous coefficient $c = 9000$ N·s/m, determine the maximum recoil deflection x_{max} of the cannon unit.

Ans. $x_{max} = 86.4$ mm

Problem 8/41

8/42 The owner of a 1600-kg pickup truck tests the action of his rear-wheel shock absorbers by applying a steady 450-N force to the rear bumper and measuring a static deflection of 75 mm. Upon sudden release of the force, the bumper rises and then falls to a maximum of 12 mm below the unloaded equilibrium position of the bumper on the first rebound. Treat the action as a one-dimensional problem with an equivalent mass of half the car mass. Find the viscous damping factor ζ for the rear end and the viscous damping coefficient c for each shock absorber assuming its action to be vertical.

Problem 8/42

8/43 Determine the damping ratio ζ of the system depicted in the figure. The mass and friction of the pulleys are negligible, and the cable remains taut at all times.

$$Ans. \ \zeta = \frac{c}{4\sqrt{k(m_1 + 4m_2)}}$$

Problem 8/43

8/44 Derive the differential equation of motion for the system shown in its equilibrium position. Neglect the mass of link AB and assume small oscillations.

Problem 8/44

8/45 Develop the equation of motion in terms of the variable x for the system shown. Determine an expression for the damping ratio ζ in terms of the given system properties. Neglect the mass of the crank AB and assume small oscillations about the equilibrium position shown.

$$Ans. \ \ddot{x} + \frac{b^2}{a^2}\frac{c}{m}\dot{x} + \frac{k}{m}x = 0$$

$$\zeta = \frac{1}{2}\frac{b^2}{a^2}\frac{c}{\sqrt{km}}$$

Problem 8/45

►**8/46** Investigate the case of Coulomb damping for the block shown, where the coefficient of kinetic friction is μ_k and each spring has a stiffness $k/2$. The block is displaced a distance x_0 from the neutral position and released. Determine and solve the differential equation of motion. Plot the resulting vibration and indicate the rate r of decay of the amplitude with time.

$$Ans. \ r = \frac{2\mu_k g}{\pi}\sqrt{\frac{m}{k}}$$

Problem 8/46

8/3 *FORCED VIBRATION OF PARTICLES*

Although there are many significant applications of free vibrations, the most important class of vibration problems is that where
the motion is continuously excited by a disturbing force. The force
may be externally applied or may be generated within the system
by such means as unbalanced rotating parts. Forced vibrations may
also be excited by the motion of the system foundation.

Various forms of forcing functions $F = F(t)$ and foundation
displacements $x_B = x_B(t)$ are depicted in Fig. 8/8. The harmonic

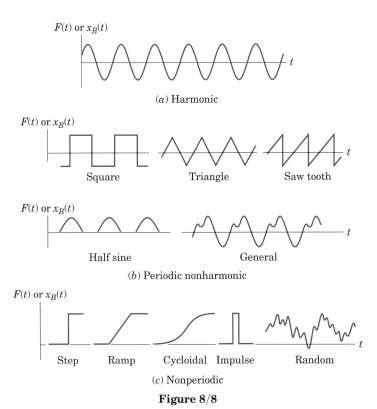

$F(t)$ or $x_B(t)$

(*a*) Harmonic

$F(t)$ or $x_B(t)$

Square   Triangle   Saw tooth

$F(t)$ or $x_B(t)$

Half sine   General

(*b*) Periodic nonharmonic

$F(t)$ or $x_B(t)$

Step   Ramp   Cycloidal   Impulse   Random

(*c*) Nonperiodic

Figure 8/8

force shown in the *a*-part of the figure occurs frequently in engineering practice, and the understanding of the analysis associated
with harmonic forces is a necessary first step in the study of more
complex forms. For this reason, we will focus our attention on harmonic excitation.

We first consider the system of Fig. 8/9*a*, where the body is
subjected to the external harmonic force $F = F_0 \sin \omega t$, in which F_0
is the force amplitude and ω is the driving frequency (in radians per

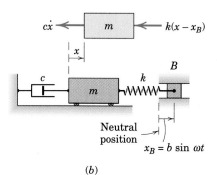

(a)

(b)

Figure 8/9

second). The student should take care to distinguish between $\omega_n = \sqrt{k/m}$, which is a property of the system, and ω, which is a property of the force applied to the system. We also note that for a force $F = F_0 \cos \omega t$, one merely substitutes cos ωt for sin ωt in the results about to be developed. From the free-body diagram of Fig. 8/9a, we may apply Newton's second law to obtain

$$-kx - c\dot{x} + F_0 \sin \omega t = m\ddot{x}$$

In standard form, with the same variable substitutions made in Art. 8/2, the equation of motion becomes

$$\ddot{x} + 2\zeta\omega_n\dot{x} + \omega_n^2 x = \frac{F_0 \sin \omega t}{m} \qquad (8/13)$$

In many cases, the excitation of the mass is due not to a directly applied force but to the movement of the base or foundation to which the mass is connected by springs or other compliant mountings. Examples of such applications are seismographs, vehicle suspensions, and structures shaken by earthquakes. Harmonic movement of the base is equivalent to the direct application of a harmonic force. To show this, consider the system of Fig. 8/9b where the spring is attached to the movable base. The free-body diagram shows the mass displaced a distance x from the neutral or equilibrium position it would have if the base were in its neutral position. The base, in turn, is assumed to have a harmonic movement $x_B = b \sin \omega t$. Note that the spring deflection is the difference between the inertial displacements of the mass and the base. From the free-body diagram, Newton's second law gives

$$-k(x - x_B) - c\dot{x} = m\ddot{x}$$

or
$$\ddot{x} + 2\zeta\omega_n\dot{x} + \omega_n^2 x = \frac{kb \sin \omega t}{m} \qquad (8/14)$$

We see immediately that Eq. 8/14 is exactly the same as our basic equation of motion, Eq. 8/13, in that F_0 is replaced by kb. Consequently, all the results about to be developed apply to either Eq. 8/13 or 8/14.

(a) Undamped forced vibration. First, we treat the case where damping is negligible ($c = 0$). Our basic equation of motion, Eq. 8/13, becomes

$$\ddot{x} + \omega_n^2 x = \frac{F_0}{m} \sin \omega t \qquad (8/15)$$

The complete solution to Eq. 8/15 is a sum of the complementary solution x_c, which is the general solution of Eq. 8/15 with the right side set to zero, and the particular solution x_p, which is *any* solution to the complete equation. Thus, $x = x_c + x_p$. We developed the complementary solution in section (*a*) of Art. 8/2. A particular solution is investigated by assuming that the response to the force should resemble the form of the force term. To that end, we assume

$$x_p = X \sin \omega t \qquad (8/16)$$

where X is the amplitude (in units of length) of the particular solution. Substituting this expression into Eq. 8/15 and solving for X yield

$$X = \frac{F_0/k}{1 - (\omega/\omega_n)^2} \qquad (8/17)$$

Thus, the particular solution becomes

$$x_p = \frac{F_0/k}{1 - (\omega/\omega_n)^2} \sin \omega t \qquad (8/18)$$

The complementary solution, known as the *transient solution*, is of no special interest here since, with time, it dies out with the small amount of damping which can never be completely eliminated. The particular solution x_p describes the continuing motion and is called the *steady-state solution*. Its period is $\tau = 2\pi/\omega$, the same as that of the forcing function. Of primary interest is the amplitude X of the motion. If we let δ_{st} stand for the magnitude of the static deflection of the mass under a static load F_0, then $\delta_{st} = F_0/k$, and we may form the ratio

$$M = \frac{X}{\delta_{st}} = \frac{1}{1 - (\omega/\omega_n)^2} \qquad (8/19)$$

The ratio M is known as the *amplitude ratio* or *magnification factor* and is a measure of the severity of the vibration. We especially note that M *approaches infinity* as ω approaches ω_n. Consequently, if the system possesses no damping and is excited by a harmonic force whose frequency ω approaches the natural frequency ω_n of the system, then M, and hence X, increases without limit. Physically, this means that the motion amplitude would reach the limits of the attached spring, which is a condition to be avoided. The value ω_n is known as the *resonant* or *critical frequency* of the system, and the condition of ω being close in value to ω_n with the resulting large displacement amplitude X is known as *resonance*. For $\omega < \omega_n$, the magnification factor M is positive, and the vibration is in phase with the force F. For $\omega > \omega_n$, the magnification factor is negative, and the vibration is 180° out of phase with F. Figure 8/10 shows a plot of the absolute value of M as a function of the driving-frequency ratio ω/ω_n.

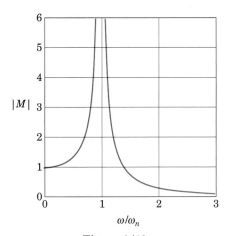

Figure 8/10

(b) Damped forced vibration. We now reintroduce damping in our expressions for forced vibration. Our basic differential equation of motion is

$$\ddot{x} + 2\zeta\omega_n\dot{x} + \omega_n^2 x = \frac{F_0 \sin \omega t}{m} \qquad [8/13]$$

As in section (*a*), the complete solution is the sum of the complementary solution x_c, which is the general solution of Eq. 8/13 with the right side equal to zero, and the particular solution x_p, which is *any* solution to the complete equation. We have already developed the complementary solution x_c in section (*b*) of Art. 8/2. By virtue of the influence of damping, we find that a single sine or cosine expression, such as we were able to use in section (*a*), is not sufficiently general for the particular solution in the case of damping. So we try

$$x_p = X_1 \cos \omega t + X_2 \sin \omega t \qquad \text{or} \qquad x_p = X \sin (\omega t - \phi)$$

Substitution of the latter expression into Eq. 8/13, matching coefficients of sin ωt and cos ωt, and solving the resulting two equations give

$$X = \frac{F_0/k}{\{[1 - (\omega/\omega_n)^2]^2 + [2\zeta\omega/\omega_n]^2\}^{1/2}} \qquad (8/20)$$

$$\phi = \tan^{-1}\left[\frac{2\zeta\omega/\omega_n}{1 - (\omega/\omega_n)^2}\right] \qquad (8/21)$$

The complete solution is now known, and for underdamped systems it can be written as

$$x = Ce^{-\zeta\omega_n t} \sin (\omega_d t + \psi) + X \sin (\omega t - \phi) \qquad (8/22)$$

Because the first term on the right side diminishes with time, it is known as the *transient solution*. The particular solution x_p is known as the *steady-state solution* and is the part of the solution in which we are primarily interested. All quantities on the right side of Eq. 8/22 are properties of the system and the applied force, except for C and ψ (which are determinable from initial conditions) and the running time variable t.

Near resonance the magnitude X of the steady-state solution is a strong function of the damping ratio ζ and the nondimensional frequency ratio ω/ω_n. It is again convenient to form the nondimensional ratio $M = X/(F_0/k)$, which is known as the *amplitude ratio* or *magnification factor*

$$M = \frac{1}{\{[1 - (\omega/\omega_n)^2]^2 + [2\zeta\omega/\omega_n]^2\}^{1/2}} \qquad (8/23)$$

An accurate plot of the magnification factor M versus the frequency ratio ω/ω_n for various values of the damping ratio ζ is shown

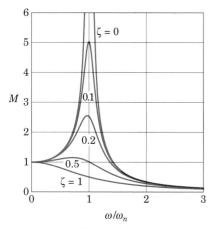

Figure 8/11

in Fig. 8/11. This figure reveals the most essential information pertinent to the forced vibration of a single-degree-of-freedom system under harmonic excitation. It is clear from the graph that, if a motion amplitude is excessive, two possible remedies would be to (a) increase the damping (to obtain a larger value of ζ) or (b) alter the driving frequency so that ω is farther from the resonant frequency ω_n. The addition of damping is most effective near resonance. Figure 8/11 also shows that, except for $\zeta = 0$, the magnification-factor curves do not actually peak at $\omega/\omega_n = 1$. The peak for any given value of ζ can be calculated by finding the maximum value of M from Eq. 8/23.

The phase angle ϕ, given by Eq. 8/21, can vary from 0 to π and represents the part of a cycle (and hence the time) by which the response x_p lags the forcing function F. Figure 8/12 shows how the phase angle ϕ varies with the frequency ratio for various values of the damping ratio ζ. Note that the value of ϕ when $\omega/\omega_n = 1$ is 90° for all values of ζ. To further illustrate the phase difference between the response and the forcing function, we show in Fig. 8/13 two examples of the variation of F and x_p with ωt. In the first example, $\omega < \omega_n$ and ϕ is taken to be $\pi/4$. In the second example, $\omega > \omega_n$ and ϕ is taken to be $3\pi/4$.

As a frequently encountered application of harmonic excitation, we consider vibration-measuring instruments such as seismometers and accelerometers. The elements of this class of instruments are

Figure 8/12

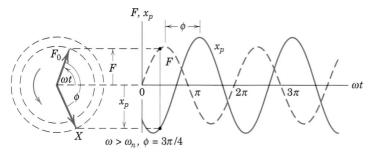

Figure 8/13

shown in Fig. 8/14a. We note that the entire system is subjected to the motion x_B of the frame. Letting x denote the position of the mass *relative* to the frame, we may apply Newton's second law and obtain

$$-c\dot{x} - kx = m\frac{d^2}{dt^2}(x + x_B) \qquad \text{or} \qquad \ddot{x} + \frac{c}{m}\dot{x} + \frac{k}{m}x = -\ddot{x}_B$$

where $(x + x_B)$ is the inertial displacement of the mass. If $x_B = b \sin \omega t$, then our equation of motion with the usual notation is

$$\ddot{x} + 2\zeta\omega_n\dot{x} + \omega_n^2 x = b\omega^2 \sin \omega t$$

which is the same as Eq. 8/13 if $b\omega^2$ is substituted for F_0/m. Again, we are interested only in the steady-state solution x_p. Thus, from Eq. 8/20, we have

$$x_p = \frac{b(\omega/\omega_n)^2}{\{[1 - (\omega/\omega_n)^2]^2 + [2\zeta\omega/\omega_n]^2\}^{1/2}} \sin (\omega t - \phi)$$

If X represents the amplitude of the relative response x_p, then the nondimensional ratio X/b is

$$X/b = (\omega/\omega_n)^2 M$$

where M is the magnification ratio of Eq. 8/23. A plot of X/b as a function of the driving-frequency ratio ω/ω_n is shown in the b-part of Fig. 8/14. The similarities and differences between the magnification ratios of Figs. 8/14(b) and 8/11 should be noted.

If the frequency ratio ω/ω_n is large, then $X/b \cong 1$ for all values of the damping ratio ζ. Under these conditions, the displacement of the mass relative to the frame is approximately the same as the absolute displacement of the frame, and the instrument acts as a *displacement meter*. To obtain a high value of ω/ω_n, we need a small

(a)

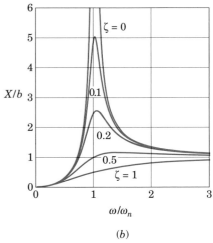

(b)

Figure 8/14

value of $\omega_n = \sqrt{k/m}$, which means a soft spring and a large mass. With such a combination, the mass will tend to stay inertially fixed. Displacement meters generally have very light damping.

On the other hand, if the frequency ratio ω/ω_n is small, then M approaches unity (see Fig. 8/11) and $X/b \cong (\omega/\omega_n)^2$ or $X \cong b(\omega/\omega_n)^2$. But $b\omega^2$ is the maximum acceleration of the frame. Hence, X is proportional to the maximum acceleration of the frame, and the instrument may be used as an *accelerometer*. The damping ratio is generally selected so that M approximates unity over the widest possible range of ω/ω_n. From Fig. 8/11, we see that a damping factor somewhere between $\zeta = 0.5$ and $\zeta = 1$ would meet this criterion.

An important analogy exists between electric circuits and mechanical spring-mass systems. Figure 8/15 shows a series circuit consisting of a voltage E which is a function of time, an inductance L, a capacitance C, and a resistance R. If we denote the charge by the symbol q, the equation which governs the charge is

$$L\ddot{q} + R\dot{q} + \frac{1}{C}q = E \qquad (8/24)$$

Figure 8/15

This equation has the same form as the equation for the mechanical system. Thus, by a simple interchange of symbols, the behavior of the electrical circuit may be used to predict the behavior of the mechanical system, or vice versa. The mechanical and electrical equivalents in the following table are worth noting:

MECHANICAL–ELECTRICAL EQUIVALENTS

MECHANICAL			ELECTRICAL		
QUANTITY	SYMBOL	SI UNIT	QUANTITY	SYMBOL	SI UNIT
Mass	m	kg	Inductance	L	H	henry
Spring stiffness	k	N/m	1/Capacitance	$1/C$	$1/F$	1/farad
Force	F	N	Voltage	E	V	volt
Velocity	$\dot{x}$	m/s	Current	I	A	ampere
Displacement	x	m	Charge	q	C	coulomb
Viscous damping constant	c	N·s/m	Resistance	R	Ω	ohm

Sample Problem 8/4

A 50-kg instrument is supported by four springs, each of stiffness 7500 N/m. If the instrument foundation undergoes harmonic motion given in meters by $x_B = 0.002 \cos 50t$, determine the amplitude of the steady-state motion of the instrument. Damping is negligible.

x_B

Solution. For harmonic oscillation of the base, we substitute kb for F_0 in our particular-solution results, so that, from Eq. 8/17, the steady-state amplitude becomes

①
$$X = \frac{b}{1 - (\omega/\omega_n)^2}$$

The resonant frequency is $\omega_n = \sqrt{k/m} = \sqrt{4(7500)/50} = 24.5$ rad/s, and the impressed frequency $\omega = 50$ rad/s is given. Thus,

②
$$X = \frac{0.002}{1 - (50/24.5)^2} = -6.32(10^{-4}) \text{ m} \quad \text{or} \quad -0.632 \text{ mm} \quad Ans.$$

Note that the frequency ratio ω/ω_n is approximately 2, so that the condition of resonance is avoided.

① Note that either $\sin 50t$ or $\cos 50t$ can be used for the forcing function with this same result.

② The minus sign indicates that the motion is 180° out of phase with the applied excitation.

Sample Problem 8/5

The spring attachment point B is given a horizontal motion $x_B = b \cos \omega t$. Determine the critical driving frequency ω_c for which the oscillations of the mass m tend to become excessively large. Neglect the friction and mass associated with the pulleys. The two springs have the same stiffness k.

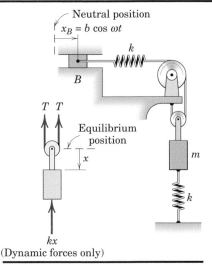

Neutral position
$x_B = b \cos \omega t$

k

B

T T

Equilibrium position

x

m

k

kx
(Dynamic forces only)

Solution. The free-body diagram is drawn for arbitrary positive displacements x and x_B. The motion variable x is measured downward from the position of static equilibrium defined as that which exists when $x_B = 0$. The additional stretch in the upper spring, beyond that which ① exists at static equilibrium, is $2x - x_B$. Therefore, the *dynamic* spring force in the upper spring, and hence the *dynamic* tension T in the cable, ② is $k(2x - x_B)$. Summing forces in the x-direction gives

$$[\Sigma F_x = m\ddot{x}] \qquad -2k(2x - x_B) - kx = m\ddot{x}$$

which becomes

$$\ddot{x} + \frac{5k}{m}x = \frac{2kb \cos \omega t}{m}$$

The natural frequency of the system is $\omega_n = \sqrt{5k/m}$. Thus,

$$\omega_c = \omega_n = \sqrt{5k/m} \qquad Ans.$$

① If a review of the kinematics of constrained motion is necessary, see Art. 2/9.

② We learned from the discussion in Art. 8/2 that the equal and opposite forces associated with the position of static equilibrium may be omitted from the analysis. Our use of the terms *dynamic* spring force and *dynamic* tension stresses that only the force increments in addition to the static values need be considered.

Sample Problem 8/6

The 45-kg piston is supported by a spring of modulus $k = 35$ kN/m. A dashpot of damping coefficient $c = 1250$ N·s/m acts in parallel with the spring. A fluctuating pressure $p = 4000 \sin 30t$ in Pa acts on the piston, whose top surface area is $50(10^{-3})$ m^2. Determine the steady-state displacement as a function of time and the maximum force transmitted to the base.

$p = p_0 \sin \omega t$

Solution. We begin by computing the system natural frequency and damping ratio:

$$\omega_n = \sqrt{\frac{k}{m}} = \sqrt{\frac{35(10^3)}{45}} = 27.9 \text{ rad/s}$$

$$\zeta = \frac{c}{2m\omega_n} = \frac{1250}{2(45)(27.9)} = 0.498 \text{ (underdamped)}$$

The steady-state amplitude, from Eq. 8/20, is

$$X = \frac{F_0/k}{\{[1 - (\omega/\omega_n)^2]^2 + [2\zeta\omega/\omega_n]^2\}^{1/2}}$$

$$= \frac{4000(50)(10^{-3})/[35(10^3)]}{\{[1 - (30/27.9)^2]^2 + [2(0.498)(30/27.9)]^2\}^{1/2}}$$

① $= 0.00528$ m or 5.28 mm

The phase angle, from Eq. 8/21, is

② $$\phi = \tan^{-1}\left[\frac{2\zeta\omega/\omega_n}{1 - (\omega/\omega_n)^2}\right]$$

$$= \tan^{-1}\left[\frac{2(0.498)(30/27.9)}{1 - (30/27.9)^2}\right]$$

$$= 1.716 \text{ rad}$$

The steady-state motion is then given by the second term on the right side of Eq. 8/22:

$$x_p = X \sin(\omega t - \phi) = 5.28 \sin(30t - 1.716) \text{ mm} \qquad Ans.$$

The force F_{tr} transmitted to the base is the sum of the spring and damper forces, or

$$F_{tr} = kx_p + c\dot{x}_p = kX \sin(\omega t - \phi) + c\omega X \cos(\omega t - \phi)$$

The maximum value of F_{tr} is

$$(F_{tr})_{max} = \sqrt{(kX)^2 + (c\omega X)^2} = X\sqrt{k^2 + c^2\omega^2}$$

$$= 0.00528\sqrt{(35\ 000)^2 + (1250)^2(30)^2}$$

① $= 271$ N $\qquad Ans.$

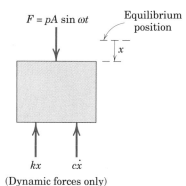

$F = pA \sin \omega t$ — Equilibrium position

$kx \qquad c\dot{x}$

(Dynamic forces only)

① The student is encouraged to repeat these calculations with the damping coefficient c set to zero so as to observe the influence of the relatively large amount of damping present.

② Note that the argument of the inverse tangent expression for ϕ has a positive numerator and a negative denominator for the case at hand, thus placing ϕ in the second quadrant. Recall that the defined range of ϕ is $0 \le \phi \le \pi$.

PROBLEMS

(Unless otherwise instructed, assume that the damping is light to moderate so that the amplitude of the forced response is a maximum at $\omega/\omega_n \cong 1$.)

Introductory problems

8/47 Determine the amplitude X of the steady-state motion of the 10-kg mass if (a) $c = 500$ N·s/m and (b) $c = 0$.

> *Ans.* (a) $X = 1.344(10^{-2})$ m
> (b) $X = 2.27(10^{-2})$ m

$k = 100$ kN/m

c

$m = 10$ kg

$F = 1000 \cos 120t$ N

Problem 8/47

8/48 A viscously damped spring-mass system is excited by a harmonic force of constant amplitude F_0 but slowly-varying frequency ω. If the amplitude of the steady-state motion is observed to decrease by a factor of 10 as the frequency ratio ω/ω_n is varied from 1 to 2, determine the damping ratio ζ of the system.

8/49 The 30-kg cart is acted upon by the harmonic force shown in the figure. If $c = 0$, determine the range of the driving frequency ω for which the amplitude of the steady-state response is less than 75 mm.

> *Ans.* $\omega < 4.99$ rad/s or $\omega > 6.86$ rad/s

$k = 1080$ N/m

c

30 kg

$F = 25 \cos \omega t$ N

Problem 8/49

8/50 If the viscous damping coefficient of the damper in the system of Prob. 8/49 is $c = 36$ N·s/m, determine the range of the driving frequency ω for which the amplitude of the steady-state response is less than 75 mm. Justify your answer in comparison with that given for Prob. 8/49.

8/51 If the driving frequency for the system of Prob. 8/49 is $\omega = 6$ rad/s, determine the required value of the damping coefficient c if the steady-state amplitude is not to exceed 75 mm.

> *Ans.* $c = 55.6$ N·s/m

8/52 The block of mass $m = 45$ kg is suspended by two springs each of stiffness $k = 3$ kN/m and is acted upon by the force $F = 350 \cos 15t$ N where t is the time in seconds. Determine the amplitude X of the steady-state motion if the viscous-damping coefficient c is (a) 0 and (b) 900 N·s/m. Compare these amplitudes to the static spring deflection δ_{st}.

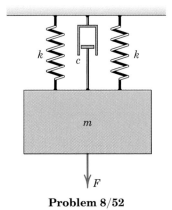

k c k

m

F

Problem 8/52

8/53 The 2-kg body is attached to two springs, each of which has a stiffness of 1.2 kN/m. The body is mounted on a shake table which vibrates harmonically in the horizontal direction with an amplitude of 12 mm and a frequency f which can be varied. Power to the shake table is turned off when electrical contact is made at A or B. Determine the maximum value of the frequency f at which the shake table may be operated without turning the power off as it starts from rest and increases its frequency gradually. Damping may be neglected. The equilibrium position is centered between the fixed contacts.

Ans. $f = 3.98$ Hz

Problem 8/53

8/54 An external force $F = F_0 \sin \omega t$ is applied to the cylinder as shown. What value ω_c of the driving frequency would cause excessively large oscillations of the system?

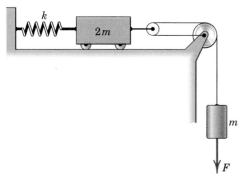

Problem 8/54

Representative problems

8/55 A viscously damped spring-mass system is forced harmonically at the undamped natural frequency $(\omega/\omega_n = 1)$. If the damping ratio ζ is doubled from 0.1 to 0.2, compute the percentage reduction R_1 in the steady-state amplitude. Compare with the result R_2 of a similar calculation for the condition $\omega/\omega_n = 2$. Verify your results by inspecting Fig. 8/11.

Ans. $R_1 = 50\%, R_2 = 2.52\%$

8/56 A single-cylinder four-stroke gasoline engine with a mass of 90 kg is mounted on four stiff spring pads, each with a stiffness of $30(10^3)$ kN/m, and is designed to run at 3600 rev/min. The mounting system is equipped with viscous dampers which have a large enough combined viscous damping coefficient c so that the system is critically damped when it is given a vertical displacement and then released while not running. When the engine is running, it fires every other revolution causing a periodic vertical displacement modeled by 1.2 cos ωt mm with t in seconds. Determine the magnification factor M and the overall damping coefficient c.

8/57 A linear spring-mass oscillator has a viscous damping factor $\zeta = 0.2$ and has an undamped natural frequency $f_n = 6$ Hz. By referring to Fig. 8/11, estimate the range of frequencies f of the periodic applied force for which the amplitude of the oscillator will not exceed twice the static deflection which would be caused by applying a static force equal in magnitude to that of the periodic force. Check your estimate by applying Eq. 8/23.

Ans. $f \leq 4.68$ Hz or $f \geq 6.66$ Hz

8/58 The collar *A* is given a harmonic oscillation along the fixed horizontal shaft with a frequency of 3 Hz. The ball is attached to the collar by the elastic rod of negligible mass and deflects 60 mm when a horizontal force of 40 N is applied statically to the ball with the collar clamped in place. If the double amplitude $2x_0$ of the ball is twice the double amplitude $2\delta_0$ of the collar with the two motions 180° out of phase, compute the mass *m* of the ball.

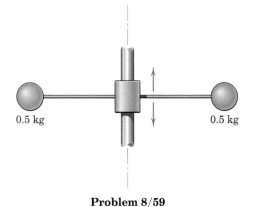

Problem 8/58

8/59 Each 0.5-kg ball is attached to the end of the light elastic rod and deflects 4 mm when a 2-N force is statically applied to the ball. If the central collar is given a vertical harmonic movement with a frequency of 4 Hz and an amplitude of 3 mm, find the amplitude y_0 of vertical vibration of each ball.
Ans. $y_0 = 8.15$ mm

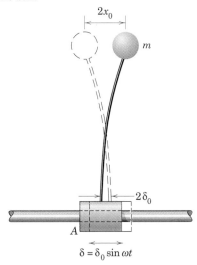

Problem 8/59

8/60 The motion of the outer cart *B* is given by $x_B = b \sin \omega t$. For what range of the driving frequency ω is the magnitude of the motion of the mass *m* relative to the cart less than $2b$?

Problem 8/60

8/61 The 20-kg variable-speed motorized unit is restrained in the horizontal direction by two springs, each of which has a stiffness of 2.1 kN/m. Each of the two dashpots has a viscous damping coefficient $c = 58$ N·s/m. At what limiting speeds *N* can the motor be run for which the magnification factor *M* will not exceed 2?
Ans. $N \leq 108.1$ rev/min or $N \geq 153.5$ rev/min

Problem 8/61

8/62 When the person stands in the center of the floor system shown, he causes a static deflection δ_{st} of the floor under his feet. If he walks (or runs quickly!) in the same area, how many steps per second would cause the floor to vibrate with the greatest vertical amplitude?

Problem 8/62

8/63 Attachment B is given a horizontal motion $x_B = b \cos \omega t$. Derive the equation of motion for the mass m and state the critical frequency ω_c for which the oscillations of the mass become excessively large.

$$Ans. \quad m\ddot{x} + c\dot{x} + (k_1 + k_2)x = k_2 b \cos \omega t$$

$$\omega_c = \sqrt{\frac{k_1 + k_2}{m}}$$

Problem 8/63

8/64 Attachment B is given a horizontal motion $x_B = b \cos \omega t$. Derive the equation of motion for the mass m and state the critical frequency ω_c for which the oscillations of the mass become excessively large. What is the damping ratio ζ for the system?

Problem 8/64

8/65 A 30-kg machine mounted on an elastic foundation is observed to vibrate harmonically in the vertical direction with an amplitude $X = 0.4$ mm under the action of a vertical force which varies harmonically from F_0 to $-F_0$ with a frequency of 4 cycles per second. Damping is negligible. When a static force of magnitude F_0 is applied vertically to the machine, the foundation is observed to deflect 0.8 mm. Calculate the equivalent spring constant k of the elastic mounting.

$$Ans. \quad k = 6320 \text{ N/m}$$

8/66 Derive an expression for the *transmission ratio* T for the system of the figure. This ratio is defined as the maximum force transmitted to the base divided by the amplitude F_0 of the forcing function. Express your answer in terms of ζ, ω, ω_n, and the magnification factor M.

Problem 8/66

8/67 A device to produce vibrations consists of the two counter-rotating wheels, each carrying an eccentric mass $m_0 = 1$ kg with a center of mass at a distance $e = 12$ mm from its axis of rotation. The wheels are synchronized so that the vertical positions of the unbalanced masses are always identical. The total mass of the device is 10 kg. Determine the two possible values of the equivalent spring constant k for the mounting which will permit the amplitude of the periodic force transmitted to the fixed mounting to be 1500 N due to the imbalance of the rotors at a speed of 1800 rev/min. Neglect damping.

Ans. $k = 227$ kN/m or 823 kN/m

Problem 8/67

8/68 The seismic instrument shown is attached to a structure which has a horizontal harmonic vibration at 3 Hz. The instrument has a mass $m = 0.5$ kg, a spring stiffness $k = 20$ N/m, and a viscous damping coefficient $c = 3$ N·s/m. If the maximum recorded value of x in its steady-state motion is $X = 2$ mm, determine the amplitude b of the horizontal movement x_B of the structure.

Problem 8/68

8/69 Derive and solve the equation of motion for the mass m in terms of the variable x for the system shown. Neglect the mass of the lever AOC and assume small oscillations.

Ans. $m\ddot{x} + c\dot{x} + \left(k_1 + k_2\dfrac{b^2}{a^2}\right)x = k_2\dfrac{b}{a}b_0 \cos \omega t$

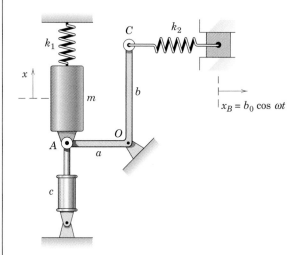

Problem 8/69

8/70 The seismic instrument is mounted on a structure which has a vertical vibration with a frequency of 5 Hz and a double amplitude of 18 mm. The sensing element has a mass $m = 2$ kg, and the spring stiffness is $k = 1.5$ kN/m. The motion of the mass relative to the instrument base is recorded on a revolving drum and shows a double amplitude of 24 mm during the steady-state condition. Calculate the viscous damping constant c.

Problem 8/70

▶8/71 Determine the amplitude of vertical vibration of the spring-mounted trailer as it travels at a velocity of 25 km/h over the corduroy road whose contour may be expressed by a sine or cosine term. The mass of the trailer is 500 kg and that of the wheels alone may be neglected. During the loading each 75 kg added to the load caused the trailer to sag 3 mm on its springs. Assume that the wheels are in contact with the road at all times and neglect damping. At what critical speed v_c is the vibration of the trailer greatest?

Ans. $X = 14.75$ mm, $v_c = 15.23$ km/h

50 mm

1.2 m

Problem 8/71

▶8/72 Derive the expression for the power loss P averaged over a complete steady-state cycle due to the frictional dissipation of energy in a viscously damped linear oscillator. The forcing function is $F_0 \sin \omega t$, and the displacement-time relation for steady-state motion is $x_P = X \sin(\omega t - \varphi)$ where the amplitude X is given by Eq. 8/20. (*Hint:* The frictional energy loss during a displacement dx is $c\dot{x}\,dx$, where c is the viscous damping coefficient. Integrate this expression over a complete cycle and divide by the period of the cycle.)

Ans. $P = cX^2\omega^2/2$

8/4 VIBRATION OF RIGID BODIES

The subject of planar rigid-body vibrations is entirely analogous to that of particle vibrations. In particle vibrations, the variable of interest is one of translation (x), while in rigid-body vibrations, the variable of primary concern is one of rotation (θ). Thus, the principles of rotational dynamics play a central role in the development of the equation of motion. We shall see that the equation of motion for rotational vibration of rigid bodies has a mathematical form identical to that developed in Arts. 8/2 and 8/3 for translational vibration of particles. As was the case with particles, it is convenient to draw the free-body diagram for an arbitrary positive value of the displacement variable, because a negative displacement value easily leads to sign errors in the equation of motion. The practice of measuring the displacement from the position of static equilibrium rather than from the position of zero spring deflection continues to simplify the formulation for linear systems because the equal and opposite forces and moments associated with the static equilibrium position cancel from the analysis.

Rather than individually treating the cases of (a) free vibration, undamped and damped, and (b) forced vibrations, undamped and damped, as was done with particles in Arts. 8/2 and 8/3, we shall go directly to the damped, forced problem.

As an illustrative example, consider the rotational vibration of the uniform slender bar of Fig. 8/16a. Figure 8/16b depicts the free-body diagram associated with the horizontal position of static equilibrium. Equating to zero the moment sum about O yields

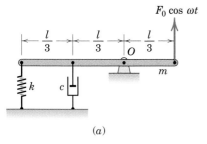

$$-P\left(\frac{l}{2} + \frac{l}{6}\right) + mg\left(\frac{l}{6}\right) = 0 \qquad P = \frac{mg}{4}$$

where P is the magnitude of the static spring force.

Figure 8/16c depicts the free-body diagram associated with an arbitrary positive angular displacement θ. Using the rotational equation of motion $\Sigma M_O = I_O\ddot{\theta}$ as treated in Chapter 6, we write

$$(mg)\left(\frac{l}{6}\cos\theta\right) - \left(\frac{cl}{3}\dot{\theta}\cos\theta\right)\left(\frac{l}{3}\cos\theta\right) - \left(P + k\frac{2l}{3}\sin\theta\right)\left(\frac{2l}{3}\cos\theta\right)$$

$$+ (F_0\cos\omega t)\left(\frac{l}{3}\cos\theta\right) = \frac{1}{9}ml^2\ddot{\theta}$$

where $I_O = \bar{I} + md^2 = ml^2/12 + m(l/6)^2 = ml^2/9$ is obtained from the parallel-axis theorem for mass moments of inertia. For small angular deflections, the approximations $\sin\theta \cong \theta$ and $\cos\theta \cong 1$ may be used. With $P = mg/4$, the equation of motion, upon rearrangement and simplification, becomes

$$\ddot{\theta} + \frac{c}{m}\dot{\theta} + 4\frac{k}{m}\theta = \frac{(F_0l/3)\cos\omega t}{ml^2/9} \tag{8/25}$$

(a)

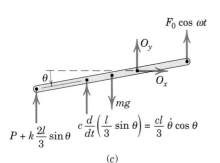

(b)

(c)

Figure 8/16

The right side has been left unsimplified in the form $M_0(\cos \omega t)/I_O$, where $M_0 = F_0 l/3$ is the magnitude of the moment about point O of the externally applied force. Note that the two equal and opposite moments associated with static equilibrium forces canceled on the left side of the equation of motion. Hence, it is not necessary to include the static-equilibrium forces and moments in the analysis.

At this point, we observe that Eq. 8/25 is identical in form to Eq. 8/13 for the linear case, so we may write

$$\ddot{\theta} + 2\zeta\omega_n\dot{\theta} + \omega_n{}^2\theta = \frac{M_0 \cos \omega t}{I_O} \qquad (8/26)$$

Thus, we may use all of the relations developed in Arts. 8/2 and 8/3 merely by replacing the linear quantities with their rotational counterparts. The following table shows the results of this procedure as applied to the rotating bar of Fig. 8/16:

LINEAR	ANGULAR (for current problem)
$\ddot{x} + \dfrac{c}{m}\dot{x} + \dfrac{k}{m}x = \dfrac{F_0 \cos \omega t}{m}$	$\ddot{\theta} + \dfrac{c}{m}\dot{\theta} + \dfrac{4k}{m}\theta = \dfrac{M_0 \cos \omega t}{I_O}$
$\omega_n = \sqrt{k/m}$	$\omega_n = \sqrt{4k/m} = 2\sqrt{k/m}$
$\zeta = \dfrac{c}{2m\omega_n} = \dfrac{c}{2\sqrt{km}}$	$\zeta = \dfrac{c}{2m\omega_n} = \dfrac{c}{4\sqrt{km}}$
$\omega_d = \omega_n\sqrt{1 - \zeta^2} = \dfrac{1}{2m}\sqrt{4km - c^2}$	$\omega_d = \omega_n\sqrt{1 - \zeta^2} = \dfrac{1}{2m}\sqrt{16km - c^2}$
$x_c = Ce^{-\zeta\omega_n t} \sin (\omega_d t + \psi)$	$\theta_c = Ce^{-\zeta\omega_n t} \sin (\omega_d t + \psi)$
$x_p = X \cos (\omega t - \phi)$	$\theta_p = \Theta \cos (\omega t - \phi)$
$X = M\left(\dfrac{F_0}{k}\right)$	$\Theta = M\left(\dfrac{M_0}{k_\theta}\right) = M\,\dfrac{F_0(l/3)}{\frac{4}{9}kl^2} = \dfrac{3F_0}{4kl}$

In the above table, the variable k_θ represents the equivalent torsional spring constant of the system of Fig. 8/16 and is determined by writing the restoring moment of the spring. For a small angle θ, this moment about O is

$$M_k = -[k(2l/3) \sin \theta][(2l/3) \cos \theta] \cong -(\tfrac{4}{9}kl^2)\theta$$

Thus, $k_\theta = \tfrac{4}{9}kl^2$. Note that M_0/k_θ is the static angular deflection which would be produced by a constant external moment M_0.

We conclude that there exists an exact analogy between the linear vibration of particles and the small angular vibration of rigid bodies. Furthermore, the utilization of this analogy can save the labor of complete rederivation of the governing relationships for a given problem of general rigid-body vibration.

Sample Problem 8/7

A simplified version of a pendulum used in impact tests is shown in the figure. Derive the equation of motion and determine the period for small oscillations about the pivot. The mass center G is located a distance $\bar{r} = 0.9$ m from O, and the radius of gyration about O is $k_O = 0.95$ m. The friction of the bearing is negligible.

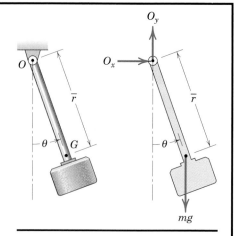

Solution. We draw the free-body diagram for an arbitrary, positive value of the angular-displacement variable θ, which is measured counterclockwise for the coordinate system chosen. Next we apply the governing equation of motion to obtain

① $[\Sigma M_O = I_O\ddot{\theta}]$ $\qquad -mg\bar{r}\sin\theta = mk_O^2\ddot{\theta}$

or $\qquad\qquad \ddot{\theta} + \dfrac{g\bar{r}}{k_O^2}\sin\theta = 0$ $\qquad$ *Ans.*

Note that the governing equation is independent of the mass. When θ is small, $\sin\theta \cong \theta$, and our equation of motion may be written as

$$\ddot{\theta} + \dfrac{g\bar{r}}{k_O^2}\theta = 0$$

② The frequency in cycles per second and the period in seconds are

$$f_n = \frac{1}{2\pi}\sqrt{\frac{g\bar{r}}{k_O^2}} \qquad \tau = \frac{1}{f_n} = 2\pi\sqrt{\frac{k_O^2}{g\bar{r}}} \qquad \textit{Ans.}$$

For the given properties: $\qquad \tau = 2\pi\sqrt{\dfrac{(0.95)^2}{(9.81)(0.9)}} = 2.01$ s $\qquad$ *Ans.*

① With our choice of point O as the moment center, the bearing reactions O_x and O_y never enter the equation of motion.

② For large angles of oscillation, determining the period for the pendulum requires the evaluation of an elliptic integral.

Sample Problem 8/8

The uniform bar of mass m and length l is pivoted at its center. The spring of constant k at the left end is attached to a stationary surface, but the right-end spring, also of constant k, is attached to a support which undergoes a harmonic motion given by $y_B = b\sin\omega t$. Determine the driving frequency ω_c which causes resonance.

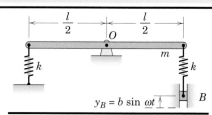

Solution. We use the moment equation of motion about the fixed point O to obtain

① $-\left(k\dfrac{l}{2}\sin\theta\right)\dfrac{l}{2}\cos\theta - k\left(\dfrac{l}{2}\sin\theta - y_B\right)\dfrac{l}{2}\cos\theta = \dfrac{1}{12}ml^2\ddot{\theta}$

Assuming small deflections and simplifying give us

$$\ddot{\theta} + \frac{6k}{m}\theta = \frac{6kb}{ml}\sin\omega t$$

② The natural frequency should be recognized from the now-familiar form of the equation to be

$$\omega_n = \sqrt{6k/m}$$

Hence, $\omega_c = \omega_n = \sqrt{6k/m}$ will result in resonance (as well as violation of our small-angle assumption!). $\qquad$ *Ans.*

① As previously, we consider only the changes in the forces due to a movement away from the equilibrium position.

② The standard form here is $\ddot{\theta} + \omega_n^2\theta = \dfrac{M_0\sin\omega t}{I_O}$, where $M_0 = \dfrac{klb}{2}$ and $I_O = \frac{1}{12}ml^2$. The natural frequency ω_n of a system does not depend on the external disturbance.

Sample Problem 8/9

Derive the equation of motion for the homogeneous circular cylinder, which rolls without slipping. If the cylinder mass is 50 kg, the cylinder radius 0.5 m, the spring constant 75 N/m, and the damping coefficient 10 N·s/m, determine

    (a) the undamped natural frequency
    (b) the damping ratio
    (c) the damped natural frequency
    (d) the period of the damped system.

In addition, determine x as a function of time if the cylinder is released from rest at the position $x = -0.2$ m when $t = 0$.

    Solution. We have a choice of motion variables in that either x or ① the angular displacement θ of the cylinder may be used. Since the problem statement involves x, we draw the free-body diagram for an arbitrary, positive value of x and write the two motion equations for the cylinder as

② $[\Sigma F_x = m\ddot{x}]$ $-c\dot{x} - kx + F = m\ddot{x}$

    $[\Sigma M_G = \bar{I}\ddot{\theta}]$ $-Fr = \frac{1}{2}mr^2\ddot{\theta}$

The condition of rolling with no slip is $\ddot{x} = r\ddot{\theta}$. Substitution of this condition into the moment equation gives $F = -\frac{1}{2}m\ddot{x}$. Inserting this expression for the friction force into the force equation for the x-direction yields

$$-c\dot{x} - kx - \frac{1}{2}m\ddot{x} = m\ddot{x} \quad \text{or} \quad \ddot{x} + \frac{2}{3}\frac{c}{m}\dot{x} + \frac{2}{3}\frac{k}{m}x = 0$$

Comparing the above equation with that for the standard damped oscillator, Eq. 8/9, allows us to state directly

(a) $\omega_n{}^2 = \frac{2}{3}\frac{k}{m}$ $\omega_n = \sqrt{\frac{2}{3}\frac{k}{m}} = \sqrt{\frac{2}{3}\frac{75}{50}} = 1$ rad/s   *Ans.*

(b) $2\zeta\omega_n = \frac{2}{3}\frac{c}{m}$ $\zeta = \frac{1}{3}\frac{c}{m\omega_n} = \frac{10}{3(50)(1)} = 0.0667$ *Ans.*

Hence, the damped natural frequency and the damped period are

(c) $\omega_d = \omega_n\sqrt{1 - \zeta^2} = (1)\sqrt{1 - (0.0667)^2} = 0.998$ rad/s   *Ans.*

(d) $\tau_d = 2\pi/\omega_d = 2\pi/0.998 = 6.30$ s   *Ans.*

From Eq. 8/12, the underdamped solution to the equation of motion is

$$x = Ce^{-\zeta\omega_n t}\sin(\omega_d t + \psi) = Ce^{-(0.0667)(1)t}\sin(0.998t + \psi)$$

The velocity is $\dot{x} = -0.0667Ce^{-0.0667t}\sin(0.998t + \psi)$
$$+ 0.998Ce^{-0.0667t}\cos(0.998t + \psi)$$

At time $t = 0$, x and $\dot{x}$ become

$$x_0 = C\sin\psi = -0.2$$
$$\dot{x}_0 = -0.0667C\sin\psi + 0.998C\cos\psi = 0$$

The solution to the two equations in C and ψ gives

$$C = -0.200 \text{ m} \quad \psi = 1.504 \text{ rad}$$

Hence, the motion is given by

$$x = -0.200e^{-0.0667t}\sin(0.998t + 1.504) \text{ m} \quad \text{Ans.}$$

---

① The angle θ is taken positive clockwise to be kinematically consistent with x.

② The friction force F may be assumed in either direction. We will find that the *actual* direction is to the right for $x > 0$ and to the left for $x < 0$; $F = 0$ when $x = 0$.

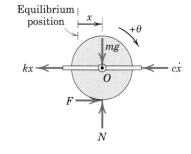

PROBLEMS

Introductory problems

8/73 The light rod and attached small spheres of mass m each are shown in the equilibrium position, where all four springs are equally precompressed. Determine the natural frequency ω_n and period τ for small oscillations about the frictionless pivot O.

$$Ans. \quad \omega_n = \sqrt{\frac{2k}{m}}, \quad \tau = \pi\sqrt{\frac{2m}{k}}$$

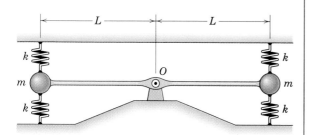

Problem 8/73

8/74 Derive the differential equation for small oscillations of the spring-loaded pendulum and find the period τ. The equilibrium position is vertical as shown. The mass of the rod is negligible.

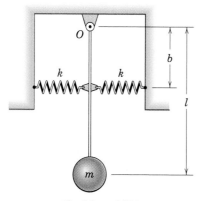

Problem 8/74

8/75 The thin square plate is suspended from a socket (not shown) which fits the small ball attachment at O. If the plate is made to swing about axis A-A, determine the period for small oscillations. Neglect the small offset, mass, and friction of the ball.

$$Ans. \quad \tau = 2\pi\sqrt{\frac{2b}{3g}}$$

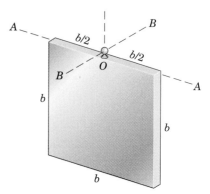

Problem 8/75

8/76 If the square plate of Prob. 8/75 is made to oscillate about axis B-B, determine the period of small oscillations and compare it with the answer for Prob. 8/75.

8/77 The rectangular frame is formed of a uniform slender rod and is suspended from a socket (not shown) which fits the small ball attachment at O. If the rectangle is made to swing about axis A-A, determine the natural frequency for small oscillations. Neglect the small offset, mass, and friction of the ball.

$$Ans. \quad \omega_n = \frac{3}{2}\sqrt{\frac{g}{2b}}$$

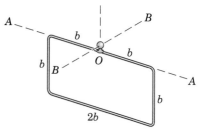

Problem 8/77

8/78 If the rectangular frame of Prob. 8/77 is made to oscillate about axis *B-B*, determine the natural frequency of small oscillations and compare it with the answer for Prob. 8/77.

8/79 The circular ring of radius *r* is suspended from a socket (not shown) which fits the small ball attachment at *O*. Determine the ratio *R* of the period of small oscillations about axis *B-B* to that about axis *A-A*. Neglect the small offset, mass, and friction of the ball.

$$\text{Ans. } R = \frac{2}{\sqrt{3}}$$

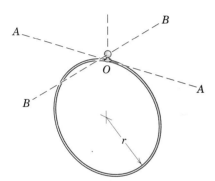

Problem 8/79

8/80 The thin-wall cylindrical shell of radius *r* and height *h* is welded to the small shaft at its upper end as shown. Determine the natural circular frequency ω_n for small oscillations of the shell about the *y*-axis.

Problem 8/80

8/81 The mass of the uniform slender rod is 3 kg. Determine the position *x* for the 1.2-kg slider such that the system period is 1 s. Assume small oscillations about the horizontal equilibrium position shown.

$$\text{Ans. } x = 0.558 \text{ m}$$

Problem 8/81

Representative problems

8/82 The uniform rod of mass *m* is freely pivoted about a horizontal axis through point *O*. Assume small oscillations and determine an expression for the damping ratio ζ. For what value c_{cr} of the damping coefficient *c* will the system be critically damped?

Problem 8/82

8/83 The compound pendulum with mass center at G and radius of gyration $\bar{k}$ about G is suspended first from O_1 and next from O_2. If the periods for small oscillations about the two points are equal and if $\bar{r}_1 \neq \bar{r}_2$, determine the equation relating $\bar{r}_1$, $\bar{r}_2$, and $\bar{k}$.

Ans. $\bar{k}^2 = \bar{r}_1 \bar{r}_2$

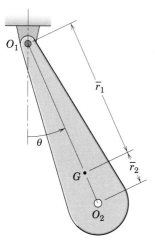

Problem 8/83

8/84 The mechanism shown oscillates in the vertical plane about the pivot O. The springs of equal stiffness k are both compressed in the equilibrium position $\theta = 0$. Determine an expression for the period τ of small oscillations about O. The mechanism has a mass m with mass center G, and the radius of gyration of the assembly about O is k_O.

Problem 8/84

8/85 The uniform square plate is suspended in a horizontal plane by its four corner cables from fixed points A and B on a horizontal line a distance b above the plate. Determine an expression for the frequency f_n of small oscillations of the plate about the axis AB.

Ans. $f_n = \dfrac{1}{\pi}\sqrt{\dfrac{3g}{13b}}$

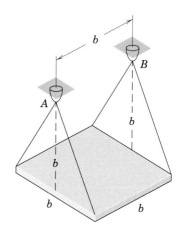

Problem 8/85

8/86 When the motor is slowly brought up to speed, a rather large vibratory oscillation of the entire motor about O-O occurs at a speed of 360 rev/min, which shows that this speed corresponds to the natural frequency of free oscillation of the motor. If the motor has a mass of 43 kg and a radius of gyration of 100 mm about O-O, determine the stiffness k of each of the four identical spring mounts.

Problem 8/86

8/87 The center of mass G of the ship may be assumed to be at the center of the equivalent 15-m square section. The metacentric height h, determined by the intersection M of the ship's center line with the line of action of the total buoyancy force acting through the center of buoyancy B, is 0.9 m. Determine the period τ of one complete roll of the ship if the amplitude is small and the resistance of the water is neglected. Neglect also the change in cross section of the ship at the bow and stern, and treat the loaded ship as a uniform solid block of square cross section.

Ans. $\tau = 12.95$ s

Problem 8/87

8/88 Determine the natural frequency f_n for small oscillations of the composite body in the vertical plane about the bearing O. Approximate the body as a slender bar of mass $m/5$ and a semicircular disk of mass m, both with the dimension r as shown.

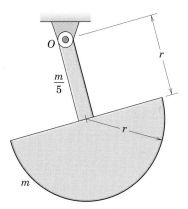

Problem 8/88

8/89 The system of Prob. 8/44 is repeated here. If the link AB now has mass m_3 and radius of gyration k_O about point O, determine the equation of motion in terms of the variable x. Assume small oscillations. The damping coefficient for the dashpot is c.

$$Ans. \ \left[m_1 + \frac{a^2}{b^2} m_2 + \frac{k_O^2}{b^2} m_3 \right] \ddot{x}$$
$$+ \left[\frac{a^2}{b^2} c \right] \dot{x} + kx = 0$$

Problem 8/89

8/90 Determine the period τ of small oscillations of the semicylinder of mass m and radius r as it rolls without slipping on the horizontal surface.

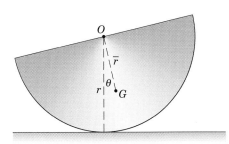

Problem 8/90

8/91 The circular sector of mass m is cut from steel plate of uniform thickness and mounted in a bearing at its center O so that it can swing freely in the vertical plane. If the sector is released from rest with $\theta = 0$, derive its differential equation of motion assuming negligible damping. Determine its period τ for small oscillations about the position $\theta = \dfrac{\pi}{2}$.

$$\text{Ans.} \quad \tau = \pi\sqrt{\frac{3r\beta}{g \sin \beta}}$$

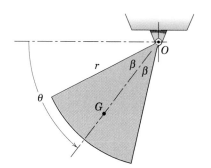

Problem 8/91

8/92 Two identical uniform bars are welded together at a right angle and are pivoted about a horizontal axis through point O as shown. Determine the critical driving frequency ω_c of the block B which will result in excessively large oscillations of the assembly. The mass of the welded assembly is m.

Problem 8/92

8/93 The uniform solid cylinder of mass m and radius r rolls without slipping during its oscillation on the circular surface of radius R. If the motion is confined to small amplitudes $\theta = \theta_0$, determine the period τ of the oscillations. Also determine the angular velocity ω of the cylinder as it crosses the vertical center line. (*Caution:* Do not confuse ω with $\dot{\theta}$ or with ω_n as used in the defining equations. Note also that θ is not the angular displacement of the cylinder.)

$$\text{Ans.} \quad \tau = 2\pi\sqrt{\frac{3(R-r)}{2g}}, \quad \omega = \frac{\theta_0}{r}\sqrt{2g(R-r)/3}$$

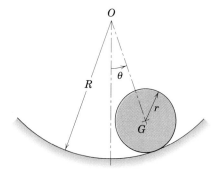

Problem 8/93

8/94 The homogeneous solid cylindrical pulley has mass m_1 and radius r. If the attachment at B undergoes the indicated harmonic displacement, determine the equation of motion of the system in terms of the variable x. The cord which connects mass m_2 to the upper spring does not slip on the pulley.

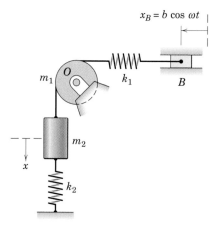

Problem 8/94

8/95 The circular disk of mass m and moment of inertia I about its central axis is welded to the steel shaft which, in turn, is welded to the fixed block. The disk is given an angular displacement θ_0 and then released, causing a torsional vibration of the disk with θ changing between $+\theta_0$ and $-\theta_0$. The shaft resists the twist with a moment $M = JG\theta/L$, where J is the polar moment of inertia of the cross section of the shaft about the rotation axis, G is the shear modulus of elasticity of the shaft (resistance to shear stress), θ is the angle of twist in radians, and L is the length of the twisted shaft. Derive the expression for the natural frequency f_n of the torsional vibration.

$$\textit{Ans. } f_n = \frac{1}{2\pi}\sqrt{\frac{JG}{IL}}$$

8/96 The two flywheels and shaft form an integral unit and are rotating at a constant angular velocity ω when a sudden impulsive torque is applied momentarily to one of the wheels causing an angular harmonic oscillation between them as shown on the diagram. The moments of inertia of the wheels are I_1 and I_2, and the shaft has a shear modulus of elasticity G and a polar moment of inertia J of its circular cross section. Determine the relationship between L_1 and L_2 which locates the nodal point in the shaft where the angular velocity does not fluctuate. Also express the natural period of torsional vibration in terms of L. (*Hint:* The oscillations may be treated as motion relative to the nodal-point section as nonrotating. Use the results of Prob. 8/95.)

Problem 8/95

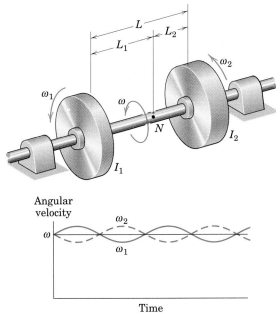

Angular velocity

Time

Problem 8/96

8/97 The segmented "dummy" of Prob. 6/105 is repeated here. The hip joint O is assumed to remain fixed to the car, and the torso above the hip is treated as a rigid body of mass m. The center of mass of the torso is at G and the radius of gyration of the torso about O is k_O. Assume that muscular response acts as an internal torsional spring which exerts a moment $M = K\theta$ on the upper torso, where K is the torsional spring constant and θ is the angular deflection from the initial vertical position. If the car is brought to a sudden stop with a constant deceleration a, derive the differential equation for the motion of the torso prior to its impact with the dashboard.
Ans. $mk_O^2\ddot{\theta} + K\theta - m\bar{r}(g \sin\theta + a \cos\theta) = 0$

Problem 8/97

▶**8/98** The elements of the "swing-axle" type of independent rear suspension for automobiles are depicted in the figure. The differential D is rigidly attached to the car frame. The half-axles are pivoted at their inboard ends (point O for the half-axle shown) and are rigidly attached to the wheels. Suspension elements not shown constrain the wheel motion to the plane of the figure. The mass of the wheel–tire assembly is $M = 45$ kg, and its mass moment of inertia about a diametral axis passing through its mass center G is 1.4 kg·m². The mass of the half-axle is negligible. The spring rate and shock-absorber damping coefficient are $k = 8.75$ kN/m and $c = 2600$ N·s/m, respectively. If a static tire imbalance is present, as represented by the additional concentrated mass $m = 0.25$ kg as shown, determine the angular velocity ω which results in the suspension system being driven at its undamped natural frequency. What would be the corresponding vehicle speed v? Determine the damping ratio ζ. Assume small angular deflections and neglect gyroscopic effects and any car frame vibration. In order to avoid the complications associated with the varying normal force exerted by the road on the tire, treat the vehicle as being on a lift with the wheels hanging free.
Ans. $\omega = 10.21$ rad/s, $v = 12.87$ km/h
$\zeta = 1.517$

Problem 8/98

8/5 ENERGY METHODS

In Arts. 8/2 through 8/4 we derived and solved the equations of motion for vibrating bodies by applying Newton's second law of motion to the body isolated with its free-body diagram. With this approach, we were able to account for the actions of all forces acting on the body, including frictional damping forces. There are many problems where the effect of damping is small and may be neglected, so that the total energy of the system is essentially conserved. For such systems, we find that the principle of conservation of energy may frequently be applied with considerable advantage in establishing the equation of motion and, when the motion is simple harmonic, in determining the frequency of vibration.

To illustrate this alternative approach, consider first the simple case of the body of mass m attached to the spring of stiffness k and vibrating in the vertical direction without damping, Fig. 8/17. As previously, we find it convenient to measure the motion variable x from the equilibrium position. With this datum, the total potential energy of the system, elastic plus gravitational, becomes

Figure 8/17

$$V = V_e + V_g = \tfrac{1}{2}k(x + \delta_{st})^2 - \tfrac{1}{2}k\delta_{st}^2 - mgx$$

where $\delta_{st} = mg/k$ is the initial static displacement. Substituting $k\delta_{st} = mg$ and simplifying give

$$V = \tfrac{1}{2}kx^2$$

Thus, the total energy of the system becomes

$$T + V = \tfrac{1}{2}m\dot{x}^2 + \tfrac{1}{2}kx^2$$

Because $T + V$ is constant for a conservative system, its time derivative is zero. Consequently,

$$\frac{d}{dt}(T + V) = m\dot{x}\ddot{x} + kx\dot{x} = 0$$

Canceling $\dot{x}$ gives us our basic differential equation of motion

$$m\ddot{x} + kx = 0$$

which is identical to Eq. 8/1 derived in Art. 8/2 for the same system of Fig. 8/3.

Conservation of energy may also be used to determine the period or frequency of vibration for a linear conservative system, without having to derive and solve the equation of motion. For a system which oscillates with simple harmonic motion about the equilibrium position, from which the displacement x is measured, the energy changes from maximum kinetic and zero potential at the equilibrium position $x = 0$ to zero kinetic and maximum potential at the position of maximum displacement $x = x_{max}$. Thus, we may write

$$T_{max} = V_{max}$$

The maximum kinetic energy is $\frac{1}{2}m(\dot{x}_{max})^2$, and the maximum potential energy is $\frac{1}{2}k(x_{max})^2$. For the harmonic oscillator of Fig. 8/17, we know that the displacement may be written $x = x_{max} \sin(\omega_n t + \psi)$ so that the maximum velocity is $\dot{x}_{max} = \omega_n x_{max}$. Thus, we may write

$$\tfrac{1}{2}m(\omega_n x_{max})^2 = \tfrac{1}{2}k(x_{max})^2$$

where x_{max} is the maximum displacement at which the potential energy is a maximum. From this energy balance, we easily obtain

$$\omega_n = \sqrt{k/m}$$

This direct determination of the frequency may be used for any linear undamped vibration.

The main advantage of the energy approach for the free vibration of conservative systems is that it becomes unnecessary to dismember the system and account for all of the forces which act on each member. In Art. 3/7 of Chapter 3 and in Arts. 6/6 and 6/7 of Chapter 6, we learned for a system of interconnected bodies that an active-force diagram of the complete system enabled us to evaluate the work U' of the external active forces and to equate it to the change in the total mechanical energy $T + V$ of the system. So for a conservative mechanical system of interconnected parts with a single degree of freedom where $U' = 0$, we may obtain its equation of motion simply by setting the time derivative of its constant total mechanical energy to zero, giving

$$\frac{d}{dt}(T + V) = 0$$

Here $V = V_e + V_g$ is the sum of the elastic and gravitational potential energies of the system. Also, for an interconnected mechanical system as for a single body, the natural frequency of vibration is obtained by equating the expression for its maximum total kinetic energy to the expression for its maximum potential energy, where the potential energy is taken to be zero at the equilibrium position. This approach to the determination of natural frequency is valid only if it can be determined that the system vibrates with simple harmonic motion.

Sample Problem 8/10

The small sphere of mass m is mounted on the light rod pivoted at O and supported at end A by the vertical spring of stiffness k. End A is displaced a small distance y_0 below the horizontal equilibrium position and released. By the energy method, derive the differential equation of motion for small oscillations of the rod and determine the expression for its natural frequency ω_n of vibration. Damping is negligible.

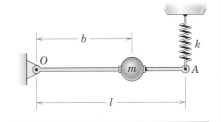

Solution. With the displacement y of the end of the bar measured from the equilibrium position, the potential energy in the displaced position for small values of y becomes

①
$$V = V_e + V_g = \frac{1}{2}k(y + \delta_{st})^2 - \frac{1}{2}k\delta_{st}^2 - mg\left(\frac{b}{l}y\right)$$

where δ_{st} is the static deflection of the spring at equilibrium. But the force in the spring in the equilibrium position, from a zero moment sum about O, is $(b/l)mg = k\delta_{st}$. Substituting this value in the expression for V and simplifying yield

②
$$V = \frac{1}{2}ky^2$$

The kinetic energy in the displaced position is

$$T = \frac{1}{2}m\left(\frac{b}{l}\dot{y}\right)^2$$

where we see that the vertical displacement of m is $(b/l)y$. Thus, with the energy sum constant, its time derivative is zero, and we have

$$\frac{d}{dt}(T + V) = \frac{d}{dt}\left[\frac{1}{2}m\left(\frac{b}{l}\dot{y}\right)^2 + \frac{1}{2}ky^2\right] = 0$$

which yields

$$\ddot{y} + \frac{l^2}{b^2}\frac{k}{m}y = 0 \qquad\qquad Ans.$$

when $\dot{y}$ is canceled. By analogy with Eq. 8/2, we may write the motion frequency directly as

$$\omega_n = \frac{l}{b}\sqrt{k/m} \qquad\qquad Ans.$$

Alternatively, we can obtain the frequency by equating the maximum kinetic energy, which occurs at $y = 0$, to the maximum potential energy, which occurs at $y = y_0 = y_{max}$, where the deflection is a maximum. Thus,

$$T_{max} = V_{max} \qquad gives \qquad \frac{1}{2}m\left(\frac{b}{l}\dot{y}_{max}\right)^2 = \frac{1}{2}ky_{max}^2$$

Knowing that we have a harmonic oscillation, which can be expressed as $y = y_{max}\sin \omega_n t$, we have $\dot{y}_{max} = y_{max}\omega_n$. Substituting this relation into our energy balance gives us

$$\frac{1}{2}m\left(\frac{b}{l}y_{max}\omega_n\right)^2 = \frac{1}{2}ky_{max}^2 \qquad so\ that \qquad \omega_n = \frac{l}{b}\sqrt{k/m} \quad Ans.$$

as before.

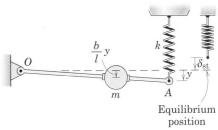

Equilibrium position

① For large values of y, the circular motion of the end of the bar would cause our expression for the deflection of the spring to be in error.

② Here again, we note the simplicity of the expression for potential energy when the displacement is measured from the equilibrium position.

Sample Problem 8/11

Determine the natural frequency ω_n of vertical vibration of the 3-kg collar to which are attached the two uniform 1.2-kg links, which may be treated as slender bars. The stiffness of the spring, which is attached to both the collar and the foundation, is $k = 1.5$ kN/m, and the bars are both horizontal in the equilibrium position. A small roller on end B of each link permits end A to move with the collar. Frictional retardation is negligible.

Solution. In the equilibrium position, the compression P in the spring equals the weight of the 3-kg collar, plus half the weight of each link or $P = 3(9.81) + 2(\frac{1}{2})(1.2)(9.81) = 41.2$ N. The corresponding static deflection of the spring is $\delta_{st} = P/k = 41.2/1.5(10^3) = 27.5(10^{-3})$ m. With the displacement variable y measured downward from the equilibrium position, which becomes the position of zero potential energy, the potential energy for each member in the displaced position is

$$\text{(Spring)} \quad V_e = \frac{1}{2}k(y + \delta_{st})^2 - \frac{1}{2}k\delta_{st}^2 = \frac{1}{2}ky^2 + k\delta_{st}y$$

$$= \frac{1}{2}(1.5)(10^3)y^2 + 1.5(10^3)(27.5)(10^{-3})y$$

$$= 750y^2 + 41.2y \text{ J}$$

$$\text{(Collar)} \quad V_g = -m_cgy = -3(9.81)y = -29.4y \text{ J}$$

$$\text{(Each link)} \quad V_g = -m_lg\frac{y}{2} = -1.2(9.81)\frac{y}{2} = -5.89y \text{ J}$$

The total potential energy of the system then becomes

$$V = 750y^2 + 41.2y - 29.4y - 2(5.89)y = 750y^2 \text{ J}$$

The maximum kinetic energy occurs at the equilibrium position, where the velocity $\dot{y}$ of the collar has its maximum value. For small motion of the collar, end B of each link has essentially no motion, so that each link may be treated as though it were rotating about B as a fixed point with an angular velocity $\dot{y}/0.3$. Thus, the kinetic energy of each part is

$$\text{(Collar)} \quad T = \frac{1}{2}m_c\dot{y}^2 = \frac{3}{2}\dot{y}^2 \text{ J}$$

$$\text{(Each link)} \quad T = \frac{1}{2}I_B\omega^2 = \frac{1}{2}\left(\frac{1}{3}m_ll^2\right)(\dot{y}/l)^2 = \frac{1}{6}m_l\dot{y}^2$$

$$\frac{1}{6}(1.2)\dot{y}^2 = 0.2\dot{y}^2$$

Hence, the kinetic energy of the collar and both links is

$$T = \frac{3}{2}\dot{y}^2 + 2(0.2\dot{y}^2) = 1.9\dot{y}^2$$

With the harmonic motion expressed by $y = y_{max}\sin\omega_n t$, we have $\dot{y}_{max} = y_{max}\omega_n$, so that the energy balance $T_{max} = V_{max}$ with $\dot{y} = \dot{y}_{max}$ becomes

$$1.9(y_{max}\omega_n)^2 = 750y_{max}^2 \quad \text{or} \quad \omega_n = \sqrt{750/1.9} = 19.87 \text{ Hz} \quad \textit{Ans.}$$

① Note that the mass center of each link moves down only half as far as the collar.

② We note again that measurement of the motion variable y from the equilibrium position results in the total potential energy being simply $V = \frac{1}{2}ky^2$.

③ Our knowledge of rigid-body kinematics is essential at this point.

④ To appreciate the advantage of the work-energy method for this and similar problems of interconnected systems, the student is encouraged to explore the steps required for solution by the force and moment equations of motion of the separate parts.

⑤ If the oscillations were large, the links could no longer be considered bodies rotating about fixed points. In this event, we would find that the angular velocity of each link would equal $\dot{y}/\sqrt{0.09 - y^2}$, which would cause a nonlinear response no longer described by $y = y_{max}\sin\omega t$.

PROBLEMS

(Solve the following problems by the energy method of Art. 8/5.)

Introductory problems

8/99 The potential energy V of a linear spring-mass system is given in joules by $64x^2$, where x is the displacement in meters measured from the neutral equilibrium position. The kinetic energy T of the system in joules is given by $8\dot{x}^2$. Determine the differential equation of motion for the system and find the period τ of its oscillation. Neglect energy loss.

Ans. $\ddot{x} + 8x = 0$, $\tau = 2.22$ s

8/100 Determine the period τ for the uniform circular hoop of radius r as it oscillates with small amplitude about the horizontal knife edge.

Problem 8/100

8/101 The spoked wheel of radius r, mass m, and centroidal radius of gyration $\bar{k}$ rolls without slipping on the incline. Determine the natural frequency of oscillation and explore the limiting cases of $\bar{k} = 0$ and $\bar{k} = r$.

$$Ans. \quad \omega_n = \sqrt{\frac{k}{m(1 + \bar{k}^2/r^2)}}$$

$$\bar{k} = 0: \omega_n = \sqrt{k/m}$$

$$\bar{k} = r: \omega_n = \sqrt{k/2m}$$

Problem 8/101

8/102 The 1.5-kg bar OA is suspended vertically from the bearing O and is constrained by the two springs each of stiffness $k = 120$ N/m and both equally precompressed with the bar in the vertical equilibrium position. Treat the bar as a uniform slender rod and compute the natural frequency f_n of small oscillations about O.

Problem 8/102

8/103 A uniform rod of mass m and length l is welded at one end to the rim of a light circular hoop of radius l. The other end lies at the center of the hoop. Determine the period τ for small oscillations about the vertical position of the bar if the hoop rolls on the horizontal surface without slipping.

$$Ans. \ \tau = 2\pi\sqrt{\frac{2l}{3g}}$$

Problem 8/103

8/104 Calculate the frequency f_n of vertical oscillation of the system shown. The 40-kg pulley has a radius of gyration about its center O of 200 mm.

$k = 2$ kN/m

300 mm

O — 40 kg $k_O = 200$ mm

30 kg

Problem 8/104

8/105 The homogeneous circular cylinder of Prob. 8/93, repeated here, rolls without slipping on the track of radius R. Determine the period τ for small oscillations.

$$Ans. \ \tau = \pi\sqrt{\frac{6(R - r)}{g}}$$

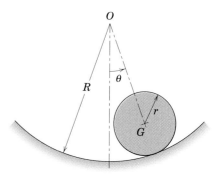

Problem 8/105

8/106 The ends of the uniform slender bar of mass m and length L move freely in the vertical and horizontal slots under the action of the two precompressed springs each of stiffness k as shown. If the bar is in static equilibrium when $\theta = 0$, determine the natural frequency ω_n of small oscillations.

Problem 8/106

8/107 Determine the period τ of small oscillations of the cylindrical shell of Prob. 8/80, repeated here, about the y-axis.

$$Ans. \quad \tau = 2\pi \sqrt{\frac{2}{gh}} \sqrt{\frac{r^2}{2} + \frac{h^2}{3}}$$

Problem 8/107

8/108 Develop an expression for the natural circular frequency ω_n of the system of Prob. 8/27, repeated here. The mass and friction of the pulleys are negligible.

Problem 8/108

8/109 The rotational axis of the turntable is inclined at an angle α from the vertical. The turntable shaft pivots freely in bearings which are not shown. If a small block of mass m is placed a distance r from point O, determine the natural frequency ω_n for small rotational oscillations through the angle θ. The mass moment of inertia of the turntable about the axis of its shaft is I.

$$Ans. \quad \omega_n = \sqrt{\frac{mgr \sin \alpha}{I + mr^2}}$$

Problem 8/109

8/110 Determine the period τ of vertical oscillations for the system composed of the 140-kg frame and two 80-kg pulleys, each of which has a radius of gyration $k_O = 400$ mm. The flexible wires do not slip on the pulleys.

Problem 8/110

8/111 Derive the natural frequency f_n of the system composed of two homogeneous circular cylinders, each of mass M, and the connecting link AB of mass m. Assume small oscillations.

$$\text{Ans. } f_n = \frac{1}{2\pi}\sqrt{\frac{mgr_0}{3Mr^2 + m(r - r_0)^2}}$$

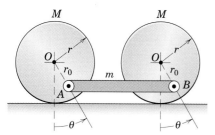

Problem 8/111

8/112 Each of the two uniform 1.5-kg slender bars is hinged freely at A with its small upper-end guide roller free to move in the horizontal guide. The bars are supported in their 45° equilibrium positions by the vertical spring of stiffness 1050 N/m. If point A is given a very small vertical displacement and then released, calculate the natural frequency of the resulting motion.

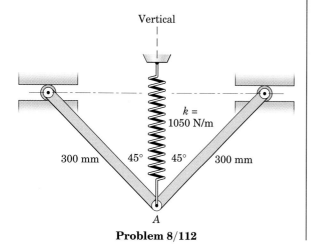

Problem 8/112

8/113 The thin homogeneous panel of mass m is hinged to swing freely about a fixed axis which makes an angle α with the vertical. Determine the period of small oscillations.

$$\text{Ans. } \tau = 2\pi\sqrt{\frac{2b}{3g \sin \alpha}}$$

Problem 8/113

8/114 The 12-kg block is supported by the two 5-kg links with two torsion springs, each of constant $K = 500$ N·m/rad, arranged as shown. The springs are sufficiently stiff so that stable equilibrium is established in the position shown. Determine the natural frequency f_n for small oscillations about this equilibrium position.

Problem 8/114

8/115 The semicylinder of mass m and radius r, repeated here from Prob. 8/90, rolls without slipping on the horizontal surface. By the method of this article, determine the period τ of small oscillations.

Ans. $\tau = 7.78 \sqrt{r/g}$

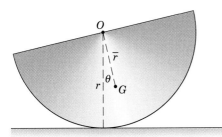

Problem 8/115

8/116 The front-end suspension of an automobile is shown. Each of the coil springs has a stiffness of 46.8 kN/m. If the mass of the front-end frame and equivalent portion of the body attached to the front end is 800 kg, determine the natural frequency f_n of vertical oscillation of the frame and body in the absence of shock absorbers. (*Hint:* To relate the spring deflection to the deflection of the frame and body, consider the frame fixed and let the ground and wheels move vertically.)

Problem 8/116

8/117 If the spring-loaded frame is given a slight vertical disturbance from its equilibrium position shown, determine its natural frequency f_n of vibration. The mass of the upper member is 24 kg, and that of the lower members is negligible. Each spring has a stiffness of 9 kN/m.

Ans. $f_n = 2.62$ Hz

Problem 8/117

8/118 The uniform slender rod of length $2b$ is supported in the horizontal plane by a bifilar suspension. The rod is set into small angular oscillation about the vertical axis through its center O. Derive the expression for the period τ of oscillation. (*Hint:* From the auxiliary sketch note that the rod rises a distance h corresponding to an angular twist θ. Also note that $l\beta \cong b\theta$ for small angles and that $\cos \beta$ may be replaced by the first two terms of its series expansion. A simple harmonic solution of the form $\theta = \theta_0 \sin \omega_n t$ may be used for small angles.)

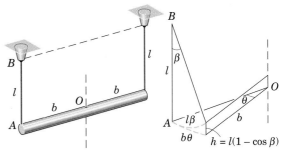

Problem 8/118

8/119 The semicircular cylindrical shell of radius r with small but uniform wall thickness is set into small rocking oscillation on the horizontal surface. If no slipping occurs, determine the expression for the period τ of each complete oscillation.

$$Ans. \quad \tau = 2\pi \sqrt{\frac{(\pi - 2)r}{g}}$$

▶8/120 A hole of radius $R/4$ is drilled through a cylinder of radius R to form a body of mass m as shown. If the body rolls on the horizontal surface without slipping, determine the period τ for small oscillations.

$$Ans. \quad \tau = 41.4 \sqrt{\frac{R}{g}}$$

Problem 8/119

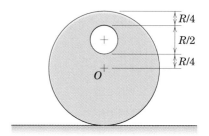

Problem 8/120

8/6 *PROBLEM FORMULATION AND REVIEW*

In studying the vibrations of particles and rigid bodies in Chapter 8, we have observed that the subject is simply a direct application of the fundamental principles of dynamics as derived in Chapters 3 and 6. In these previous chapters, we established the dynamic behavior of a body at an instant of time and also found the changes in motion resulting from finite intervals of displacement or time. Chapter 8, on the other hand, has dealt with the solution of the defining differential equations of motion, so that the linear or angular displacement can be expressed as a function of time. This information is of critical importance in many engineering problems.

Our study of the time response of particles has been divided into the two categories of free and forced motion, with the further subdivisions of negligible and significant damping. It has been seen that the damping ratio ζ is a convenient variable for determining the nature of unforced but viscously damped vibrations. The prime lesson associated with harmonic forcing is that driving a lightly damped system with a force whose frequency is near the natural frequency can cause motion of excessively large amplitude—a condition called resonance that usually must be carefully avoided.

In our study of rigid-body vibrations, we observed that the equation of motion has a form identical to that for particle vibrations. Whereas particle vibrations may be described completely by the equations governing linear motion, rigid-body vibrations usually require the equations of rotational dynamics.

In the final article of Chapter 8, we learned how the energy method can facilitate the determination of the natural frequency ω_n in free vibration problems where damping may be neglected. Here the total mechanical energy of the system is assumed to be constant. Setting its first time derivative to zero leads directly to the differential equation of motion for the system. The energy approach also permits the analysis of a conservative system of interconnected parts without dismembering the system. When mechanical energy dissipation cannot be neglected, we must use the laws of motion in terms of force, mass, and acceleration.

Throughout the chapter, we have restricted our attention to systems having one degree of freedom, where the position of the system can be specified by a single variable. One motion variable suffices for many engineering problems. If a system possesses n degrees of freedom, it has n natural frequencies. Thus, if a harmonic force is applied to such a system which is lightly damped, there are n driving frequencies which can cause motion of large amplitude. By a process called modal analysis, a complex system with n degrees of freedom can be reduced to n single-degree-of-freedom systems. For this reason, the thorough understanding of the material of this chapter is vital for the further study of vibrations.

REVIEW PROBLEMS

8/121 Determine the value of the damping coefficient c for which the system is critically damped if $k = 70$ kN/m and $m = 100$ kg.

Ans. $c = 4580$ N·s/m

Problem 8/121

8/122 The uniform circular disk is suspended by a socket (not shown) which fits over the small ball attachment at O. Determine the period of small motion if the disk swings freely about (a) axis A-A and (b) axis B-B. Neglect the small offset, mass, and friction of the ball.

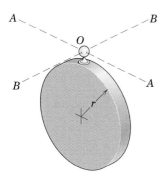

Problem 8/122

8/123 The block of mass M is suspended by the two uniform slender rods each of mass m. Determine the natural frequency ω_n of small oscillations for the system shown.

$$Ans. \ \omega_n = \sqrt{\frac{g}{l}\left(\frac{M+m}{M+\frac{2}{3}m}\right)}$$

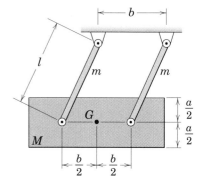

Problem 8/123

8/124 The triangular frame is constructed of uniform slender rod and pivots about a horizontal axis through point O. Determine the critical driving frequency ω_c of the block B which will result in oscillations of the assembly which would tend to become excessively large. The total mass of the frame is m.

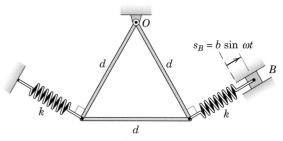

Problem 8/124

8/125 Determine the period τ for small oscillations of the assembly composed of two light bars and two particles, each of mass m. Investigate your expression as the angle α approaches values of 0 and 180°.

$$Ans.\ \ \tau = 2\pi\sqrt{l/(g\ \cos\ \alpha/2)}$$

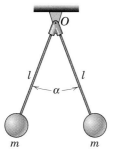

Problem 8/125

8/126 The uniform triangular plate pivots freely about a horizontal axis through point O. Determine the natural frequency of small oscillations.

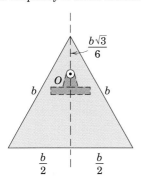

Problem 8/126

8/127 Determine the largest amplitude x_0 for which the uniform circular disk will roll without slipping on the horizontal surface.

$$Ans.\ \ x_0 = \frac{3\mu_s mg}{k}$$

Problem 8/127

8/128 Calculate the damping ratio ζ of the system shown if the mass and radius of gyration of the stepped cylinder are $m = 8$ kg and $\overline{k} = 135$ mm. Also, the spring constant is $k = 2.6$ kN/m, and the damping coefficient of the hydraulic cylinder is $c = 30$ N·s/m. The cylinder rolls without slipping on the radius $r = 150$ mm and the spring can support tension as well as compression.

Problem 8/128

8/129 The cylinder A of radius r, mass m, and radius of gyration $\overline{k}$ is driven by a cable-spring system attached to the drive cylinder B, which oscillates as indicated. If the cables do not slip on the cylinders, and if both springs are stretched to the degree that they do not go slack during a motion cycle, determine an expression for the amplitude θ_{max} of the steady-state oscillation of cylinder A.

$$Ans.\ \ \theta_{max} = \phi_0 \frac{r_0/r}{1 - (\omega/\omega_n)^2}\ \text{where}\ \omega_n = \frac{r}{\overline{k}}\sqrt{\frac{2k}{m}}$$

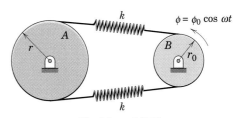

Problem 8/129

8/130 A linear oscillator with mass m, spring constant k, and viscous damping coefficient c is set into motion when released from a displaced position. Derive an expression for the energy loss Q during one complete cycle in terms of the amplitude x_1 at the start of the cycle. (See Fig. 8/7.)

8/131 With collar A held in position, a static horizontal force of 14 N applied to the 2.2-kg sphere B gives it a deflection of 15 mm against the elastic resistance of the slender rod of negligible mass to which it is attached. If the collar A is given a horizontal harmonic oscillation with a frequency of 2 cycles per second and an amplitude of 7.5 mm, calculate the amplitude X of the horizontal vibration of the sphere. Assume negligible damping.

Ans. $X = 11.95$ mm

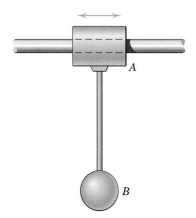

Problem 8/131

8/132 The seismic instrument shown is secured to a ship's deck near the stern where propeller-induced vibration is most pronounced. The ship has a single three-bladed propeller which turns at 180 rev/min and operates partly out of water, thus causing a shock as each blade breaks the surface. The damping ratio of the instrument is $\zeta = 0.5$, and its undamped natural frequency is 3 Hz. If the measured amplitude of A relative to its frame is 0.75 mm, compute the amplitude δ_0 of the vertical vibration of the deck.

Problem 8/132

8/133 A 220-kg experimental engine is mounted on a test stand with spring mounts at A and B, each with a stiffness of 105 kN/m. The radius of gyration of the engine about its mass center G is 115 mm. With the motor not running, calculate the natural frequency $(f_n)_y$ of vertical vibration and $(f_n)_\theta$ of rotation about G. If vertical motion is suppressed and a slight rotational imbalance occurs, at what speed N should the engine not be run?

Ans. $(f_n)_y = 4.92$ Hz, $(f_n)_\theta = 10.69$ Hz,
$N = 641$ rev/min

Problem 8/133

8/134 The uniform bar of mass M and length l has a small roller of mass m with negligible bearing friction at each end. Determine the period τ of the system for small oscillations on the curved track.

Problem 8/134

▶8/135 The system composed of four light rigid rods and the cylinder of mass m is shown in its equilibrium position. Determine the natural frequency ω_n of small-amplitude vertical oscillations of the cylinder. (*Hint:* Retain only first-order terms.)

$$Ans. \quad \omega_n = \sqrt{\frac{k}{m} + \frac{g}{b}\sqrt{2}}$$

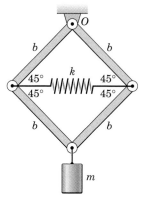

Problem 8/135

Computer-oriented problems

*8/136 The mass of the system shown is released with the initial conditions $x_0 = 0.1$ m and $\dot{x}_0 = -5$ m/s at $t = 0$. Plot the response of the system and determine the time(s) (if any) at which the displacement $x = -0.05$ m.

Problem 8/136

*8/137 The 10-kg oscillator contains an unbalanced motor whose speed N in revolutions per minute can be varied. The oscillator is restrained in its horizontal motion by a spring of stiffness $k = 1080$ N/m and by a viscous damper whose piston is resisted by a force of 30 N when moving at a speed of 0.5 m/s. Determine the viscous damping factor ζ and plot the magnification factor M for motor speeds from zero to 300 revolutions per minute. Determine the maximum value of M and the corresponding motor speed.

$$Ans. \quad M_{max} = 1.809 \text{ at } N = 90.6 \text{ rev/min}$$
$$\zeta = 0.289$$

Problem 8/137

*8/138 Plot the response x of the 20-kg body over the time interval $0 \leq t \leq 1$ second. Determine the maximum and minimum values of x and their respective times. The initial conditions are $x_0 = 0$ and $\dot{x}_0 = 2$ m/s.

Problem 8/138

*8/139 Shown in the figure are the elements of a displacement meter used to study the motion $y_B = b \sin \omega t$ of the base. The motion of the mass relative to the frame is recorded on the rotating drum. If $l_1 = 360$ mm, $l_2 = 480$ mm, $l_3 = 600$ mm, $m = 0.9$ kg, $c = 1.4$ N·s/m, and $\omega = 10$ rad/s, determine the range of the spring constant k over which the magnitude of the recorded relative displacement is less than $1.5b$. It is assumed that the ratio ω/ω_n must remain greater than unity. *Ans.* $0 < k < 27.4$ N/m

Problem 8/139

*8/140 The 4-kg cylinder is attached to a viscous damper and to the spring of stiffness $k = 800$ N/m. If the cylinder is released from rest at time $t = 0$ from the position where it is displaced a distance $y = 100$ mm from its equilibrium position, plot the displacement y as a function of time for the first second for the two cases where the viscous damping coefficient is (*a*) $c = 124$ N·s/m and (*b*) $c = 80$ N·s/m.

Problem 8/140

*8/141 Determine and plot the response $x = f(t)$ for the undamped linear oscillator subjected to the force F which varies linearly with time for the first ¾ second as shown. The mass is initially at rest with $x = 0$ at time $t = 0$. *Ans.* $x = 0.0926(t - 0.0913 \sin 10.95t)$ m

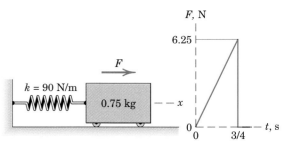

Problem 8/141

*8/142 The damped linear oscillator of mass $m = 4$ kg, spring constant $k = 200$ N/m, and viscous damping factor $\zeta = 0.1$ is initially at rest in a neutral position when it is subjected to a sudden impulse loading F over a very short period of time as shown. If the impulse $I = \int F \, dt = 8$ N·s, determine the resulting displacement x as a function of time and plot it for the first two seconds following the impulse.

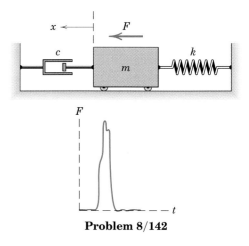

Problem 8/142

AREA MOMENTS OF INERTIA

See Appendix A of *Vol. 1 Statics* for a treatment of the theory and calculation of area moments of inertia. Because this quantity plays an important role in the design of structures, especially those dealt with in statics, we present only a brief definition in this *Dynamics* volume so that the student can appreciate the basic differences between area and mass moments of inertia.

The moments of inertia of a plane area A about the x- and y-axes in its plane and about the z-axis normal to its plane, Fig. A/1, are defined by

$$I_x = \int y^2 \, dA \qquad I_y = \int x^2 \, dA \qquad I_z = \int r^2 \, dA$$

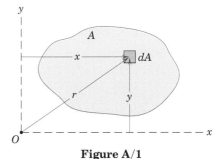

Figure A/1

where dA is the differential element of area and $r^2 = x^2 + y^2$. Clearly, the polar moment of inertia I_z equals the sum $I_x + I_y$ of the rectangular moments of inertia. For thin flat plates, the area moment of inertia is useful in the calculation of the mass moment of inertia, as explained in Appendix B.

The area moment of inertia is a measure of the distribution of area about the axis in question and, for that axis, is a constant property of the area. The dimensions of area moment of inertia are (distance)^4 expressed in m^4 or mm^4 in SI units. In contrast, mass moment of inertia is a measure of the distribution of mass about the axis in question, and its dimensions are (mass)(distance)^2 which are expressed in kg·m^2 in SI units.

MASS MOMENTS OF INERTIA

B

B/1 MASS MOMENTS OF INERTIA ABOUT AN AXIS

The rotational equation of motion about an axis normal to the plane of motion for a rigid body in plane motion contains an integral which depends on the distribution of mass with respect to the moment axis. This integral occurs whenever a rigid body has an angular acceleration about its axis of rotation.

Consider a body of mass m, Fig. B/1, rotating about an axis O-O with an angular acceleration α. All particles of the body move in parallel planes which are normal to the rotation axis O-O. We may choose any one of the planes as the plane of motion, although the one containing the center of mass is usually the one so designated. An element of mass dm has a component of acceleration tangent to its circular path equal to $r\alpha$, and by Newton's second law of motion the resultant tangential force on this element equals $r\alpha\,dm$. The moment of this force about the axis O-O is $r^2\alpha\,dm$, and the sum of the moments of these forces for all elements is $\int r^2\alpha\,dm$. For a rigid body, α is the same for all radial lines in the body and we may take it outside the integral sign. The remaining integral is known as the mass moment of inertia I of the body about the axis O-O and is

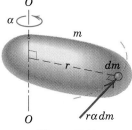

Figure B/1

$$I = \int r^2\,dm \qquad \textbf{(B/1)}$$

This integral represents an important property of a body and is involved in the analysis of any body which has rotational acceleration about a given axis. Just as the mass m of a body is a measure of the resistance to translational acceleration, the moment of inertia I is a measure of resistance to rotational acceleration of the body.

The moment-of-inertia integral may be expressed alternatively as

$$I = \Sigma r_i^2 m_i \qquad (B/1a)$$

where r_i is the radial distance from the inertia axis to the representative particle of mass m_i and where the summation is taken over all particles of the body.

If the density ρ is constant throughout the body, the moment of inertia becomes

$$I = \rho \int r^2 \, dV$$

where dV is the element of volume. In this case, the integral by itself defines a purely geometrical property of the body. When the density is not constant but is expressed as a function of the coordinates of the body, it must be left within the integral sign and its effect accounted for in the integration process.

In general, the coordinates that best fit the boundaries of the body should be used in the integration. It is particularly important that we make a good choice of the element of volume dV. To simplify the integration, an element of lowest possible order should be chosen, and the correct expression for the moment of inertia of the element about the axis involved should be used. For example, in finding the moment of inertia of a solid right-circular cone about its central axis, we may choose an element in the form of a circular slice of infinitesimal thickness, Fig. B/2a. The differential moment

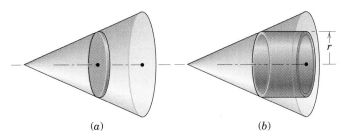

(a)   (b)

Figure B/2

of inertia for this element is the expression for the moment of inertia of a circular cylinder of infinitesimal altitude about its central axis. (This expression will be obtained in Sample Problem B/1.) Alternatively, we could choose an element in the form of a cylindrical shell of infinitesimal thickness as shown in Fig. B/2b. Since all of the mass of the element is at the same distance r from the inertia axis, the differential moment of inertia for this element is merely $r^2 \, dm$ where dm is the differential mass of the elemental shell.

From the definition of mass moment of inertia, its dimensions are (mass)(distance)^2 and are expressed in the units kg$\cdot$m^2 in SI units.

(a) *Radius of gyration.* The radius of gyration k of a mass m about an axis for which the moment of inertia is I is defined as

$$\boxed{k = \sqrt{\frac{I}{m}} \quad \text{or} \quad I = k^2 m} \qquad \textbf{(B/2)}$$

Thus, k is a measure of the distribution of mass of a given body about the axis in question, and its definition is analogous to the definition of the radius of gyration for area moments of inertia. If all the mass m of a body could be concentrated at a distance k from the axis, the moment of inertia would be unchanged. The moment of inertia of a body about a particular axis is frequently indicated by specifying the mass of the body and the radius of gyration of the body about the axis. The moment of inertia is then calculated from Eq. B/2.

(b) *Transfer of axes.* If the moment of inertia of a body is known about an axis passing through the mass center, it may be determined easily about any parallel axis. To prove this statement, consider the two parallel axes in Fig. B/3, one being an axis through the mass center G and the other a parallel axis through some other point C. The radial distances from the two axes to any element of mass dm are r_0 and r, and the separation of the axes is d. Substituting the law of cosines $r^2 = r_0{}^2 + d^2 + 2r_0 d \cos \theta$ into the definition for the moment of inertia about the axis through C gives

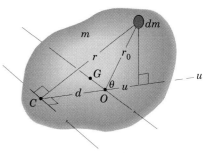

Figure B/3

$$I = \int r^2 \, dm = \int (r_0{}^2 + d^2 + 2r_0 d \cos \theta) \, dm$$

$$= \int r_0{}^2 \, dm + d^2 \int dm + 2d \int u \, dm$$

The first integral is the moment of inertia $\bar{I}$ about the mass-center axis, the second term is md^2, and the third integral equals zero, since the u-coordinate of the mass center with respect to the axis through G is zero. Thus, the parallel-axis theorem is

$$\boxed{I = \bar{I} + md^2} \qquad \textbf{(B/3)}$$

It must be remembered that the transfer cannot be made unless one axis passes through the center of mass and unless the axes are parallel. When the expressions for the radii of gyration are substituted in Eq. B/3, there results

$$\boxed{k^2 = \bar{k}^2 + d^2} \qquad \textbf{(B/3a)}$$

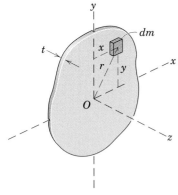

Figure B/4

Equation B/3a is the parallel-axis theorem for obtaining the radius of gyration k about an axis a distance d from a parallel axis through the mass center, for which the radius of gyration is $\bar{k}$.

For plane-motion problems where rotation occurs about an axis normal to the plane of motion, a single subscript for I is sufficient to designate the inertia axis. Thus, if the plate of Fig. B/4 has plane motion in the x-y plane, the moment of inertia of the plate about the z-axis through O is designated I_O. For three-dimensional motion, however, where components of rotation may occur about more than one axis, we use a double subscript to preserve notational symmetry with product-of-inertia terms, which are described in Art. B/2. Thus, the moments of inertia about the x-, y-, and z-axes are labeled I_{xx}, I_{yy}, and I_{zz}, respectively, and from Fig. B/5 we see that they become

$$
\begin{aligned}
I_{xx} &= \int r_x^2\, dm = \int (y^2 + z^2)\, dm \\
I_{yy} &= \int r_y^2\, dm = \int (z^2 + x^2)\, dm \\
I_{zz} &= \int r_z^2\, dm = \int (x^2 + y^2)\, dm
\end{aligned}
\qquad \textbf{(B/4)}
$$

These integrals are cited in Eqs. 7/10 of Art. 7/7 on angular momentum in three-dimensional rotation.

The similarity between the defining expressions for mass moments of inertia and area moments of inertia is easily observed. An exact relationship between the two moment-of-inertia expressions exists in the case of flat plates. Consider the flat plate of uniform thickness in Fig. B/4. If the constant thickness is t and the density is ρ, the mass moment of inertia I_{zz} of the plate about the z-axis normal to the plate is

$$
I_{zz} = \int r^2\, dm = \rho t \int r^2\, dA = \rho t I_z \qquad \textbf{(B/5)}
$$

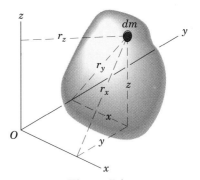

Figure B/5

Thus, the mass moment of inertia about the z-axis equals the mass per unit area ρt times the polar moment of inertia I_z of the plate area about the z-axis. If t is small compared with the dimensions of the plate in its plane, the mass moments of inertia I_{xx} and I_{yy} of the plate about the x- and y-axes are closely approximated by

$$I_{xx} = \int y^2 \, dm = \rho t \int y^2 \, dA = \rho t I_x$$
$$I_{yy} = \int x^2 \, dm = \rho t \int x^2 \, dA = \rho t I_y$$

$$(B/6)$$

Hence, the mass moments of inertia equal the mass per unit area ρt times the corresponding area moments of inertia. The double subscripts for mass moments of inertia distinguish these quantities from area moments of inertia.

Inasmuch as $I_z = I_x + I_y$ for area moments of inertia, we have

$$I_{zz} = I_{xx} + I_{yy} \qquad (B/7)$$

which holds *only* for a thin flat plate. This restriction is observed from Eqs. B/6, which do not hold true unless the thickness t or z-coordinate of the element is negligible compared with the distance of the element from the corresponding x- or y-axis. Equation B/7 is very useful when dealing with a differential mass element taken as a flat slice of differential thickness, say, dz. In this case, Eq. B/7 holds exactly and becomes

$$dI_{zz} = dI_{xx} + dI_{yy} \qquad (B/7a)$$

(c) Composite bodies. The defining integral, Eq. B/1, involves the square of the distance from the axis to the element and so is always positive. Thus, as in the case of area moments of inertia, the mass moment of inertia of a composite body is the sum of the moments of inertia of the individual parts about the same axis. It is often convenient to treat a composite body as defined by positive volumes and negative volumes. The moment of inertia of a negative element, such as the material removed to form a hole, must be considered a negative quantity.

Thorough familiarity with the calculation of mass moments of inertia for rigid bodies is an absolute necessity for the study of the dynamics of rotation.

A summary of some of the more useful formulas for mass moments of inertia of various masses of common shape is given in Table D/4, Appendix D.

Sample Problem B/1

Determine the moment of inertia and radius of gyration of a homogeneous right-circular cylinder of mass m and radius r about its central axis O-O.

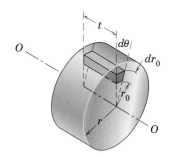

Solution. An element of mass in cylindrical coordinates is $dm =$ ① $\rho\,dV = \rho t r_0\,dr_0\,d\theta$, where ρ is the density of the cylinder. The moment of inertia about the axis of the cylinder is

②
$$I = \int r_0{}^2\,dm = \rho t \int_0^{2\pi}\!\!\int_0^r r_0{}^3\,dr_0\,d\theta = \rho t\,\frac{\pi r^4}{2} = \tfrac{1}{2}mr^2 \quad Ans.$$

The radius of gyration is

$$k = \sqrt{\frac{I}{m}} = \frac{r}{\sqrt{2}} \qquad Ans.$$

① If we had started with a cylindrical shell of radius r_0 and axial length t as our mass element dm, then $dI = r_0{}^2\,dm$ directly. The student should evaluate the integral.

② The result $I = \tfrac{1}{2}mr^2$ applies *only* to a solid homogeneous circular cylinder and cannot be used for any other wheel of circular periphery.

Sample Problem B/2

Determine the moment of inertia and radius of gyration of a homogeneous solid sphere of mass m and radius r about a diameter.

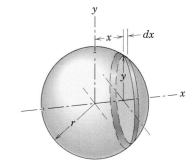

Solution. A circular slice of radius y and thickness dx is chosen as the volume element. From the results of Sample Problem B/1, the moment of inertia about the x-axis of the elemental cylinder is

①
$$dI_{xx} = \tfrac{1}{2}(dm)y^2 = \tfrac{1}{2}(\pi\rho y^2\,dx)y^2 = \frac{\pi\rho}{2}(r^2 - x^2)^2\,dx$$

where ρ is the constant density of the sphere. The total moment of inertia about the x-axis is

$$I_{xx} = \frac{\pi\rho}{2}\int_{-r}^{r}(r^2 - x^2)^2\,dx = \tfrac{8}{15}\pi\rho r^5 = \tfrac{2}{5}mr^2 \qquad Ans.$$

The radius of gyration is

$$k = \sqrt{\frac{I}{m}} = \sqrt{\frac{2}{5}}\,r \qquad Ans.$$

① Here is an example where we utilize a previous result to express the moment of inertia of the chosen element, which in this case is a right-circular cylinder of differential axial length dx. It would be foolish to start with a third-order element, such as $\rho\,dx\,dy\,dz$, when we can easily solve the problem with a first-order element.

Sample Problem B/3

Determine the moments of inertia of the homogeneous rectangular parallelepiped of mass m about the centroidal x_0- and z-axes and about the x-axis through one end.

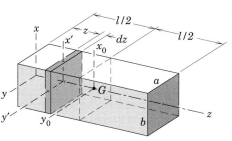

Solution. A transverse slice of thickness dz is selected as the element of volume. The moment of inertia of this slice of infinitesimal thickness equals the moment of inertia of the area of the section times the mass per unit area $\rho\, dz$. Thus, the moment of inertia of the transverse slice about the y'-axis is

$$dI_{y'y'} = (\rho\, dz)(\tfrac{1}{12}ab^3)$$

and that about the x'-axis is

$$dI_{x'x'} = (\rho\, dz)(\tfrac{1}{12}a^3b)$$

① As long as the element is a plate of differential thickness, the principle given by Eq. B/7a may be applied to give

$$dI_{zz} = dI_{x'x'} + dI_{y'y'} = (\rho\, dz)\frac{ab}{12}(a^2 + b^2)$$

These expressions may now be integrated to obtain the desired results.
    The moment of inertia about the z-axis is

$$I_{zz} = \int dI_{zz} = \frac{\rho ab}{12}(a^2 + b^2)\int_0^l dz = \tfrac{1}{12}m(a^2 + b^2) \qquad Ans.$$

where m is the mass of the block. By interchange of symbols, the moment of inertia about the x_0-axis is

$$I_{x_0x_0} = \tfrac{1}{12}m(a^2 + l^2) \qquad Ans.$$

The moment of inertia about the x-axis may be found by the parallel-axis theorem, Eq. B/3. Thus,

$$I_{xx} = I_{x_0x_0} + m\left(\frac{l}{2}\right)^2 = \tfrac{1}{12}m(a^2 + 4l^2) \qquad Ans.$$

This last result may be obtained by expressing the moment of inertia of the elemental slice about the x-axis and integrating the expression over the length of the bar. Again, by the parallel-axis theorem

$$dI_{xx} = dI_{x'x'} + z^2\, dm = (\rho\, dz)(\tfrac{1}{12}a^3b) + z^2\rho ab\, dz = \rho ab\left(\frac{a^2}{12} + z^2\right) dz$$

Integrating gives the result obtained previously:

$$I_{xx} = \rho ab \int_0^l \left(\frac{a^2}{12} + z^2\right) dz = \frac{\rho abl}{3}\left(l^2 + \frac{a^2}{4}\right) = \tfrac{1}{12}m(a^2 + 4l^2)$$

The expression for I_{xx} may be simplified for a long prismatic bar or slender rod whose transverse dimensions are small compared with the length. In this case, a^2 may be neglected compared with $4l^2$, and the moment of inertia of such a slender bar about an axis through one end normal to the bar becomes $I = \tfrac{1}{3}ml^2$. By the same approximation, the moment of inertia about a centroidal axis normal to the bar is $I = \tfrac{1}{12}ml^2$.

① Refer to Eqs. B/6 and recall the expression for the area moment of inertia of a rectangle about an axis through its center parallel to its base.

PROBLEMS

Introductory problems

B/1 Follow the suggestion of Note 1 of Sample Problem B/1 and use the differential element shown in the figure to show that the mass moment of inertia of the homogeneous right-circular cylinder about its central axis O-O is $I = \frac{1}{2}mr^2$.

Problem B/1

B/2 The two small spheres of mass m each are connected by the light rigid rod which lies in the x-z plane. Determine the mass moments of inertia of the assembly about the x-, y-, and z-axes.

Problem B/2

B/3 Determine the mass moments of inertia about the x-, y-, and z-axes of the slender rod of length L and mass m which makes an angle β with the x-axis as shown.

$$Ans.\ I_{xx} = \tfrac{1}{12}mL^2 \sin^2 \beta$$
$$I_{yy} = \tfrac{1}{12}mL^2 \cos^2 \beta$$
$$I_{zz} = \tfrac{1}{12}mL^2$$

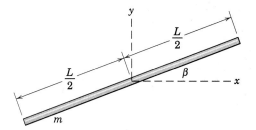

Problem B/3

B/4 Every "slender" rod has a finite radius r. Refer to Table D/4 and derive an expression for the percentage error e which results if one neglects the radius r of a homogeneous solid cylindrical rod of length l when calculating its moment of inertia I_{zz}. Evaluate your expression for the ratios $r/l = 0.01$, 0.1, and 0.5.

Problem B/4

B/5 The moment of inertia of a solid homogeneous cylinder of radius r about an axis parallel to the central axis of the cylinder may be obtained approximately by multiplying the mass of the cylinder by the square of the distance d between the two axes. What percentage error e results if (a) $d = 10r$ and (b) $d = 2r$?

$$Ans.\ (a)\ e = 0.498\%$$
$$(b)\ e = 11.11\%$$

Problem B/5

B/6 From the results of Sample Problem B/2, state without computation the moments of inertia of the solid homogeneous hemisphere of mass m about the x- and z-axes.

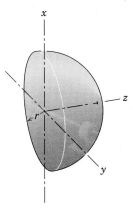

Problem B/6

B/7 In order to better appreciate the greater ease of integration with lower-order elements, determine the mass moment of inertia I_{xx} of the homogeneous thin plate by using the square element (a) and then by using the rectangular element (b). The mass of the plate is m. Then by inspection state I_{yy}, and finally, determine I_{zz}.

$$Ans. \quad I_{xx} = \tfrac{1}{12}ma^2$$
$$I_{yy} = \tfrac{1}{12}mb^2$$
$$I_{zz} = \tfrac{1}{12}m(a^2 + b^2)$$

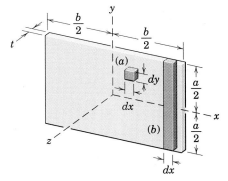

Problem B/7

B/8 Determine the moment of inertia of the thin equilateral triangular plate of mass m about the z-z axis normal to the plate through its mass center G. Solve by using the results for the triangular area in Table D/3, the relations developed for thin flat plates, and the transfer-of-axis theorem.

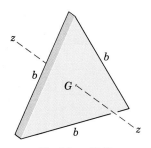

Problem B/8

B/9 Determine I_{xx} for the cylinder with a centered circular hole. The mass of the body is m.

$$Ans. \quad I_{xx} = \tfrac{1}{2}m(r_2{}^2 + r_1{}^2)$$

Problem B/9

B/10 Calculate the radius of gyration about axis O-O for the steel disk with the hole.

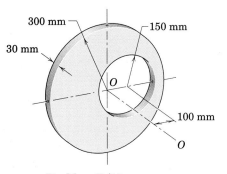

Problem B/10

B/11 The homogeneous bar of mass m has a uniform cross section of area A throughout its length. If the dimensions of A are small compared with the length of the bar, the bar is referred to as a "slender bar." Derive the expressions for the moments of inertia of the bar about the y'-axis and about the y-axis through the mass center. The mass density is ρ. (Commit these much-used results to memory.)

$$\text{Ans. } I_{y'y'} = \tfrac{1}{3}mL^2, I_{yy} = \tfrac{1}{12}mL^2$$

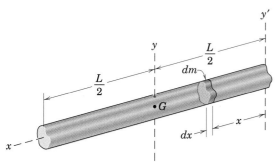

Problem B/11

Representative problems

B/12 Determine the length L of each of the slender rods of mass $m/2$ which must be centrally attached to the faces of the thin homogeneous disk of mass m in order to make the mass moments of inertia of the unit about the x- and z-axes equal.

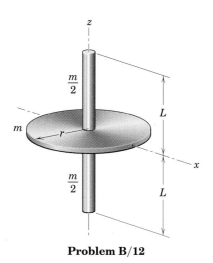

Problem B/12

B/13 A badminton racket is constructed of uniform slender rods bent into the shape shown. Neglect the strings and the built-up wooden grip and estimate the mass moment of inertia about the y-axis through O, which is the location of the player's hand. The mass per unit length of the rod material is ρ.

$$\text{Ans. } I_{yy} = (\tfrac{43}{192} + \tfrac{83}{128}\pi)\rho L^3$$

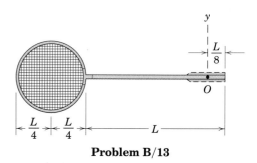

Problem B/13

B/14 Calculate the moment of inertia of the steel control wheel, shown in section, about its central axis. There are eight spokes, each of which has a constant cross-sectional area of 200 mm². What percent n of the total moment of inertia is contributed by the outer rim?

Dimensions in millimeters

Problem B/14

B/15 The uniform rod of length $4b$ and mass m is bent into the shape shown. The diameter of the rod is small compared with its length. Determine the moments of inertia of the rod about the three coordinate axes.

Ans. $I_{xx} = \frac{3}{4}mb^2$, $I_{yy} = \frac{1}{6}mb^2$, $I_{zz} = \frac{3}{4}mb^2$

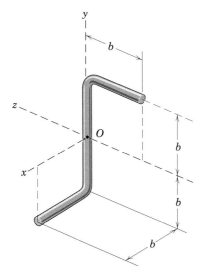

Problem B/15

B/16 The clock pendulum consists of the slender rod of length l and mass m and the bob of mass $7m$. Neglect the effects of the radius of the bob and determine I_O in terms of the bob position x. Calculate the ratio R of I_O evaluated for $x = \frac{3}{4}l$ to I_O evaluated for $x = l$.

Problem B/16

B/17 Determine the mass moments of inertia of the thin parabolic plate of mass m about the x-, y-, and z-axes.

Ans. $I_{xx} = \frac{1}{7}mh^2$, $I_{yy} = \frac{3}{5}mb^2$

$$I_{zz} = m\left(\frac{3b^2}{5} + \frac{h^2}{7}\right)$$

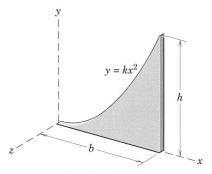

Problem B/17

B/18 Determine the moment of inertia of the half-ring of mass m about its diametral axis a-a and about axis b-b through the midpoint of the arc normal to the plane of the ring. The radius of the circular cross section is small compared with r.

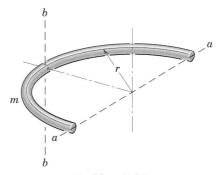

Problem B/18

B/19 Calculate the moment of inertia of the tapered steel rod of circular cross section about an axis normal to the rod through O. Note that the rod diameter is small compared with its length.

Ans. $I_O = 158.9$ kg·mm^2

100 mm

10 mm

O

5 mm

Problem B/19

B/20 The varying radius y of the solid of revolution is proportional to the square of its x-coordinate. If the mass of the body is m, determine I_{xx}.

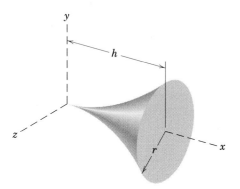

Problem B/20

B/21 Determine the radius of gyration about the z-axis of the paraboloid of revolution shown. The mass of the homogeneous body is m.

$$Ans. \quad k_z = \frac{r}{\sqrt{3}}$$

Problem B/21

B/22 Calculate the moment of inertia of the homogeneous right-circular cone of mass m, base radius r, and altitude h about the cone axis x and about the y-axis through its vertex.

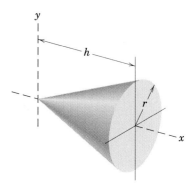

Problem B/22

B/23 In the study of high-speed reentry into the earth's atmosphere, small solid cones are fired at high velocities into low-density gas. A condition of critical stability occurs when the moment of inertia of the cone about its axis of generation a-a equals that about a transverse axis b-b through the mass center. Determine the critical value of the cone angle α for this condition.

$$Ans. \quad \alpha = 26.6°$$

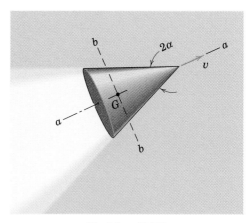

Problem B/23

B/24 Determine the moment of inertia about the x-axis of the homogeneous solid semiellipsoid of revolution having mass m.

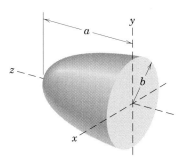

Problem B/24

B/25 Determine the mass moment of inertia about the x-axis of the thin elliptical plate of mass m.

$$Ans. \ I_{xx} = \tfrac{1}{4}mb^2$$

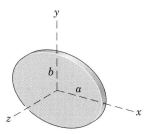

Problem B/25

B/26 The S-shaped piece of mass m is formed from a rod of diameter d and bent into the two semicircular shapes. Determine the moments of inertia of the piece about the x-, y-, and z-axes. Assume that d is small compared with r.

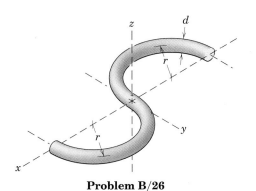

Problem B/26

B/27 Solve Prob. B/26 for the case where $d = r/4$. Make use of Table D/4.

$$Ans. \ I_{xx} = 0.510mr^2$$
$$I_{yy} = 1.510mr^2$$
$$I_{zz} = 2.01mr^2$$

B/28 In the design of a sorting machine, the steel half-cylinder is subject to rapid angular acceleration and deceleration about the y-axis, and its moment of inertia about this axis is required. Calculate I_{yy}. Use Tables D/1 and D/4 as needed.

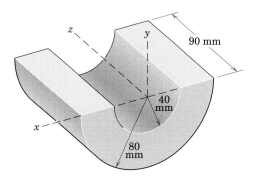

Problem B/28

B/29 Determine the mass moment of inertia about the x-axis of the solid spherical segment of mass m.

$$Ans. \ I_{xx} = \tfrac{53}{200}mR^2$$

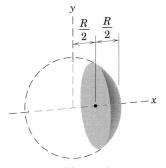

Problem B/29

B/30 Determine the moment of inertia about the x-axis of the portion of the homogeneous sphere shown. The mass of the sphere portion is m.

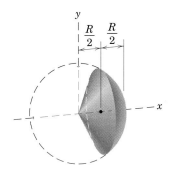

Problem B/30

B/31 The uniform coiled spring has a mass of 2 kg. Approximate its moments of inertia about the x-, y-, and z-axes from the analogy to the properties of a cylindrical shell.

$$Ans. \; I_{xx} = I_{yy} = 11\,820 \; \text{kg} \cdot \text{mm}^2$$
$$I_{zz} = 2810 \; \text{kg} \cdot \text{mm}^2$$

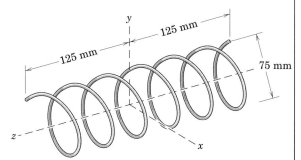

Problem B/31

B/32 The welded assembly shown is made from steel rod which has a mass of 0.993 kg per meter of length. Calculate the moment of inertia of the assembly about the x-x axis.

Dimensions in millimeters

Problem B/32

B/33 Determine the moments of inertia of the half-spherical shell with respect to the x- and z-axes. The mass of the shell is m, and its thickness is negligible compared with the radius r.

$$Ans. \; I_{xx} = I_{zz} = \tfrac{2}{3}mr^2$$

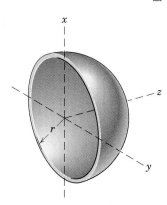

Problem B/33

B/34 A square plate with a quarter-circular sector removed has a net mass m. Determine its moment of inertia about axis A-A normal to the plane of the plate.

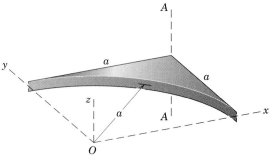

Problem B/34

B/35 Determine the moment of inertia, about the generating axis, of the hollow circular tube of mass m obtained by revolving the thin ring shown in the sectional view completely around the generating axis.

$$Ans. \ I = \tfrac{1}{2}m(2R^2 + 3a^2)$$

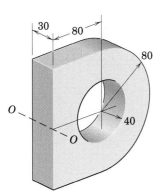

Problem B/35

B/36 The machine element is made of steel and is designed to rotate about axis O-O. Calculate its radius of gyration k_O about this axis.

Dimensions in millimeters

Problem B/36

B/37 The thickness of the homogeneous triangular plate of mass m varies linearly with the distance from the vertex toward the base. The thickness a at the base is small compared with the other dimensions. Determine the moment of inertia of the plate about the y-axis along the center line of the base.

$$Ans. \ I_{yy} = \tfrac{1}{10}mh^2$$

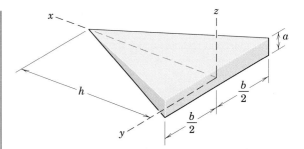

Problem B/37

B/38 Determine the moment of inertia of the triangular plate described in Prob. B/37 about the z-axis.

B/39 Determine I_{xx} for the cone frustum, which has base radii r_1 and r_2 and mass m.

$$Ans. \ I = \tfrac{3}{10}m \ \frac{r_2{}^5 - r_1{}^5}{r_2{}^3 - r_1{}^3}$$

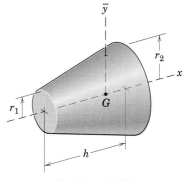

Problem B/39

▶B/40 The cube with semicircular grooves in two opposite faces is cast of lead. Calculate the moment of
inertia of the solid about the a-a axis.
                           Ans. $I_{aa} = 0.367$ kg·m^2

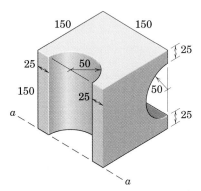

Dimensions in millimeters

Problem B/40

▶B/41 The figure shows one of the three blades of a var-
iable-pitch propeller for a small commercial aircraft. The figure includes a table of the measured
cross-sectional area A of each blade as a function
of its radial distance r from the rotation axis. A
design analysis of the torsional vibration of the
propeller shaft requires a calculation of the moment of inertia of the propeller assembly. Determine the moment of inertia I of each blade extending from $r = 150$ mm to $r = 1500$ mm. The
blade is a forged aluminum alloy with a density of
2850 kg/m^3. (*Suggestion:* Plot A vs. r and select
at least $n = 6$ increments of blade length for determining the contribution of each to the total
value.) Would you anticipate any significant
change in I as the pitch of the blade is changed?
                      Ans. $I = 4.22$ kg·m^2 ($n = 6$)
                            $I = 4.15$ kg·m^2 ($n = 11$)

r, mm	A, mm^2
150	3750
250	3420
500	2690
750	2050
1000	1494
1250	1019
1500	625

Problem B/41

▶B/42 A shell of mass m is obtained by revolving the circular section about the z-axis. If the thickness of
the shell is small compared with a and if $r = a/3$,
determine the radius of gyration of the shell about
the z-axis.
                           Ans. $k_z = 0.890a$

Problem B/42

▶**B/43** By direct integration, determine the moment of inertia about the Z-axis of the thin semicircular disk of mass m and radius R inclined at an angle θ from the X-y plane.

$$\text{Ans. } I_{ZZ} = \tfrac{1}{4}mR^2(1 + \cos^2\theta)$$

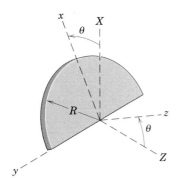

Problem B/43

****B/44** A preliminary design model to ensure rotational stability for a spacecraft consists of the cylindrical shell and the two square panels as shown. The shell and panels have the same thickness and density. It can be shown that rotational stability about the z-axis can be maintained if I_{zz} is less than I_{xx} and I_{yy}. For a given value of r, determine the limitation for L.

$$\text{Ans. } L > 4.54r$$

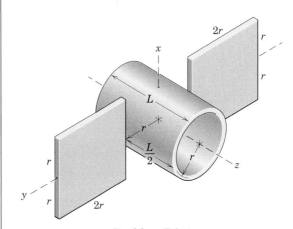

Problem B/44

B/2 PRODUCTS OF INERTIA

For problems in the rotation of three-dimensional rigid bodies, the expression for angular momentum contains, in addition to the moment-of-inertia terms, *product-of-inertia* terms defined as

$$
\begin{aligned}
I_{xy} &= I_{yx} = \int xy \, dm \\[4pt]
I_{xz} &= I_{zx} = \int xz \, dm \\[4pt]
I_{yz} &= I_{zy} = \int yz \, dm
\end{aligned}
\tag{B/8}
$$

These expressions were cited in Eqs. 7/10 in the expansion of the expression for angular momentum, Eq. 7/9.

The calculation of products of inertia involves the same basic procedure which we have followed in calculating moments of inertia and in evaluating other volume integrals as far as the choice of element and the limits of integration are concerned. The only special precaution we need to observe is to be doubly watchful of the algebraic signs in the expressions. Whereas moments of inertia are always positive, products of inertia may be either positive or negative. The units of products of inertia are the same as those of moments of inertia.

We have seen that the calculation of moments of inertia is often simplified by using the transfer-of-axis theorem. A similar theorem exists for transferring products of inertia, and we prove it easily as follows. In Fig. B/6 is shown the *x-y* view of a rigid body with parallel axes x_0-y_0 passing through the mass center G and located from the *x-y* axes by the distances d_x and d_y. The product of inertia about the *x-y* axes by definition is

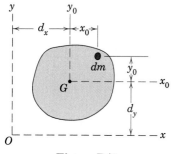

Figure B/6

$$
\begin{aligned}
I_{xy} &= \int xy \, dm = \int (x_0 + d_x)(y_0 + d_y) \, dm \\[4pt]
&= \int x_0 y_0 \, dm + d_x d_y \int dm + d_x \int y_0 \, dm + d_y \int x_0 \, dm \\[4pt]
&= I_{x_0 y_0} + m d_x d_y
\end{aligned}
$$

The last two integrals vanish since the first moments of mass about the mass center are necessarily zero. Similar relations exist for the remaining two product-of-inertia terms. Dropping the zero subscripts and using the bar to designate the mass-center quantity give

$$
\begin{aligned}
I_{xy} &= \bar{I}_{xy} + m d_x d_y \\[4pt]
I_{xz} &= \bar{I}_{xz} + m d_x d_z \\[4pt]
I_{yz} &= \bar{I}_{yz} + m d_y d_z
\end{aligned}
\tag{B/9}
$$

These transfer-of-axis relations are valid *only* for transfer to or from *parallel axes* through the *mass center*.

With the aid of the product-of-inertia terms, we can calculate the moment of inertia of a rigid body about any prescribed axis through the origin of coordinates. For the rigid body of Fig. B/7, let it be required to determine the moment of inertia about axis *OM*. The direction cosines of *OM* are l, m, n, and a unit vector $\boldsymbol{\lambda}$ along *OM* may be written $\boldsymbol{\lambda} = l\mathbf{i} + m\mathbf{j} + n\mathbf{k}$. The moment of inertia about *OM* is

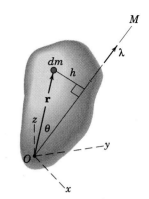

Figure B/7

$$I_M = \int h^2\, dm = \int (\mathbf{r} \times \boldsymbol{\lambda}) \cdot (\mathbf{r} \times \boldsymbol{\lambda})\, dm$$

where $|\mathbf{r} \times \boldsymbol{\lambda}| = r \sin \theta = h$. The cross product is

$$(\mathbf{r} \times \boldsymbol{\lambda}) = (yn - zm)\mathbf{i} + (zl - xn)\mathbf{j} + (xm - yl)\mathbf{k}$$

and, after we collect terms, the dot-product expansion gives

$$(\mathbf{r} \times \boldsymbol{\lambda}) \cdot (\mathbf{r} \times \boldsymbol{\lambda}) = h^2 = (y^2 + z^2)l^2 + (x^2 + z^2)m^2 + (x^2 + y^2)n^2$$
$$- 2xylm - 2xzln - 2yzmn$$

Thus, with the substitution of the expressions of Eqs. B/4 and B/8, we have

$$\boxed{I_M = I_{xx}l^2 + I_{yy}m^2 + I_{zz}n^2 - 2I_{xy}lm - 2I_{xz}ln - 2I_{yz}mn} \quad \textbf{(B/10)}$$

This expression gives the moment of inertia about any axis *OM* in terms of the direction cosines of the axis and the moments and products of inertia about the coordinate directions.

    (a) Principal axes of inertia.   As noted in Art. 7/7, the array

$$\begin{bmatrix} I_{xx} & -I_{xy} & -I_{xz} \\ -I_{yx} & I_{yy} & -I_{yz} \\ -I_{zx} & -I_{zy} & I_{zz} \end{bmatrix}$$

whose elements appear in the expansion of the angular-momentum expression, Eq. 7/11, for a rigid body with attached axes, is known as the *inertia matrix* or *inertia tensor*. If we examine the moment- and product-of-inertia terms for all possible orientations of the axes with respect to the body for a given origin, we will find in the general case an orientation of the *x-y-z* axes for which the product-of-inertia terms vanish and the array takes the diagonalized form

$$\begin{bmatrix} I_{xx} & 0 & 0 \\ 0 & I_{yy} & 0 \\ 0 & 0 & I_{zz} \end{bmatrix}$$

Such axes *x-y-z* are called the *principal axes of inertia*, and I_{xx}, I_{yy}, and I_{zz} are called the *principal moments of inertia* and represent the maximum, minimum, and intermediate values of the moments of inertia for the particular origin chosen.

It may be shown* that for any given orientation of axes x-y-z the solution of the determinant equation

$$\begin{vmatrix} I_{xx} - I & -I_{xy} & -I_{xz} \\ -I_{yx} & I_{yy} - I & -I_{yz} \\ -I_{zx} & -I_{zy} & I_{zz} - I \end{vmatrix} = 0 \qquad (B/11)$$

for I yields three roots I_1, I_2, and I_3 of the resulting cubic equation which are the three principal moments of inertia. Also, the direction cosines l, m, and n of a principal inertia axis are given by

$$(I_{xx} - I)l - I_{xy}m - I_{xz}n = 0$$

$$-I_{yx}l + (I_{yy} - I)m - I_{yz}n = 0 \qquad (B/12)$$

$$-I_{zx}l - I_{zy}m + (I_{zz} - I)n = 0$$

These equations along with $l^2 + m^2 + n^2 = 1$ will enable a solution for the direction cosines to be made for each of the three I's .

To assist with the visualization of these conclusions, consider the rectangular block, Fig. B/8, having an arbitrary orientation

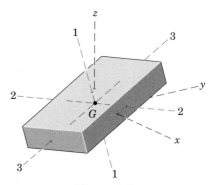

Figure B/8

with respect to the x-y-z axes. For simplicity, the mass center G is located at the origin of the coordinates. If the moments and products of inertia for the block about the x-y-z axes are known, then solution of Eq. B/11 would give the three roots, I_1, I_2, and I_3, which are the principal moments of inertia. Solution of Eq. B/12 using each of the three I's, in turn, along with $l^2 + m^2 + n^2 = 1$ would give the direction cosines l, m, and n for each of the respective principal axes, which are always mutually perpendicular. From the proportions of the block as drawn, we see that I_1 is the maximum moment of inertia, I_2 is the intermediate value, and I_3 is the minimum value.

*See, for example, the first author's *Dynamics, SI Version*, 1975, John Wiley & Sons, Art. 41.

Sample Problem B/4

The bent plate has a uniform thickness t which is negligible compared with its other dimensions. The density of the plate material is ρ. Determine the products of inertia of the plate with respect to the axes as chosen.

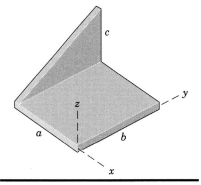

Solution. Each of the two parts is analyzed separately.

① *Rectangular part.* In the separate view of this part, we introduce parallel axes x_0-y_0 through the mass center G and use the transfer-of-axis theorem. By symmetry, we see that $\bar{I}_{xy} = I_{x_0 y_0} = 0$ so that

$$[I_{xy} = \bar{I}_{xy} + m d_x d_y] \qquad I_{xy} = 0 + \rho t a b \left(-\frac{a}{2}\right)\left(\frac{b}{2}\right) = -\frac{1}{4}\rho t a^2 b^2$$

Since the z-coordinate of all elements of the plate is zero, it follows that $I_{xz} = I_{yz} = 0$.

Triangular part. In the separate view of this part, we locate the mass center G and construct x_0-, y_0-, and z_0-axes through G. Since the x_0-coordinate of all elements is zero, it follows that $\bar{I}_{xy} = I_{x_0 y_0} = 0$ and $\bar{I}_{xz} = I_{x_0 z_0} = 0$. The transfer-of-axis theorems then give us

$$[I_{xy} = \bar{I}_{xy} + m d_x d_y] \qquad I_{xy} = 0 + \rho t \frac{b}{2} c(-a)\left(\frac{2b}{3}\right) = -\frac{1}{3}\rho t a b^2 c$$

$$[I_{xz} = \bar{I}_{xz} + m d_x d_z] \qquad I_{xz} = 0 + \rho t \frac{b}{2} c(-a)\left(\frac{c}{3}\right) = -\frac{1}{6}\rho t a b c^2$$

We obtain I_{yz} by direct integration, noting that the distance a of the plane of the triangle from the y-z plane in no way affects the y- and z-coordinates. With the mass element $dm = \rho t \, dy \, dz$, we have

$$\left[I_{yz} = \int yz \, dm\right] \qquad I_{yz} = \rho t \int_0^b \int_0^{cy/b} yz \, dz \, dy = \rho t \int_0^b y \left[\frac{z^2}{2}\right]_0^{cy/b} dy$$

$$= \frac{\rho t c^2}{2b^2} \int_0^b y^3 \, dy = \frac{1}{8}\rho t b^2 c^2$$

Adding the expressions for the two parts gives

$$I_{xy} = -\tfrac{1}{4}\rho t a^2 b^2 - \tfrac{1}{3}\rho t a b^2 c = -\tfrac{1}{12}\rho t a b^2 (3a + 4c) \qquad Ans.$$

$$I_{xz} = 0 \qquad -\tfrac{1}{6}\rho t a b c^2 = -\tfrac{1}{6}\rho t a b c^2 \qquad Ans.$$

$$I_{yz} = 0 \qquad +\tfrac{1}{8}\rho t b^2 c^2 = +\tfrac{1}{8}\rho t b^2 c^2 \qquad Ans.$$

① We must be careful to preserve the same sense of the coordinates. Thus, plus x_0 and y_0 must agree with plus x and y.

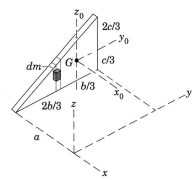

② We choose to integrate with respect to z first, where the upper limit is the variable height $z = cy/b$. If we were to integrate first with respect to y, the limits of the first integral would be from the variable $y = bz/c$ to b.

Sample Problem B/5

The angle bracket is made from aluminum plate with a mass of 13.45 kg per square meter. Calculate the principal moments of inertia about the origin O and the direction cosines of the principal axes of inertia. The thickness of the plate is small compared with the other dimensions.

Dimensions in millimeters

Solution. The masses of the three parts are

$$m_1 = 13.45(0.21)(0.1) = 0.282 \text{ kg}$$

$$m_2 = -13.45\pi(0.035)^2 = -0.0518 \text{ kg}$$

$$m_3 = 13.45(0.12)(0.11) = 0.1775 \text{ kg}$$

① Note that the mass of the hole is treated as a negative number.

Part 1

$$I_{xx} = \tfrac{1}{3}mb^2 = \tfrac{1}{3}(0.282)(0.1)^2 = 9.42(10^{-4}) \text{ kg} \cdot \text{m}^2$$

$$I_{yy} = \tfrac{1}{3}m(a^2 + b^2) = \tfrac{1}{3}(0.282)[(0.21)^2 + (0.1)^2] = 50.9(10^{-4}) \text{ kg} \cdot \text{m}^2$$

$$I_{zz} = \tfrac{1}{3}ma^2 = \tfrac{1}{3}(0.282)(0.21)^2 = 41.5(10^{-4}) \text{ kg} \cdot \text{m}^2$$

$$I_{xy} = 0 \qquad I_{yz} = 0$$

$$I_{xz} = 0 + m\frac{a}{2}\frac{b}{2} = 0.282(0.105)(0.05) = 14.83(10^{-4}) \text{ kg} \cdot \text{m}^2$$

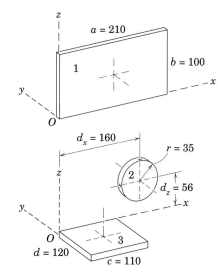

Part 2

$$I_{xx} = \tfrac{1}{4}mr^2 + md_z^2 = -0.0518\left[\frac{(0.035)^2}{4} + (0.050)^2\right]$$

$$= -1.453(10^{-4}) \text{ kg} \cdot \text{m}^2$$

$$I_{yy} = \tfrac{1}{2}mr^2 + m(d_x^2 + d_z^2)$$

$$= -0.0518\left[\frac{(0.035)^2}{2} + (0.16)^2 + (0.05)^2\right]$$

$$= -14.86(10^{-4}) \text{ kg} \cdot \text{m}^2$$

$$I_{zz} = \tfrac{1}{4}mr^2 + md_x^2 = -0.0518\left[\frac{(0.035)^2}{4} + (0.16)^2\right]$$

$$= -13.41(10^{-4}) \text{ kg} \cdot \text{m}^2$$

$$I_{xy} = 0 \qquad I_{yz} = 0$$

$$I_{xz} = 0 + md_xd_z = -0.0518(0.16)(0.05) = -4.14(10^{-4}) \text{ kg} \cdot \text{m}^2$$

② You can easily derive this formula. Also check Table D/4.

Sample Problem   B/5 (Continued)

Part 3

$$I_{xx} = \tfrac{1}{3}md^2 = \tfrac{1}{3}(0.1775)(0.12)^2 = 8.52(10^{-4}) \text{ kg} \cdot \text{m}^2$$

$$I_{yy} = \tfrac{1}{3}mc^2 = \tfrac{1}{3}(0.1775)(0.11)^2 = 7.16(10^{-4}) \text{ kg} \cdot \text{m}^2$$

$$I_{zz} = \tfrac{1}{3}m(c^2 + d^2) = \tfrac{1}{3}(0.1775)[(0.11)^2 + (0.12)^2]$$

$$= 15.68(10^{-4}) \text{ kg} \cdot \text{m}^2$$

$$I_{xy} = m\frac{c}{2}\left(\frac{-d}{2}\right) = 0.1775(0.055)(-0.06) = -5.86(10^{-4}) \text{ kg} \cdot \text{m}^2$$

$$I_{yz} = 0 \qquad I_{xz} = 0$$

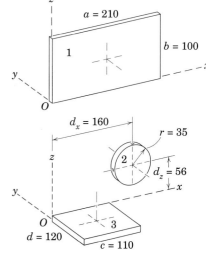

a = 210

b = 100

1

$d_x = 160$

r = 35

2

$d_z = 56$

d = 120

3

c = 110

(Figure repeated)

Totals

$$I_{xx} = 16.48(10^{-4}) \text{ kg} \cdot \text{m}^2 \qquad I_{xy} = -5.86(10^{-4}) \text{ kg} \cdot \text{m}^2$$

$$I_{yy} = 43.2(10^{-4}) \text{ kg} \cdot \text{m}^2 \qquad I_{yz} = 0$$

$$I_{zz} = 43.8(10^{-4}) \text{ kg} \cdot \text{m}^2 \qquad I_{xz} = 10.69(10^{-4}) \text{ kg} \cdot \text{m}^2$$

Substitution into Eq. B/11, expansion of the determinant, and simplification yield

$$I^3 - 103.5(10^{-4})I^2 + 3180(10^{-8})I - 24\,800(10^{-12}) = 0$$

③ Solution of this cubic equation yields the following roots, which are the principal moments of inertia

$$I_1 = 48.3(10^{-4}) \text{ kg} \cdot \text{m}^2$$

$$I_2 = 11.82(10^{-4}) \text{ kg} \cdot \text{m}^2$$

$$I_3 = 43.4(10^{-4}) \text{ kg} \cdot \text{m}^2$$

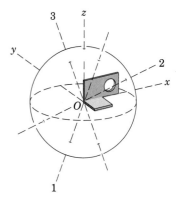

The direction cosines of each principal axis are obtained by substituting each root, in turn, into Eq. B/12 and using $l^2 + m^2 + n^2 = 1$. The results are

$$l_1 = 0.357 \qquad l_2 = 0.934 \qquad l_3 = 0.01830$$

$$m_1 = 0.410 \qquad m_2 = -0.1742 \qquad m_3 = 0.895$$

$$n_1 = -0.839 \qquad n_2 = 0.312 \qquad n_3 = 0.445$$

The bottom figure shows a pictorial view of the bracket and the orientation of its principal axes of inertia.

③ A computer program for the solution of a cubic equation may be used, or an algebraic solution using the formula cited in item 4 of Art. C/4, Appendix C, may be employed.

PROBLEMS

Introductory problems

B/45 Determine the products of inertia about the co-ordinate axes for the unit which consists of three small spheres, each of mass m, connected by the light but rigid slender rods.

$$Ans. \ I_{xy} = 0, \ I_{xz} = I_{yz} = -2ml^2$$

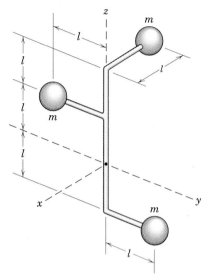

Problem B/45

B/46 Determine the products of inertia about the co-ordinate axes for the unit which consists of four small particles, each of mass m, connected by the light but rigid slender rods.

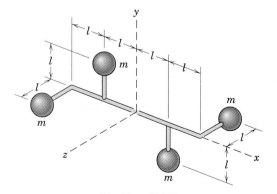

Problem B/46

B/47 Determine the products of inertia about the co-ordinate axes for the uniform slender rod, which has a mass of 0.60 kg.

$$Ans. \ I_{xy} = 4.05(10^{-3}) \ kg \cdot m^2$$
$$I_{yz} = 8.10(10^{-3}) \ kg \cdot m^2$$
$$I_{xz} = 2.43(10^{-3}) \ kg \cdot m^2$$

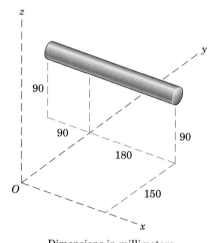

Dimensions in millimeters

Problem B/47

B/48 Determine the products of inertia about the co-ordinate axes for the thin square plate with two circular holes. The mass of the plate material per unit area is ρ.

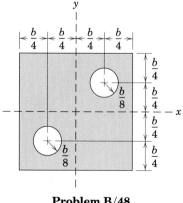

Problem B/48

B/49 Determine the products of inertia for the rod of Prob. B/15, repeated here.

Ans. $I_{xy} = -\frac{1}{8}mb^2$, $I_{xz} = 0$, $I_{yz} = -\frac{1}{8}mb^2$

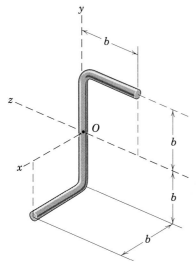

Problem B/49

B/50 Determine the products of inertia about the coordinate axes for the thin plate of mass m which has the shape of a circular sector of radius a and angle β as shown.

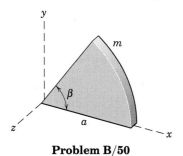

Problem B/50

B/51 Determine the products of inertia about the coordinate axes for the triangular plate of mass m.

Ans. $I_{xz} = \frac{1}{3}mab$, $I_{yz} = \frac{1}{3}mah$

$I_{xy} = \frac{1}{12}mbh$

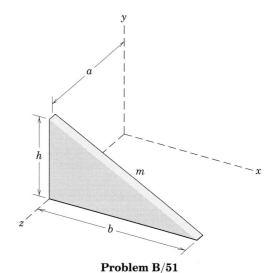

Problem B/51

B/52 Determine the products of inertia of the solid homogeneous half-cylinder of mass m for the axes shown.

Problem B/52

Representative problems

B/53 The semicircular disk of mass m and radius R, inclined at an angle θ from the X-y plane, of Prob. B/43 is repeated here. By the methods of this article, determine the moment of inertia about the Z-axis.

$$\text{Ans. } I_{ZZ} = \tfrac{1}{4}mR^2(1 + \cos^2 \theta)$$

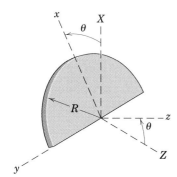

Problem B/53

B/54 Prove that the moment of inertia of the rigid assembly of three identical balls, each of mass m and radius r, has the same value for all axes through O. Neglect the mass of the connecting rods.

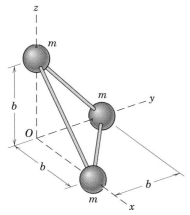

Problem B/54

B/55 Determine the products of inertia for the S-shaped rod of Prob. B/26, repeated here, where d is small compared with r.

$$\text{Ans. } I_{xy} = 2mr^2/\pi, I_{xz} = I_{yz} = 0$$

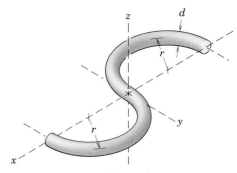

Problem B/55

B/56 For the slender rod of mass m bent into the configuration shown, determine its products of inertia I_{xy}, I_{xz}, and I_{yz}.

Problem B/56

B/57 Determine the moment of inertia of the solid cube of mass m about the diagonal axis A-A through opposite corners.

$$\text{Ans. } I_{AA} = \frac{ma^2}{6}$$

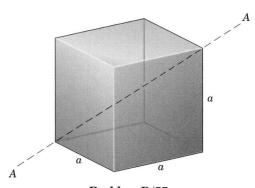

Problem B/57

B/58 The steel plate with two right-angle bends and a central hole has a thickness of 15 mm. Calculate its moment of inertia about the diagonal axis through the corners A and B.

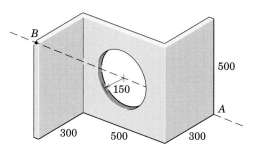

500

150

A

300   500   300

Dimensions in millimeters

Problem B/58

Computer-oriented problems

*B/59** The assembly of three small spheres connected by light rigid bars of Prob. B/45 is repeated here. Determine the principal moments of inertia and the direction cosines associated with the axis of maximum moment of inertia.

$$Ans.\ I_1 = 9ml^2,\ I_2 = 7.37ml^2$$
$$I_3 = 1.628ml^2$$
$$l_1 = 0.816,\ m_1 = 0.408,\ n_1 = 0.408$$

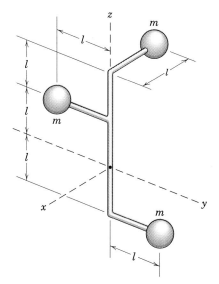

Problem B/59

*B/60** The L-shaped piece is cut from steel plate having a mass per unit area of 160 kg/m². Determine and plot the moment of inertia of the piece about axis A-A as a function of θ from $\theta = 0$ to $\theta = 90°$ and find its minimum value.

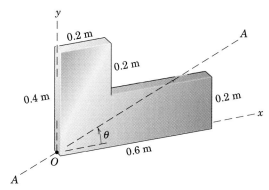

Problem B/60

*B/61** Determine the moment of inertia I about axis OM for the uniform slender rod bent into the shape shown. Plot I versus θ from $\theta = 0$ to $\theta = 90°$ and determine the minimum value of I and the angle α which its axis makes with the x-direction. (*Note:* Because the analysis does not involve the z-coordinate, the expressions developed for area moments of inertia, Eqs. A/9, A/10, and A/11 in Appendix A of *Vol. 1 Statics*, may be utilized for this problem in place of the three-dimensional relations of Appendix B.) The rod has a mass ρ per unit length.

$$Ans.\ I_{min} = 0.1870\rho r^3,\ \alpha = 38.6°$$

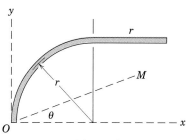

Problem B/61

*B/62 The bent rod of Probs. B/15 and B/49 is repeated here. Its mass is m, and its diameter is small compared with its length. Use the results of Probs. B/15 and B/49 and determine the principal moments of inertia of the rod about the origin O. Also find the direction cosines for the axis of minimum moment of inertia.

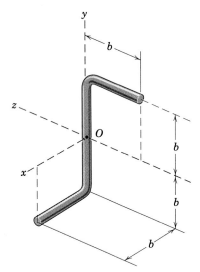

Problem B/62

*B/63 The thin plate has a mass ρ per unit area and is formed into the shape shown. Determine the principal moments of inertia of the plate about axes through O.

$$\text{Ans. } I_1 = 3.78\rho b^4$$
$$I_2 = 0.612\rho b^4$$
$$I_3 = 3.61\rho b^4$$

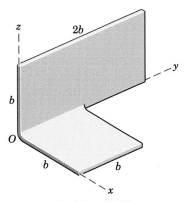

Problem B/63

*B/64 The slender rod has a mass ρ per unit length and is formed into the shape shown. Determine the principal moments of inertia about axes through O and calculate the direction cosines of the axis of minimum moment of inertia.

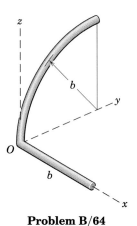

Problem B/64

SELECTED TOPICS OF MATHEMATICS

C

C/1 INTRODUCTION

Appendix C contains an abbreviated summary and reminder of selected topics in basic mathematics which find frequent use in mechanics. The relationships are cited without proof. The student of mechanics will have frequent occasion to use many of these relations, and he or she will be handicapped if they are not well in hand. Other topics not listed will also be needed from time to time.

As the reader reviews and applies mathematics, he or she should bear in mind that mechanics is an applied science descriptive of real bodies and actual motions. Therefore, the geometric and physical interpretation of the applicable mathematics should be kept clearly in mind during the development of theory and the formulation and solution of problems.

C/2 PLANE GEOMETRY

1. When two intersecting lines are, respectively, perpendicular to two other lines, the angles formed by the two pairs are equal.

$\theta_1 = \theta_2$

4. Circle

Circumference $= 2\pi r$
Area $= \pi r^2$
Arc length $s = r\theta$
Sector area $= \frac{1}{2}r^2\theta$

2. Similar triangles

$$\frac{x}{b} = \frac{h - y}{h}$$

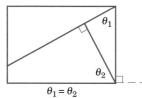

5. Every triangle inscribed within a semicircle is a right triangle.

$\theta_1 + \theta_2 = \pi/2$

3. Any triangle

Area $= \frac{1}{2}bh$

6. Angles of a triangle

$\theta_1 + \theta_2 + \theta_3 = 180°$
$\theta_4 = \theta_1 + \theta_2$

C/3 SOLID GEOMETRY

1. Sphere

   Volume $= \frac{4}{3}\pi r^3$
   Surface area $= 4\pi r^2$

2. Spherical wedge

   Volume $= \frac{2}{3}r^3\theta$

3. Right-circular cone

   Volume $= \frac{1}{3}\pi r^2 h$
   Lateral area $= \pi r L$
   $L = \sqrt{r^2 + h^2}$

4. Any pyramid or cone

   Volume $= \frac{1}{3}Bh$
   where B = area of base

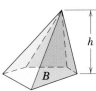

C/4 ALGEBRA

1. Quadratic equation

   $$ax^2 + bx + c = 0$$

   $$x = \frac{-b \pm \sqrt{b^2 - 4ac}}{2a}, \quad b^2 \geq 4ac \text{ for real roots}$$

2. Logarithms

   $b^x = y, x = \log_b y$

   Natural logarithms

   $b = e = 2.718\ 282$
   $e^x = y, x = \log_e y = \ln y$

   $\log (ab) = \log a + \log b$
   $\log (a/b) = \log a - \log b$
   $\log (1/n) = -\log n$
   $\log a^n = n \log a$
   $\log 1 = 0$
   $\log_{10} x = 0.4343 \ln x$

3. Determinants
   2nd order

   $$\begin{vmatrix} a_1 & b_1 \\ a_2 & b_2 \end{vmatrix} = a_1 b_2 - a_2 b_1$$

   3rd order

   $$\begin{vmatrix} a_1 & b_1 & c_1 \\ a_2 & b_2 & c_2 \\ a_3 & b_3 & c_3 \end{vmatrix} = \begin{matrix} +a_1 b_2 c_3 + a_2 b_3 c_1 + a_3 b_1 c_2 \\ -a_3 b_2 c_1 - a_2 b_1 c_3 - a_1 b_3 c_2 \end{matrix}$$

4. Cubic equation

   $$x^3 = Ax + B$$

   Let $p = A/3, q = B/2$.

   Case I: $q^2 - p^3$ negative (three roots real and distinct)

   $$\cos u = q/(p\sqrt{p}), \quad 0 < u < 180°$$
   $$x_1 = 2\sqrt{p} \cos (u/3)$$
   $$x_2 = 2\sqrt{p} \cos (u/3 + 120°)$$
   $$x_3 = 2\sqrt{p} \cos (u/3 + 240°)$$

   Case II: $q^2 - p^3$ positive (one root real, two roots imaginary)

   $$x_1 = (q + \sqrt{q^2 - p^3})^{1/3} + (q - \sqrt{q^2 - p^3})^{1/3}$$

   Case III: $q^2 - p^3 = 0$ (three roots real, two roots equal)

   $$x_1 = 2q^{1/3}, x_2 = x_3 = -q^{1/3}$$

   For general cubic equation

   $$x^3 + ax^2 + bx + c = 0$$

   Substitute $x = x_0 - a/3$ and get $x_0^3 = Ax_0 + B$. Then proceed as above to find values of x_0 from which $x = x_0 - a/3$.

C/5 *ANALYTIC GEOMETRY*

1. Straight line

$$y = a + mx$$

$$\frac{x}{a} + \frac{y}{b} = 1$$

3. Parabola

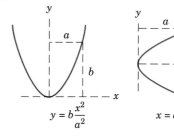

$$y = b\frac{x^2}{a^2}$$

$$x = a\frac{y^2}{b^2}$$

2. Circle

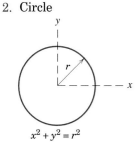

$$x^2 + y^2 = r^2$$

$$(x - a)^2 + (y - b)^2 = r^2$$

4. Ellipse

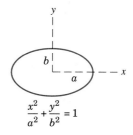

$$\frac{x^2}{a^2} + \frac{y^2}{b^2} = 1$$

5. Hyperbola

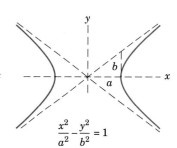

$$xy = a^2$$

$$\frac{x^2}{a^2} - \frac{y^2}{b^2} = 1$$

C/6 *TRIGONOMETRY*

1. Definitions

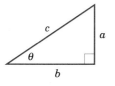

$$\sin \theta = a/c \quad \csc \theta = c/a$$
$$\cos \theta = b/c \quad \sec \theta = c/b$$
$$\tan \theta = a/b \quad \cot \theta = b/a$$

2. Signs in the four quadrants

	I	II	III	IV
$\sin \theta$	+	+	−	−
$\cos \theta$	+	−	−	+
$\tan \theta$	+	−	+	−
$\csc \theta$	+	+	−	−
$\sec \theta$	+	−	−	+
$\cot \theta$	+	−	+	−

3. Miscellaneous relations

$$\sin^2 \theta + \cos^2 \theta = 1$$
$$1 + \tan^2 \theta = \sec^2 \theta$$
$$1 + \cot^2 \theta = \csc^2 \theta$$

$$\sin \frac{\theta}{2} = \sqrt{\tfrac{1}{2}(1 - \cos \theta)}$$

$$\cos \frac{\theta}{2} = \sqrt{\tfrac{1}{2}(1 + \cos \theta)}$$

$$\sin 2\theta = 2 \sin \theta \cos \theta$$
$$\cos 2\theta = \cos^2 \theta - \sin^2 \theta$$
$$\sin (a \pm b) = \sin a \cos b \pm \cos a \sin b$$
$$\cos (a \pm b) = \cos a \cos b \mp \sin a \sin b$$

4. Law of sines

$$\frac{a}{b} = \frac{\sin A}{\sin B}$$

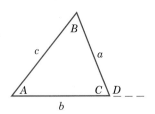

5. Law of cosines

$$c^2 = a^2 + b^2 - 2ab \cos C$$
$$c^2 = a^2 + b^2 + 2ab \cos D$$

C/7 VECTOR OPERATIONS

1. **Notation.**   Vector quantities are printed in boldface type, and scalar quantities appear in lightface italic type. Thus, the vector quantity **V** has a scalar magnitude V. In longhand work vector quantities should always be consistently indicated by a symbol such as $\underline{V}$ or $\vec{V}$ to distinguish them from scalar quantities.

2. **Addition**

   Triangle addition   **P** + **Q** = **R**
   Parallelogram addition   **P** + **Q** = **R**
   Commutative law   **P** + **Q** = **Q** + **P**
   Associative law   **P** + (**Q** + **R**) = (**P** + **Q**) + **R**

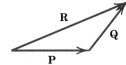

3. **Subtraction**

$$\mathbf{P} - \mathbf{Q} = \mathbf{P} + (-\mathbf{Q})$$

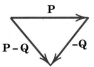

4. **Unit vectors i, j, k**

$$\mathbf{V} = V_x \mathbf{i} + V_y \mathbf{j} + V_z \mathbf{k}$$

   where $|\mathbf{V}| = V = \sqrt{V_x^2 + V_y^2 + V_z^2}$

5. **Direction cosines** l, m, n are the cosines of the angles between **V** and the x-, y-, z-axes. Thus,

$$l = V_x/V \quad m = V_y/V \quad n = V_z/V$$

   so that $\mathbf{V} = V(l\mathbf{i} + m\mathbf{j} + n\mathbf{k})$

   and $l^2 + m^2 + n^2 = 1$

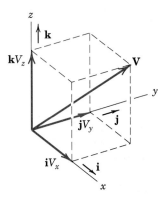

6. *Dot or scalar product*

$$\mathbf{P \cdot Q} = PQ \cos \theta$$

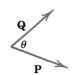

This product may be viewed as the magnitude of **P** multiplied by the component $Q \cos \theta$ of **Q** in the direction of **P**, or as the magnitude of **Q** multiplied by the component $P \cos \theta$ of **P** in the direction of **Q**.

Commutative law $\mathbf{P \cdot Q} = \mathbf{Q \cdot P}$

From the definition of the dot product

$$\mathbf{i \cdot i} = \mathbf{j \cdot j} = \mathbf{k \cdot k} = 1$$

$$\mathbf{i \cdot j} = \mathbf{j \cdot i} = \mathbf{i \cdot k} = \mathbf{k \cdot i} = \mathbf{j \cdot k} = \mathbf{k \cdot j} = 0$$

$$\mathbf{P \cdot Q} = (P_x \mathbf{i} + P_y \mathbf{j} + P_z \mathbf{k}) \cdot (Q_x \mathbf{i} + Q_y \mathbf{j} + Q_z \mathbf{k})$$
$$= P_x Q_x + P_y Q_y + P_z Q_z$$

$$\mathbf{P \cdot P} = P_x^2 + P_y^2 + P_z^2$$

It follows from the definition of the dot product that two vectors **P** and **Q** are perpendicular when their dot product vanishes, $\mathbf{P \cdot Q} = 0$.

The angle θ between two vectors $\mathbf{P}_1$ and $\mathbf{P}_2$ may be found from their dot product expression $\mathbf{P}_1 \cdot \mathbf{P}_2 = P_1 P_2 \cos \theta$, which gives

$$\cos \theta = \frac{\mathbf{P}_1 \cdot \mathbf{P}_2}{P_1 P_2} = \frac{P_{1_x} P_{2_x} + P_{1_y} P_{2_y} + P_{1_z} P_{2_z}}{P_1 P_2} = l_1 l_2 + m_1 m_2 + n_1 n_2$$

where l, m, n stand for the respective direction cosines of the vectors. It is also observed that two vectors are perpendicular to each other when their direction cosines obey the relation $l_1 l_2 + m_1 m_2 + n_1 n_2 = 0$.

Distributive law $\mathbf{P \cdot (Q + R)} = \mathbf{P \cdot Q} + \mathbf{P \cdot R}$

7. *Cross or vector product.*   The cross product $\mathbf{P \times Q}$ of the two vectors **P** and **Q** is defined as a vector with a magnitude

$$|\mathbf{P \times Q}| = PQ \sin \theta$$

and a direction specified by the right-hand rule as shown. Reversing the vector order and using the right-hand rule give $\mathbf{Q \times P} = -\mathbf{P \times Q}$.

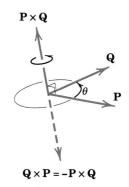

Distributive law $\mathbf{P \times (Q + R)} = \mathbf{P \times Q} + \mathbf{P \times R}$

From the definition of the cross product, using a *right-handed coordinate system*, we get

$$\mathbf{i \times j} = \mathbf{k} \qquad \mathbf{j \times k} = \mathbf{i} \qquad \mathbf{k \times i} = \mathbf{j}$$

$$\mathbf{j \times i} = -\mathbf{k} \qquad \mathbf{k \times j} = -\mathbf{i} \qquad \mathbf{i \times k} = -\mathbf{j}$$

$$\mathbf{i \times i} = \mathbf{j \times j} = \mathbf{k \times k} = 0$$

With the aid of these identities and the distributive law, the vector product may be written

$$\mathbf{P} \times \mathbf{Q} = (P_x\mathbf{i} + P_y\mathbf{j} + P_z\mathbf{k}) \times (Q_x\mathbf{i} + Q_y\mathbf{j} + Q_z\mathbf{k})$$
$$= (P_yQ_z - P_zQ_y)\mathbf{i} + (P_zQ_x - P_xQ_z)\mathbf{j} + (P_xQ_y - P_yQ_x)\mathbf{k}$$

The cross product may also be expressed by the determinant

$$\mathbf{P} \times \mathbf{Q} = \begin{vmatrix} \mathbf{i} & \mathbf{j} & \mathbf{k} \\ P_x & P_y & P_z \\ Q_x & Q_y & Q_z \end{vmatrix}$$

8. *Additional relations*

Triple scalar product $(\mathbf{P} \times \mathbf{Q})\cdot\mathbf{R} = \mathbf{R}\cdot(\mathbf{P} \times \mathbf{Q})$. The dot and cross may be interchanged as long as the order of the vectors is maintained. Parentheses are unnecessary since $\mathbf{P} \times (\mathbf{Q}\cdot\mathbf{R})$ is meaningless because a vector $\mathbf{P}$ cannot be crossed into a scalar $\mathbf{Q}\cdot\mathbf{R}$. Thus, the expression may be written

$$\mathbf{P} \times \mathbf{Q}\cdot\mathbf{R} = \mathbf{P}\cdot\mathbf{Q} \times \mathbf{R}$$

The triple scalar product has the determinant expansion

$$\mathbf{P} \times \mathbf{Q}\cdot\mathbf{R} = \begin{vmatrix} P_x & P_y & P_z \\ Q_x & Q_y & Q_z \\ R_x & R_y & R_z \end{vmatrix}$$

Triple vector product $(\mathbf{P} \times \mathbf{Q}) \times \mathbf{R} = -\mathbf{R} \times (\mathbf{P} \times \mathbf{Q})$
$$= \mathbf{R} \times (\mathbf{Q} \times \mathbf{P})$$

Here we note that the parentheses must be used since an expression $\mathbf{P} \times \mathbf{Q} \times \mathbf{R}$ would be ambiguous because it would not identify the vector to be crossed. It may be shown that the triple vector product is equivalent to

$$(\mathbf{P} \times \mathbf{Q}) \times \mathbf{R} = \mathbf{R}\cdot\mathbf{P}\mathbf{Q} - \mathbf{R}\cdot\mathbf{Q}\mathbf{P}$$

or $$\mathbf{P} \times (\mathbf{Q} \times \mathbf{R}) = \mathbf{P}\cdot\mathbf{R}\mathbf{Q} - \mathbf{P}\cdot\mathbf{Q}\mathbf{R}$$

The first term in the first expression, for example, is the dot product $\mathbf{R}\cdot\mathbf{P}$, a scalar, multiplied by the vector $\mathbf{Q}$.

9. *Derivatives of vectors* obey the same rules as they do for scalars.

$$\frac{d\mathbf{P}}{dt} = \dot{\mathbf{P}} = \dot{P}_x\mathbf{i} + \dot{P}_y\mathbf{j} + \dot{P}_z\mathbf{k}$$

$$\frac{d(\mathbf{P}u)}{dt} = \mathbf{P}\dot{u} + \dot{\mathbf{P}}u$$

$$\frac{d(\mathbf{P}\cdot\mathbf{Q})}{dt} = \mathbf{P}\cdot\dot{\mathbf{Q}} + \dot{\mathbf{P}}\cdot\mathbf{Q}$$

$$\frac{d(\mathbf{P} \times \mathbf{Q})}{dt} = \mathbf{P} \times \dot{\mathbf{Q}} + \dot{\mathbf{P}} \times \mathbf{Q}$$

10. *Integration of vectors.* If $\mathbf{V}$ is a function of x, y, and z and an element of volume is $d\tau = dx\ dy\ dz$, the integral of $\mathbf{V}$ over the volume may be written as the vector sum of the three integrals of its components. Thus,

$$\int \mathbf{V}\,d\tau = \mathbf{i}\int V_x\,d\tau + \mathbf{j}\int V_y\,d\tau + \mathbf{k}\int V_z\,d\tau$$

C/8 SERIES

(Expression in brackets following series indicates range of convergence.)

$$(1 \pm x)^n = 1 \pm nx + \frac{n(n-1)}{2!}x^2 \pm \frac{n(n-1)(n-2)}{3!}x^3 + \cdots\ [x^2 < 1]$$

$$\sin x = x - \frac{x^3}{3!} + \frac{x^5}{5!} - \frac{x^7}{7!} + \cdots \qquad\qquad [x^2 < \infty]$$

$$\cos x = 1 - \frac{x^2}{2!} + \frac{x^4}{4!} - \frac{x^6}{6!} + \cdots \qquad\qquad [x^2 < \infty]$$

$$\sinh x = \frac{e^x - e^{-x}}{2} = x + \frac{x^3}{3!} + \frac{x^5}{5!} + \frac{x^7}{7!} + \cdots \qquad [x^2 < \infty]$$

$$\cosh x = \frac{e^x + e^{-x}}{2} = 1 + \frac{x^2}{2!} + \frac{x^4}{4!} + \frac{x^6}{6!} + \cdots \qquad [x^2 < \infty]$$

$$f(x) = \frac{a_0}{2} + \sum_{n=1}^{\infty} a_n \cos \frac{n\pi x}{l} + \sum_{n=1}^{\infty} b_n \sin \frac{n\pi x}{l}$$

$$\text{where } a_n = \frac{1}{l}\int_{-l}^{l} f(x) \cos \frac{n\pi x}{l}\,dx, \qquad b_n = \frac{1}{l}\int_{-l}^{l} f(x) \sin \frac{n\pi x}{l}\,dx$$

$$[\text{Fourier expansion for } -l < x < l]$$

C/9 DERIVATIVES

$$\frac{dx^n}{dx} = nx^{n-1}, \qquad \frac{d(uv)}{dx} = u\frac{dv}{dx} + v\frac{du}{dx}, \qquad \frac{d\left(\dfrac{u}{v}\right)}{dx} = \frac{v\dfrac{du}{dx} - u\dfrac{dv}{dx}}{v^2}$$

$$\lim_{\Delta x \to 0} \sin \Delta x = \sin dx = \tan dx = dx$$

$$\lim_{\Delta x \to 0} \cos \Delta x = \cos dx = 1$$

$$\frac{d\sin x}{dx} = \cos x, \qquad \frac{d\cos x}{dx} = -\sin x, \qquad \frac{d\tan x}{dx} = \sec^2 x$$

$$\frac{d\sinh x}{dx} = \cosh x, \qquad \frac{d\cosh x}{dx} = \sinh x, \qquad \frac{d\tanh x}{dx} = \operatorname{sech}^2 x$$

C/10 INTEGRALS

$$\int x^n \, dx = \frac{x^{n+1}}{n+1}$$

$$\int \frac{dx}{x} = \ln x$$

$$\int \sqrt{a + bx} \, dx = \frac{2}{3b} \sqrt{(a + bx)^3}$$

$$\int x\sqrt{a + bx} \, dx = \frac{2}{15b^2} (3bx - 2a)\sqrt{(a + bx)^3}$$

$$\int x^2\sqrt{a + bx} \, dx = \frac{2}{105b^3} (8a^2 - 12abx + 15b^2x^2)\sqrt{(a + bx)^3}$$

$$\int \frac{dx}{\sqrt{a + bx}} = \frac{2\sqrt{a + bx}}{b}$$

$$\int \frac{\sqrt{a + x}}{\sqrt{b - x}} \, dx = -\sqrt{a + x}\sqrt{b - x} + (a + b) \sin^{-1} \sqrt{\frac{a + x}{a + b}}$$

$$\int \frac{x \, dx}{a + bx} = \frac{1}{b^2} [a + bx - a \ln (a + bx)]$$

$$\int \frac{x \, dx}{(a + bx)^n} = \frac{(a + bx)^{1-n}}{b^2} \left(\frac{a + bx}{2 - n} - \frac{a}{1 - n} \right)$$

$$\int \frac{dx}{a + bx^2} = \frac{1}{\sqrt{ab}} \tan^{-1} \frac{x\sqrt{ab}}{a} \quad \text{or} \quad \frac{1}{\sqrt{-ab}} \tanh^{-1} \frac{x\sqrt{-ab}}{a}$$

$$\int \frac{x \, dx}{a + bx^2} = \frac{1}{2b} \ln (a + bx^2)$$

$$\int \sqrt{x^2 \pm a^2} \, dx = \tfrac{1}{2}[x\sqrt{x^2 \pm a^2} \pm a^2 \ln (x + \sqrt{x^2 \pm a^2})]$$

$$\int \sqrt{a^2 - x^2} \, dx = \tfrac{1}{2}\left(x\sqrt{a^2 - x^2} + a^2 \sin^{-1} \frac{x}{a} \right)$$

$$\int x\sqrt{a^2 - x^2} \, dx = -\tfrac{1}{3}\sqrt{(a^2 - x^2)^3}$$

$$\int x^2\sqrt{a^2 - x^2} \, dx = -\frac{x}{4} \sqrt{(a^2 - x^2)^3} + \frac{a^2}{8} \left(x\sqrt{a^2 - x^2} + a^2 \sin^{-1} \frac{x}{a} \right)$$

$$\int x^3\sqrt{a^2 - x^2} \, dx = -\tfrac{1}{5}(x^2 + \tfrac{2}{3}a^2)\sqrt{(a^2 - x^2)^3}$$

$$\int \frac{dx}{\sqrt{a + bx + cx^2}} = \frac{1}{\sqrt{c}} \ln \left(\sqrt{a + bx + cx^2} + x\sqrt{c} + \frac{b}{2\sqrt{c}} \right) \quad \text{or} \quad \frac{-1}{\sqrt{-c}} \sin^{-1} \left(\frac{b + 2cx}{\sqrt{b^2 - 4ac}} \right)$$

$$\int \frac{dx}{\sqrt{x^2 \pm a^2}} = \ln (x + \sqrt{x^2 \pm a^2})$$

$$\int \frac{dx}{\sqrt{a^2 - x^2}} = \sin^{-1} \frac{x}{a}$$

$$\int \frac{x\ dx}{\sqrt{x^2 - a^2}} = \sqrt{x^2 - a^2}$$

$$\int \frac{x\ dx}{\sqrt{a^2 \pm x^2}} = \pm \sqrt{a^2 \pm x^2}$$

$$\int x\sqrt{x^2 \pm a^2}\ dx = \tfrac{1}{3}\sqrt{(x^2 \pm a^2)^3}$$

$$\int x^2\sqrt{x^2 \pm a^2}\ dx = \frac{x}{4}\sqrt{(x^2 \pm a^2)^3} \mp \frac{a^2}{8} x\sqrt{x^2 \pm a^2} - \frac{a^4}{8} \ln (x + \sqrt{x^2 \pm a^2})$$

$$\int \sin x\ dx = -\cos x$$

$$\int \cos x\ dx = \sin x$$

$$\int \sec x\ dx = \frac{1}{2} \ln \frac{1 + \sin x}{1 - \sin x}$$

$$\int \sin^2 x\ dx = \frac{x}{2} - \frac{\sin 2x}{4}$$

$$\int \cos^2 x\ dx = \frac{x}{2} + \frac{\sin 2x}{4}$$

$$\int \sin x \cos x\ dx = \frac{\sin^2 x}{2}$$

$$\int \sinh x\ dx = \cosh x$$

$$\int \cosh x\ dx = \sinh x$$

$$\int \tanh x\ dx = \ln \cosh x$$

$$\int \ln x\ dx = x \ln x - x$$

$$\int e^{ax}\ dx = \frac{e^{ax}}{a}$$

$$\int xe^{ax}\ dx = \frac{e^{ax}}{a^2} (ax - 1)$$

$$\int e^{ax} \sin px \, dx = \frac{e^{ax}(a \sin px - p \cos px)}{a^2 + p^2}$$

$$\int e^{ax} \cos px \, dx = \frac{e^{ax}(a \cos px + p \sin px)}{a^2 + p^2}$$

$$\int e^{ax} \sin^2 x \, dx = \frac{e^{ax}}{4 + a^2}\left(a \sin^2 x - \sin 2x + \frac{2}{a}\right)$$

$$\int e^{ax} \cos^2 x \, dx = \frac{e^{ax}}{4 + a^2}\left(a \cos^2 x + \sin 2x + \frac{2}{a}\right)$$

$$\int e^{ax} \sin x \cos x \, dx = \frac{e^{ax}}{4 + a^2}\left(\frac{a}{2} \sin 2x - \cos 2x\right)$$

$$\int \sin^3 x \, dx = -\frac{\cos x}{3}(2 + \sin^2 x)$$

$$\int \cos^3 x \, dx = \frac{\sin x}{3}(2 + \cos^2 x)$$

$$\int \cos^5 x \, dx = \sin x - \tfrac{2}{3} \sin^3 x + \tfrac{1}{5} \sin^5 x$$

$$\int x \sin x \, dx = \sin x - x \cos x$$

$$\int x \cos x \, dx = \cos x + x \sin x$$

$$\int x^2 \sin x \, dx = 2x \sin x - (x^2 - 2) \cos x$$

$$\int x^2 \cos x \, dx = 2x \cos x + (x^2 - 2) \sin x$$

Radius of curvature
$$
\begin{cases}
\rho_{xy} = \dfrac{\left[1 + \left(\dfrac{dy}{dx}\right)^2\right]^{3/2}}{\dfrac{d^2y}{dx^2}} \\[4ex]
\rho_{r\theta} = \dfrac{\left[r^2 + \left(\dfrac{dr}{d\theta}\right)^2\right]^{3/2}}{r^2 + 2\left(\dfrac{dr}{d\theta}\right)^2 - r\dfrac{d^2r}{d\theta^2}}
\end{cases}
$$

C/11   NEWTON'S METHOD FOR SOLVING INTRACTABLE EQUATIONS

Frequently, the application of the fundamental principles of mechanics leads to an algebraic or transcendental equation which is not solvable (or easily solvable) in closed form. In such cases, an iterative technique, such as Newton's method, can be a powerful tool for obtaining a good estimate to the root or roots of the equation.

Let us place the equation to be solved in the form $f(x) = 0$. The a-part of the accompanying figure depicts an arbitrary function $f(x)$ for values of x in the vicinity of the desired root x_r. Note that x_r is merely the value of x at which the function crosses the x-axis. Suppose that we have available (perhaps via a hand-drawn plot) a rough estimate x_1 of this root. Provided that x_1 does not closely correspond to a maximum or minimum value of the function $f(x)$, we may obtain a better estimate of the root x_r by extending the tangent to $f(x)$

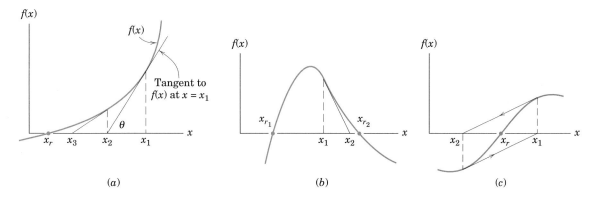

(a)   (b)   (c)

at x_1 so that it intersects the x-axis at x_2. From the geometry of the figure, we may write

$$\tan \theta = f'(x_1) = \frac{f(x_1)}{x_1 - x_2}$$

where $f'(x_1)$ denotes the derivative of $f(x)$ with respect to x evaluated at $x = x_1$. Solving the above equation for x_2 results in

$$x_2 = x_1 - \frac{f(x_1)}{f'(x_1)}$$

The term $-f(x_1)/f'(x_1)$ is the correction to the initial root estimate x_1. Once x_2 is calculated, we may repeat the process to obtain x_3, and so forth.

Thus, we generalize the above equation to

$$x_{k+1} = x_k - \frac{f(x_k)}{f'(x_k)}$$

where

$$x_{k+1} = \text{the } (k+1)\text{th estimate of the desired root } x_r$$

$$x_k = \text{the } k\text{th estimate of the desired root } x_r$$

$$f(x_k) = \text{the function } f(x) \text{ evaluated at } x = x_k$$

$$f'(x_k) = \text{the function derivative evaluated at } x = x_k$$

This equation is repeatedly applied until $f(x_{k+1})$ is sufficiently close to zero and $x_{k+1} \cong x_k$. The student should verify that the equation is valid for all possible sign combinations of x_k, $f(x_k)$, and $f'(x_k)$.

Several cautionary notes are in order:

1. Clearly, $f'(x_k)$ must not be zero or close to zero. This would mean, as restricted above, that x_k exactly or approximately corresponds to a minimum or maximum of $f(x)$. If the slope $f'(x_k)$ is zero, then the tangent to the curve never intersects the x-axis. If the slope $f'(x_k)$ is small, then the correction to x_k may be so large that x_{k+1} is a worse root estimate than x_k. For this reason, experienced engineers usually limit the size of the correction term; that is, if the absolute value of $f(x_k)/f'(x_k)$ is larger than a preselected maximum value, that maximum value is used.

2. If there are several roots of the equation $f(x) = 0$, we must be in the vicinity of the desired root x_r in order that the algorithm actually converges to that root. The *b*-part of the figure depicts the condition when the initial estimate x_1 will result in convergence to x_{r_2} rather than x_{r_1}.

3. Oscillation from one side of the root to the other can occur if, for example, the function is antisymmetric about a root which is an inflection point. The use of one-half of the correction will usually prevent this behavior, which is depicted in the *c*-part of the accompanying figure.

Example: Beginning with an initial estimate of $x_1 = 5$, estimate the single root of the equation $e^x - 10 \cos x - 100 = 0$.

The table below summarizes the application of Newton's method to the given equation. The iterative process was terminated when the absolute value of the correction $-f(x_k)/f'(x_k)$ became less than 10^{-6}.

k	x_k	$f(x_k)$	$f'(x_k)$	$x_{k+1} - x_k = -\dfrac{f(x_k)}{f'(x_k)}$
1	5.000 000	45.576 537	138.823 916	$-0.328\ 305$
2	4.671 695	7.285 610	96.887 065	$-0.075\ 197$
3	4.596 498	0.292 886	89.203 650	$-0.003\ 283$
4	4.593 215	0.000 527	88.882 536	$-0.000\ 006$
5	4.593 209	$-2(10^{-8})$	88.881 956	$2.25(10^{-10})$

C/12 *SELECTED TECHNIQUES FOR NUMERICAL INTEGRATION*

    1. Area determination.   Consider the problem of determining the shaded area under the curve $y = f(x)$ from $x = a$ to $x = b$, as depicted in the *a*-part of the figure, and suppose that analytical integration is not feasible. The function may be known in tabular form from experimental measurements or it may be known in analytical form. The function is taken to be continuous within the interval $a < x < b$. We may divide the area into n vertical strips, each of width $\Delta x = (b - a)/n$, and then add the areas of all strips to obtain $A = \int y\,dx$. A representative strip of area A_i is shown with darker shading in the figure. Three useful numerical approximations are cited. In each case the greater the number of strips, the more accurate becomes the approximation geometrically. As a general rule, one can begin with a relatively small number of strips and increase the number until the resulting changes in the area approximation no longer improve the desired accuracy.

(a)

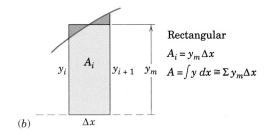

(b)

Rectangular
$$A_i = y_m \Delta x$$
$$A = \int y\,dx \cong \Sigma\, y_m \Delta x$$

    I. *Rectangular* [Figure (b)]   The areas of the strips are taken to be rectangles, as shown by the representative strip whose height y_m is chosen visually so that the small cross-hatched areas are as nearly equal as possible. Thus, we form the sum Σy_m of the effective heights and multiply by Δx. For a function known in analytical form, a value for y_m equal to that of the function at the midpoint $x_i + \Delta x/2$ may be calculated and used in the summation.

II. *Trapezoidal* [Figure (c)] The areas of the strips are taken to be trapezoids, as shown by the representative strip. The area A_i is the average height $(y_i + y_{i+1})/2$ times Δx. Adding the areas gives the area approximation as tabulated. For the example with the curvature shown, clearly the approximation will be on the low side. For the reverse curvature, the approximation will be on the high side.

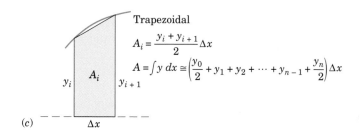

Trapezoidal

$$A_i = \frac{y_i + y_{i+1}}{2}\Delta x$$

$$A = \int y\, dx \cong \left(\frac{y_0}{2} + y_1 + y_2 + \cdots + y_{n-1} + \frac{y_n}{2}\right)\Delta x$$

(c)

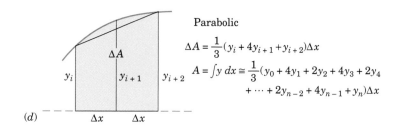

Parabolic

$$\Delta A = \frac{1}{3}(y_i + 4y_{i+1} + y_{i+2})\Delta x$$

$$A = \int y\, dx \cong \frac{1}{3}(y_0 + 4y_1 + 2y_2 + 4y_3 + 2y_4$$
$$+ \cdots + 2y_{n-2} + 4y_{n-1} + y_n)\Delta x$$

(d)

III. *Parabolic* [Figure (d)] The area between the chord and the curve (neglected in the trapezoidal solution) may be accounted for by approximating the function by a parabola passing through the points defined by three successive values of y. This area may be calculated from the geometry of the parabola and added to the trapezoidal area of the pair of strips to give the area ΔA of the pair as cited. Adding all of the ΔA's produces the tabulation shown, which is known as Simpson's rule. To use Simpson's rule, the number n of strips must be even.

Example: Determine the area under the curve $y = x\sqrt{1 + x^2}$ from $x = 0$ to $x = 2$. (An integrable function is chosen here so that the three approximations can be compared with the exact value, which is $A = \int_0^2 x\sqrt{1 + x^2}\, dx = \frac{1}{3}(1 + x^2)^{3/2}\big|_0^2 = \frac{1}{3}(5\sqrt{5} - 1) = 3.393\,447.$)

NUMBER OF SUBINTERVALS	AREA APPROXIMATIONS		
	RECTANGULAR	TRAPEZOIDAL	PARABOLIC
4	3.361 704	3.456 731	3.392 214
10	3.388 399	3.403 536	3.393 420
50	3.393 245	3.393 850	3.393 447
100	3.393 396	3.393 547	3.393 447
1000	3.393 446	3.393 448	3.393 447
2500	3.393 447	3.393 447	3.393 447

Note that the worst approximation error is less than 2 percent, even with only four strips.

  2. Integration of first-order ordinary differential equations.   The application of the fundamental principles of mechanics frequently results in differential relationships. Let us consider the first-order form $dy/dt = f(t)$, where the function $f(t)$ may not be readily integrable or may be known only in tabular form. We may numerically integrate by means of a simple slope-projection technique, known as Euler integration, which is illustrated in the figure.

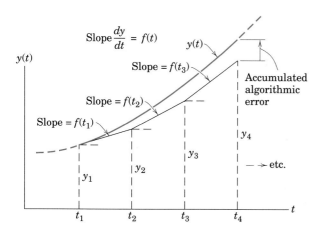

Beginning at t_1, at which the value y_1 is known, we project the slope over a horizontal subinterval or step $(t_2 - t_1)$ and see that $y_2 = y_1 + f(t_1)(t_2 - t_1)$. At t_2, the process may be repeated beginning at y_2, and so forth until the desired value of t is reached. Hence, the general expression is

$$y_{k+1} = y_k + f(t_k)(t_{k+1} - t_k)$$

If y versus t were linear, i.e., if $f(t)$ were constant, the method would be exact, and there would be no need for a numerical approach in that case. Changes in the slope over the subinterval introduce error. For the case shown in the figure, the estimate y_2 is clearly less than the true value of the function $y(t)$ at t_2. More accurate integration techniques (such as Runge-Kutta methods) take into account changes in the slope over the subinterval and thus provide better results.

As with the area-determination techniques, experience is helpful in the selection of a subinterval or step size when dealing with analytical functions. As a rough rule, one begins with a relatively large step size and then steadily decreases the step size until the corresponding changes in the integrated result are much smaller than the desired accuracy. A step size which is too small, however, can result in increased error due to a very large number of computer operations. This type of error is generally known as "round-off error," while the error which results from a large step size is known as algorithm error.

Example: For the differential equation $dy/dt = 5t$ with the initial condition $y = 2$ when $t = 0$, determine the value of y for $t = 4$.

Application of the Euler integration technique yields the following results:

NUMBER OF SUBINTERVALS	STEP SIZE	y at $t = 4$	PERCENT ERROR
10	0.4	38	9.5
100	0.04	41.6	0.95
500	0.008	41.92	0.19
1000	0.004	41.96	0.10

This simple example may be integrated analytically. The result is $y = 42$ (exactly).

USEFUL TABLES

D

TABLE D/1 PHYSICAL PROPERTIES

Density (kg/m^3) *and specific weight* (lb/ft^3)

	kg/m^3	lb/ft^3		kg/m^3	lb/ft^3
Air*	1.2062	0.07530			
Aluminum	2 690	168	Iron (cast)	7 210	450
Concrete (av.)	2 400	150	Lead	11 370	710
Copper	8 910	556	Mercury	13 570	847
Earth (wet, av.)	1 760	110	Oil (av.)	900	56
(dry, av.)	1 280	80	Steel	7 830	489
Glass	2 590	162	Titanium	3 080	192
Gold	19 300	1205	Water (fresh)	1 000	62.4
Ice	900	56	(salt)	1 030	64
*At 20°C (68°F) and atmospheric			Wood (soft pine)	480	30
pressure			(hard oak)	800	50

Coefficients of friction

(The coefficients in the following table represent typical values under normal working conditions. Actual coefficients for a given situation will depend on the exact nature of the contacting surfaces. A variation of 25 to 100 percent or more from these values could be expected in an actual application, depending on prevailing conditions of cleanliness, surface finish, pressure, lubrication, and velocity.)

CONTACTING SURFACE	TYPICAL VALUES OF COEFFICIENT OF FRICTION	
	STATIC, μ_s	KINETIC, μ_k
Steel on steel (dry)	0.6	0.4
Steel on steel (greasy)	0.1	0.05
Teflon on steel	0.04	0.04
Steel on babbitt (dry)	0.4	0.3
Steel on babbitt (greasy)	0.1	0.07
Brass on steel (dry)	0.5	0.4
Brake lining on cast iron	0.4	0.3
Rubber tires on smooth pavement (dry)	0.9	0.8
Wire rope on iron pulley (dry)	0.2	0.15
Hemp rope on metal	0.3	0.2
Metal on ice		0.02

TABLE D/2   SOLAR SYSTEM CONSTANTS

Universal gravitational constant	$G = 6.673(10^{-11})$ m^3/(kg·s^2)
	$= 3.439(10^{-8})$ ft^4/(lbf-s^4)
Mass of Earth	$m_e = 5.976(10^{24})$ kg
	$= 4.095(10^{23})$ lbf-s^2/ft
Period of Earth's rotation (1 sidereal day)	$= 23$ h 56 min 4 s
	$= 23.9344$ h
Angular velocity of Earth	$\omega = 0.7292(10^{-4})$ rad/s
Mean angular velocity of Earth–Sun line	$\omega' = 0.1991(10^{-6})$ rad/s
Mean velocity of Earth's center about Sun	$= 107\ 200$ km/h
	$= 66,610$ mi/h

BODY	MEAN DISTANCE TO SUN km (mi)	ECCENTRICITY OF ORBIT e	PERIOD OF ORBIT solar days	MEAN DIAMETER km (mi)	MASS RELATIVE TO EARTH	SURFACE GRAVITATIONAL ACCELERATION m/s^2 (ft/s^2)	ESCAPE VELOCITY km/s (mi/s)
Sun	—	—	—	1 392 000 (865 000)	333 000	274 (898)	616 (383)
Moon	384 398* (238 854)*	0.055	27.32	3 476 (2 160)	0.0123	1.62 (5.32)	2.37 (1.47)
Mercury	57.3 × 10^6 (35.6 × 10^6)	0.206	87.97	5 000 (3 100)	0.054	3.47 (11.4)	4.17 (2.59)
Venus	108 × 10^6 (67.2 × 10^6)	0.0068	224.70	12 400 (7 700)	0.815	8.44 (27.7)	10.24 (6.36)
Earth	149.6 × 10^6 (92.96 × 10^6)	0.0167	365.26	12 742[†] (7 918)[†]	1.000	9.821[‡] (32.22)[‡]	11.18 (6.95)
Mars	227.9 × 10^6 (141.6 × 10^6)	0.093	686.98	6 788 (4 218)	0.107	3.73 (12.3)	5.03 (3.13)

* Mean distance to Earth (center-to-center)

[†] Diameter of sphere of equal volume, based on a spheroidal Earth with a polar diameter of 12 714 km (7900 mi) and an equatorial diameter of 12 756 km (7926 mi)

[‡] For nonrotating spherical Earth, equivalent to absolute value at sea level and latitude 37.5°

TABLE D/3 PROPERTIES OF PLANE FIGURES

FIGURE	CENTROID	AREA MOMENTS OF INERTIA
Arc Segment	$\bar{r} = \dfrac{r \sin \alpha}{\alpha}$	—
Quarter and Semicircular Arcs	$\bar{y} = \dfrac{2r}{\pi}$	—
Circular Area	—	$I_x = I_y = \dfrac{\pi r^4}{4}$ $I_z = \dfrac{\pi r^4}{2}$
Semicircular Area	$\bar{y} = \dfrac{4r}{3\pi}$	$I_x = I_y = \dfrac{\pi r^4}{8}$ $\bar{I}_x = \left(\dfrac{\pi}{8} - \dfrac{8}{9\pi} \right) r^4$ $I_z = \dfrac{\pi r^4}{8}$
Quarter-circular Area	$\bar{x} = \bar{y} = \dfrac{4r}{3\pi}$	$I_x = I_y = \dfrac{\pi r^4}{16}$ $\bar{I}_x = \bar{I}_y = \left(\dfrac{\pi}{16} - \dfrac{4}{9\pi} \right) r^4$ $I_z = \dfrac{\pi r^4}{8}$
Area of Circular Sector	$\bar{x} = \dfrac{2}{3} \dfrac{r \sin \alpha}{\alpha}$	$I_x = \dfrac{r^4}{4} \left(\alpha - \dfrac{1}{2} \sin 2\alpha \right)$ $I_y = \dfrac{r^4}{4} \left(\alpha + \dfrac{1}{2} \sin 2\alpha \right)$ $I_z = \dfrac{1}{2} r^4 \alpha$

TABLE D/3 **PROPERTIES OF PLANE FIGURES** *Continued*

FIGURE	CENTROID	AREA MOMENTS OF INERTIA
Rectangular Area	—	$I_x = \dfrac{bh^3}{3}$ $\bar{I}_x = \dfrac{bh^3}{12}$ $\bar{I}_z = \dfrac{bh}{12}(b^2 + h^2)$
Triangular Area	$\bar{x} = \dfrac{a+b}{3}$ $\bar{y} = \dfrac{h}{3}$	$I_x = \dfrac{bh^3}{12}$ $\bar{I}_x = \dfrac{bh^3}{36}$ $I_{x_1} = \dfrac{bh^3}{4}$
Area of Elliptical Quadrant	$\bar{x} = \dfrac{4a}{3\pi}$ $\bar{y} = \dfrac{4b}{3\pi}$	$I_x = \dfrac{\pi ab^3}{16}, \ \bar{I}_x = \left(\dfrac{\pi}{16} - \dfrac{4}{9\pi}\right)ab^3$ $I_y = \dfrac{\pi a^3 b}{16}, \ \bar{I}_y = \left(\dfrac{\pi}{16} - \dfrac{4}{9\pi}\right)a^3 b$ $I_z = \dfrac{\pi ab}{16}(a^2 + b^2)$
Subparabolic Area $y = kx^2 = \dfrac{b}{a^2}x^2$ Area $A = \dfrac{ab}{3}$	$\bar{x} = \dfrac{3a}{4}$ $\bar{y} = \dfrac{3b}{10}$	$I_x = \dfrac{ab^3}{21}$ $I_y = \dfrac{a^3 b}{5}$ $I_z = ab\left(\dfrac{a^3}{5} + \dfrac{b^2}{21}\right)$
Parabolic Area $y = kx^2 = \dfrac{b}{a^2}x^2$ Area $A = \dfrac{2ab}{3}$	$\bar{x} = \dfrac{3a}{8}$ $\bar{y} = \dfrac{3b}{5}$	$I_x = \dfrac{2ab^3}{7}$ $I_y = \dfrac{2a^3 b}{15}$ $I_z = 2ab\left(\dfrac{a^2}{15} + \dfrac{b^2}{7}\right)$

TABLE D/4   PROPERTIES OF HOMOGENEOUS SOLIDS

(m = mass of body shown)

BODY		MASS CENTER	MASS MOMENTS OF INERTIA
	Circular Cylindrical Shell	—	$I_{xx} = \frac{1}{2}mr^2 + \frac{1}{12}ml^2$ $I_{x_1x_1} = \frac{1}{2}mr^2 + \frac{1}{3}ml^2$ $I_{zz} = mr^2$
	Half Cylindrical Shell	$\bar{x} = \dfrac{2r}{\pi}$	$I_{xx} = I_{yy}$ $\qquad = \frac{1}{2}mr^2 + \frac{1}{12}ml^2$ $I_{x_1x_1} = I_{y_1y_1}$ $\qquad = \frac{1}{2}mr^2 + \frac{1}{3}ml^2$ $I_{zz} = mr^2$ $\bar{I}_{zz} = \left(1 - \dfrac{4}{\pi^2}\right)mr^2$
	Circular Cylinder	—	$I_{xx} = \frac{1}{4}mr^2 + \frac{1}{12}ml^2$ $I_{x_1x_1} = \frac{1}{4}mr^2 + \frac{1}{3}ml^2$ $I_{zz} = \frac{1}{2}mr^2$
	Semicylinder	$\bar{x} = \dfrac{4r}{3\pi}$	$I_{xx} = I_{yy}$ $\qquad = \frac{1}{4}mr^2 + \frac{1}{12}ml^2$ $I_{x_1x_1} = I_{y_1y_1}$ $\qquad = \frac{1}{4}mr^2 + \frac{1}{3}ml^2$ $I_{zz} = \frac{1}{2}mr^2$ $\bar{I}_{zz} = \left(\dfrac{1}{2} - \dfrac{16}{9\pi^2}\right)mr^2$
	Rectangular Parallelepiped	—	$I_{xx} = \frac{1}{12}m(a^2 + l^2)$ $I_{yy} = \frac{1}{12}m(b^2 + l^2)$ $I_{zz} = \frac{1}{12}m(a^2 + b^2)$ $I_{y_1y_1} = \frac{1}{12}mb^2 + \frac{1}{3}ml^2$ $I_{y_2y_2} = \frac{1}{3}m(b^2 + l^2)$

TABLE D/4 **PROPERTIES OF HOMOGENEOUS SOLIDS** *Continued*

(m = mass of body shown)

BODY		MASS CENTER	MASS MOMENTS OF INERTIA
	Spherical Shell	—	$I_{zz} = \frac{2}{3}mr^2$
	Hemispherical Shell	$\bar{x} = \dfrac{r}{2}$	$I_{xx} = I_{yy} = I_{zz} = \frac{2}{3}mr^2$ $\bar{I}_{yy} = \bar{I}_{zz} = \frac{5}{12}mr^2$
	Sphere	—	$I_{zz} = \frac{2}{5}mr^2$
	Hemisphere	$\bar{x} = \dfrac{3r}{8}$	$I_{xx} = I_{yy} = I_{zz} = \frac{2}{5}mr^2$ $\bar{I}_{yy} = \bar{I}_{zz} = \frac{83}{320}mr^2$
	Uniform Slender Rod	—	$I_{yy} = \frac{1}{12}ml^2$ $I_{y_1y_1} = \frac{1}{3}ml^2$

TABLE D/4   **PROPERTIES OF HOMOGENEOUS SOLIDS**   *Continued*

(m = mass of body shown)

BODY	MASS CENTER	MASS MOMENTS OF INERTIA
Quarter-Circular Rod	$\bar{x} = \bar{y}$ $= \dfrac{2r}{\pi}$	$I_{xx} = I_{yy} = \frac{1}{2}mr^2$ $I_{zz} = mr^2$
Elliptical Cylinder	—	$I_{xx} = \frac{1}{4}ma^2 + \frac{1}{12}ml^2$ $I_{yy} = \frac{1}{4}mb^2 + \frac{1}{12}ml^2$ $I_{zz} = \frac{1}{4}m(a^2 + b^2)$ $I_{y_1y_1} = \frac{1}{4}mb^2 + \frac{1}{3}ml^2$
Conical Shell	$\bar{z} = \dfrac{2h}{3}$	$I_{yy} = \frac{1}{4}mr^2 + \frac{1}{2}mh^2$ $I_{y_1y_1} = \frac{1}{4}mr^2 + \frac{1}{6}mh^2$ $I_{zz} = \frac{1}{2}mr^2$ $\bar{I}_{yy} = \frac{1}{4}mr^2 + \frac{1}{18}mh^2$
Half Conical Shell	$\bar{x} = \dfrac{4r}{3\pi}$ $\bar{z} = \dfrac{2h}{3}$	$I_{xx} = I_{yy}$ $\quad = \frac{1}{4}mr^2 + \frac{1}{2}mh^2$ $I_{x_1x_1} = I_{y_1y_1}$ $\quad = \frac{1}{4}mr^2 + \frac{1}{6}mh^2$ $I_{zz} = \frac{1}{2}mr^2$ $\bar{I}_{zz} = \left(\dfrac{1}{2} - \dfrac{16}{9\pi^2} \right) mr^2$
Right-Circular Cone	$\bar{z} = \dfrac{3h}{4}$	$I_{yy} = \frac{3}{20}mr^2 + \frac{3}{5}mh^2$ $I_{y_1y_1} = \frac{3}{20}mr^2 + \frac{1}{10}mh^2$ $I_{zz} = \frac{3}{10}mr^2$ $\bar{I}_{yy} = \frac{3}{20}mr^2 + \frac{3}{80}mh^2$

TABLE D/4   PROPERTIES OF HOMOGENEOUS SOLIDS *Continued*
(m = mass of body shown)

BODY	MASS CENTER	MASS MOMENTS OF INERTIA
Half Cone	$\bar{x} = \dfrac{r}{\pi}$ $\bar{z} = \dfrac{3h}{4}$	$I_{xx} = I_{yy}$ $\quad = \frac{3}{20}mr^2 + \frac{3}{5}mh^2$ $I_{x_1 x_1} = I_{y_1 y_1}$ $\quad = \frac{3}{20}mr^2 + \frac{1}{10}mh^2$ $I_{zz} = \frac{3}{10}mr^2$ $\bar{I}_{zz} = \left(\dfrac{3}{10} - \dfrac{1}{\pi^2}\right)mr^2$
$\dfrac{x^2}{a^2} + \dfrac{y^2}{b^2} + \dfrac{z^2}{c^2} = 1$ Semiellipsoid	$\bar{z} = \dfrac{3c}{8}$	$I_{xx} = \frac{1}{5}m(b^2 + c^2)$ $I_{yy} = \frac{1}{5}m(a^2 + c^2)$ $I_{zz} = \frac{1}{5}m(a^2 + b^2)$ $\bar{I}_{xx} = \frac{1}{5}m(b^2 + \frac{19}{64}c^2)$ $\bar{I}_{yy} = \frac{1}{5}m(a^2 + \frac{19}{64}c^2)$
$\dfrac{x^2}{a^2} + \dfrac{y^2}{b^2} = \dfrac{z}{c}$ Elliptic Paraboloid	$\bar{z} = \dfrac{2c}{3}$	$I_{xx} = \frac{1}{6}mb^2 + \frac{1}{2}mc^2$ $I_{yy} = \frac{1}{6}ma^2 + \frac{1}{2}mc^2$ $I_{zz} = \frac{1}{6}m(a^2 + b^2)$ $\bar{I}_{xx} = \frac{1}{6}m(b^2 + \frac{1}{3}c^2)$ $\bar{I}_{yy} = \frac{1}{6}m(a^2 + \frac{1}{3}c^2)$
Rectangular Tetrahedron	$\bar{x} = \dfrac{a}{4}$ $\bar{y} = \dfrac{b}{4}$ $\bar{z} = \dfrac{c}{4}$	$I_{xx} = \frac{1}{10}m(b^2 + c^2)$ $I_{yy} = \frac{1}{10}m(a^2 + c^2)$ $I_{zz} = \frac{1}{10}m(a^2 + b^2)$ $\bar{I}_{xx} = \frac{3}{80}m(b^2 + c^2)$ $\bar{I}_{yy} = \frac{3}{80}m(a^2 + c^2)$ $\bar{I}_{zz} = \frac{3}{80}m(a^2 + b^2)$
Half Torus	$\bar{x} = \dfrac{a^2 + 4R^2}{2\pi R}$	$I_{xx} = I_{yy} = \frac{1}{2}mR^2 + \frac{5}{8}ma^2$ $I_{zz} = mR^2 + \frac{3}{4}ma^2$

PHOTO CREDITS

INDEX